THE BASAL GANGLIA III

ADVANCES IN BEHAVIORAL BIOLOGY

Recent Volumes in this Series

A Continuation Order Plan is available for this series. A continuation order will bring delivery of each new volume immediately upon publication. Volumes are billed only upon actual shipment. For further information please contact the publisher.

THE BASAL GANGLIA III

Edited by

Giorgio Bernardi

Second University of Rome
Rome, Italy

Malcolm B. Carpenter

Uniformed Services University of the Health Sciences
Bethesda, Maryland

Gaetano Di Chiara

University of Cagliari
Cagliari, Italy

Micaela Morelli

University of Cagliari
Cagliari, Italy

and

Paolo Stanzione

Second University of Rome
Rome, Italy

SPRINGER SCIENCE+BUSINESS MEDIA,LLC

Library of Congress Cataloging in Publication Data

International Basal Ganglia Society. Symposium (3rd: 1989: Cagliari, Italy)
  The basal ganglia III / edited by Giorgio Bernardi . . . [et al.].
    p.   cm.—(Advances in behavioral biology; v. 39)
  "Proceedings of the Third Triennial Meeting of the International Basal Ganglia Society,
held June 10–13, 1989, in Cagliari, Italy"—T.p. verso.
  Includes bibliographical references and index.
  ISBN 978-1-4684-5873-2   ISBN 978-1-4684-5871-8 (eBook)
  DOI 10.1007/978-1-4684-5871-8
  1. Basal ganglia—Physiology—Congresses. 2. Basal ganglia—Diseases—Congres-
ses. I. Bernardi, Giorgio. II. Title. III. Title: Basal ganglia 3. IV. Series.
  [DNLM: 1. Basal Ganglia—congresses. WL 307 I605b 1989]
QP383.3.I58 1989
599′.0188—dc20
DNLM/DLC   90-14329
for Library of Congress   CIP

Proceedings of the Third Triennial Meeting of the International Basal
Ganglia Society, held June 10–13, 1989, in Cagliari, Italy

© Springer Science+Business Media New York 1991
Originally published by Plenum Press, New York 1991
Softcover reprint of the hardcover 1st edition 1991

PREFACE

This volume represents the collected papers presented at the Third Triennial Symposium of the International Basal Ganglia Society (IBAGS) held at Capo Boi, Italy, June 10-13, 1989. About 300 members of the Society and participants attended the symposium which was held in a delightful environment conducive to the formal and informal exchange of scientific thought. The interdisciplinary nature of the symposium was unique in its coverage of the neurosciences from molecular biology to clinical and behavioural studies. The 80 papers collected here reflect the wide spectrum and the depth of studies on virtually all aspects of the basal ganglia.
Unfortunately, this book does not capture the cordial and congenial atmosphere which has characterized this, and all prior symposia of the Society. Any cooperative endeavour of this kind requires a tremendous effort and dedication, usually by a small number of individuals. The Society is especially pleased to acknowledge the support and encouragement of the "Italian Ministry of University and Scientific Research" and the "Italian National Research Council". In addition the Society received financial support from numerous Foundations and Corporations, which are listed separately under acknowledgements.
Finally the Editors are pleased that Plenum Press, which has published the two previous symposia, has accepted this program for publication. It is our hope that vast scientific efforts reflected in these pages will be widely disseminated and further encourage every kind of research related to the basal ganglia.

Giorgio Bernardi
Malcolm Carpenter
Gaetano Di Chiara
Micaela Morelli
Paolo Stanzione

ACKNOWLEDGEMENTS

The following Organizations and Institutions are gratefully acknowledged:

Comune di Cagliari (Italy)
Comune di Villasimius (Italy)
Ente Provinciale per il Turismo (Italy)
Ente Sardo Industrie Turistiche (Italy)
Assessorato al Turismo, Regione Sardegna (Italy)
Ass. Pubblica Istruzione, Regione Sardegna (Italy)
Ministero Pubblica Istruzione (Italy)
Consiglio Nazionale delle Ricerche (Italy)
Credito Industriale Sardo (Italy)
Banca d'Italia (Italy)

American Parkinson Disease Association (USA)
National Parkinson Foundation (USA)
Parkinson's Disease Foundation (USA)
United Parkinson Foundation (USA)

Aziende Chimiche Riunite Angelini Francesco (Italy)
Du Punt de Nemours (USA)
E.Lilly & Co. (USA)
FIDIA Research Laboratories (Italy)
Hoffman-La Roche (Switzerland)
Idma (Italy)
ISF S.p.a. (Italy)
Merck Sharp and Dohme Research Laboratories (USA)
Polifarma S.p.a. (Italy)
Proter S.p.a. (Italy)
Sandoz (Italy)
Sandoz (Switzerland)
Sigma-Tau S.p.a. (Italy)
Zambon S.p.a. (Italy)

CONTENTS

ANATOMICAL ORGANIZATION AND CHEMICAL NEUROANATOMY OF THE BASAL GANGLIA

PHARMACOLOGICAL EFFECTS UPON BEHAVIOUR

MODELS OF BASAL GANGLIA PATHOLOGY

ANATOMICAL ORGANIZATION AND CHEMICAL
NEUROANATOMY OF THE
BASAL GANGLIA

STRIOSOMES AND MATRISOMES

A. M. Graybiel, A.W. Flaherty and J.-M.Giménez-Amaya*

Dept. of Brain and Cognitive Sciences
Massachusetts Institute of Technology
Cambridge, Massachusetts, USA

*Departamento de Morfología
Facultad de Medicina
Universidad Autónoma de Madrid, Spain

In the past twelve years it has become clear that most neurotransmitter systems in the striatum are differentially distributed with respect to striosomes ("striatal bodies") and their surrounding extrastriosomal matrix (Graybiel and Ragsdale, 1978, 1983; Herkenham and Pert, 1981; Gerfen, 1984; Gerfen et al., 1985a). The boundaries between striosomes and matrix govern the distributions of neurotransmitter-related compounds ranging from cholinergic and monoaminergic receptors and uptake sites to neuropeptides, benzodiazepine receptors, and calcium-binding proteins. The neurochemical differences between striosomes and matrix reflect differential distributions of striatal afferent fibers, interneurons and projection neurons in striosomes and matrix (see Graybiel, 1990 for review). The different neurochemistry, inputs, and outputs of striosomes and matrix all suggest that the two striatal compartments have different functional roles.

This neurochemical and connectional compartmentalization of the striatum, although internally consistent, has nevertheless created a puzzle. The striosomes make up only about 20% of the volume of the striatum (Graybiel et al., 1981; Groves et al., 1988; Desban et al., 1989). Are we to think that the remaining 80% of the striatum, the matrix, is homogeneous? Large parts of the cerebral cortex, including parieto--temporo-occipital association cortex, sensorimotor cortex, and parts of lateral frontal cortex project preferentially to the matrix (Ragsdale and Graybiel, 1981, 1990; Dono-ghue and Herkenham, 1986), and the matrix is the main source of striatal outputs to the globus pallidus and substantia nigra pars reticulata (Graybiel et al., 1979; Gerfen, 1985b; Jimenez-Castellanos and Graybiel, 1989; Sadikot and Parent, 1989 Gimenez--Amaya and Graybiel, 1990a). Compartmental clustering of these inputs and outputs within the matrix would impose controls over their interactions similar to the constraints on the relative input-output interactions of striosomes and matrix.

The Basal Ganglia III, Edited by G. Bernardi *et al.*
Plenum Press, New York

Growing evidence indicates that there is, in fact, considerable heterogeneity in the striatal matrix. First, many inputs to the matrix do not innervate it uniformly, but innervate discrete zones of matrix that are similar in size and shape to striosomes. The presence of this extrastriosomal clustering of striatal afferents was first documented in the cat, for the projection to the striatum from the somatosensory cortex (Malach and Graybiel, 1986). Patchy terminations in explicitly identified matrix have since been shown for fiber systems from other major forebrain regions that project to the striatum -- those from the amygdala (Ragsdale and Graybiel, 1988a), from the thalamus (Ragsdale and Graybiel, 1988b), and from other areas of cortex (Alexander et al., 1988; Fotuhi et al., 1989). Studies in the cat have also demonstrated clustering, within the matrix, of output neurons projecting to the substantia nigra and to the globus pallidus (Desban et al., 1989; Jimenez-Castellanos and Graybiel, 1989). Thus, at least in the cat, the efferent architecture of the striatal matrix is as modular as its afferent architecture.

Single unit recording studies suggest that there is physiological as well as anatomical patchiness in the striatal matrix. Clusters of neurons with similar somatosensory receptive fields and similar relations to active movements have been observed (Schneider and Lidsky, 1981; Crutcher and DeLong, 1984; Liles and Updyke, 1985) and striatal microexcitable zones have been found in which electrical stimulation elicits movements (Alexander and DeLong, 1985). In none of these studies have the locations of the physiological zones been established in relation to striosomal boundaries. Anatomical studies, however, have shown that inputs from somatosensory cortex and motor cortex are distributed to the matrix, and in one physiological study the clusters of arm movement-related cells in the striatum have been shown to correspond to corticostriatal projection-clusters from the arm area of motor cortex (Liles and Updyke, 1985). It thus is likely that these functionally defined clusters are in the striatal matrix.

In this review, we summarize new experimental evidence (Flaherty et al., 1989; Gimenez-Amaya and Graybiel, 1990a, 1990b) that the extrastriosomal matrix of the primate striatum contains both input and output modules. These afferent and efferent subsystems have an organization sufficiently systematic to suggest that modularity is a major feature of the organization of the striatal matrix. To differentiate these subsystems from the striosome/matrix compartmentalization of the striatum, we refer provisionally to the compartments within the matrix as "matrisomes".

CLUSTERING OF AFFERENT FIBERS IN THE MATRIX

Our evidence for modules of afferent fibers in the matrix comes from experiments in the squirrel monkey in which we injected anterograde tracers into the somatosensory cortex. We then studied the distributions of labeled fibers in the striatum with respect to the distribution of striosomes detected by enkephalin immunohistochemistry or butyrylcholinesterase histochemistry. Our observations were of two types.

First, we examined the corticostriatal projections issuing from electrophysiologically defined loci within the three cortical somatosensory maps found in Brodmann areas 3a (containing cells with primarily deep receptive fields), 3b (primarily cutaneous receptive fields), and 1 (primarily cutaneous receptive fields). We concentrated on map regions containing the most detailed representations: the hand (especially the digit representations), the face (especially the representation of the lips), and the foot (especially the toes). The cortical injection sites had diameters on the order of 1 mm, and in most animals all cortical layers were injected. Regardless of both the cortical area injected and the body part represented at the injection site, single injections in somatosensory cortex produced a pattern of striatal labeling that was characterized by multiple corticostriatal fiber clusters in the somatosensory sector of the putamen.

The somatosensory projections to the putamen followed the familiar foot-dorsal, head-ventral somatotopy first described by Künzle (1975, 1977). Within each of these regions, however, the map of the body was broken into a dispersed set of modules. For example, Fig. 1A shows five prominent clusters of afferent fibers in the anterior putamen, labeled by a single injection of radioactive amino acid into cortical area 3a. Farther posteriorly in the putamen, many -- but not all -- of the afferent clusters tended to fuse, forming branching diagonal bands that extended dorsomedially from the lateral border of the putamen.

The afferent clusters seen in single cross-sections had sizes and orientations that were similar to those of striosomes. In part of the putamen it is difficult to demonstrate striosomes with any known striosomal marker, but in our cases there was an extensive region of the putamen in which we could compare the distributions of somatosensory afferent clusters to striosomes in serial sections. In every case the labeled clusters were located within the extrastriosomal matrix. Fig. 1 shows an example of such a serial-section comparison. Although the afferent clusters sometimes abutted striosomes, often they shared no common boundary with them, suggesting that the locations of the striosomes and fiber clusters do not depend on each other.

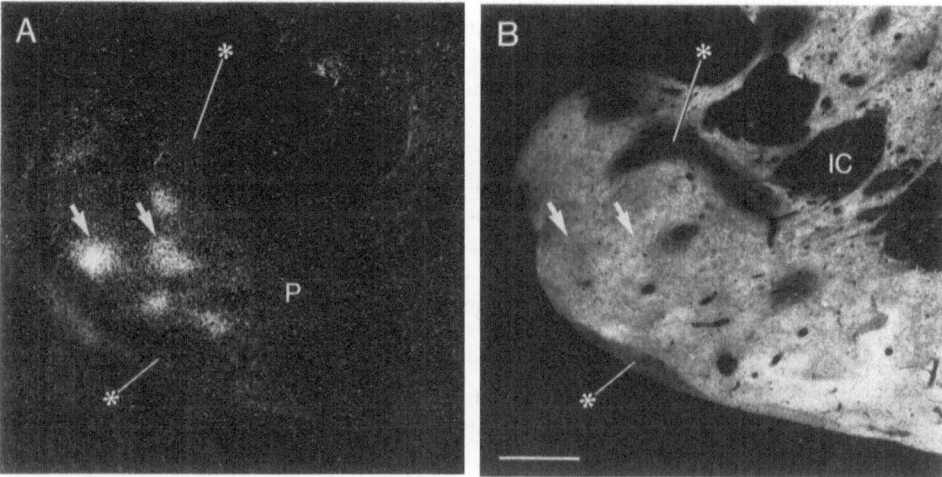

Fig. 1. Afferent-fiber clusters in the striatal matrix. A. Transverse section through the forebrain of a squirrel monkey , showing multiple patches of anterograde label (examples at arrows) in the putamen produced by a single injection (diameter ca. 1.5 mm) of [^{35}S]methionine into the foot region of cortical area 3a. B. Serially adjoining section, stained for met-enkephalin-like immunoreactivity, in which striosomes are visible as regions of low enkephalin immunostaining (Graybiel and Chesselet, 1984, see examples marked by asterisks). Comparison of A and B (see matching fiduciary arrows and asterisks) demonstrates that all of the labeled fiber-clusters are contained within the extrastriosomal matrix, and that their clustering does not simply result from an avoidance of striosomes. P = putamen; IC = internal capsule. Scale bar = 1 mm.

In a second set of experiments, we compared the distributions of corticostriatal projections from two different somatosensory areas by injecting distinguishable anterograde tracers into regions representing matched body parts -- for example, the index finger regions of cortical areas 3a and 1. We found that the projections from matched regions of all three cortical areas overlapped extensively, especially in the region of the putamen that received the strongest somatosensory cortical input. The overlap was not always complete at anterior levels, but in all cases the two sets of corticostriatal clusters were similar in shape and were in close proximity, suggesting that the organizations of the cortical projections were not independent of one another. These results point to the existence of a truly modular corticostriatal projection system: the fibers from the cortex do not merely form randomly arranged clusters in the striatum, they are systematically organized with respect to each other. It is interesting that equivalent body part representations in different cortical areas send projections that overlap in the putamen of primates to a greater extent than has been seen in the cat (Malach and Graybiel, 1986). This suggests that in the primate the somatosensory matrisomes have more to do with integrating body-part information coming from different cortical areas than they do with sorting out somatosensory submodalities.

CLUSTERING OF PROJECTION NEURONS IN THE MATRIX

The striatal matrix sends a massive fiber projection to the globus pallidus. Evidence from dual-retrograde tracer experiments suggests that in cats and monkeys, the projections to the external and internal segments of the globus pallidus (GPe and GPi) mainly arise from different neurons (Beckstead and Cruz, 1986; Koliatsos et al., 1988; Parent et al., 1989; Gimenez-Amaya and Graybiel, 1990a) although single neurons that send processes to both segments have been observed in rats (Chang et al., 1982). Because findings in the cat suggested that at least some matrix projection neurons form clusters (Desban et al., 1989; Jimenez-Castellanos and Graybiel, 1989), we injected retrograde tracers into the globus pallidus of the squirrel monkey to determine whether there is a comparable mosaicism in the primate's striatopallidal projection.

The evidence indicates that extensive clustering of striatal output neurons does exist. When injection sites involved one pallidal segment more than the other, retrogradely labeled neurons were not only concentrated in a main field, but also appeared in discrete clusters at the edges of the main field and even at distances of several millimeters from it (Gimenez-Amaya and Graybiel, 1990a). Such projection-neuron clusters have also been noted in experiments in macaques (Koliatsos et al., 1988; Selemon and Goldman-Rakic, 1989).

To study this clustering phenomenon, we made more restricted injections of retrograde tracer into the globus pallidus, attempting to infiltrate either the GPi or the GPe alone (Gimenez-Amaya and Graybiel, 1990b). Though none of the injection sites obtained was perfectly confined to one segment, many were nearly so. In these cases there was very distinct clustering of retrogradely labeled neurons in the matrix of the putamen and, in some cases, the caudate nucleus. Figure 2A illustrates a typical field of patchy retrograde labeling produced by an injection of wheatgerm-conjugated horseradish peroxidase (WGA-HRP) that was confined almost completely to the GPi. The majority of the retrogradely labeled pallidostriatal neurons were collected in clusters and bands that in size and orientation resembled striosomes. In the caudate nucleus, where the striosomal system is easy to demonstrate histochemically, it was clear that the clusters and bands lay in the extrastriosomal matrix. In the putamen, in all instances in which we could see striosomes in the regions of retrograde labeling, the clusters of retrogradely labeled neurons were similarly excluded from striosomes. Fig. 2B shows such a region. Finally, even within fields of more continuous retrograde labeling, in which there were no zones in the matrix devoid of labeled neurons, there nevertheless was commonly a systematic waxing and waning in the intensity of retrograde labeling, as though there were adjacent cell clusters within the main field.

Current neuroanatomical techniques cannot produce injection sites that fill exactly all and only all of a single pallidal segment with a retrograde tracer (nor all and only all of each of the two segments with two distinguishable tracers). Within the limits of the experimental methods available, however, our results suggest that the clusters may represent a sorting out of neurons projecting, respectively, to the two segments of the globus pallidus. Alternate possibilities -- for example, that the clusters are targeted to other subdivisions within the globus pallidus -- are not excluded by the results. In fact, it would be surprising if only one sorting principle held for the ordering of these striatal output neurons. What is clear from our experiments is that aggregation of projection neurons into clusters and bands is a general phenomenon in the striatal matrix of the primate.

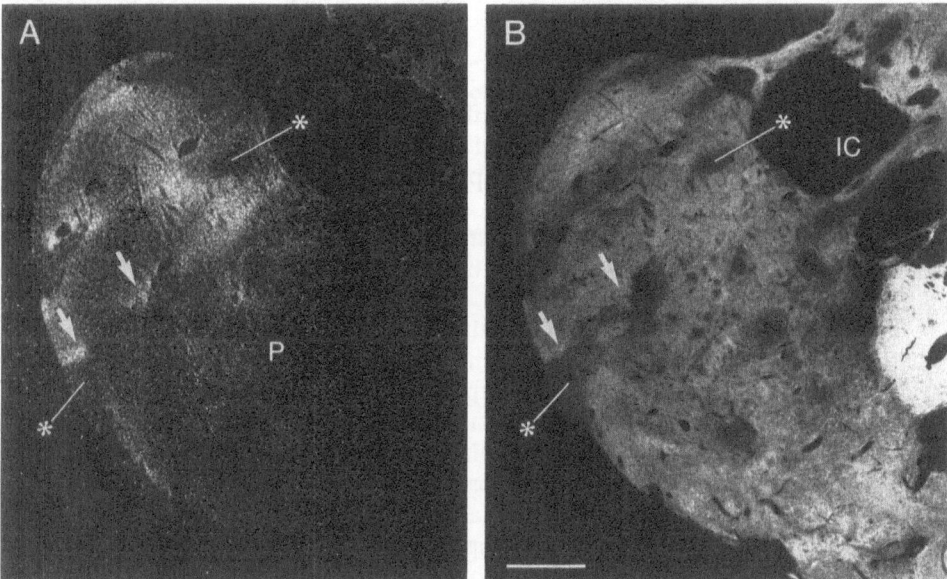

Fig. 2. Efferent-neuron clusters in the striatal matrix. A. Transverse section through the putamen of a squirrel monkey in which a single injection (diameter ca. 2.0 mm) of wheatgerm-conjugated horseradish peroxidase was placed in the GPi. In the putamen, there are multiple clusters of retrogradely labeled neurons (arrows point to examples). B. Serially adjoining section, stained for met-enkephalin-like immunoreactivity, demonstrates the locations of striosomes (regions of low enkephalin immunostaining; see examples marked by asterisks). Comparison of A and B (see matching fiduciary arrows and asterisks) shows that the cell-clusters are contained within the extrastriosomal matrix, and that their clustering does not simply result from an avoidance of striosomes. P = putamen; IC = internal capsule. Scale bar = 1 mm.

DIRECTIONS FOR THE FUTURE: MATRISOMES AND MODULAR PROCESSING IN THE STRIATUM

The existence in the matrix of modules containing afferent fibers and efferent neurons raises questions about how these modules relate to each other and to striosomes. Are there many afferent-fiber clusters abutting and overlapping one another in different patterns, so that already at the input side of striatal processing there are modular constraints on intrastriatal processing? Are the locations of these afferent matrisomes related in regular ways to the striosomes that lie spaced at intervals through the matrix? Are the afferent modules targeted to particular output-cell modules? Or do they spread information from a given source through several output systems? Are the output matrisomes divided mainly in relation to the three principal targets of the matrix (the GPe, the GPi, and the pars reticulata of the substantia nigra), or are there many finer divisions within these structures that receive individually specific inputs from subsets of clustered output cells in the matrix? And, within the striatum, how are the local axonal arbors of the projection neurons, and the processes of interneurons, organized with respect to these input and output modules? Answers to these questions would be important guides for further work on information processing in the extrapyramidal motor system.

Figure 3 illustrates three of the simplest possibilities for how striatal afferents and efferents might be coupled within matrisomes. In Fig. 3A, afferent-fiber clusters and efferent-neuron clusters in the matrix are shown as being randomly distributed with respect to each other, constrained only to avoid overlapping with striosomes. This arrangement would allow each output system to sample the activity of many input systems. Our evidence that matched parts of different somatosensory cortical areas all project to the same "matrisomes" suggests that random overlap of afferent-fiber clusters is not likely. Preliminary reports that the motor cortex projects to the same clusters in the putamen that are innervated by somatosensory cortex, and avoids those areas of the putamen innervated by supplementary cortex (Alexander et al., 1988; Fotuhi et al., 1989), also makes random overlap of inputs unlikely. Whether such inputs to the matrix are randomly or systematically distributed with respect to the output modules of the matrix remains to be determined.

Evidence that a single small region in the cortex projects to multiple clusters in the striatum raises a second possibility, shown in Fig. 3B, in which inputs to the matrix from a given region split to innervate output matrisomes that have different targets. In Fig. 3C, individual input and output clusters are shown as being constrained to form functional "channels" through the matrix, similar to the channeling that obtains for the inputs and outputs of the matrix and striosomes. Such an arrangement would, for instance, allow projections from specific areas of cortex to innervate parts of the matrix that project to only one of the three principal targets of the matrix, and to avoid parts of the matrix that project to the other two.

Of course, these three schemes do not exhaust the possibilities for modular organization of the striatum. Nor do they address the important question of how adjacent modules within the striatum interact. If, for example, matrisomes form parallel or branching channels through the striatum, what roles do the interactions at borders between adjacent matrisomes (and between matrisomes and striosomes) play in integrating information going through adjacent channels?

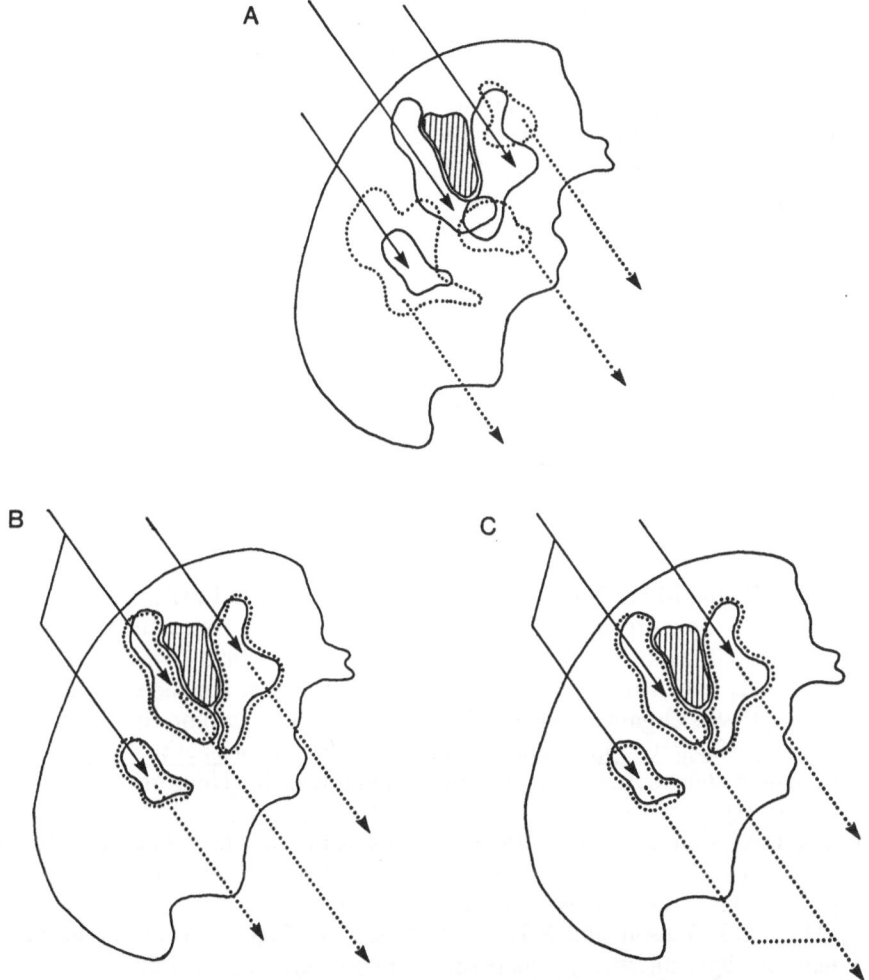

Fig. 3. Three possible input-output organizations in the striatal matrix. Solid lines surround input matrisomes, and solid arrows show their inputs; dotted lines surround output matrisomes, and dotted arrows show their outputs; and shading denotes striosomes (for which inputs and outputs are not indicated). A. Random overlap of input and output matrisomes, permitting extensive mixing of different classes of inputs and outputs. B. Systematic overlapping of input and output matrisomes, with one class of input diverging to two different categories of output matrisome. C. Coupled matrisomal inputs and outputs forming trans-striatal channels.

The form of interactions among striatal inputs, outputs, and interneurons clearly will have important implications for the functional organization of the basal ganglia. For instance, if matrisomes coordinate the inputs and outputs of the matrix into distinct channels, they could allow primary targeting of certain cortical inputs to the GPe (and thus to the subthalamic control loop) and targeting of other cortical inputs to the GPi (the main source of pallidal input to the thalamus). They could also restrict convergence of different types of input -- for example, those from thalamus and neocortex. Understanding matrisomal clustering should also help in probing the cellular basis of the different vulnerabilities now being uncovered in the striatum in extrapyramidal disease states. In Huntington's disease, for example, neurons projecting to the two pallidal segments and to the substantia nigra pars reticulata appear to be affected to different degrees at different stages of the disease (Reiner et al., 1988). If the neurons progressively lost in this disease are related to the matrisomal compartments described here, we may gain insight into the etiology of the patterns of cell death and functional impairment that occur during the evolution of the disease.

ACKNOWLEDGEMENTS

This work was supported by NIH Javits Award R01 25529, NIH EY02866, and the Seaver Institute.

REFERENCES

Alexander, G.E. and M.R. DeLong (1985). Microstimulation of the primate striatum: II. Somatotopic organization of striatal microexcitable zones and their relation to neuronal response properties. J.Neurophysiol. *53*:1433-1446.

Alexander, G.E., V.E. Koliatsos, L.J. Martin, J. Hedreen, I. Hamada and M.R. De-Long (1988). Organization of primate basal ganglia "motor circuit": 1. Motor cortex (MC) and supplementary motor area (SMA) project to complementary regions within matrix compartment of putamen. Soc.Neurosci.Abstr. *14*:720. (Abstract)

Beckstead, R.M. and C.J. Cruz (1986). Striatal axons to the globus pallidus, entopeduncular nucleus and substantia nigra come mainly from separate cell populations in cat. Neuroscience. *19*:147-158.

Chang, H.T., C.J. Wilson and S.T. Kitai (1982). A golgi study of rat neostriatal neurons: light microscopic analysis. J.Comp.Neurol. *208*:107-126.

Crutcher, M.D. and M.R. DeLong (1984). Single cell studies of the primate putamen. I. Functional organization. Exp.Brain Res. *53*:233-243.

Desban, M., C. Gauchy, M.L. Kemel, M.J. Besson and J. Glowinski (1989). Three-dimensional organization of the striosomal compartment and patchy distribution of striatonigral projections in the matrix of the cat caudate nucleus. Neuroscience. *29*:551-566.

Donoghue, J.P. and M. Herkenham (1986). Neostriatal projections from indiviual cortical fields conform to histochemically distinct striatal compartments in the rat. Brain Res. *365*:397-403.

Flaherty, A.W., A.M. Graybiel, M. Sur and P.E. Garraghty (1989). Distinctive patterns of projections to striatum from physiologically-mapped somatosensory representations in primate cortex. Soc.Neurosci.Abstr. *15*:659. (Abstract)

Fotuhi, M., V.E. Koliatsos, G.E. Alexander and M.R. DeLong (1989). Patterns of sensorimotor integration in the primate neostriatum: primary somatosensory cortex (SC) and motor cortex (MC) project to coextensive territories in the putamen. Soc.Neurosci.Abstr. *15*:285. (Abstract)

Gerfen, C.R. (1984). The neostriatal mosaic: compartmentalization of corticostriatal input and striatonigral output systems. Nature. *311*:461-464.

Gerfen, C.R., K.G. Baimbridge and J.J. Miller (1985a). The neostriatal mosaic: Compartmental distribution of calcium-binding protein and parvalbumin in the basal ganglia of the rat and monkey. Proc.Natl.Acad.Sci.USA. *82*:8780-8784.

Gerfen, C.R. (1985b). The neostriatal mosaic. I. Compartmental organization of projections from the striatum to the substantia nigra in the rat. J.Comp.Neurol. *236*:454-476.

Gimenez-Amaya, J.-M. and A.M. Graybiel (1990a). Compartmental origins of the striatopallidal projection in the primate. Neuroscience. *34*:111-126.

Gimenez-Amaya, J.-M. and A.M. Graybiel (1990b). Modular organization of the matrix-compartment of the primate's striatum: clustering of striatopallidal projection neurons. (Submitted).

Graybiel, A.M. and M.-F. Chesselet (1984). Compartmental distribution of striatal cell bodies expressing met-enkephalin-like immunoreactivity. Proc.Natl.Acad.Sci.-USA. *81*:7980-7984.

Graybiel, A.M. and C.W. Ragsdale (1978). Histochemically distinct compartments in the striatum of human, monkey, and cat demonstrated by acetyl-thiocholinesterase staining. Proc.Natl.Acad.Sci.USA. *75*:5723-5726.

Graybiel, A.M., C.W. Ragsdale and S. Moon Edley (1979). Compartments in the striatum of the cat observed by retrograde cell-labelling. Exp.Brain Res. *34*:189-195.

Graybiel, A.M., C.W. Ragsdale, E.S. Yoneoka and R.P. Elde (1981). An immunohistochemical study of enkephalins and other neuropeptides in the striatum of the cat with evidence that the peptides are arranged to form mosaic patterns in register with striosomal compartments visible by acetylcholinesterase staining. Neuroscience. *6*:377-397.

Graybiel, A.M. (1990). Neurotransmitters and neuromodulators in the basal ganglia. TINS. (in press)

Graybiel, A.M. and C.W. Ragsdale (1983) Biochemical Anatomy of the Striatum. In P.C. Emson (ed): Chemical Neuroanatomy. New York: Raven Press, pp. 427-504.

Groves, P.M, M. Martone, S.J. Young and D.M. Armstrong (1988). Three-dimensional pattern of enkephalin-like immunoreactivity in the caudate nucleus of the cat. J.Neurosci. *8*:892-900.

Herkenham, M. and C.B. Pert (1981). Mosaic distribution of opiate receptors, parafascicular projections and acetylcholinesterase in rat striatum. Nature. *291*:415-418.

Jimenez-Castellanos, J. and A.M. Graybiel (1989). Compartmental origins of striatal efferent projections in the cat. Neuroscience. *32*:297-321.

Koliatsos, V.E., L.J. Martin, J. Hedreen, G.E. Alexander, I. Hamada, D.L. Price and M.R. DeLong (1988). Organization of primate basal ganglia "motor circuit": 2. putaminal projections to internal (GPi) and external (GPe) globus pallidus originate in distinct neuronal populations within the matrix compartment. Soc.Neurosci.Abstr. *14*:720. (Abstract)

Kunzle, H. (1975). Bilateral projections from precentral motor cortex to the putamen and other parts of the basal ganglia. Brain Res. *88*:195-210.

Kunzle, H. (1977). Projections from the primary somatosensory cortex to basal ganglia and thalamus in the monkey. Exp.Brain Res. *30*:481-492.

Liles, S.L. and B.V. Updyke (1985). Projection of the digit and wrist area of precentral gyrus to the putamen: relation between topography and physiological properties of neurons in the putamen. Brain Res. *339*:245-255.

Malach, R. and A.M. Graybiel (1986). Mosaic architecture of the somatic sensory-recipient sector of the cat's striatum. J.Neurosci. *6*:3436-3458.

Parent, A., Y. Smith, M. Filion and J. Dumas (1989). Distinct afferents to internal and external pallidal segments in the squirrel monkey. Neurosci Lett. *96*:140-144.

Ragsdale, C.W. and A.M. Graybiel (1981). The fronto-striatal projection in the cat and monkey and its relationship to inhomogeneities established by acetyl-cholinesterase histochemistry. Brain Res. *208*:259-266.

Ragsdale, C.W. and A.M. Graybiel (1988a). Fibers from the basolateral nucleus of the amygdala selectively innervate striosomes in the caudate nucleus of the cat. J.Comp.Neurol. *269*:506-522.

Ragsdale, C.W. and A.M. Graybiel (1988b) Multiple patterns of thalamostriatal innervation in the cat. In M. Bentivoglio, and R. Spreafico (eds): Cellular Thalamic Mechanisms. Amsterdam: Elsevier, pp. 261-267.

Ragsdale, C.W. and A.M. Graybiel (1990). Novel ordering of neocortical areas established by the compartmental organization of their striatal projections. Proc.-Natl.Acad.Sci.USA. (in press)

Reiner, A., R.L. Albin, K.D. Anderson, C.J. D'Amato and J.B. Penney (1988). Differential loss of striatal projection neurons in Huntington disease. Proc.Natl.-Acad.Sci.USA. *85*:5733-5737.

Sadikot, A.F. and A. Parent (1989). Differential projections of centre median and parafascicular nuclei in squirrel monkey. Soc.Neurosci.Abstr. *15*:288. (Abstract)

Schneider, J.S. and T.I. Lidsky (1981). Processing of somatosensory information in the striatum of behaving cats. J.Neurophysiol. *45*:841-851.

Selemon, L.D. and P.S. Goldman-Rakic (1989). Retrograde labeling of striatonigral and striatopallidal neurons in the rhesus monkey. Soc.Neurosci.Abstr. *15*:659. (Abstract)

THE DEVELOPMENT OF STRIATAL COMPARTMENTALIZATION:

THE ROLE OF MITOTIC AND POSTMITOTIC EVENTS

Janice G. Johnston, Gord Fishell, Leslie A. Krushel
and Derek van der Kooy

Neurobiology Research Group, Department of Anatomy
University of Toronto, Toronto, Ontario, Canada, M5S 1A8

An intriguing question in the evolution of the mammalian brain is how pattern can be conserved despite enormous changes in size and functional capacities. This phenomenon is dramatically demonstrated in the striatum where compartmental pattern is conserved across different mammalian species (Johnston et al., 1990)(Fig.1). Although the majority of the mammalian striatum is comprised of a seemingly homogeneous population of medium spiny neurons (Kemp and Powell, 1970), it can be compartmentalized into two subdivisions, the patch and matrix compartments, on the basis of different distributions of neurotransmitters, receptors and connections (Graybiel et al. 1981; Gerfen et al, 1984; van der Kooy et al., 1987). Compartmentalization of the striatum into small patches in a matrix background is restricted to mammals (Reiner et al., 1989) and has been shown in rodents (Herkenham and Pert, 1981), cats (Graybiel et al., 1981), monkeys (Goldman-Rakic, 1982) and humans (Haber, 1986). Despite the large differences in total striatal size, the ratio of patch/matrix area and the number of individual patches is maintained in different mammalian species (Johnston et al., 1990; Fig. 1). The maintenance of striatal compartmentalization suggests that common developmental mechanisms are involved in establishing this pattern and raises many questions concerning how this is achieved. We hypothesize, first, that cell lineages are committed to becoming members of compartments while still proliferating in the ventricular zone and that these precursors share common properties with the precursors which produce the compartments of the other major portion of the telencephalon, the cerebral cortex. Second, we suggest that the adult striatal pattern of small patches embedded within a matrix background is produced through cellular interactions in the postmitotic region that position compartmentally committed cells.

1) Proliferative commitment to compartmental identity

a. Birthdate predicts compartmental membership. The development of the brain is unique in that neuronal proliferation takes place in a circumscribed area adjacent to the cerebral ventricles (called the ventricular zone) that is separate from the final position that postmitotic

The Basal Ganglia III, Edited by G. Bernardi *et al.*
Plenum Press, New York

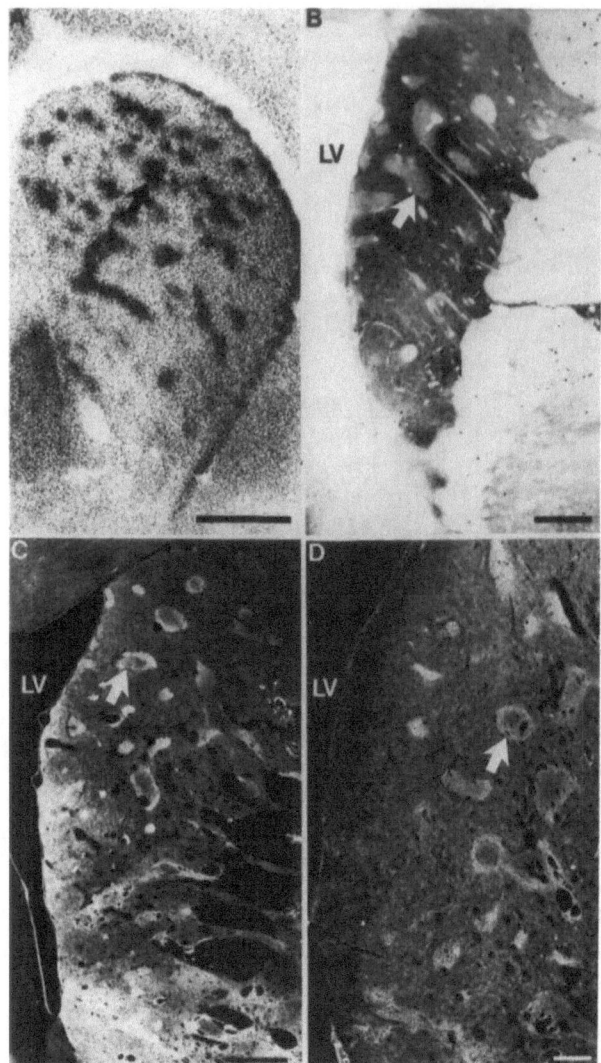

Figure 1. Compartmental organization in coronal
sections through the adult rat (A), rhesus monkey (B and C)
and human striatum (D) showing opiate receptor binding
(A), calcium-binding protein staining (B) and enkephalin
staining (C and D). The rat section (A) shows the entire
striatum. The monkey sections (B and C) show the entire
caudate. The human section (D) shows the dorsomedial
quarter of the caudate. Reverse images are shown of the
enkephalin stained sections (C and D) for better resolution
of patches. Arrows indicate individual patches.
LV=lateral ventricle. Scale bars=1mm. (From Johnston et
al, 1990)

cells will adopt. 'Birthdating' studies utilizing pulses of DNA synthesis
markers, such as ^{3}H-thymidine (or bromodeoxyuridine), have been used to
identify the time when neurons become postmitotic (Angevine and Sidman,
1961; Fishell and van der Kooy, 1987). Striatal birthdating studies (van der

14

Kooy and Fishell, 1987) have shown that the majority of neurons comprising the patch compartment become postmitotic before the majority of matrix cells leave the cell cycle. Similarly, the deep layers of the cerebral cortex (infragranular) are born prior to the superficial cortical layers (supragranular)(Angevine and Sidman, 1961), and the peaks of striatal and cortical compartmental neurogenesis occur concomitantly (van der Kooy and Fishell, 1987). More specifically, early born cells in the forebrain become striatal patch neurons and deep cortical neurons, whereas late born cells become striatal matrix neurons and superficial cortical neurons. Although consistent with the idea of separate stem cells for the different compartments, this evidence does not rule out the possibility that temporally regulated environmental factors influence the differentiation of newly postmitotic cells into specific compartmental phenotypes.

b. Intraocular transplants suggest early irreversible commitment. In order to discern between these two contrasting views of compartmental determination, one can challenge the cellular commitment of postmitotic cells by exposure to a different environment. Intraocular transplants of embryonic striatal tissue (Johnston et al., 1987) test whether striatal cells are committed to a certain phenotype independent of their connections (afferent and efferent) and exposure to certain local growth factors. The expression of phenotypic markers representative of the patch and matrix compartments (opiate receptor binding and acetylcholinesterase staining, respectively) in a complementary fashion in intraocular transplants of E16 striatal tissue, suggests that striatal cells can differentiate independent of nearby brain regions (Johnston et al, 1987). The transplants were composed of approximately 50% patch staining cells and 50% matrix staining cells, indicating that the striatal cells have maintained their commitment to compartments in a ratio expected from the time of removal of the graft.

Heterochronic transplants performed by McConnell (1988) also suggest that cortical precursors are committed to certain compartmental phenotypes. Early born ventricular cells will migrate to deep cortical lamina and extend normal phenotypic subcortical projections, despite being transplanted to animals at an age when superficial cortical neurons are becoming postmitotic (McConnell, 1988; McConnell, 1989). Further evidence for mitotic commitment to specific compartmental phenotype comes from the finding that neurons displaced from their natural adult laminar position (eg. in *reeler* mice [Drager, 1981; Lemmon and Pearlman, 1981] and x-radiated embryos [Jensen and Killackey, 1984]) will still send out axonal projections predicted from their times of birth. These studies have looked primarily for the presence of subcortical projections, rather than at specific sites of termination. However, transplants performed by Stanfield and O'Leary (1985) have revealed that cortical neurons will maintain in the adult the very specific axonal projections appropriate for their new host cortical location, suggesting that the environment is important in determining precise cellular phenotype. Moreover, columns of neurons in the somatosensory cortex of the rodent that represent the vibrissae (barrel fields) have been found to be modified in shape following deprivation of sensory information to the cortex (Simons et al., 1984). Therefore, intrinsic (perhaps ventricular zone) factors may determine laminar position in the cortex and gross axonal projections, but adult columnar locations and precise axonal terminal field may depend on cell interactions with nearby brain cells. Similarly, the intraocular striatal transplants suggest that environmental factors (such as afferents from or efferents to the substantia nigra [Lanca et al.,1986; Johnston et al.,1990; van der Kooy et al., 1987; Johnston et al, 1990]) play a role in the precise formation of pattern, since the transplants did not display the normal small patches embedded within a matrix background. Again, the gross commitment of cells to a compartmental identity may be cell autonomous from early embryonic times, but the precise topographic positioning of cells in the striatum may be a result of environmental interactions with other brain structures.

Therefore, the commitment to become a member of a compartment appears to be an early event in the striatum. However, in order to determine whether commitment to a compartment takes place within proliferating cell lineages, we must look for evidence of heterogeneity within the proliferative region itself.

c. Retroviral clonal analysis suggests separate compartmental precursors. Evidence for specialization within the ventricular zone has been hampered by the fact that most markers either are not expressed in the ventricular zone (eg. neurotransmitters, receptors) or do so in a homogeneous fashion (D1.1 [Levine et al.,1984], Rat 401 [Frederikson and McKay, 1988]). Although heterogeneity in the ventricular zone has been found which differentiates between neuronal and glial precursors (Levitt et al.,1981) and between radial glial populations (Johnston and van der Kooy, 1989), there is as of yet no marker that differentiates between neuronal precursors. Ventricular commitment to a postmitotic neuronal phenotype, however, can be inferred through the permanent labeling of precursors in lineage studies which utilize replication deficient retroviral vectors, and any subsequent restriction of their progeny (Price and Thurlow, 1988; Walsh and Cepko, 1988; Lushkin et al, 1988; Krushel et al.,1989). Retroviral lineage studies provide some of the strongest support for the existence of ventricular zone events that determine the forebrain compartmental identity of proliferating neuronal populations. Clonally related neurons in the striatum fall into three categories. Those restricted to either the patch or matrix compartment and a small population which span the boundary between the patch and matrix compartments (Krushel et al., 1989). Similarly, cortical clones are either restricted to deep or superficial laminae or are distributed throughout all layers of the cortex (Price and Thurlow, 1988; Krushel et al.,1989). Quantitative evidence (Krushel et al., 1989) suggests that there are at least two types of neuronal precursors within the proliferating ventricular zone: 1) a population committed to producing early born neurons (striatal patch or deep cortical layers), 2) a population committed to producing both early and late born neurons (striatal patch or matrix and deep or superficial cortical layers).

d. Chimeric animals suggest separate precursor populations for early (patch or deep layers) versus late (matrix or superficial layers) born neurons. Support for the existence of separate precursor populations for different forebrain compartments also comes from the study of genotype ratios in chimeric mice (Fishell et al.,1989, Herrup and Crandall, 1989). These studies show that different genotype ratios exist for early versus late born cells in both the cortex (Fishell et al.,1989; Herrup and Crandall, 1989) and striatum (Fishell et al.,1989). It is surprising then, that the cortex and striatum are not more structurally similar, since their respective compartments (deep cortical layers and striatal patch versus superficial cortical layers and striatal matrix) seem to share similar birthdates (Angevine and Sidman, 1961; van der Kooy and Fishell, 1987), adhesiveness (Krushel et al.,1989) and precursor populations (Fishell et al.,1989; Herrup and Crandall, 1989). Their differences in structure (laminae versus patches) may reflect the mechanisms of migration to their positions in the two regions. A possible explanation is that radial glial cells, which are thought to be important in guiding neuronal migration (Rakic, 1972), may not have as strong of an influence in pattern formation in the striatum, as compared to the cortex (Johnston et al.,1989).

2. Postmitotic cell interactions pattern groups of compartmentally committed cells

a. Intraocular transplants maintain the compartmental commitment of single striatal cells but not the exact positioning of patch cells into distinct patches. Although cells appear to be committed to compartmental fate while still proliferating, the exact positioning of neurons into their adult compartmental pattern appears to be a function of the cells' interactions with the environment. This is evident in the intraocular transplants (Johnston et al., 1987) in which compartmental phenotype was maintained in the ratio expected from time of removal of the graft. However, instead of the normal striatal compartmental pattern of small patches embedded within a matrix background, the transplants consisted of one large patch and one large matrix. Co-grafting experiments (eg. striatum and substantia nigra) would test the influence of connections with distal targets (eg. afferents and efferents) on the formation of the adult multicellular compartmental organization.

b. Early born cells are selectively adhesive. The self-aggregation of striatal patch cells in the intraocular transplants into one large patch suggests that these cells may be adhesive to one another. Indeed, *in vitro* reaggregation studies (Krushel et al.,1989) suggest that patch cells (and not striatal matrix cells) are selectively self-adhesive. Furthermore, co-reaggregation studies of striatum with cortex suggest that it is the early born cells that are selectively adhesive, since cells destined to become members of striatal patches or cortical deep lamina(e) will reaggregate together in these cultures (Krushel et al., 1988). The preferential adhesiveness of cells of similar birthdates (ie. deep cortical cells and striatal patch cells) over cells of similar tissue type (ie. deep and superficial cortical cells) further supports the idea that compartmental commitments may be similar across structures throughout the telencephalon. The selective adhesion of early born striatal patch cells would ensure the maintenance of striatal patch aggregates in the face of the massive influx of later born matrix cells. Indeed, selective cell adhesion has been suggested to be a basis for pattern formation throughout the central nervous system (Edelman, 1987).

c. Early lesions of the afferent and efferent connections of the striatum may test their role in pattern formation. Selective adhesion of early born cells does not, however, explain why the patch compartment is finally composed of small islands of cells distributed in a sea of matrix cells. The separation of the patch compartment into small groups of cells (the number of groups or patches is maintained throughout mammals (Johnston et al., 1990) is hypothesized to occur through the action of bundles of striatal efferents and/or afferents. The earliest fibers to innervate the striatum contain dopamine (Voorn et al., 1988), and the earliest striatal efferents project prenatally to the globus pallidus and substantia nigra (Fishell and van der Kooy, 1987). Furthermore, the formation of early striatal efferent projections appears important in cells selectively surviving the cell death period (Fishell and van der Kooy, 1990). We suggest that these striatal afferents or efferents might possess positional information that allows committed patch cells to form distinct multicellular patches. The influence of the connections of the striatum has been tested by intraocular transplantation (Johnston et al., 1987), knife cut lesions of the early postnatal or adult nigrostriatal pathway (Lanca et al., 1986), early postnatal decortication (Lanca et al., 1986) and embryonic day 19 lesions of the substantia nigra (Fishell and van der Kooy, unpublished observations). Severing the connections with the cortex in the neonate or with the substantia nigra in the adult had no effect on the striatal patterning of compartments (Lanca et al., 1986). However, destroying the connection with the substantia nigra, either embryonically or early postnatally, decreased the level of expression of phenotypic patch markers (ie. opiate receptors)(Lanca et al., 1986; Johnston et al., 1987; Fishell and van der Kooy, unpublished observations). These results suggest three possible neonatal roles for striatal afferents or efferents in striatal pattern formation: 1) maintaining the survival of striatal patch cells, 2) positioning patch aggregates in specific striatal locations, and/or 3) allowing full phenotypic expression of adult markers such as neurotransmitter receptors. By labeling patch neurons prior to the lesions, it should be possible to discriminate among these suggestions.

Conclusions

Commitment of single cells to a striatal compartment occurs within the proliferating ventricular zone cell lineages. Precursors in the ventricular zone are divided into populations that are committed to producing either the early born neurons of the striatum (patch) and cortex (deep layers) or the late born neurons of the striatum (matrix) and cortex (superficial layers)(Fishell et al., 1989; Herrup and Crandall, 1989; Krushel et al., 1989). These compartmentally committed cells may be positioned into the final adult location and induced to full phenotypic expression by interactions with neurons in distant brain areas (Johnston et al., 1987).

References

Angevine, J.B. and R. L. Sidman, Autoradiographic study of cell migration during histogenesis of cerebral cortex in mice (1961) Nature **192**: 766-768.

Drager, U.C., Observations on the organization of the visual cortex in the reeler mouse (1981) J. Comp. Neurol. **201**: 555-570.

Edelman, G.M., Adult maps: stabilized competition with fixed circuitry (1987) in Neural Darwinism (Basic Books Inc., New York) pp.126-138.

Fishell, G. and D. van der Kooy, Neurons with early projections to the substantia nigra survive the cell death period (1990) J. Neurosci., in press.

Fishell, G. and D. van der Kooy, Pattern formation in the striatum: Developmental changes in the distribution of striatonigral neurons (1987) J. Neurosci. **7**: 1969-1978.

Fishell, G., Rossant, J. and D. van der Kooy, Neuronal lineages in chimeric mouse forebrain are segregated between compartments and in the rostrocaudal and radial planes (1989) Develop. Biol., in press.

Gerfen, C.R., The neostriatal mosaic: compartmentalization of corticostriatal input and striatonigral input and output systems (1984) Nature **311**: 461-464.

Goldman-Rakic, P.S., Prenatal formation of cortical input and development of cytoarchitectonic compartmentalization of the monkey (1981) J. Neurosci. 1:721-735.

Graybiel, A.M., C.W. Ragsdale Jr., E.S. Yoneka and R.P. Elde, An immunohistochemical study of enkephalins and other neuropeptides in the striatum of the cat with evidence that the opiate peptides are arranged to form mosaic patterns in register with the striosomal compartments visible by acetylcholinesterase staining (1981) Neuroscience 6: 377-397.

Haber, S.N., Transmitter systems of the human and nonhuman primate basal ganglia (1986) Human Neurobiology 5: 159-168.

Herkenham, M. and C.B. Pert, Mosaic distribution of opiate receptors, parafasicular projections and acetylcholinesterase in the rat striatum (1981) Nature 291: 415-418.

Herrup, K. and J.E. Crandall, Cell lineage in the mouse somatosensory cortex: analysis of chimeras (1989) Neurosci Abstr., 15: 599

Jensen, K.F. and H.P. Killackey, Subcortical projections from ectopic neocortical neurons (1984) Proc. Natl. Acad. Sci. USA 81: 964-968.

Johnston, J.G., Boyd, S.R. and D. van der Kooy, Compartmentalization of the embryonic striatum after intraocular transplantation (1987) Dev. Brain Res. 33: 310-314.

Johnston, J.G. and D. van der Kooy, Protooncogene expression identifies a transient columnar organization of the forebrain within the late embryonic ventricular zone (1989) Proc. Natl. Acad. Sci. USA 86: 1066-1070.

Johnston, J.G., Krushel, L.A. and D. van der Kooy, Maintenance of striatal compartments in the reeler mouse suggests pattern formation independent of glial guidance (1989) Neurosci. Abstr., 15: 101.

Johnston, J.G., S. Haber, C.R. Gerfen and D. van der Kooy, Mechanisms of striatal pattern formation: Conservation of mammalian compartmentalization (1990), submitted.

Kemp, J.M. and T.P.S. Powell, The structure of the caudate nucleus of the cat: light and electron microscopy (1971) Phil. Trans B. 262: 383-401.

Krushel, L.A., J. Connolly, and D. van der Kooy, Pattern formation in the mammalian forebrain: patch neurons from the rat striatum selectively reassociate in vitro (1989) Dev. Brain Res. 47: 137-142.

Krushel, L.A. and D. van der Kooy, Birthdate is more important than tissue type for the in vitro reassociation of early postmitotic cortical and striatal neurons (1988) Neurosci. Abstr., 14: 91.

Krushel, L.A., Johnston, J.G., Fishell, G. and D. van der Kooy, Retroviral vectors reveal compartmental specific cell lineages in the developing rat forebrain (1989) Neurosci Abstr., 15: 598.

Lanca, A.J., S. Boyd, B.E. Kolb and D. van der Kooy, The development of a patchy organization of the rat striatum (1986) Dev. Brain Res. 27: 1-10.

Lemmon, V. and A.L. Pearlman, Does laminar position determine field properties of cortical neurons? A study of corticotectal cells in area 17 of the normal mouse and the reeler mutant (1981) J. Neurosci. 1: 83-93.

Levine, J.M., Beasley, L. and W.B. Stallcup, The D1.1 antigen: a cell surface marker for germinal cells of the central nervous system (1984) J. Neurosci. 4: 820-831.

Levitt, P, Cooper, M.L. and P. Rakic, Coexistence of neuronal and glial precursors cells in the cerebral ventricular zone of the fetal monkey: an ultrastructural immunoperoxidase analysis (1981) J. Neurosci. 1: 27-39.

Lushkin, M.B., Pearlman, A.L. and J. R. Sanes, Cell lineage in the cerebral cortex of the mouse studied in vivo and in vitro with a recombinant retrovirus (1988) <u>Neuron</u> 1: 635-647.

McConnell, S.K., Fates of visual cortical neurons in the ferret after isochronic and heterochronic transplantation (1988) <u>J. Neurosci.</u> 8: 945-974.

McConnell, S. K., The determination of cell fate in the cerebral cortex (1989) <u>Trends Neurosci.</u> 12: 342-349.

Price, J. and L. Thurlow, Cell lineage in the rat cerebral cortex: a study using retroviral-mediated gene transfer (1988) <u>Development</u> 104: 473-482.

Rakic, P., Mode of cell migration to the superficial layers of the fetal monkey neocortex (1972) <u>J. Comp. Neurol.</u> 145: 61-84.

Reiner, A., S.E. Brauth, C.A. Kitt and R. Quiron, Distribution of mu, delta, and kappa opiate receptor types in the forebrain and midbrain of pigeon (1989) <u>J. Comp. Neurol.</u> 280: 359-382.

Simons, D.J., Durham, D. and T.A. Woosley, Functional organization of mouse and rat SmI barrel cortex following vibrissal damage on different postnatal days (1984) <u>Somatosens. Res.</u> 1:207-245.

Stanfield, B.B. and D. D.M. O'Leary, Fetal occipital cortical neurons transplants to the rostral cortex can extend and maintain a pyramidal tract axon (1985) <u>Nature</u> (London) 313: 135-137.

van der Kooy, D. and G. Fishell, Neuronal birthdate underlies the development of striatal compartments (1987) <u>Brain Res.</u> 401: 155-161.

van der Kooy, D., G. Fishell, L.A. Krushel, and J. Johnston, The development of striatal compartments: from proliferation to patches (1987) <u>In</u> M.B. Carpenter, and A. Jayaraman (eds) The Basal Ganglia II, Plenum Publishing Corp., New York, pp. 81-98.

Voorn, P., A. Kalsbeek, B. Jorrisam-Byham and H.J. Groenewegen, The pre- and postnatal development of the dopamine cell groups in the ventral mesencephalon and dopaminergic innervation of the striatum of the rat (1988) <u>J. Neurosci.</u> 25: 857-888.

Walsh C. and C.L. Cepko, Clonally related cortical cells show several migration patterns (1988) <u>Science</u> 241: 1342-1345.

CORRELATED LIGHT AND ELECTRON MICROSCOPY OF GOLGI-IMPREGNATED NEOSTRIATAL NEURONS AFTER 6-HYDROXYDOPAMINE LESIONS IN THE RAT

C.A. Ingham, S.H. Hood and G.W. Arbuthnott

Department of Preclinical Veterinary Sciences
R(D)SVS, University of Edinburgh, Summerhall
Edinburgh, EH9 1QH, UK

INTRODUCTION

The removal of the dopamine input to the neostriatum by 6-hydroxydopamine lesions is a useful method of studying possible neuronal responses which may occur in the neostriatum in Parkinson's disease. We have shown that medium size spiny neurons have a lower spine density in the neostriatum which has lost its dopamine input compared with the control unlesioned neostriatum (Ingham et al., 1989). This result persists to the same degree at relatively short and long time periods after the lesion. The present study expands on these results and takes the same material to the electron microscope for ultrastructural evaluation.

METHODS

Anaesthetised, male Han Wistar rats (200–250 g) received injections of 6-hydroxydopamine (6-OHDA; 4 µg in 2 ul saline with 0.4 µg ascorbic acid) into the right medial forebrain bundle (stereotaxic coordinates: 5.2 mm anterior of the interaural line; 1.6 lateral to Bregma and 8.2 mm ventral to the surface of the brain). The injections were made 30 min after intraperitoneal injections of pargyline (50 mg/kg) and desmethylimipramine (25 mg/kg) with a Hamilton syringe attached to a 30 gauge needle. Nine days after the lesion the animal's turning behaviour was tested in response to apomorphine (0.25 mg/kg). Those animals turning more than 200 times in a direction contralateral to the lesion, during a 45 min testing period were retained for future use and could be expected to have a 90% reduction of dopamine in the lesioned neostriatum (Hefti,1980). At short (12–26 days) or long (7–13 months) times after the lesion (see Table 1) the rats were anaesthetised and perfused through the aorta with either calcium free Tyrode's solution or heparinized saline for 1-2 min followed by either:-

1.   0.5-4% paraformaldehyde and 2.5% glutaraldehyde in 0.1 M sodium phosphate buffer, pH 7.4 (PB) for 30 min at a rate of about 13 mls/min or
2.   4% paraformaldehyde, 0.1% glutaraldehyde in PB for 10 min, followed by 3% paraformaldehyde in PB for 5 min at a rate of 20 mls/min.

    Rats perfused according to protocol 2. were also used for immunohistochemical experiments. Two old (equivalent to 10 months post lesion) control animals were similarly prepared although these rats were unoperated. Sections (70 μm) were cut on a Vibrating microtome (Vibratome) into PB and after thorough washing were treated with 1% osmium tetroxide in PB for 40 min. The section Golgi-impregnation procedure of Izzo et al. (1987) was then followed. Briefly, sections were immersed in 3.5% potassium dichromate immediately after osmium treatment for 1-3 h, they were then placed between glass slides, taped together at one end and placed vertically into a 1% silver nitrate solution for 6-18 h. Sections containing Golgi-impregnated neurons were illuminated, gold-toned, dehydrated (including 1% uranyl acetate in 70% ethanol for 40 min) and mounted on light microscope slides in resin (Durcupan, ACM, Fluka) for light microscopical evaluation (Bolam and Ingham, 1990). The resin was polymerized for 48 h at 60° and the slides were coded so that subsequent analysis could be carried out blind. Twenty five well impregnated medium size spiny neurons were chosen from both the control and lesioned neostriata of all the animals; the position of each neuron was noted as was the order of each dendrite studied. An estimate of spine density was made as described previously (Ingham et al., 1989). In brief, spines were counted over a 28 μm length of dendrite 14 μm from the start of the spines at 1000X magnification. The mean spine density of each sample of 25 was calculated and statistical analysis was carried out on these means. One neuron from each side of each animal was selected for further study. These neurons were drawn using a light tube attached to a microscope at 1000x magnification. Those portions of some of these neurons that were used for the spine density estimate were rembedded in resin and ultrathin serial sections through them were examined in the electron microscope after staining with lead citrate (Reynolds, 1963).

RESULTS

Light Microscopy

    Most of the Golgi-impregnated neurons were of the medium size spiny type although non-spiny neurons were also infrequently observed. Mean spine density estimates are shown on Table 1. A two way analysis of variance carried out on the mean spine density estimates revealed that spine density was significantly lower in the lesioned neostriata of both groups of rats regardless of length of time after the lesion (F-ratio=13.45, p>0.002). Similarly, longterm animals had a significantly lower spine density than short term animals (F-ratio=45.53, p>>0.0001. However, there was no interaction between the effect caused by the lesion and the effect caused by length of time after the lesion (F-ratio=1.27, p>0.81).

Table 1. Showing mean spine density/28 μm dendrite, for individual rats which rotated in response to apomorphine 9 days after the 6-OHDA lesion (1-10), and for control rats (11-12) which did not receive a lesion.

Rat	Number of turns contralateral to lesion	Survival time, post-lesion	Mean count/28 μm (n=25) ± SD	
			Control	Lesioned
1	237	12 days	53.12 ± 8.7	53.60 ± 10.2
2	457	12 days	59.80 ± 9.0	52.36 ± 8.6
3	400	19 days	56.64 ± 8.4	45.60 ± 9.0
4	222	26 days	57.56 ± 10.1	45.68 ± 15.5
5	461	26 days	63.84 ± 10.5	54.88 ± 11.8
6	255	7 months	45.92 ± 9.2	43.52 ± 11.9
7	325	7.5 months	48.48 ± 14.4	37.52 ± 11.3
8	440	8 months	40.96 ± 8.7	42.00 ± 10.8
9	367	12 months	40.96 ± 9.8	29.28 ± 8.7
10	321	13 months	45.28 ± 14.0	35.48 ± 8.4
11	control	10 months	45.64 ± 14.7	42.60 ± 9.4
12	control	10 months	40.80 ± 9.7	41.12 ± 6.9

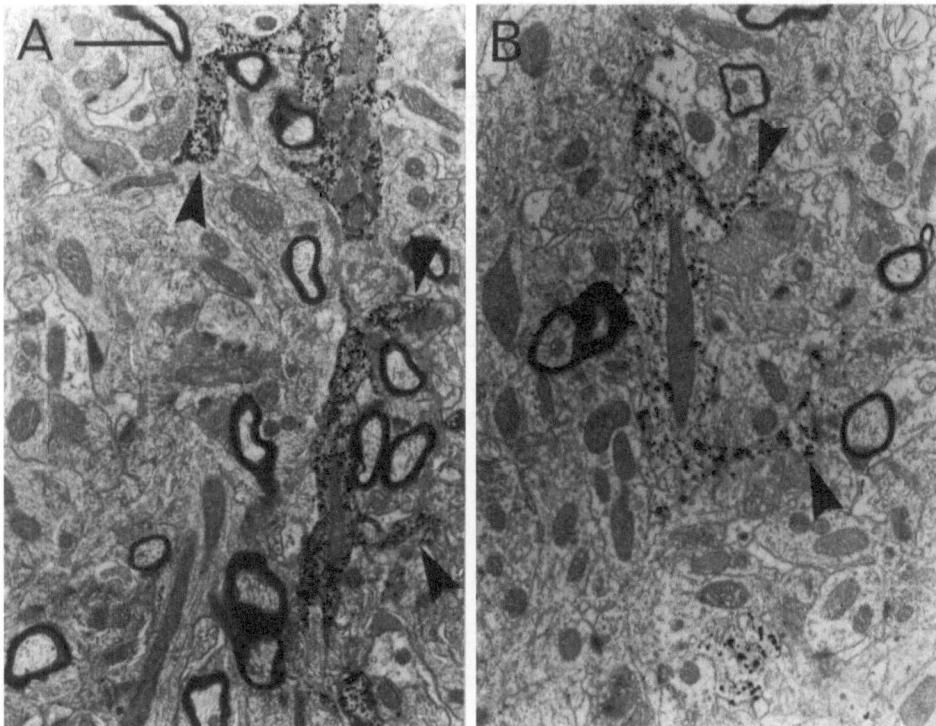

Fig. 1. Electron micrographs of Golgi-impregnated dendrites of medium size spiny neurons from the control (A) and lesioned (B) neostriatum. Arrowheads show spines emerging from the dendritic shaft. Scales: A,B, 1 μm.

Electron Microscopy

The general features of the ultrastructure were similar throughout all the animals although on both the control and lesioned sides of older rats features of aging were observed.

These included a large number of lysosomes in the neuronal cytoplasm and the occasional degenerating process. In the lesioned neostriatum of 12 day lesioned rats, small degenerating processes were occasionally observed although most of these were engulfed in glia and their structure could not be determined.

The Golgi-impregnated neurons that were examined in the electron microscope were recognised by correlation from light micrographs and by the presence of electron dense particles of the gold-toned Golgi precipitate (Figure 1,2). The precipitate was scattered throughout the cytoplasm of the dendrites and spines. Spines which were free of the Golgi-precipitate were never seen emerging from these identified neurons. Synaptic specializations were found along the lengths of dendrite studied and consisted of four different types. The most commonly observed synaptic interaction was that formed by presynaptic boutons which made asymmetric contact with the heads of spines (Figure 2D). These synapses typically stretched over 5-8 serial sections (aprox. 0.5 μm). Symmetrical synapses were also observed onto the necks of spines and onto dendrites. Those onto the necks of spines were always small and usually could only be observed in one or two consecutive sections (Figure 2C). Symmetrical synapses onto dendritic shafts were usually observed on 2-5 consecutive sections. Asymmetrical synaptic contacts were occasionally observed onto dendritic shafts. In the 5-10 μm lengths of dendrite observed in serial sections from the different animals, no obvious differences in synaptic arrangements were observed.

Fig. 2. Light and electron micrographs of a Golgi-impregnated medium size spiny neuron from the neostriatum of a control short term animal. A. Light micrograph of the dendrite used for spine density estimation. Arrowheads point to spines which are shown in subsequent electron micrographs. B. Electron micrograph of part of the same dendrite shown in A. where spines 1. and 2. are clearly visible. C. Electron micrograph of part of spine 3. shown in A. where a presynaptic bouton is making symmetrical synaptic contact (asterisks) with the neck of the spine. D. Electron micrograph of spine 2 shown in A. and B. where a presynaptic bouton makes asymmetrical contact (asterisks) with the head of the spine.

Scales: A, 10 μm; B, 1 μm; C,D, 0.2 μm.

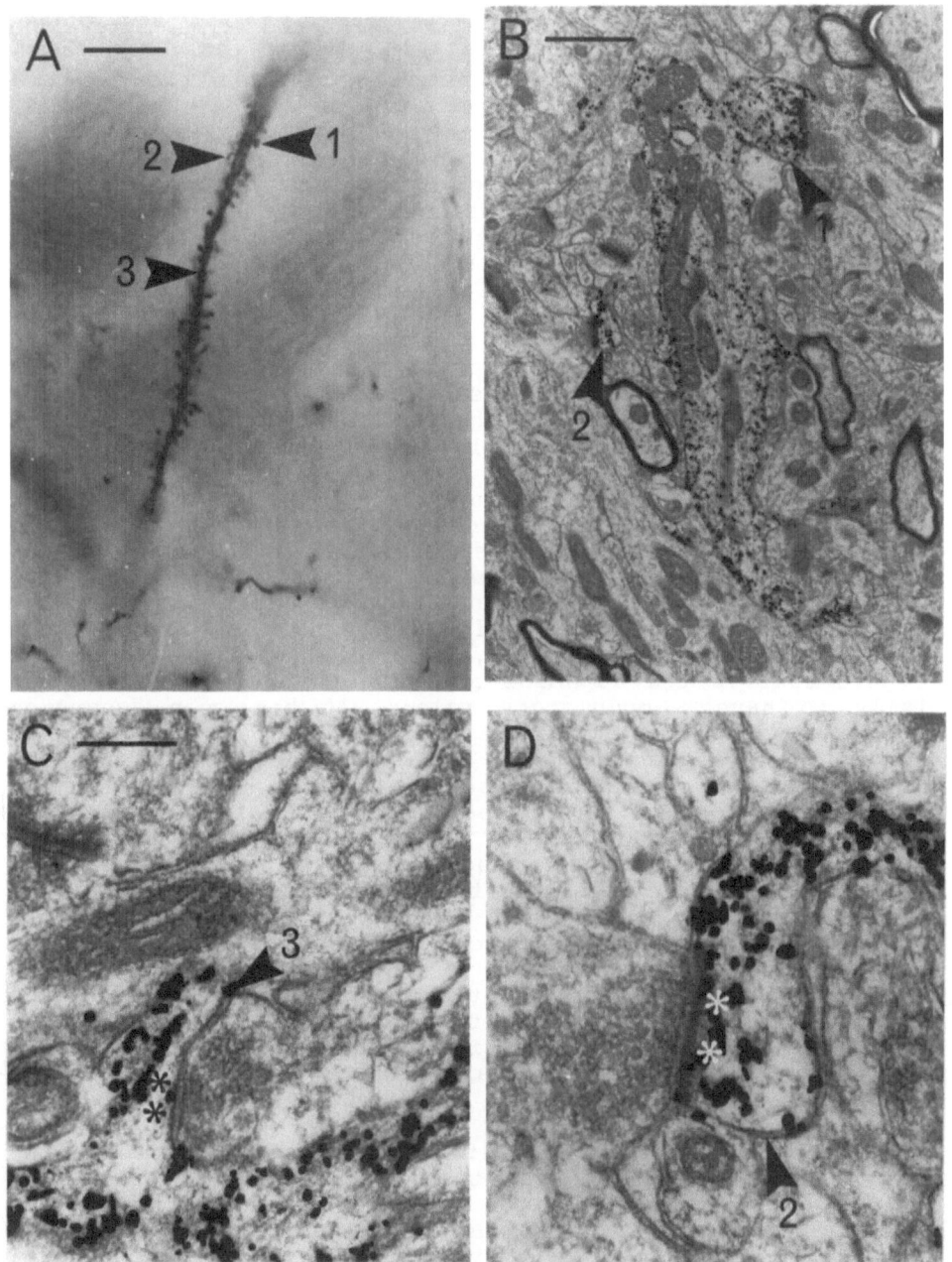

Fig. 2.

DISCUSSION

This study expands on previous results and confirms that spine density is reduced on medium size spiny neurons when the dopamine input is removed (Ingham et al., 1989). It also confirms the loss of spines with increased time after the lesion which is probably due to aging (Levine et al., 1986; Rafols et al., 1989). The neostriata from two control animals showed similar spine densities on both sides. Insufficient numbers of control animals means that statistical analysis has not yet been carried out, but the spine density from these control neostriata is similar to the control neostriata from the lesioned animals of the same age group. Thus it seems likely that the lesion has changed spine density only on the operated side of the brain and any compensation occurring contralateral to the lesion does not involve loss of spines.

The ultrastructural part of this investigation confirmed the fact that spine density loss was due to an actual loss of spines and not simply due to the failure of some spines to be filled with the Golgi deposit in the experimental conditions.

Manipulation of afferent inputs has been shown to affect the spine density previously (Kemp and Powell, 1971). In the cat caudate, lesions in the cortex or thalamus resulted in lower spine density counts on medium spiny neurons. These inputs are presumed to be excitatory and form asymmetrical contacts with spines or dendritic shafts (Kemp and Powell, 1971; Dube et al., 1988). In the present study we have removed the dopamine afferent pathway which is more likely to have a modulatory role. It has been suggested (Freund et al., 1984) that the dopamine terminals make small symmetrical synaptic contacts with the necks of spines but that the latter always also receive a more prominent asymmetrical contact. Our results suggest that spines may disappear simply by removing the small but possibly vital symmetrical dopamine containing input. Indeed Hattori and Fibiger (1982) show ultrastructural evidence for degenerating spines and their presynaptic boutons within glial elements two or three days after either a cortical lesion or a large intraventricular injection of 6-OHDA. The question of the fate of afferent synapses onto the heads of spines which disappear is then raised. Whether they themselves degenerate or whether they find alternative postsynaptic targets needs detailed electron microscopical analysis before coming to a conclusion.

Electrophysiological studies have shown an increase in spontaneous activiy of neostriatal neurons after 6-OHDA lesions which is apparent from 1-3 weeks after the lesion but which subsequently disappears (Schultz, 1982). Any compensatory mechanism underlying this electrophysiological change probably does not involve regrowth of spines for this would lead to a smaller side difference in spine density in the long term animals which was clearly not evident in our study. It is more likely that biochemical changes such as D2 receptor supersensitivity (Ungerstedt, 1971) or changes in neurotransmitter levels and turnover underly this mechanism (Kerkerian et al., 1988; Lindefors et al., 1989; Voorn et al., 1987; Young et al., 1986).

26

ACKNOWLEDGEMENTS

The authors would like to thank Drs Paul Bolam, David Maxwell
and Professor Alan Brown for helpful discussions throughout
the course of this study. This work was supported by the
MRC.

REFERENCES

Bolam J.P. and Ingham C.A., 1990. Combined morphological and
histochemical techniques for the study of neuronal
microcircuits. "Handbook of Chemical Neuroanatomy,
Volume 8, Neuronal Microcircuits - Combined
morphological, immunocytochemical and
electrophysiological techniques for the study of
synaptic interactions between identified CNS neurons."
Van den Pol A. and Wouterlood F. eds, Elsevier,
Amsterdam.

Dube L., Smith A.D., and Bolam J.P., 1988. Identification of
synaptic terminals of thalamic or cortial origin in
contact with distinct medium size spiny neurons in the
rat neostriatum. J. Comp. Neurol. 267; 455-471.

Freund T.F., Powell J.F., and Smith A.D., 1984. Tyrosine
hydroxlase-immunoreactive boutons in synaptic contact
with identified striatonigral neurons, with particular
reference to dendritic spines. Neuroscience 13; 1198-
1215.

Hattori T. and Fibiger H.C., 1982. On the use of lesions of
afferents to localize neurotransmitter receptor sites in
the striatum. Brain Res. 238; 245-250.

Hefti F., Melamed E. and Wurtum R.J., 1980. Partial lesions
of the dopaminergic nigrostriatal system in rat brain:
biochemical characterization. Brain Res. 195; 123-137.

Ingham C.A., Hood S.H. and Arbuthnott G.W., 1989. Spine
density on neostriatal neurons changes with 6-
hydroxydopamine lesions and with age. Brain Res. 503;
334-338.

Izzo P.N., Graybiel A.M. and Bolam J.P., 1977.
Characterization of substance P- and [met]enkphalin-
immunoreactive neurons in the caudate nucleus of cat and
ferret by single section Golgi procedure. Neuroscience
20; 577-587.

Kemp J.M. and Powell T.P.S., 1971. The termination of fibres
from the cerebal cortex and thalamus upon dendritic
spines in the caudate nucleus: a study with the Golgi
method. Phil. Trans. R. Soc. Lond. B 262; 429-439.

Kerkerian L., Salin P. and Nieoullon A., 1988.
Pharamcological characterization of dopaminergic
influence on expression of neuropeptide Y
immunoreactivity by rat striatal neurons. Neuroscience
26; 809-817.

Levine M.S., Adinolfi A.M., Fisher R.S., Hull C.D., Buchwald N.A. and McAllister J.P., 1986. Quantative morphology of medium-size caudate spiny neurons in aged cats. Neurobiol. of Ageing 7; 277-286.

Lindefors N., Brodin E., Tossman U., Segovia J. and Ungerstedt U., 1989. Tissue levels and in vivo release of tachykinins and GABA in striatum and substantia nigra of rat brain after unilateral striatal dopamine denervation. Exp. Brain Res. 74; 527-534.

Rafols J.A., Wei Chang H., and McNeill T.H., 1989. Golgi study of the mouse striatum: age-related dendritic changes in different neuronal populations.J. comp. Neurol. 279; 212-227.

Reynolds E.S., 1963. The use of lead citrate at high pH as an electron opaque stain in electron microscopy. J. Cell. Biol. 17; 208-212.

Schultz W., 1982. Depletion of dopamine in the striatum as an experimental model of parkinsonism: direct effects and adaptive mechanisms. Prog. Neurobiol. 18; 121-166.

Ungerstedt U., 1971. Postsynaptic supersenstivity after 6-hydroxydopamine induced degeneration of the nigrostriatal dopamine system. Acta physiol. scand. Suppl. 367; 69-93.

Voorn P., Roest G. and Groenewegen H.J., 1987. Increase of enkephalin and decrease of substance P-immunoreactivity in the dorsal and ventral striatum of the rat after midbrain 6-hydroxydopamine lesions. Brain Res. 412; 391-396.

Young W.S. III, Bonner T.I. and Brann M.R., 1986. Mesencephalic dopamine neurons regulate expression of neuropeptide mRNAs in the rat forebrain. Proc. Natn. Acad. Sci. U.S.A. 83; 9827-9831.

COEXPRESSION OF NEUROPEPTIDES AND GLUTAMIC ACID DECARBOXYLASE IN CAT STRIATAL NEURONS: DEPENDENCE STRIOSOMAL COMPARTMENTATION

Marie-Jo Besson*, Ann M. Graybiel**, and Bruce Quinn**

*Laboratoire de Neurochimie-Anatomie
IDN, CNRS-Universite P. et M. Curie, Paris, France

**Laboratory of Neuroanatomy
Department of Brain & Cognitive Sciences
M.I.T., Cambridge, Mass. USA

INTRODUCTION

In the last two decades, there has been a sharp increase in information about the input-output systems that link the striatum with other parts of the basal ganglia and their allied nuclei as well as with the cerebral cortex and thalamus. During the same time, many of the newly-discovered neuropeptide and amino acid neuromodulators have been found in the caudate-putamen complex, joining the "classical" neurotransmitters of this region: dopamine, acetylcholine, and gamma-aminobutyric acid (GABA).

An organizing principle for studying these extrapyramidal pathways and their neuromodulatory systems has emerged through investigation of the compartmentation of the striatum. The striosome-matrix compartmentalization of the striatum was initially suggested on the basis of the heterogeneous distribution of acetylcholinesterase staining (Graybiel & Ragsdale, 1978), and the two main compartments (striosomes and matrix) defined by this histochemistry have been found to correlate strongly with the distributions of a very wide array of other neuronal markers, including the opioid receptor patches delineated by Pert and colleagues (1975), and the islandic dopamine afferent organization first detected in the developing brain (Olson et al., 1972; Tennyson et al., 1972). In immunohistochemical studies, fibers containing neuropeptides display prominently heterogenous, patchy distributions, which for a given neuropeptide may be selectively more intense in either the striosome or matrix compartments. Similarly, striosome-selective and matrix-selective differential distributions are (depending on the neuropeptide) characteristic of neuronal perikarya

The Basal Ganglia III, Edited by G. Bernardi *et al.*
Plenum Press, New York

immunoreactive for the neuropeptides expressed in striatal projection neurons and interneurons. Binding sites labelled by a wide variety of specific radiolabeled receptor ligands also show high and low densities that differ in the striosomes and the matrix. This neurochemical compartmentalization is parallelled by striosome-matrix ordering observed by retrograde-tracer labeled perikarya of striatal efferent pathways, and by anterograde tracer-labeled fibers of striatal afferent pathways. The cumulative weight of many different studies has thus firmly established striosomal compartmentalization as a fundamental organizing plan imposing neurochemical specialization on the output and input connectivities of the basal ganglia and thus on the interactions of the basal ganglia with the thalamus, cerebral cortex, and limbic system.

In this chapter we summarize new immunohistochemical studies (Besson et al., 1990) focused on striatal projection neurons that express three neuropeptides, dynorphin B, substance P, and enkephalin (DYNB, SP, and ENK). We analyzed, in the cat's caudate nucleus, the differential patterns of coexpression of these neuropeptides in single striatal neurons identified by their location in either the striosomal or matrix compartment. We further examined the coexpression of immunoreactivity for glutamic acid decarboxylase (GAD, a marker for the inhibitory neurotransmitter GABA) in the neuropeptide-positive perikarya we identified.

From earlier work, we knew that there are marked differences in the compartmental expression of both SP-like and DYNB-like immunoreactivity in the cat's striatum, particularly in the caudate nucleus. Neurons with high SP-like immunoreactivity and neurons with high DYNB-like immunoreactivity appear clustered in striosomes and are scattered at much lower densities in the matrix (Beckstead, 1978; Chesselet & Graybiel, 1983; Graybiel & Chesselet, 1984; Penny et al., 1988). By contrast, neurons with strong ENK-like immunoreactivity are concentrated in the matrix (Graybiel & Chesselet, 1984). This compartmental distribution of peptide-containing neurons can be strikingly altered by pretreatment with colchicine, a drug that interferes with axonal transport processes and can thus be administered intracerebroventricularly in order to reveal perikarya that were formerly immunonegative or only very weakly immunopositive for the neuropeptide of interest. In the caudate nucleus of the colchicine-treated cats, SP- and DYNB-immunoreactive neurons are still clustered in striosomes, but many also appear in the matrix, yielding a less strongly differentiated compartmentalization of immunopositive neurons. In the same animals, many more ENK-immunoreactive neurons appear in striosomes than in untreated cats (Graybiel & Chesselet, 1984).

The similar compartmental distribution of SP- and DYNB-immunoreactive neurons in untreated cats, and the similar compartmental blurring for both types of immunoreactive neuron following colchicine treatment, suggested the possibility that these two peptides could coexist in striatal neurons, as already reported for non-mammalian striatal neurons in turtles and pigeons (Reiner, 1986). Such colocalization, in mammalian species, would help to account for the similarity of immunostaining for SP- and DYNB-positive fibers in known target regions of the medium-sized striatal neurons (the internal pallidum and the pars compacta and pars reticulata of the substantia nigra). The presence of many ENK-positive neurons in both compartments, at least in the colchicine-treated cat, opened the further possibility that individual neurons might express ENK in addition to DYNB or SP. However, the terminal patterns for ENK-immunoreactive fibers largely differ from those of DYNB- and SP-positive fibers in the pallidal targets of the striatum (where ENK is prominent in the external segment) and also in the substantia nigra. Finally, because it had been suggested that medium-sized neurons expressing GAD immunoreactivity were uniformly distributed in the striatum (Ribak et al., 1979; Oertel et al., 1984; Penny et al., 1986) it seemed likely that GAD-containing medium-sized neurons--and hence GABA--might coexist extensively with ENK, DYNB, and SP peptides, regardless of the compartment in which the striatal neurons were located.

The methods we followed for these experiments are described elsewhere (Besson, Graybiel, & Quinn, 1990), but it is important to mention two technical details. First, we carried out our immunohistochemical studies primarily on the head of the caudate nucleus, a region where the striosomal compartmentation of SP, ENK, and DYNB is particularly prominent, and treated the cats with bilateral intraventricular injections of colchicine 2-3 days before perfusion with phosphate-buffered 4% paraformaldehyde. Second, we carried out studies both with the two-fluorochrome fluorescence method for detecting different antigens in the same sections, and with a thin (3 micron) serial-section cryostat method using polyethylene glycol that allows reliable serial colocalization studies of neuronal antigens even in such relatively small neurons as the striatal peptidergic neurons examined here (Quinn & Weber, 1987). Because many but not all labeled perikarya would appear on any given pair of serial sections, we studied triplets of thin sections stained with the same primary antibody on the first and third sections, and made a quantitative analysis of disjunct and colocalized peptide expression with respect to the middle section. We also quantitatively studied sequential thin sections stained with the same antisera, to assess the percentage of colocalization actually achieved on paired serial sections, where the theoretical colocalization would be 100%. Because the medium-sized neurons we studied are relatively homogeneous in size, these measurements provided a 'correction factor' for the proportion of neurons appearing immunopositive on paired 3-4 micron sections. We used a rat monoclonal SP antibody (Seralab), and rabbit polyclonal DYNB and ENK antisera. Immunohistochemical controls indicated specificity among the antisera/antigen combinations used, but we did not test whether the SP antiserum might label other tachykinins, whether the DYNB antiserum might label other dynorphin-family peptides, or whether the ENK antiserum was fully specific for met- rather than leu-enkephalin under the experimental staining conditions.

COLOCALIZATION OF SP-LIKE AND DYNB-LIKE NEUROPEPTIDES IN STRIOSOMAL AND EXTRASTRIOSOMAL NEURONS

Each of the methods we followed indicated a very high level of coexistence of SP and DYNB medium-sized neurons of the caudate nucleus. For the most part, the intensities of SP and DYNB immunostaining in single neurons were similar, but occasionally they were discordant. Of 208 SP-positive neurons studied by the dual fluorochrome method, 93% were double-labeled. Analysis of PAP-stained thin sections in triplets (DYNB-SP-DYNB, or SP-DYNB-SP) yielded a mean DYNB-SP coexistence of 92% in striosomes and 78% in matrix; paired sections (with correction factors) showed that about 96% of DYNB-immunoreactive neurons in striosomes contained SP immunoreactivity, and that 89% of all DYNB-positive neurons in the matrix co-expressed SP immunoreactivity. We also compared the populations of peptide-immunoreactive neurons to all Nissl-stained neurons in each compartment: SP/DYNB positive neurons corresponded to about 60% of all striosomal and 40% of all matrix neurons. From these experiments, we estimate all but about 16% of striosomal and 13% of matrix neurons contain both SP- and DYNB-immunoreactivity. This high level of SP/DYNB colocalization in striatal neurons in the cat is remarkably similar to that found in the turtle and pigeon (Reiner, 1986), which suggests that expression patterns for these opioid and tachykinin peptides are phylogenetically conserved in striatal tissue.

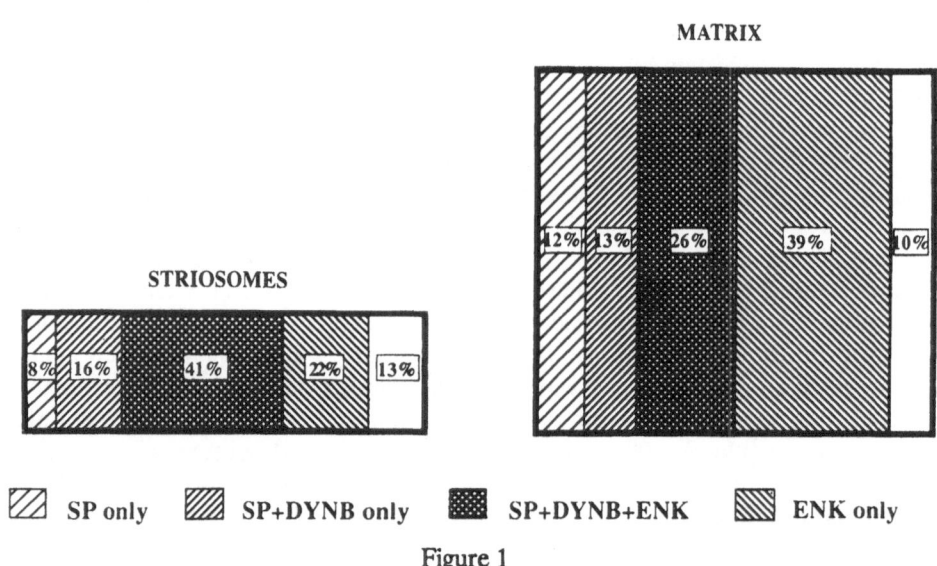

Figure 1

Summary of patterns of peptide colocalization in medium-sized neurons of the cat's striatum.

COEXPRESSION OF ENK-LIKE IMMUNOREACTIVITY IN SP/DYNB-POSITIVE NEURONS

There was a striking range of neuronal immunoreactivity for ENK in the colchicine-treated material, ranging from extremely high to extremely low in both striatal compartments. When all ENK-immunoreactive neurons were counted, they corresponded to about 65% of all Nissl-stained neurons in either compartment. With the PAP method on pairs of serial sections, we found that about 65% of all ENK-positive neurons overlapped with SP- or DYNB- positive neurons in striosomes; the coexistence figure was only about 30% for ENK-positive neurons in the matrix. When triplets of stained sections were examined, similar figures were obtained. Conversely, among SP or DYNB-positive neurons, the corrected estimate of ENK coexpression was about 76% in both compartments. Extrapolation from the number of all Nissl-stained neurons suggested an estimate for triple-peptide coexistence of 41% in striosomes and 26% in matrix, taking into account the very high colocalization of SP/DYN-immunoreactivity described above. Because even weakly ENK-immunostained neurons were counted, these figures for ENK-coexpression might be overestimates. For example, the ENK antiserum might have weakly cross-reacted with leu-ENK cleaved from the prodynorphin precursor itself in DYNB-positive striatal neurons (Zamir et al., 1984). When only strongly ENK-immunolabeled neurons were counted, about 30% and 40% of SP/DYN neurons were labeled in striosomes and matrix, respectively. However, the higher figures may be valid; the large total percentage of ENK-positive neurons we found in both striosomes and matrix of colchicine-treated cats does correspond to the high values obtained for ENK mRNA-positive neurons in rat striatum by in situ hybridization (Gerfen & Young, 1988).

COEXPRESSION OF NEUROPEPTIDES IN GAD-POSITIVE STRIATAL NEURONS

In the thin sections immunostained for ENK and for GAD, striosomal compartments could not be delineated, but we estimated that at least 90% of all ENK-positive neurons counted contained GAD-like immunoreactivity, whereas only 55% of GAD-positive neurons were ENK-immunoreactive. If we refer to the percentage of ENK-positive neurons among all Nissl-stained neurons (obtained on other sections), this suggests that about 60% of medium-sized striatal neurons contain both ENK and GAD (whereas the great majority of all Nissl-stained, medium-sized neurons are GAD-positive.) If, in fact, 90% of ENK-positive neurons contain GAD, and there is a roughly 75% co-occurrence of ENK in SP/DYNB neurons, then GAD must be present in many SP/DYNB/ENK neurons in striosomes. Only a very limited number of GAD-immunostained sections serial to SP-immunostained sections were analyzed. From these, however, we obtained lower estimates (about 60%) for SP-GAD coexistence. Further work on these patterns is needed.

COMMENT

Understanding the expression, and the selective coexpression, of opioid and tachykinin peptides in striosomal and matrix neurons of the striatum must underlie analysis of the striatal output pathways modulating functionally distinct regions of the globus pallidus and substantia nigra. Further, such a detailed cellular analysis is crucial to interpreting the results of studies demonstrating changes in neuropeptide and neuropeptide mRNA levels after pharmacological or other manipulations of striatal afferents, including the nigrostriatal dopaminergic system. Our observations suggest that there are both single-cell level and compartment (striosome/matrix) level constraints on peptide expression in the striatum. These findings point to the promise of a new understanding of the pharmacological specialization of functional subsystems within the basal ganglia.

Striatal projection neurons co-expressing both SP and DYNB comprise about one-half of neurons in the striosomal compartment, and about one-half of all neurons in the extrastriosomal matrix. Neurons expressing DYNB alone were rare in either compartment, but about 7-12% of SP-positive neurons did not stain for DYNB. From an anatomical viewpoint, this cross-compartment coexpression of DYNB and SP is in good accord with the similarity of SP and DYNB terminal staining patterns in the globus pallidus and substantia nigra pars reticulata, both of which receive dense projections from the matrix, and it also is consistent with the parallel distribution of SP- and DYNB-immunoreactive innervations seen in the region of the densocellular zone of the nigral pars compacta, a region that appears to be innervated by striosomal projection neurons (Jimenez-Castellanos & Graybiel, 1989). The additional colocalization of ENK immunoreactivity in many SP/DYNB neurons is more puzzling, however, because there is more disjunction than overlap in the pallidal and nigral terminal fields of ENK-immunoreactive and SP- and DYNB- immunoreactive fibers.

Selective agonists and antagonists for dopamine D1 and D2 receptors have differential effects on the peptides SP, DYNB, and ENK both at the level of peptide expression, determined by radioimmunoassay, and at the level of mRNA transcription (for review see Graybiel, 1990). This is remarkable in view of the widespread colocalizations we have described, and implies that a given drug, acting on D1 or D2 receptors, can have a differential cascade of effects causing suppression or enhancement of the promotor sites of the genes encoding these different neuropeptides. On the other hand, even in striosomes, which have the highest density of SP/DYNB positive neurons, and even after colchicine treatment, approximately 40% of medium-sized neurons remained SP-DYNB immunonegative by the methods used here. It is unknown whether the density of D1 and D2 receptors differs consistently between the SP/DYNB-negative and -positive neurons within striosomes, or whether the dopamine-related terminals and receptors are similar and it is, instead, intracellular differentiation that accounts for the discrepancies in SP/DYNB expression among neurons of the striosomal compartment. The latter explanation of our results would predict that only a subset of medium-sized striatal neurons is capable of expressing each neuropeptide, and, that, within this theoretical fraction, the percentage expressing detectable peptide immunoreactivity varies with the dopaminergic milieu. An interesting model for studying these issues would be in the developing striatum, where dopamine-containing fibers form islands during an extended period of development. Interestingly, in the cat, Baylon et al. (1990) have reported that a subset of striatal neurons begin differentiation to express tachykinin immunoreactivity already during the migratory phase of neuronal development.

Of further interest, the effects of colchicine treatment may be more complex than originally believed. Ceccatelli et al. (1989) have shown that colchicine can induce the expression of the proto-oncogene c-*fos* in neurons, and c-fos has been directly associated with the up-regulation of ENK transcription by its actions on the ENK promotor region (Sonnenberg et al., 1989). Conceivably, the increase in expression of ENK in matrix neurons following colchicine pretreatment could reflect c-*fos*-mediated intracellular events at the transcriptional level, over and above the accumulation of peptide in neuronal perikarya due to the blockage of axonal transport (Graybiel et al., in preparation). Other factors affecting the ultimate detection of immunoreactive peptide under pharmacological manipulations include post-transcriptional variables such as regulation of either the degradation rate of the peptides or the translation rate of cellular mRNA. A first approach to sorting out these variables would be cellular quantitation of both peptide immunoreactivity and mRNA after administration of colchicine or dopamimetic agents.

The very high colocalization of SP and DYNB we have described is also important for interpreting the normal physiological role of these peptides as neuromodulators in the striatopallidal and striatonigral pathways. SP has been reported to increase (Collingridge & Davies, 1982; Pinnock & Dray, 1982) but ENK and DYN to decrease (Collingridge & Davies, 1982; Lavin & Garcia-Muñoz, 1986) the firing rate of neurons in the substantia nigra pars reticulata of the rat. The dopaminergic neurons of the pars compacta have been reported to be relatively insensitive to DYN (Lavin & Garcia-Munoz, 1986), and to show variable activation by SP (Walker et al., 1976; Collingridge & Davies, 1982; Pinnock & Dray, 1982), but half of such pars compacta neurons are efficiently stimulated by substance K (Innis et al., 1985). On the other hand, behavioral studies suggest that SP and DYN, applied to the substantia nigra, produce similar motor behaviors such as contralateral turning, and indicate that such behaviors are potentiated when SP and DYN are simultaneously applied (Tan & Tsou, 1988). The evidence we present for extensive coexpression of these peptides suggests that future studies on differential or concordant release of SP (or SK) and DYN in the substantia nigra, under different physiological or pharmacologic conditions, could shed light on the link between the cellular colocalization of these peptides and the electrophysiological effects of exogenously administered peptides.

Finally, our findings can also be viewed from the perspective of the GABAergic projection systems of the striatum, for nearly all medium spiny projection neurons are thought to be immunoreactive for GAD (Oertel & Mugnaini, 1984; Penney et al., 1986) or to contain GAD mRNA transcripts (Chesselet et al., 1987), as our immunohistochemical findings also suggest. If one takes GABA as the 'primary' or classical neurotransmitter in these projection systems, then our findings point to the presence of at least four potential subclasses of GABAergic neurons, containing ENK, SP, SP/DYN, and all three peptides. GABAergic neurons likely contain other neuropeptides as well, such as leu-enkephalin from the enkephalin precursor, dynorphin A and alpha-neoendorphin from the dynorphin precursor, and substance K from the tachykinin precursor (Horikawa et al., 1983; Kakidani et al., 1982; Nawa et al., 1984). They are known to contain neurotensin: apparently ENK-positive medium cells but not SP-positive medium cells can contain neurotensin as a cotransmitter (Sugimoto and Mizuno, 1987). If co-release of these neuromodulators is as frequent as their anatomical coexistence, then under physiological conditions, neither GABA nor the individual peptides would often act alone at postsynaptic sites of the striatopallidal and striatonigral pathways.

ACKNOWLEDGMENTS

The research summarized in this chapter was supported by a Javits Neuroscience Investigator Award from the National Institutes of Health, the Seaver Institute, and the National Parkinson Foundation.

REFERENCES

Boylan, M.K., Levine, M.S., Buchwald, N.A., and Fisher, R.S., 1990, Patterns of tachykinin expression and localization in developing feline neostriatum, J. Comp. Neurol., 293:151.

Beckstead, R.M., 1987, Striatal substance P cell clusters coincide with the high density terminal zones of the discontinuous nigrostriatal dopaminergic projection system in the cat: a study by combined immunohistochemistry and autoradiographic axon-tracing, Neuroscience, 20:557.

Besson, M.J., Graybiel, A.M., and Quinn, B., 1990, Patterns of coexpression of neuropeptides and glutamic acid decarboxylase in neurons of the striatum: an immunohistochemical study in the cat, Neuroscience, in press.

Ceccatelli, S., Villar, M.J., Goldstein, M., and Hokfelt, T., 1989, Expression of c-fos immunoreactivity in transmitter-characterized neurons after stress, Proc. Natl. Acad. Sci., 86:9569.

Chesselet, M.F., and Graybiel, A.M., 1983, Met-enkephalin-like and dynorphin-like immunoreactivities of the basal ganglia of the cat, Life Sci., 33:37.

Chesselet, M.F., Weiss, L., Wuenschell, C., Tobin, A.J., and Affolter, H.U., 1987, Comparative distribution of mRNAS for glutamic acid decarboxylase, tyrosine hydroxylase and tachykinins in the basal ganglia: an in situ hybridization study in the rodent brain, J. Comp. Neurol., 262:125.

Collingridge, G.L., and Davies, J., 1982, Actions of substance P and opiates in the rat substantia nigra, Neuropharmacology, 21:715.

Gerfen, C.R., and Young, W.S., 1988, Distribution of striatonigral and striatopallidal peptidergic neurons in both patch and matrix compartments: an in situ hybridization histochemistry and fluorescent retrograde tracing study, Brain Res., 460:161.

Graybiel, A.M., and Chesselet, M.F., 1984, Compartmental distribution of striatal cell bodies expressing [Met]enkephalin-like immunoreactivity, Proc. Nat. Acad. Sci., 81:7980.

Graybiel, A.M., and Ragsdale, C.W., 1978, Histochemically distinct compartments in the striatum of human, monkey and cat demonstrated by acetylthiocholinesterase staining. Proc. Nat. Acad. Sci., 75:5723.

Graybiel, A.M., Ragsdale, C.W., Yoneoka, E.S., and Elde, R.P., 1981, An immunohistochemical study of enkephalins and other neuropeptides in the striatum of the cat with evidence that the opiates peptides are arranged to form mosaic patterns in register with the striosomal compartments visible by acetylcholinesterase staining, Neuroscience, 6:377.

Graybiel, A.M., Pickel, V.M., Joh, T.H., Reis, D.J., and Ragsdale, C.W., 1981, Direct demonstration of a correspondence between the dopamine islands and acetylcholinesterase patches in the developing striatum, Proc. Natl. Acad. Sci., 78:5871.

Graybiel, A.M., 1990, Neurotransmitters and neuromodulators in the basal ganglia, Trends in Neurosciences, in press.

Horikawa, S., Takai, T., Toyosato, M., Takahashi, H., Noda, M., Kakidana, H., Kubo, T., Hirose, T., Inayama, S., Hayashida, H., Miyata, T., and Numa, S., 1983, Isolation and structural organization of the human preproenkephalin B gene, Nature 306:611.

Innis, R.B., Andrade, R., and Aghajanian, G.K., 1985, Substance K excite dopaminergic and non-dopaminergic neurons in rat substantia nigra, Brain Res., 335:381.

Izzo, P.N., Graybiel, A.M., and Bolam, J.P., 1987, Characterization of substance P and [Met]enkephalin immunoreactive neurons in the caudate nucleus of cat and ferret by a single section Golgi procedure, Neuroscience 20:577.

Jimenez-Castellanos, J., and Graybiel, A.M., 1989, Compartmental origins of striatal efferent projections in the cat, Neuroscience, 32:297.

Kakidani, H., Furutani, Y., Takahashi, H., Noda, M., Morimoto, Y., Mirose, T., Asai, M., Inayama, S., Nakanishi, S. and Numa, S., 1982, Cloning and sequence analysis of CDNA for porcine a-neo-endorphin/dynorphin precursor, Nature, 298:245.

Lavin, A., and Garcia-Muñoz, M., 1986, Electrophysiological changes in substantia nigra after dynorphin administration, Brain Res. 369:298.

Nawa, H., Kotani, H., and Nakanishi, S., 1984, Tissue-specific generation of two preprotachykinin mRNAS from one gene by alternative RNA splicing, Nature, 312:729.

Oertel, W.H., and Mugnaini, E., 1984, Immunocytochemical studies of GABAergic neurons in rat basal ganglia and their relations to other neuronal systems. Neurosci. Lett., 47:233.

Olson, L., Seiger, A, and Fuxe, K, 1972, Heterogeneity of striatal and limbic dopamine innervation: highly fluorescent islands in developing and adult rat, Brain Res., 44:283.

Penny, G.R., Afsharpour, S., and Kitai, S.T., 1986, The glutamate decarboxylase-, leucine enkephalin-, methionine enkephalin- and substance P-immunoreactive neurons in the neostriatum of the rat and cat: envidence for partial population overlap, Neuroscience, 17:1011.

Penny, G.R., Wilson, C.J., and Kitai, S.T., 1988, Relationship of the axonal and dendritic geometry of spiny projection neurons to the compartmental organization of the neostriatum, J. Comp. Neurol., 269:275.

Pert, C.B., Kuhar, M.J., and Snyder, S., 1975, Autoradiographic localization of the opiate receptor in rat brain, Proc. Natl. Acad. Sci., 72:1849.

Pinnock, R.D., and Dray, A., 1982, Differential sensitivity of presumed dopaminergic neurons in rat substantia nigra to electrophoretically applied substance P, Neurosci. Lett., 29:153.

Quinn, B., and Weber, E., 1987, Serial 2 micron cryostat sections: a modified colocalization technique for immunohistochemistry. Soc. Neurosci. Abst., 13:777.

Reiner, A., 1986, The co-occurrence of substance P-like immunoreactivity and dynorphin-like immunoreactivity in striatopallidal and striatonigral projection neurons in birds and reptiles, Brain Res., 371:155.

Ribak, C.E., Vaughn, J.E., and Roberts, E., 1979, The GABA neurons and their axon terminals in rat corpus striatum as demonstrated by GAD immunocytochemistry. J. Comp. Neurol., 187:281.

Sonnenberg, J.L., Rauscher, R.J., Morgan, J.I., and Curran, T., 1989, Regulation of proenkephalin by Fos and Jun, Science, 246:1622.

Sugimoto, T., and Mizuno, N., 1987, Neurotensin in projection neurons of the striatum and nucleus accumbens with reference to coexistence with enkephalin and GABA: an immunohistochemical study in the cat, J. Comp. Neurol., 257:383.

Tan, D.P., and Tsou, K., 1988, Differential effects of tachykinins injected intranigrally on striatal dopamine metabolism, J. Neurochem., 51:1333.

Tennyson, V.M., Barret, R.E., Cohen, G., Cote, L., Heikkila, R., Mytilneous, C., 1972, The developing neostriatum of the rabbit: correlation of fluorescence histochemistry, electron microscopy, endogenous dopamine levels, and 3H dopamine uptake, Brain Res., 46:251.

Walker, R.J., Kemp, J.A., Yajima, H., Kitagawa, K., and Woodruff, G.N., 1976, The action of substance P on mesencephalic reticular and substantia nigral neurones of the rat, Experientia, 32:214.

Zamir, N., Palkovits, M., Weber, E., Mezey, E., and Brownstein, M.J., 1984 A dynorphinergic pathway of leu-enkephalin production in the substantia nigra, Nature 307:643.

ORGANIZATION OF CHOLINERGIC PERIKARYA IN THE CAUDATE NUCLEUS OF THE CAT

WITH RESPECT TO HETEROGENEITIES IN ENKEPHALIN AND SUBSTANCE P STAINING

Maryann E. Martone, Stephen J. Young, David M. Armstrong*and
Philip M. Groves

Departments of Psychiatry and Neuroscience, University of California
San Diego, San Diego, California, USA
*FIDIA, Georgetown Institute of Neurosciences, Washington, D. C.

We present here a series of studies examining the organization of cholinergic perikarya in the caudate nucleus of the cat and their relationship to heterogeneities observed in enkephalin and substance P staining. Cholinergic interneurons comprise only a small proportion of the total number of striatal cells (Phelps et al., 1985). Most studies have not reported any obvious organization in the distribution of these cells in single sections (Fibiger, 1982) although some authors have commented that they did not appear to be evenly distributed but were organized into poorly defined clusters (Everitt et al., 1988; Mesulam et al., 1984). In addition, no consistent relationship has been demonstrated between the cholinergic perikarya and the striosomal compartment as delineated by acetylcholinesterase. While cholinergic neuropil and the high affinity choline uptake system are distributed primarily within the matrix (Graybiel et al., 1986; Rhodes et al., 1987), cholinergic cell bodies have been reported to occur both within patch and matrix in equal densities (Brand, 1980) or to be slightly increased in density within the matrix compartment (Graybiel et al., 1983, 1986).

In an earlier study from this laboratory, we used three-dimensional computer reconstruction techniques to examine the organization of patches of intense enkephalin immunoreactivity in the caudate nucleus of the cat (Groves et al., 1988). Enkephalin-rich patches have been found previously to correspond with acetylcholinesterase-poor zones seen in adjacent sections (Graybiel et al., 1981). We found that the enkephalin rich patches seen in individual sections were actually part of highly structured three dimensional networks distributed throughout the head of the caudate nucleus, as illustrated in figure 1. The major structural feature was the presence of regularly spaced diagonal bands radiating from the ventricular edge across the width of the caudate. These bands curved in the rostral caudal dimension so that they were generally not visible in a single plane of section. Since the enkephalin patches are cytoarchitecturally distinct from the rest of the striatum (Penny et al., 1988), it follows that the organization of the enkephalin network also describes the distribution of a subpopulation of striatal cells.

Although some differences existed in the details of the distribution between animals, there was a remarkable overall similarity in the patch networks from animal to animal. Since the striatum possesses no obvious cytoarchitectural organization analogous to the lamination visible in the cortex, we reasoned that the patch network could be used as an organizational framework for the examination of the distribution of other neurotransmitter markers, cell types and afferent input in the cat striatum. As with the enkephalin patches, such organization might not be evident in two-dimensional

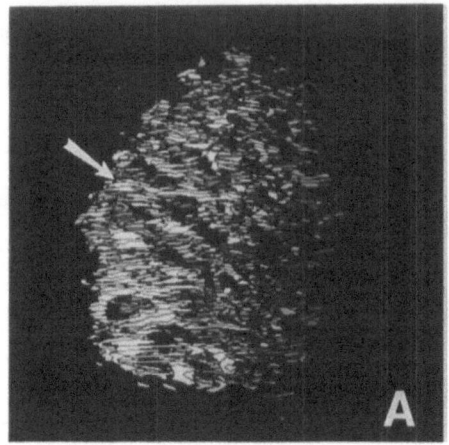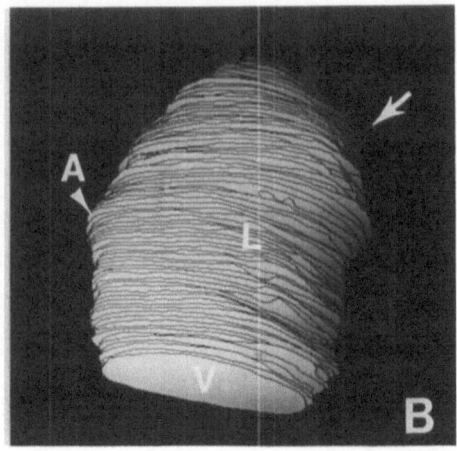

Figure 1. Three-dimensional computer reconstruction of enkephalin patches (A) based on a series of 80 horizontal sections (every other section). Enkephalin patches appear to be distributed in regularly spaced diagonal bands or "fingers" (arrow) radiating from the ventricular edge. A reconstruction of the outlines of the caudate nucleus in the same rotation is shown in B. The image is rotated so that a lateral view of the caudate nucleus is obtained: the arrowhead points to the anterior tip of the caudate. The tail of the caudate is just barely visible (arrow). A=anterior, L=lateral, V=ventral.

sections. In the present series of studies, three dimensional computer reconstruction and immuncytochemical techniques are used to examine the organization of the cholinergic interneurons of the striatum and compare it to the mosaic patterns of enkephalin and substance P staining.

THREE DIMENSIONAL ORGANIZATION OF CHOLINERGIC PERIKARYA AND THEIR RELATIONSHIP TO ENKEPHALIN-RICH PATCHES

The following experiments were carried out on adult mongrel cats. Series of 50 μm sections were cut on a sliding microtome through the head of the caudate nucleus along the coronal, horizontal or sagittal plane. Details of tissue preparation are given in Groves et al. (1988). Alternate sections were processed for choline acetyl transferase (antibody kindly provided by D. L. Hersh) and leu-enkephalin (Incstar). Series were also prepared that were stained in sequence for enkephalin, substance P (Sera Labs) and ChAT respectively. Prior to sectioning, a Pasteur pipette was passed perpendicular to the plane of section through areas of white matter lateral and dorsal to the striatum. The resulting holes were later used to aid in alignment of the sections.

Camera lucida drawings were prepared of each section showing the position of the pipette holes, the edge of the tissue block, the outline of the caudate nucleus, the position of major blood vessels and either the outlines of the enkephalin patches or the position of ChAT+ cell bodies. Drawings of consecutive sections were overlain on a light box and fiducial marks chosen so as to obtain the best fit between the pipette holes, the edge of the section, the outlines of the caudate and the major blood vessels. These drawings along with the fiducial marks for each section were traced into an IBM-PC computer using a high resolution digitizing tablet. The programs used to input and display the data have been described elsewhere (Groves et al., 1988; Young et al., 1988). Reconstructions could be rotated on the x, y and z axes and were photographed directly off the computer screen.

Cholinergic perikarya were distributed throughout the extent of the head of the caudate nucleus although they appeared to decrease in density at more caudal levels. These cells did not appear to be evenly distributed within the caudate and local regions of increased and decreased cell density were apparent. Although the patterning was subtle, areas of increased cell density were visible which appeared similar in shape to the enkephalin

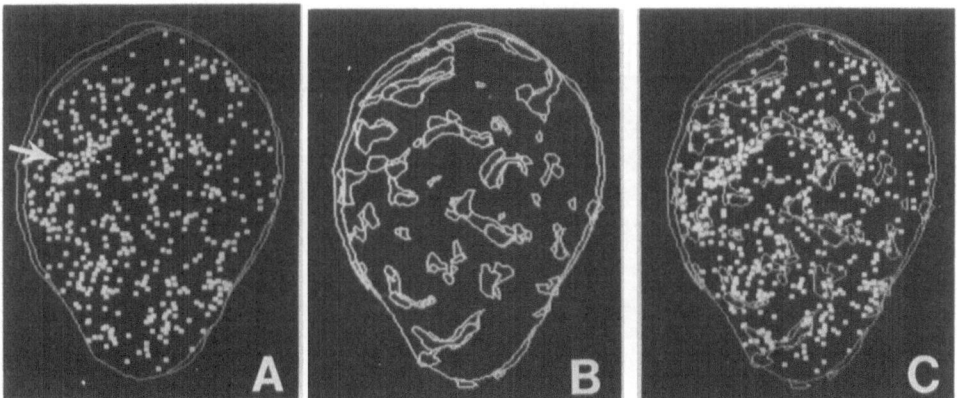

Figure 2. Reconstruction of sagittal series of sections viewed from the lateral edge of the caudate nucleus. The distribution of ChAT+ cell bodies is shown in A (1 square = 1 cell body) and that of the enkephalin patches in B. The inset (B') shows the outlines of the caudate nucleus in the same rotation. The arrowhead in B' points to the tail of the caudate. Alternating regions of increased and decreased density of ChAT+ cells appear to be visible in approximate correspondence to regions of the enkephalin network. L=lateral.

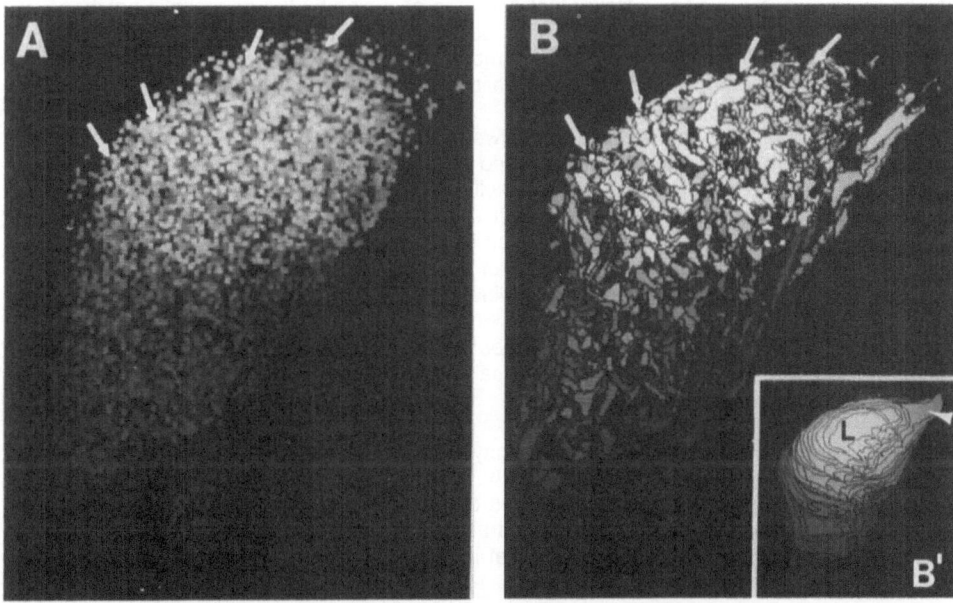

Figure 3. Reconstructions of limited series of horizontal sections from the dorsal third of the caudate showing the ChAT+ cells and enkephalin patches alone (A and B) and superimposed (C). Medial is to the left and anterior to the top in all reconstructions. Areas of increased ChAT+ cell density (arrow) appear to occur in regions where there are also enkephalin patches. When the two are superimposed, these regions of increased ChAT+ cell density often overlap with enkephalin patches or appear to group around them.

patch networks visible in similar reconstructions. For example, in figure 2A, regularly alternating bands of high and low density are seen lined up in the rostral caudal dimension. A similar organization is present in the organization of enkephalin patches (the "fingers" of Groves et al., 1988). When reconstructions of limited series of sections were prepared which showed both the enkephalin patches and the cholinergic cell bodies superimposed, these areas of increased cell density appeared to be localized in and around the enkephalin fingers (fig. 3). The cholinergic perikarya were not exclusively contained

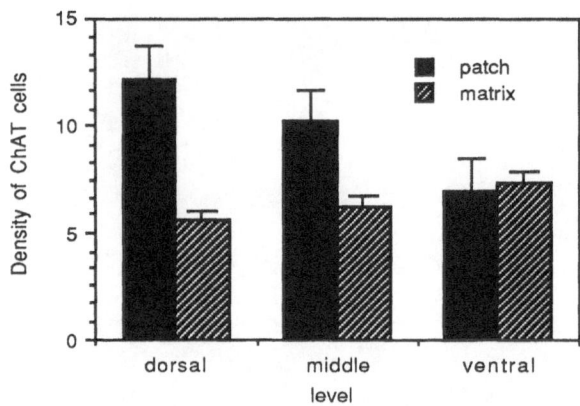

Fig. 4. Density of cholinergic perikarya (per mm^2) lying within the enkephalin patches and the matrix for three dorsal-ventral levels of the caudate nucleus (N=4). Each level corresponded to approximately one third of the caudate nucleus. *=p< .05.

within the enkephalin patches but seemed to group in and around them so that the area of increased density tended to be larger than the area occupied by the enkephalin patches. This pattern held true for the dorsal third to dorsal half of the caudate nucleus. At more ventral levels, cholinergic perikarya appeared to be distributed uniformly between patch and matrix or possibly to avoid the enkephalin patches (not shown).

In order to quantitate the relationship between the enkephalin patches and the cholinergic perikarya, sections were prepared that were double-labeled for ChAT and enkephalin and the density of cholinergic cells in the patch and matrix compartments determined. To avoid using two primary antisera raised in the same species, a mouse monoclonal antibody to enkephalin was used in the double-labeling experiments (Sera Labs). Sections were labeled sequentially for ChAT and enkephalin using nickel-intensified and unintensified diaminobenzidine to distinguish between the two labels. Because it appeared from the three-dimensional images that the density of cholinergic perikarya lying within the patch compartment differed along the dorsal-ventral axis, the caudate nucleus was divided into three dorsal-ventral levels The density within patch and matrix was determined for each level.

In the double-labeled sections, the density of cholinergic perikarya within the enkephalin patches was over twice that of the matrix in the dorsal third of the caudate nucleus (Tukey's HSD, p < .05; fig. 4). This difference in density gradually decreased from dorsal to ventral levels and in the ventral third of the caudate there was no difference in density between patch and matrix.

Discussion

The three-dimensional images of the distribution of the cholinergic perikarya suggest that they are not evenly distributed within the caudate nucleus. These results are consistent with observations in the primate that clustering of striatal cholinergic cells does occur although this effect is subtle in two dimensions (Mesulam et al., 1984; Everitt et al., 1988). A previous study examining the three-dimensional organization of cholinergic neurons in the forebrain of the rat did not find any evidence of organization in the caudoputamen (Schwaber et al., 1987). However the patterning observed in the present study may have been obscured by the numerous fiber fascicles running through the rat neostriatum. Comparison of the distribution of cholinergic perikarya with enkephalin-rich patches suggests that these clusters of cholinergic cells are related to the striosomal compartment in the dorsal striatum where areas of increased ChAT+ cell density occur in and around the enkephalin patches. This is not true at more ventral levels of the striatum where there appeared to be decreased density of cholinergic cells within the patches or no difference in density between patch and matrix. These relationships between ChAT cells and enkephalin patches could also be seen in single sections double-labeled for ChAT and enkephalin, although clustering in and around the enkephalin patches was not as apparent as in the three-dimensional images. This may be due to the small number of ChAT+ positive cells on any given section: overlapping of sections emphasized regions where consistently greater or lesser numbers of cells were found from section to section. Quantitative analysis confirmed that the density of ChAT cells is significantly higher in the enkephalin patches compared to the matrix in the dorsal striatum.

Enkephalin-rich patches have been shown to correspond to AChE-poor zones seen in adjacent sections (Graybiel et al., 1981). Thus, the increased density of cholinergic cells in the enkephalin patches in the dorsal caudate nucleus seen in the present study are not consistent with other reports in the cat which suggest that cholinergic perikarya tend to avoid the striosomes (Graybiel et al., 1983; 1986) or which report no difference in density between patch and matrix (Brand, 1980). This discrepancy is likely due to the regional differences observed in the present study. Cholinergic perikarya were most likely to be present within the enkephalin patches in the dorsal caudate nucleus where AChE-poor zones are not well defined. We did find at more ventral levels that there was no difference in density between patch and matrix as has been reported by others using AChE as a striosomal marker (Brand, 1980).

Although there did appear to be a relationship between the cholinergic perikarya and the enkephalin patches, this relationship was not overwhelming compared to other markers that obey striosomal ordering. Even though most striosomal markers are found both within patch and matrix, the distribution of most, e.g. enkephalin, show a sharp transition between patch and matrix while the cholinergic cells appeared to be grouped in and around the patches rather than exclusively within them. Thus the areas of increased cholinergic cell density were larger than the corresponding enkephalin patch. This latter configuration is also characteristic of the relationship between patches of substance P fibers and enkephalin patches in the dorsal caudate nucleus of the cat (Graybiel et al., 1981; personal observations). Substance P staining exhibits a complex staining pattern in the caudate nucleus of the cat. In the dorsal caudate, substance P rich zones are visible against a lighter staining background These zones tend to be somewhat larger and more diffuse in appearance than do enkephalin patches (Graybiel et al., 1981; personal observations). Because it appeared that the relationship of cholinergic perikarya to the enkephalin patches in the dorsal striatum was similar to that seen for substance P and enkephalin, we compared directly the distribution of cholinergic perikarya to the pattern of substance P staining.

RELATIONSHIP OF CHOLINERGIC PERIKARYA TO HETEROGENEITIES IN SUBSTANCE P STAINING

Tissue from adult mongrel cats for computer reconstructions and double-labeling procedures was prepared as above. Substance P was localized using a monoclonal antibody raised in rat (Sera Labs). The procedures for three-dimensional computer reconstructions and double-labeling immunocytochemistry were the same as described for the above experiments.

The pattern of SP staining obtained in the present study was similar to that described in previous studies (Graybiel et al., 1981; Beckstead, 1987). In approximately the dorsal third of the caudate, SP rich zones were present against a lighter staining background. When the pattern of SP staining was compared to that of enkephalin seen in adjacent sections, SP rich zones in the dorsal striatum generally corresponded to enkephalin-rich patches (fig.5) although they were usually larger and less discretely defined. In addition, the pale rims of reduced staining which outline many of the enkephalin patches were not generally visible for SP patches. However, as has been reported by others, the appearance of these patches differed somewhat depending upon

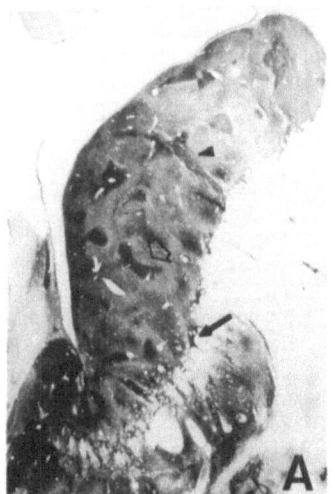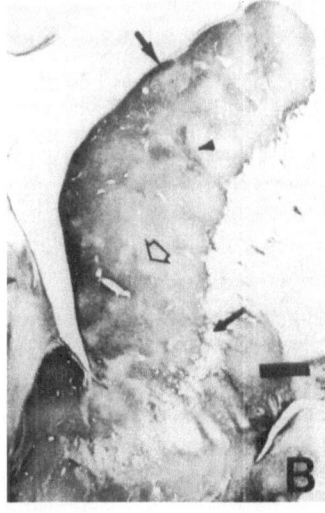

Figure 5. Adjacent sections stained for enkephalin (A) and substance P (B). A general correspondence is seen between enkephalin- and substance P-rich patches in the caudate nucleus (solid arrowheads and small arrows) although the substance P patches tend to be slightly larger and more diffuse than corresponding enkephalin patches. It was often difficult to find precise matches between the sets of patches. In the medial caudate, correspondence also occurs between substance P-poor areas and enkephalin rich patches (large arrow and open arrow). Scale bar in B=1µm.

whether cell body or fiber staining predominated (Graybiel and Chesselet, 1984; Bolam et al., 1988). When SP+ cell bodies were visible, they were seen to be clustered within SP rich patches in the dorsal striatum (Beckstead and Kersey, 1985). The borders of such patches were fairly crisp and good correspondence was seen in size and shape to corresponding enkephalin patches. When fiber staining predominated, the patches took on the more diffuse appearance described above.

In the ventral caudate nucleus, the pattern of dark patches on a light background reversed itself. In the medial aspects of this region, SP-poor zones were seen embedded within a SP rich matrix. Substance P poor zones were also visible in the most medial aspect of the dorsal caudate nucleus (fig. 5, large arrow). The matrix itself was not homogeneously stained but showed both intense SP staining and a more moderate level that was still distinguishable from that in the SP poor zones. The SP poor zones in the medial region appeared to correspond to enkephalin-rich patches seen in adjacent sections. However, in the ventral-lateral striatum, SP rich zones were seen which also overlapped with enkephalin patches (see arrow in fig. 5). In fact, single enkephalin patches were seen which overlapped with a SP poor zone in the medial region and a SP rich zone in the lateral caudate nucleus.

Because the distribution of substance P was not very discrete, camera lucida drawings did not quite capture the subtleties of its distribution. Regardless of whether cell body or fiber staining predominated, areas of intense SP were present both dorsally and ventrally which did not appear to correspond to enkephalin patches in adjacent sections and which lacked clearly definable borders. Such regions were difficult to render in the camera lucida drawings and so were not included in the three-dimensional reconstructions. Nevertheless, as would be expected, there was an overall similarity in the pattern of SP patch distribution and that seen for enkephalin in dorsal regions.

Comparisons between the distribution of SP patches and cholinergic neurons in both the three dimensional images (not shown) and the double-labeled sections showed a striking correspondence between regions of increased cholinergic cell density and SP rich patches. Unlike the sections double-labeled for enkephalin and

ChAT, a clear relationship between areas of heightened SP immunoreactivity and cholinergic perikarya was apparent in individual sections double-labeled for SP and ChAT (fig.6). Our impressions were that cholinergic cells tended to be found in areas which stained intensely for SP. Thus, cholinergic neurons tended to be found within the substance P rich patches in the dorsal striatum and to avoid substance P poor regions, including the SP-poor zones seen in the ventral caudate nucleus (fig. 6). Quantitative measurements of the density of cholinergic cells in the SP rich areas of the dorsal striatum showed that the density of cholinergic cells was significantly higher in SP rich areas compared to regions which stained less intensely for SP in both the dorsal (14.87 ± 0.84 cells/mm^2 vs. 5.35 ± 0.32, $p<.05$) and in the ventral third of the caudate (12.87 ± 1.73 cells/mm^2 vs. 6.75 ± 0.54, $p<.05$). Density measurements for the middle third of the caudate have not been performed yet. Although not analyzed in detail, there appeared to be a rather striking correlation between substance P staining and cholinergic perikarya in other ventral striatal and basal forebrain areas (fig. 6). As in the caudate nucleus, cholinergic cells appeared to be concentrated within substance P rich regions.

Discussion

The results of the second experiment appear similar to those found in the first experiment for enkephalin in that cholinergic perikarya were found concentrated within the substance P rich areas of the dorsal striatum. Since both enkephalin and substance P are considered markers of the striosomal compartment in the dorsal striatum (Graybiel and Chesselet, 1984; Malach and Graybiel, 1986), our results might suggest that the cholinergic perikarya are preferentially associated with the striosomal compartment in the dorsal striatum. However, cholinergic perikarya did not appear to be associated with the striosomal compartment in the ventral caudate nucleus. Many other markers also exhibit regional variation in their association with the striosomes in the cat. In addition to

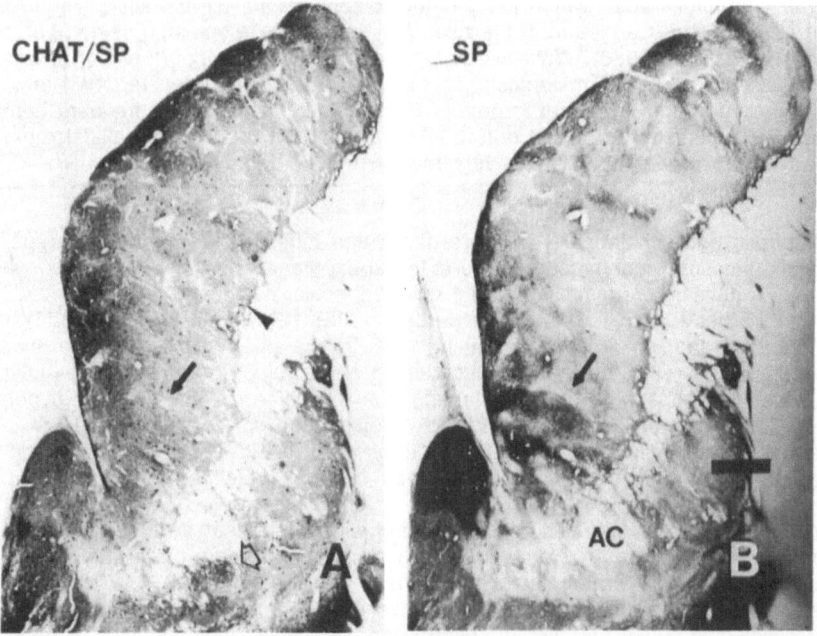

Figure 6. Low power photomicrograph of a section double-labeled for ChAT and substance P (A). The adjacent section stained for substance P alone is shown in B for comparison. ChAT+ neurons appeared to be found in areas which stained intensely for substance P in both the dorsal and the ventral caudate nucleus (arrowhead). They generally were not found in substance P poor areas (arrows in A and B). This pattern also appeared to be true for ventral striatal regions below the level of the anterior commissure (open arrow). AC=anterior commissure, scale bar in B=1μm.

substance P, afferents from certain prefrontal cortical areas and from subregions of the substantia nigra are distributed within the striosomes dorsally but tend to avoid them ventrally (Ragsdale and Graybiel, 1981; Beckstead, 1987; Jiminez-Castellanos and Graybiel, 1987). During development, the dopamine rich islands in the cat striatum are most clearly defined in the dorsal-lateral striatum and are less well delineated at more ventral levels (Graybiel, 1984).

Given that this regional variation exists in the distribution of several markers, an alternative account for the above results is that the cholinergic perikarya are distributing along with one or more of these markers and are only circumstantially related to the striosomal compartment. In the present experiments, cholinergic perikarya appeared to distribute with substance P immunoreactivity, regardless of whether it occurred in patch or matrix. Our impression of the double-labeled material was that the distribution of cholinergic perikarya was more closely related to the distribution of substance P rich areas than to that of enkephalin rich patches when both the dorsal and ventral regions of the caudate were considered together. Substance P and enkephalin are found in largely separate populations of striatal neurons similar in size and morphology to the common spiny neurons identified in Golgi studies (Penny et al., 1986; Gerfen and Young, 1988; Izzo et al., 1987), the major projection cell of the neostriatum. While cholinergic neurons in the striatum are known to interact with many intrinsic and extrinsic systems (DiFiglia and Carey, 1986; Phelps et al., 1985; Chang, 1988; Kubota et al., 1987), the patterns observed in the present study suggest that perhaps cholinergic interneurons interact predominantly with substance P containing elements compared to those which contain enkephalin in the caudate nucleus. We have begun to examine this hypothesis in the rat using a double-labeling electron microscopic procedure (Martone et al., 1989). Thus far we have found very few synaptic contacts between enkephalin containing terminals and ChAT+ perikarya and dendrites while such contacts between substance P terminals and cholinergic elements were commonly observed by us and by others (Bolam et al., 1986). The majority of substance P and enkephalin terminals seen within the striatum displayed the ultrastructural features of a type of terminal which arises from the axon collaterals of the common spiny neuron (Wilson and Groves, 1980). While we recognize the inherent opportunities for false negative errors in electron microscopy, a similar situation has been reported for cholinergic neurons in the basal forebrain. Bolam et al. (1986) found frequent contacts of substance P containing terminals onto cholinergic cells in the ventral pallidal region, whereas Chang et al. (1987) found very few enkephalin inputs onto peripallidal cholinergic cells.

Enkephalin and substance P also are distributed differentially in striatal targets. Enkephalin staining predominates in the external segment of the globus pallidus while substance P staining predominates in the entopeduncular nucleus and the substantia nigra (Haber and Elde, 1981; Beckstead and Kersey, 1985). Gerfen and Young (1988) have shown recently that enkephalin containing cells in the striatum project preferentially to the globus pallidus while substance P containing cells project to the entopeduncular nucleus and the substantia nigra confirming that the differential distribution of peptides in these structures can be accounted for by projections from chemically distinct populations of striatal cells. Our electron microscopic results imply that these two populations of striatal cells may differ in their intrinsic projections as well.

The relationship between substance P and cholinergic perikarya demonstrated in the present experiments does not preclude the possibility that some other intrinsic or extrinsic system is governing the distribution of cholinergic cells. Ultrastructural analyses suggest that cholinergic cells receive input from multiple intrinsic and extrinsic sources including GABA and tyrosine hydroxylase containing elements (Bolam and Izzo, 1987; Chang et al., 1988; Kubota et al., 1987). Beckstead (1987) showed that injections of radioactive tracers in the substantia nigra pars compacta resulted in patches of heavy labeling that coincided with substance P rich zones in the dorsal striatum. At more ventral striatal levels, labeling was more evenly distributed with zones containing little or no labeling apparent. In a separate study, he described zones of heightened D2 receptor binding in the dorsal striatum which were similar in size and shape to the substance P rich patches in this area (Beckstead et al., 1988). Dopaminergic inputs to the striatum are thought to exert a tonic inhibitory effect upon striatal cholinergic neurons and this effect may be mediated through the D2 receptor (Fage and Scatton, 1986). The coincidence of dense projections from the nigra with increased densities of substance P staining, cholinergic perikarya and perhaps D2 receptor binding suggest a possible preferential interaction between these elements in the caudate nucleus. Given the historical and clinical interest in acetylcholine-dopamine interactions in the striatum, these relationships are deserving of further study.

ACKNOWLEDGEMENTS

This work supported by grants from the Office of Naval Research and the National Institute on Drug abuse to P. M. G.

REFERENCES

Beckstead, R. M., 1987,Striatal substance P cell clusters coincide with the high density terminal zones of the discontinuous nigrostriatal dopaminergic projection system in the cat: A study by combined immunohistochemistry and autoradiographic axon-tracing, Neuroscience. 20: 557.

Beckstead, R. M. and Kersey, K. S., 1985, Immunohistochemical demonstration of differential substance P-, Met-enkephalin-, and glutamic-acid-decarboxylase-containing cell body and axon distributions in the corpus striatum of the cat, J. Comp. Neurol., 232: 481.

Beckstead, R. M., Wooten, G.F. and Trugman, J. M., 1988, Distribution of D1 and D2 dopamine receptors in the basal ganglia of the cat determined by quantitative autoradiography, J. Comp. Neurol., 268: 131.

Bolam, J. P., Ingham, C. A., Izzo, P. N., Levey, A. I., Rye, D. B., Smith, A. D. and Wainer, B.H.,1986, Substance P-containing terminals in synaptic contact with cholinergic neurons in the neostriatum and basal forebrain: A double immunocytochemical study in the rat, Brain Res., 397: 279.

Bolam, J. P. and Izzo, P. N., 1987, Possible sites of transmitter interaction in the neostriatum: An anatomical approach, in Sandler, M., Feuerstein, C. and Scatton, B. (eds) Neurotransmitter interactions in the basal ganglia, New York: Raven Press, p. 47.

Bolam, J. P., Izzo, P. N. and Graybiel, A. M., 1988, Cellular substrate of the histochemically defined striosome/matrix system of the caudate nucleus: A combined Golgi and immunocytochemical study in cat and ferret, Neuroscience. 24: 853.

Brand, S., 1980, A comparison of the distribution of acetylcholinesterase and muscarinic cholinergic receptors in the feline neostriatum, Neurosci.Lett.,17:113.

Chang, H. T., 1988, Dopamine-acetylcholine interaction in the rat striatum: A dual labeling immunocytochemical study, Brain Res. Bull., 21: 295.

Chang, H. T., Penny, G. R. and Kitai, S. T., 1987, Enkephalinergic-cholinergic interaction in the rat globus pallidus: A pre-embedding double-labeling immunocytochemistry study, Brain Res., 426: 197.

DiFiglia, M . and Carey, J., 1986, Large neurons in the primate neostriatum examined with the combined Golgi-electron microscopic method, J. Comp. Neurol., 244: 36.

Everitt, B. J., Sirkia, T. E., Roberts, A. C., Jones, G. H. and Robbins, T. W., 1988, Distribution and some projections of cholinergic neurons in the brain of the common marmoset, Callithrix jacchus, J. Comp. Neurol., 271: 533.

Fage, D. and Scatton, B., 1986, Opposing effects of D-1 and D-2 receptor antagonists on acetylcholine levels in the rat striatum, European. J. Pharmacol., 129: 359.

Fibiger, H. C., 1982, The organization and some projections of cholinergic neurons of the mammalian forebrain, Brain Res. Rev., 4: 327.

Gerfen, C. R. and Young, W. S., 1988, Distribution of striatonigral and striatopallidal peptidergic neurons in both patch and matrix compartments: An in situ hybridization histochemistry and fluorescent retrograde tracing study, Brain Res., 460: 161.

Graybiel, A.M., 1984, Correspondence between the dopamine islands and striosomes of the mammalian striatum, Neuroscience. 13: 1157.

Graybiel, A. M., Baugham, R. W. and Eckenstein, F., 1986, Cholinergic neuropil of the striatum observes striosomal boundaries, Nature. 323: 625.

Graybiel, A. M. and Chesselet, M.-F., 1984, Compartmental distribution of striatal cell bodies expressing [Met] enkephalin-like immunoreactivity, Proc. Natl. Acad. Sci. USA. 81: 7980.

Graybiel, A. M., Chesselet, M.-F., Wu, J.-Y., Eckenstein, F. and Joh, T. E., 1983,The relation of striosomes in the caudate nucleus of the cat to the organization of early developing dopaminergic fibers, GAD-positive neuropil, and CAT-positive neurons, Soc. Neurosci. Abstr.,9: 14.

Graybiel, A. M., Ragsdale, C. W., Yoneoka, E. S. and Elde, R. P., 1981, An immunohistochemical study of enkephalins and other neuropeptides in the striatum of the cat with evidence that the opiate peptides are arranged to form mosaic patterns in register with the striosomal compartments visible by acetylcholinesterase staining, Neuroscience. 6: 377.

Groves, P. M. , Martone, M., Young, S. J. and Armstrong, D. M., 1988, Three-dimensional pattern of enkephalin-like immunoreactivity in the caudate nucleus of the cat, J. Neurosci., 8: 892.

Haber, S. and Elde, R., 1981, Correlation between met-enkephalin and substance P immunoreactivity in the primate globus pallidus, Neuroscience. 6: 1291.

Izzo, P. N., Graybiel, A. M. and Bolam, J. P., 1987, Characterization of substance P- and [met]enkephalin-immunoreactive neurons in the caudate nucleus of cat and ferret by a single section Golgi procedure, Neuroscience. 20: 577.

Jimenez-Castellanos, J. and Graybiel, A. M., 1987, Subdivisions of the dopamine-containing A8-A9-A10 complex identified by their differential mesostriatal innervation of striosomes and extrastriosomal matrix, Neuroscience, 23: 223.

Kubota, Y., Inagaki, S., Shimada, S., Kito, S., Eckenstein, F. and Tohyama, M., 1987, Neostriatal cholinergic neurons receive direct synaptic inputs from dopaminergic axons, Brain Res., 413: 179.

Malach, R. and Graybiel, A. M., 1986, Mosaic architecture of the somatic sensory-recipient sector of the cat's striatum. J. Neurosci. 6: 3436.

Martone, M., Young, S. J., Armstrong, D. M. and Groves, P. M., 1989, Ultrastructural examination of enkephalin and substance P input to cholinergic neurons in the rat neostriatum. Soc. Neurosci. Abstr. 15: 911.

Mesulam, M.-M., Mufson, E. J., Levey, A. I. and Wainer,B. H., 1984, Atlas of cholinergic neurons in the forebrain and upper brainstem of the macaque based on monoclonal choline acetyltransferase immunohistochemistry and acetylcholinesterase histochemistry, Neuroscience. 3: 669.

Penny, G. R., Afsharpour and Kitai, S. T.,1986, The glutamate decarboxylase, leucine enkephalin-, methionine enkephalin- and substance P-immunoreactive neurons in the neostriatum of the rat and cat: Evidence for partial population overlap, Neuroscience. 17:1011.

Penny, G. R., Wilson, C. J. and Kitai, S. T., 1988, Relationship of the axonal and dendritic geometry of spiny projection neurons to the compartmental organization of the neostriatum, J. Comp. Neurol.. 269: 275.

Phelps, P. E., Houser, C. R. and Vaughn, J. E., 1985, Immunocytochemical localization of choline acetyltransferase within the rat neostriatum: A correlated light and electron microscopic study of cholinergic neurons and synapses, J. Comp. Neurol. , 238: 286.

Ragsdale, C. W. and Graybiel, A. M., 1981,The fronto-striatal projections in the cat and monkey and its relationship to inhomogeneities established by acetylcholinesterase histochemistry, Brain Res., 208: 259.

Rhodes, K. J., Joyce, J. N., Sapp, D. W. and Marshall, J. F., 1987, [3H] Hemicholinium-3 binding in rabbit striatum: Correspondence with patchy acetylcholinesterase staining and a method for quantifying striatal compartments, Brain Res., 412: 400.

Schwaber, J. S., Rogers, W. T., Satoh, K. and Fibiger, H. C., 1987, Distribution and org-anization of cholinergic neurons in the rat forebrain demonstrated by computer- aided data acquisition and three-dimensional reconstruction, J. Comp. Neurol., 263: 309.

Wilson, C. J. and Groves, P. M., 1980, Fine structure and synaptic connections of the common spiny neuron of the rat neostriatum: A study employing intracellular injection of horseradish peroxidase, J. Comp. Neurol. 194: 599.

Young, S. J., Royer, S. M., Groves, P. M. and Kinnamon, J. C., 1987, Three- dimensional reconstructions from serial micrographs using the IBM PC, J. Electron Microsc. Tech., 6: 207.

NEUROPEPTIDE Y NEURONS IN THE STRIATAL NETWORK. FUNCTIONAL ADAPTIVE RESPONSES TO IMPAIRMENT OF STRIATAL INPUTS

Lydia Kerkerian-Le Goff, Pascal Salin, Jacqueline Vuillet* and André Nieoullon

Unité de Neurochimie - Laboratoire de Neurosciences Fonctionnelles du CNRS - BP 71 - 13402 Marseille cedex 09, France
Laboratoire associé à l'Université Aix-Marseille II
* *Faculté de Médecine de Marseille*

Neuropeptide Y (NPY), a 36 amino acid peptide enriched with tyrosine residues first isolated from porcine brain extracts (Tatemoto, 1982; Tatemoto et al, 1982), is thought to be the neuroactive member of the pancreatic polypeptide family (Di Maggio et al, 1985). It is one of the most abundant and widely distributed peptides isolated so far within the mammalian central nervous system and it shows a distinctive pattern of distribution as compared with other neuropeptides. The most salient feature of its distribution is the high concentration centred on the striatum, especially in the human brain (Adrian et al, 1983; Allen et al, 1983). Immunohistochemical studies have shown that NPY is present in numerous cell bodies and fibers in both the dorsal striatum or caudate putamen (CP) and the ventral striatum or nucleus accumbens (NAcc) (Adrian et al, 1983; Allen et al, 1983; De Quidt and Emson,1986; Smith et al, 1985). Within these structures and in related cortical areas NPY has been reported to coexist extensively with somatostatin and the enzyme nicotinamide adenine dinucleotide phosphate-diaphorase (NADPH-D) (Chronwall et al, 1984; Gaspar et al, 1987; Kowall et al, 1987; Smith and Parent, 1986; Vincent et al, 1982; 1983). The presence of high levels of NPY in the basal ganglia nuclei has suggested that this peptide may play a fundamental role in the control of sensorimotor function and has led to detailed investigations on the morphological features and cellular relationships of the NPY neurons within these structures. The purpose of the present report is to briefly review data from biochemical and immunohistochemical studies in an attempt to clarify the anatomical and functional position of NPY elements in the striatal network (dorsal striatum).

I - MORPHOLOGY AND SYNAPTIC INVOLVEMENT OF NPY-CONTAINING NEURONS IN STRIATAL CIRCUITRY. LIGHT AND ELECTRON MICROSCOPIC IMMUNOCYTOCHEMICAL ANALYSIS

Morphology, distribution and synaptic relationships of NPY neurons in the striatum

NPY-immunoreactive cell bodies and fibers are abundant in the striatum of the rat, cat and monkey but nearly absent from the pallidum and the substantia nigra which are the main targets of striatal output neurons (Smith et al, 1985; Kerkerian et al, 1986).In agreement with these findings, concentrations of NPY were found to be substantially higher in the striatum than in the pallidum and the substantia nigra. In those striatal target structures, the NPY levels were found to be unaffected by excitotoxic lesion of the striatum (Christie et al, 1986). All in all these data suggest that the NPY-containing neurons are short axoned local circuit neurons (interneurons) within the striatum. In addition to being functionally characterized as interneurons, the striatal NPY neurons have been morphologically classified in the rat, cat and monkey, as belonging to the medium-sized aspiny subtype (Smith and Parent, 1986; Vuillet et al, 1989b). These NPY-positive neurons were observed in the rat to receive numerous synaptic inputs mainly of the symmetrical type on their perikarya and proximal dendrites and of the asymmetrical type on their distal dendrites (Vuillet et al,1989b). Since striatal asymmetrical and symmetrical junctions are thought to be primarily of extrinsic and intrinsic origin, respectively (Hassler et al, 1977), NPY neurons were presumed to constitute targets for both striatal afferent pathways and intrastriatal neurons. The NPY terminals in the striatum may primarily belong to the intrinsic NPY neuronal population. The possibility cannot been ruled out that some of these terminals may be of extrinsic origin. NPY nerve terminals were found to form synaptic contacts of the symmetrical type with dendrites, spines, and perikarya. Most of the neurons contacted by NPY axons are of the spiny type (Vuillet et al, 1989b). In the monkey, NPY neurons and fibers were found to be homogeneously distributed within the striatum (Smith and Parent, 1986). In the rat, a fairly even distribution was also observed although a higher neuronal density was noted in the middle striatal areas than in the dorsolateral areas (Kerkerian et al, in the press), whereas an uneven distribution was observed in the cat striatum where the NPY immunoreactivity was concentrated in patches surrounded by poorly immunostained areas (Smith and Parent, 1986).

Relationships with striatal afferent fibers originating in the cerebral cortex and the substantia nigra

Using dual labeling methods and combining deafferentation procedures with NPY immunocytochemistry, we recently investigated whether two of the main striatal afferent pathways, namely the nigrostriatal dopaminergic and the corticostriatal presumably glutamatergic pathway, might provide inputs to striatal NPY neurons. After 6-hydroxydopamine (6-OHDA)-induced lesion of nigral dopaminergic neurons, degenerating boutons were observed to form straight appositions with NPY-positive cell bodies or dendrites (Vuillet et al, 1989b). Moreover, by means of dual labeling in intact animals, morphologically defined synaptic contacts were visualized between tyrosine hydroxylase (TH)-labeled nerve terminals and NPY-positive neurons (Vuillet et al, 1989a). After hemidecortication, degenerating corticostriatal terminals were found to form asymmetrical synaptic contacts with NPY-positive neurons (Vuillet et al, 1989a). Taken as a whole, these data show that NPY neurons constitute targets for nigrostriatal dopaminergic and corticostriatal inputs and provide ultrastructural support for the idea that these two afferent pathways may participate in regulating striatal NPY system. Some pictures suggestive of axo-axonic contacts between TH- and NPY-containing profiles were also obtained : these contacts might constitute the morphological basis of the reported reciprocal action of NPY on the striatal dopamine system (Beal et al, 1986b).

Relationships with intrinsic striatal elements

Using dual labeling methods to simultaneously locate NPY and the GABA synthetic enzyme glutamate decarboxylase (GAD) immunoreactivities in striatal sections, we observed numerous synaptic associations of the symmetrical type between GAD-positive axonal boutons and NPY-positive cell bodies or dendrites. Pictures of axo-axonic contacts between GAD and NPY-labeled terminals were also obtained (Vuillet et al, submitted). These data show that NPY neurons constitute targets for intrinsic GABAergic striatal neurons and support the view that functional presynaptic interactions may occur between GABA and NPY neuronal systems within the striatum.

The postsynaptic targets of NPY neurons still remain to be identified. The finding that NPY-positive terminals contact dendritic spines and spiny type perikarya, together with data showing that a large proportion of striatal spiny neurons contain GABA, suggests that conversely, GABAergic elements may constitute a postsynaptic target for NPY interneurons in the striatum. Interestingly, NPY-positive axonal processes were also frequently observed in apposition with perivascular cells, which may provide the morphological basis of the potent vasoconstrictor effects of NPY on cerebral vessels.

Conclusions

The above data show that NPY neurons constitute targets both for terminals originating outside the structure and for intrinsic neurons, which suggests that the NPY system may play a fundamental integrative role in the striatal circuitry. The fact that NPY terminals were found to be involved in axo-axonic and neurovascular associations supports the view that NPY may play a major neuromodulatory role within the striatum. It is worth noting the similarities observed between the cytological features and connectivity of NPY neurons in the CP and the NAcc. As reported above in connection with the CP, NPY immunoreactivity within the NAcc has also been found in aspiny type neurons (Massari et al, 1988). These neurons were shown to receive dopaminergic extrinsic and GABAergic intrinsic inputs (Aoki and Pickel, 1988; Massari et al, 1988) and a prevalent association of NPY processes with vascular cells was observed. These data provide further support for the view that the dorsal and ventral striatum have similar although separate connectivity patterns and functions.

II - INTRANEURONAL COLOCALIZATION OF NPY WITH OTHER NEUROTRANSMITTERS IN THE STRIATUM

Biochemical studies have shown the distribution of NPY and somatostatin immunoreactivities to be highly correlated in the basal ganglia nuclei (Beal et al, 1986a; 1987). Immunocytochemical investigations have confirmed that NPY is frequently colocalized with somatostatin in neurons of the rat, monkey and human striatum (Kowall et al, 1987; Smith and Parent, 1986; Vincent et al, 1982). On the other hand, several reports have mentioned the presence of the enzyme NADPH-D in NPY/somatostatin-positive striatal neurons in the same species (Kowall et al, 1987; Villalba et al, 1988; Vincent et al; 1983). A similar triple colocalization may also exist at cortical level.

By means of dual labeling approaches we recently investigated the possibility that NPY may also be associated with GABA in rat striatal neurons

(Vuillet et al, submitted). Our results showed that all the aspiny type cells undoubtedly displaying NPY immunoreactivity also exhibited GAD immunoreactivity. Some other neurons of the aspiny or spiny type were found to be immunoreactive to GAD alone. GAD/NPY colocalization was observed not only in cell bodies and dendrites but also in axon terminals. These data suggest that, as previously reported at cortical level (Hendry et al, 1984), the majority of NPY neurons in the striatum also contain GABA, although many more neurons contain GABA alone. Moreover, since the NPY neurons in the striatum are presumably interneurons, our data confirm the existence of short-axoned intrinsic GABAergic neurons in the striatum.

Consequently, these data suggest that a multiple colocalization of putative neurotransmitters may occur in NPY-containing interneurons of the aspiny type. It is furthermore possible that different intraneuronal associations of these various neuroactive substances may divide the striatal NPY population into neuronal subsets with diverse functional specificities.

III - ADAPTIVE RESPONSES OF STRIATAL NPY NEURONS TO IMPAIRMENTS IN THE TRANSMISSION OF STRIATAL INPUTS. IMMUNOCYTOCHEMICAL EVIDENCE SUPPORTING MODULATION OF THE PEPTIDE GENOMIC EXPRESSION

We recently investigated the contribution of the three main striatal pathways originating in the substantia nigra, the cerebral cortex and the thalamus, respectively, to striatal NPY regulation in the rat. This was achieved by analyzing, at the light microscopic level, the effects of impairments in the transmission of these inputs on striatal NPY immunostaining by the peroxidase anti-peroxidase method using a specific rabbit anti-NPY antiserum. Since control experiments showed that the cellular labeling obtained was specific to NPY, any changes in NPY immunostaining observed in our experiments might therefore be assumed to primarily reflect variations in the intraneuronal levels of the peptide.

Impairment of nigrostriatal dopaminergic transmission

Non disruptive impairment of dopaminergic transmission by α-methylparatyrosine (α-MPT) or haloperidol administration was found to markedly reduce the number and staining intensity of striatal NPY-positive cells. By contrast, unilateral infusion of 6-OHDA into the substantia nigra resulted, 12 to 90 days later, in an increase in the same parameters in the ipsilateral striatum. The effects of 6-OHDA lesion were completely reversed when an apomorphine treatment was applied prior to perfusion of the animals, and can be thus assumed to be directly linked to changes in striatal DA receptor activation resulting from DA depletion induced by the lesion (Kerkerian et al, 1986; 1988). The changes in NPY immunostaining occuring in response to these manipulations of DA transmission suggest that striatal NPY interneurons are under the influence of nigrostriatal dopaminergic afferent fibers. The above reported visualization of synaptic contacts between DA terminals and NPY neurons further suggests that the action of DA upon NPY neurons is direct. The apparently contradictory results obtained on the one hand with α-MPT or haloperidol treatment and on the other hand with 6-OHDA lesions raises the question as to the nature of the DA control. Among the various hypothesis put

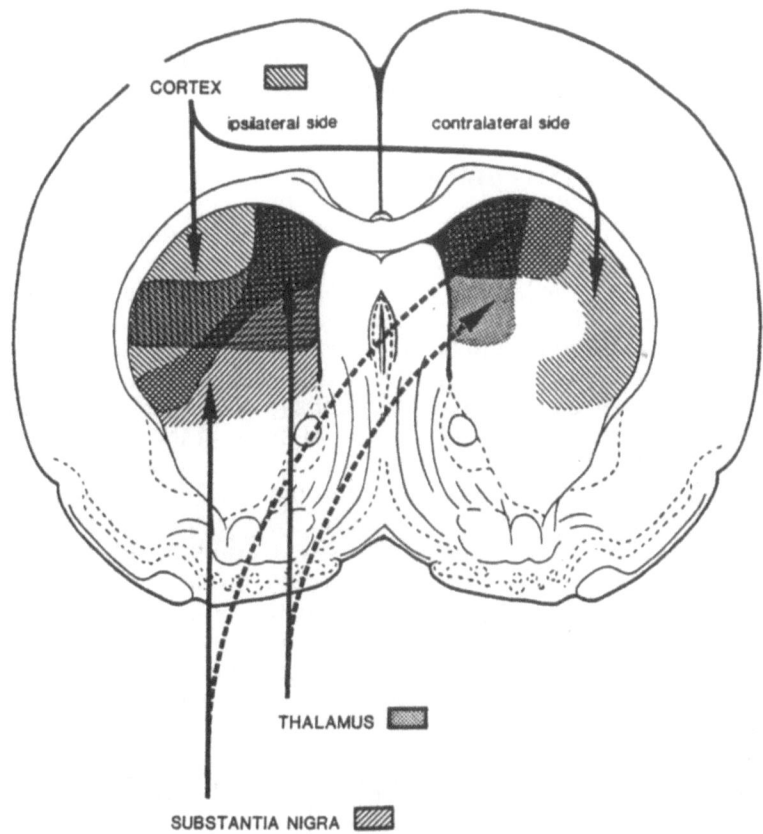

Fig. 1. Topography of the increases in NPY-immunoreactive neuron density in the rat striatum observed after selective unilateral lesion of the nigrostriatal dopaminergic, the corticostriatal and the thalamostriatal pathways (survival time : 15-30 days). The reported changes are > 15% as compared to control values (P at least < 0.05, Student's t-test).

forward to explain this discrepancy, the most plausible seems to be that DA exerts a dual action over NPY neurons through D1 and D2 receptors, which may be differentially affected from one experimental situation to another. Indeed this hypothesis is supported by the finding that selective blockade of D2 receptors by sulpiride administration reduces striatal NPY immunostaining, as does α-MPT or haloperidol treatment, whereas blokade of D1 receptors by means of SCH 23390 treatment induces a slight increase in the peptide neuronal immunostaining that partly mimicked the effects of 6-OHDA lesion (see Kerkerian et al, 1988). Further support for this hypothesis is provided by the topography of the changes induced by the 6-OHDA lesion : the changes in the density of NPY neurons in response to dopaminergic denervation were found to be preferentially localized in the ventral and medial areas of the deafferented striatum. This topography does not agree with the distribution pattern of DA projections, of D1 and D2 receptors, or of the changes in the density of these receptors induced by similar 6-OHDA lesions (Savasta et al, 1988). Interestingly, it overlaps striatal sectors which, in response to lesions of this kind, do not manifest any D2 receptor proliferation and display an increase in D1 receptor sensitivity, as indicated by increases in the maximal

DA stimulation of striatal adenylate cyclase activity (Hervé et al, 1988). These data therefore support the view that NPY neuronal responses to nigrostriatal dopaminergic deafferentation result primarily from changes in D1 receptor subtype sensitivity.

Corticostriatal deafferentation

Hemidecortication was found to induce, 20 to 30 days later, a bilateral increase in the density and staining intensity of striatal NPY-positive neurons (Kerkerian et al, in the press). This result, together with the electron microscopic finding that synaptic contacts existed between corticostriatal terminals and NPY neurons, suggests that corticostriatal afferents may monosynaptically regulate striatal NPY function. From topographical analysis of data, it was found in addition that the increases in NPY neuron density observed after corticostriatal deafferentation are prevalent in the dorsolateral areas of the striatum, which are the main site of terminating corticostriatal projections (Cospito et al, 1981), whereas increases in response to dopaminergic denervation occured preferentially in ventromedial areas (figure 1). Comparisons between the regional changes in NPY neuron density detected after nigrostriatal dopaminergic versus corticostriatal deafferentation then revealed that the cerebral cortex and the substantia nigra may primarily influence topographically separate NPY neuron populations. However, a partial overlapping of cortical and nigral influences was seen to occur from the increase in the staining intensity of labeled cells observed after both types of deafferentation in the same ventral and middle striatal areas.

Combined nigrostriatal and corticostriatal deafferentations

Hemidecortication was performed one week after unilateral intranigral 6-OHDA injection on the same brain side and the effects of the combined lesions were estimated 20 to 30 days later. The main result of this experiment was the striking disappearance of the ventromedial increase in NPY neuron density related to selective nigrostriatal dopaminergic deafferentation. In addition, the onset of the dorsolateral increase in NPY neuron density related to selective corticostriatal deafferentation was delayed and the amplitude of this increase was lower in comparison with the effects of hemidecortication alone. Furthermore, the increase in the staining intensity observed in ventromedial areas disappeared (Kerkerian et al, in the press).These data suggest that the response of NPY neurons to striatal DA deafferentation may be dependent on corticostriatal transmission, and hence that functional interactions between the nigrostriatal and corticostriatal pathways are involved in the regulation of striatal NPY function. We have reported above that NPY responses to DA deafferentation may be attributable to an increase in D1 receptor sensitivity. The first possible explanation for the suppression of this specific NPY response after combined deafferentation might be that the increase in D1 receptor sensitivity is abolished after corticostriatal deafferentations. This hypothesis is supported by the recent finding that alteration of corticostriatal transmission abolished the increase in adenylate cyclase activity observed in response to selective unilateral destruction of the nigrostriatal dopaminergic pathway (Hervé et al, 1988). Moreover, cortical projections have been previously shown to contribute to the regulation of D1-mediated DA transmission at the level of the NAcc (Reibaud et al., 1984). On the other hand, it was shown that the up-regulations

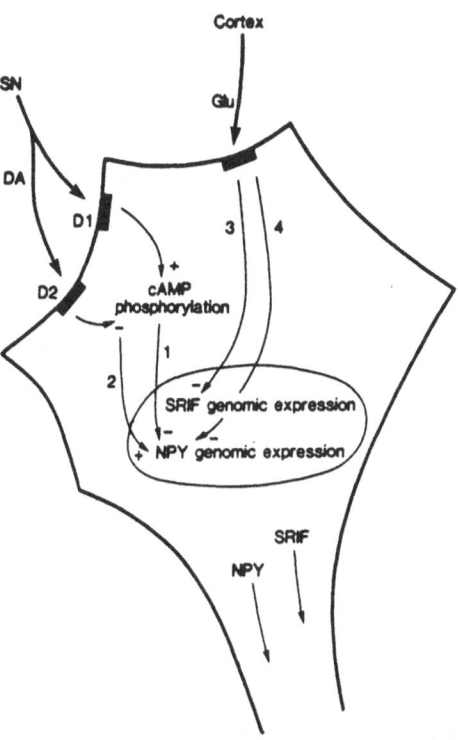

Fig. 2. Hypothetical mechanisms involved in the control of NPY/somatostatin-immunoreactive (SRIF) striatal neurons by the nigrostriatal dopaminergic and the corticostriatal glutamatergic pathways. The nigrostriatal dopaminergic input may exert a dual action upon NPY genomic expression, inhibitory throught D1(1) and activatory throught D2(2) receptor subtypes, but may not influence SRIF genomic expression in the same striatal neurons. On the other hand, the cerebral cortex may exert a parallel inhibitory influence on genomic expression of both SRIF (3) and NPY (4).

of D2 receptor observed in response to 6-OHDA denervation and to receptor blockade by haloperidol are additive, which suggests that a regulatory factor other than DA may be involved in the mechanisms of D2 receptor regulation at striatal level (Reches et al, 1983). Assuming that the corticostriatal input contributes to the regulation of D2 receptors and that DA exerts a dual action over both NPY and adenylate cyclase activities, an alternative interpretation might be that cortical lesion may have induced the development of D2 receptor sensitivity in medial striatal areas, thus counteracting the effects of D1 receptor supersensitivity on NPY and adenylate cyclase activities. In conclusion, whatever the exact mechanism responsible for the NPY neuron responses, our data suggest that the corticostriatal pathway has permissive effects on the expression of DA-NPY striatal interactions through a putative DA postsynaptic receptors regulation by receptors mediating corticostriatal transmission. Recent biochemical data have suggested that a reciprocal heteroregulation of a subtype of excitatory aminoacid receptors (NMDA type) may be effected by DA afferents (Samuel et al, in the press). Taken as a whole, the available data support the view that besides previously reported presynaptic interactions, postsynaptic receptor-receptor interactions may be involved in the functional reciprocal relationships between nigrostriatal and corticostriatal transmissions.

Finally, we have also shown that kainic acid-induced unilateral lesion of the thalamostriatal pathway induced significant bilateral increases in NPY-positive neuron density in very restricted medial striatal areas (Kerkerian et al, in the press). The topography of these changes (figure 1) did not resemble the unilateral widespread distribution of thalamostriatal afferent fibers (Jones and Leavitt,1974). At the present stage, the NPY responses to thalamostriatal deafferentation are therefore not attributable to a direct thalamic influence over striatal NPY neurons.

Are striatal NPY and somatostatin functions correlated?

In view of the presumably frequent neuronal colocalization of NPY with somatostatin at striatal level, one would expect a similar adaptation of the cell populations containing these peptides to occur in response to experimentally induced modifications of their biochemical environment. It thus appeared to be of interest to compare the effects induced by interruption of nigrostriatal and corticostriatal transmission on striatal somatostatin activity with those reported on NPY activity under similar conditions. These data were obtained on the rat. The functional hypothesis raised are illustrated in figure 2. Hemidecortication was found to induce parallel changes in the expression of somatostatin and NPY immunoreactivities by striatal neurons (Salin et al, submitted), which suggests that the corticostriatal pathway may influence both striatal somatostatin and NPY function in a similar way. The increase observed in both NPY and somatostatin-positive neuron density within the ipsilateral striatum was prevalent in the same dorsolateral areas of the structure. However, changes in NPY immunostaining were bilateral whereas changes in somatostatin labeling were mainly ipsilateral. The adaptation of the two peptide functions in response to haloperidol treatment also appeared to be correlated. Indeed, in agreement with the decreased striatal NPY immunostaining possibly reflecting decreased intraneuronal NPY concentrations that we reported above, haloperidol was found to decrease striatal somatostatin concentrations measured by radioimmunoassay (Beal and Martin, 1984) as well as preprosomatostatin mRNA levels (Salin et al, in the press). In contrast, α-MPT administration and 6-OHDA nigrostriatal deafferentation were both found to have no effect on somatostatin immmunostaining, whereas both treatments were previously reported to induce marked changes in NPY immunostaining. These data raise two questions :

First, the discrepancy between the effects of haloperidol and the lack of effect of α-MPT- or 6-OHDA-induced DA depletion on the somatostatin contents. Since haloperidol has been reported to preferentially block D2 receptors, this apparent conflict may be accounted for by an opposite action of DA on somatostatin function through D1 and D2 receptors as previously hypothesized in connection with the dopaminergic control of striatal NPY or enkephalin functions (see Salin et al, in the press). On the other hand, the peptidergic response to haloperidol treatment cannot be said, as in the case of 6-OHDA intranigral injection, to have involved the nigrostriatal tract. Therefore at this stage of the experiments, there is no convincing evidence that somatostatin function may be regulated by nigrostriatal dopaminergic projections.

The second question concerns the discrepancy between the somatostatin and NPY modifications which occur in response to striatal DA depletion. A similar conflict was also evidenced by data from human neuropathology showing that changes in NPY and somatostatin levels are not always parallel in brain regions where their neuronal colocalization has been found to be most predominant (see Gaspar et al, 1987). For instance, in subjects with Parkinson's or Alzheimer's disease, cortical levels of somatostatin were found to have decreased, whereas no change in the NPY levels was observed. However in the latter pathology, parallel increases in both peptide contents were detected at the level of the substantia innominata. Likewise, in choreic subjects, the concentrations of both peptides are known to increase considerably and neurons immunoreactive to these peptides are spared and even increase in number, whereas the whole striatal neuronal population is depleted by more than 50%. In a recent report,

Gaspar et al (1989) showed that the apparent conflict between the somatostatin and NPY alterations at cortical level in senile dementia of the Alzheimer type is due to the selective vulnerability of the neurons in which somatostatin is not colocalized with NPY. In our experiments, however, the hypothesis that NPY responses to α- MPT and 6-OHDA treatments may selectively involve the neurons containing only NPY and not those containing both peptides is unlikely. Indeed, besides changes in the density of NPY cells that may be restricted to a neuronal subset in which NPY is not colocalized with somatostatin, marked changes in the staining intensity of the vast majority of NPY-positive cells were observed after both treatments in the absence of any change in somatostatin immunostaining, which suggests that virtually all of the striatal NPY-containing neuron populations are affected. Assuming that NPY/somatostatin neuronal colocalization occurs in the striatum, these data suggest rather that the nigrostriatal dopaminergic pathway may differentially influence NPY and somatostatin functions in the same neurons. It is therefore likely that the differential responses of coexisting peptides may not necessarily be mediated by the same mechanisms in every cases and may be attributable either to incomplete overlapping in the distribution of the two peptides or to the fact that differential control is exerted on each of the coexisting peptides in a single neuronal population. Since so many neurons exhibit extensive colocalization in various brain structures, the possibility of an independant regulation of coexisting neuroactive substances may be a general feature providing a restricted number of neuronal elements with great functional diversity.

Cellular mechanisms underlying changes in striatal NPY or somatostatin immunostaining. Genomic or metabolic responses ?

Changes in the expression of NPY or somatostatin immunoreactivities by rat striatal neurons were observed after disruption of striatal afferent inputs. These changes reflected an adaptation of the peptide function and provide evidence that functional relationships may exist between NPY/somatostatin interneurons and striatal afferent pathways. The question then arises as to the mechanisms underlying the changes in peptide immunoreactivity detected by means of immunocytochemical techniques. An increase in the density of immunolabeled cells may indicate that some neurons normally not containing NPY or somatostatin have newly acquired the ability to express the peptide. This suggests that a derepression of the gene encoding the peptide prohormones may have occured. Conversely, a decrease in the neuron density may mean that some cells have stopped the peptide synthesizing. In view of the detection limits of the immunocytochemical method, another possibility is that changes in the number of labeled neurons may be due to increases or decreases in the peptide intraneuronal levels which favour or prevent, respectively, the immunodetection of a neuronal subset containing the peptide. The finding that changes in the density of labeled neurons are generally accompanied by changes in the staining intensity of the majority of the labeled cells in the same regions argues in favour of the latter proposal. One might indeed expect that, depending on the initial peptide contents of the cells, similar modifications in the peptide levels may induce either changes in the number of immunodetectable cells or changes in the labeling intensity of the cells. Similar changes in both parameters may be observed when cell populations containing the same peptide concentrations are subject to variable levels of regulation. This

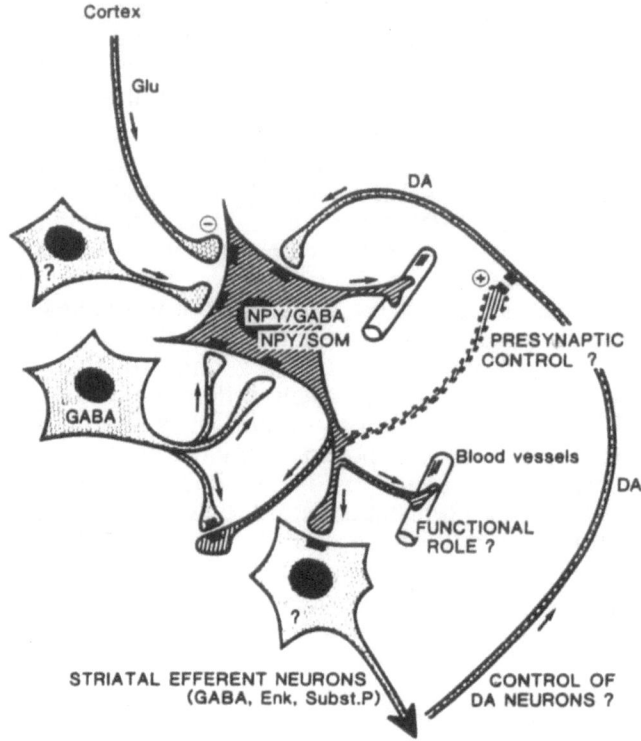

Fig. 3. The neuropeptide Y-containing neurons in the striatal network. Anatomical and possible functional cellular interactions.

kind of changes in the intraneuronal peptide concentrations may reflect an adaptive change in the propeptide gene expression, in the processing from prohormone to final peptide or in the release of the peptide. It may be significant that the peptide responses we observed by immunocytochemistry are frequently correlated to changes in the peptide mRNA levels detected by in situ hybridization under similar experimental conditions. This correlation suggests that changes in immunostaining may reflect an adaptation of peptide gene expression. Whatever the mechanisms involved, one major interest of these immunocytochemical data was to reveal that, although NPY is to be found in a single population of aspiny type neurons quite evenly distributed in the striatum, this neuronal striatal population is functionally heterogeneous.

CONCLUSIONS

In the rat striatum, NPY is to be found in a single population of aspiny type interneurons which are fairly evenly distributed and is frequently colocalized with somatostatin and GABA. The NPY-containing neurons may constitute targets for striatal afferents originating in the substantia nigra and the cerebral cortex as well as for intrinsic GABAergic neurons, and are involved in the functional interactions between corticostriatal and nigrostriatal inputs. Conversely, the NPY neurons may contribute to the regulation of striatal DA and GABA functions. Indeed intrastriatal NPY administration has been found to induce a dose dependent increase in DA turnover as reflected by an increase in the DOPAC/DA ratio (Beal et al, 1986b). Accordingly, in preliminary experiments conducted by our group, an increase in the striatal release of DA was measured using a voltammetric method after intraventricular injection of the peptide. In parallel, we observed a decrease in high affinity GABA uptake reflecting an alteration of striatal GABAergic function in response to intracerebral NPY administration. The effects of NPY on striatal DA and GABA functions may involve either direct presynaptic interactions, as suggested by the axo-axonic contacts detected or postsynaptic interactions with striatal local circuit and output neurons (see figure 3). The specific regional changes in NPY-positive neuron density and staining intensity which occured in response to striatal deafferentations revealed that the striatal NPY neuron population can be divided into subsets with different functional specificities. This kind of functional heterogeneity may result from:

1) a differential connectivity of NPY neuron populations depending on their position in the striatal network (as regards the patterning of striatal afferent pathways and related receptors, and the organisation of other intrinsic neurons for instance)

2) differential intracellular associations of NPY with other neuroactive substances (somatostatin, GABA)

3) a differential control exerted either on the metabolism or on the gene expression of the coexisting peptides.

Aknowledgements. The authors are grateful to Dr. G. Pelletier for kindly providing NPY antisera, to Dr. H. Vaudry and Dr. J. Epelbaum for allowing the measurements of peptides levels and to Dr. O. Bosler for his contribution to the development of the immunocytochemical electron microscopic methods. This work was supported by grants from CNRS, Université Aix-Marseille II, INSERM (CRE : 876 016), La Fondation pour la Recherche Médicale and from DRET, (convention n° 88 049).

BIBLIOGRAPHY

Adrian, T.E., Allen, J.M., Bloom, S.R., Ghatei, M.A., Rossor, M.N., Roberts, G.W., Crow, T.J., Tatemoto, K. and Polak, J.M., 1983, Neuropeptide Y distribution in humain brain. Nature, 306: 584-586.

Allen, Y.S., Adrian, T.E., Allen, J.M., Tatemoto, K., Crow, T.J., Bloom, S.R. and Polak, J.M., 1983, Neuropeptide Y distribution in the rat brain, Science, 221: 877-879.

Aoki, C. and Pickel, V.M., 1988, Neuropeptide Y-containing neurons in the rat striatum : ultrastructure and cellular relations with tyrosine hydroxylase-containing terminals and with astrocytes, Brain Res., 459: 205-225.

Beal, M.F., Chattha, G.K. and Martin, J.B., 1986a, A comparison of regional somatostatin and neuropeptide Y distribution in rat striatum and brain. Brain Res., 377: 240-245.

Beal, M.F., Franck, R.C., Ellison, D.W. and Martin, J.B., 1986b, The effects of neuropeptide Y on striatal catecholamines, Neurosci.Lett., 71: 118-123.

Beal, M.F. and Martin, J.B., 1984, Effects of neuroleptic drugs on brain somatostatin-like immunoreactivity, Neurosci.Lett., 47: 125-130.

Beal, M.F., Mazurek, M.F. and Martin, J.B., 1987, A comparison of somatostatin and neuropeptide Y distribution in monkey brain. Brain Res., 405: 213-219.

Christie, M.J., Beart, P.M., Jarrott, B. and Maccarone, C., 1986, Distribution of neuropeptide Y immunoreactivity in the rat basal ganglia : effects of excitotoxin lesions to caudate-putamen, Neurosci.Lett., 63: 305-309.

Chronwall, B.M., Chase, T.N. and O'Donohue, T.L.,1984, Coexistence of neuropeptide Y and somatostatin in rat and human cortical and rat hypothalamic neurons, Neurosci.Lett., 52: 213-217.

Cospito, J.A. and Kultas-Ilinsky, K., 1981 Synaptic organization of motor corticostriatal projections in the rat, Exp.Neurol., 72: 257-266.

De Quidt, M.E. and Emson, P.C., 1986, Distribution of neuropeptide Y-like immunoreactivity in the rat central nervous system - II. Immunohistochemical analysis, Neurosci., 18: 545-618.

Di Maggio, D.A., Chronwall, B.M., Buchanan, K. and O'Donohue, T.L., 1985, Pancreatic polypeptide immunoreactivity in rat brain is actually neuropeptide Y, Neurosci., 15: 1149-1157.

Gaspar, P., Berger, B., Lesur, A., Borsotti, J.P. and Febvret, A., 1987, Somatostatin 28 and neuropeptide Y innervation in the septal area and related cortical and subcortical structures of the human brain. Distribution, relationships and evidence for differential coexistence. Neurosci., 22: 49-73.

Gaspar, P., Duyckaerts, C., Febvret, A., Benoit, R., Beck, B. and Berger, B., 1989, Subpopulations of somatostatin 28-immunoreactive neurons display different vulnerability in senile dementia of the Alzheimer type, Brain Res., 490: 1-13.

Hassler, R., Chung, J.W. and Rinne, U., 1977, Experimental demonstration of intrinsic synapse in cat's caudate nucleus, Neurosci.Lett., 5: 117-121.

Hendry, S.H.C., Jones, E.G., De Felipe, J., Schmechel, D., Brandon, C. and Emson, P.C., 1984, Neuropeptide-containing neurons of the cerebral cortex are also GABAergic. Proc.Natl.Acad.Sci. USA, 81: 6526-6530.

Hervé, D., Trovero, F., Blanc, G., Studler, J.M., Thierry, A.M., Glowinski, J. and Tassin, J.P., 1988, Heterologous regulation of dopaminergic D_1 receptors in the rat striatum : involvement of cortico-subcortical relationships. Abst. Europ. Neurosci. Association. 11th annual meeting. Europ. J. Neurosci. Suppl., 338 : Abst.n°88.3.

Jones, E.G. and Leavitt, R.Y., 1974, Retrograde axonal transport and the demonstration of   non-specific projections to the cerebral cortex and striatum from thalamic intralaminar nuclei in the rat, cat and monkey, J.Comp.Neurol., 154: 349-378.

Kerkerian, L., Bsler, O., Pelletier, G. and Nieoullon, A.,1986, Striatal neuropeptide Y neurons are under the influence of the nigrostriatal dopaminergic pathway : immunohistochemical evidence. Neurosci. Lett., 66 : 106-112.

Kerkerian, L., Salin, P. and Nieoullon, A.,1988, Pharmacological characterization of dopaminergic influence on expression of neuropeptide Y immunoreactivity by rat striatal neurons. Neuroscience, 26 : 809-817.

Kerkerian, L., Salin, P. and Nieoullon A., Cortical regulation of striatal neuropeptide Y (NPY)-containing neurons in the rat, Europ.J.Neurosci., in the press.

Kowall, N.W., Ferrante, R.J., Beal, M.F., Richardson, Jr. E.P., Sofroniew, M.V., Cuello, A.C. and Martin, J.B., 1987, Neuropeptide Y, somatostatin and reduced nicotinamide adenine dinucleotide phosphate-diaphorase in the human striatum : a combined immunocytochemical and enzyme histochemical study, Neurosci., 20: 817-828.

Massari, V.J., Chan, J., Chronwall, B.M., O'Donohue, T.L., Oertel, W.H. and Pickel, V.M., 1988, Neuropeptide Y in the rat nucleus accumbens : ultrastructural localization in aspiny neurons receiving synaptic input from GABAergic terminals, J.Neurosci.Res., 19: 171-186.

Reches,A., Wagner R.H., Jackson V., Yablonskaya-Alter, E. and Fahn, S., 1983, Dopamine receptors in the denervated striatum : further supersensitivity by chronic haloperidol treatment, Brain Res., 275 : 183-185.

Reibaud, M., Blanc, G., Studler, J.M., Glowinski, J. and Tassin J.P., 1984, Non-DA prefronto-cortical efferents modulate D_1 receptors in the nucleus accumbens, Brain Res.,305 : 43-50.

Salin, P., Mercugliano, M. and Chesselet, M.F, Differential effects of chronic treatment with haloperidol and clozapine on the level of preprosomatostatin mRNA in the striatum, nucleus accumbens and frontal cortex of the rat,
Mol.Cell. Neurobiol., in the press.

Samuel, D., Errami, M. and Nieoullon, A., Localization of NMDA receptors in the rat striatum : effects of specific lesions on the ^{3}H-CPP binding, J.Neurochem., in the press.

Savasta, M., Dubois, A., Benavidès, J. and Scatton, B. ,1988, Different plasticity changes n D1 and D2 receptors in rat striatal subregions following impairment of dopaminergic transmission. Neurosci. Lett., 85 : 119-124.

Smith, Y. and Parent, A., 1986, Neuropeptide Y-immunoreactive neurons in the striatum of cat and monkey : morphological characteristics, intrinsic organization and co-localization with somatostatin, Brain Res., 372 : 241-252.

Smith, Y., Parent, A., Kerkerian, L. and Pelletier, G., 1985, Distribution of neuropeptide Y immunoreactivity in the basal forebrain and upper brainstem of the squirrel monkey (Saimiri sciureus), J. Comp. Neurol., 236: 71-89.

Tatemoto, K., 1982, Neuropeptide Y : complete amino acid sequence of the brain peptide, Proc.Natl.Acad.Sci., USA, 79: 5485-5489.

Tatemoto, K., Carlquist, M. and Mutt, V., 1982, Neuropeptide Y- a novel brain peptide with structural similarities to peptide YY and pancreatic polypeptide, Nature, 296: 659-660.

Villalba, R.M., Martinez-Murillo, R., Blasco, I., Alvarez, F.J. and Rodrigo, J., 1988, C-Pon containing neurons in the rat striatum are also positive for NADPH-diaphorase activity. A light microscopic study, Brain Res., 462, 359-362.

Vincent, S.R., Johansson, O., Hökfelt, T., Skirboll, L., Elde, R.P., Terenius, L., Kimmel, J. and Goldstein, M., 1983, NADPH-diaphorase : a selective histochemical marker for striatal neurons containing both somatostatin- and avian pancreatic polypeptide (APP)-like immunoreactivities, J.Comp.Neurol., 217, 252-263.

Vincent, S.R., Skirboll, L., Hökfelt, T., Johansson, O., Lündberg, J.M., Elde, R.P., Terenius, L. and Kimmel, J., 1982, Coexistence of somatostatin- and avian pancreatic polypeptide (APP)-like immunoreactivities in some forebrain neurons, Neurosci., 7: 439-446.

Vuillet, J., Kerkerian, L., Kachidian, P., Bosler, O. and Nieoullon, A.,1989a, Ultrastructural correlates of functional relationships between nigral dopaminergic or cortical afferent fibers and neuropeptide Y-containing neurons in the rat striatum, Neurosci. Lett., 100 : 99-104.

Vuillet, J., Kerkerian, L., Salin, P. and Nieoullon, A., 1989b, Ultrastructural features of NPY-containing neurons in the rat striatum, Brain Res., 477 : 241-251.

TOPOGRAPHICAL ORGANIZATION OF THE THALAMOSTRIATAL PROJECTION

IN THE JAPANESE MONKEY, MACACA FUSCATA, WITH SPECIAL REFERENCE TO

THE CENTROMEDIAN-PARAFASCICULAR AND MOTOR THALAMIC NUCLEI

Katsuma Nakano[1], Yasuo Hasegawa[2], Tetsuro Kayahara[1]
and Yoshihiro Kuga[1]

[1]Department of Anatomy, School of Medicine, Mie University
Tsu, Mie 514, Japan
[2]Department of Anatomy, School of Medicine
Kagoshima University, Kagoshima 890, Japan

ABSTRACT

Thalamostriatal projection was studied in monkeys (Macaca fuscata) using the axonal transport techniques of WGA-HRP and autoradiography to clarify the topographical organization. Major findings were as follows: The dorsomedial part of the centromedian nucleus (CM) projects to the dorsolateral strip in the leg territory of the putamen (Put); the ventromedial CM projects to the ventromedial strip in the face area; and the lateral CM projects to the intermediate strip in the arm region of the Put. The lateral part of the parafascicular nucleus (Pf) and nucleus ventralis anterior pars parvicellularis project to the lateral part of the head of the caudate nucleus (CN) and to the dorsolateral part of the rostral Put. Whereas the medial Pf and nucleus ventralis anterior pars magnocellularis project to the medial and ventral parts of the CN and to the ventromedial part of the rostral Put. The nucleus ventralis lateralis pars oralis projects to the Put. The medial part of the nucleus subthalamicus (STN) projects to the head of CN and rostral Put, while the lateral STN projects to the remaining whole part of the Put.

INTRODUCTION

The neostriatum appears to have a heterogeneous structural organization, in spite of a fairly homogeneous cytoarchitecture. On an anatomical and functional basis the striatum has been divided into limbic and nonlimbic portions (Kelley et al., 1982). Heterogeneity in the neostriatum is expressed by the large number of different sources of afferents, by their distinct topographical distribution and by their unique neurotransmitters (Selemon & Goldman-Rakic, 1985; see reviews by Carpenter, 1984 and Graybiel & Ragsdale, 1983). Because the neostriatum is the most important receptive portion of the basal ganglia, there is a need for considerably greater detail concerning inputs. In contrast to the organization of the cortical afferents, part of which is somatotopic, papers dealing with the thalamostriatal projections in monkeys are limited, and lack specific detail (Powell & Cowan, 1956; Jones & Leavitt, 1974; Kalil, 1978; Parent et al., 1983; Smith & Parent, 1986; Parent & Smith, 1987).

The Basal Ganglia III, Edited by G. Bernardi *et al.*
Plenum Press, New York

The purpose of this paper was to investigate cellular origins of the thalamic afferents to the striatum in the macaque monkey, as the distinct projections from the well developed centromedian-parafascicular complex and subdivisions of the motor thalamic nuclei to the heterogeneous striatum are presumed.

METHODS

Retrograde and anterograde axonal transport methods were used. Experiments were performed on 39 Japanese monkeys (<u>Macaca fuscata</u>) ranging in weight 4.2-12.4 kg. Each monkey was anesthetized with ketamine hydrochloride (Ketalar, 10mg/kg intramuscular injection) followed by an intraperitoneal injection of sodium pentobarbital (15-30mg/kg) for aseptic surgery. The animal's heads were held in a sterotaxic apparatus (David Kopf). Then a tracer was injected into the striatum or centromedian-parafascicular complex (CM-Pf) following craniotomy.

<u>Retrograde axonal tracing method</u>: Wheat germ agglutinin conjugated horseradish peroxidase (WGA-HRP) was used as a marker. Single injections of 2% WGA-HRP (Toyobo) were placed ipsilaterally or bilaterally into the various portions of the caudate nucleus (CN) or the putamen (Put) by manual pressure (0.03-0.06 μl) from a glass micropipette or by electrophoresis (3-5 μA, 200-ms duration, 2.5 Hz, for 20-30 minutes). After a survival time of 24-48h, all animals were perfused under deep anesthesia with 500 ml of 0.9 % saline followed by 3000 ml of freshly prepared fixative (1.25 % glutaraldehyde-0.1 % paraformaldehyde mixture in a 0.1 M phosphate buffer, pH 7.4 or 9 % formalin in 0.1 M phosphate buffer, pH 7.2-7.3), and finally by 1000 ml of the same buffer containing 10 % sucrose at 4 °C. Serial frozen sections of the brain were processed by diaminobenzidine (DAB) or tetramethyl benzidine (TMB) for HRP histochemistry.

<u>Anterograde axonal tracing method of an autoradiography</u>: The ^{3}H-leucine was evaporated under nitrogen gas, and then reconstituted in concentration of 20-100 μci/μl with sterile saline. Also, ^{3}H-leucine was injected stereotaxically into the CM through a glass micropipette attached to a needle of a 1 μl Hamilton syringe following craniotomy. After 7 days of survival, the monkeys were perfused with 10 % buffered neutral formalin and processed for autoradiography using Sakura NR-M2 emulsion and a 4-20 week exposure time.

Retrogradely labeled neurons or anterogradely labeled terminals and fibers were observed microscopically under bright and darkfield illumination. The nomenclature of the thalamic nuclei was taken from Olszewski (1952).

RESULT

<u>WGA-HRP studies</u>. WGA-HRP was injected into the various parts of the head of the CN in 10 cases, and of the Put in 27 cases. These experiments indicated topographical arrangement between the striatum and the CM-Pf complex. These results will be published in a separate paper (Nakano et al.).

When large injection of WGA-HRP was made into the central portion of the dorsal part of the Put at the levels of the rostral portion of the thalamus, labeled neurons were observed in the dorsomedial part of the CM, except for its dorsomedial corner, and in the lateral part of the subthalamic nucleus (STN), nucleus ventralis lateralis pars oralis (VLo), and in the nucleus tegmenti pedunculopontinus (PPN). A lesser number of labeled neurons were seen in the intralaminar thalamic nuclei, midline thalamic nuclei, dorsal raphe nuclei and hypothalamic area.

After large injection was placed into the ventromedial portion of the Put at the anterior commmissure level, the greatest number of labeled neurons were seen in the ventromedial part of the CM and in the lateral part of the STN. Also, a moderate number of labeled neurons were observed in the nucleus ventralis anterior pars parvicellularis (VApc), midline thalamic nuclei, intralaminar nuclei and the PPN.

When WGA-HRP was injected obliquely in ventromedial direction into the intermediate portion of the Put at the levels of the rostral end of the thalamus, the number of the labeled neurons were the greatest in the lateral portion of the CM; moderate in the VApc, lateral part of the STN and the PPN; and lesser in the intralaminar nuclei, hypothalamic area and the dorsal raphe nuclei.

When WGA-HRP was injected into the lateral part of the head of the CN with partial involvement of adjacent white matter, well labeled neurons were found from the dorsal to the ventrolateral parts of the nucleus parafascicularis (Pf), VApc, and in the medial part of the STN. There were also labeled neurons in the anterior thalamic nuclei, the intralaminar nuclei and the nucleus medialis dorsalis (MD).

Following WGA-HRP injection into the dorsomedial part of the head of the CN, a great number of labeled neurons were found in the medial part of the Pf. A moderate number of the labeled neurons were detected in the medial part of the STN, in addition to the PPN, the midline thalamic nuclei and the nucleus ventralis anterior pars magnocellularis (VAmc).

When WGA-HRP injection was placed in the ventral part of the head of the CN, well labeled neurons were seen from the most medial part of the Pf to the midline thalamic nuclei. Labeled neurons were also observed in the VAmc, PPN, dorsal raphe nuclei and the STN.

<u>Autoradiographic studies</u>. In these experiments, ^{3}H-leucine injections were made into the various portions of the CM-Pf complex in 14 cases to confirm the topographical arrangement, which has been traced by the retrograde transport of WGA-HRP. These injection sites were divided into four groups; the dorsomedial, lateral, ventromedial and the medial parts of CM. The latter part involves portions of the Pf (Fig. 1).

Four injections were made in the dorsomedial part of the CM. In M 134L, a ^{3}H-leucine injection was made in the middle third of the dorsal half of the CM (Fig. 1B). The labeled terminals were observed in the

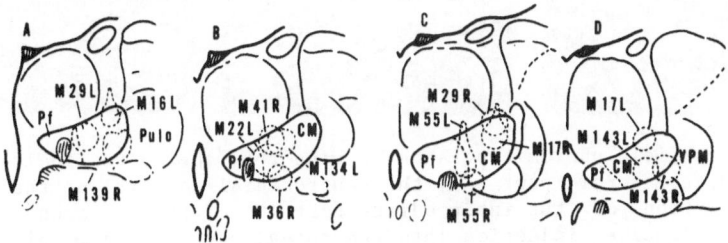

Fig. 1. Diagrams of transverse sections through the center of isotope deposits (dotted line circle) in cetromedian-parafascicular complex (CM-Pf).

dorsolateral part of the Put caudally to the anterior commissural level
(Fig. 2A). These labeled terminals were also well localized in its dorsolateral portion of the caudal pole of the Put. More rostrally to the
anterior commissure, the labeled terminals were seen in the lateral margin
of the Put. There were also labeled terminals in the lateral end of the
head of the CN. In M 41R and M 17L, the injection sites were located more
rostrally than that of M 134L, and involved dorsomedial part of the CM and
adjacent ventrolateral part of the MD (Fig. 1B,D). Labeled terminals were
observed in the dorsolateral part of the Put and the lateral part of the
head of the CN. At the levels rostral to the anterior commissure, the
labeled terminals were well localized in the lateral zone of the Put. In
M 29L the injection site was shifted more medially, and involved the
dorsomedial part of the CM and the dorsolateral part of the Pf (Fig. 1A).
Labeled terminals were present in the dorsal part of the Put and the
lateral part of the CN.

The lateral part of the CM received five ^{3}H-leucine injections. The
injection sites in M 16L, M 17R and M 29R were located more dorsally, and
those in M 139R and M 143R located more ventrally. In M 17R, the injection was placed in the dorsolateral part of the CM (Fig. 1C). Labeled
terminals were present in the intermediate oblique strip, extending ventrolaterally, and in the lateral margin of the Put (Fig. 3). More rostrally, these labelings tended to localize more laterally in the lateral
margin of the Put (Fig. 3E, F) and no terminal labelings were seen in the
rostral tip of the Put. A small amount of the terminal labelings were
seen in the ventrolateral portion of the head of CN and were adjacent to
the internal capsule (Fig. 3C-F). There were also labeled terminals in
the lateral part of the STN (Fig. 3B). In this case patches of the
labeled terminals were seen in the Put. In other two cases, M 29R and M
16L, the injection sites were located in the dorsolateral part of the CM

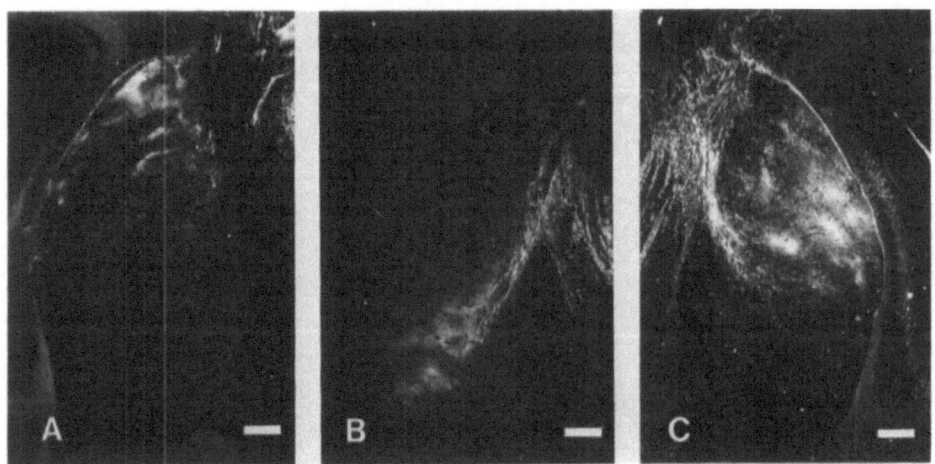

Fig. 2. Darkfield photomicrographs showing the labeled terminals in the
dorsolateral part (A), the ventromedial part (B) or the lateral
portion of the intermediate part (C) of the putamen following
^{3}H-leucine injection into the dorsal part of the middle third of
the CM (M 134L), the ventral part of the middle third of the CM
(M 143L), or into the ventrolateral part of the CM (M 139R),
respectively. Scale bars = 1 mm.

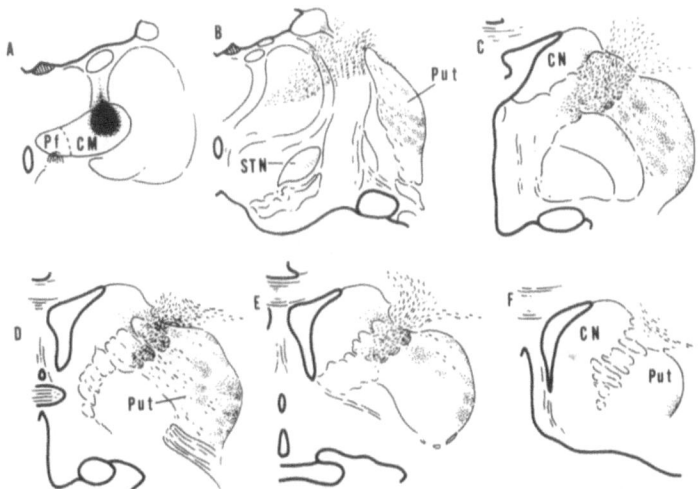

Fig. 3 Diagrams of selected transverse sections to show the striatal
labeling (B-F) from an isotope deposit in the dorsolateral CM (A).
Labeled terminals and fibers are indicated by dots and dashes,
respectively (17R).

and in the ventral part of the nucleus centralis lateralis of the intralaminar nuclei (Fig. 1A, C). Labeled terminals were present dorsally in
the dorsal and intermediate strips of the Put, and in the lateral part of
the head and body of the CN. More rostrally, the labeled terminals were
seen in the lateral zone of the Put. In M 139R and M 143R the injection
sites were located in the ventrolateral part of the CM, and in adjacent
parts of the oral part of the pulvinar nucleus or of the nucleus ventralis
posteromedialis (VPM), respectively (Fig. 1A, D). The injection site of
the former case was located more caudally. The labeled terminals in the
Put were observed in the lateral part of the intermediate strip in M 139R
(Fig. 2C). These labelings shifted more ventromedially in M 143R.

In three cases, an injection into the ventromedial part of the CM was
attempted. In M 143L, the injection site was located in the middle third
of the ventral half of the CM (Fig. 1D). Labeled terminals were observed
in the ventromedial portion of the Put (Fig. 2B), and in the small area of
the ventrolateral part of the CN. In M 55R an injection was made in the
ventromedial corner of the CM and the medial part of the VPMpc (Fig. 1C).
Labeled terminals were seen in the medial zone of the Put, parallel to the
lateral medullary lamina. No labeled terminals were seen in the most-
rostral portion of the Put. In M 36R, the injection site was centrally
located in the medial part of the VPMpc, and the ventromedial part of the
CM also resulted in faint labeling (Fig. 1B). Labeled terminals were
observed in the medial to ventromedial zone of the Put, and in the ventral
margin of the lateral half of the head and body of the CN.

In M 22L and M 55L, the injection sites were placed in the medial part
of the CM and encroached upon the lateral part of the Pf (Fig. 1B, C).
Labeled terminals were observed in the dorsal half of the Put and aggregated into several strips extending ventrolaterally. These labelings were
more dense in the dorsal or dorsomedial portions of the Put caudally to
the anterior commissure level. More rostrally, the labelings in the Put
tended to localize more dorsolaterally. Labeled terminals were present
also in the lateral part of CN from the head to the tail in M 22L. There
were also labeled terminals in the lateral margin of the STN.

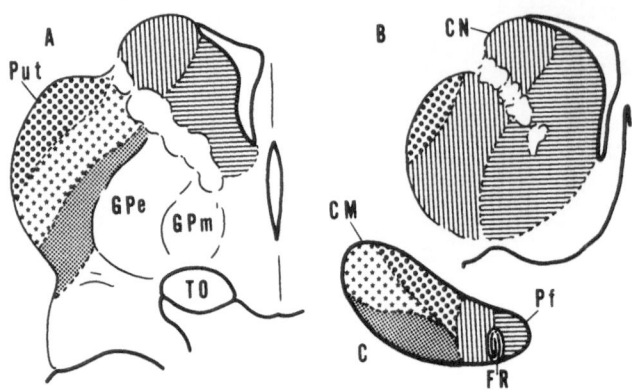

Fig. 4. Summary diagrams showing the topographical arrangement of the
striatal projections from the CM-Pf complex. The dorsomedial part
of CM projects to the dorsolateral strip corresponding to the leg
territory in the putamen (Put)(dotted area); the lateral part of
CM projects to the intermediate strip corresponding to the arm
territory in the Put (star area); or the ventromedial part
projects to the ventromedial strip corresponding to the face
territory (stippled area). The lateral part of Pf projects to the
lateral portions in the caudate nucleus (CN) and to the most
rostral part of the Put (vertical line area), and from the medial
part of the Pf to the medial portions in the CN and the most
rostral part of the Put (horizontal line area). GPe: lateral
pallidal segment; GPm: medial pallidal segment; TO: tractus opticus; FR: fasciculus retroflexus. Modified from Nakano et al..

DISCUSSION

There appear to be species differences between cat and monkey concerning the projections of the CM-Pf complex to the CN and Put among several
authors (Cowan & Powell, 1955; Powell & Cowan, 1956; Johnson, 1961; Sato
et al., 1979; Royce, 1978a,b; Royce & Mourey, 1985). Selective CM projection to the Put has been reported by some researchers in the monkey
(Powell & Cowan, 1956; Mehler, 1966; Parent et al., 1983; Smith & Parent,
1986).

In the present findings, the CM projects mainly to the Put with only a
few additional terminations in the dorsolateral corner of the CN adjacent
to the internal capsule in the macaque monkey. In the cat both projections to the Put and CN from the CM were observed (Nakano et al., 1983).
These findings indicate that the CM projects preferentially to the Put in
the monkey and to both the Put and CN in the cat. According to Beckstead
(1984), the CM projects to the laterodorsal portions of both the Put and
CN, whereas the Pf projects to the ventromedial portions of the same two
structures in the cat. A clear differential projection of the CM-Pf
complex to the striatum was reported in the squirrel monkey; the CM
projects exclusively to the Put, and the Pf only to the CN (Parent et al.,
1983; Smith & Paren, 1986). Our findings indicate that the Pf projects to
the CN and to the ventromedial region of the rostral-most Put. The injections of WGA-HRP, which were placed ventrally in the head CN, resulted in
the occurrence of labeled neurons in the more dorsomedial region of the
Pf. The dorsomedial Pf and midline thalamic nuclei project to the ventral
striatum (Groenewegen et al.,1980; Beckstead, 1984). This finding was

supported in our study. The ventral striatum, midline thalamic nuclei and the dorsomedial Pf may be engaged in the limbic connections.

WGA-HRP injections in the present experiments and detailed observations in a separate paper (Nakano et al.) indicate the topographical arrangement; the dorsomedial portion of the CM projects to the dorsolateral area in the Put, the ventromedial portion of the CM projects to the ventromedial area, and the lateral CM projects to the intermediate area in the Put (Fig. 4). Autoradiographic findings in the present experiments supported this topographical organization. These areas appear to correspond to the cortical territories indicated by Künzle's (1975) in the Put.

The striatal projections from the VA were described (Royce, 1978a,1983; Beckstead, 1984; Macchi et al.,1984; Smith & Parent, 1986; Tanaka et al.,1986), as were those from the VL (Royce, 1983; Jayaraman, 1985; Smith & Parent, 1986; Tanaka et al., 1986). In the squirrel monkey, the lateral third of the VA-VL complex projects to the Put, whereas the medial third of the nuclei projects to the CN (Smith & Parent, 1986). Distinct striatal projections from subdivisions of the VA-VL complex was indicated in our experiments. The VAmc projects to the medial CN, and the VApc projects to the lateral CN and rostral Put, while the VLo projects to the Put except for its rostral tip (Fig. 4). Neurons in the VA have a branching axon projecting to both the CN and motor cortex (Jinnai & Matsuda, 1981; Macchi et al., 1984). A more dense projection of the VAmc, VApc and VLo fibers to the cerebral cortex were seen than to the striatum, based on the autoradiographic studies (in preparation). The thalamostriatal projection arising from these nuclei appears to be axon collaterals of the thalamocortical fibers.

Subthalamic neurons were labeled retrogradely after striatal injections of tracers (Beckstead, 1983; Royce & Laine, 1984; Smith & Parent, 1986; Parent & Smith, 1987). This subthalamostriatal projection was confirmed in the present experiments. Our data support the medio-lateral topography of the subthalamo-striatal projection by Beckstead (1983). The medial STN projects to the CN and rostral-most Put, whereas the lateral STN projects to the remaining whole Put. These projection neurons were observed most frequently in the middle third, less often in the rostral third and least frequently in the caudal third of the STN.

On the basis of our findings, the Put receives afferents from the VLo, CM and lateral STN, whereas the CN receives afferents from the VAmc, VApc, Pf and medial STN. Functional and anatomical relationships seem to coincide with afferent connections from these thalamic and subthalamic nuclei, and from the prefrontal, premotor and sensorimotor areas. The VLo, CM, and lateral STN afferents from the motor cortex have a somatotopical arrangement (Künzle, 1976; Hartmann-von Monakow et al., 1978). The Put seems to be modulated indirectly by the motor cortex via the VLo, CM, and lateral STN in addition to the direct motor activation.

The lateral portion of the CN receives afferents from the premotor cortex (Künzle, 1978). In addition, our sutdies show the projection to the lateral part of CN from the VApc, lateral Pf and the medial STN. The premotor cortex projects to the VApc, the lateral Pf and the medial STN (Hartmann-von Monakow et al., 1978; Akert & Hartmann-von Monakow, 1980). The lateral portion of CN seems to be modulated directly by premotor activation and indirectly by the premotor cortex via the VApc, lateral Pf and the medial STN.

Other authors reported that the central and dorsomedial portions of the CN receive afferents from the prefrontal cortex (Goldman & Nauta, 1977; Künzle & Akert, 1977; Künzle, 1978). Our data indicate that afferents to these portions originate also from the VAmc, the medial Pf and the medial STN. The ventromedial portion of the CN receives afferents from the orbital cortex, superior temporal and anterior cingulate gyri (Yeterian & Hoesen, 1978; Selemon & Goldman-Rakic, 1985), as well as the VAmc, medial Pf and medial STN. The prefrontal cortex projects also to the VAmc, medial Pf and medial STN (Hartmann-von Monakow et al., 1978; Kunzle, 1978; Akert & Hartmann-von Monakow, 1980). Also, the VAmc receives afferents from the SN and projects to the prefrontal and orbital cortices (Ilinsky et al., 1985). The dorsomedial, central and ventro-medial portions of the CN seem to be modulated directly by prefrontal and orbital cortical activations, and indirectly by these cortices via the VAmc, medial Pf and the medial STN.

The Put projects to the ventrolateral part of the pallidum and caudal SN, whereas the CN projects to the dorsomedial part of the pallidum and rostral SN (Smith & Parent, 1986; Percheron et al., 1987). It seems that the CN and rostral Put connect with the association cortex, Pf, VAmc, VApc, medial STN, dorsomedial pallidum and the rostral SN, whereas the Put, except for its rostral and adjacent ventromedial portions, connects with sensorimotor cortex, CM, VLo, lateral STN and ventrolateral pallidum, and the caudal SN. The former seems to contribute to complex loops and the latter to motor loops. The functional significance of such a complex connectional system remains to be investigated, especially on the basis of the highly heterogeneous organization in the neostriatum.

ACKNOWLEDGEMENTS

The authors wish to express their appreciation to Mr. K. Yamaguchi, Mr. S. Imamura, Mrs. C. Imamura and Miss. S. Ogata for their technical assistance. This work was supported, in part, by Grants-in-Aid for Scientific Research from the Ministry of Education, Science and Culture of Japan (56570022, 57570023, 61570030).

LITERATURE CITED

Akert, K., and Hartmann-von Monakow, K., 1980, Relationships of precentral, premotor and prefrontal cortex to the mediodorsal and intralaminar nuclei of the monkey thalamus, Acta Neurobiol. Exp., 40:7-25.
Beckstead, R. M., 1983, A reciprocal axonal connection between the subthalamic nucleus and the neostriatum in the cat, Brain Res., 275:137-142.
Beckstead, R. M., 1984, The thalamostriatal projection in the cat, J. Comp. Neurol., 223:313-346.
Carpenter, M. B., 1984, Interconnections between the corpus striatum and brain stem nuclei, in: "The Basal Ganglia; Structure and Function", Advances in Behavioral Biology, Vol. 27, pp.1-68, J. S. McKenzie, R. E. Kemm, and C. N. Wilcock, eds., Plenum Press, New York.
Cowan, W. M., and Powell, T. P. S., 1955, The projection of the midline and intralaminar nuclei of the thalamus of the rabbit, J. Neurol. Neurosurg. Psychiat., 18:266-279.
Goldman, P. S., and Nauta, W. J. H., 1977, An intricately patterned prefronto-caudate projection in the rhesus monkey, J. Comp. Neurol., 171:369-386.
Graybiel, A. M., and Ragsdale, C. W. Jr., 1983, Biochemical anatomy of the striatum, in "Chemical Neuroanatomy", pp. 427-504, P. C. Emson, ed., Reven Press, New York.
Groenewegen, H. J., Becker, N. E. H. M., and Lohman, A. H. M., 1980, Sub-cortical afferents of the nucleus accumbens septi in the cat, studied with retrograde axonal transport of horseradish peroxidase and bisben-zimid, Neuroscience, 5:1903-1916.

Hartmann-von Monakow, K., Akert, K., and Kunzle, H., 1978, Projections of the precentral motor cortex and other cortical areas of the frontal lobe to the subthalamic nucleus in the monkey, Exp. Brain Res., 33:395-403.

Ilinsky, I. A., Jouandest, M. L., and Goldman-Rakic, P. S., 1985, Organization of the nigrothalamocortical system in the rhesus monkey, J. Comp. Neurol., 236:315-330.

Jayaraman, A., 1985, Organization of thalamic projections in the nucleus accumbens and the caudate nucleus in cats and its relation with hippocampal and other subcortical afferents, J. Comp. Neurol., 231:396-420.

Jinnai, K., and Matsuda, Y., 1981, Thalamocaudate projection neurons with a branching axon to the cerebral motor cortex, Neurosci. Lett., 26:95-99.

Johnson, T. N., 1961, Fiber connections between the dorsal thalamus and corpus striatum in the cat, Exp. Neurol., 3:556-569.

Jones, E. G., and Leavitt, R. Y., 1974, Retrograde axonal transport and the demonstration of non-specific projections to the cerebral cortex and striatum from thalamic intralaminar nuclei in the rat, cat and monkey, J. Comp. Neurol., 154:349-378.

Kalil, K. 1978, Patch-like termination of thalamic fibers in the putamen of the rhesus monkey: an autoradiographic study, Brain Res. 140:333-339.

Kelley, A. E., Domesik, V. B., J. H. Nauta, W. J. H., 1982, The amygdalo-striatal projection in the rat - an anatomical study by anterograde and retrograde tracing methods, Neuroscience, 7:615-630.

Künzle, H. 1975, Bilateral projections from precentral motor cortex to the putamen and other parts of the basal ganglia. An autoradiographic study in Macaca fascicularis, Brain Res., 88:195-209.

Künzle, H., 1976, Thalamic projections from the precentral motor cortex in macaca fascicularis, Brain Res., 105:253-267.

Künzle, H., 1978, An autoradiographic analysis of the efferent connections from premotor and adjacent prefrontal regions (areas 6 and 9) in Macaca fascicularis, Brain Behav. Evol., 15:185-234.

Künzle, H., and Akert, K., 1977, Efferent connections of cortical area 8 (frontal eye field) in macaca fascicularis. A reinvestigation using the autoradiographic technique, J. Comp. Neurol., 173:147-164.

Macchi, G., Bentivoglio, M., Molinari, M., and Minciacchi, D., 1984, The thalamo-caudate versus thalamo-cortical projections as studied in the cat with fluorescent retrograde double labeling, Exp. Brain Res., 54:225-239.

Mehler, W. R., 1966, Further notes on the center median nucleus of Luys, in: "The Thalamus", pp. 109-122, D. P. Purpura, and M. D. Yahr, eds, Columbia University Press, New York.

Nakano, K., Kohno, M., Hasegawa, Y., and Tokushige, A., 1983, Efferent projections of thalamic centromedian nucleus in the cat and monkey as studied by an autoradiographic technique, Acta Anat. Nippon. Abstr., 58:253.

Nakano, K., Hasegawa, Y., Tokushige, A., Nakagawa, S., Kayahara, T., and Mizuno, N., Direct projections from the thalamus, subthalamic nucleus and pedunculopontine tegmental nucleus to the striatum in the Japanese monkey, Macaca fuscata, submitted.

Olszewski, J., "The Thalamus of the Macaca Mulatta. An Atlas for Use with the Stereotaxic Instrument", Karger, Basel (1952).

Parent, A., and Smith, Y., 1987, Organization of efferent projections of the subthalamic nucleus in the squirrel monkey as revealed by retrograde labeling methods, Brain Res., 436: 296-310.

Parent, A., Mackey, A., and De Bellefeuille, L., 1983, The subcortical afferents to caudate nucleus and putamen in primate: A fluorescence retrograde double labeling study, Neuroscience, 10:1137-1150.

Percheron, G., François, C., and Yelnik, J., 1987, Spatial organization and information processing in the core of the basal ganglia. in: "The Basal Ganglia II; Structure and Function - Current Concepts", Advances in Behavioral Biology, Vol. 32, pp. 205-226. M. B. Carpenter, and A. Jayaraman, eds., Plenum Press, New York and London.

Powell, T. P. S., and Cowan, W. M., 1956, A study of thalamo-striate relations in the monkey, Brain, 79:364-390.

Royce, G. J., 1978a, Cells of origin of subcortical afferents to the caudate nucleus: a horseradish peroxidase study in the cat, Brain Res., 153:465-475.

Royce, G. J., 1978b, Autoradiographic evidence for a discontinuous projection to the caudate nucleus from the centromedian nucleus in the cat, Brain Res., 146:145-150.

Royce, G. J., 1983, Single thalamic neurons which project to both the rostral cortex and caudate nucleus studied with the fluorescent double labeling method, Exp. Neurol., 79:773-784.

Royce, G. J., and Laine, E. J., 1984, Efferent connections of the caudate nucleus, including cortical projections of the striatum and other basal ganglia: an autoradiographic and horseradish peroxidase investigation in the cat, J. Comp. Neurol., 226:28-49.

Royce, G. J., and Mourey, R. J., 1985, Efferent connections of the centromedian and parafascicular thalamic nuclei: An autoradiographic investigation in the cat, J. Comp. Neurol.,235:277-300.

Sato, M., Itoh, K., and Mizuno, N., 1979, Distribution of thalamo-caudate neurons in the cat as demonstrated by horseradish peroxidase, Exp. Brain Res., 34:143-153.

Selemon, L. D., and Goldman-Rakic, P. S., 1985 Longitudinal topography and interdigitation of corticostriatal projections in the rhesus monkey, J. Neurosci., 5:776-794.

Smith, Y., and Parent, A., 1986, Differential connections of caudate nucleus and putamen in the squirrel monkey (saimiri sciureus), Neuroscience, 18:347-371.

Tanaka, D., Jr., Isaacson, L. G., and Trosko, B. K., 1986, Thalamostriatal projections from the ventral anterior nucleus in the dog. J. Comp. Neurol., 247:56-68.

Yeterian, E. H., and Van Hoesen, G. W., 1978 Cortico-striate projections in the rhesus monkey: The organization of certain cortico-caudate connections, Brain Res., 139:43-63.

COMPLEMENTARITY OF THE TWO PALLIDAL SEGMENTS

IN THE PRIMATE

Michel Filion, Léon Tremblay and Vijayakumar Chockkan

Centre de Recherche en Neurobiologie
Université Laval et Hôpital de l'Enfant-Jésus
1401, 18e Rue, Québec, Qué., G1J 1Z4, Canada

INTRODUCTION

In the present volume, Parent and collaborators demonstrate that the neurons of the internal (GPi) and external (GPe) segments of the primate globus pallidus differ fundamentally with respect to their afferent and efferent connections, in spite of their virtually identical morphology and chemospecificity. To develop further this anatomical point of view, the present paper reviews recent findings of physiological differences between the two pallidal segments in monkeys. Those differences are clearly revealed when the dopaminergic innervation of the basal ganglia is destroyed, quite specifically, by the neurotoxin 1-methyl-4-phenyl-1,2,3,6-tetrahydropyridine (MPTP), which renders the animal parkinsonian. We submit that GPi and GPe constitute a matching, contrasting and complementary pair of structures.

EXPERIMENTAL PROCEDURES

The procedures have been described in details elsewhere[1,2]. Five young female cynomolgus monkeys (*Macaca fascicularis*) were studied. One or two intravenous injections of MPTP (0.6 to 1.5 mg/kg) were used to render the animals parkinsonian. The first two monkeys (A and B) were studied both before and after MPTP. The others two (C and D) provided data only in the intact state, since they did not tolerate the neurotoxin. The last monkey (E) had been parkinsonian for 13 months when the electrophysiological studies were begun. To evaluate the MPTP-induced dopaminergic denervation, we counted the number of nigral neurons of the compacta-type, using Nissl stain and criteria published by Poirier et al.[3]. The percentages of decrease of dopaminergic neurons were 86, 90, and 99 in the parkinsonian monkeys.

For extracellular unit recording, the animal sat in a primate chair with the head fixed in the stereotaxic planes by means of a rigid platform, previously anchored to the skull of the animal under general anesthesia. The animals were free to move body and limbs and rapidly adapted to the painless head restraint. Microelectrodes were lowered vertically on either side of the brain. The neurons were identified by their characteristic firing patterns and rates and by the estimated position of the microelectrode tip, which was later verified by histological study. Neuronal activity was recorded when the animal was immobile and waking.

Six bipolar concentric electrodes were used for electrical stimulation of the striatum. They were implanted bilaterally: one in the head of the caudate nucleus, one in the putamen at the same frontal plane, and a third also in the putamen but more caudally. The responses were studied in the form of peristimulus interval histograms constructed during unit recording. In a number of cases in parkinsonian animals, the records of spontaneous activity and the histograms were repeated during and after the effects of apomorphine: a short-acting (30-45 minutes) dopamine agonist injected subcutaneously in doses of 20-40 µg/kg.

RESULTS AND COMMENTS

Spontaneous activity

The mean firing rates of GPi and GPe neurons were approximately equal in intact monkeys. In MPTP-treated monkeys, however, this similarity disappeared: the mean firing rate increased in GPi from 78/s (±26, n= 104) to 95/s (±32, n= 167), whereas it inversely decreased in GPe from 76/s (±28, n= 108) to 51/s (±27, n= 175). In both cases the difference is significant at the 0.001 level (Student's t-test). Identical results were obtained independently by Miller and DeLong[4].

Those abnormal firing rates could be normalized by the injection of apomorphine. However, if the dose of the dopaminergic agonist was sufficient, the abnormal rates were not normalized but inverted: GPi neurons being completely silenced and GPe neurons becoming three to four times more active[5]. Some years ago, we had obtained identical results in monkeys rendered parkinsonian by electrolytic midbrain lesions[6]. It is as if dopamine maintained an equilibrium between the activity of the two pallidal segments in the intact animal, the balance being shifted to one side when dopamine is lacking, as in the MPTP-treated animal, and shifted to the other side when apomorphine is given to the latter animal.

Evoked activity

The great majority of responses of pallidal neurons to electrical stimulation of the striatum in intact monkeys consisted of an initial inhibition followed by excitation[2]. The responses were not different in GPi and GPe, except for a slight preponderance of excitation in GPe. In the MPTP-treated animals, however, striatal stimulation characteristically induced late inhibitory responses in GPi and, inversely, late excitatory responses in GPe[7]. Apomorphine restored normal responses, similar in GPi and GPe. Thus, again, the lack of dopamine in the MPTP-treated animal brings the two pallidal segments to react in opposite directions.

Morphology

In the parkinsonian monkeys (A and B) sacrificed 5 weeks after MPTP and in the intact monkeys (C and D), GPe and GPi neurons, as well as those of the pars reticulata of the substantia nigra (SNr), all displayed similar shapes and their granular and evenly distributed Nissl bodies stained with similar intensities (Fig. 1, left side). However, in parkinsonian monkey E, sacrificed as much as 15 months after MPTP, the great majority of GPe neurons were pale, whereas inversely most GPi and SNr neurons were darkly stained (Fig. 1, right side). The changes in staining intensity could not be explained by changes in cell volume. Moreover, the strikingly specific localization of the pale neurons in GPe and of the dark neurons in the two output structures of the basal ganglia, GPi and SNr, rules out a general staining artifact. Furthermore, the apparently normal number and size of astrocytes in the affected areas appear to exclude degenerative conditions.

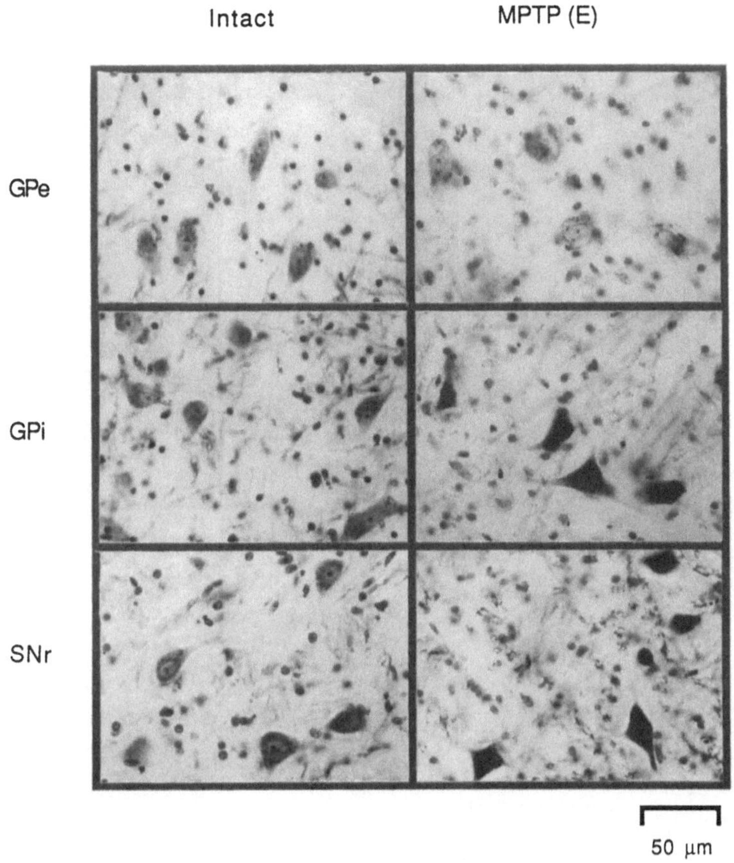

Intact   MPTP (E)

GPe

GPi

SNr

50 µm

Fig. 1. Photomicrographs of frontal sections through the globus pallidus (GPe
and GPi) and substantia nigra pars reticulata (SNr) of one intact monkey
and of a parkinsonian monkey (E) sacrificed 15 months after receiving
MPTP. The sections were stained with cresyl violet and fast blue.

Interestingly, the morphological and electrophysiological results strongly suggest a
correlation between dark neurons and high firing rates and between pale neurons and low
firing rates. The correlation even holds in the case of a few pallidal neurons displaying the
inverse abnormal morphological characteristics. In fact, a number of GPe neurons were dark
and a number of GPi neurons were pale. The latter neurons were mostly grouped in particular areas of the nuclei, where recording tracks indeed revealed low firing rates in the vicinity
of pale neurons and high firing rates in the vicinity of dark neurons (Fig. 2 A-C).
Abnormally high firing rates certainly require synthesis of large quantities of neurotransmitters, whereas inversely, abnormally low firing rates are likely to relieve the synthesis apparatus. The changes in Nissl stain, already illustrated, suggest corresponding alterations in the
endoplasmic reticulum. Moreover, we stained carbohydrates with the periodic acid Schiff
reaction (P.A.S.) and nucleic acids with the gallocyanin method on alternate sections. In
both cases, GPe neurons were paler and GPi neurons darker in monkey E compared to the
intact animal (Fig. 2 D-E).

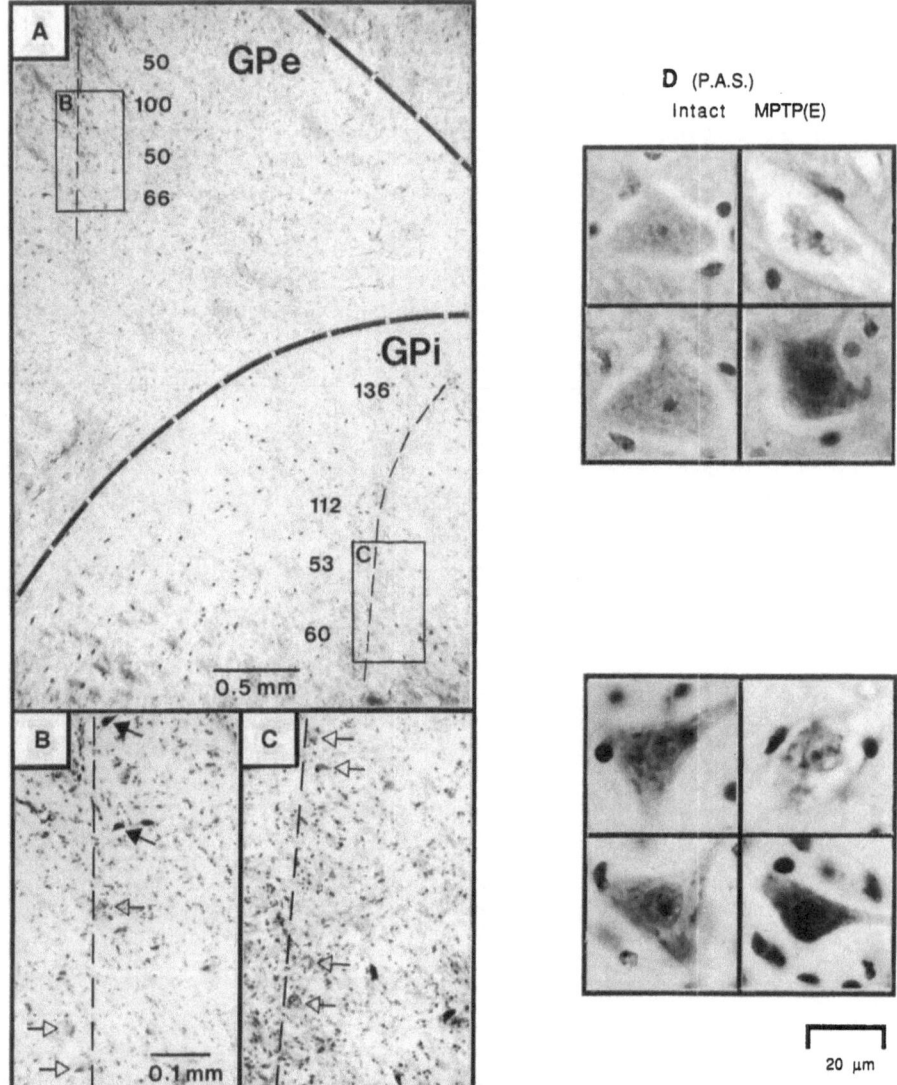

Fig. 2. (A-C) Photomicrographs of a frontal section through the globus pallidus of parkinsonian monkey E (cresyl violet and fast blue). (A) Note the numerous darkly stained neurons located almost exclusively in GPi and contrasting with the pale neurons in the adjacent GPe. Exceptions (dark neurons in GPe and pale neurons in GPi) are shown in the rectangles (B and C), which have been reproduced below at higher magnification. The numbers along the tracks (dashed lines) indicate the location and firing rate per second of units recorded in the tracks. Note the correspondance between high rates and dark neurons (dark arrows in B), and between low rates and pale neurons (empty arrows in B and C). (D-E) Photomicrographs from the same monkeys as in Fig. 1. However, the sections were stained for carbohydrates with the periodic acid Schiff reaction (D), and for nucleic acids with the gallocyanin method (E).

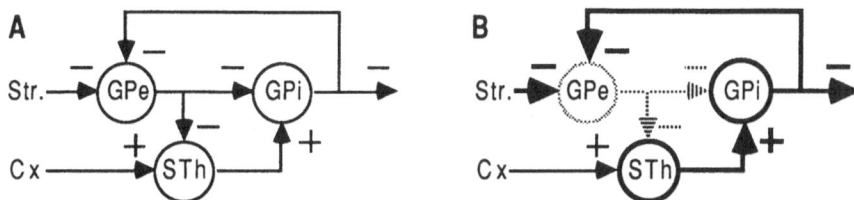

Fig. 3. Diagram of the direct and indirect connections between the pallidal segments in an intact (A) and in a parkinsonian monkey (B). The plus signs indicate excitatory and the minus signs inhibitory effects. Thick and striped lines indicate increased and decreased activity, respectively. Abbreviations: Cx, cortex; GPe, external segment of globus pallidus; GPi, internal segment of globus pallidus; STh, subthalamic nucleus; Str., striatum.

Parkinsonian monkeys A and B also displayed abnormal firing rates, but no morphological abnormalities. This lack of change cannot be attributed to less severe dopaminergic lesions, since the percentage of degeneration of compacta-type neurons in monkey E (90%) was intermediate between those of monkeys A (86%) and B (99%). Moreover, all monkeys were of the same sex and of nearly the same age. However, the morphological alterations may be associated with the exceptionally long duration of parkinsonism in monkey E: 15 months compared to 5 weeks for monkeys A and B. MPTP-treated monkeys are generally sacrificed much earlier, which may explain the uniqueness of the present results. Nevertheless, those results again reveal reactions opposite between the pallidal segments.

Role of intersegmental connections

In the present volume, Parent and collaborators demonstrate direct, reciprocal connections between the two pallidal segments. Those connections are very probably inhibitory, since all pallidal neurons appear to use GABA as neurotransmitter[8]. On the other hand, GPe can also influence GPi indirectly, through the subthalamic nucleus (STh) (Fig. 3A). This indirect influence is likely to be of the same polarity as the direct influence, since GPe has inhibitory GABAergic projections to STh, which has a non-inverting, excitatory glutamatergic influence on GPi[9,10]. Consequently, a decrease in the activity of GPe neurons, as occurs in MPTP-treated animals, is likely to decrease the inhibition that they exert directly on GPi neurons and to increase simultaneously the excitation that they exert indirectly on the same neurons, through STh (Fig. 3B). The resulting increased activity of GPi neurons can be fed back on GPe neurons, through the intersegmental connections, thus further decreasing the activity of the latter neurons. Therefore, the direct and indirect intersegmental connections could at least partly explain the opposite reactions of the pallidal segments described above. Moreover, the scheme suggests that pallidal imbalance could be triggered by acting only at the GPe level.

Role of striatopallidal afferents

Another mechanism could also partly explain the pallidal imbalance described above. On the basis of metabolic, behavioral and biochemical studies, it has been suggested that dopamine excites the striatal neurons projecting to GPi but inversely inhibits those projecting to GPe (see Ref. 11 for a review). In this respect, it has been shown recently that the striatal neurons projecting to GPi and those projecting to GPe constitute two populations that are not only distinct but also topographically segregated in the monkey: the striatal neurons

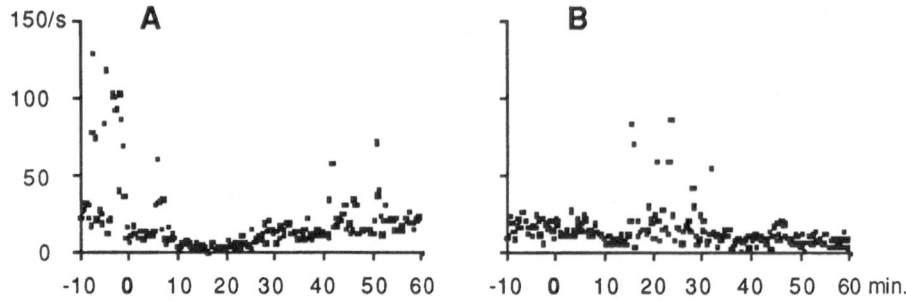

Fig. 4. Effects of apomorphine (30 μg/kg, sc), injected at time zero, on the firing rate of striatal neurons primed by a continuous local perfusion of glutamate (10^{-2} M, 1 μl/min.) in an MPTP-treated monkey. Decreases (A) and increases (B) of firing were found both in the striatal area projecting to GPi and in that projecting to GPe. The changes of firing rates occurred some 10 minutes after the injection and corresponded with the disappearance of the signs of parkinsonism and occurrence of dyskinesia.

projecting to GPi being grouped ventromedially to those projecting to GPe[12]. This suggested a direct test of the above hypothesis. In an MPTP-treated monkey, a level of activity was induced in striatal neurons by local perfusion of glutamate (10^{-2} M, 1 μl/min.) and was recorded extracellularly by a microelectrode located within and protruding 0.5 mm from the tip of the cannula. Apomorphine was then injected systemically to activate dopaminergic mechanisms. Preliminary results have shown excitatory, inhibitory, and null effects (Fig. 4). Apparently in contradiction with the above hypothesis, excitatory and inhibitory effects were obtained both in the striatal region projecting to GPi and in that projecting to GPe. More data are required to test the less stringent hypothesis that dopamine excites a majority of neurons in the striatal region projecting to GPi and inhibits a majority of neurons in the striatal region projecting to GPe, the computation of the predominant effect being preformed on the long pallidal dendrites, where there is convergence of a large number of striatopallidal fibers[13].

Role of dopaminergic afferents

In the present volume, Parent and collaborators recall their demonstration of a pallidal innervation originating from a population of nigral dopaminergic neurons that is distinct from that projecting to the striatum and appears to be at least partly resistant to the neurotoxic effects of MPTP. Interestingly, this dopaminergic innervation is more abundant in GPi than in GPe. This unequal dopaminergic influence on the pallidum, in addition to a complete or partial absence of dopaminergic influence on the striatum, may be related to the pallidal imbalance described above. The mechanisms involved, however, are unknown.

CONCLUSIONS

Dopamine appears to maintain the activities of the two pallidal segments of the primate in a state of equilibrium. Indeed, when dopaminergic mechanisms are abnormally weak or strong the neuronal reactions are opposite in each pallidal segment. Explanations for this imbalance are suggested by the recent anatomical demonstrations of reciprocal connections between the pallidal segments and of distinct striatopallidal and nigropallidal projections.

ACKNOWLEDGEMENTS

The study was supported by the Medical Research Council of Canada (Grant MT-5750), the Fonds de la Recherche en Santé du Québec (FRSQ), and the Fonds pour la Formation de Chercheurs et l'Aide à la Recherche. L.T. holds a Studentship from the FRSQ.

REFERENCES

1   Filion, M., Tremblay, L. and Bédard, P.J., Abnormal influences of passive limb movement on the activity of globus pallidus neurons in parkinsonian monkeys, *Brain Research*, 444 (1988) 165-176.

2   Tremblay, L. and Filion, M., Responses of pallidal neurons to striatal stimulation in intact waking monkeys, *Brain Research*, 498 (1989) 1-16.

3   Poirier, L.J., Giguère, M. and Marchand R., Comparative morphology of the substantia nigra and ventral tegmental area in the monkey, cat and rat, *Brain Res. Bull.*, 11 (1983) 371-397.

4   Miller, W.C. and DeLong, M.R., Altered tonic activity of neurons in the globus pallidus and subthalamic nucleus in the primate MPTP model of parkinsonism. In M.B. Carpenter and A. Jayaraman (Eds.), *The Basal Ganglia II, Advances in behavioral biology, Vol. 32*, Plenum, New York, 1987, pp. 415-427.

5   Filion, M. and Tremblay L., Electrophysiological studies in MPTP-treated monkeys. In J.S. Schneider and M. Gupta (Eds.), *Current Concepts in Parkinson's Disease Research*, Hogrefe & Huber, Toronto, 1990 (in press).

6   Filion, M., Effects of interruption of the nigrostriatal pathway and of dopaminergic agents on the spontaneous activity of globus pallidus neurons in the awake monkey, *Brain Research*, 178 (1979) 425-441.

7   Tremblay, L., Filion, M. and Bédard, P.J., Responses of pallidal neurons to striatal stimulation in monkeys with MPTP-induced parkinsonism, *Brain Research*, 498 (1989) 17-33.

8   Smith, Y., Parent, A., Séguéla, P. and Descarries, L., Distribution of GABA-immunoreactive neurons in the basal ganglia of the squirrel monkey (*Saimiri sciureus*), *J. Comp. Neurol.*, 129 (1987) 50-64.

9   Smith, Y. and Parent A., Neurons of the subthalamic nucleus in primates display glutamate but not GABA immunoreactivity, *Brain Research*, 453 (1988) 353-356.

10   Féger, J., Vezole, I., Renwart, N. and Robledo, P., The rat subthalamic nucleus: electrophysiological and behavioral data. In A.R. Crossman and M.A. Sambrook (Eds.), *Neural Mechanisms in Disorders of Movement*, John Libbey, London, 1989, pp. 37-43.

11   Young, A.B., Albin, R.L. and Penney J.B., Neuropharmacology of basal ganglia functions: relationship to pathophysiology of movement disorders. In A.R. Crossman and M.A. Sambrook (Eds.), *Neural Mechanisms in Disorders of Movement*, John Libbey, London, 1989, pp. 17-27.

12   Parent, A., Smith, Y., Filion, M. and Dumas, J., Distinct afferents to internal and external pallidal segments in the squirrel monkey, *Neurosci. Lett.*, 96 (1989) 140-144.

13   Percheron, G., Yelnik, J. and François, C. A Golgi analysis of the primate globus pallidus. III. Spatial organization of the striato-pallidal complex, *J. Comp. Neurol.*, 227 (1984) 214-227.

THE PALLIDUM AS A DUAL STRUCTURE IN PRIMATES

André Parent, Lili-Naz Hazrati and Brigitte Lavoie

Centre de recherche en neurobiologie
Hôpital de l'Enfant-Jésus
1401, 18e rue, Québec, Québec G1J 1Z4 Canada

INTRODUCTION

This paper reviews some of our most recent findings on the organization of the afferent and efferent connections of the globus pallidus (GP) in the squirrel monkey (*Saimiri sciureus*). Its main purpose is to demonstrate that even though all pallidal neurons express the same phenotype in terms of both morphology and chemospecificity, the analysis of afferent and efferent connections of GP reveals that this major component of the basal ganglia is in fact a dual entity in primates.

EXPERIMENTAL PROCEDURES

The data presented here are based on experiments undertaken by means of a combination of axonal transport neuroanatomical methods and immunohistochemistry. For the tracing of anatomical connections we used various retrograde tracers, such as WGA-HRP, fast blue, nuclear yellow and propidium iodide, as well as the anterograde tracer PHA-L[1,2]. Our analysis of the chemospecific neuronal systems projecting to the pallidum was restricted to dopaminergic (DA) and serotoninergic (5-HT) afferents. The DA fibers were visualized by using antibodies directed against tyrosine hydroxylase (TH), whereas the 5-HT innervation of GP was revealed by means of antibodies raised against serotonin itself[3,4].

RESULTS AND COMMENTS

Anatomical Findings

Pallidal afferents. The two major sources of afferents to GP are the striatum and the subthalamic nucleus[5,6]. Other less massive projections arising from midbrain dopaminergic and serotoninergic neurons will be dealt with in detail below. The striatal and subthalamopallidal afferents terminate in both the external (GPe) and internal (GPi) segment of the pallidum where they arborize in the form of elongated bands aligned parallel to the medullary laminae[2]. This band-like pattern is one of the most characteristic features of the pallidal afferents. It most likely corresponds to the arrangement of the large and discoidal dendritic arborization of the pallidal neurons, which also lie parallel to the medullary laminae[7,8].

In an attempt to disclosed the cellular origin and degree of collateralization of striatal and subthalamopallidal afferents, one fluorescent tracer was injected in GPi and another in GPe. This study revealed that GPe and GPi receive projections from different cell populations in both the striatum and the subthalamic nucleus. Striatal neurons projecting to either GPe or GPi formed wide and non-overlapping bands oriented obliquely in the putamen and parts of the caudate nucleus, whereas subthalamic nucleus neurons projecting to GPe were more abundant and more laterally located than those projecting to GPi. Few striatal and subthalamic nucleus neurons were found to project to both GPe and GPi[9]. Striatal neurons projecting to the pallidum use GABA as inhibitory neurotransmitter, but also contain a neuropeptide. Striatal neurons projecting to GPe are enriched with met-enkephalin and those projecting to GPi express substance P[10]. In contrast, subthalamic nucleus neurons projecting in a topographic manner to the entire pallidum use glutamate as excitatory neurotransmitter[11]. Up to now no specific neuropeptide has been found to be co-localized with glutamate in the subthalamic nucleus neurons.

Pallidal efferents. In primates the major efferent projections of the GPe is back to the subthalamic nucleus. Other less massive efferents terminate in the striatum and the substantia nigra[6]. In contrast, GPi is a major output structure of the basal ganglia. It has long been known to project to the ventral anterior (VA)/ventral lateral (VL) thalamic nuclei, the centre médian (CM) nucleus, the habenula (HB) and the pedunculopontine (TPP) nucleus of the midbrain-pontine tegmentum[12]. The projections to VA/VL, CM and TPP nuclei were shown to arise mostly from axon collaterals of the same GPi neurons, whereas the projection to HB originated principally from a distinct cell population located peripherally[13,14].

In order to further our knowledge of GP efferent projections in primates, PHA-L was injected either in GPe or GPi of squirrel monkeys. Injection of PHA-L in GPe (Fig. 1) led to massive terminal labeling into the subthalamic nucleus (Fig. 3A). Before entering the subthalamic nucleus, some fibers were seen to emit collaterals that swept ventrally to reach the substantia nigra where they arborized principally in the lateral portion of the pars reticulata. Other anterogradely labeled fibers terminate massively in GPi (Fig. 1B), whilst a small contingent of fibers could be followed laterally in the putamen (Fig. 1A-C). The labeled fibers in the striatum were characteristically long, they followed a linear course and branched only infrequently (Fig. 3E).

In contrast, PHA-L injections in GPi produced anterograde labeling of axons that arborized profusely within VA/VL nuclei, and less massively in CM, HB and TPP (Fig. 2). The patterns of fiber terminations were markedly different in these four sites. The GPi efferents arborized into numerous small plexuses in VA/VL nuclei (Fig. 3B), whereas fibers reaching HB formed a relatively dense field of very small terminals in the central portion of the lateral habenular nucleus (Fig. 3D). Numerous labeled fibers that appeared to make contacts of the *en passage* type with cell bodies were visualized in CM (Figs. 2D, 3C), whereas relatively short and sinuous labeled fibers that did not make pericellular contacts were found in the TPP area.

In addition, numerous labeled fibers were seen to cross the midline after GPi injections. These decussating fibers abounded particularly at the level of nucleus centralis medialis and midline nuclei located rostral to CM. These data agrees with our earlier findings obtained after retrograde double-labeling studies, which revealed that a significant proportion (15 to 20%) of pallidal neurons were retrogradely labeled in GPi located contralateral to the VA/VL injection site[14]. Other anterogradely labeled fibers were seen in significant number in the ipsilateral GPe and less abundantly in the ipsilateral putamen (Fig. 2A-C). On the other hand, since virtually all pallidal neurons display GABA immunoreactivity in the squirrel monkey[15], it may be presumed that all the efferent projections of GPe and GPi utilize GABA as their main neurotransmitter.

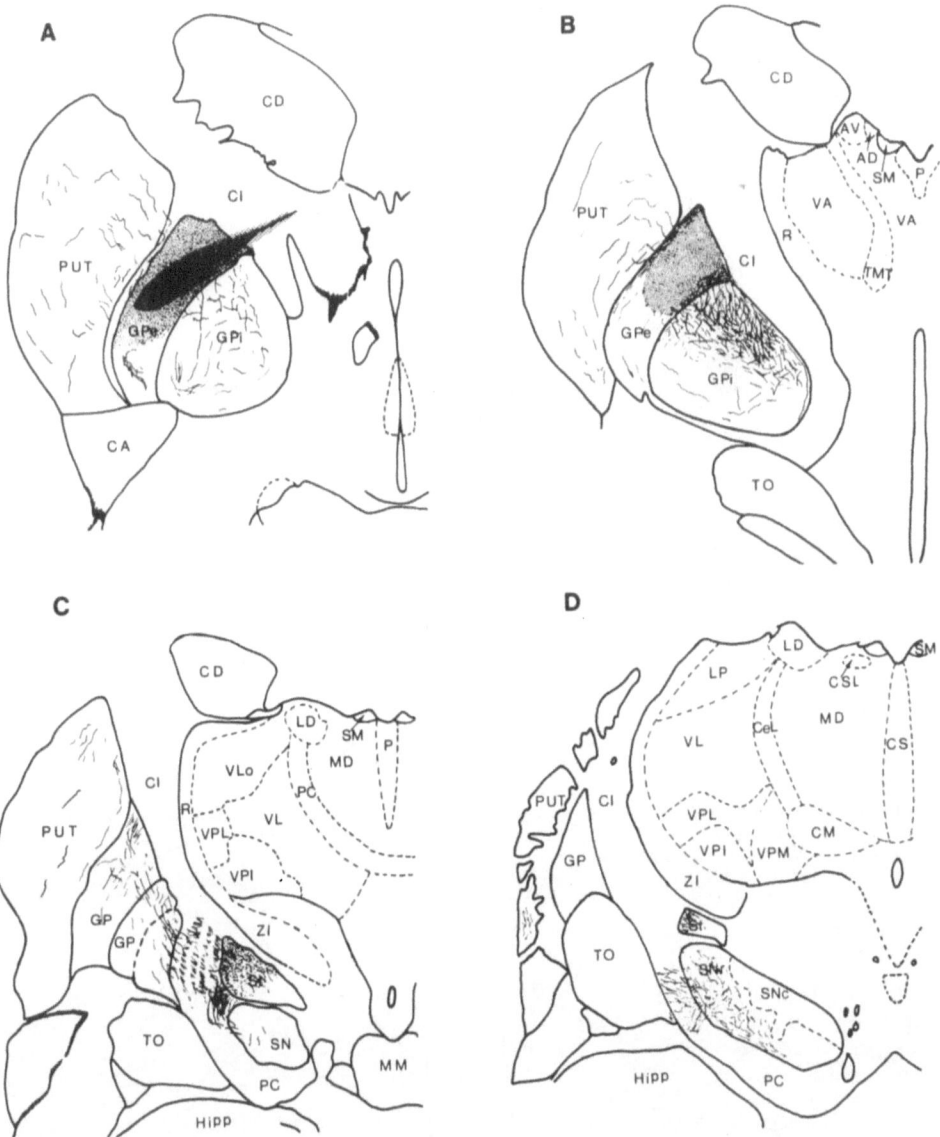

Fig. 1. Camera lucida drawings of transverse hemi-sections through the basal ganglia of a squirrel monkey that received a PHA-L injection into the GPe. The injection site is shown in dark and stippled areas, whereas the efferent fibers are represent as sinuous lines. The drawings are set out in a rostrocaudal order.

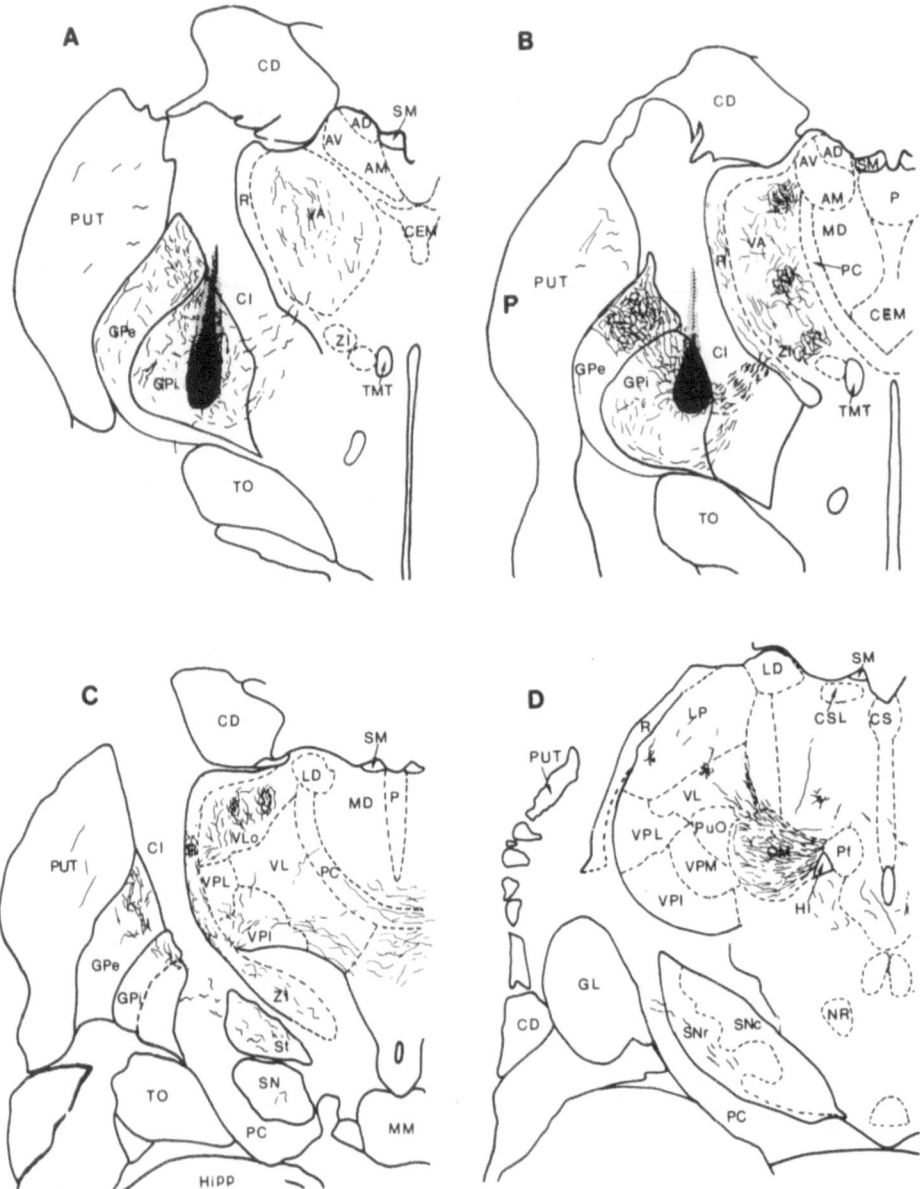

Fig. 2. Camera lucida drawings of transverse hemi-sections through the basal ganglia of a squirrel monkey that received a PHA-L injection into the GPi. The symbols are the same as in figure 1.

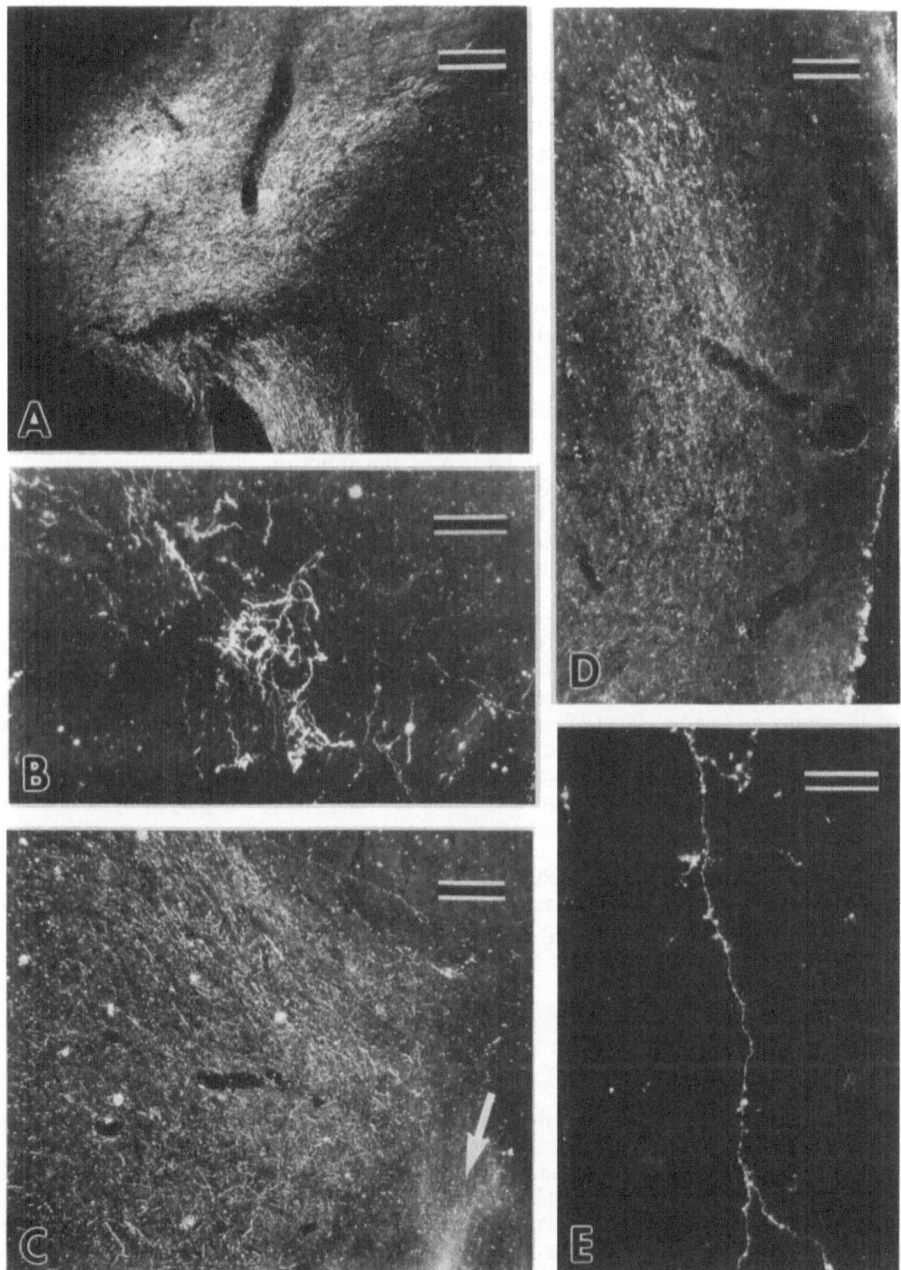

Fig. 3. Darkfield photomicrographs showing examples of PHA-L terminal fields visualized after pallidal injections. (a) Dense field of fine axon terminals in the ventromedial portion of the subthalamic nucleus and fibers descending toward the substantia nigra after GPe injection. (b) Typical fiber plexuses in ventral lateral thalamic nucleus after GPi injection. (c) Pattern of innervation of the centre médian after GPi injection. Note the absence of labeling in the parafascicular nucleus, lateral to the habenulo-interpeduncular tract. (d) Terminal field in the lateral nucleus of the habenula after GPi injection. (e) Typical long and unbranched fibers seen in the putamen after GPe injection. Scale bars: 200 μm (A,C,D) and 50 μm (B,E).

85

Immunohistochemical Findings

Dopaminergic pallidal innervation. The pallidum of the squirrel monkey receives a particularly dense DA innervation, which arborizes heavily within GPi and much less so in GPe[16]. Retrograde double-labeling studies involving injections of one fluorescent tracer in the putamen and/or the caudate nucleus and the other in GPi showed that the DA neurons projecting to GPi are distributed throughout the substantia nigra (SN)/ventral tegmental area (VTA) complex, but abounded particularly at the junction between VTA and SN[17]. More importantly, neurons projecting to GPi were found to be largely distinct from those projecting to the striatum, indicating that the nigropallidal dopaminergic projection is a separate subsystem in primates. This is at variance with the finding obtained by means of histofluorescence method in the rat where GP was reported to receive a DA input largely made up of collaterals of the nigrostriatal pathway[18]. The DA fibers in the globus pallidus of rodents were said to make synapses of the *en passant* type with pallidal neurons.

The DA nature of the nigropallidal projection in the squirrel monkey was confirmed by experiments combining the use of propidium iodide as retrograde marker and TH immunohistochemistry[17]. Little is known of the function of this nigropallidal DA projection in primates, but clues could come from recent observations in cynomolgus monkeys rendered parkinsonian by the administration of the neurotoxin MPTP. In such cases the nigropallidal DA projection was found to be selectively spared, in contrast to the nigrosriatal DA pathway which was massively destroyed[19].

Serotoninergic pallidal afferents. The pallidum in the squirrel monkey receives a relatively dense 5-HT input that is also differentially distributed: most of the serotoninergic fibers arborize within GPi and much less so in GPe[3]. The 5-HT innervation of GP derives from fascicles that detach themselves from the main 5-HT bundle ascending within the lateral hypothalamic area. These fascicles reach the pallidum by coursing ventrally within the sublenticular gray or by penetrating more directly through the ventromedial aspect of GPi. Within the pallidum 5-HT fibers were aligned parallel to medullary laminae (Fig 4). Their arrangement was thus very similar to the band-like pattern displayed by striatal and subthalamic nucleus pallidal afferents.

Taken together, our immunohistochemical findings reveal that, in addition to their well-established influence upon the striatum, the midbrain dopaminergic and serotoninergic neurons can act directly onto the output neurons of the basal ganglia at pallidal levels.

CONCLUSIONS

Because virtually all pallidal neurons display the same phenotype, detailed analyses of their morphology and chemospecificity reveal little of the role that GP plays in the organization of the basal ganglia. In contrast, studies of the organization and neurotransmitter content of pallidal afferent and efferent connections allow a better understanding of the function of each pallidal segment. Our findings in the squirrel monkey have revealed that the two pallidal segments in primates are differentially modulated by various chemospecific neuronal systems, that they are reciprocally linked with one another, and that the information processed by each segment is forwarded to different target structures along distinct fiber systems with specific termination patterns. The differential modulation of the two pallidal segments in primates could play a crucial role in the development of hyper- and hypokinetic disorders[20].

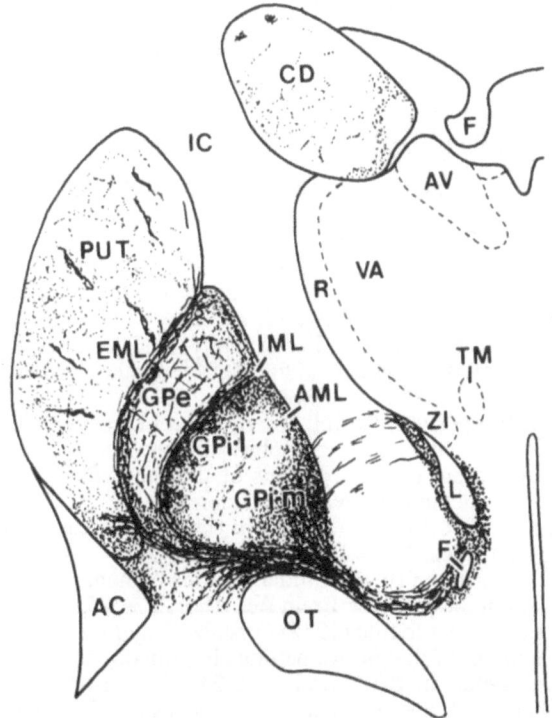

Fig. 4. Schematic drawing of a transverse half-section through the middle third of the basal ganglia in the squirrel monkey showing the distribution of the serotonin-immunoreactive fibers (sinuous lines) and axon terminals (dots). Note the dense and differential innervation of the pallidum.

REFERENCES

1. Parent, A., A. Mackey and L. De Bellefeuille (1983) The subcortical afferents to caudate nucleus and putamen in primates: a fluorescence retrograde double labeling study. Neuroscience 10: 1137-1150.
2. Smith, Y., L.-N. Hazrati and A. Parent (1989) Efferent projections of the subthalamic nucleus in the squirrel monkey as revealed by the PHA-L anterograde tracing method. J. Comp. Neurol. (in press).
3. Lavoie, B. and A. Parent (1989) Immunohistochemical study of the serotoninergic innervation of the basal ganglia in the squirrel monkey. J. Comp. Neurol. (submitted).
4. Lavoie, B., Y. Smith and A. Parent (1989) Dopaminergic innervation of the basal ganglia in the squirrel monkey as revealed by tyrosine hydroxylase immunohistochemistry. J. Comp. Neurol. 289: 36-52.
5. Carpenter, M.B. (1984) Interconnections between the corpus striatum and brain stem nuclei. In J.S. McKenzie, R.E. Kemm and L.N. Wilcock (eds): The Basal Ganglia: Structure and Function. Adv. Behav. Biol., Vol 27. Plenum Press: New York, pp. 1-68.
6. Parent, A. (1986) Comparative Neurobiology of the Basal Ganglia. John Wiley & Sons, New York, pp. 335.
7. Park, M.R., W.M. Falls and S.T. Kitai (1982) An intracellular HRP study of the rat globus pallidus. I. Responses and light microscopic analysis. J. Comp. Neurol., 211: 284-294.
8. Yelnik, J., G. Percheron and C. François (1984) A Golgi analysis of the primate globus pallidus. II. Quantitative morphology and spatial orientation of dendritic arborization. J. Comp. Neurol. 227: 200-213.
9. Parent, A., Y. Smith, M. Filion and J. Dumas (1989) Distinct afferents to internal and external pallidal segments in the squirrel monkey. Neurosci. Lett. 96: 140-144.
10. Haber, S. and R. Elde (1981) Correlation between met-enkephalin and substance P immunoreactivity in the primate globus pallidus. Neuroscience 6: 1291-1297.
11. Smith, Y. and A. Parent (1988) Neurons of the subthalamic nucleus in primates display glutamate but not GABA immunoreactivity. Brain Res. 453: 353-356.
12. Nauta, W.J.H. and W.R. Mehler (1966) Projections of the lentiform nucleus in the monkey. Brain Res. 1: 3-42.
13. Harnois, C. and M. Filion (1980) Pallidal neurons branching to the thalamus and to the midbrain in the monkey. Brain Res. 186: 222-225.
14. Parent, A. and L. De Bellefeuille (1982) Organization of efferent projection from the internal segment of the globus pallidus in primates as revealed by fluorescent retrograde labeling method. Brain Res. 245: 201-214.
15. Smith, Y., A. Parent, P. Séguéla and L. Descarries (1987) Distribution of GABA-immunoreactive neurons in the basal ganglia of the squirrel monkey (*Saimiri sciureus*). J. Comp. Neurol. 259: 50-65.
16. Parent, A. and Y. Smith (1987) Differential dopaminergic innervation of the two pallidal segments in the squirrel monkey (*Saimiri sciureus*). Brain Res. 426: 397-400.
17. Smith, Y. B. Lavoie, J. Dumas and A. Parent (1989) Evidence for a separate nigropallidal dopaminergic projection in the squirrel monkey. Brain Res. 482: 381-386.
18. Lindvall, O. and A. Björklund (1979) Dopaminergic innervation of the globus pallidus by collaterals from the nigrostriatal pathway. Brain Res. 172: 169-173.
19. Parent, A. and B. Lavoie (1989) Dopaminergic innervation of the basal ganglia in normal and parkinsonian monkeys. In J.S. Schneider and M.Gupta (eds): Current Concepts in Parkinson's Disease Research. Hans Huber: Toronto (in press).
20. Young, A.B., R.L. Abin and J.B. Penney (1989) Neuropharmacology of basal ganglia functions: relationship to pathophysiology of movement disorders. In A.R. Crossman and M.A. Sambrook (eds): Neural Mechanisms in Disorders of Movement. John Libbey: London, pp. 17-27.

THE CONNECTIONS OF THE MEDIAL PART OF THE SUBTHALAMIC NUCLEUS IN THE RAT:

EVIDENCE FOR A PARALLEL ORGANIZATION

Henk W. Berendse and Henk J. Groenewegen

Department of Anatomy and Embryology, Vrije Universiteit
Van der Boechorststraat 7, 1081 BT, Amsterdam
The Netherlands

SUMMARY

The anatomical relationships of the subthalamic nucleus (STh) with ventral striatopallidal parts of the basal ganglia and prefrontal cortical areas were studied in the rat, using anterograde and retrograde tracing methods. The anterograde tracer Phaseolus vulgaris-leucoagglutinin was injected in the STh, the prefrontal cortex, the globus pallidus, the ventral pallidum, and the parafascicular thalamic nucleus. Injections of the retrograde tracer Fluoro-Gold were placed in the globus pallidus and the ventral pallidum. The results of these experiments reveal that the medial part of the STh and the adjacent lateral hypothalamic area are reciprocally connected with ventral parts of the basal ganglia and receive projections from the prefrontal cortex and the medial part of the parafascicular nucleus. The connections are organized in parallel to those of the lateral part of the STh. Furthermore, three zones can be identified in the medial part of the STh and the adjacent lateral hypothalamic area, each having parallel connections with distinct parts of the pallidum and receiving projections from different prefrontal cortical areas. The present findings suggest that the subthalamic nucleus is also involved in non-motor functions.

INTRODUCTION

The subthalamic nucleus is implicated in movement control since damage to this nucleus is accompanied by a violent form of hemichorea (Martin, 1927; Whittier and Mettler, 1949; Carpenter et al., 1950; Crossman, 1987). In all mammals studied, substantial projections from the subthalamic nucleus (STh) are directed towards the external and internal (the entopeduncular nucleus of cats and rodents) segments of the globus pallidus, the substantia nigra, and the pedunculopontine tegmental nucleus (Nauta and Cole, 1978; Ricardo, 1980; Carpenter et al., 1981; Beckstead, 1983; Hammond et al., 1983; Kita and Kitai, 1987). The major sources of afferents to the STh are the motor and premotor cortical areas, the external segment of the globus pallidus, the parafascicular thalamic nucleus and the pedunculopontine tegmental nucleus (Hartmann-Von Monakow et al., 1978; Nomura et al., 1980; Carpenter et al., 1981; Hammond et al., 1983; Sugimoto et al., 1983; Afsharpour, 1985). Within the STh of primates three subsectors with different efferent connections have recently been distinguished (Parent and Smith, 1987; Parent et al., 1989).

The Basal Ganglia III, Edited by G. Bernardi *et al.*
Plenum Press, New York

In the last decade, it has become clear that within the circuitry of the basal ganglia parallel dorsal and ventral striatopallidothalamic pathways exist (Heimer and Wilson, 1975; Haber et al., 1985). Moreover, it has been postulated that in primates at least five parallel basal ganglia-thalamocortical circuits can be recognized, each focusing on a functionally distinct part of the frontal cortex and including anatomically distinct sectors of the striatum, the pallidum, and the thalamus (Alexander et al, 1986). The results of our tracing experiments demonstrate a similar parallel organization of the connections of these forebrain structures in the rat, with additional parallel circuits involving the midline and intralaminar thalamic nuclei and the amygdala (Groenewegen et al., 1990). Apart from its association with the ventral pallidum (Ricardo, 1980; Haber et al., 1985; Kita and Kitai, 1987), little is known of the connections of the STh with the limbic-innervated parts of the basal ganglia nor of its relationship with the prefrontal cortical areas that project to these ventral parts of the basal ganglia. In the present report the connections of the STh with the prefrontal cortex and the limbic-related parts of the basal ganglia in the rat, studied by means of anterograde and retrograde tracing techniques, will be described. In addition, data will be provided on the afferents of the STh from the thalamus.

MATERIALS AND METHODS

Anterograde Tracing Experiments

In order to study the afferent connections of the STh, the anterograde tracer Phaseolus vulgaris-leucoagglutinin (PHA-L) was injected into several forebrain structures of female Wistar rats. The injections were placed in the medial and lateral prefrontal cortices (n=35), the ventral pallidum (n=8), the medial part of the globus pallidus (n=2), the olfactory tubercle (n=3), and the parafascicular thalamic nucleus (n=4). The efferents of the STh were studied following PHA-L injections in the medial part of the STh (n=2). The methods used to deliver the tracer and the subsequent histological and immunohistochemical procedures have been described in detail elsewhere (Berendse and Groenewegen, 1990; Groenewegen and Wouterlood, 1990).

Retrograde Tracing Experiments

In a separate series of experiments the fluorescent tracer Fluoro-Gold (FG; 2% in acetate buffer; Schmued and Fallon, 1986) was injected iontophoretically in several parts of the pallidum of 6 rats. Following a survival time of 7-17 days, the animals were perfused transcardially with a fixative containing 4% paraformaldehyde in 0,1 M phosphate buffer pH 7,4. The brains were postfixated for 1-2 hours in the same fixative, stored overnight at 4°C in 30% phosphate buffered sucrose or in a mixture of 20% glycerol and 2% dimethylsulfoxide in distilled water, and cut on a freezing microtome. Sections were then mounted, xylene cleared and coverslipped with Entellan. They were examined under a Zeiss IV F fluorescence microscope using filter-mirror system 0,1 (365 nm).

RESULTS AND DISCUSSION

Pallidal Connections

PHA-L injections in the ventral pallidum result in an area of dense termination in the dorsomedial part of the STh throughout its rostrocaudal extent and in the adjacent part of the lateral hypothalamus (LH; Fig. 1A). Following an injection of PHA-L in the medial part of the globus pallidus

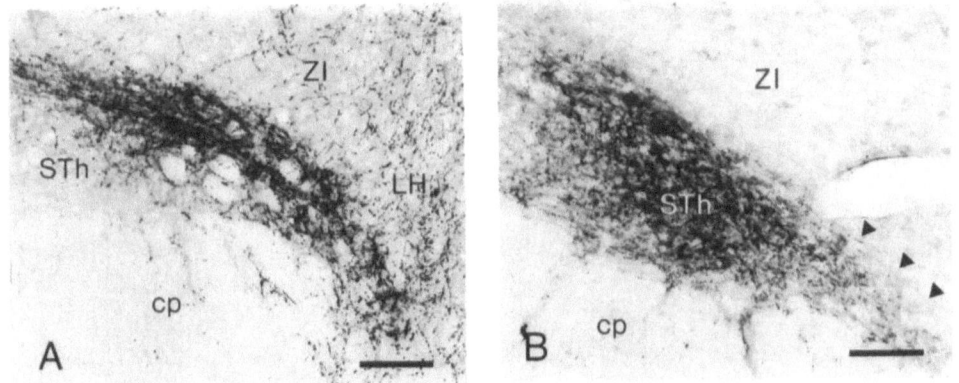

Fig. 1. Photomicrographs illustrating the pallidosubthalamic projections.
A, PHA-L labelled fibres terminating in the dorsomedial part of the
medial STh and the adjacent part of the LH following an injection in
the ventral pallidum. B, The position of the label in the medial
part of the STh after a PHA-L injection in the medial part of the
globus pallidus is more ventral and lateral. Arrowheads in (B) mark
the dorsomedial border of the STh. Bars = 100 μm. For abbreviations,
see legend to figure 3.

the majority of terminals are observed in the medial part of the STh also,
but in a more lateral and ventral position (Fig. 1B). In cases with injections in the olfactory tubercle, anterogradely labelled axons terminate in
the LH at the level of the rostral part of the STh. The density of terminations in the LH is rather sparse compared to that in the medial STh after
the injections in the ventral pallidum and the globus pallidus.

The projections from the STh back to the pallidal areas were studied
by means of PHA-L injections in the STh and FG injections in several parts
of the dorsal and ventral pallidum. PHA-L injections in the medial part of
the STh result in dense terminal fields in the lateral part of the subcommissural ventral pallidum (Fig. 2A; cf. Zahm, 1989), the medial part of
the globus pallidus (Fig. 2B), and the lateral hypothalamic area just medial
to and extending into the medial part of the entopeduncular nucleus. The
dense projections to this part of the LH stress the idea that the area is
comparable to the entopeduncular nucleus and probably constitutes part of
the internal segment of the ventral pallidum (Heimer et al., 1985; Groenewegen and Berendse, 1990). Another contingent of fibres from the medial STh
terminates in the ventrolateral part of the globus pallidus bordering the
caudate-putamen (Fig. 2B). A similar strip of subthalamic terminals in the
globus pallidus was reported following injections in the lateral part of the
STh (Kita and Kitai, 1987), which suggests that there is convergence of
subthalamopallidal fibres in this lateral band of the globus pallidus. The
projections from the STh to more medial parts of the globus pallidus, on the
other hand, appear to be topographically organized (Kita and Kitai, 1987;
Groenewegen and Berendse, 1990). The results obtained with FG can be summarized as follows. Injections in the subcommissural part of the ventral
pallidum result in retrograde labelling of a large number of intensely
fluorescent neurons in the dorsomedial part of the STh (Fig. 2D). In
ventral and lateral directions the labelled neurons rapidly diminish in
number. Scattered labelled cells lie between the fascicles of the cerebral
peduncle and in the medially adjacent lateral hypothalamic area. After an

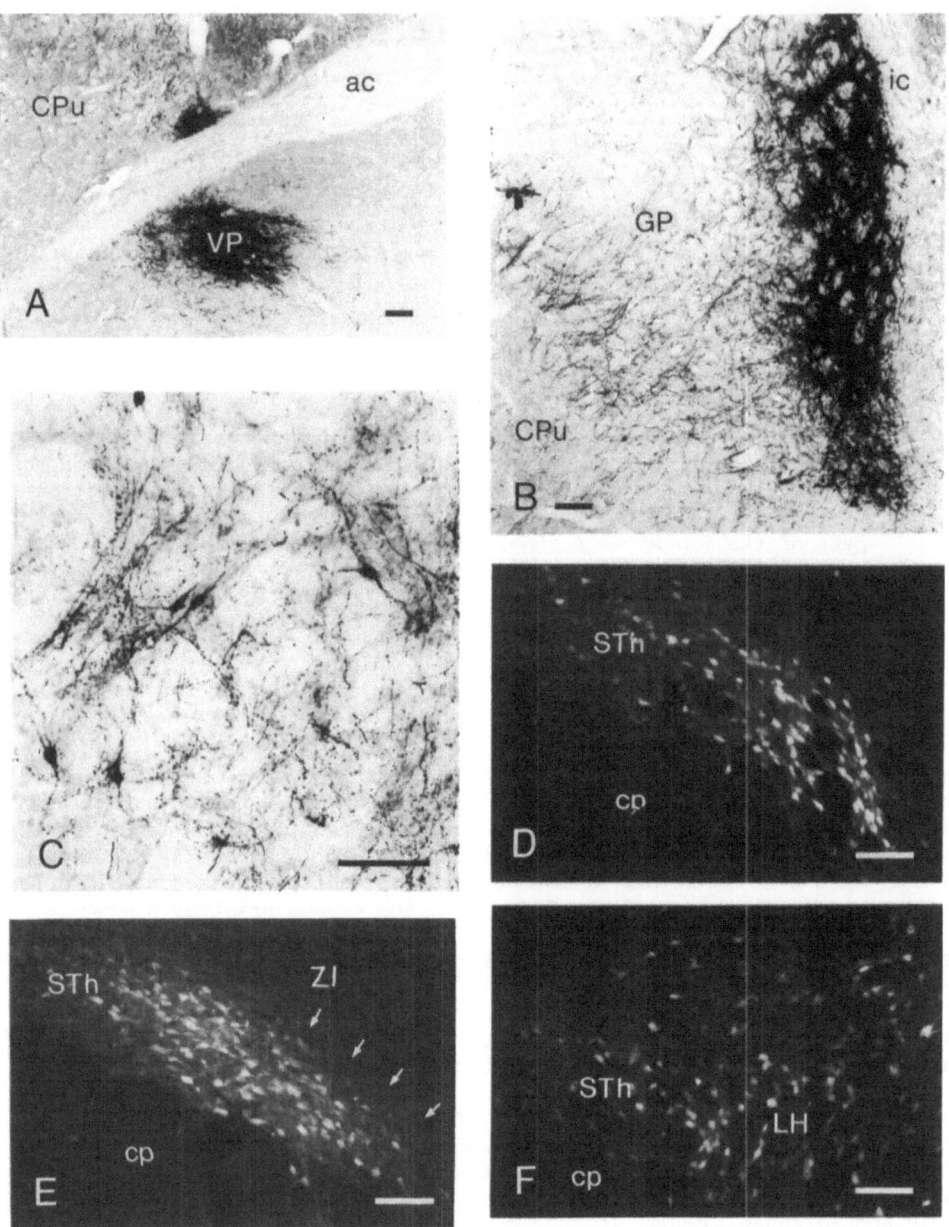

Fig. 2. A-C, PHA-L labelling in the lateral part of the subcommissural
    ventral pallidum (A) and the medial part of the globus pallidus (B)
    after an injection in the medial part of the STh. Note in (B) the
    ventrolateral strip of labelling in the globus pallidus. Within the
    terminal field in the medial part of the globus pallidus there is
    concomitant retrograde labelling of neurons (C). D-F, FG-labelled
    neurons in the medial part of the STh and the adjacent part of the
    LH following injections in the ventral pallidum (D), the medial
    part of the globus pallidus (E), and the olfactory tubercle (F).
    Arrows in (E) mark the dorsomedial border of the STh. Bars = 100 μm.
    Abbreviations: ac: anterior commissure; GP: globus pallidus; ic:
    internal capsule; VP: ventral pallidum; see also legend to figure 3.

injection of FG in the medial part of the globus pallidus retrograde labelling is present more ventrally and laterally in the STh than in the ventral pallidal cases, thus leaving the extreme dorsomedial part of the nucleus free of labelling (Fig. 2E). Following an injection involving the deep layers of the caudal part of the olfactory tubercle only a few cells are retrogradely labelled in the extreme dorsomedial corner of the STh, whereas most retrogradely labelled neurons are located in the adjacent part of the LH (Fig. 2F). The neurons in this area closely resemble the labelled neurons in the STh in size and shape but they are less densely packed.

The present data confirm that the most prominent fibre connections of the STh are established with pallidal structures (e.g. Nauta and Cole, 1978; Ricardo, 1980; Carpenter et al., 1981; Kita and Kitai, 1987) and further indicate that distinct parts of the pallidal complex are reciprocally connected with distinct parts of the medial STh and the adjacent lateral hypothalamic area (Fig. 5B). The reciprocity of these connections is illustrated most clearly by the concomitant anterograde and retrograde labelling in the same pallidal area following PHA-L injections in the STh (Fig. 2C). Moreover, the labelled subthalamopallidal fibres were found to be in close contact with the retrogradely labelled neurons. It could further be demonstrated that a population of neurons outside the borders of the STh in the adjacent lateral hypothalamic area have connectional characteristics comparable to those of STh neurons (cf. Heimer et al., 1985). An argument in favour of the notion that neurons lying outside the STh proper belong to the same neuronal system is the similarity in morphology of the retrogradely labelled neurons in the LH and in the STh following injections in the olfactory tubercle and the ventral pallidum, respectively. Further knowledge of the synaptology and physiological characteristics of these lateral hypothalamic neurons is necessary to classify them with certainty as outlying STh neurons.

Projections from the Prefrontal Cortex

In rats and primates topographically organized projections to the subthalamic nucleus have been demonstrated from motor, premotor, and adjacent cortical areas (Hartmann-Von Monakow et al., 1978; Afsharpour, 1985). These cortical areas also project to dorsal parts of the striatum. Since the prefrontal cortex (for subdivisions see Groenewegen, 1988) is intricately involved in the ventral striatopallidal system (Groenewegen et al., 1990), we speculated that this part of the cortex would project to the medial part of the STh and the adjacent lateral hypothalamic area which are associated with ventral parts of the pallidum (see above), the ventral striatum, and the medial part of the ventral mesencephalon (Groenewegen and Berendse, 1990).

PHA-L injections in the prefrontal cortex (see for example Fig. 3C) all result in labelling in the ipsilateral STh and/or the adjacent part of the LH. An injection of PHA-L in the dorsal anterior cingular cortex gives rise to terminal labelling ventrally and laterally in the medial part of the STh, most densely at caudal levels (Fig. 3A). After injections into the dorsal part of the prelimbic cortex labelling is present caudally in the most medial part of the STh, with only a few fibres in the adjacent LH (Fig. 3B). The projection of the ventral part of the prelimbic cortex is confined to the LH (Fig. 3C,D). Injections in the dorsal agranular insular cortex result in terminations in the rostral part of the STh. At the rostral extreme of the nucleus, the labelled fibres occupy a ventral position, whereas somewhat more caudally they shift to a dorsomedial position (Fig. 3E). A few fibres are present in the LH. The projections from the medial orbital, infralimbic and ventral agranular insular cortices are confined to the lateral hypothalamic area adjacent to the STh (Fig. 3F).

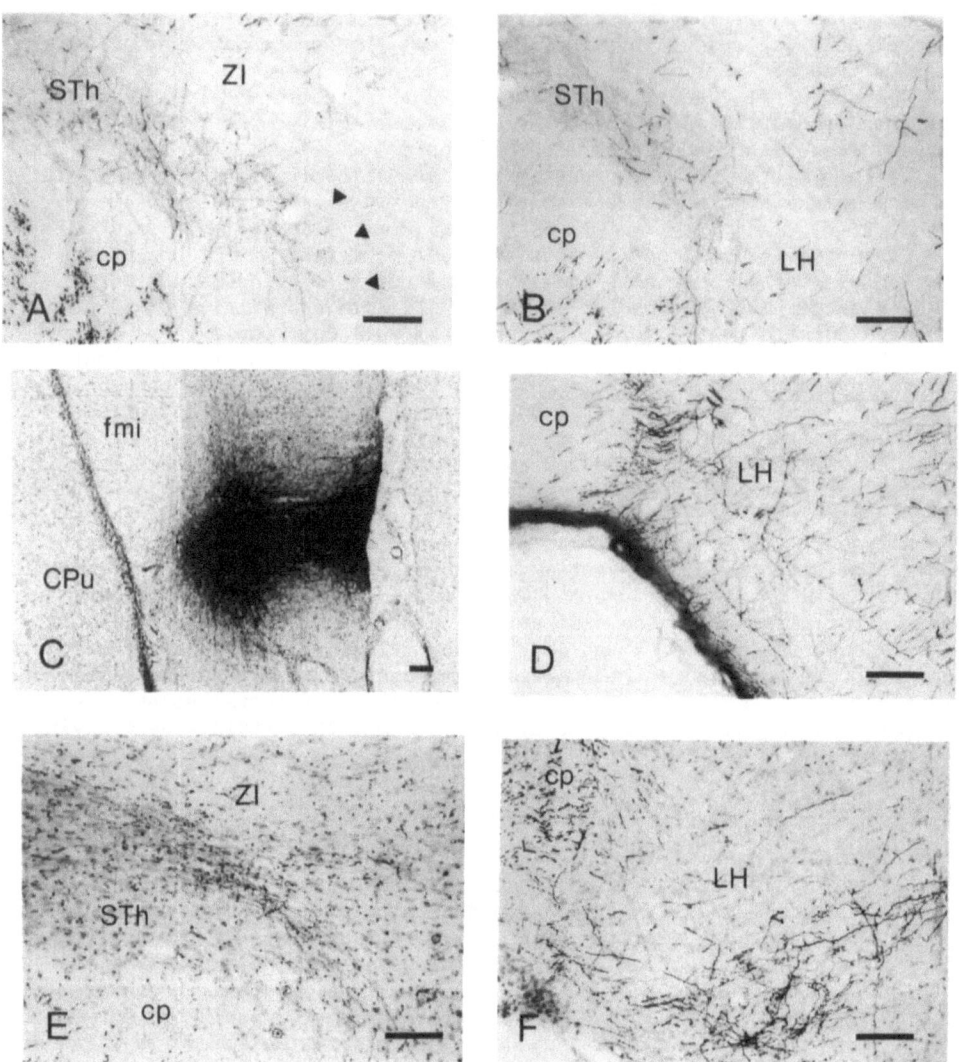

Fig. 3. Photomicrographs illustrating the organization of the prefrontal
corticosubthalamic projections. A, PHA-L labelled fibres in the
medial part of the STh after an injection in the dorsal anterior
cingular cortex. Arrowheads mark the dorsomedial border of the STh.
Note that the extreme dorsomedial part of the STh remains free of
labelling. B, A PHA-L injection in the dorsal part of the prelimbic
cortex results in labelling in the extreme medial part of the STh at
caudal levels. A few fibres are present in the adjacent part of the
LH. C, PHA-L injection site in the ventral part of the prelimbic
cortex. D, PHA-L labelling resulting from the injection illustrated
in (C) in the part of the LH adjacent to the STh. E, PHA-L labelled
fibres in the dorsomedial part of the STh following an injection in
the dorsal agranular insular cortex. F, PHA-L labelling in the LH
after an injection of PHA-L in the ventral agranular insular cortex.
Note the difference in position of the labelling compared to (D).
Bars = 100 μm. *Abbreviations*: cp: cerebral peduncle; CPu: caudate-
putamen; fmi: forceps minor of the corpus callosum; LH: lateral
hypothalamus; STh: subthalamic nucleus; ZI: zona incerta.

These results demonstrate that the projections from the prefrontal cortex are topographically distributed over the medial part of the STh and the adjacent lateral hypothalamic area (Fig. 5B). As indicated in figure 5B, the various prefrontal areas can also indirectly influence these areas by way of relays through distinct parts of the striatum and the pallidum (see also Groenewegen et al., 1990). The projections from the prefrontal cortex appear to overlap to some extent with projections from directly adjacent cortical areas (Afsharpour, 1985). However, in the experiments of Afsharpour (1985), the dorsomedial part of the STh and the adjacent part of the LH, where most prefrontal cortical afferent fibres terminate, allways remained free of labelling.

Projections from the Thalamus

The projections from the thalamus to the STh were examined in a series of experiments with small PHA-L injections in individual nuclei of the midline and intralaminar thalamic complex (see also: Berendse and Groenewegen, 1990). Anterograde labelling in the STh was observed following injections of PHA-L in the parafascicular thalamic nucleus only. A projection from this nucleus to the STh has been reported previously with anterograde autoradiographic tracing techniques in rats and cats (Sugimoto et al., 1983; Royce and Mourey, 1985). According to our data the thalamosubthalamic pathway is topographically organized. The most medial part of the parafascicular nucleus, medial to the fasciculus retroflexus, projects to the medial part of the STh (Fig. 4A). The same part of the parafascicular nucleus innervates the lateral part of the nucleus accumbens and the medial part of the caudate-putamen (Berendse and Groenewegen, 1990), which are related to the medial part of the STh by routes through the lateral part of the subcommissural ventral pallidum and the medial part of the globus pallidus, respectively (Fig. 5B). The part of the parafascicular nucleus located lateral to the fasciculus retroflexus projects to more lateral parts of the STh (Fig. 4B) and to extensive lateral parts of the caudate-putamen (Berendse and Groenewegen, 1990), the latter of which can influence the lateral part of the STh indirectly by way of the main body of the globus pallidus (Fig. 5A; Ricardo, 1980; Gerfen, 1985; Kita and Kitai, 1987).

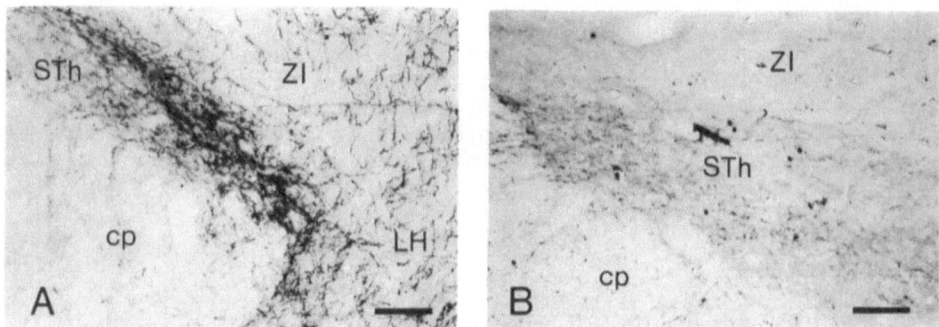

Fig. 4. PHA-L labelled thalamosubthalamic fibres in the STh following
injections in the medial (A) and lateral (B) parts of the parafascicular nucleus. Note that the projection originating from the
lateral part of the parafascicular nucleus is sparser than that
arising from the medial part. Bars = 100 μm. For abbreviations see
legend to figure 3.

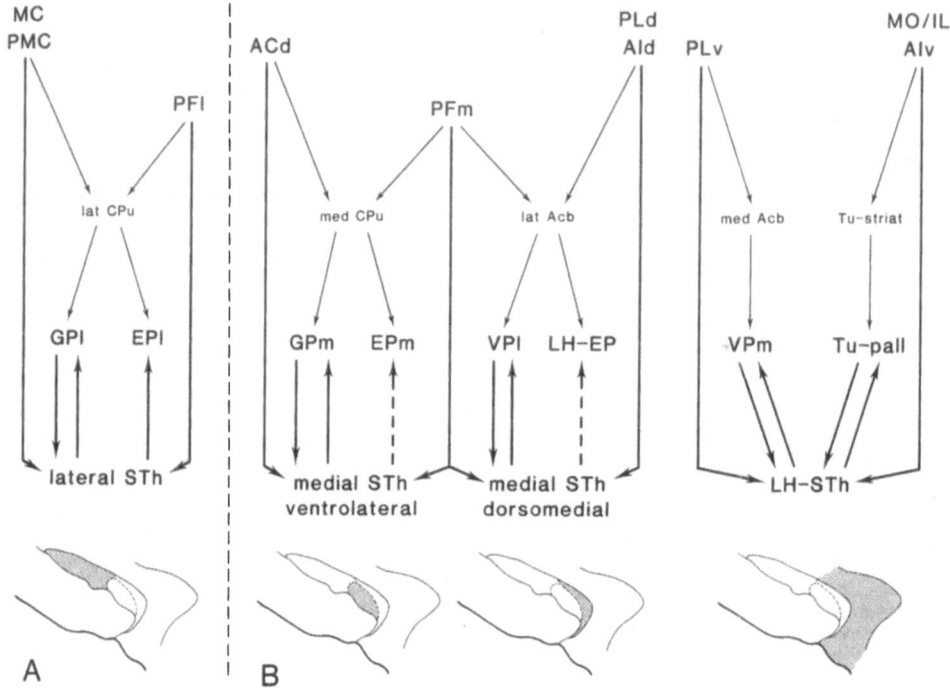

Fig. 5   Summary diagram of the connections of the lateral (A) and medial (B)
parts of the STh and the adjacent part of the LH with the cerebral
cortex, the thalamus, and the pallidum based on data from the
literature, the present report and unpublished observations. The
indirect pathways from the cortex and thalamus to the STh via the
striatum and the pallidum are drawn in thin lines. The projections
from the medial part of the STh to the EPm and the LH-EP are indicated by broken lines, as they cannot be distinguished from each
other on the basis of the available data. *Abbreviations*: Acb:
nucleus accumbens; ACd: dorsal anterior cingular cortex; AId: dorsal
agranular insular cortex; AIv: ventral agranular insular cortex;
CPu: caudate-putamen; EP: entopeduncular nucleus; GP: globus
pallidus; IL: infralimbic cortex; l: lateral; LH-EP: lateral hypothalamic area adjacent to the EP; LH-STh: lateral hypothalamic area
adjacent to the STh; MC: motor cortex; m: medial; MO: medial
orbital cortex; PF: parafascicular thalamic nucleus; PLd: dorsal
part of prelimbic cortex; PLv: ventral part of prelimbic cortex;
PMC: premotor cortex; STh: subthalamic nucleus; Tu-pall: pallidal
elements of the olfactory tubercle; Tu-striat: striatal elements of
the olfactory tubercle; VP: ventral pallidum.

CONCLUDING REMARKS

The medial part of the STh and the adjacent lateral hypothalamic area
are reciprocally connected with ventral parts of the pallidum and receive
projections from the medial and lateral parts of the prefrontal cortex and
the medial part of the parafascicular thalamic nucleus, similar and parallel
to the connections of the lateral part of the STH with motor-related parts
of the pallidum and its input from motor and premotor cortical areas and the
lateral part of the parafascicular nucleus (Fig. 5). Furthermore, the
present data provide evidence for the existence of at least three zones

within the medial part of the STh and the adjacent part of the LH that each maintain parallelly organized connections with distinct parts of the prefrontal cortex and the pallidum (Fig. 5B). The organization of the above-described connections of the STh and of those with the striatum and the substantia nigra (Groenewegen and Berendse, 1990) fit in with the topographical pattern of connections in ventral parts of the basal ganglia (Groenewegen et al., 1990).

The present findings as well as the results of a previous study on the connections of the STh in primates (Parent and Smith, 1987; Parent et al., 1989) indicate that the STh, in addition to its well-established role in motor processing, probably subserves non-motor functions as well.

ACKNOWLEDGEMENTS

Supported in part by Medigon/NWO Program Grant No. 900-550-093. The authors would like to thank Prof. Dr. A.H.M. Lohman for critically reading the manuscript and Mr. Dirk de Jong for photographical assistance.

REFERENCES

Afsharpour, S., 1985, Topographical projections of the cerebral cortex to the subthalamic nucleus, J. Comp. Neurol., 236:14.

Alexander, G.E., DeLong, M.R., and Strick, P.L., 1986, Parallel organization of functionally segregated circuits linking basal ganglia and cortex, Ann. Rev. Neurosci., 9:357.

Beckstead, R.M., 1983, A reciprocal axonal connection between the subthalamic nucleus and the neostriatum in the cat, Brain Res., 275:137.

Berendse, H.W., and Groenewegen, H.J., 1990, The organization of the thalamostriatal projections in the rat, with special emphasis on the ventral striatum, Submitted.

Carpenter, M.B., Carleton, S.C., Keller, J.T., and Conte, P., 1981, Connections of the subthalamic nucleus in the monkey, Brain Res., 224:1.

Carpenter, M.B., Whittier, J.R., and Mettler, F.A., 1950, Analysis of choreoid hyperkinesia in the rhesus monkey. Surgical and pharmacologic analysis of hyperkinesia resulting from lesions in the subthalamic nucleus of Luys, J. Comp. Neurol., 92:293.

Crossman, A.R., 1987, Primate models of dyskinesia: The experimental approach to the study of basal ganglia-related involuntary movement disorders, Neuroscience, 21:1.

Gerfen, C.R., 1985, The neostriatal mosaic. I. Compartmental organization of projections from the striatum to the substantia nigra in the rat, J. Comp. Neurol., 236:454.

Groenewegen, H.J., 1988, Organization of the afferent connections of the mediodorsal thalamic nucleus in the rat, related to the mediodorsal-prefrontal topography, Neuroscience, 24:379.

Groenewegen, H.J., and Berendse, H.W., 1990, Connections of the subthalamic nucleus with ventral striatopallidal parts of the basal ganglia in the rat, J. Comp. Neurol., in press.

Groenewegen, H.J., and Wouterlood, F.G., 1990, Light and electron microscopic tracing of neuronal connections with Phaseolus vulgaris-leucoagglutinin (PHA-L), and combinations with other neuroanatomical techniques, in: "Handbook of Chemical Neuroanatomy", Vol. 8, A. Björklund, T. Hökfelt, F.G. Wouterlood, and A. Van den Pol, eds., Elsevier, Amsterdam, p 47.

Groenewegen, H.J., Berendse, H.W., Wolters, J.G., and Lohman, A.H.M., 1990, The anatomical relationship of the prefrontal cortex with the striato-pallidal system, the thalamus and the amygdala: evidence for a parallel organization, in: "The Prefrontal Cortex: its Structure, Function, and Pathology" (Prog. Brain Res.), H.B.M. Uylings, C.G. Van Eden, J.P.C. De Bruin, M.A. Corner, and M.G.P. Feenstra, eds., Elsevier, Amsterdam, in press.

Haber, S.N., Groenewegen, H.J., Grove, E.A., and Nauta, W.J.H., 1985, Efferent connections of the ventral pallidum: evidence of a dual striatopallidofugal pathway, J. Comp. Neurol., 235:322.

Hammond, C., Rouzaire-Dubois, B., Féger, J., Jackson, A., and Crossman, A.R., 1983, Anatomical and electrophysiological studies on the reciprocal projections between the subthalamic nucleus and nucleus tegmenti pedunculopontinus in the rat, Neuroscience, 9:41.

Hartmann-Von Monakow, K., Akert, K., and Künzle, H., 1978, Projections of the precentral motor cortex and other cortical areas of the frontal lobe to the subthalamic nucleus in the monkey, Exp. Brain Res., 33:395.

Heimer, L., and Wilson, R.D., 1975, The subcortical projections of the allocortex: similarities in the neural associations of the hippo-campus, the piriform cortex, and the neocortex, in: "Golgi Centennial Symposium", M. Santini, ed., Raven Press, New York, p 177.

Heimer, L., Alheid, G.F., and Zaborszky, L., 1985, Basal ganglia, in: "The Rat Nervous System", G.W. Paxinos, ed., Academic Press, Sydney, p 37.

Kita, H., and Kitai, S.T., 1987, Efferent projections of the subthalamic nucleus in the rat: Light and electron microscopic analysis with the PHA-L method, J. Comp. Neurol., 260:435.

Martin, J.P., 1927, Hemichorea resulting from a focal lesion of the brain (The syndrome of the body of Luys), Brain, 50:637.

Nauta, H.J.W., and Cole, M., 1978, Efferent projections of the subthalamic nucleus: an autoradiographic study in monkey and cat, J. Comp. Neurol., 180:1.

Nomura, S., Mizuno, N., and Sugimoto, T., 1980, Direct projections from the pedunculopontine tegmental nucleus to the subthalamic nucleus in the cat, Brain Res., 196:223.

Parent, A., and Smith, Y., 1987, Organization of efferent projections of the subthalamic nucleus in the squirrel monkey as revealed by retro-grade labeling methods, Brain Res., 436:296.

Parent, A., Hazrati, L.-N., and Smith, Y., 1989, The subthalamic nucleus in primates. A neuroanatomical and immunohistochemical study, in: "Neural Mechanisms in Disorders of Movement", A.R. Crossman, and M.A. Sambrook, eds., John Libbey, London, p 29.

Ricardo, J.A., 1980, Efferent connections of the subthalamic region in the rat. I. The subthalamic nucleus of Luys, Brain Res., 202:257.

Royce, G.J., and Mourey, R.J., 1985, Efferent connections of the centro-median and parafascicular thalamic nuclei: an autoradiographic investigation in the cat, J. Comp. Neurol., 235:277.

Schmued, L.C., and Fallon, J.H., 1986, Fluoro-Gold: a new fluorescent retrograde axonal tracer with numerous unique properties, Brain Res., 377:147.

Sugimoto, T., Hattori, T., Mizuno, N., Itoh, K., and Sato, M., 1983, Direct projections from the centre median-parafascicular complex to the subthalamic nucleus in the cat and rat. J. Comp. Neurol., 214:209.

Whittier, J.R., and Mettler, F.A., 1949, Studies on the subthalamus of the rhesus monkey. II. Hyperkinesia and other physiological effects of subthalamic lesions, with special reference to the subthalamic nucleus of Luys, J. Comp. Neurol., 90:319.

Zahm, D.S., 1989, The ventral striatopallidal parts of the basal ganglia in the rat - II. Compartmentation of ventral pallidal efferents, Neuroscience, 30:33.

THE SUBTHALAMIC NUCLEUS : NEW DATA, NEW QUESTIONS

J. Féger, P. Robledo and N. Renwart

Laboratoire de Pharmacologie
Faculté des Sciences Pharmaceutiques et Biologiques de l'Université
R.Descartes, F-75270 Paris Cedex 06

INTRODUCTION

Since the first "Basal Ganglia " meeting, there has been a significant evolution in the knowledges concerning the subthalamic nucleus specially in relation to its importance and function. This new viewpoint is demonstrated in those diagrammatic representations so passionately used by basal ganglia addict people. For example, in 1983, the subthalamic nucleus was absent from the diagramms of basal ganglia function proposed by Penney and Young. In 1989, this structure is present in a central position (Albin et al.,1989). Today, the subthalamic nucleus is a well recognized structure among the basal ganglia . However, most diagramms are still published with a lot of arrows turning around a empty place where the subthalamic nucleus is like a ghost. Functionnaly, as it was claimed in a bright and premonitory formulation, now the subthalamic nucleus appears as an important driving force onto the basal ganglia (Kitai and Kita, 1987).

This proposition reversed the previously and commonly prevailing assumption which ascribed an inhibitory function to the subthalamo-pallidal efference, in parallel to the striato-pallidal gabaergic pathway (see ref. in De Long and Georgopoulos, 1981). This new trend was forced by the growing set of morphological data obtained with light and electronmicroscopic observation (Chang et al. 1984, Kitai and Kita,1987)as well as immunocytochemical results (Parent et al. 1989) and electrophysiological data recorded in *in vitro* experiments (Kitai and Kita, 1987, Nakanishi et al. 1987,1988).

In a first part of this paper, we report electrophysiological and metabolic data obtained in the rat, showing that the level of neuronal activity in the pars reticulata of the substantia nigra and in the pallidal complex (globus pallidus and nucleus entopeduncular) is largely dependent on the excitatory and probably glutamatergic subthalamic efferences ending in these structures. The second part deals with the relationship between the dopaminergic system and subthalamic activity. In the third part we provide behavioral data obtained in rats receiving micro-injections at subthalamic level of drugs able to induce an activation or a depression of neuronal activity. These results were reported in the doctoral thesis by N.Renwart and P.Robledo and are or will be published (Robledo et al.1988, Féger et al.1989, Robledo and Féger 1990, Renwart et al. 1990).

RESULTS

1° Electrophysiological and metabolic arguments for the driving activity ensured by the subthalamic nucleus on the substantia nigra,pars reticulata, and the pallidal complex.

In the rat, most subthalamic neurons project to both the substantia nigra and the pallidal complex via collaterals running in a caudal and rostral directions (Deniau et al. 1978, Van Der Kooy and Hattori, 1980). In the previous *in vivo* experiments, electrical stimulation within the subthalamic nucleus leads to opposite results depending on whether records were performed respectively in substantia nigra (Hammond et al. 1978, 1983,) or in the pallidal complex (Perkins and Stone, 1980, Hammond et al. 1983), even though it was attempted to be in best conditions for a selective action. This unexpected result was minutiously discussed by Kitai and Kita (1987) and a number of argument were put forward against the inhibitory effect attributed to the subthalamo-pallidal projection. Briefly,

- there are no immunocytochemical evidence for GAD or GABA positive neurons in subthalamic nucleus,

- a subthalamic lesion did not induce a change in GABA concentration in the pallidum,

- the synaptic organization of identified subthalamic terminals at pallidal and nigral level have the same characteristics and correspond to the asymetric type,

- electrophysiological experiments performed on brain slices preparation unquestionably confirm the excitatory activity of the subthalamo-nigral efferences (Nakanishi et al. 1987).

All these arguments were an incentive for us to perform new *in vivo* experiments. Moreover, we also found interesting to compare electrophysiological results to those obtained in a set of behavioural experiments showing the production of a dyskinesia in rat by injection of the gaba antagonist, bicuculline, into the subthalamic nucleus. The chemical stimulation by microinjection of bicuculline (volume injected : 0,2 µl, concentration : 0,39 mmol/l)in the subthalamic nucleus avoids unspecific activation of fibers running through or in the vicinity of the subthalamic nuleus. At this concentration and volume, a strong and long duration activation was always obtained. The excitation induced by this drug is related to the release of tonic gabaergic inhibition (Rouzaire-Dubois et al.1980). The mean increase in the firing rate of recorded subthalamic neurons was of 358 % (basal activity : $15 \pm 2,1$ spikes per second , level at plateau : $69,9 \pm 5,8/$ s, this level was quite stable until loss of the cell). In the target structures, the changes in rate occured within 30 to 60 s after the end of injection. In the substantia nigra, pars reticulata, the activity was enhanced in 20 out 21 cells (baseline : $32,4 \pm 1,5$ spikes/s to $54,6 \pm 3,7$ spikes/s). In the entopeduncular nucleus, an increase of activity was observed in 11 out 14 neurons, but three cells were not retained since the record was suddenly interrupted before determination of the mean activity . One cell showed a decrease in firing rate. The observed variation was from $15,6 \pm 3,1$ spike/s to $30,7 \pm 3,8$ spikes/s. In the globus pallidus, similar results were obtained since the activity of 9 out 13 neurons was increased by 112 % whereas the rate remains at baseline level for 2 neurons and decreases in two others. Then, the main and statistically significant response to subthalamic stimulation was a powerfull enhancement of the neuronal activity (Figure 1). This excitatory effect seems to be based on glutamatergic transmission since enhancement of nigral activity is blocked by previous injection of kynurenic acid ($0,1$ µg/ $0,2$ µl) in the substantia nigra (Robledo and Féger, 1990). For dopaminergic nigral cells, identified on the basis of electrophysiological characteristics, the responses to subthalamic activation were heterogeneously distributed : the activity of 9 out 19 neurons goes from mean $\pm$SEM $4,6 \pm 0,7$ to $5,6 \pm 0,6$ spikes per second ($+ 22\%$, $p < 0,03$), the activity of ten cells decreases from $6,3 \pm 0,8$ to $4,6 \pm 0,8$ spikes per second ($- 27$ %, $p < 0,01$). In the striatum, a total number of 14 neurons were tested and the following variations were observed : a slight but significant increase in 5 cells is observed ($1,1 \pm 0,5$ to $1,52 \pm 0,7$ sp/s), a weak non significant decrease was recorded in 4 cells ($1,9 \pm 1,2$ to $0,5 \pm 0,3$ sp/s), an increase followed by a decrease in two others and no changes in the last two cells. In all these structures, no changes were recorded on the contralateral side of the injection.

Thus, the microinjection of bicuculline into the subthalamic nucleus enhances the activity of subthalamic neurons as well as the activity of those neurons localised in the target structures like the pallidal complex and the pars reticulata of the substantia nigra. Still, another question comes to light. The baseline activity in these structure is it an intrinsic processus or is it the result of two main exogenous and opposite influences : inhibitory by the striatal and pallidal efferences, activatory by the subthalamo-nigral and subthalamo-pallidal projections ? Indeed, the subthalamic neurons are characterized by a quite regular activity ranging between 10 to 35 spikes per second with a mean around 20 spikes per second. Is the "spontaneous " activity of pallidal and nondopaminergic nigral cells dependant on this subthalamic discharge rate? We attempted to answer this question using the microinjection technique since it is possible to inject a compound able to depress the neuronal activity with again an effect restricted to the somatodendritic part of neurons without involvement of axonal fibers.

Indeed, muscimol, a GABA agonist, at a concentration of 0,95 mmol/l (25 ng in 0, 2 µl), injected into the subthalamic nucleus induced a complete and long lasting inhibition of all recorded subthalamic neurons (Figure 1). In the target structures, the decrease in firing rate begun 30 to 60 seconds after the end of the injection but the maximal effect is obtained progressively in 10 minutes: in the pars reticulata of substantia nigra, the depression reached 50 % (n= 10, mean ±SEM changes from 28,9 ± 2,9 to 14,1 ± 5,2 spikes per second), in the entopeduncular nucleus there is quite a total arrest (n = 9, baseline activity 21,9 ± 3,2 to 0,49 ± 0,3 sp/s), in the globus pallidus, 7 out 8 neurons showed a 47 % decrease (23,9 ±4,5 to 12,9 ±3 sp/s). Similar to results obtained in experiments with bicuculline, muscimol injected into the subthalamic nucleys induced heterogenous and non significant changes in the striatum and in the pars compacta of the substantia nigra.

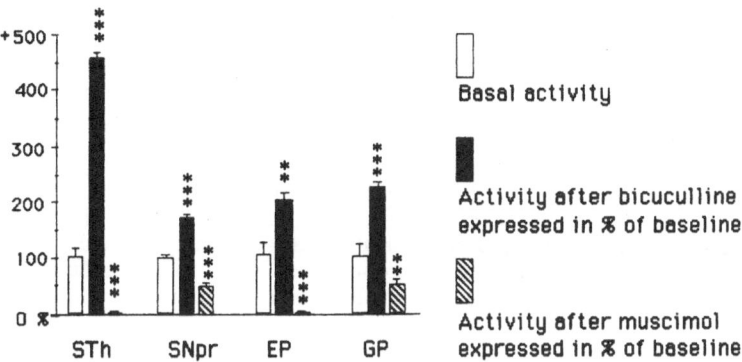

Figure 1. Histogramm showing the effect of a microinjection of bicuculline (0,2 μl of a 0,39 mol/l solution) or muscimol (0,2μl of a 0,95 mmol/l solution) on the neuronal activity in the subthalamic nucleus (STh), the substantia nigra pars reticulata (SNpr), the entopeduncular nucleus (EP) and the globus pallidus (GP).

Another way to check constant and long lasting changes in neuronal activities lies in the determination of local cerebral glucose utilization as an index of neuronal energetic metabolism. Briefly, experiments with [14 C] 2-deoxyglucose were performed in the same conditions (anesthesia with ketamine, 100 mg/kg), as in electrophysiological experiments. Injection of the [14 C] 2-deoxyglucose (12,5 μCi/ 100g body weight, diluted to a final volume of 1,2 ml saline solution 0,9% injected in 30 second) was carried out through a catheter placed in the femoral vein. Timed blood samples were collected through a catheter placed in femoral artery in order to determine the plasmatic concentration of glucose and [14 C] 2-deoxyglucose. Autoradiogrammes were prepaired as classicaly described. Regional concentrations of [14 C] 2-deoxyglucose were determined in at least four consecutive sections, using a computerized image processing system (Starwise, IMSTAR or Histopericolor, NUMELEC, France). The local cerebral glucose utlization was calculated from this regional concentration , and from the plasmatic values and appropriate constants according to the operational equation for the method (Sokoloff et al. 1977). Since the experimental conditions did not allow a comparable level of brain metabolism from one animal to the other, we chose to compare the difference in the local cerebral glucose utilization for each structure of interest between the ispsilateral and contralateral sides.

The following results were obtained. In control group (5 animals receiving only a microinjection of vehicle, saline with 1% of eosine), the difference in local cerebral glucose utilization between the ispilateral and the contralateral sides were weak : a slight increase in the ipsilateral substantia nigra pars reticulata (+ 2,5 %) and in the entopeduncular nucleus (+1,9 %), a decrease in the globus pallidus (- 0,3 %). None of these variations were statistically significant. In the group receiving bicuculline (4 animals retained out of 5), the local cerebral glucose utilization was of 54 ± 4,3 μmol/100g/mn in the contralateral pars reticulata of substantia nigra and of 69,4 ± 3,3 in the ipsilateral side. In the contralateral entopeduncular nucleus, the LCGU was found to be of 51,2 ±3,8 and of 66 ± 6,5 in the ipsilateral side. In the contralateral globus pallidus, the LCGU is of 58,05 ± 5,5 and of 71,8 ± 6,3 in the ipsilateral structure. In the group receiving muscimol (4 animals) opposite variations were observed : a decrease from 56,7 ±4,6 μmol/100 g/mn to 38,4 ± 2 in the case of the substantia nigra pars

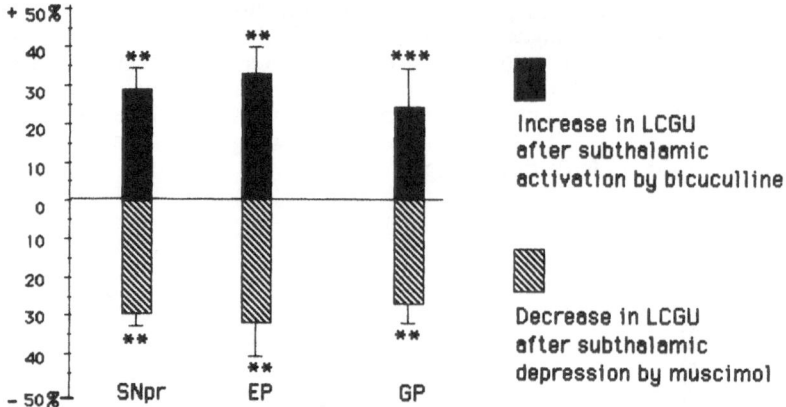

Figure 2. Changes in local cerebral glucose utilization (LCGU) after a microinjection of bicuculline (40 ng/0,2 μl)or muscimol (25 ng/0,2 μl) into the subthalamic nucleus. The variations in the substantia nigra pars reticulata (SNpr) , entopeduncular nucleus (EP) and globus pallidus (GP) are expressed in percentage of the difference between the ipsi and contralateral side to the injected subthalamic nucleus side.

reticulata, from 49,2 ± 4,4 to 33,3 ± 3,1 in the entopeduncular nucleus and from 57,5 ± 3,7 to 41,9 ± 3 for the globus pallidus (Figure 2).

Since electrophysiological results, obtained under the same conditions, provide strong arguments for an excitatory synaptic relation between subthalamic terminals and neurons localized in these different structures, the activity of terminals and of somato-dendritic parts in substantia nigra and in pallidal complex varies in the same direction. No inference can thus be made, as to which neuronal element is retaining most of the [14 C] 2-deoxyglucose, i.e. the terminals or the somato-dendritic part . However, in the ventro-medial thalamic nucleus, VM, which receives inhibitory inputs from the substantia nigra and the entopeduncular nucleus, the activity of the efferent terminals and of the somato-dendritic part varies in an opposite direction. Since activation or reduction of the afferent terminals by bicuculline and by muscimol produces reduced and increased local cerebral glucose utilization respectively, it seems that the accumulation of [14 C] 2-deoxyglucose is related to changes in activity at post-synaptic level. However, this preliminary conclusion needs to be confirmed in further experiments.

In conclusion, electrophysiologiocal data and metabolic results with the [14 C] 2-deoxy-glucose technique provides obvious arguments permitting to assign an essential function to the subthalamic nucleus as a generator of neuronal activity in the substantia nigra, pars reticulata and in the pallidal complex. The subthalamic nucleus itself receives excitatory cortical projections (Kitai and Deniau, 1981). They originate bilaterally from parietal and frontal areas, and in addition from the ipsilateral temporal and occipital areas (Rouzaire-Dubois and Scarnati, 1985). Part of these cortical projections send back peripheral somesthesic activations to the subthalamic nucleus (Hammond et al. 1978, Canteras et al. 1988). Therefore, subthalamic activity is in itself maintened by cortical influence, and maybe by thalamic projections, originating from CM-PF and recently identified (Sugimoto et al. 1983).

2° Interactions between subthalamic nucleus and its efferences with the dopaminergic system

The relationship between the subthalamic nucleus and the dopaminergic system involves two main aspects:

On the one side, attention was first turned to the dopaminergic effect on this structure. Several studies, using the [14 C] 2-deoxyglucose method, emphasize the intensity of increase in local cerebral glucose utilization at the level of the subthalamic nucleus after systemic administration of amphetamine or apomorphine . This effect could either be due to an indirect action, through the striatum or to a direct effect on the subthalamic neurons. A direct effect is supported by available data on the dopaminergic innervation of subthalamic nucleus (Brown et al. 1979,1987 Meibach and Katzman, 1979). Moreover, the activity of subthalamic neurons is enhanced by iontophoretic applications of dopamine (Campbell et al.1985, Mintz et al. 1986a,) or by systemic administration of apomorphine or amphetamine. In this last situation, an indirect effect mediated by a primary involvement at striatal or pallidal level can be rejected since this response is still obtained in rats with subthalamic nucleus disconnected from the striatum and globus pallidus (Figure 3 - our unpublished results).

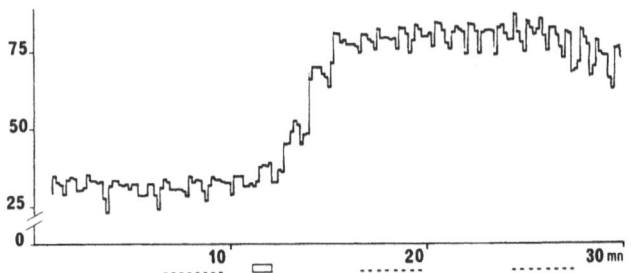

Figure 3 . Record of integrated activity enregistred in the subthalamic nucleus in rats with acute hemitransection just rostral to this structure. The rate of unit activity, counted during three minutes (----) increases from 32,3 ±2,8 spikes per second to 80,1±5 and 79,3±4 after the injection of amphetamine (⊂⊃ , 5mgKg / IV).

On the other side, electrical or chemical stimulation of the subthalamic nucleus enhance significantly the release of [^{3}H] dopamine in the substantia nigra (Mintz et al. 1986b). These data imply potential interactions between the subthalamic nucleus and the dopaminergic system. Therefore we found it interesting to pose the following questions :
a- How does the dopaminergic innervation of the subthalamic nucleus contribute to the tonic discharge rate of subthalamic neurons ?
b- Is the release of dopamine after subthalamic stimulation involved in the intensity of the nigral response to the subthalamic excitatory input ?
   The simplest experimental design in order to obtain a response to these questions consists on the impairment of the dopaminergic system by pretreatment with reserpine and with an inhibitor of tyrosine hydroxylase(alpha-mpt) . This double treatment induces a strong decrease (94%) in striatal dopamine level. Since reserpine can deplete serotoninergic terminals, abundant in the substantia nigra, control experiments were performed in this structure using only alpha-mpt and provide the same results. After inactivation of the dopaminergic system, the firing rate of subthalamic neurons was significantly increased by 53 % compared to untreated animals (15,2 ± 2,1 and 23 ±3 spikes/s respectively). After injection of bicuculline , the increase in activity was relatively weaker in treated animals as compared to the baseline, but the maximal discharge rate was the same for the two groups (68,1 ± 8,8 instead of 69,9 ± 5,8). In the pallidal complex,the mean effect of subthalamic activation remains the same in treated animals as in normal rats. However, the pattern of spontaneous activity was modified but a detailed analysis cannot be reported briefly. The main differences observed between normal rats and rats with low dopamine level appear at nigral level. No changes in the spontaneous activity were recorded, but activation of subthalamic nucleus induced a stronger increase in nigral neuronal activity in the pars reticulata : 217 % instead of 69 %. Consequently, the level of maximal discharge rate is higher in treated rats than in control animals : mean ±SEM 91 ±31 instead of 55,8 ±3,8 (p< 0,01). This enhanced response of nigral cells in the pars reticulata to the excitatory subthalamic input was unexpected since :
- iontophoretic application of dopamine produces an excitatory effect on this neuronal population (Matthews and German, 1986),
- release of endogenous dopamine or iontophoretic application reduce the inhibitory effect of GABA (Waszack and Walters, 1986),
- there is a spontaneous release of dopamine at nigral level (Chéramy et al., 1981) and this release is enhanced after subthalamic stimulation (Mintz et al. 1986b).
   Then it can be assumed that an impairment of the dopaminergic system will throw the balance toward a depression of activity. In our experiments, no changes in spontaneous activity was

observed presumably due to a compensatory action from the subthalamic activatory influence. As for the potentiation of the subthalamic activatory effect in dopamine depleted animals one possible explanation could be suggested,based on data obtained in the striatum, which of course need further experimental support. In this last structure, the release of glutamate in response to activation of cortico-striatal projections induces a release of dopamine . Conversely dopamine interacts with cortico-striatal terminals and inhibits the release of glutamate (Nieoullon et al. 1983, Chesselet 1984). Our results show that a similar processus might be present in the substantia nigra. Alternatively, the released dopamine could interact with gabaergic terminals and therefore reduce the firing rate through an increase of the strength of the inhibitory pressure on the pars reticulata neurons. Is it possible that such opposite effects on subthalamic glutamatergic or gabaergic terminals is only localized at nigral level, but not at the level of the entopedoncular nucleus ? Another interesting question is wether or not this potentiation is present in primates and if so, could it be involved in the paradoxical dyskinesia observed in parkinsonian patients ?

3 °-Behavioural data

The purposes of these experiments were :
- first, to determine if it is possible to obtain, in the rat, dyskinetic activities reminiscent of those obtained in monkey, both induced by microinjection of bicuculline into the subthalamic nucleus.
- Second, to analyse the neuronal mechanisms related to the induction of these dyskinesia by comparison of behavioural and electrophysiological results.

The microinjections were performed using chronically implanted tubes as guide for injection needless. An injection of 0,1 µg of bicuculline by side in a volume of 0,3 µl induces several changes in motor activities : repeated head movement and different forelimb movements such an alternance of flexion and extension and rubbing or clapping with extended limbs. However, with this dose, only a minority of the tested rat exhibited these movements. Moreover, abnormal movements were discontinue throughout the observation. With higher doses, the frequency and/or the duration of theses responses were enhanced and the proportion of animal shoving them was increased. Only the dose of 0,4 µg induced a real dyskinetic state, characterized by continuous abnormal movements of head and forelimbs. At this concentration, the majority of animals are dyskinetic and presented an abundant salivation (Table I).

In electrophysiological experiments, bicuculline was injected in the subthalamic nucleus of ketamine anesthetized rats in a range of 0,1 to 0,4 µg in 0,3 µl. As already reported, these microinjections enhance the subthalamic neuronal activity extracellularly recorded. When the injected dose increases, the electrophysiological records of neuronal activity include progressive changes marked by increase in slow oscillations of the baseline, reduction of spikes amplitudes and at the end disapearances of spikes. These changes are explained as being the results of an intense excitatory activity leading to a depolarization blockade. The percentage of depolarizing blocks increases with the dose injected: 0 % after 0,1 µg, 29 % with 0,2 µg and 86 % with 0,4 µg. Consequently, with bicuculline in doses equal to 0,4 µg, a production of important dyskinetic movements can be related to a functionnal inactivation of a largest part of subthalamic neurons (Table I).

However, abnormal motor activities are observed with a statistically significant intensity when the doses used induce mainly a pure increase in the firing rate of subthalamic neurons and anyone or a restricted frequency of depolarization blockades. A possible explanation on this discrepancy between the production of dyskinetic movements and the relative lack of firing blockade lies on the use of ketamine as anesthetic drug. It is known thats this molecule acts as a non-competitive antagonist on the NMDA subtype among the excitatory aminoacid receptors (Martin et al. 1985). Then a weaker involvement of these receptor reduce the permanent activatory effect induced by the cortico-subthalamic projection and finally the excitability of subthalamic neurons. Thus, the activation induced by the bicuculline is insufficient to reach the state of depolarization blockade in anesthetized rats compared to conscious animals used in behavioural experiments. Obvious explanations need electrophysiological records performed in chronic conditions on awake animals. On the other hand, is the production of these dyskinetic activities linked to an arrest of firing due to a depolarization blockade ?

There are several arguments against the coupling of abnormal movements and a simple blockade of the subthalamic neuronal activity. First of all, a micro-injection of 25 ng of muscimol into the subthalamic nucleus induces a complete inhibition of all recorded subthalamic neurons, but no changes were observed in motor activities which can be compared to a dyskinetic state. Even with 400 ng, no particular motor activities were observed but only less frequent grooming. Moreover, a catalepsy and a backward motion was observed.

Therefore the reduction of firing rate in subthalamic nucleus and the simple abolition of excitatory influx to the target structures is by itself insufficient to produce abnormal forelimb activity.

Table I allowing a comparison between intensity and frequence in different abnormal movements observed after a bilateral microinjection of bicuculline in the subthalamic nucleus,the proportion (%) of animals reaching a dyskinetic state and the proportion of recorde neurons exhibiting a block of depolarisation (statistical tests : Mann-Whitney and Kruskall-Wallis).

BICUCULLINE	saline	0,1 µg/0,3 µl	0,2 µg/0,3 µl	0,4 µg/0,3 µl
Observed movements :				
Head movements	0	3,5 ±1,2** (33%)	5,3 ±1,4** (69 %)	5,3 ±0,8** (90 %)
Body movements:				
- torsion	0	0,4 ±0,4	2,6 ±1,4 (23 %)	4,9 ±2,6 (40 %)
- jumps	0	0,7 ±0,7 (8 %)	0	2 ±0,9 (30 %)
Forelimbs movements alternate flex/extens.	0	2,5 ±1,3 (25%)	1,9 ±1,9 (23%)	9±2,7** (70 %)
mvts. with extended forelimb	0,1± 0,1	4 ±1,9* (41%)	4 ±1,5* (54%)	12 ±2,6 (80%)
% of recorded cells with arrest by block	0%	O%	29%	86%

This is confirmed by application of kynurenic acid in the substantia nigra and in entopeduncular nucleus, in a range of doses from 0,33 to 3,33 µg. It is expected that this glutamatergic antagonist blocks all the excitatory influence driven by the subthalamic nucleus. Nevertheless, no statistically changes were observed in the motor activities. Moreover, muscimol, even at a dose of 400 ng injected in the entopeduncular nucleus is again unable to challenge increase in head or forelimb movement. Thus,muscimol or kynurenate were without effect in regard to the production of abnormal movement, although each of them induce a reduction or an arrest of neuronal activities mimicking the depression produced by the disappearance of excitatory influx originating from the subthalamic nucleus. On the contrary, injection of an excitatory drugs into the entopeducular nucleus, kainic acid at concentrations below those which produce a lesion, induces biphasic responses : during the first fifteen minutes after injection of 36 ng animals exhibit only an increase in chewing but in a discontinuous manner. After this delay, the rats show body and forelimb movements reminiscent of those observed after injection of bicuculline into the subthalamic nucleus but only during a short period alternating with normal motor activities. It is of interest to note that a microinjection of the same dose of kainic acid into the substantia nigra induces also enhancement of alternate flexion/extension and movements with extended forelimbs.

All these results suggest that the abnormal movements observed after a microinjection of bicuculline into the subthalamic nucleus did not originates from a plain and unimodal modification in the neuronal activity of subthalamic neurons. It seems necessary to produce successively and/or simultaneously an increase of activity mixed with an arrest of activity in a part of the neuronal population. It is possible that the needs for an activation followed by a depression in rate discharge can explain the latency and the progressive intensity of hemiballismus observed in monkey after microinjection with bicuculline. Nevertheless, how accurate are our comparisons between changes in behaviour observed in conscious animal and changes in unit activities recorded on anesthetized animals? Are results obtained in rats transposable to primates ? In their experiments, Crossman and colleagues (1987, Mitchell et al. 1985, Sambrook et al. 1989) produced hemichorea and hemiballismus by use of large quantities of bicuculline : 30 µg in 2 µl, but the linear dimension of subthalamic nucleus in primates is larger than in rats in a ratio of 4 or 5 in the antero-posterior direction,width and thickness. In addition the subthalamic nucleus projects to the globus pallidus and the substantia nigra through separate population. In the rat, on the other hand, a homogenous population projects by branched axons to these two structures. In their metabolic studies, Crossman's group describes a diminishing intake of 2-deoxyglucose in the two segments of globus pallidus.These data together with our electrophysiological and metabolic results lead to the conclusion that subthalamic terminals activity is reduced as well as pallidal activity since the excitatory subthalamo-pallidal projection seems necessary to drive the spiking activity in this structure. However if hemiballismus was related to a simple arrest of pallidal activity, it is difficult to understand how ablative lesion of medial pallidal or of its efferences to the thalamus could abolish hyperkinesia produced by primary lesion at subthalamic level. How can suppression of an unactivated structure be favourable ? On the contrary, if it is assumed that the lesionned subthalamic nucleus resembles an irritative focus producing a disorganized activity in the medial pallidum, the abolition of abnormal movement by secondary pallidal lesion is more likely explainable. Metabolic studies provide information about the mean activity of neurons during the time allowed for the accumulation of 2-deoxyglucose but no information is obtained on the occurence of alternate periods of paroxystic activities and silence. Accurate explanations may be provided by chronique recordings of neuronal activities in the thalamic nuclei receiving pallidal afferences during hemiballistic behavior in monkeys.

CONCLUSION

The subthalamic nucleus fulfills a particular role within the basal ganglia. It sends excitatory projections to the entopeduncular nucleus and to the pars reticulata of the substantia nigra. This structure appears as a counterweight to the main inhibitory influence provided by the striatum on the outputs of the basal ganglia. But how are these two opposite influence organized? Are they? Is there competition between the subthalamic excitatory influence and the linear, hierarchical concept of desinhibition as the expression of striatal functions (Deniau and Chevalier, 1985) or is there alternance of striatal and subthalamic influences ? If so,what is there the sequential timing during preparation and execution of a motor activity ?

The similarity of the subthalamic nucleus and the striatum is emphasized by the presence of afferences originating from various cortical area and from the CM-Pf, a non specific thalamic nucleus at level of both these structures. But the subthalamic nucleus receives in addition an abundant projection from the lateral pallidum, thus indirectly it is also controlled by the striatum. It is of interest to note that, in primates, if the direct cortico-subthalamic projections are provided by motor areas, the indirect projections via the striatum and the pallidum are from the associative cortex with the subthalamic nucleus (G. Percheron, personnal communication). Moreover, Berendse and Groenenwegen have pointed out the existence of a topographically organized relation between the limbic part of the basal ganglia and the subthalamic nucleus. How is this connection functionnaly important ? Can this input be related to a motivationnal or emotionnal tone of the subthalamic nucleus. Striatal, pallidal and nigral lesions are not solely related to disorders of motor behaviour but also to disturbances of mental activities and affective states. How is the emotional component of behaviour affected in hemiballismus?

ACKNOWLEDGEMENTS

The authors wish to aknowledge financial support of INSERM (CRE 846008 and 876010). This story on the subthalamic nucleus start with C.Ohye, in "Institut Marey" and was carryed out owing to the collaboration of several foreign and french neurophysiologists. It is a pleasure for J.F. to aknowledge their friend and clever collaboration and contribution to all our reports in this field.

REFERENCES

Albin,R.L., Young,A.B. and Penney,J.B., 1989, The functionnal anatomy of basal ganglia disorders, T.I.N.S., 12: 366-375.
Berendse, H. W. and Groenenwegen, H. J.,1989, Anatomical evidence for a limbic part of the subthalamic nucleus in the rat. Basal Ganglia Soc. III rd meet. Abstr. N° 15.

Brown, L. L., Wolfson, L. I. and Feldman, S. M., 1987, Functional neuroanatomic mapping of the rat striatum : regional differences in glucose utilization in normal controls and after treatment with apomorphine. Brain Research, 411:65-71.

Campbell, G. A., Eckardt, M. J. and Weight, F. F., 1985, Dopaminergic mecanisms in the subthalamic nucleus of rat: Analysis using horseradish peroxydase and microionto-phoresis. Brain Research, 333: 261-270.

Chang, H. T., Kita, H. and Kitai, S. T., 1984, The ultrastructural morphology of the subthalamo-nigral axon terminals intracellularly labeled with horseradish peroxidase. Brain Res., 229: 182-185.

Cheramy, A., Leviel, V. and Glowinski, J., 1981, Dendritic release of dopamine in the substantia nigra. Nature, 289: 537-542.

Chesselet, M. F., 1984, Presynaptic regulation of neurotransmitter release in the brain : facts and hypothesis. Neuroscience, 12: 347-375.

Crossman, A. R., 1987, Primates models of dyskinesia : the experimental approach to the study of basal ganglia-related involuntary movements disorders. Neuroscience, 21: 1-40.

De Long, M. R. and Georgopoulos, A. P. 1981, Motor functions of the basal ganglia. In "Handbook of Physiology, the nervous system, II, Part 2 "J. M.Brookhart and V. B. Mountcastle (eds.), , Am. Physiol. Soc., Bethesda, pp. 1017-1071.

Deniau, J. M., Hammond, C., Chevalier, G. and Feger, J. 1978a, Evidence for branched subthalamic nucleus projections to substantia nigra, entopeduncular nucleus and globus pallidus. Neurosci. Lett., 9: 117-121.

Deniau, J. M., Hammond, C., Rizk, A.and Féger, J., 1978b, Electrophysiological properties of identified output neurons of the rat substantia nigra (pars compacta and pars reticulata) Evidence for the existence of branched neurons. Exp. Brain Res. 32: 4O2-422

Deniau, J. M. and Chevalier, G., 1985, Desinhibition as a basic process in the expession of striatal fonction. II. The striato-nigral influence on thalamocortical cells of the ventromedial thalamic nucleus. Brain Research, 334: 227-233.

Féger, J., Vézole,I., Renwart, N. and Robledo, P., 1989, The rat subthalamic nucleus: electrophysiological and behavioral data. In A. R. Crossman and M. A. Sambrook (eds.), Current problems in Neurology, Neural mechanisms in disorders of movement, Vol. 9 , John Libbey, London, pp. 37-43.

Hammond, C., Deniau, J. M., Risk, A. and Féger, J. 1978a, Electrophysiological demonstration of an excitatory subthalamonigral pathway in the rat. Brain Research, 151: 235-244.

Hammond, C., Deniau, J. M., Rouzaire-Dubois, B. and Féger, J., 1978b,Peripheral input to the rat subthalamic nucleus,an electrophysiological study.Neurosc. Lett.,,9:171-176.

Hammond,C.,Shibazaki,T.and Rouzaire-Dubois,B.,1983, Branched output neurons of the rat subthalamic nucleus: electrophysiological study of the sysnaptic effects on identified cells in the two main target nuclei, the entopeduncular nucleus and the substantia nigra. Neuroscience, 9: 511-520.

Kerkerian, L., Cupo, A., Dusticier, N., Epelbaum, J., Errami, M., Ettayebi, K. Forni, C., Jarry, T.,Kumar, U., Salin, P.,Samuel, D., Danger,J.M.and Nieoullon, A. , 1989, Régulation dopaminergique de l'activité de systèmes neuronaux peptidergiques,GABAergiques et glutamatergiques au niveau du striatum. L'Encéphale 15: 143-154.

Kita, H. and Kitai, S,T., 1987, Efferent projections of the subthalamic nucleus in the rat: light and electron microscopic analysis with the PHA-L method, J. Comp. Neurol., 260: 435-452.

Kitai, S. T. and Deniau, J. M. 1981, Cortical imputs to the subthalamus: intracellular analysis. Brain Research, 214: 411-415.

Kitai, S. T. and Kita, H., 1987, Anatomy and Physiology of the subthalamic nucleus: a driving force of the basal ganglia. In " The basal ganglia II- Structure and function: Current concepts ", M. B. Carpenter and A. Jayaraman (eds.), Plenum Press, New York, pp. 357-373.

Mathews,R.T. and German,D.C.,1986, Evidence for a functionnal role of dopamine D1 receptors in the substantia nigra of rats. Eur.J.Pharmacol.,120: 87-93.

Meibach, R. C. and Katzman, R., 1979, Catecholaminergic innervation of the subthalamic nucleus : evidence for a rostral continuation of A9 (substantia nigra) dopaminergic group. Brain Research. 173: 364-368.

Mintz, I., Hammond, C. and Féger, J., 1986a, Excitatory effect of iontophoretically applied dopamine on identified neurons of the rat subthalamic nucleus. Brain Research , 375: 172-175.

Mintz, I., Hammond, C., Guibert, B. and Leviel, V., 1986b.,Stimulation of the subthalamic nucleus enhances the release of dopamine in the rat substantia nigra, Brain Research ,376: 406-408.

Mitchell, I., Sambrook, M. A. and Crossman, A. R., 1985, Subcortical changes in the regional uptake of (3H) 2-deoxyglucose in the brain of the monkey during experimental choreiform dyskinesia elicited by injection of a gamma-aminobutyric acid antagonist into the subthalamic nucleus, Brain, 108: 405-422.

Nakanishi, H., Kita, H. and Kitai, S. T., 1987, Intracellular study of rat substantia nigra pars reticulata neurons in an in vitro preparation: electrical membrane properties and response characteristics to subthalamic stimulation. Brain Research, 437: 45-55.

Nakanishi, H., Kita, H. and Kitai, S. T., 1988, Electrophysiology of entopeduncular neurons and their responses to subthalamic stimulation in the rat brain slice preparations, Soc. Neurosci. Abstr. Vol. 14 : 408.14.

Nieoullon, A., Kerkerian, L., Dusticier, N., 1983, Presynaptic controls in the neostriatum: reciprocal interactions between the nigrostriatal dopaminergic neurons and the cortico-striatal glutamatergic pathway. Exp. Brain Res., 7: 54-65.

Parent, A., 1986, " Comparative Neurobiology of the Basal Ganglia ". In R. G. Northcutt (ed.), Wiley series in neurobiology, John Wiley & Sons, New York,1986, pp. 335.

Parent,A., Hazrati,L.N. and Smith,Y.,1989, The subthalamic nucleus in primates. A neuroanatomical and immunohistochemical study. In: " Neural Mechanisms in Disorders of Movement ", Current Problems in Neurology, vol.9. Eds. A. R. Crossman and M.A.Sambrook, Vol.9, London, pp. 29-35.

Penney,J. B. and Young, A.B., 1983, Speculations on the functional anatomy of the basal ganglia disorders. Ann. Rev. Neurosci., 6: 73-94.

Perkins, M. N. and Stone, T. W., 1980, Subthalamic projections to the globus pallidus: an electrophysiological study in the rat. Exp. Neurol., 68: 500-511.

Renwart, N., 1989. Role de ganglions de la base dans la production de dyskinésies: Etude Comportementale chez le rat. Thèse de Doctorat de l'Université de Paris V.

Robledo,P.,1989, Mise en évidence de l'effet activateur tonique des efférences du noyau subthalamique : approches électrophysiologiques etmétaboliques chez le rat. Thèse de Doctorat de l'université Paris VI.

Robledo, P., Vezole, I. et Féger, J., 1988, Mise en evidence d'un effet excitateur des efferences subthalamo-nigrales et subthalamo-pallidales chez le rat, C. R. Acad. Sci. Paris, t. 307, Serie III: pp.133-138.

Robledo,P. and Féger,J., 1990, Excitatory influence of rat subthalamic nucleus to substantia nigra pars reticulata and the pallidal complex : an in vivo electrophysiological study. Brain Research, under press.

Rouzaire-Dubois, B., Hammond, C., Hamon, B. and Féger, J., 1980, Pharmacological blockade of the globus pallidus-induced inhibitory responses of the subthalamic cells in the rat. Brain Research, 200: 321-329.

Rouzaire-Dubois, B. and Scarnati, E., 1985, Bilateral corticosubthalamic nucleus projections: an electrophysiological study in rats with chronic cerebral lesions. Neuroscience,15: 69-79.

Sambrook,M.A.,Crossman,A.R.,Mitchell,I.,Robertson,R.G.,Clarke,C.E.and Boyce,S.,1989, The basal ganglia mechanisms mediating primate models of movement disorders. In " Neural Mechanisms in Disorders of Movement ", Current Problems in Neurology, vol.9. Eds. A. R. Crossman et M. A. Sambrook , Vol. 9, John Libbey, London, pp. 123-144.

Sokoloff, L., Reivich, M., Kennedy, C., Des Rosiers, M. H., Patlak, C. S., Pettigrew, D. K., Sakurada, O. and Shinahara, M., 1977, The 14 C-deoxyglucose method for the mesurement of local cerebral glucose utilization : theory, procedures and normal values in the conscious and anesthetized rat. J. Neurochem., 28: 897-916.

Sugimoto, T., Hattori, T., Mizuno, N., Itoh, K. and Sato., 1983, Direct projections from the centre median-parafascicular complex to the subthalamic nucleus in the cat and rat. J. Comp. Neurol., 214: 209-216.

Van Der Kooy, D. and Hattori, T., 1980, Single subthalamic neurons project to both the globus pallidus and the substantia nigra in the rat. J. Comp. Neurol.,192: 751-790.

Waszczak,B.L. and Walters,J.R.,1986, Endogenous dopamine can modulate inhibition of substantia nigra reticulata neurons elicited by GABA iontophoresis or striatal stimulation. J. Neurosci., 6, 120-126.

SUBTHALAMIC NUCLEUS AFFERENTS: ANATOMICAL AND

IMMUNOCYTOCHEMICAL FEATURES

M. B. Carpenter and A. Jayaraman

Uniformed Services University
Bethesda, MD   and Louisiana State
University, New Orleans, LA

INTRODUCTION

The subthalamic nucleus (STN) may be the only CNS site at which a discrete destructive or biochemical lesion consistently results in a characteristic contralateral dyskinesia in man and monkey (Whittier,1947; Carpenter et al.,1950; Hammond et al.,1979). The STN, a hypothalamic derivative, consists of a single population of cells, virtually all of which are projections neurons (Richter, 1965; Van der Kooy and Hattori,1980). The neurotransmitter of STN neurons is unknown, but GABA immunoreactive terminals surround STN neurons in the monkey (Smith et al., 1987).

The object of this study was to co-define afferents to the STN and their putative neurotransmitters in the squirrel monkey (Saimiri Sciureus) and cat using retrograde tracers in combination with immunocytochemical methods.

MATERIALS AND METHODS

In a series of squirrel monkeys and cats attempts were made to inject small volumes (80-100nl) of 1% wheat germ agglutinin-horseradish peroxidase (WGA-HRP) into the STN on one or both sides. After postoperative survivals of 24-48 hours animals were anesthetized and perfused with 4% paraformaldehyde and 1% glutaraldehyde in 0.1 M phosphate buffered saline (PBS). Brain were removed, blocked in either transverse or sagittal planes, further fixed and cryoprotected with 30% sucrose in PBS. Blocks of tissue were cut on a freezing microtome at 40 μm, divided into two or three series and collected in 0.1 M PBS. Series of sections, reacted with tetramethyl benzidine (TMB), were stabilized by rinsing in 0.01 M acetate buffer (pH 3.3) and processing in 0.02% cobalt chloride, 0.05% diaminobenzidine (DAB) and 0.01% H_2O_2 in 0.01 M phosphate buffer at pH 7.4 (Rye et al.1984). One series of sections was mounted and stained with neutral red and the other series were immunoreacted with different antisera. Stablized TMB sections were reacted with polyclonal antisera for choline acetyltransferase (ChAT), gamma-aminobutyric acid (GABA), leucine enkephalin (L-ENK) and substance P (SP). Free-floating sections were incubated with primary antibody in concentrations of 1:1000-1:3000 for 48 to 96 hrs. at 4^0 C., incubated with Avidin-Biotin peroxidase complex and reacted with DAB and H_2O_2. Sections from normal monkeys were immunoreacted with each of the antisera. Control sections were processed in the same fashion, substituting normal rabbit serum or preadsorbed antiserum for the primary antibody. Nearly serial section of discrete STN injections were projected and drawn at magnifications of X45. Measurements of the STN and densely labeled portions of the nucleus, made with a Hewlett-Packard digitizer, provided percentage estimates of the labeled part of the nucleus (Dornfeld et al.,1942).

RESULTS

WGA-HRP injections were made in parts of the STN in 7 monkeys and two cats. Unilateral injections in two monkeys selectively labeled large portions of the STN. Attempted bilateral injections in 5 monkeys labeled smaller portions of the STN on one side. One unilateral injection in a cat labeled large parts of the STN selectively; less discrete injections in the other cat labeled virtually all parts of the STN bilaterally.

Monkey

Similar injections in two monkeys labeled 63% and 47% of the lateral two-thirds of the STN selectively (Figs. 1A,F). There was no involvement of the internal capsule or substantia nigra (SN) in either animal. Injections involved smaller portions of the STN unilaterally in 5 monkeys. These STN injections labeled: (1) ventrolateral portions caudally (20%),(2) dorsomedial portions caudally (18%) (Fig.1G) ,(3) rostromedial portions (21%),(4) the medial third of the nucleus (38%) and (5) the lateral pole of the nucleus (9%).

Retrograde transport: In the caudal third of the lateral pallidal segment(LPS) only few labeled cells were seen dorsolaterally; no cells were identified ventrally or in the caudal pole of the LPS (Fig. 1B). Large STN injections labeled cells in the dorsal two-thirds of the middle third of the LPS, forming two arrays parallel to the medullary laminae. The largest cellular array was close to the lateral medullary laminae; the medial array did not extend to the dorsal border of the LPS (Fig.1C). Few label cells were seen in the ventral third of the LPS. Rostral to the anterior commissure a smaller number of cells formed a single narrow band in the LPS near to the lateral medullary lamina. No labeled neurons were found in the ventral pallidum or in the oral pole of the LPS. No labeled cells were found in any part of the striatum, the substantia innominata (SI) or the substantia nigra (SN).

Smaller injections of the STN produced varied cellular labeling in the LPS. Involvement of ventral portions of the lateral half of the nucleus produced the pattern described, but with fewer cells. Injection of the medial STN caudally, labeled an island of cells in the middle third of the LPS dorsally and an area of terminals in the medial pallidal was associated with a ventrolateral band of cells in the segment (MPS)(Figs.1G,H). Involvement of the rostromedial STN rostral LPS. Labeling of the medial third of the STN, labeled cells only in medial regions of the rostral LPS. Selective labeling of the lateral pole of the STN failed to label any cells in the LPS.

A small number of cells were labeled in the centromedian-parafascicular (CM-PF) complex and the pedunculopontine nucleus (PPN) in animals with larger STN injections. No cells in the SN or the dorsal nucleus of the raphe were labeled. Serial sections of the frontal cortex in 4 animals with selective injections of the STN revealed no retrograde transport to cortical neurons.

Anterograde transport: Multiple bundles of STN efferents traversed portions of the internal capsule and filled the entire caudal third of the MPS (Fig.1B). In this mass of heavily beaded fibers, it was not possible to identify terminal fields. Caudal portions of the LPS contained few fibers. At more rostral levels fibers from the MPS occupied successively more lateral and dorsal regions, leaving medial portions of the MPS free of fibers. In the rostral half of the MPS, virtually all fibers had projected into the LPS. In the middle third of the LPS fibers and terminals closely surrounded arrays of labeled cells (Fig. 1C). The medial array of terminals did not extend to the dorsal border of the LPS. Few fibers and no terminals were seen in the ventral third of the LPS. In the rostral LPS small number of fibers and terminals surrounded labeled neurons.

In dorsal portions of the middle third of the LPS small fascicles and individual fibers traversed the medullary lamina to enter adjacent portions of the putamen. These fibers did not penetrate the putamen for any distance and no terminal fields could be identified. Small numbers of fibers and terminal fields were seen in the SN. No STN projections were seen in any part of the SI or in the frontal cortex.

Immunocytochemistry: In 6 monkeys with WGA-HRP injections in the STN sections stabilized for TMB were immunoreacted with antisera for GABA, L-ENK, SP and ChAT. Anterograde and retrograde transport of WGA-HRP appeared identical to that in TMB sections. Similar series of sections from control animals were processed using the same antisera.

GABA: In three monkeys large numbers of GABA immunoreactive (GABA-IR) cells were present in all parts of the striatum and in both pallidal segments. In dorsal parts of the middle third of the LPS, large numbers of arrayed pallidal neurons contained WGA-HRP granules. GABA-IR could not be detected in the most densely retrogradely labeled cells, but adjacent GABA-IR pallidal neurons contained HRP granules (Fig. 1D). Comparisons with the uninjected side and with sections from normal animals suggested that virtually all large pallidal neurons were GABA-IR. Data were interpreted as indicating that virtually all LPS neurons projecting to the STN were positive for GABA.

L-ENK: Immunoreactivity (IR) to L-ENK was studied in three monkeys with WGA-HRP injections in the STN and in control animals. Only L-ENK fibers were seen in the LPS, but both cells and fibers were identified in ventrocaudal parts of the

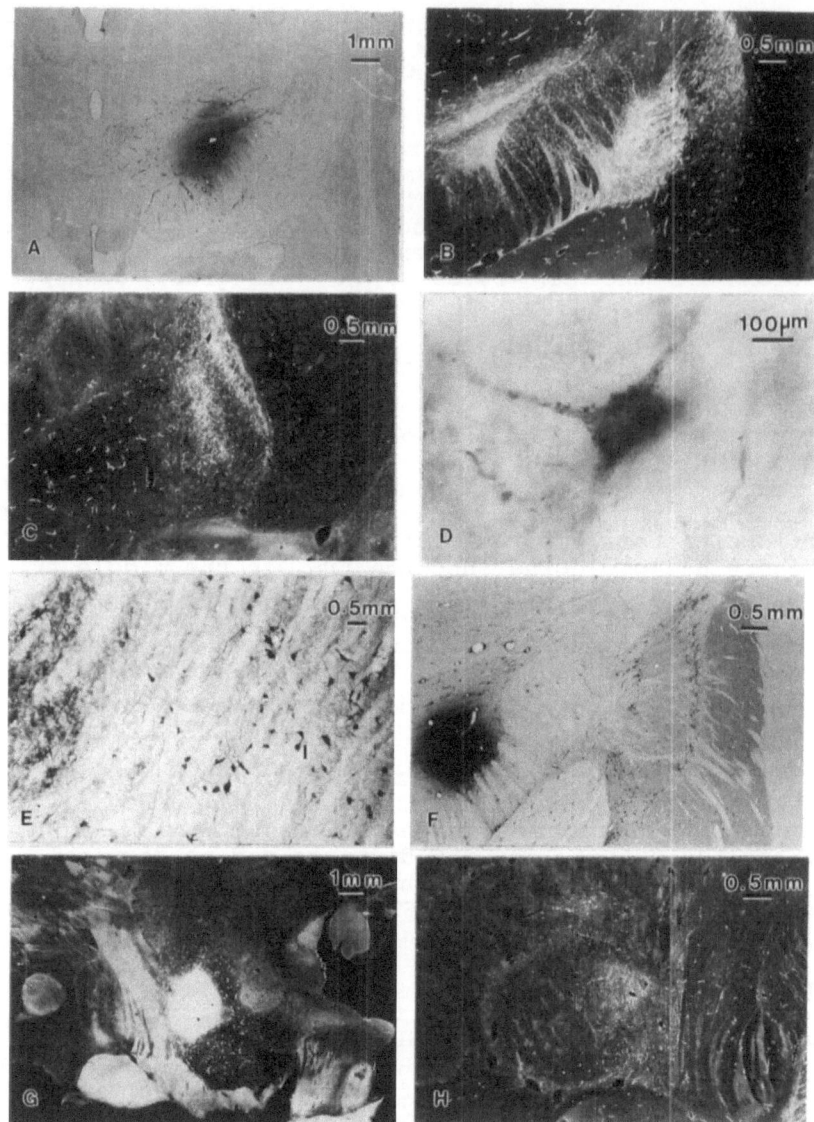

Fig. 1. Monkeys U-445(A-E),U-436(F) and U-458(G,H).
A: Photomicrograph of WGA-HRP injection labeling 63% of the
subthalamic nucleus(STN). B-C: Darkfield photomicrographs of
anterograde and retrograde transport to the globus pallidus.
Fibers filled the medial pallidal segment(MPS) caudally (B);
rostrally virtually all fibers from the MPS projected into the
lateral pallidal segment(LPS) where they terminated about
parallel arrays of cells (C). D: Cell in the LPS double-labeled
for GABA and WGA-HRP. E: Retrogradely labeled cells in the LPS
were most numerous in regions with the least dense L-ENK-IR
fibers; dense L-ENK-IR fibers and terminals appear on the left.
F: Section through the STN injection showing ChAT-IR cells in
the medullary laminae of the pallidum, none of which were
double labeled. G: Sagittal section through a WGA-HRP injection
in the medial pole of the STN. H: Sagittal section showing an
anterograde terminal field in the MPS and retrogradely labeled
cells in a localized region of the LPS.

putamen. L-ENK-IR fibers and terminals in the LPS exhibited a constant pattern with regional variations in density. The caudal third of the LPS was filled with L-ENK-IR fibers and granules, with the darkest staining ventrally. In the middle third of the LPS less dense immunostaining was seen, except in the ventral third of the nucleus and in a small dorsomedial wedge near the MPS (Fig. 1E). Dense L-ENK-IR fibers in the rostral LPS were found near the internal capsule and the margins of the anterior commissure.

Retrogradely labeled neurons, in parallel arrays in dorsal portions of the middle third of the LPS, were found in regions containing the lowest density of L-ENK-IR fibers and terminals (Fig.1E). No labeled pallidal neurons were closely surrounded by L-ENK-IR terminals. Few WGA-HRP labeled neurons were found in the ventral third of the LPS or in the dorsomedial wedged-shaped area. Identified pallidosubthalamic neurons projecting to the lateral two-third of the STN were located in regions of the LPS with the least dense L-ENK-IR fibers and terminals.

SP: Sections from two animals with WGA-HRP injections in the STN and from two control animals were immunoreacted with antiserum to SP. WGA-HRP retrogradely labeled cells in the LPS were readily identified in SP-immunostained sections. No SP-IR cells were seen in any part of the globus pallidus, though they were present ventrocaudally in the putamen. SP-IR fibers and terminals were dense in the apex of the MPS, although a few were noted ventrocaudally in the LPS. There appeared to be no relationship between labeled pallidosubthalamic neurons and SP-IR fibers in the LPS.

ChAT: Immunoreactivity (IR) to ChAT antiserum was studied in two monkeys with WGA-HRP injections of the STN and in several control animals. Large ChAT-IR neurons were distributed throughout the striatum and surrounded caudal portions of the LPS (Fig.1F). None of the ChAT-positive neurons in the striatum, the medullary laminae of the pallidum or in the SI contained HRP granules. ChAT-positive cells in the pedunculopontine nucleus (PPN) and the lateral dorsal tegmental nucleus were abundant, but no double labeled cells could be identified.

Cat

A discrete unilateral injection in one animal labeled about 75% of the STN, and larger bilateral injections in another cat labeled nearly all of the nuclei. Retrograde labeling was found in dorsal and lateral regions of the middle and rostral thirds of the globus pallidus (GP). Labeled cells occurred in clusters and never formed parallel arrays. Rostrally, labeled GP neurons extended into the internal capsule and the ventral pallidum. Anterograde fibers and terminals surrounding labeled GP neurons were not as numerous or as impressive as in the monkey. Only a small number of fibers projected to the rostral SNR. No fibers or terminals were seen in any part of the striatum or in the SI. Small numbers of cells were labeled in CM-PF and PPN in both animals. No cortical cells were labeled.

Immunocytochemistry: Sections immunoreacted for GABA and ChAT were the same as in the monkey. Less heavily labeled pallidal neurons were double-labeled for WGA-HRP and GABA. None of the ChAT-IR cells in the striatum, SI or the PPN contained HRP granules. L-ENK-IR fibers and terminals in the GP were dense, but not uniformly distributed; lower densities of ENK-IR fibers was seen in dorsolateral regions. Pallidosubthalamic neurons were found primarily in regions with low density L-ENK-IR, but some labeled cells were seen in ventral regions with dense immunoreactivity.

DISCUSSION

The most massive inputs to the STN in both monkey and cat arises from the LPS or it equivalent. In the monkey the lateral two-thirds of the STN received afferents from two well-defined cellular arrays in the dorsal two-thirds of the central part of the LPS and from a smaller cellular band in the rostral LPS. The medial third of the STN received afferents from the rostral third of the LPS. Ventral regions of the central LPS and the caudal third of the LPS contributed few STN afferents. Subthalamopallidal projections appeared reciprocal to all LPS regions providing STN afferents. STN projections to the MPS, arose only from medial regions of the nucleus and appeared small in number. No cell group of the STN was found to project exclusively to the MPS. Terminals in the MPS may be collaterals of axons projecting to the LPS. This finding differs from that of Parent et al.(1989) whom found few STN neurons projecting to both pallidal segments.

In the monkey virtually all pallidal neurons were immunoreactive to GABA (Smith et al.,1987). This observation was confirmed in the monkey and cat with the additional information that few GABAergic neurons in caudal and ventral regions of the LPS appeared to project to the STN. Identified pallidosubthalamic neurons in the monkey were most numerous in regions with the least dense L-ENK-IR fibers. Electron micorscopic (EM) studies indicating L-ENK-IR fibers ensheath and synapse upon pallidal dendrites raise questions as to whether this pattern pertains for all regions of the LPS (DiFiglia et al.,1982). This important question requires further exploration at the EM level. None of the SP-IR or ChAT-IR cells or fibers in the forebrain appeared related to STN afferent systems.

Cortical STN afferents have been demonstrated in autoradiographic and degenerations studies at the light and EM levels (Hartmann von Monakow et al.,1978; Afsharpour,1985; Moriizumi et al,1987). All retrograde studies at the light microscopic level have been unsuccessful in labeling cortical neurons (Rinvik et al.1979; McBride and Larson,1980; Carpenter et al.,1981). Failure to confirm this projection suggests the these fibers probably are collaterals of corticofugal fibers destined for other loci. It seems curious that of two structures (globus pallidus and subthalamic nucleus) derived from the same anlage (hypothalamus), only the STN should have cortical inputs. The somatotopic organization of this projection from the motor cortex suggests these fibers may be collaterals of corticostriate projections (Künzle,1975).

114

Studies of STN efferents in the monkey based upon injections in the striatum have indicated labeling of an appreciable number of STN neurons with some topographic features (Parent and Smith,1987). PHA-L studies in the rat suggest that few STN fibers entered the striatum (Kita and Kitai,1987). In this study few fibers projected into the putamen and no terminal fields were identified.

Small numbers of cells in CM-PF project to the oral pole of the STN in the rat and cat (Sugimoto et al.1983). Similar small projections were seen in monkey. Because most cells of the CM-PF complex project to the striatum, it seems likely that these fibers may be collaterals of thalamostriate projections (Sugimoto et al.,1983; Smith and Parent,1986).

Although no cells in the SN were retrogradely labeled following injections limited to the STN, small numbers of fibers projected into the SNR. Subthalamonigral fibers consistituted only a small fraction of those projecting to the LPS. About 80% of the STN efferent to the SNR arise from the lateral two-third of the STN (Parent and Smith, 1987). Our estimate of STN projections to the SNR was about 10% of fibers projecting to the LPS.

The number of cells in PPN projecting to the STN in monkey and cat appeared very small. Cells in PPN are heterogeneous, but those in the pars compacta are cholinergic and have been considered to project to the STN (Sugimoto and Hattori,1984). In the rat PPN has been defined as consisting solely of the cholinergic neurons (Lee et al.,1988). These authors have shown that ChAT-IR cells of PPN projected to the thalamus, while non-cholinergic neurons in PPN projected to extrapyramidal nuclei. The number of retrogradely labeled cells in PPN in this study was too small to draw a conclusion, but none of the ChAT-IR cells of PPN appeared to be double labeled.

ACKNOWLEDGEMENTS

Supported by Grants from DOD,USUHS (C07005) and NINDS (NS-26658).

REFERENCES

Afsharpour,S. 1985, Topographical projections of the cerebral cortex to the subthalamic nucleus. J. Comp. Neurol., 236: 14-28.

Carpenter, M.B.,Carleton,S.C.,Keller,J.T. and Conte,P. 1981, Connections of the subthalamic nucleus in the monkey. Brain Res., 224: 1-29.

Carpenter,M.B., Whittier,J.R. and Mettler,F.A. 1950, Analysis of choreoid hyperkinesia in the rhesus monkey. Surgical and pharmacological analysis of hyperkinesia resulting from lesions of the subthalamic nucleus of Luys. J. Comp. Neurol., 92: 293-331.

DiFiglia, M.,Aronin,N. and Martin,J.B. 1982, Light and electron microscopic localization of immunoreactive Leu-enkephalin in the monkey basal ganglia. J. Neurosci., 3: 303-320.

Dornfeld,E.J.,Slater,D.W. and Scheffé,H. 1942, A method for accurate determination of volume and cell numbers in small organs. Anat. Rec., 82: 255-259.

Hammond,C., Féger,J.,Bioulac,B. and Souteyrand,J.P. 1979, Experimental hemiballism in the monkey produced by unilateral kainic acid lesion in corpus Luysii. Brain Res.,171: 577-580.

Hartmann-von Monakow,K., Akert,K. and Künzle,H. 1978, Projections of the precentral motor cortex and other cortical areas of the frontal lobe to the subthalamic nucleus in the monkey. Exp. Brain Res., 33: 395-403.

Kita,H. and Kitai,S.T. 1987, Efferent projections of the subthalamic nucleus in the rat: light and electron microscopic analysis with PHA-L. J. Comp. Neurol., 260: 435-452.

Kooy,D.Van der and Hattori,T. 1980, Single subthalamic nucleus neurons project to both globus pallidus and substantia nigra in rat. J. Comp. Neurol., 192: 751-768.

Künzle,H. 1975, Bilateral projections from precentral motor cortex to the putamen and other parts of the basal ganglia. An autoradiographic study in Macaca fascicularis. Brain Res., 88: 195-210.

Lee,H.J., Rye,D.B., Hallanger,A.E., Levey,A.I. and Wainer,B.H. 1988, Cholinergic Vs noncholinergic efferents from the mesopontine tegmentum to the extrapyramidal motor systems nuclei. J. Comp. Neurol., 275: 469-492.

McBride,R.L. and Larsen,K.D. 1980, Projections of the feline globus pallidus. Brain Res., 189: 3-14.

Moriizumi,T.,Nakamura,Y.,Kitao,Y. and Kudo,M. 1987, Ultrastructural analysis of afferent terminals in the subthalamic nucleus of the cat with a combined degneration and horseradish peroxidase tracing method. J. Comp. Neurol., 265: 159-174.

Parent,A. and Smith,Y. 1987, Organization of efferent projections of the subthalamic nucleus in the squirrel monkey revealed by retrograde labeling methods. Brain Res., 436: 296-310.

Parent,A., Smith,Y., Filion,M. and Dumas,J. 1989, Distinct afferents to internal and external pallidal segments in the squirrel monkey. Neurosci. Lett., 96: 140-144.

Richter,E. 1965," Die Entwicklung des Globus Pallidus und des Corpus Subthalamicum". Springer-Verlag, Berlin, pp.133.

Rinvik,E., Grofova,I., Hammond,C., Féger,J. and Deniau,J.M. 1979, A study of the afferent connections of the subthalamic nucleus in the monkey and cat using the HRP technique. Advances in Neurology,24: 53-70.

Rye,D.B., Saper,C.B. and Wainer,B.H. 1984, Stabilization of tetramethylbenzidine (TMB) reaction product. J. Histochem. Cytochem., 32: 1145-1153.

Sugimoto,T. and Hattori,T. 1983, Confirmation of thalamo-subthalamic projections by electron microscopic autoradiography. Brain Res., 267: 335-339.

Sugimoto,T. and Hattori,T. 1984, Organization and efferent projections of nucleus tegmenti pedunculopontinus pars compacta with special reference to it cholinergic aspects. Neurosci., 11: 931-946.

Sugimoto,T., Hattori,T., Mizuno,N., Itoh,K. and Sato,M.
       1983, Direct projections from the centre median-
       parafascicularis complex to the subthalamic nucleus
       in the cat and rat. J. Comp. Neurol., 214: 209-216.
Smith,Y. and Parent,A. 1986, Differential connections of
       the caudate nucleus and putamen in the squirrel
       monkey (Saimiri sciureus). Neurosci., 18: 347-371.
Smith,Y., Parent,A., Sequela,P. and Descarries,L. 1987,
       Distribution of GABA-immunoreactive neurons in the
       basal ganglia of the squirrel monkey (Saimiri
       sciureus). J. Comp. Neurol., 259: 50-64.
Whittier,J.R. 1947, Ballism and the subthalamic nucleus
       (nucleus hypothalamicus; corpus Luysi).
       Arch. Neurol. Psychiat., 58: 672-692.

CHARACTERIZATION OF THE SYNAPTIC INPUTS TO DOPAMINERGIC

NEURONES IN THE RAT SUBSTANTIA NIGRA

Paul Bolam and Yoland Smith

MRC Anatomical Neuropharmacology Unit
University Department of Pharmacology, South
Parks Road, Oxford, U.K.

INTRODUCTION

The dopaminergic nigrostriatal pathway has long been known as crucial in the circuitry of the basal ganglia. The critical role of this projection in basal ganglia function is exemplified by the marked motor disturbances in Parkinson's disease and the models of Parkinson's disease involving selective destruction of the nigral dopaminergic neurones. This projection terminates profusely throughout the entire extent of the striatum and intrastriatal release of dopamine from these terminals represents one of the major modulatory feedback mechanisms of information flow through the basal ganglia. Electron microscopic analyses have shown that most of the nigrostriatal dopaminergic terminals form synaptic contacts with the type of striatal neurone that also receives afferent input from the cerebral cortex (Bouyer et al., '84; Freund et al., '84). This pattern of synaptic organization of the cortical and dopaminergic terminals on the striatal neurones suggests that the information flow through the cortico-striato-pallidal and cortico-striato-nigral pathways may be modulated by the dopaminergic afferents. Within the striatum the release of dopamine from the terminals in contact with striatal neurones is dependent on the activity of nigral axons and ultimately on the activity of the afferent synaptic input to the dendrites and perikarya of the parent neurones in the substantia nigra. It is thus important to know the origin, chemistry and pattern of synaptic input to the dopaminergic neurones in order to more fully understand their role in the functional organization of the basal ganglia. We have carried out a series of experiments aimed at elucidating the pattern of synaptic organization of the substance P-, gamma-aminobutyric acid (GABA)- and enkephalin-containing

The Basal Ganglia III, Edited by G. Bernardi *et al.*
Plenum Press, New York

terminals in contact with the dopaminergic neurones in the rat substantia nigra. Furthermore, the anterograde transport of various axonal tracers was combined with post-embedding immunocytochemistry for GABA in order to identify some of the sources of GABA in the rat substantia nigra.

METHODS

Double Immunocytochemistry

All immunocytochemical analyses were carried out on sections of the mesencephalon of female albino Wistar rats (160-250g). The rats were perfuse-fixed with Ca^{++}-free Tyrode's solution followed by mixtures of glutaraldehyde and paraformaldehyde in sodium phosphate buffer. The mesencephalon was dissected, sectioned at 50-70 μm on a vibrating microtome and washed several times in phosphate buffered saline. The sections were subjected to double immunocytochemical staining using either the double pre-embedding technique of Levey et al (1986) or a combination of pre-embedding and post-embedding immunocytochemistry (see Bolam and Ingham, 1990).

Double pre-embedding immunocytochemistry. After several washes the sections were incubated with the primary antibodies used to reveal populations of axon terminals in the substantia nigra. The primary antibodies that were used were directed against: glutamic acid decarboxylase (GAD, 1:1000 or 1:2000) (Oertel et al., 1981), substance P (SP, 1:1000 or 1:1500) (Cuello et al., 1979) and leucine-enkephalin (leu-enk, 1:500 or 1:1000) (Senba et al., 1982). The sections were then incubated in appropriate secondary antibodies followed by appropriate peroxidase antiperoxidase (PAP) complexes. The first antigen was then localized using diaminobenzidine (DAB) as the chromogen for the peroxidase reaction. On completion of this reaction the dopaminergic neurones were immunostained using an antiserum against tyrosine hydroxylase (TH, 1:1000) (van den Pol et al., 1984) and the PAP method but using benzidine dihydrochloride (BDHC) as the chromogen for the peroxidase reaction (for details see Levey et al.,1986; Bolam and Ingham,1990).

Pre- and post-embedding immunocytochemistry. Sections of the mesencephalon were incubated to reveal TH immunoreactivity using the PAP method and DAB as the chromogen as described above. They were then prepared for electron microscopy and mounted in resin on microscope slides. Regions of the substantia nigra containing TH-immunoreactive structures were re-embedded in blocks of resin and ultrathin sections were collected on coated single-slot gold grids. These sections were then immunostained to reveal GABA (Hodgson et al., 1985; Somogyi and Hodgson, 1985; Somogyi et al., 1985) using the immunogold method (for details see Ingham et al., 1988; Bolam and Ingham, 1990).

In order to test the possibility of the co-existence of GABA and SP in boutons in synaptic contact with dopaminergic neurones, some sections were processed to reveal first, SP and TH by the double pre-embedding method as described above, and then ultrathin sections were subjected to the immunogold staining to reveal GABA.

Anterograde Labeling of Nigral Afferent Terminals Combined
with Pre-embedding Immunostaining for Tyrosine Hydroxylase
or Post-embedding Immunostaining for GABA

Anaesthetized rats received stereotaxic pressure
injections of wheatgerm agglutinin conjugated to horseradish
peroxidase (WGA/HRP) or biotinylated lysine (biocytin; King
et al., 1989) into the striatum or iontophoretic injections
of Phaseolus vulgaris-leucoagglutinin (PHA-L) into the
globus pallidus. After appropriate survival times the
animals were perfuse-fixed using relatively high
concentrations of glutaraldehyde. Vibrating microtome
sections of the mesencephalon were incubated to reveal the
transported WGA/HRP by incubation in a solution of DAB or
dimethylbenzidine and hydrogen peroxide (Bolam and Ingham,
1990). Transported biocytin was revealed by the same method
after pre-incubation in an avidin-biotin-peroxidase complex.
The transported PHA-L was visualized using a biotinylated
anti-PHA-L antiserum and an avidin-biotin-peroxidase complex
using DAB as the chromogen. Some of the sections were then
immunostained to reveal TH immunoreactivity using BDHC as
the chromogen for the peroxidase reaction as described
above. The sections were then osmium-treated, dehydrated
and embedded in resin on microscope slides. Areas of the
substantia nigra that contained anterogradely labeled
terminals were re-embedded in blocks of resin and ultrathin
sections were immunogold stained to reveal GABA.

Analysis of Material

All the sections were examined in the light microscope
for the distribution and relationships of immunostained
structures. Areas for subsequent electron microscopic
analysis were sometimes photographed, their positions in the
sections noted and then cut out from the microscope slides
and either re-embedded in resin in the form of a block or
glued to the surface of precured blocks. The tissue was
then sectioned and the ultrathin sections examined in the
electron microscope.

RESULTS

Substance P and Tyrosine Hydroxylase

The sections of the substantia nigra that were
incubated to reveal SP-immunoreactive structures and then
TH-immunoreactive structures showed the typical distribution
of immunostaining. Thus the substantia nigra pars compacta
(SNC) contained a dense accumulation of TH-immunoreactive
neurones and dendrites (Fig. 1A), identified by the granular
BDHC reaction product, that extended into the more medial
ventral tegmental area. The pars reticulata (SNR) also
contained many TH-immunoreactive structures (Fig. 1C).
These consisted of the ventrally extended dendrites of the
SNC cells as well as the perikarya and dendrites of a
smaller population of immunoreactive cells located in the
caudo-ventral SNR. The SP immunoreactivity revealed using
DAB as the chromogen was in the form of axons and terminals
containing the typical brown amorphous reaction product
formed by DAB, the density of these structures gave an

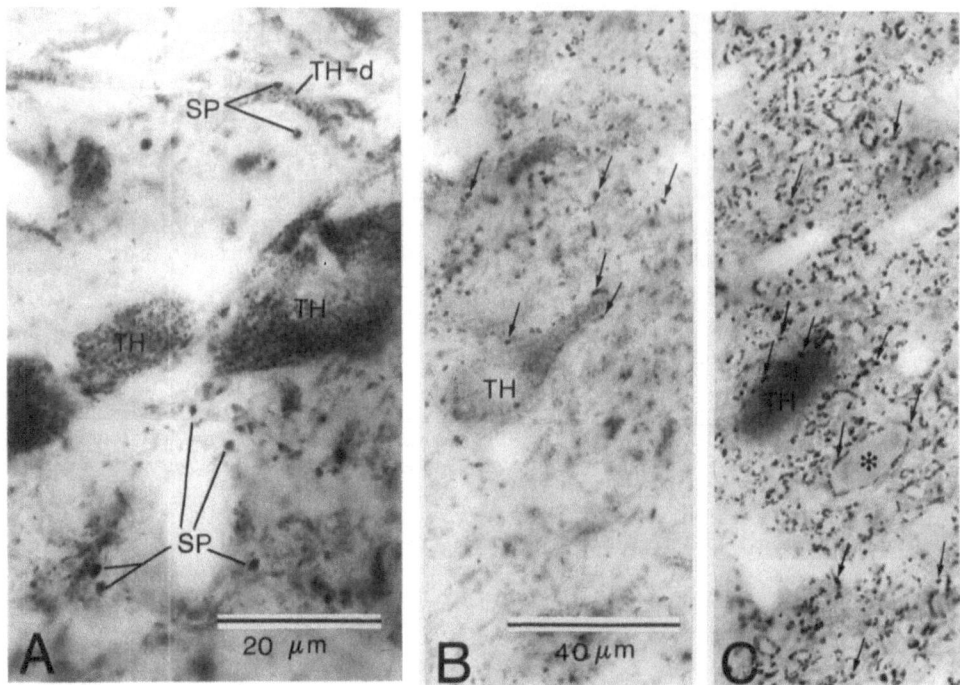

Fig.1.   Light micrographs of the substantia nigra
          incubated to reveal either substance P (SP, DAB as
          chromogen) and tyrosine hydroxylase
          immunoreactivities (TH, BDHC as chromogen) (A) or
          glutamic acid decarboxylase (GAD) and TH
          immunoreactivities (B,C). A depicts a high power
          view of TH-immunoreactive cells (TH) and SP-
          immunoreactive terminals (SP) in the substantia
          nigra compacta (SNC). Occasionally SP-positive
          boutons can be seen in association with TH-
          immunoreactive dendrites (TH-d) or perikarya. B
          shows a TH-immunoreactive perikaryon (TH)
          surrounded by many GAD-immunoreactive terminals
          in the SNC. C illustrates a TH-immunoreactive
          cell (TH) and a non-immunoreactive cell (asterisk)
          surrounded by numerous GAD-positive terminals in
          the substantia nigra reticulata (SNR). In B and C
          some of these terminals are indicated by small
          arrows. Note the difference in the density of
          GAD-immunoreactive terminals between SNC (A,B)
          and SNR (C). The scale marker in B is valid for C.

overall brown appearance to the surface of the SNR (Fig.
 1A). Many of the SP-immunoreactive structures were more
discretely stained and had the appearance of distinct
terminals or axon swellings (Fig. 1A). The SNC also
contained many SP-immunoreactive terminals albeit at a much
lower density than in the SNR. The two sets of
immunoreactive structures were interspersed with each other
both in the SNC and SNR and were easily distinguishable.
The SP-immunoreactive terminals often showed associations
with non-immunoreactive structures, dendrites and perikarya.
In addition many of the immunoreactive terminals were
closely apposed to dendrites or perikarya that were

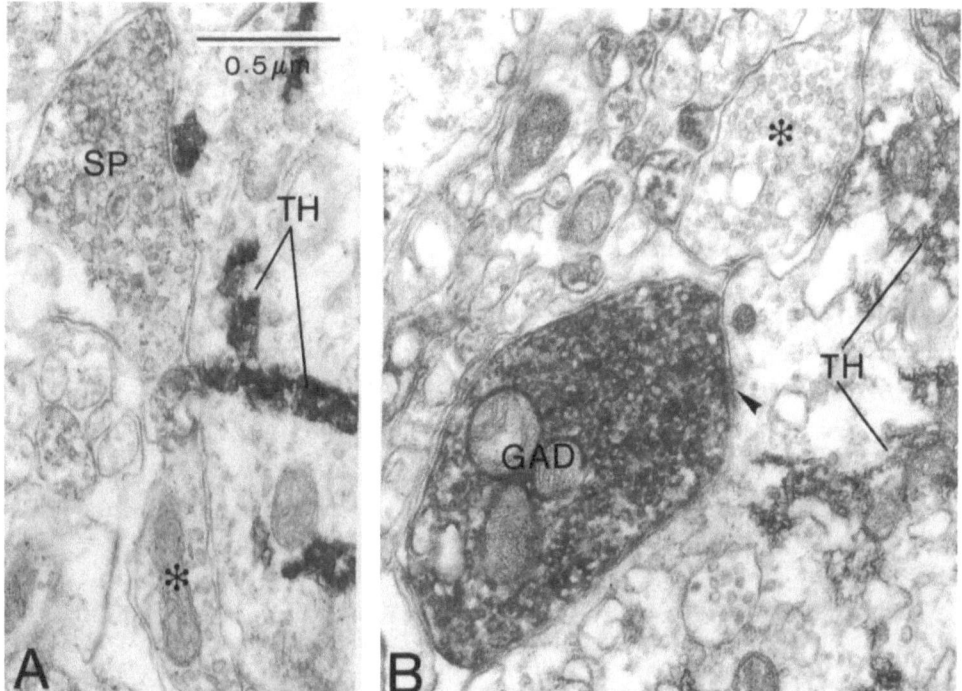

Fig. 2.   Electron micrographs showing TH-immunoreactive
           dendrites apposed by a SP-immunoreactive bouton
           (A) and a GAD-immunoreactive bouton (B). Note the
           difference between the BDHC (in TH-immunoreactive
           structures) and the DAB (in SP- and GAD-positive
           structures) peroxidase reaction products. Non-
           immunoreactive boutons (asterisks) are apposed to
           the same dendrites. In B the arrowhead indicates
           the symmetrical membrane specialization between
           the GAD-positive bouton and the TH-containing
           dendrite. The scale marker in A is valid for B.

immunoreactive for TH i.e. contained the granular BDHC
reaction product (Fig. 1A). In some cases several SP-
immunoreactive terminals were seen associated with one TH-
immunoreactive dendrite or perikaryon.

        In the electron microscope SP-immunoreactive boutons
were found throughout the SNC and SNR. They were of medium-
size and contained many vesicles and occasional
mitochondria. The vesicles were slightly pleomorphic in
shape. The synaptic membrane specializations were mainly of
the symmetrical type but many contacts were observed that
exhibited asymmetrical specializations and post-junctional
dense bodies. The majority of synaptic contacts involving
SP-positive boutons were with dendritic shafts (Figs 2A,
3B).

        Many of the SP-positive boutons formed symmetrical
synaptic contacts with structures that displayed
immunoreactivity for TH (Figs 2A, 3B). As with the overall
distribution of SP-positive boutons, the majority of
contacts identified were with TH-positive dendritic shafts

and only a small proportion in contact with TH-positive perikarya which may reflect the relative densities of dendrites and perikarya. The same TH-positive structures that received input from SP-positive boutons were also postsynaptic to non-immunoreactive boutons (Fig. 2A).

Glutamic Acid Decarboxylase and Tyrosine Hydroxylase

The distribution of BDHC-stained structures i.e. TH-immunoreactive perikarya and dendrites, was similar to that described above. The GAD-immunoreactive structures revealed using DAB consisted predominantly of axons or punctate boutons (Fig. 1B,C) and a few lightly stained neuronal perikarya. Within the SNR the stained boutons were distributed in an ordered fashion, outlining both unstained dendrites and perikarya as well as lightly immunoreactive perikarya (Fig. 1C). In contrast to this, in the SNC the GAD-immunoreactive terminals were distributed in an apparently random fashion and did not surround or outline other structures (Fig. 1B). They did show associations with other structures (dendrites and perikarya) but the density was much lower than in the reticulata (compare Figs 1B and C).

As with the sections double stained for SP and TH, the DAB-stained (ie GAD-immunoreactive) structures were easily distinguishable from the TH-immunoreactive structures on the basis of their colour, texture and location. There was a marked and consistent association between the TH-immunoreactive structures and the GAD-immunoreactive boutons. All TH-immunoreactive perikarya and proximal dendrites that were at the same depth in the tissue as the GAD-immunostained structures had many GAD-immunoreactive boutons associated with them (Fig. 1B,C). The number and density of terminals in contact with TH-immunostained structures although high, were much lower than for non-immunostained structures in the SNR.

In the electron microscope the GAD-immunoreactive boutons were variable in size, contained many packed vesicles and formed symmetrical membrane specializations (Fig. 2B). The majority formed synaptic contacts with dendritic shafts but, as mentioned above this may simply reflect the overall relative density of dendrites and perikarya. Within the region of the penetration of the immunoreagents the majority of boutons in contact with individual structures displayed GAD immunoreactivity.

Many of the GAD-positive boutons made synaptic contact with TH-immunoreactive dendrites and perikarya (Fig. 2B). In fact, within the region of the penetration of the immunoreagents for GAD, virtually all TH-immunoreactive structures received input from GAD-positive boutons. Quantitative analysis revealed that, within the depth of penetration of both sets of antisera, about 50% of the terminals in contact with TH-positive structures were immunoreactive for GAD.

GABA and Tyrosine Hydroxylase

The sections that were immunostained by the post-embedding method for GABA contained many immunoreactive terminals that displayed the morphological characteristics described above for the GAD-positive boutons. Synaptic

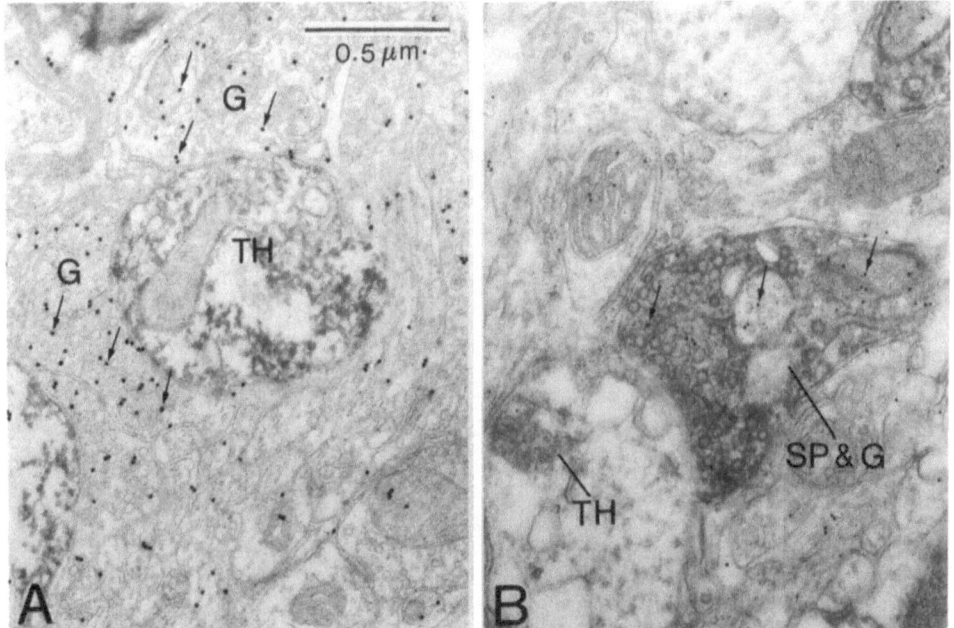

Fig. 3. Electron micrographs showing various features of
the GABA immunostaining in the substantia nigra. A
shows a TH-positive dendrite apposed by two GABA-
immunoreactive boutons (G). The TH
immunoreactivity was revealed using DAB as
chromogen. B illustrates a TH-positive dendrite
apposed by a bouton that displays
immunoreactivities for both SP and GABA (SP & G).
In this case the TH immunoreactivity was revealed
using BDHC as chromogen. The small arrows indicate
some of the immunogold particles that overlie the
GABA-immunoreactive elements. The scale marker in
A is valid for B.

specializations were of the symmetrical type and represented
a high proportion of all nigral boutons. In sections that
were both pre-embedding stained for TH using either DAB or
BDHC as the chromogen and post-embedding stained for GABA,
numerous GABA-positive boutons were identified in
symmetrical synaptic contact predominantly with dendrites
and less frequently with perikarya that displayed TH
immunoreactivity (Fig. 3A). A high proportion of the
terminals afferent to TH-positive dendrites and perikarya
were immunoreactive for GABA (Fig. 3A). In a quantitative
analysis, a total of 389 TH-positive structures were
examined and found to have boutons in synaptic contact with
or apposed to their membranes, and in the region of 70% of
these were immunoreactive for GABA.

Enkephalin and Tyrosine Hydroxylase

In sections that were stained for both TH and leu-enk
many lightly stained bouton-like structures were seen
distributed throughout the substantia nigra. The density of

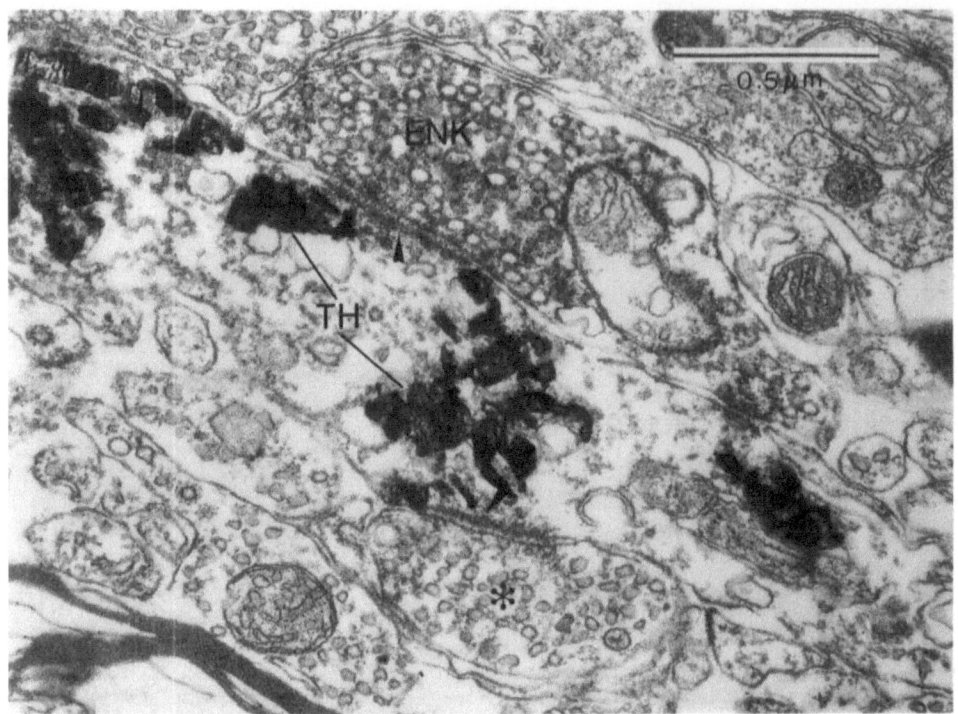

Fig. 4.   Electron micrograph showing a TH-immunoreactive
           dendrite (TH) contacted by an enkephalin-
           immunoreactive bouton (ENK) and a non-
           immunoreactive bouton (asterisk). Both boutons
           display asymmetrical membrane specializations
           (arrowheads).

immunoreactive boutons was far lower than that observed for
the GAD-immunoreactive structures and lower than that
observed for SP. Enkephalin-positive boutons were often
seen in close proximity to TH-stained structures.
       In the electron microscope the leu-enk-positive boutons
were of variable size and contained many slightly
pleomorphic vesicles. All the synaptic contacts so far
identified were with dendritic shafts and had asymmetrical
synaptic specializations. Some of the enkephalin-positive
boutons were found to form asymmetrical synaptic contact
with TH-immunoreactive dendrites (Fig. 4).

Substance P, GABA and Tyrosine Hydroxylase

       Since the fixatives that were used contained relatively
low concentrations of glutaraldehyde to retain both SP and
TH immunoreactivities, GABA immunoreactivity was often
absent or very low. However, in some cases GABA
immunoreactivity was present (Fig. 3B). In sections from
these cases TH-immunoreactive dendrites were often seen to
receive inputs from both GABA- and SP-immunoreactive
terminals. Substance P-immunoreactive terminals or preterminal boutons were sometimes seen to be also
immunoreactive for GABA. These double-stained boutons were

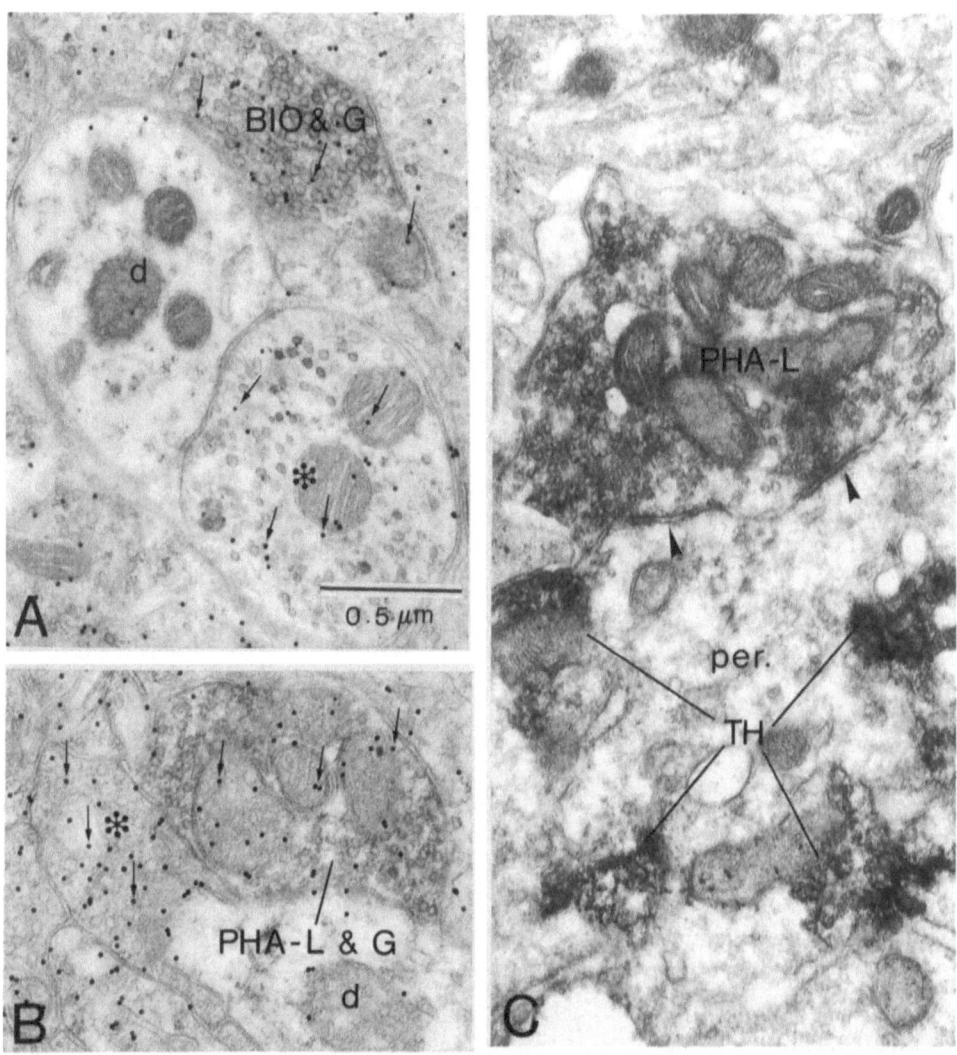

Fig. 5.   Electron micrographs illustrating various examples
of labelled structures observed in the substantia
nigra after combination of anterograde labelling
with either post-embedding immunocytochemistry for
GABA (A and B) or pre-embedding
immunocytochemistry for TH (C). A and B show a
biocytin-labelled striatal bouton (A) and a PHA-L-
labelled pallidal bouton (B) apposed to dendrites
in the substantia nigra. These boutons have
numerous immunogold particles overlying them (some
of which are indicated by small arrows) indicating
the GABA immunoreactivity. C illustrates a PHA-L-
labelled pallidal bouton that forms symmetrical
synaptic contacts (arrowheads) with a TH-
immunoreactive perikaryon. Note the presence of
the BDHC peroxidase reaction product in the TH-
positive cell. The scale marker in A is valid for
B and C.

identified in contact with non-immunoreactive dendrites or
dendrites that displayed immunoreactivity for TH (Fig. 3B).
Occasionally SP/GABA double-stained boutons and boutons that
displayed immunoreactivity for only GABA as well as non-
immunoreactive boutons were seen in contact with single TH-
immunoreactive dendrites.

Anterograde Labelling of Nigral Afferent Terminals Combined with Immunocytochemistry for GABA or Tyrosine Hydroxylase

Following injections of WGA/HRP or biocytin into the
striatum, anterogradely labelled terminal fields and axons
of passage were identified in the globus pallidus and the
SNr. The anterogradely labelled boutons formed symmetrical
synaptic specializations and had morphological
characteristics similar to GAD- or GABA-positive terminals.
In the sections in which there was a good retention of GABA
immunoreactivity, the anterogradely labelled terminals were
identified as GABA-positive (Fig. 5A).

The injection of PHA-L into the globus pallidus
resulted in the labelling of a population of large boutons
mainly in the pars reticulata of the substantia nigra. The
boutons, that were distributed according to a strict
topography, characteristically formed pericellular baskets
around the perikarya and proximal dendrites of reticulata
neurones (Smith and Bolam, 1989). The post-embedding
immunostaining for GABA revealed that the pallidonigral
terminals were immunoreactive for GABA (Fig. 5B). In
addition to those boutons found in the reticulata a small
proportion were located in the ventral aspects of the pars
compacta. In the sections that were stained to detect the
transported PHA-L and then immunostained to reveal TH,
pallidonigral terminals were seen in close association with
TH-immunoreactive neurones. Examination in the electron
microscope revealed that the pallidonigral boutons made
symmetrical synaptic contacts with TH-immunoreactive
dendrites and perikarya (Fig. 5C).

CONCLUDING REMARKS

The results of the present experiments help to
elucidate the chemical nature and pattern of synaptic input
to dopaminergic neurones in the substantia nigra of the rat.
In confirmation and extension of other observations (Chang,
1988; Kawai et al., 1987; Mahalik, 1988; Van den Pol et al.,
1985) they demonstrate first, that dopaminergic neurones
located both within the SNC and SNR receive symmetrical
synaptic input form terminals that display immunoreactivity
for substance P. Secondly, they show that dopaminergic
neurones receive a dense symmetrical synaptic input from
GABAergic terminals, and that this input represents at least
70% of all their afferent synapses. Thirdly, they reveal
that some of the substance P-immunoreactive boutons also
display GABA immunoreactivity and at least some of these
make synaptic contact with dopaminergic neurones. Fourthly,
they demonstrate that dopaminergic neurones receive
asymmetrical synaptic input from enkephalin-containing
boutons. The fact that these terminals form asymmetrical
synaptic contacts implies that they are not derived from the
striatum or the globus pallidus (see below). Finally, the

results of the experiments in which anterograde labeling was combined with GABA immunocytochemistry demonstrate that the striatum and the globus pallidus are both sources of GABA-positive terminals that make synaptic contacts in the substantia nigra. Furthermore, one of the synaptic targets of the pallidonigral terminals, like striatonigral terminals (Somogyi et al., 1981; Wassef et al., 1981), are the dopaminergic neurones.

These findings together with the demonstration of other inputs to nigral dopaminergic neurones (Nedergaard et al., 1988; Tokuno et al., 1988; Woulfe and Beaudet, 1989) suggest that these cells are under a highly complex synaptic control. The output of dopaminergic neurones and the release of dopamine in the striatum is thus likely to be dependent on the activity and topographic relationships of a large number of afferent synaptic boutons.

ACKNOWLEDGEMENTS

The authors express their gratitude to Frank Kennedy and Caroline Francis for their technical assistance and to Pierre Izzo for his advice on the use of biocytin as an anterograde tracer. Y.S. holds a fellowship from the Medical Research Council of Canada.

REFERENCES

Bolam, J.P. and Ingham, C.A., 1990, Combined morphological and histochemical techniques for the study of neural microcircuits, in: "Methods for Analysis of Neuronal Microcircuits. Handbook of Chemical Neuroanatomy, vol.8," F. Wouterlood and A. Van den Pol eds., Elsevier Biomedical Publication, Amsterdam.

Bouyer, J.J., Park, D.H., Joh, T.H. and Pickel, V.M., 1984, Chemical and structural analysis of the relation between cortical inputs and tyrosine hydroxylase-containing terminals in rat neostriatum, Brain Research, 302:267-275.

Chang, H.T., 1988, Substance P-dopamine relationship in the rat substantia nigra: a light and electron microscopy study of double immunocytochemically labeled materials, Brain Research, 448:391-396.

Cuello, A.C., Galfre, G. and Milstein, C., 1979, Detection of substance P in the central nervous system by a monoclonal antibody, Proc. Natl. Acad. sci., 76:3532-3536.

Freund, T.F., Powell, J. and Smith, A.D., 1984, Tyrosine hydroxylase-immunoreactive boutons in synaptic contact with identified striatonigral neurons, with particular reference to dendritic spines, Neuroscience, 13:1189-1215.

Hodgson, A.J., Penke, B., Erdei, A., Chubb, I.W. and Somogyi, P., 1985, Antiserum to γ-aminobutyric acid. I. Production and characterization using a new model system, J. Histochem. Cytochem., 33:229-239.

Ingham, C.A., Bolam, J.P. and Smith, A.D., 1988, GABA-immunoreactive synaptic boutons in the rat basal forebrain: comparison of neurons that project to the neocortex with pallidosubthalamic neurons, J. Comp. Neurol., 273:263-282.

Kawai, Y., Takagi, H., Kumoi, Y., Shiosaka, S. and Tohyama, M., 1987, Nigrostriatal dopamine neurons receive substance P-ergic inputs in the substantia nigra: application of the immunoelectron microscopic mirror technique to fluorescent double-staining for transmitter-specific projections, Brain Research, 401:371-376.

King, M.A., Louis, P.M., Hunter, B.E. and Walker, D.W., 1989, Biocytin: a versatile anterograde neuroanatomical tract-tracing alternative, Brain Research, 497:361-367

Levey, A.I., Bolam, J.P., Rye, D.B., Hallanger, A.E., Demuth, R.M., Mesulam, M.M. and Wainer, B.H., 1986, A light and electron microscopic procedure for sequential double antigen localization using diaminobenzidine and benzidine dihydrochloride, J. Histochem. Cytochem., 34:1449-1457.

Mahalik, T.J., 1988, Direct demonstration of interactions between substance P immunoreactive terminals and tyrosine hydroxylase immunoreactive neurons in the substantia nigra of the rat: an ultrastructural study, Synapse, 2:508-515.

Nedergaard, S., Bolam, J.P. and Greenfield, S.A., 1988, Facilitation of a dendritic calcium conductance by 5-hydroxytryptamine in the substantia nigra, Nature, 333:174-177.

Oertel, W.H., Schmechel, D.E., Tappaz, M.L. and Kopin, I.J., 1981, Production of a specific antiserum to rat brain glutamic acid decarboxylase by injection of an antigen-antibody complex, Neuroscience, 6:2689-2700.

Senba, E., Shiosaka, S., Miller, R.J., Inagaki, S., Kawai,Y., Takatsuki, K., Sakanaka, M., Iida, H., Takagi, H., Minagawa, H. and Tohyama, M., 1982, Ontogeny of the leucine-enkephalin neuron system of the rat: immunohistochemical analysis. I. Lower brain stem, J.Comp. Neurol., 205:341-359.

Smith, Y. and Bolam, J.P., 1989, Neurons of the substantia nigra reticulata receive a dense GABA-containing input from the globus pallidus in the rat, Brain Research, 493:160-167.

Somogyi, P., Bolam, J.P., Totterdell, S. and Smith, A.D., 1981, Monosynaptic input from the nucleus accumbens-ventral striatum region to retrogradely labelled nigrostriatal neurones, Brain Research, 217:245-263.

Somogyi, P. and Hodgson, A.J., 1985, Antiserum to γ-aminobutyric acid. III. Demonstration of GABA in Golgi-impregnated neurons and in conventional electron microscopic sections of cat striate cortex, J. Histochem. Cytochem., 33:249-257.

Somogyi, P., Hodgson, A.J., Chubb, I.W., Penke, B. and Erdei, A., 1985, Antisera to γ-aminobutyric acid. II. Immunocytochemical application to the central nervous system, J. Histochem. Cytochem., 33:240-248.

Tokuno, H., Moriizumi, T., Kudo, M. and Nakamura, Y., 1988,
    A morphological evidence for monosynaptic projections
    from the nucleus tegmenti pedunculopontinus pars
    compacta (TPC) to nigrostriatal projection neurons,
    Neurosci. Letts., 85:1-4.

Van den Pol, A.N., Herbst, R. and Powell, J.F., 1984,
    Tyrosine hydroxylase-immunoreactive neurons in the
    hypothalamus: a light and electron microscopic study,
    Neuroscience, 13:1117-1156.
Van den Pol, A.N., Smith, A.D. and Powell, J.F., 1985, GABA
    axons in synaptic contact with dopamine neurons in the
    substantia nigra: double immunocytochemistry with
    biotin-peroxidase and protein A-colloidal gold, Brain
    Research, 348:146-154.
Wassef, M., Berod, A. and Sotelo, C., 1981, Dopaminergic
    dendrites in the pars reticulata of the rat substantia
    nigra and their striatal input. Combined
    immunocytochemical localization of tyrosine hydroxylase
    and anterograde degeneration, Neuroscience, 6:2125-
    2139.
Woulfe, J. and Beaudet, A., 1989, Immunocytochemical
    evidence for direct connections between neurotensin-
    containing axons and dopaminergic neurons in the rat
    ventral midbrain tegmentum, Brain Research, 479:402-
    406.

MULTI-COLLATERALIZATION OF THE DOPAMINERGIC NIGROTECTAL PROJECTION

IN THE RAT

M. Takada*, K. J. Campbell and T. Hattori

Department of Anatomy, University of Toronto, Toronto, Ontario
Canada

INTRODUCTION

In view of the existence of massive efferents projecting outside the basal ganglia, the substantia nigra (especially the pars reticulata (SNr)), in conjunction with the pallidum, has been recognized as an output source of the basal ganglia (Nauta, 1979). The superior colliculus (SC) (Hopkins and Niessen, 1976; Rinvik et al., 1976; Jayaraman et al., 1977; Deniau et al., 1978; Faull and Mehler, 1978; Graybiel, 1978; Beckstead et al., 1979) is one of two major target sites of the SNr (the other is the ventromedial thalamic nucleus (VM) (Rinvik, 1975; Carpenter et al., 1976; Clavier et al., 1976; Faull and Mehler, 1978; Beckstead et al., 1979)). The fiber projection from the SNr to the SC is relatively well-organized in a topographic manner; most of projection neurons originate in the ventral portion of the SNr through its rostral 2/3 extent, and terminate in the intermediate/deep layers of the SC through its caudal 2/3 extent. This nigrotectal pathway is predominantly ipsilateral and to a lesser degree contralateral (Hopkins and Niessen, 1976; Rinvik et al., 1976; Deniau et al., 1977; Jayaraman et al., 1977; Gerfen et al., 1982). The pathway has also been implicated in the initiation of saccadic eye movements (Wurtz and Goldberg, 1972; Stryker and Schiller, 1975; Graybiel, 1978; Schiller et al., 1980; Wurtz and Albano, 1980). Indeed, a subpopulation of SNr cells attenuate their activity prior to saccadic eye movements and exert a tonic, gamma-aminobutyric acid (GABA)-ergic inhibition on saccade-related cells in the SC (Hikosaka and Wurtz, 1983, 1985a, b). Many clinical reports have shown that individuals suffering from parkinsonism often exhibit abnormal saccadic eye movements (DeJong and Melvill-Jones, 1971; Melvill-Jones and DeJong, 1971; Corin et al., 1972; Shibasaki et al., 1979; Teräväinen and Calne, 1980; Shimizu et al., 1981; White et al., 1983). In this respect, we have recently suggested in the rat that a specific population of SNr cells may provide a critical neuronal substrate for such clinical findings (Takada et al., 1988a, b). Thus, this cell group, which is localized primarily in the ventrolateral portion of the SNr at its rostral level, projects to both the striatum and SC by way of axon collaterals (Takada et al., 1988a). At least part of these bifurcating projection neurons are dopaminergic and can be eliminated by injecting a parkinsonism-inducing drug, 1-methyl-4-phenyl-1,2,3,6-tetrahydropyridine (MPTP) into the medial forebrain bundle (MFB, nigrostriatal dopamine fiber tract) (Takada et al., 1988b).

In the present report, we have investigated the possible existence of a specific nigral dopamine cell population which simultaneously innervates both the ipsilateral striatum and bilateral SC in the rat.

---

*Present adress: Department of Anatomy and Neurobiology, The University of Tennessee, Memphis, TN.

EXPERIMENTAL PROCEDURES

Twenty adult male albino rats (Wistar, 280-320 g b. wt.) were used for this study. Animals were intraperitoneally anesthetized with sodium pentobarbital (60 mg/kg b. wt.) and were each injected into the bilateral SC with the following 2 different fluorescent retrograde tracers: True blue (TB, 5% aqueous suspension) and Diamidino yellow (DY, 3% aqueous suspension). TB injections were mede stereotaxically into the SC on one side of the brain over 2 needle penetrations, while DY injections were made into the opposite side of the brain over 2 needle penetrations. A total volume of 0.2-0.4 μl of each tracer was symmetrically deposited centered on the intermediate/deep layers of the SC at its middle level (12 rats) or at its caudal level (6 rats), because nigrotectal fibers terminate mainly within the intermediate/deep layers of the nucleus over its caudal 2/3 extent (Jayaraman et al., 1977; Graybiel, 1978; Beckstead et al., 1979; Gerfen et al., 1982). Of 12 rats injected bilaterally with TB and DY at the middle level of the SC, 6 rats also received unilateral injections of a third fluorescent retrograde tracer, Propidium iodide (PI, 10% aqueous suspension), into the striatum over 2 needle penetrations. A total of 0.6 μl of PI was usually deposited ipsilaterally to the tectal DY injections. The tracers were slowly delivered through a 1-μl Hamilton microsyringe for 20 min to avoid unnecessary diffusion to neighboring regions (especially mutual tracer contamination from the bilateral SC injections). The injection needle was kept in place for an additional 20 min to minimize the leakage to overlying structures along the needle track.

The animals were allowed to survive for 2-4 days, deeply reanesthetized with an overdose of sodium pentobarbital, and fixed by transcardial perfusion with 300 ml of 10% ice-cold formalin dissolved in 0.1 M phosphate buffer (pH 7.4). The brains were immediately removed, immersed in the same buffer containing 25% sucrose at 4°C, and then cut serially into coronal sections of 30-μm thickness on a cryostat. The sections through the midbrain (including the SC injection sites and substantia nigra) obtained from the brains with the bilateral SC injections, were mounted onto clean slides and observed with a Leitz epifluorescence microscope to examine and quantify the distribution of SNr cells labeled with TB and/or DY. The nigral sections from the brains which had received the unilateral striatal injections combined with the bilateral SC injections, were divided into 2 groups. The first group was mounted onto slides and observed under the same microscope to examine and quantify the distribution of SNr cells labeled with TB, DY and/or PI. The second group was subsequently processed for immunofluorescence histochemistry for tyrosine hydroxylase (TH, 1:200 dilution). Briefly, the sections were incubated in rabbit antisera against TH (Eugene) for 48 h at 4°C, followed by purified goat anti-rabbit IgG conjugated to fluorescein isothiocyanate (FITC, Cappel, 1:50 dilution) for 1 h at room temperature, and were then observed under 3 different filters: an ultraviolet filter providing excitation light of approximately 360 nm wave-length to view the blue-emitting TB (labeling cytoplasm only)-containing and yellow-emitting DY (labeling nuclei only)-containing cells, a green filter providing excitation light of approximately 550 nm wave-length to view the red-emitting PI-containing cells, and a blue filter providing excitation light of approximately 440 nm wave-length to view the green-emitting TH-immunoreactive cells. The alternate use of these 3 filters on the same section greatly aided identification of single SNr cells quadruple-labeled with the 4 different markers.

In the remaining 2 rats, unilateral injections of TB were made into the middle SC. These animals received colchicine pretreatment (100 μg/10 μl saline) 24 h prior to sacrifice. Cryostat-cut, coronal sections through the substantia nigra were then processed for sequential double antigen immunofluorescence histochemistry for glutamic acid decarboxylase (GAD) and TH. Briefly, the sections were incubated in rabbit antisera against GAD (kindly donated by Dr. J.-Y. Wu, 1:200 dilution) for 24 h, followed by anti-rabbit IgG conjugated to FITC (Cappel, 1:200 dilution) for 1 h, and then incubated in mouse antisera against TH (Incstar, 1:700 dilution) for 48 h, followed by anti-mouse IgG conjugated to Tetramethyl rhodamine isothiocyanate (TRITC). They were observed with the same microscope under 3 different filters: an ultraviolet filter for blue-emitting TB-positive cells, a blue filter for green-emitting GAD-immunoreactive cells and a green filter for red-emitting TH-immunoreactive cells.

RESULTS

Bilateral SC Injections

Bilateral injections of TB and DY almost symmetrically involved the intermediate/deep layers of the middle 1/3 (Fig. 1A) or caudal 1/3 (Fig. 1B) of the SC on each side. However, there was no discernible encroachment of either tracer on the opposite side of the SC across the midline. These 2 types of bilateral SC injections resulted in markedly different distribution patterns of SNr cells retrogradely labeled with TB and/or DY.

Following bilateral ionjections over the middle 1/3 portion of the SC, a large number of retrogradely labeled neurons were observed in the SNr throughout its rostral 2/3 (to 3/4) extent with a rostrocaudal gradient (Fig. 1A). These labeled cells were found predominantly ipsilateral and to a lesser degree (32.3% of total labeled and 37.8% of ipsilaterally labeled cells) contralateral to each tracer injection. Thus, the SNr ipsilateral to the DY injection site contained a multitude of DY-positive and a smaller number of TB-positive cells, and vice versa. The neuronal labeling from the ipsilateral and contralateral SC occurred largely in the ventral 1/2 and far less frequently in the dorsal 1/2 of the SNr (Fig. 1A). Moreover, along with numerous single-labeled cells, bilateral SC injections of this type produced a substantial number (17.9% of the total) of cells double-labeled with both TB and DY in the SNr. The number of these double-labeled cells also amounted to 55.4% of the contralaterally labeled or 21.0% of the ipsilaterally labeled cell population. As might be predicted from each distribution pattern of sigle-labeled cells, the SNr cells containing both tracers were concentrated in the ventral 1/2 of the nucleus, and more significantly, they were most often seen in the ventrolateral portion at the level of the more rostral SNr (in the rostral 1/3) (Fig. 1A). Additionally, retrogradely labeled neurons were also found in the substantia nigra pars lateralis (SNl) ipsilateral to each tracer injection. However, the SNl was almost devoid of cell labeling from the contralateral and therefore bilateral SC.

More caudal injections (involving the caudal 1/3) of TB and DY into the bilateral SC, on the other hand, showed a striking difference in the retrograde cell labeling pattern in the SNr. Although SNr cells were, in fact, single- and double-labeled with either or both of the tracers with a rostrocaudal gradient, each number of these 3 types of (ipsilaterally, contralaterally and bilaterally) labeled cells noticeably went down. In particular, the double-labeled cell population only made up 6.2% of total labeled, 30.6% of contralaterally labeled and 7.2% of ipsilaterally labeled cells. Furthermore, the SNr cell labeling as a whole seemed to be restricted more ventrally with a slight rostrocaudal gradient. However, double-labeled cells were not localized specifically to the more lateral portion of the SNr, as compared to the case of the more rostral SC injections described above, but distributed scattered over the mediolateral plane (Fig. 1B). Additionally, fewer neurons labeled only ipsilaterally with either of the tracers were detected in the SNl.

Striatal Combined with Bilateral SC injections

In a second series of experiments, the possible existence of SNr cells simultaneously innervating both the ipsilateral striatum and bilateral SC, was examined on the basis of our recent (Takada et al., 1988a) and present data. Given a similar distribution pattern between nigrotectal cells of origin sending axon collaterals to the ipsilateral striatum (Takada et al., 1988a) and the contralateral SC (especially in the case of the more rostral SC injections as documented above), unilateral striatal injections of PI were designed to be combined with bilateral injections of TB and DY into the middle 1/3 of the SC. Striatal PI was usually deposited on the side ipsilateral to DY injections.

TB and DY injections into the middle1/3 of the bilateral SC and the resultant retrograde labeling pattern of SNr cells were almost identical to those mentioned above (see Fig. 1A). The cell population containing both tracers appeared mostly in the rostral, ventrolateral portion of the SNr and again with similar frequency (16.1% of total, 54.6% of contralateral and 18.6% of ipsilateral nigrotectal cells). PI injections into the striatum involved the major portion of the nucleus with some diffusion to the overlying frontal

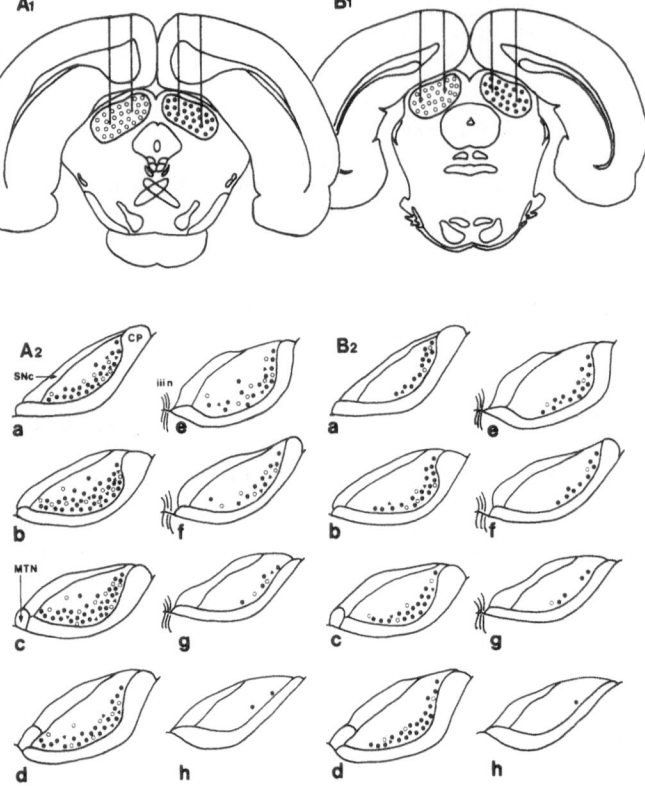

Fig. 1. (A1, B1) Line drawings of representative coronal sections through the sites of bilateral injections of DY (solid circles) and TB (open circles) into the SC at the middle (A1) and caudal (B1) levels. (A2, B2) Projection drawings made from 8 rostrocaudally (a-h) arranged coronal sections through the substantia nigra following the rostral (A2) and caudal (B2) SC injections. The drawings are of the rostral 3/4 of the substantia nigra ipsilateral to the DY injection site. DY-single-labeled SNr (ipsilaterally projecting nigrotectal) cells are represented with solid circles, TB-single-labeled SNr (contralaterally projecting nigrotectal) cells with open circles, and double-labeled SNr (bilaterally projecting nigrotectal) cells with solid triangles. Abbreviations: CP, cerebral peduncle; MTN, medial terminal nucleus of the accessory optic tract; SNc, substantia nigra pars compacta; iiin, root fibers of the oculomotor nerve.

cortex. Dense cell labeling with PI was evident in the substantia nigra pars compacta (SNc) throughout its entire extent. These labeled SNc cells were found bilaterally with an ipsilateral predominance. A considerable number of PI-positive cells were also seen exclusively ipsilaterally in the SNr. These SNr cells labeled from the striatum were located primarily in its ventral portion (ventral 1/3) and to a lesser degree in its more dorsal part. At the more rostral level (in the rostral 1/3), the ventrally situated PI-labeled neurons were observed more laterally. As the SNr level went more caudally, they became more numerous and were aggregated more medially (see Fig. 1 in a paper by Takada et al., 1988a).

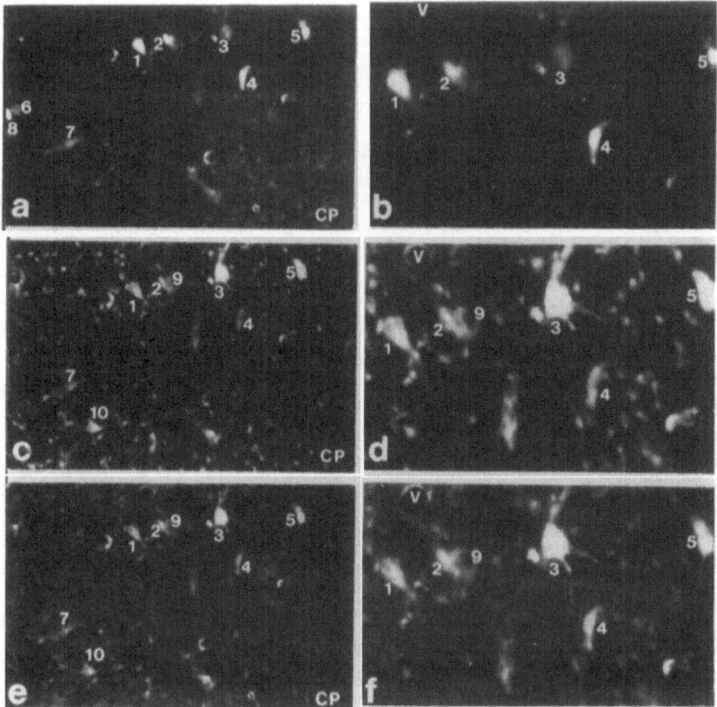

Fig. 2. Photomicrographs of TH-immunoreactive SNr cells following PI injections into the ipsilateral striatum combined with DY (ipsilateral) and TB (contralateral) injections into the bilateral SC through the middle 1/3 extent. The rostral, ventrolateral SNr ipsilateral to DY (PI) injections. (a) DY- and/or TB-positive (unilaterally or bilaterally projecting nigrotectal) cells. (c) PI-positive (ipsilaterally projecting nigrostriatal) cells in the same section as (a). (e) TH-immunoreactive (dopaminergic) cells in the same section as (a) and (c). 1, 2, 4, 5 and 7, cells quadruple-labeled with the 4 markers; 3, cell triple-labeled with TB, PI and TH antisera; 6, cell single-labeled with TB; 8, cell single-labeled with DY; 9 and 10, cells double-labeled PI and TH antisera. CP, cerebral peduncle. A cluster of quadruple-labeled cells in (a), (c) and (e) are shown with higher magnification in (b), (d) and (f), respectively. The same blood vessel is represented with V.

Together with more massive single labeling, triple tracer injections of this type often yielded the appearance of SNr cells labeled with more than one tracer. Thus, not only the above-documented double labeling with both TB and DY from the bilateral SC, but 3 other conceivable combinations of multiple labelings concurred as follows. A significant number of SNr cells were double-labeled with both TB and DY from the bilateral SC (but not with PI from the striatum) (5.9% of the total), or with both PI and DY from the striatum and ipsilateral SC (but not with TB from the contralateral SC) (4.9% of the total). On the other hand, only occasional cells (0.4% of the total) were encountered double-labeled with both PI and TB from the striatum and contralateral SC (but not with DY from the ipsilateral SC). Furthermore, 5.8% of total labeled SNr cells contained all the 3 tracers. These triple-labeled cells also constituted 27.1% of the contralateral, 9.2% of the ipsilateral and as many as 49.6% of the bilateral nigrotectal cell population. Interestingly, all these combinations of multiple labelings were again present primarily in the rostral, ventrolateral SNr (Fig. 2). In addition to cells single-labeled ipsilaterally with one of the 3 tracers, neuronal cell bodies in the SNl were relatively frequently double-labeled with both PI and DY ipsilateral to each injection site.

TH Immunohistochemistry Combined with Triple Tracer Injections

In a third series of experiments, the possible existence of dopaminergic SNr cells giving off axon collaterals to both the ipsilateral striatum and bilateral SC, was examined on the basis of our recent work (Takada et al., 1988b) showing the dopaminergic nigrotectal projection. Following triple tracer injections into the bilateral SC (TB and DY) and unilateral (ipsilateral to DY) striatum (PI) as mentioned above, nigral sections were incubated in TH antisera.

After the immunofluorescence procedure using FITC, cell bodies in the SNc were extensively double-labeled with TH antisera and PI from the striatum. A considerable number of SNr cells also exhibited the characteristic green fluorescence induced by FITC. The site of these TH-immunoreactive cells well corresponded with that of PI-positive cells in the SNr. Indeed, almost all the PI-labeled cells were double-labeled with TH antisera. In the rostral, ventrolateral SNr region, these double-labeled cells also often contained TB (from the contralateral SC) and/or DY (from the ipsilateral SC). Thus, at least 75% of cells containing PI and either or both of TB and DY were TH-immunoreactive. Interestingly, the vast majority (more than 85%) of SNr cells triple-labeled with all the 3 tracers were immunostained (quadruple-labeled) with TH-antisera as well (Fig. 2). However, SNr cells containing either or both of TB and DY only (but not PI) failed to display TH immunoreactivity. Additionally, in many instances neurons labeled with PI only or both PI and DY were TH-immunopositive in the SNl.

Sequential TH and GAD Immunohistochemistry Combined with SC Injections

The possible coexistence of dopamine and GABA in single nigrotectal cells was investigated using sequential double antigen immunofluorescence histochemistry for TH and GAD. As shown in Fig. 3, a certain population of TB-positive SNr (ipsilateral nigrotectal) cells were also immunohistochemically stained with both TH and GAD antisera. These triple-labeled cells were again localized in the rostral, ventrolateral portion of the SNr adjacent to the cerebral peduncle. A significant number of cells triple-labeled with all the 3 markers were also evident in the SNl.

DISCUSSION

The present double labeling study demonstrates that the rostral (middle 1/3) versus caudal (caudal1/3) SC injections produce a great diversity in the distribution pattern of double-labeled SNr cells, resulting in a topography that the majority of bilaterally projecting nigrotectal cells arise from the rostral, ventrolateral SNr and terminate in the SC at the more rostral levels. These bifurcating cells are equivalent to more than 1/6 of the total, 1/2 of the contralateral and 1/5 of the ipsilateral nigrotectal cell population. It should be added here that crossed nigrotectal fiber termination (especially in the rat) is confined to the lateral portion of the deep layers of the middle 1/3 SC (Gerfen et al., 1982).

Our recent study (Takada et al., 1988a) has elucidated that single cells in the rostral, ventrolateral SNr send axon collaterals ipsilaterally to both the striatum and SC. The present data confirm and extend these reports with the demonstration that such bifurcating SNr cells also issue third collateral branches to the contralateral SC. This multi-collateralized cell population constitutes almost 1/2 of bilaterally projecting nigrotectal cells. The overall results from our triple tracer labeling experiments indicate that a substantial number of cells in the specific (rostral, ventrolateral) SNr region innervate simultaneously more than one structure of the striatum and bilateral SC. In contrast, it might be of some importance to note the paucity of the branching pattern to both the striatum and contralateral SC only (but not to the ipsilateral SC). In view of the fact that no available evidence has so far supported the existence of either collateral SNr projections to both the striatum and thalamus (Bentivoglio et al., 1979; Takada et al., 1988a), or crossed SNr projections to the striatum (Takada et al., 1987), the possibility

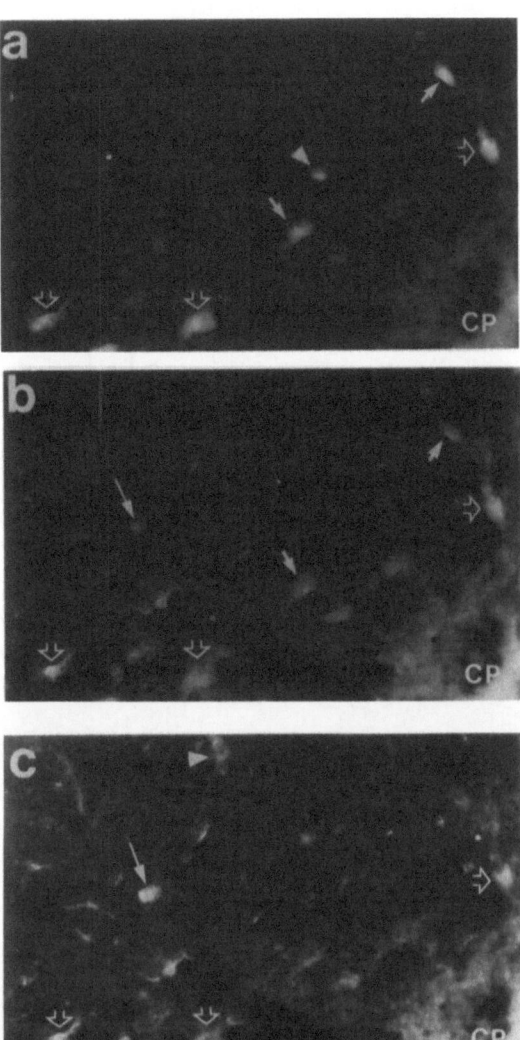

Fig. 3. Photomicrographs of GAD- and/or TH-immunoreactive SNr cells following TB injections into the SC through the middle 1/3 extent. The rostral, ventrolateral SNr on the side ipsilateral to TB injections. (a) TB-positive (ipsilaterally projecting nigrotectal) cells. (b) GAD-immunoreactive (GABAergic) cells in the same section as (a). (c) TH-immunoreactive (dopaminergic) cells in the same section as (a) and (b). Cells single-labeled with either TB or TH antisera are represented with arrow heads, cells double-labeled with both TB and GAD antisera or both GAD and TH antisera with solid arrows, and cells triple-labeled with all markers with open arrows. CP, cerebral peduncle.

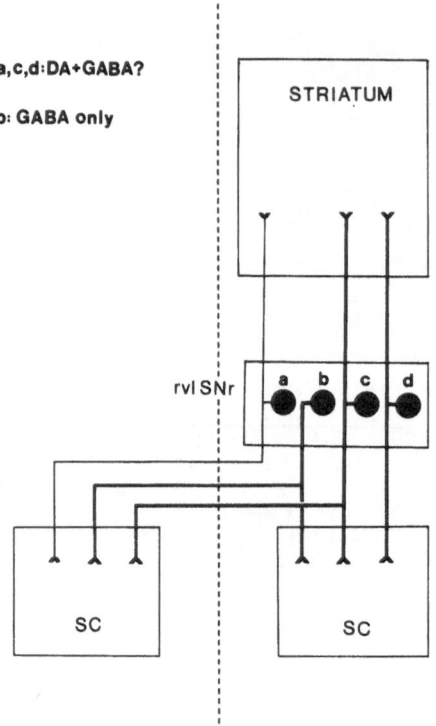

Fig. 4. Schematic diagram demonstrating four different patterns of collateral projections from the rostral ventrolateral (rvl) SNr to the ipsilateral striatum and bilateral SC. Cell (a) is far less frequently observed than cells (b), (c) and (d). Cells (a), (c) and (d) possibly contain both dopamine and GABA, while cell (b) contains GABA only.

that the specifically collateralized cells under study issue additional branches to the thalamus and/or contralateral striatum, would not be respected.

Furthermore, our TH immunofluorescence histochemistry combined with the triple tracer labeling shows that the majority (more than 75%) of SNr cells with collaterals to both the striatum (ipsilaterally) and SC (unilaterally or bilaterally), express dopamine. Above all, the dopamine-containing population of SNr cells giving rise to multicollateral projections to both the ipsilateral striatum and bilateral SC, remarkably exceeds 85%. It should also be emphasized here that even in the rostral, ventrolateral SNr neurons projecting to the SC (unilaterally or bilaterally) never displayed dopaminergic traits, if they lacked axon collaterals innervating the striatum. This finding suggests that dopamine expression in SNr cells may closely be associated with the outgrowth of their axons towards the striatum or the accomplishment of their connection with the striatum. However, the causal relation between these 2 events remains to be established definitively. It is generally accepted that the nigrotectal pathway exerts GABAergic inhibitory transmission (Vincent et al., 1978; Di Chiara et al., 1979; Chevalier et al., 1981). Accordingly, the existence of dopaminergic nigrotectal projection cells recently (Takada et al., 1988b) and presently identified favors the possible colocalization of dopamine and GABA in these specific SNr cells. Our preliminary investigation indeed reveals that the coexistence of both substances does occur in single nigrotectal cells.

The nigrotectal pathway has been suggested to play a crucial role in the onset of saccadic responses by cells in the deep layers of the SC (Wurtz and Goldberg, 1972; Stryker and Schiller, 1975; Graybiel, 1978; Schiller et al., 1980; Wurtz and Albano, 1980). In fact, Hikosaka and Wurtz (1983, 1985a, b) have clearly shown in the monkey that a certain population of SNr cells exert a tonic, GABAergic inhibition on saccade-related cells in the SC, and that these saccadic eye movements are severely disrupted by muscimol infused into the SNr as well as into the SC. In many clinical cases, abnormalities in saccadic eye movements are manifested by parkinsonian patients (DeJong and Melvill-Jones, 1971; Melvill-Jones and DeJong, 1971; Corin et al., 1972; Shibasaki et al., 1979; Teräväinen and Calne, 1980; Shimizu et al., 1981; White et al., 1983). This oculomotor disorder has been proposed to result from imbalances of GABAergic transmission to the SC caused by serial damage to putative indirect pathways, the nigro-striato-nigro-tectal and/or visual cortico/frontal eye field-striato-nigro-tectal links (Bizzi, 1968; Künzle and Akert, 1977; Schiller et al., 1980; Melis and Gale, 1983; White et al., 1983; Faull et al., 1986). However, our SNr dopamine cells simultaneously innervating both the striatum (ipsilaterally) and SC (unilaterally or bilaterally) (Fig. 4), might constitute an alternative and presumably more prominent neuronal basis for such a clinical phenomenon. Indeed, the fact that MPTP treatment of the MFB induces dopaminergic nigrotectal cell degeneration (Takada et al., 1988b), supports the notion of their vulnerability in parkinsonism. Pharmacological manipulation of this multi-collateralized dopaminergic nigrotectal system would be expedient to evaluate the viability of our hypothesis.

ACKNOWLEDGEMENTS

This work was supported by the Medical Research Council of Canada. K.J.C. is a recipient of an MRC Studentship.

REFERENCES

Beckstead, R. M., Domesick, V. B., and Nauta, W. J. H., 1979, Efferent connections of the substantia nigra and ventral tegmental area in the rat, Brain Res., 175:191.
Bentivoglio, M., van der Kooy, D., and Kuypers, H. G. J. M., 1979, The organization of the efferent projections of the substantia nigra in the rat. A retrograde fluorescent double labeling study, Brain Res., 174:1.
Bizzi, E., 1968, Discharges of frontal eye field neurons during saccadic and following eye movements in unanesthetized monkeys, Exp. Brain Res., 6:69.
Carpenter, M. B., Nakano, K., and Kim, R., 1976, Nigrothalamic projections in the monkey demonstrated by autoradiographic technics, J. Comp. Neurol., 165:401.
Chevalier, G., Thierry, A. M., Shibazaki, T., and Féger, J., 1981, Evidence for a GABAergic inhibitory pathway in the rat, Neurosci. Lett., 21:67.
Clavier, R. M., Atmadja, S., and Fibiger, H. C., 1976, Nigrothalamic projections in the rat as demonstrated by orthograde and retrograde tracing techniques, Brain Res. Bull., 1:379.
Corin, M. S., Elizan, T. S., and Bender, M. B., 1972, Oculomotor function in patients with Parkinson's disease, J. Neurol. Sci., 15:251.
DeJong, J. D., and Melvill-Jones, G., 1971, Akinesia, hypokinesia, and bradykinesia in the oculomotor system of patients with Parkinson's disease, Exp. Neurol., 32:58.
Deniau, J. M., Chevalier, G., and Féger, J., 1978, Electrophysiological study of the nigro-tectal pathway in the rat, Neurosci. Lett., 10:215.
Deniau, J. M., Hammond-LeGuyader, C., Féger, J., and McKenzie, J. S., 1977, Bilateral projection of nigro-collicular neurons: an electrophysiological analysis in the rat, Neurosci. Lett., 5:45.
Di Chiara, G., Porceddu, M. L., Morelli, M., Mulas, M. L., and Gessa, G. L., 1979, Evidence for a GABAergic projection from the substantia nigra to the ventromedial thalamus and to the superior colliculus of the rat, Brain Res., 176:173.
Faull, R. L. M., and Mehler, W. R., 1978, The cells of origin of nigrotectal, nigrothalamic and nigrostriatal projections in the rat, Neuroscience, 3:989.
Faull, R. L. M., Nauta, W. J. H., and Domesick, V. B., 1986, The visual cortico-striato-nigral pathway in the rat, Neuroscience, 19:1119.

Gerfen, C. R., Staines, W. A., Arbuthnott, G. W., and Fibiger, H. C., 1982, Crossed connections of the substantia nigra in the rat, J. Comp. Neurol., 207:283.

Graybiel, A. M., 1978, Organization of the nigrotectal connection: an experimental tracer study in the cat, Brain Res., 143:339.

Hikosaka, O., and Wurtz, R. H., 1983, Visual and oculomotor functions of monkey substantia nigra pars reticulata. IV. Relation of substantia nigra to superior colliculus, J. Neurophysiol., 49:1285.

Hikosaka, O., and Wurtz, R. H., 1985a, Modification of saccadic eye movements by GABA-related substances. I. Effects of muscimol and bicucullin in monkey superior colliculus, J. Neurophysiol., 53:266.

Hikosaka, O., and Wurtz, R. H., 1985b, Modification of saccadic eye movements by GABA-related substances. II. Effects of muscimol in monkey substantia nigra pars reticulata, J. Neurophysiol., 53:292.

Hopkins, D. A., and Niessen, L. W., 1976, Substantia nigra projections to the reticular formation, superior colliculus and central gray in the rat, cat and monkey, Neurosci. Lett., 2:253.

Jayaraman, A., Batton, R. R., III, and Carpenter, M. B., 1977, Nigrotectal projections in the monkey: an autoradiographic study, Brain Res., 135:147.

Künzle, H., and Akert, K., 1977, Efferent connections of cortical area 8 (frontal eye field) in Macaca fascicularis. A reinvestigation using the autoradiographic technique, J. Comp. Neurol., 173:147.

Melis, R. M., and Gale, K., 1983, Effects of dopamine agonists on gamma-aminobutyric acid (GABA) turnover in the superior colliculus: evidence that nigrotectal GABA projections are under the influence of dopaminergic transmission, J. Pharmacol., 226:425.

Melvill-Jones, G., and DeJong, J. D., 1971, Dynamic characteristics of saccadic eye movements in Parkinson's disease, Exp. Neurol., 31:17.

Nauta, H. J. W., 1979, A proposed conceptual reorganization of the basal ganglia and telencephalon, Neuroscience, 4:1875.

Rinvik, E., 1975, Demonstration of nigrothalamic connections in the cat by retrograde axonal transport of horseradish peroxidase, Brain Res., 90:313.

Rinvik, E., Grofová, I., and Ottersen, O. P., 1976, Demonstration of nigrotectal and nigroreticular projections in the cat by axonal transport of proteins, Brain Res., 112:388.

Schiller, P. H., True, S. D., and Conway, J. L., 1980, Deficits in eye movements following frontal eye-field and superior colliculus ablations, J. Neurophysiol., 44:1175.

Shibasaki, H., Tsuji, S., and Kuroiwa, Y., 1979, Oculomotor abnormalities in Parkinson's disease, Archs Neurol., 36:360.

Shimizu, N., Naito, M., and Yoshida, M., 1981, Eye-head co-ordination in patients with Parkinsonism and cerebellar ataxia, J. Neurol. Neurosurg. Psychiat., 43:509.

Stryker, M. P., and Schiller, P. H., 1975, Eye and head movements evoked by electrical stimulation of monkey superior colliculus, Exp. Brain Res., 23:103.

Takada, M., Li, Z. K., and Hattori, T., 1987, A note on the projections of pars compacta neurons within pars reticulata of the substantia nigra in the rat, Brain Res. Bull., 18:285.

Takada, M., Li, Z. K., and Hattori, T., 1988a, Collateral projection from the substantia nigra to the striatum and superior colliculus in the rat, Neuroscience, 25:563.

Takada, M., Li, Z. K., and Hattori, T., 1988b, Dopaminergic nigrotectal projection in the rat, Brain Res., 457:165.

Teräväinen, H., and Calne, D. B., 1980, Studies of parkinsonian movement: I. Programming and execution of eye movements, Acta Neurol. Scand., 62:137.

Vincent, S. R., Hattori, T., and McGeer, E. G., 1978, The nigrotectal projection: a biochemical and ultrastructural characterization, Brain Res., 151:159.

White, O. B., Saint-Cyr, J. A., Tomlinson, R. D., and Sharpe, J. A., 1983, Ocular motor deficits in Parkinson's disease. II. Control of the saccadic and smooth pursuit systems, Brain, 106:571.

Wurtz, R. H., and Albano, J. E., 1980, Visual-motor function of the primate superior colliculus, Ann. Rev. Neurosci., 3:189.

Wurtz, R. H., and Goldberg, M. E., 1972, Activity of superior colliculus in behaving monkey. III. Cells discharging before eye movements, J. Neurophysiol., 35:575.

CONVERGENT PROJECTIONS FROM SUBSTANTIA NIGRA AND CEREBELLUM

ON PONTINE RETICULAR FORMATION OF RAT

Sabina Berretta and Vincenzo Perciavalle

Institute of Human Physiology, University of Catania
Viale Andrea Doria 6, 95125, Catania, Italy

INTRODUCTION

Although basal ganglia and cerebellum represent two important motor structures, few data exist in literature regarding their cooperation in motor control. The only recent papers on reciprocal influences between these structures regard the control exerted by paleostriatum on neocerebellar neurons (Perciavalle et al., 1987) and the projections from intracerebellar nuclei to the dopaminergic structures of ventral midbrain tegmentum (Perciavalle et al., 1989). Cerebellum and basal ganglia could cooperate by sending their outputs on common motor targets, as thalamic nuclei controlling cortical motor areas or spinal-projecting brainstem structures. At thalamic level, no evidence has been found for convergence of cerebellar inputs on those neurons controlled by basal ganglia (Ueki et al., 1977; Uno and Yoshida, 1975). On the other hand, it has been demonstrated that cerebellar lateral nucleus (LN; Bantli and Bloedel, 1975) and pars reticulata of substantia nigra (SN; Perciavalle, 1987) are capable of exerting monosynaptic influences on single spinal-projecting neurons of pontomedullary reticular formation (RF). However, it is not yet known whether these cerebellar and nigral projections reach separate populations of RF cells or control the same RF neurons.

The present study was undertaken 1) to identify, with neuroanatomical approach, the RF area reached by both cerebellar and nigral axons and the location of neurons giving off these projections, and 2) to verify and quantify, with electrophysiological approach, the possibility that nigral and cerebellar influences converge on the same RF cells.

METHODS

The present experiments were carried out on 27 male Wistar rats (230 to 280 g), 22 of which anesthetized with urethane (1.2 g/kg i.p.) and 5 with ketamine hydrochloride (100 mg/kg i.p.).

Neuroanatomical approach

It has been used a 4% solution of horseradish peroxidase conjugated to wheat germ agglutinin (WGA-HRP; Sigma) dissolved in 0.9% saline. All the injections (15-25 nl) were hydraulically made by using glass micropipettes (tips: 40-60 μm) connected with a 1 ul Hamilton syringe. In the experiments using the retrograde transport of the tracer, 6 rats were submitted to unilateral hydraulic injections of WGA-HRP within pontine RF at planes A 0.7 - P 1.0, L 0.5 - 1.0 and H 0.5 - 1.5 (cf. atlas of Paxinos and Watson, 1986). For anterograde HRP experiments, 7 rats received a unilateral hydraulic injection of WGA-HRP in the SN (4 rats) at planes A 3.2, L 2.0 and H 1.6 or in LN (3 rats) at planes P 2.3, L 3.5 and H 3.7 (cf. atlas of Paxinos and Watson, 1986). To reduce the spread of tracer along the injection track, the micropipettes were left in place for 15 min before injection and for an additional 15 min before being removed from the brain.

After a 40-48 h survival period, the rats were reanesthetized with urethane (2 g/kg i.p.) and perfused sequentially with 500 ml of 0.9% saline at 42 °C, 500 ml of a 1.25% glutaraldehyde - 1% paraformaldehyde in 0.1 M phosphate buffer (pH 7.4) solution at room temperature and 500 ml of an ice-cold 10% sucrose solution in 0.1 M phosphate buffer (pH 7.4). The brains were removed and placed in the same 10% sucrose in phosphate buffer solution at° 4 C until sectioned.

Frontal frozen sections (20-40 μm) were cut, collected in 0.1 M pH 7.4 phosphate buffer, processed with tetramethylbenzidine (Sigma) and incubated with hydrogen peroxide, according to the method of Mesulam (1979). Sections through the injection sites were reacted with diamino benzidine (Sigma), in order to determine the core uptake area. The sections were transferred into a sodium acetate buffer solution at 4 °C and mounted onto glass slides coated with chrome alum. To facilitate the identification of cytoarchitectonic details, half of the sections processed for HRP were counterstained with 1% neutral red (pH 4.8), while the other half was left uncounterstained to reduce the loss of reaction-product. Finally, all the sections, whether counterstained or not, were dehydrated with ethanol, cleared with xylene and coverslipped with Entellan (Merck).

The slides were examined under both bright- and dark-field illumination. The effective injected area was considered the zone surrounding the HRP injection site where deposition of reaction-product was very dense and where individual labeled axons and perikarya cannot be detected (Horton et al., 1979). Injected areas were quantitatively evaluated by adapting the method of Wolf and Gollob (1980). The pattern of orthogradely transported label, judged by conventional criteria (Mesulam, 1982), was plotted onto camera lucida drawings. Photographs were taken of salient features.

Electrophysiological approach

In 14 rats, extracellular responses of single pontine reticular cells to stimulation of SN and LN were recorded with glass micropipettes (7-18 MOhm resistance) filled with a 4% solution of pontamine sky blue in KCl 1M, the indifferent electrode being a screw positioned in the skull. Ketamine hydrochloride anesthesia was done in one group of rats (n=5) to assess differences in responses, if any, between the rats anesthetized with urethane and those with ketamine hydrochloride anesthesia. Action potentials arising from the cell somata were those which appeared initially negative-positive and then, as the electrode advanced, positive-negative with a clear inflection on the ascending phase (Hubel, 1960). The RF explored regions were enclosed between planes A 2.0 and P 1.5 (cf. atlas of Paxinos and Watson, 1986). The end of each electrode penetration was marked with a small (50-100 um) deposit of pontamine by ionophoresis (10 μA for 20 min), for subsequent histological reconstruction. Stimulation of SN and LN was bilaterally carried out through monopolar nickel-chrome wires (0.6-1.2 MOhm resistance). The SN and LN were activated with 1-5 cathodal shocks (0.05-0.2 ms, 100-700 Hz) delivered at intervals of 1.5 - 5 s. Stimulation electrodes were inserted within SN and LN at the same stereotaxic coordinates used for WGA-HRP injections.

After the recording sessions, the stimulating sites were marked and the animals were perfused, under deep urethane (2 g/kg i.p.) anesthesia, with a 4% paraformaldehyde solution. Reconstructions of stimulation sites and electrode penetrations were made from frozen sections of 40 μm thickness, stained with neutral red.

The evaluation of unitary discharges was performed by converting 60-90 responses into post-stimulus time histograms and cumulative frequency distributions. Excitatory and inhibitory responses were sequences of at least 3 bins, which showed frequency values of more than twice the standard deviation (SD) above or below the mean value during spontaneous activity. Solitary bins between these sequences were not considered. Latency was the time interval between the last shock of the stimulating train and the first bin of the sequence, and the duration was the distance between first and last bin (Li Volsi et al., 1982).

RESULTS

Neuroanatomical experiments

Fig. 1 combines the results from two experiments in which the enzyme injection involved, in one rat, nearly all the left LN and in the other rat the central part of the right SN. Orthogradely labeled axons from WGA-HRP placement sites in LN reached the pons. In this area, an evident terminal labeling was present in some contralateral ventral structures of pons, as pontine nuclei, raphe magnus nucleus, reticulotegmental nucleus and pontine reticular nucleus oral part (PnO) and caudal part (PnC). Following unilateral hydraulic injection of WGA-HRP in the SN area, an ipsilateral dust-like terminal labeling (often detectable only on uncounterstained sections) was present in the raphe magnus nucleus, pedunculopontine nucleus, reticulotegmental nucleus, PnO and PnC. It is evident in this figure that a wide overlap of terminal labeling from SN and LN is present in the more ventral and medial portions of PnO and PnC. Anterograde labeling appeared as a network of fibers branching irregularly with bouton-like swellings at their endings and that fibers coming from SN were less numerous and thicker than those from LN.

Following WGA-HRP injections in the ventral part of pontine reticular formation, retrogradely labeled cells were found in

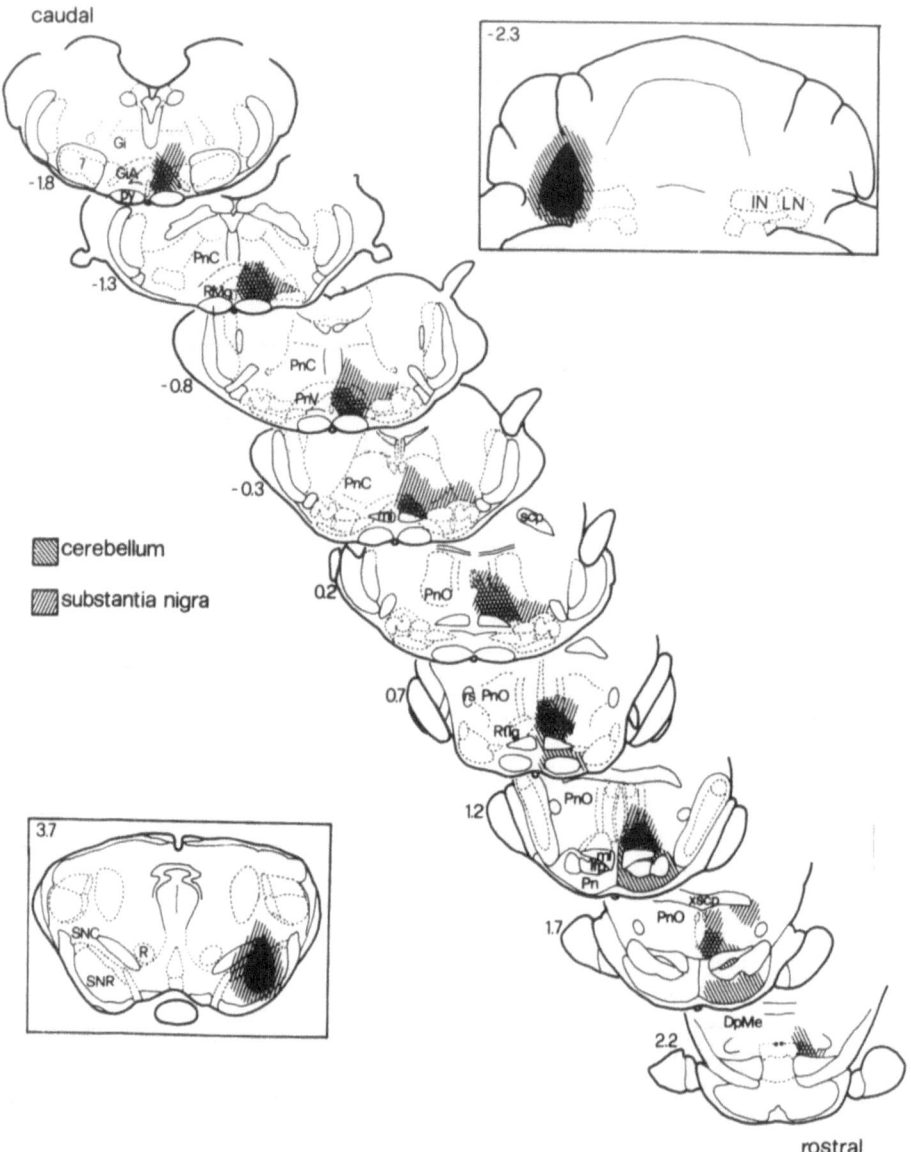

Fig.1. Camera lucida drawings showing the anterograde terminal labeling observed in two rats following a single WGA-HRP injection by pressure in the lateral aspect of cerebellum (left-to-right etched area) and in the central part of substantia nigra (right-to-left etched area). Black area, injection site; surrounding etched area, zone of passive diffusion around the injection site. Abbreviations: see list.

ipsilateral SN pars reticulata and in contralateral LN. Fig. 2 illustrates results of an experiment. The injection of 20 nl of 4% WGA-HRP involves almost exclusively the medial part of PnO. In this experiment, various-sized labeled neurons were found in LN and, although with a lower density, in lateral parts of interpositus nucleus (IN) of the contralateral side; the majority of these labeled cells were medium or small-size. No retrogradely labeled neurons were found in the fastigial nucleus of both sides. Within SN, labeled cells were observed throughout the pars reticulata of the ipsilateral side; most of these cells were large-size. Examples of retrogradely labeled cells of varying size and intensity in LN and SN of the rat presented in Fig. 2 are illustrated in Fig. 3.

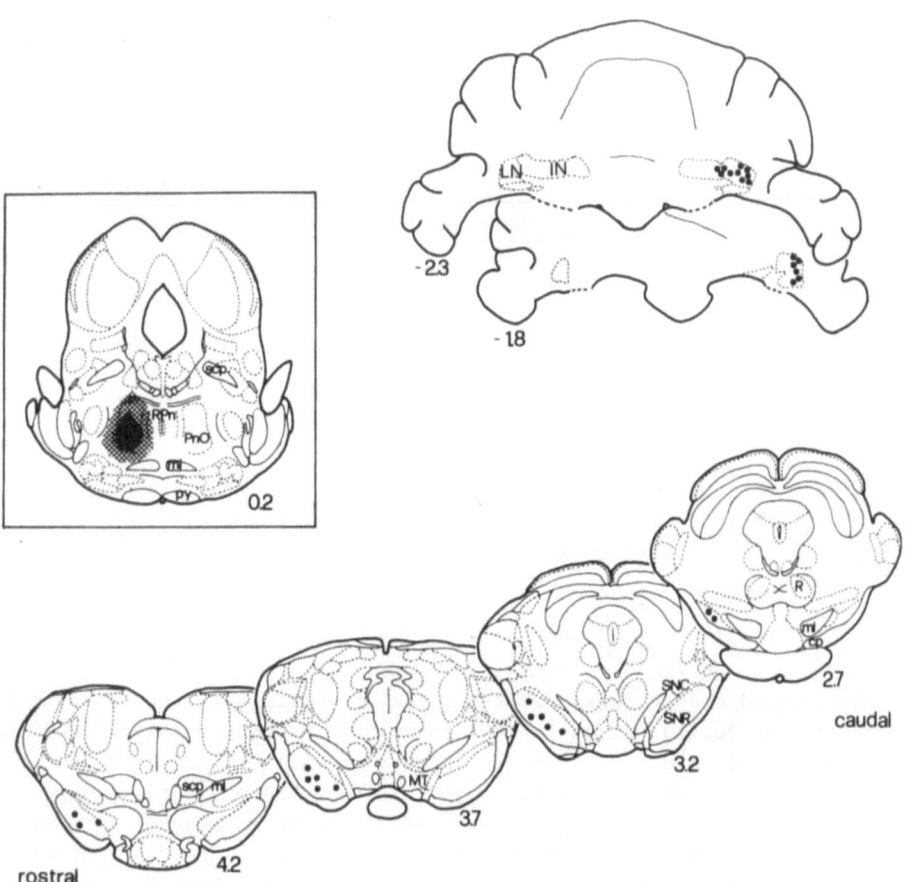

Fig.2. Camera lucida drawings showing the injection locus and distribution of retrogradely labeled cells (filled circles) within the cerebellar nuclei and the substantia nigra following a WGA-HRP injection involving the medial portion of PnO. Black area, injection site; surrounding etched area, zone of passive diffusion around the injection site. Abbreviations: see list.

Electrophysiological experiments

Unitary activity was recorded from 135 pontine reticular cells, 79 of which belonged to PnO and remaining 56 to PnC. About 80% (108/135) of these cells were influenced by SN and/or LN stimulation. In particular, 50/135 neurons (37%) were influenced only by SN stimulation, 9/135 neurons (6.7%) only by LN stimulation, whereas the remaining 49 neurons (36.3%) were influenced by both SN and LN activation (convergence cells); these latter cells were found with about the same incidence in both PnO (29/79; 36.7%) and PnC (20/56; 35.7%). Concerning the 99 reticular cells responsive to SN stimulation, 65/99 cells (65.7%) were influenced only by ipsilateral SN, 3/99 cells (3%) only by contralateral SN and the remaining 31/99 cells (31%) by SN of both sides.

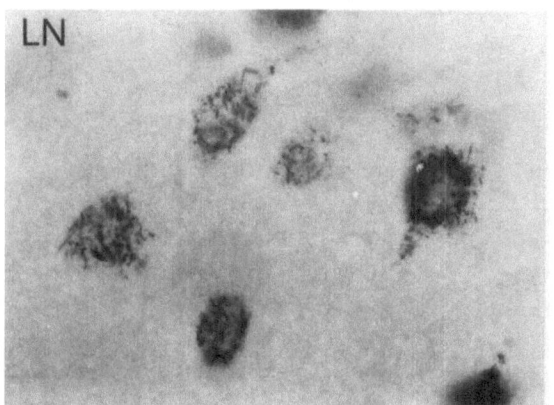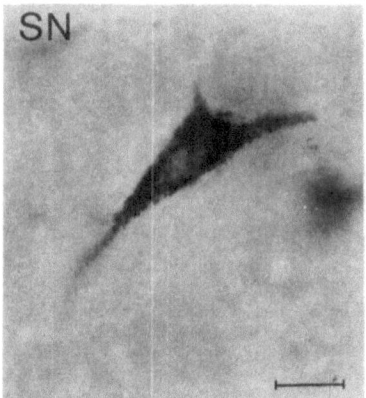

Fig.3. Retrogradely labeled cells of varying size and intensity in SN and LN of the animal presented in Fig. 2. Scale line: 20 um.

As regard the response patterns, the prevailing SN-induced response (70/99 cells; 70.7%) was an inhibition with an average duration of 62 ms (+/- 38.53 SD) and an average latency of 4.1 ms (+/- 2.87 SD), the remaining 29/99 cells (29.3%) being excited. The excitatory responses evoked by stimulation of ipsilateral SN displayed an average latency of 3.3 ms (+/- 2.55 SD) while excitations from contralateral SN appeared with an average latency of 9.7 ms (+/- 5.61 SD). Following LN stimulation 23/58 cells (39.7%) displayed a short latency (mean value: 2.8 ms +/- 1.92 SD) excitation and 35/58 neurons (60.3%) an inhibition with a average duration of 41.6 ms (+/- 21.78 SD) and an average latency of 4.4 ms (+/- 3.05 SD). Both SN- and LN-induced inhibitions were usually evoked with a single shock and followed stimulation frequencies up to 50 Hz.

Figure 4 illustrates the effects elicited by SN and LN stimulation on two pontine reticular neurons, the first located within PnO and the second within PnC. It can be seen by post-stimulus time histograms and cumulative frequency distributions that these neurons were inhibited upon stimulation of ipsi- as well as contralateral SN whereas activation of contralateral LN evoked in the first cell a short latency excitation and in the second one a short latency inhibition. In Fig. 5 are reported, on two sagittal sections of brainstem, in A the location within pontine RF of cells influenced only by SN, only by LN, or not influenced at all, and in B location and response patterns displayed by convergence cells. It can be detected that convergence cells are distributed throughout the pontine RF without any particular segregation.

In the present experiments, no significant differences in latency, duration as well as incidence of the different pattern responses were observed between rats anesthetized with urethane and those with ketamine hydrochloride anesthesia.

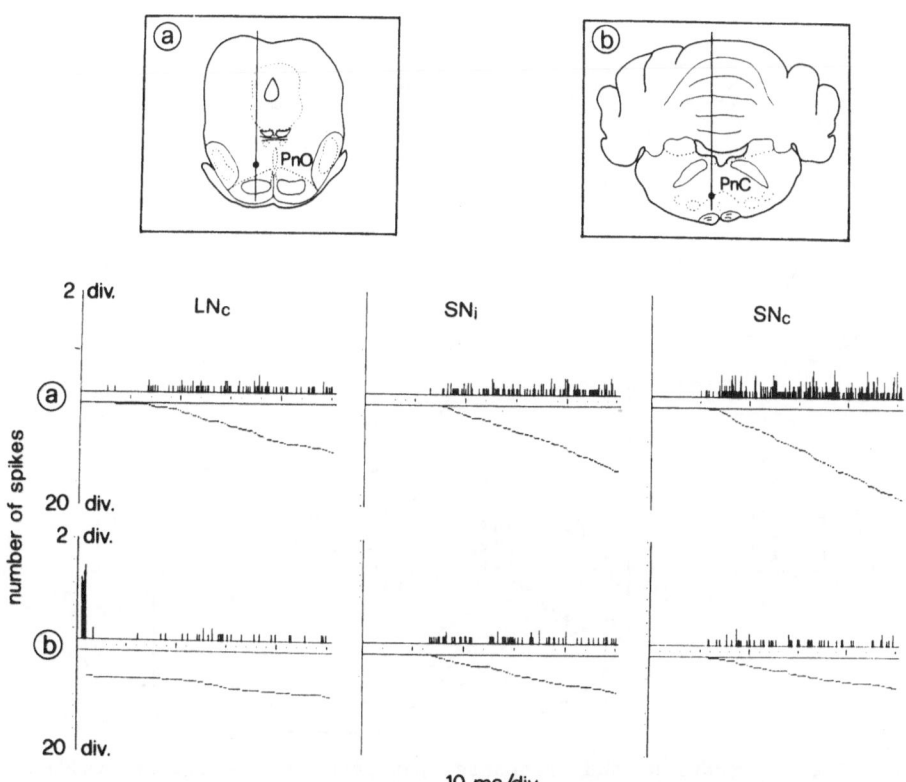

Fig.4. Post-stimulus time histograms and cumulative frequency distributions showing the effects elicited by stimulation of contralateral LN and of both ipsi- and contralateral SN on two pontine reticular neurons, the first (a) located within PnO and the second (b) within PnC (see inset diagrams).

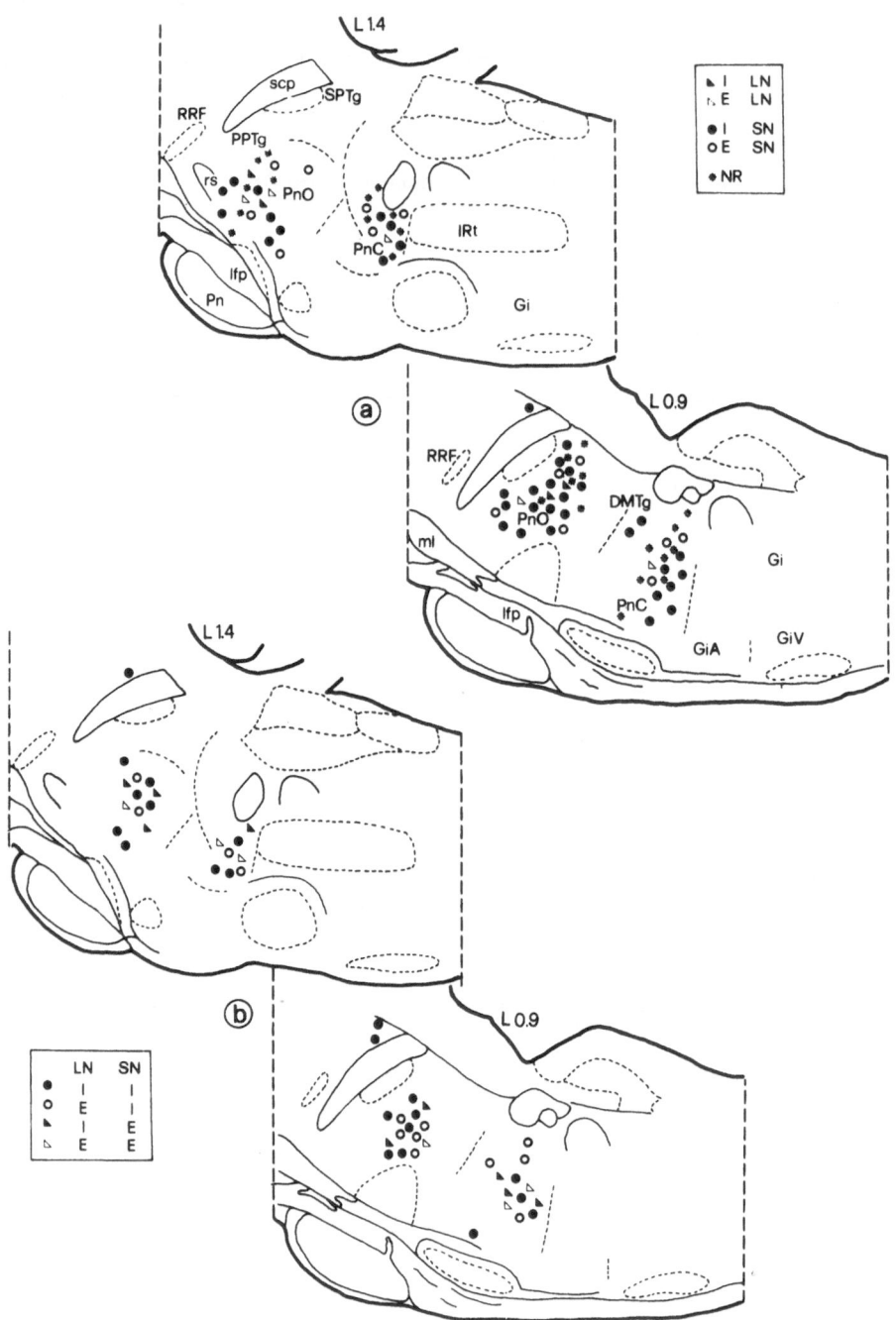

Fig.5. Location and response patterns of reticular cells influenced only by SN, only by LN or not influenced (a) as well as location and response patterns of reticular cells influenced by both SN and LN (b). Cells are plotted onto two sagittal sections of brain stem redrawn from Paxinos and Watson (1986). Abbreviations: see list.

CONCLUSIONS

The present study has demonstrated that 1) the same zones of PnO and PnC are reached by fibers coming from large-sized neurons of ipsilateral SN pars reticulata and various-sized cells of contralateral LN and (although with a lower density) IN, and 2) more than one third (36.3%) of neurons belonging these pontine nuclei integrates influences coming from SN and LN. It is noteworthy that in addition to these convergence cells, a similar percentage (37%) of pontine reticular cells was influenced exclusively by SN, but only a small fraction of them (6.7%) was controlled by LN alone.

It has been observed that in the majority of cases both LN-induced and SN-induced effects are inhibitory in nature and appear with latency values compatible with a monosynaptic linkage. Monosynaptic inhibitions induced by LN stimulation on reticular cells have not been previously observed. However, several small cell bodies containing gamma-aminobutirric acid (GABA) have been found scattered within the deep cerebellar nuclei, with the high incidence in the LN nucleus (Mugnaini and Oertel, 1985; Kumoi et al., 1988), and GABA is an inhibitory neurotransmitter in various brain region. It can be, therefore, suggested that LN-induced inhibitions are due to the activation of cerebellopontine GABAergic neurons; in fact, in the present study most of LN cells retrogradely labeled following a WGA-HRP injection in the pontine RF were small-sized neurons. An alternative possibility is that the LN-induced inhibitory responses are presynaptic inhibitions due to excitatory cerebellofugal axons establishing axo-axonic synapses on other excitatory reticulopetal axons (corticoreticular ?).

GABAergic inhibitions induced by SN stimulation on reticular neurons have been previously described (Perciavalle, 1987). In fact, it is well demonstrated that neurons of SN pars reticulata give off axons that appear to be GABAergic and exert on their target neurons an inhibitory effect (for review, cf. Dray, 1980). In the present experiments the SN-induced inhibitions display latencies that are similar to those elicited by SN stimulation on thalamocortical (Deniau et al., 1978) and reticulospinal neurons (Perciavalle, 1987). This is in good agreement with the observation, confirmed in the present study, that following HRP injection in the pontomedullary RF labeled neurons were found only in this SN portion (Grofova et al., 1978; Rinvik et al., 1976). Moreover, electric stimulation of pontine reticular area evokes antidromic action potentials only in neurons of SN pars reticulata, with latencies similar to those obtained for antidromic invasion of the same cells from the thalamus (Guyenet and Aghajanian, 1978). These findings allow us to conclude that inhibitions induced at short latency by SN stimulation are due to the activation of GABAergic fibers arising from the SN pars reticulata. Concerning the SN-induced excitatory responses, it must be emphasized that excitations from ipsilateral SN display latency values similar to those evoked on reticulospinal cells by stimulating the cerebral peduncle or the SN area pre-treated with kainic acid (Perciavalle, 1987). It is, therefore, reasonable that these excitations may be due to current spread which activates corticoreticular fibers running in close proximity to the SN.

Excitations from contralateral SN, instead, appear with latencies compatible with the involvement of intercalated structures. On the basis of these observations, of the 49 pontine reticular neurons influenced by both SN and LN stimulation, only those inhibited upon SN activation (34/49; 69.4%) should be considered convergence cells.

It is well known that cells belonging to pontine RF are influenced, besides SN and LN, from several areas of cerebral cortex (Fukushima et al., 1981, Magni and Willis, 1964; Pilyavsky and Gokin, 1978), as well as from fastigial nucleus (Batton et al., 1977; Voogt, 1964; Walberg et al., 1962), vestibular nerve (Fukushima et al., 1980), superior colliculus and neck muscle afferents (Fukushima et al., 1981). Lesions involving pontine RF interfere with voluntary movements and righting reflexes (Sirkin et al., 1980), and stimulation experiments have suggested that reticulospinal neurons belonging to this area act on motoneurons by way of indirect linkages including segmental or propriospinal pathways (Peterson, 1979). These data agree with the findings from experiments with SN and LN electrical activation. SN stimulation, although is not capable of eliciting movements, enhances reflex responses in both flexor and extensor motoneurons independently of corticofugal influences (Anden et al., 1971; Hassler and Wagner, 1975; York, 1972; 1973). LN stimulation, in rats with lesions eliminating corticospinal and rubrospinal pathways, activates only muscles of the scapular or the pelvic belt, presumably through the reticulospinal pathway (Cicirata et al., 1989). It is worth noting that basal ganglia and neocerebellum receive copy of the ongoing activity from the cerebral motor areas through collateral axons of pyramidal tract (Giuffrida et al., 1983; 1986).

The whole of these observations supports the functional view that the integration on the same pontine RF cells of information issued by cerebellum as well as basal ganglia, shown in the present study, could play an indirect role in the motor control, as the coordination between posture and movement (cf. Massion and Dufosse, 1988), rather than directly controlling limb movements.

ACKOWLEDGEMENTS

This work was partly supported by a grant of Ministero della Pubblica Istruzione.

ABBREVIATIONS USED IN FIGURES

7,	facial nerve
c,	contralateral
DpMe,	deep mesencephalic nucleus
E,	excitation
Gi,	gigantocellular reticular nucleus
GiA,	gigantocellular reticular nucleus, alpha
GiV,	gigantocellular reticular nucleus, ventral
I,	inhibition
i,	ipsilateral

Ifp,	longitudinal fasciculus pons
IN,	cerebellar interpositus nucleus
IRt,	intermediate reticular nucleus
LN,	cerebellar lateral nucleus
ml,	medial lemniscus
NR,	non-responsive
Pn,	pontine nuclei
PnC,	pontine reticular nucleus, caudal
PnO,	pontine reticular nucleus, oral
PnV,	pontine reticular nucleus, ventral
py,	pyramidal tract
PPTg,	pedunculopontine tegmental nucleus
R,	red nucleus
RMg,	raphe magnus nucleus
RPn,	raphe pontis nucleus
RtTg,	reticulotegmental nucleus of pons
RRF,	retrorubral field
scp,	superior cerebellar peduncle
SN,	substantia nigra
SNC,	pars compacta of substantia nigra
SNR,	pars reticulata of substantia nigra
xscp,	decussatio superior cerebellar peduncle

REFERENCES

Anden N.E., Larsson K., and Steg G., 1971, The influence of the nigro-neostriatal dopamine pathway on spinal motoneuron activity. Acta Physiol. Scand., 82: 268-271.

Bantli H., and Bloedel J.R., 1975, Monosynaptic activation of a direct reticulospinal pathway by the dentate nucleus. Pflugers Arch. ges. Physiol., 357: 237-242.

Batton R.R. III, Jayaraman A., Ruggiero D., and Carpenter M.B., 1977, Fastigial efferent projections in the monkey: an autoradiographic study. J. Comp. Neurol., 174: 281-306.

Cicirata F., Angaut P., Pantò M.R., and Serapide M.F., 1989, Neocerebellar control of motor activity: experimental analysis in the rat. Comparative aspects. Brain Res. Rev. 14: 117-141.

Deniau J.M., Lackner D., and Feger J. , 1978, Effect of substantia nigra stimulation on identified neurons in the VA-VL thalamic complex: comparison between intact and chronically decorticated cats. Brain Res. 145, 27-35.

Dray A., 1980, The physiology and pharmacology of mammalian basal ganglia. Prog. Neurobiol. 14, 221-335.

Fukushima K., Murakami S., Ohno M., and Kato M., 1980, Properties of mesencephalic reticulospinal neurons in the cat. Exp. Brain Res., 41: 75-78.

Fukushima K., Ohno M., and Kato M., 1981, Responses of cat mesencephalic reticulospinal neurons to stimulation of superior colliculus, pericruciate cortex, and neck muscle afferents. Exp. Brain Res., 44: 441-444.

Giuffrida R., Licata F., Li Volsi G., Perciavalle V., and Urbano A., 1983, Pyramidal input to the intracerebellar nuclei. <u>Neuroscience</u>, 9: 421-427.

Giuffrida R., Li Volsi G., Maugeri G., and Perciavalle V., 1986, Pyramidal input to the basal ganglia in the cat. <u>Exp. Brain Res.</u>, 61: 645-648.

Grofova I., Ottersen O.P., and Rinvik E., 1978, Mesencephalic and diencephalic afferents to the superior colliculus and periaqueductal gray substance demonstrated by retrograde axonal transport of horseradish peroxidase in the cat. <u>Brain Res.</u> 146, 205-220.

Guyenet P.G., and Aghajanian G.K., 1978, Antidromic identification of dopaminergic and other output neurons of the rat substantia nigra. <u>Brain Res.</u> 150, 69-84.

Hassler R., and Wagner A., 1975, Locomotor activity and speed of movements in relation to monoamine-acting drugs. <u>Int. J. Neurol.</u> 10, 80-97.

Horton J.C., Greenwood M.M., and Hubel D.H., 1979, Non-retinotopic arrangement of fibers in cat optic nerve. Nature, 282: 720-722.

Hubel D.H., 1960, Single unit activity in lateral geniculate body and optic tract of unrestrained cats. <u>J. Physiol. (Lond.)</u>, 150: 91-104.

Kumoi K., Saito N., Kuno T., and Tanaka C., 1988, Immunohistochemical localization of gamma-aminobutirric acid- and aspartate-containing neurons in the rat deep cerebellar nuclei, <u>Brain Res.</u>, 439: 302-310.

Li Volsi G., Pacitti C., Perciavalle V., Sapienza S., and Urbano A., 1982, Interpositus nucleus influences on pyramidal tract neurons in the cat. <u>Neuroscience</u>, 7: 1929-1936.

Magni F., and Willis W.D., 1964, Cortical control of brain stem reticular neurons. <u>Arch. Ital. Biol.</u>, 102: 418-433.

Massion J., and Dufosse M., 1988, Coordination between posture and movement: why and how? <u>NIPS</u>, 3: 88-93.

Mesulam M.M., 1978, Tetramethyl benzidine for horseradish peroxidase neurohistochemstry: a non-carcinogenic blue reaction-product with superior sensivity for visualizing neural afferents and efferents. <u>J. Histochem. Cytochem.</u>, 26: 106-117.

Mesulam M.M., 1982, Principles of horseradish peroxidase neurochemstry and their application for tracing neural pathways - axonal transport enzyme histochemstry and light microscopic analysis. In: "<u>Tracing Neural Connections with Horseradish Peroxidase</u>", M.M. Mesulam, ed., pp 1-152, Wiley, New York.

Mugnaini E., and Oertel W.H., 1985, An atlas of the distribution of GABAergic neurons and terminals in the rat CNS as revealed by GAD immunohistochemstry. In: "<u>Handbook of Chemical Anatomy, Vol. 4, GABA and Neuropeptides in the CNS</u>", A. Bjorklund and T. Hokfelt, eds, pp. 436-608, Elsevier, Amsterdam.

Paxinos G., and Watson C., 1986, "<u>The Rat Brain in Stereotaxic Coordinates</u>", 2nd edn, Academic Press, Sydney - New York - London.

Perciavalle V., 1987, Substantia nigra influences on the reticulospinal neurons: an electrophysiological and ionophoretic study in cats and rats. <u>Neuroscience</u>, 23: 243-251.

Perciavalle V., Berretta S., Li Volsi G., and Polizzi M.C., 1987, Basal ganglia influences on the cerebellum of the cat. <u>Arch. Ital. Biol.</u>, 125: 29-35.

Perciavalle V., Berretta S., and Raffaele R., 1989, Projections from the intracerebellar nuclei to the ventral midŭrain tegmentum in the rat. <u>Neuroscience</u>, 29: 109-119.

Peterson B.W., 1979, Reticulospinal projection to spinal motor nuclei. <u>Ann. Rev. Physiol.</u>, 41: 127-140.

Pilyavsky A.L., and Gokin A.P., 1978, Investigation of the cortico-reticulo-spinal connections in cats. <u>Neuroscience</u>, 3: 99-103.

Rinvik E., Grofova I., and Ottersen O.P., 1976, Demonstration of nigrotectal and nigroreticular projections in the cat by axonal transport of proteins. <u>Brain Res</u>. 112, 388-394.

Sirkin D.W., Schallert T., and Ottersen O.P., 1976, Demonstration of nigrotectal and nigroreticular projections in the cat by axonal transport of proteins. <u>Brain Res.</u>, 112: 435-457.

Ueki A., Uno M., Anderson M., and Yoshida M., 1977, Monosynaptic inhibition of thalamic neurons produced by stimulation of substantia nigra. <u>Experientia</u>, 33: 1480-1481.

Uno M., and Yoshida M., 1975, Monosynaptic inhibition of thalamic neurons produced by stimulation of pallidal nucleus in cats. <u>Brain</u> Res., 99: 377-380.

Voogd J., 1964, "<u>The Cerebellum of the Cat: Structure and Fibre Connexions</u>", Van Gorcum, Assen.

Walberg F., Pompeiano O., Westrum L.E., and Hauglie-Hanssen E., 1962, Fastigioreticular fibers in cat: an experimental study with silver methods. J. Comp. Neurol., 119: 187-199.

Wolf G., and Gollob H.F., 1980, Quantitative assessment of brain lesions. <u>Physiol. Behav.</u>, 24: 1195-1199.

York D.H., 1972, Alterations in monosynaptic reflex produced by stimulation of the substantia nigra. In: "<u>Corticothalamic Projections and Sensorimotor Activities</u>", T.L. Frigyesi and E. Rinvik, eds, pp. 445-447, Raven Press, New York.

York D.H., 1973, Motor responses induced by stimulation of the substantia nigra.<u>Exp. Neurol.</u>, 41: 323-330.

INVOLVEMENT OF THE NUCLEUS TEGMENTI PEDUNCULOPONTINUS IN

THE DESCENDING PATHWAYS OF THE BASAL GANGLIA IN THE RAT

Irena Grofova, Bryan M. Spann, and Kathy Bruce

Department of Anatomy
Michigan State University
E. Lansing, MI 48823, U.S.A.

INTRODUCTION

Modern concept regarding the role of the basal ganglia in control of movement emphasizes the importance of the pallido/nigro-thalamo-cortical pathway. In addition, several lines of evidence have indicated that some aspects of motor behavior may be controlled through direct pathways linking the basal ganglia to brainstem structures such as the superior colliculus[1] and nucleus tegmenti pedunculopontinus[2].

The nucleus tegmenti pedunculopontinus (PPN) contains a mixture of cholinergic and non-cholinergic neurons[3] that surround the superior cerebellar peduncle in the pontomesencephalic junction. It consists of two subdivisions, a smaller pars compacta (PPNc) and a larger pars dissipata (PPNd) that comprises the entire rostral part of the PPN in addition to the medial two-thirds of the caudal PPN[4]. This nucleus is of particular interest since its position corresponds to the core of the mesencephalic locomotor region in non-primate species[5]. Furthermore, it has been recently suggested that the PPN in the rat mediates striatally induced inhibition of muscle activity of some neck and shoulder muscles[6]. Since direct PPN projections to the spinal cord are modest[4], it is assumed that the motor effects are mediated through an indirect pathway involving the pontomedullary reticular formation.

The present report summarizes our recent studies on the anatomical organization of pathways linking the pars reticulata of the substantia nigra (SNR) to the PPN and the PPN to the brainstem reticular formation. The efferent projections of the SNR and the descending connections of the PPN were investigated in two separate series of experiments utilizing anterograde transport of Phaseolus vulgaris - leucoagglutinin (PHA-L). In the third experimental series, a double-labeling technique combining anterograde transport of PHA-L and retrograde transport of Cholera toxin subunit B (CTB) was employed in order to ascertain whether the nigropedunculopontine fibers terminate around the PPN-reticular neurons.

MATERIALS AND METHODS

In the first and second series of experiment, 2.5% solution of PHA-L (Vector) was iontophoretically injected into the SNR or into the PPN. The stereotaxic coordinates were derived from the Paxinos and Watson's atlas of the rat brain[7]. Following 7-14 days survival, deeply anesthetized animals were perfused transcardially with a saline solution followed by a fixative consisting of 4% paraformaldehyde and 0.2% glutaraldehyde in sodium phosphate buffer. The brains were divided into a rostral block including the forebrain, midbrain and

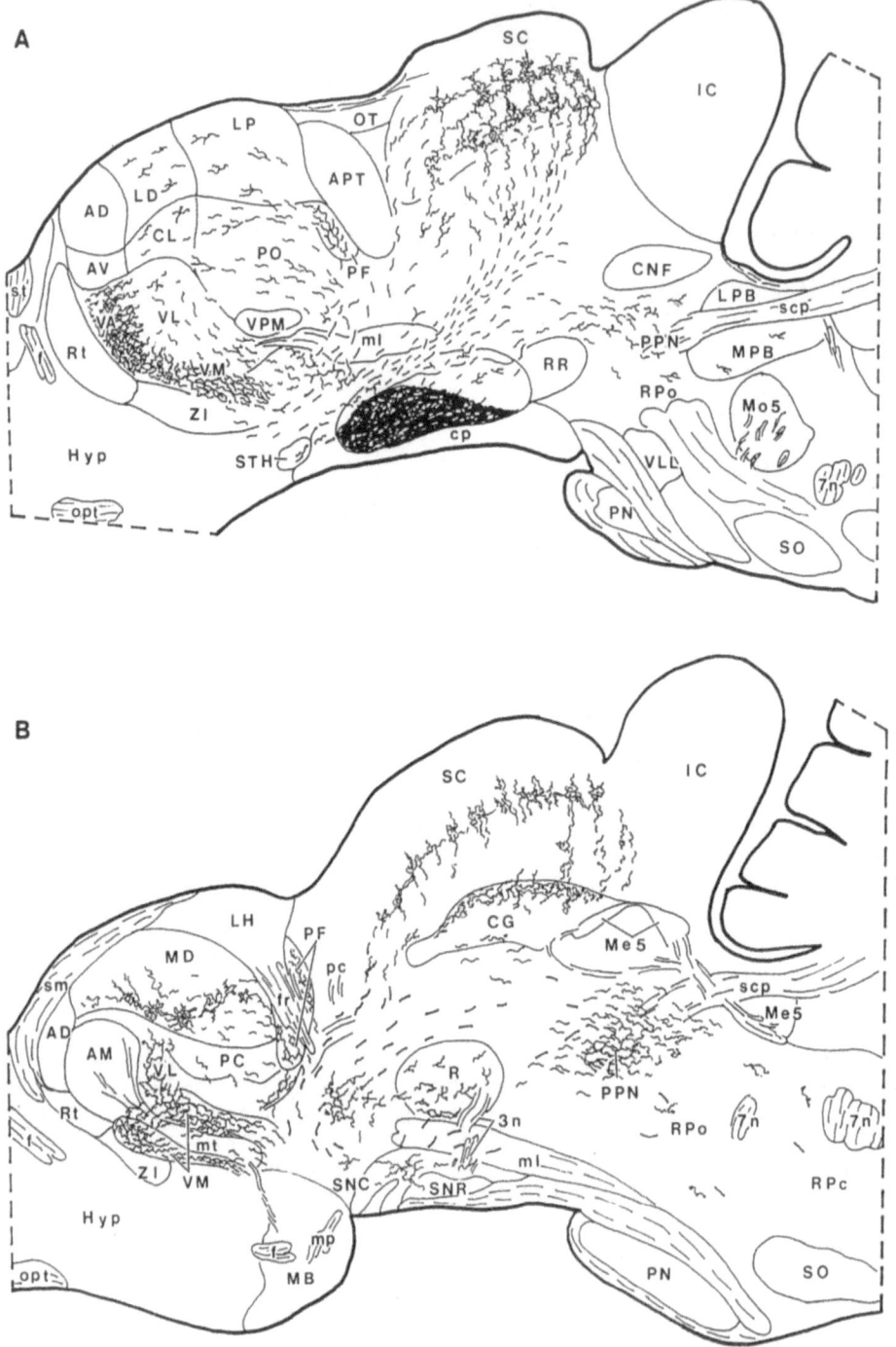

Fig. 1. Line drawings of lateral (A) and medial (B) sagittal sections through the diencephalon, midbrain and pons illustrating the course and distribution of PHA-L labeled SNR efferents; IC, inferior colliculus; LH, lateral habenular nu; MB, mammillary body. For the remaining abbreviations see Ref. 7.

rostral pons, and a caudal block containing the caudal pons and medulla. Serial 30 μm thick sections were cut on a vibratome in a sagittal (rostral block) or coronal (caudal block) plane and processed for PHA-L immunohistochemistry using a modified protocol by Gerfen and Sawchenko[8]. Alternating sections were postfixed in OsO_4, dehydrated in graded acetone and section-embedded in plastic for electron microscopic analyses. A complete EM protocol as well as detailed description of the ultrastructural analysis of the nigro-PPN projection will be published elsewhere[9].

In the double-labeling experiments, the anterograde tracer PHA-L was iontophoretically injected to the SNR, while pressure injections of the retrograde tracer CTB were made bilaterally in the ventral subdivisions of the gigantocellular nucleus. Fixation, blocking of the brains and cutting were the same as for the single labeling experiments. The sections were further processed for PHA-L immunohistochemistry using DAB as a chromogen followed by CTB immunohistochemistry with 0.04% nickel chloride added to the DAB solution in order to yield blue-black reaction product contrasting with brown staining of PHA-L labeled fibers.

RESULTS

Efferent projections of the SNR

The distribution of PHA-L labeled fibers and terminal plexuses following an injection involving almost the entire mediolateral and dorsoventral extents of the SNR are depicted in Fig. 1. In addition to the well-known nigral projections to the ventromedial (VM), mediodorsal (MD) and parafascicular (PF) thalamic nuclei, and to the superior colliculus (SC), pontomesencephalic reticular formation and central gray (CG), these experiments disclosed other hitherto unknown targets of SNR efferents. In the thalamus, the densest termination of nigral fibers was seen in the rostromedial cap of the ventral lateral nucleus (VL) in all cases with the injection involving the ventral portion of the SNR. This thalamic region has been tentatively labeled ventral anterior (VA) nucleus in Fig. 1 A. Furthermore, dense patches of labeled terminal plexuses were consistently observed in the dorsal part of the zona incerta (ZI), around the cells in the Forel's field, in the red nucleus (R), and in the nuclei associated with the medial longitudinal fasciculus, particularly the rostral interstitial nucleus (not illustrated). In the PPN, the nigral fibers terminated preferentially in the medial two-thirds of the PPNd (Figs. 1 B and 2), while the smaller PPNc received only occasional nigral fibers exhibiting preterminal and terminal varicosities. A remarkable feature of the distribution of nigral fibers in the PPN was a patchy organization of the terminal plexuses. Thus within the same region of the PPN, the nigral fibers were concentrated around some cells or groups of cells while avoiding the others (Fig. 2). Electron microscopic analysis of the material revealed that the majority of the PHA-L labeled nigral terminals established synapses with medium-sized dendrites although a few synapsed on large primary dendrites (Fig. 3) and cell soma.

Descending PPN-reticular projections

The PHA-L injections into the PPN were intended to involve mainly or exclusively the medial two-thirds of the PPNd (Fig. 5 A) which, as determined in the first series of experiments, received the bulk of nigral afferents. In addition to labeled fibers streaming rostrally from the injection site, a substantial number of PPN efferents were seen to descend through the pontine reticular nuclei to the ventromedial portions of the nucleus gigantocellularis, GiA and GiV (Figs. 4 B, C and 5 C). Although many fibers appeared to cross the midline in the caudal pons and rostral medulla, the terminal arborizations within the ventromedial portions of the gigantocellular nucleus were somewhat richer ipsilaterally (Figs. 4 B, C and 5 C). During the course through the pontine reticular nuclei, numerous varicose branches of the descending PPN fibers were seen to terminate around large reticular neurons, particularly in the nucleus reticularis pontis caudalis (RPc; Figs. 4 A and 5 B). A relatively small number of labeled fibers continued their course to the lower medulla and to the spinal cord. In ultrathin sections, the GiA contained a number of labeled thin myelinated fibers (≤1 μm in diameter) as well as darkly labeled boutons that were found in contacts with large dendrites (Fig. 5 D).

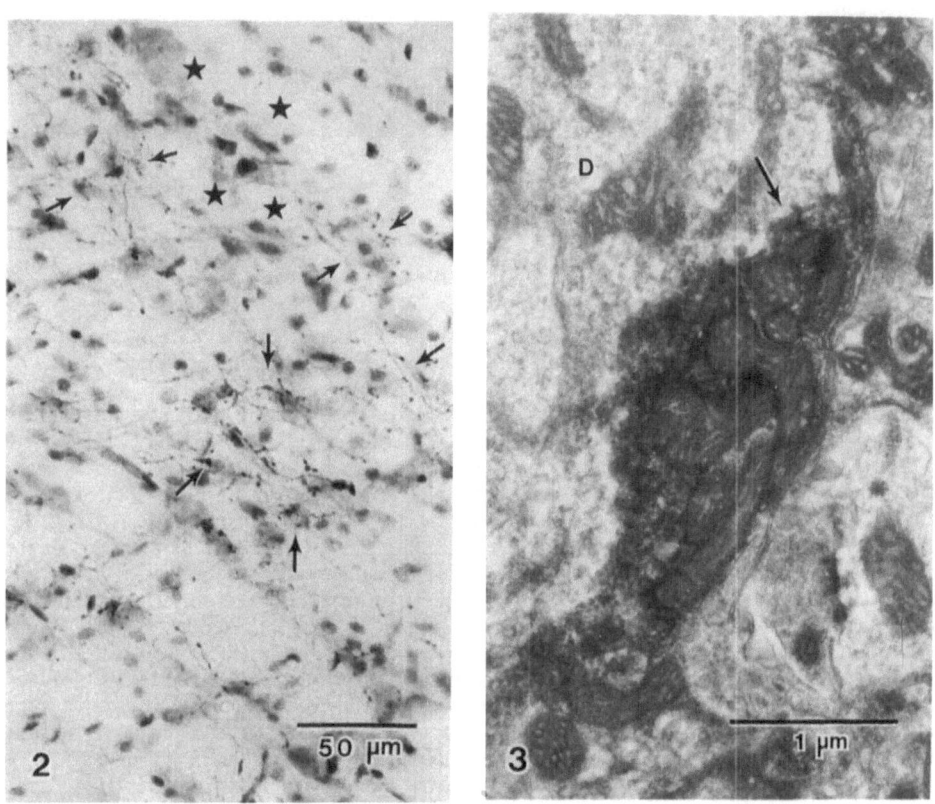

Fig. 2. Bright field photomicrograph showing the terminal plexus of PHA-L labeled SNR fibers in the medial PPN in the case illustrated in Fig. 1. Arrows point to dense patches of label, stars indicate neighbouring regions that seem to be avoided by nigral fibers.

Fig. 3. Electron micrograph of a PHA-L labeled SNR terminal synapsing (arrow) on a large proximal dendrite (D) in the medial portion of the PPN.

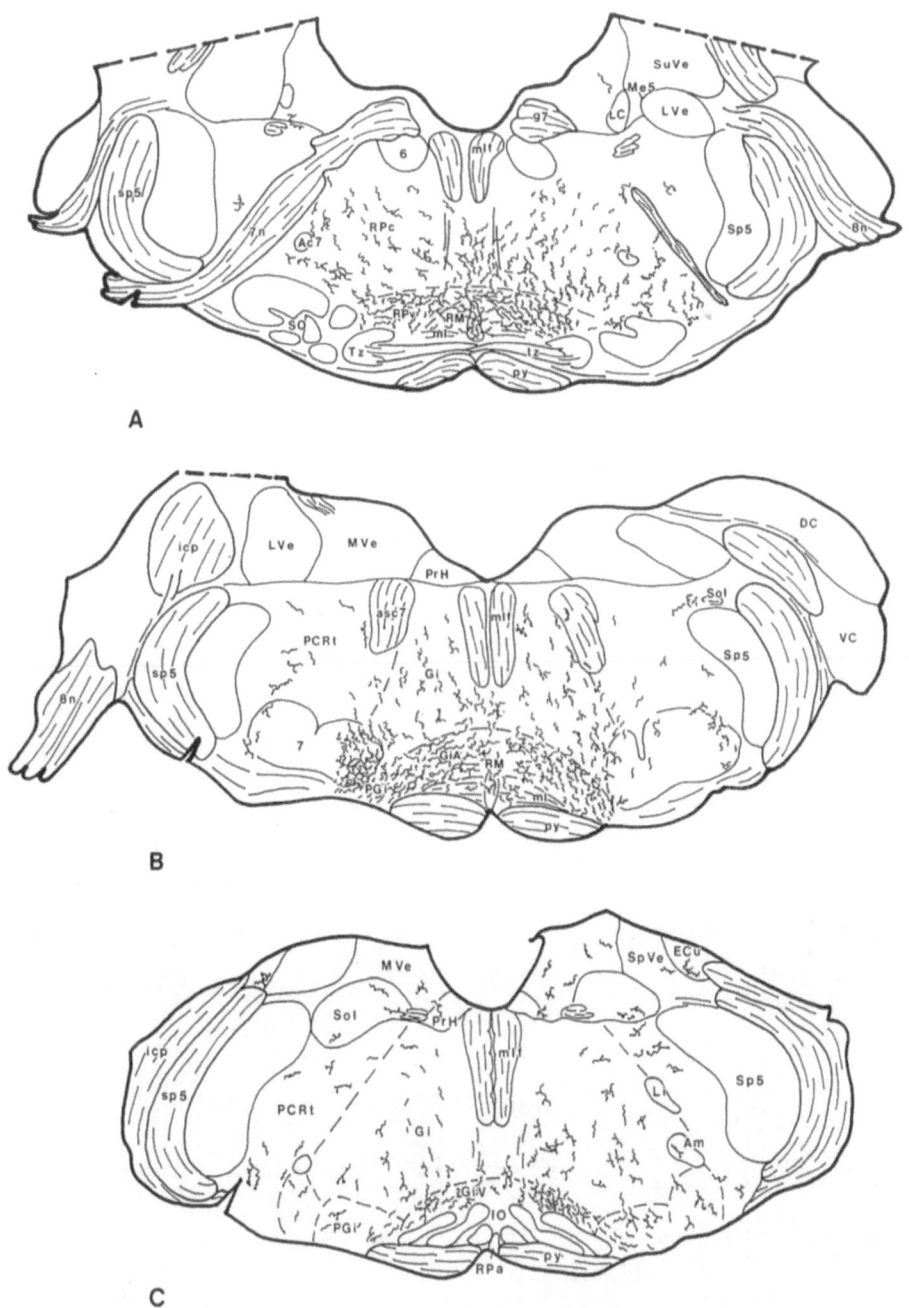

Fig. 4. Line drawings of transverse sections through the brainstem are arranged in a rostro-caudal sequence (A-C) and show the distribution of labeled fibers following a PHA-L injection involving the medial half of the PPN. The right half of the drawings is ipsilateral to the injection site; Am, ambiguus nu; Ac7, accessory facial nu; IO, inferior olivary nu; PGi, paragigantocellular reticular nu; RM, raphe magnus nu; RPc, reticularis pontis caudalis nu; RMv, reticularis pontis ventralis nu; SO, superior olivary nu. For remaining abbreviations see Ref. 7.

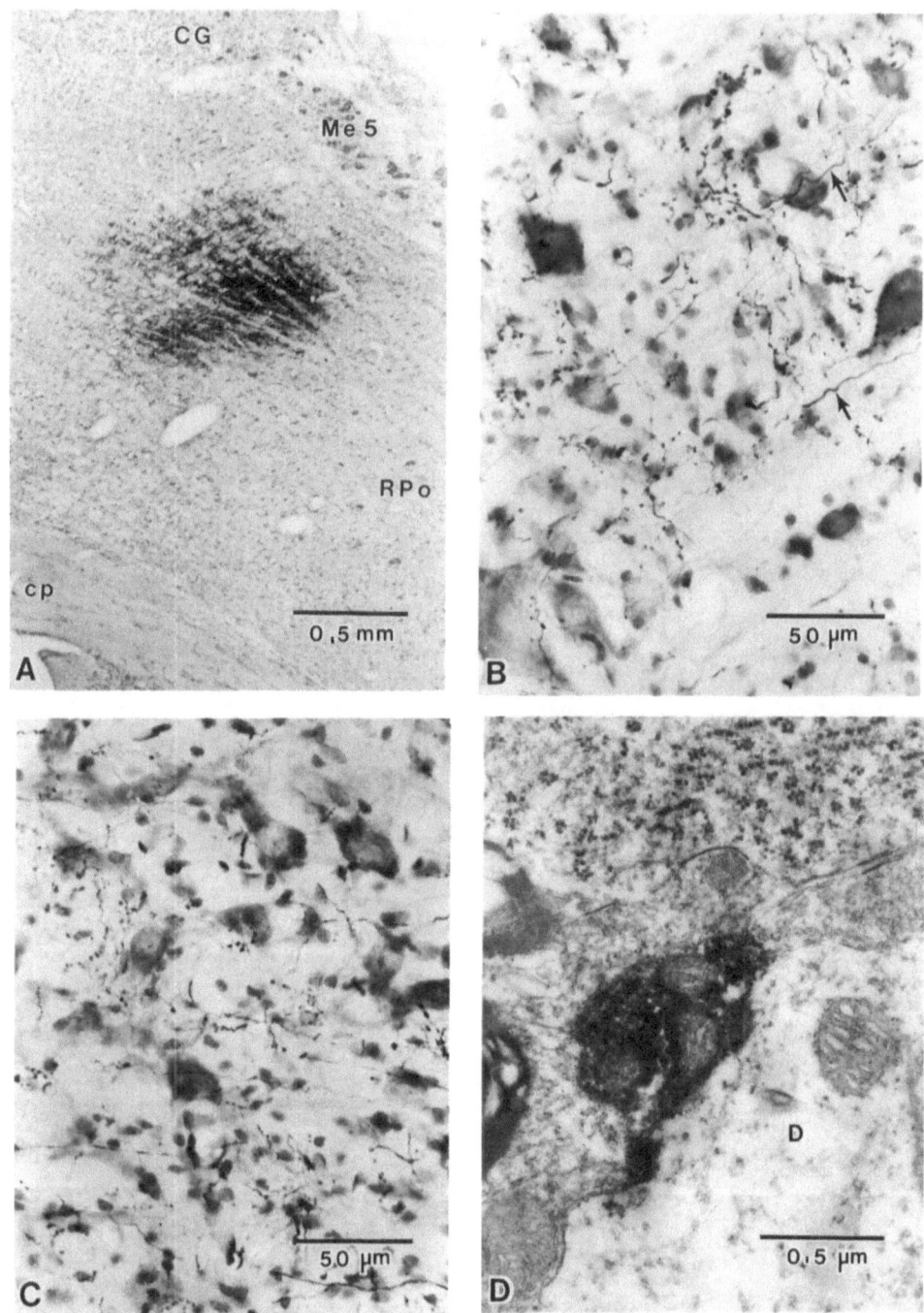

Fig. 5. A-D Photomicrographs illustrating the results of the second series of experiments. A. PHA-L injection in the medial portion of the PPN. B & C. Labeled fibers exhibiting terminal varicosities in the RPc and GiA respectively. Arrows in B point to thick fibers descending through the pons. D. Electron micrograph of a PHA-L labeled terminal in the GiA apposed to a large dendrite (D).

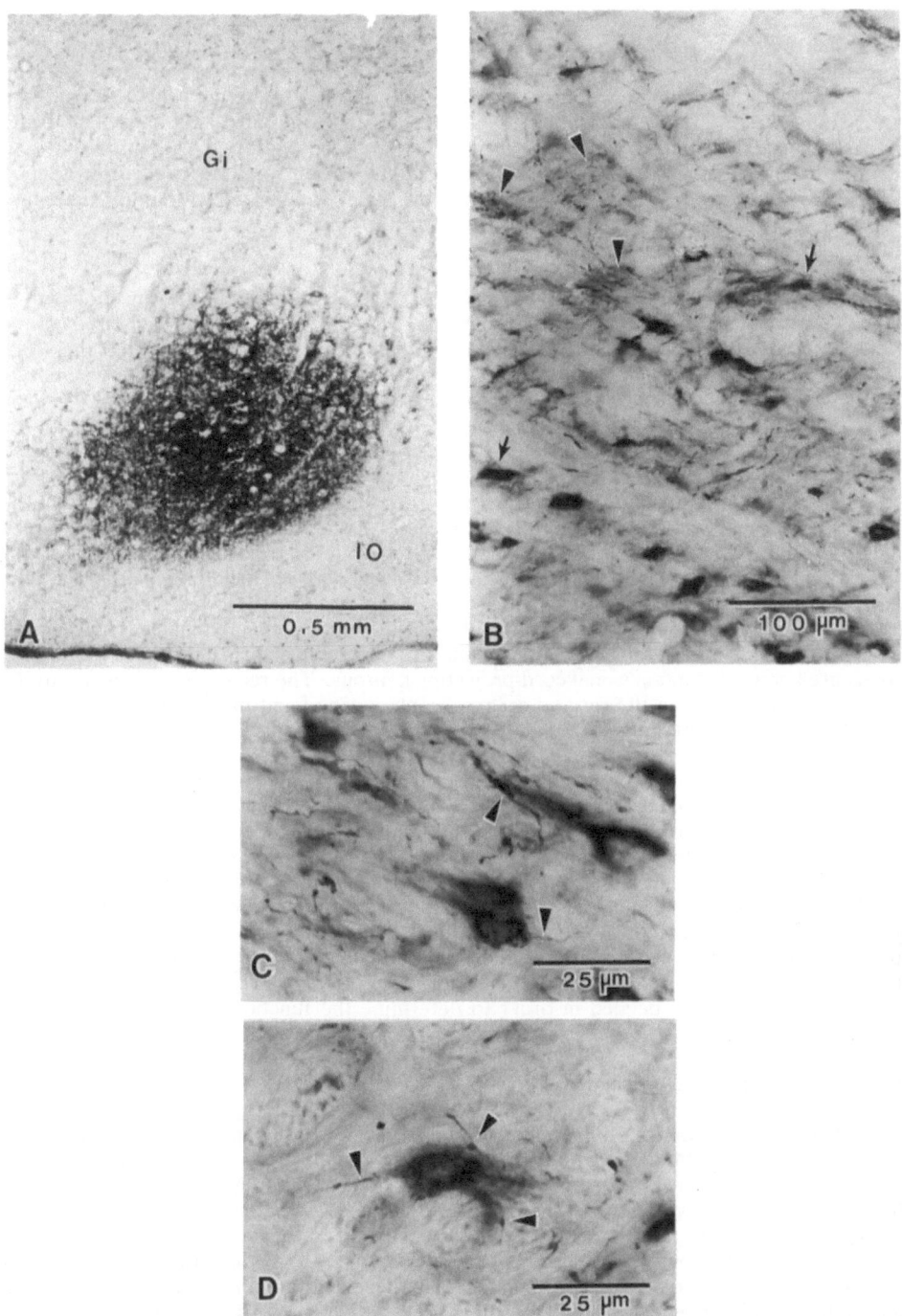

Fig. 6 A-D Photomicrographs illustrating the results of the double-labeling expeiments. A. CTB injection in the GiV. B. PHA-L labeled fibers in the PPN are present in the proximity of both unlabeled (arrowheads) as well as labeled (arrows) cells. C & D. PHA-L labeled nigral fibers with terminal varicosities (arrowheads) applied to the dendrites and somata of the retrogradely labeled PPN-reticular neurons.

DISCUSSION AND CONCLUSIONS

The present report represents a part of our ongoing studies aimed at the demonstration of anatomical pathways linking the basal ganglia to the brainstem structures that are involved in the control of motor functions. Examination of nigral efferents labeled with the anterograde tracer PHA-L confirmed the existence of the nigrothalamic, nigrotectal and nigroreticular fibers and provided complementary observations on their distribution and organization of their terminal plexuses. Furthermore, these experiments revealed several hereto unknown brainstem targets of nigral output that are associated with motor functions.

Double-labeling experiments

The results of these experiments are illustrated in Fig. 6. The PHA-L injections in the SNR were similar in size to that shown in Fig. 1 A, while bilateral deposits of the retrograde tracer CTB always involved portions of both GiA and GiV (Fig. 6 A). Within the PPN, retrogradely labeled neurons exhibited black granules (Fig. 6 C and D) or homogeneously distributed black reaction product extending into the proximal dendrites (Fig. 6 B). The labeled cells were intermingled with unlabeled neurons in all portions of the nucleus. Dense terminal arborizations of the PHA-L labeled nigral fibers were seen around both, unlabeled as well as labeled PPN cells (Figs. 6 B, C and D).

These include the red nucleus, nucleus of the Forel's field, rostral interstitial nucleus of the medial longitudinal fasciculus and interstitial nucleus of Cajal. Since we have not observed labeling of the corticospinal and corticopontine fibers following PHA-L injections involving the cerebral peduncle, or labeling of the cerebellorubral and cerebellothalmic fibers in case of PHA-L injections in the PPN which inevitably included the superior cerebellar peduncle, we feel confident that the labeled terminal ramifications in the red nucleus and nuclei associated with the medial longitudinal fasciculus are of nigral origin. All the afore-mentioned targets are known to contain spinal cord-projecting neurons. The red nucleus gives rise to the rubrospinal tract which in the rat and cat represents a prominent motor pathway extending throughout the length of the spinal cord. Physiological experiments in the cat suggest that the red nucleus and the motor cortex complement each other in the control of movement of individual muscles[10]. The rostral interstitial nucleus of the medial longitudinal fasciculus including cells in the caudal field of Forel as well as the interstitial nucleus of Cajal in the cat contain neurons that participate in the control of both movement and position of the head and eyes[11]. In the light of these findings it is tempting to speculate that via nigral projections to the afore-mentioned brainstem nuclei, the basal ganglia may indeed control movement (particularly head and eye movements) through a direct pathway, circumventing the cerebral cortex. However, based on the present observation such speculation would be premature since both the red nucleus as well as the nuclei associated with the medial longitudinal fasciculus also contain neurons projecting to the inferior olive[11]. Further refined anatomical and physiological studies are needed in order to determine the functional significance of these nigral connections.

Nigro-pedunculopontine-reticular pathway

The possibility that the PPN may participate in the extrapyramidal motor control was raised by Nauta and Mehler[12] who were the first to demonstrate a prominent projection linking the primate internal segment of the globus pallidus to the PPN. However, subsequent studies on the connectivity and chemistry of the PPN have accumulated data indicating that the connections and possible functions of this nucleus are much more complex. Numerous studies have described prominent ascending projections of the PPN that distribute to the basal ganglia nuclei, thalamus, hypothalamus and basal forebrain (for review see Ref. 4). Consequently, it has been suggested that the PPN may be involved in the reticular activation system, in the modulation of rapid eye movement sleep, and feedback loops between the basal ganglia nuclei (for review see Ref. 3). However, descending connections to both the medullary reticular formation and spinal cord have also been documented. It is likely that these projections may be involved in some aspects of motor behavior, since the PPN had been identified as the central component of the mesencephalic locomotor region. Since the PPN neurons appear to lack any substantial axonal collateralization[4], and are randomly intermingled within the nucleus, it remained unclear which of them actually receive the nigral input.

The present observation that the terminal ramifications of nigral fibers within the PPN are prominent around some groups of cells while avoiding the others suggest that the basal ganglia input may be directed towards specific subpopulations of PPN neurons. Our double-labeling experiments support recent findings in the cat[13] demonstrating a monosynaptic nigral input to the pedunculopontine neurons projecting to the medullary reticular formation. However, our observations also suggest that the descending PPN neurons may not be the exclusive targets of the nigral input.

ACKNOWLEDGEMENT

This work was supported by NIH grant NS25744

REFERENCES

1. M. N. Williams, and R. L. M. Faulls, The nigrotectal projection and tectospinal neurons in the rat. A light and electron microscopic study demonstrating monosynaptic nigral input to identified tectospinal neurons, Neuroscience 25:533-562 (1988).
2. A. Jackson and A. R. Crossman, Nucleus tegmenti pedunculopontinus: Efferent connections with special references to the basal ganglia, studied in the rat by anterograde and retrograde transport of horseradish peroxidase, Neuroscience 10:725-765 (1983).
3. M. -M. Mesulam, C. Geula, M. A. Bothwell and L. B. Hersch, Human Reticular Formation: Cholinergic neurons at the pedunculopontine and laterodorsal tegmental nuclei and some cytochemical comparisons to forebrain cholinergic neurons, J. Comp. Neurol. 281:611-633 (1989).
4. B. M. Spann and I. Grofova, Origin of ascending and spinal pathways from the nucleus tegmenti pedunculopontinus in the rat, J. Comp. Neurol. 283:13-27 (1989).
5. E. Garcia-Rill, The basal ganglia and locomotor regions, Brain Res. Rev. 11:47-63 (1986).
6. M. D. Kelland and D. Asdourian, Pedunculopontine tegmental nucleus-induced inhibition of muscle activity in the rat, Behavioral Brain Res. 34:213-234 (1989).
7. G. Paxinos and C. Watson, "The Rat Brain in Stereotaxic Coordinates," Academic Press, New York (1986).
8. C. R. Gerfen and P. E. Sawchenko, An anterograde neuroanatomical tracing method that shows the detailed morphology of neurons, their axons and terminals: Immunohistochemical localization of an axonally transported plant lectin, Phaseolus vulgaris Leucoagglutinin (PHA-L), Brain Res. 290:219-238 (1984).
9. B. M. Spann and I. Grofova, A light and electron microscopic study on the substantia nigra afferents of the nucleus tegmenti pedunculopontinus in the rat, Submitted to J. Comp. Neurol. (1990).
10. C. Ghez, Input-output relations of the red nucleus in the cat, Brain Res. 98:93-108 (1975).
11. G. Holstege and R. J. Cowie, Projections from the rostral mesencephalic reticular formation to the spinal cord, Exp. Brain Res. 75:265-279 (1989).
12. H. J. W. Nauta and W. R. Mehler, Projections of the lentiform nucleus in the monkey, Brain Res. 1:3-42 (1966).
13. Y. Nakamura, H. Tokuno, T. Moriizumi, Y. Kitao and M. Kudo, Monosynaptic nigral inputs to the pedunculopontine tegmental nucleus neurons which send their axons to the medial reticular formation in the medulla oblongata. An electron microscopic study in the rat, Neuroscience Let. 103:145-150 (1989).

SYNAPTIC ORGANIZATION OF NUCLEUS ACCUMBENS (VENTRAL STRIATUM)

Gloria E. Meredith and Floris G. Wouterlood

Department of Anatomy and Embryology
Faculty of Medicine, Vrije Universiteit
P.O. Box 7161, 1007 MC Amsterdam, THE NETHERLANDS

INTRODUCTION

Experimental studies have produced many data over the past few years which implicate nucleus accumbens in the motivational and attentional processes that are important for goal-directed behaviors (Willner and Scheel-Krüger, 1990). Moreover, disorders, such as schizophrenia, which involve the disruption of these behaviors, are thought to be specifically linked to this nucleus (Matthysse, 1981). There is experimental evidence that supports a role for dopamine (DA) in these behaviors in nucleus accumbens but the importance of other transmitter systems such as those containing gamma aminobutyric acid (GABA) and acetylcholine, has also become apparent (Costa et al., 1978; Jones et al., 1981; Matthysse, 1981; Willner and Scheel-Krüger, 1990). However, our understanding of the differential role of each of these neurochemicals is limited, especially with respect to their interactions. A major reason for this limitation is the absence of relevant data on the synaptic relationships in accumbens circuitry. We know that nucleus accumbens receives extensive projections from limbic-related parts of the cortex, the amygdala, the midline thalamic nuclei, and the ventral tegmental area (VTA) and projects to major targets in the ventral pallidum and the midbrain i.e., the substantia nigra, VTA, and retrorubral field (see Groenewegen et al., 1990, for review) but we have yet to delineate the synaptic relationships underlying this flow of information into and through the nucleus.

The primary objective of this report is to review briefly the morphology and connectivity of accumbens neurons, with special reference to the cholinergic population, in order to shed some light on the anatomical connections important to accumbens circuitry. The data are based entirely on the rat ventral striatum and are derived primarily from recent studies in our laboratory (Groenewegen et al., 1989; Meredith et al., 1989; Meredith and Wouterlood, 1990; Meredith et al., 1990).

MORPHOLOGICALLY IDENTIFIED NEURONS

According to Golgi studies conducted by Danner and Pfister (1981), nucleus accumbens contains three types of spiny and two types of aspiny neurons. Of the spiny neurons, the most commonly encountered cells are

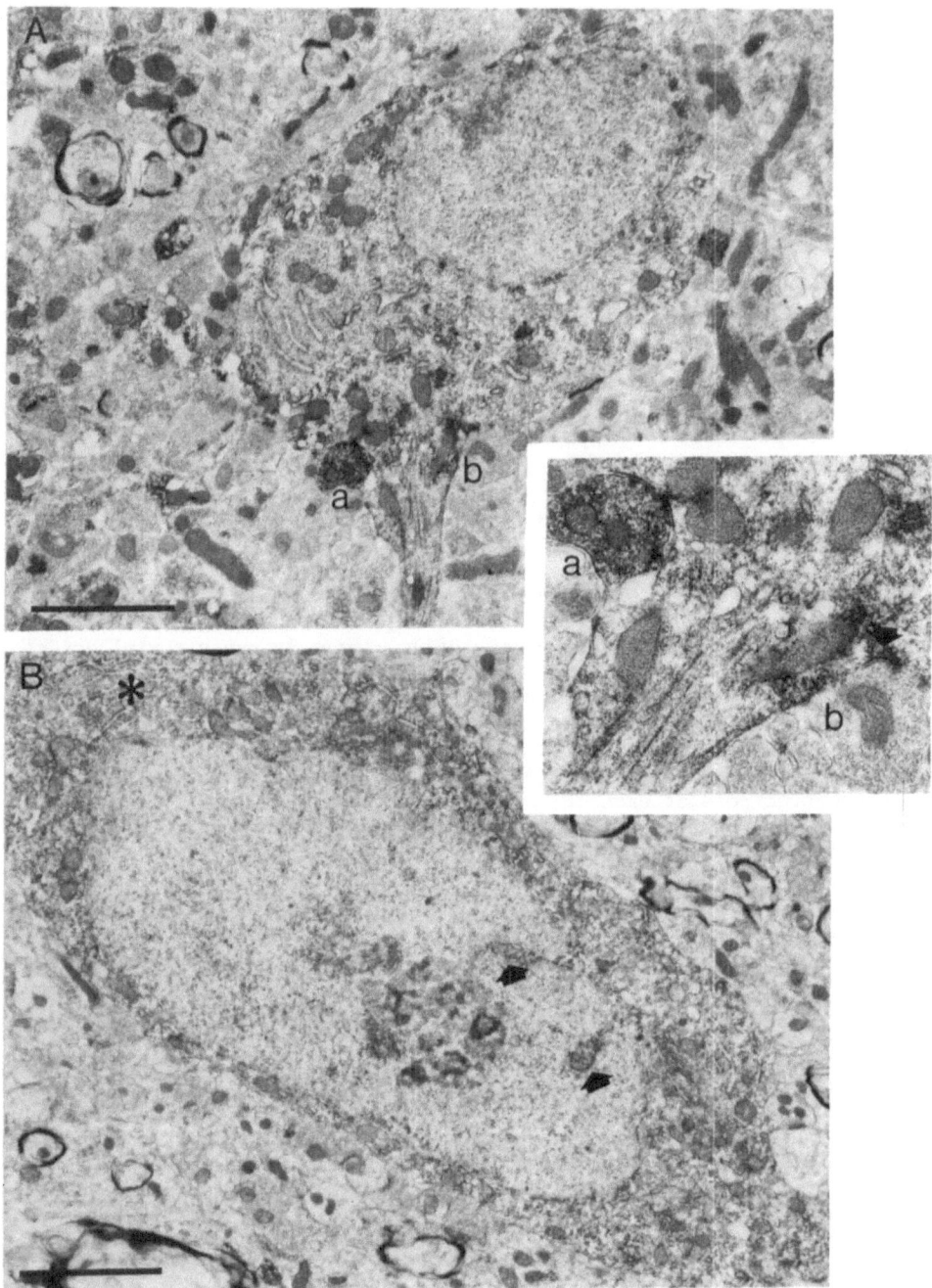

Figure 1

medium-sized, i.e., somata of 10-20 µm (Chang and Kitai, 1985) and have round or slightly elongated somata. They have four to six primary dendrites, which are covered in spines except for small spine-free proximal segments, and which spread out in the surrounding neuropil (Danner and Pfister, 1981). The axon arises from the soma or from an initial spine-free part of the dendrite and arborizes extensively before leaving the nucleus (Chang and Kitai, 1985). A second type of neuron is similar to the first except that the soma is polygonal and the four to five dendrites are less densely spined (Danner and Pfister, 1981). The third type of cell is bi- or tri- polar with very sparsely spined dendrites and an axon that branches frequently. These cells are often found in close association with medium-sized spiny neurons (type 1 above, Danner and Pfister, 1981).

The first type of aspiny cell is very large ("macro") and polymorphous with three smooth, sometimes varicose, dendrites that extend several hundred microns into the neuropil; the axon courses ventrocaudally (Danner and Pfister, 1981). The second type of neuron is much smaller with delicate, aspiny dendrites which spread out in the surrounding neuropil in a spidery fashion (Danner and Pfister, 1981).

NEUROCHEMICALLY IDENTIFIED NEURONS AND THEIR TERMINALS

Gamma Aminobutyric Acid- and Glutamate Decarboxylase-immunoreactive Neurons

The vast majority of neurons in nucleus accumbens are immunoreactive for GABA or its synthetic enzyme, glutamate decarboxylase (GAD). These cells, which are presumably GABAergic, are small to medium-sized and resemble spiny projection neurons (Pickel et al., 1988b; Meredith et al., 1990).

As expected, both the GABA- and the GAD-positive cells are similar in form and, when examined with the electron microscope, two neuronal types can be distinguished. These are small to medium-sized and their perikarya are round to oval. The more commonly encountered type (Fig. 1A) contains a central nucleus, moderate amounts of cytoplasm, and an occasional spine that protrudes directly from the perikaryal membrane. The nucleus, which is round to oval and unindented, contains one or occasionally two nucleoli, one of which is eccentrically located. Dendrites are initially free of spines but have spines more distally. The less common cell type contains a large, deeply indented nucleus with a single central nucleolus and a shallow rim of cytoplasm. Pickel et al. (1988b) noted a bar-shaped inclusion in the nucleus. The synaptic input to both cell types is sparse and the synaptic specializations of these terminals are typically symmetric. Occasionally, one or more terminals can be seen in contact with the axon hillock (Fig. 1A), where they presumably exert a powerful influence. The GAD-immunoreactive terminal seen in Fig. 1A lies opposite an ending with an asymmetric specialization which may be derived from the cortex or thalamus (see below and Hassler and Chung, 1976; Meredith and Wouterlood, 1990).

Numerous immunolabelled spines (Fig. 2A) can be identified both in colchicine-treated and untreated animals using anti-GAD antibodies. These

Fig. 1. Photomicrographs of (A) the more common type of GAD-immunoreactive cell. Note the axon arising from the cell body, and the GAD-immunoreactive ending (a) and the unlabelled terminal (b) in contact with the axon hillock (See inset for higher power view). (B) typical ChAT-immunoreactive perikaryon. Note the infolded nuclear envelope (arrows) and the stacks of rough endoplasmic reticulum (*). Scale bars (A,B) are 2.0 µm.

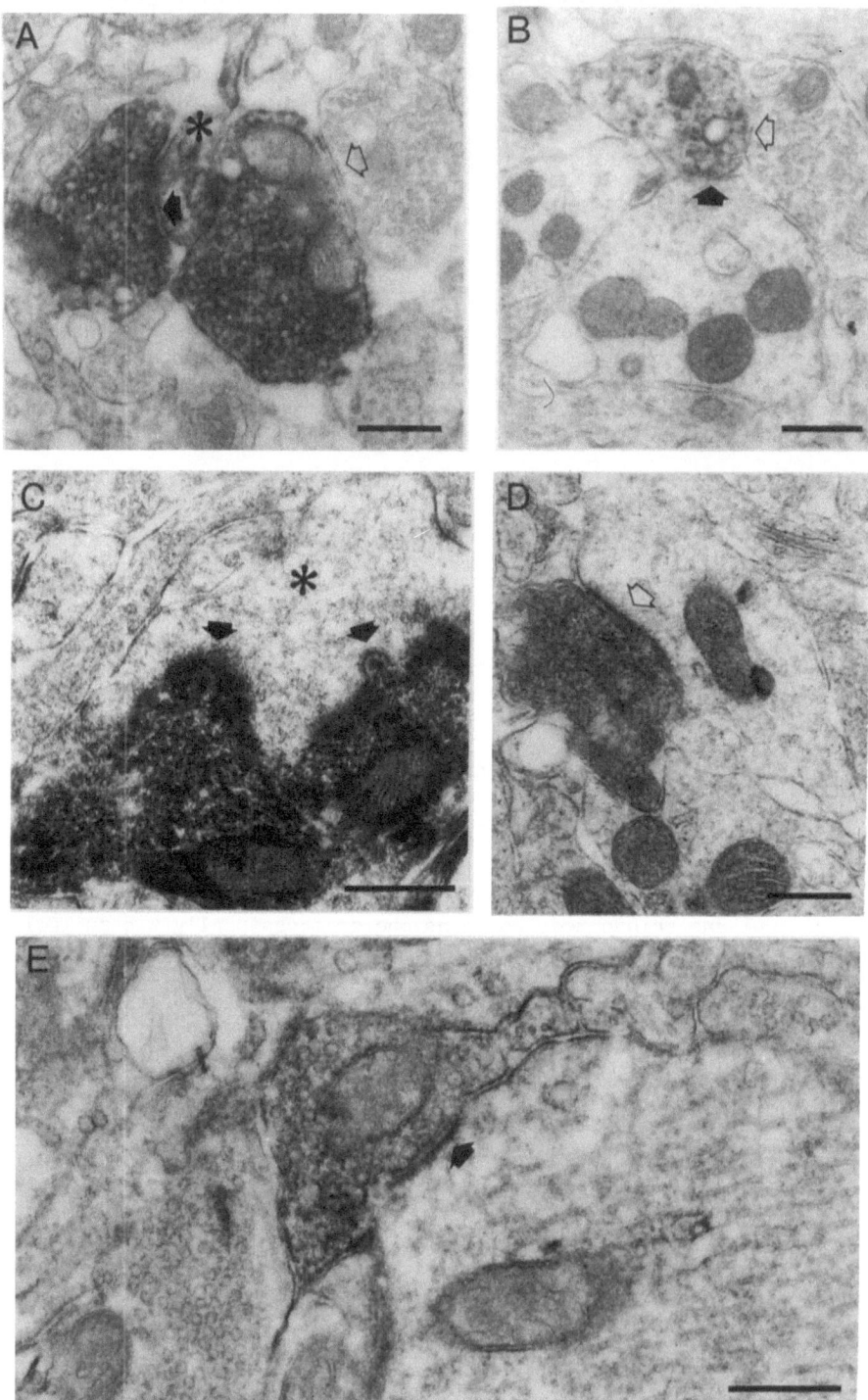

Figure 2

spines often receive one or more synapses, i.e., usually an asymmetric input to the head and a symmetric synapse at the neck. The asymmetric inputs are thought to be cortical or thalamic in origin (Hassler and Chung, 1976; Meredith and Wouterlood, 1990).

The axon collaterals of medium-sized, spiny projection neurons branch extensively within nucleus accumbens (Chang and Kitai, 1985) and presumably, many of the numerous GABA- (Pickel et al., 1988b) and GAD- (Meredith et al., 1990) immunoreactive terminal boutons seen in the nucleus with the electron microscope are endings of these axons. The synaptic specializations of these terminals are primarily symmetric, although we found a few that are asymmetric. Our recent results (Meredith et al., 1990) show that 35% of all GAD-immunoreactive terminals contact dendritic shafts, 15% spines and 6% perikarya (Meredith et al., 1990). Some immunoreactive terminals make contact with GABA- or GAD-positive elements i.e., 7% of all GABA-immunoreactive endings (Pickel et al., 1988b) and 37% of GAD-immunoreactive terminal boutons. This difference in percentage can probably be attributed to the inclusion of labelled spines in the targets of GAD-immunoreactive terminals (Fig. 2A); spines remain unlabelled in material immunoreacted for GABA (Pickel et al., 1988b). The presence of the numerous GABAergic endings on GABAergic elements (Fig. 2A) possibly forms the basis for the IPSPs recorded *in vitro* or *in vivo* from accumbens neurons following stimulation of the excitatory pathway from the cortex (Chang and Kitai, 1985; Boeijinga et al., 1990).

According to Pickel et al. (1988b), many GABAergic terminals which fail to exhibit well-defined membrane specializations, are closely apposed to the plasmalemma of adjacent unlabelled axon terminals that form contacts with neighboring dendrites or spines. These axo-axonic associations are also evident in our GAD-immunoreactive material (Fig. 2A). Inputs from the cortex or thalamus, which utilize an excitatory amino acid as neuro-transmitter, presumably establish many of the unlabelled asymmetric synapses in the striatum (see below, and Bolam and Izzo, 1987; Meredith and Wouterlood, 1990). In light of recent evidence that receptors can occupy binding sites at axo-axonic appositions which lack the morphological specializations that characterize synapses (Hamel and Beaudet, 1987), the association of GAD-positive and unlabelled terminals could reflect nonsynaptic sites of interaction (Pickel et al., 1988b).

Peptide-immunoreactive Neurons

Dense fiber plexuses which are immunoreactive for neuropeptides, e.g., substance P (SP) and enkephalin are found throughout the nucleus. These substances have been seen in fibers (Groenewegen et al., 1989), and terminals (Pickel et al., 1988a). Neurotensin and SP have been localized to perikarya (Zahm and Heimer, 1988). There is evidence from the dorsal

---

Fig. 2. Photomicrographs of synaptic terminals in nucleus accumbens. (A) Two GAD-immunoreactive boutons apposed to a GAD-immunoreactive spine (*). Note the symmetric junction (filled arrow) of one GAD-immunoreactive terminal and the close apposition of the other (open arrow) with an unlabelled terminal. (B) ChAT-immunoreactive terminal in symmetric synapse (filled arrow) with an unlabelled dendrite. Note the close apposition (open arrow) with an unlabelled terminal. (C) A terminal, labelled anterogradely from the ventral subiculum with PHA-L, in asymmetric contact (filled arrows) with a spine (*). (D) Degenerating bouton of a lesioned fiber from the paraventricular thalamic nucleus forming an asymmetric junction (open arrow) with a dendrite. (E) DA-immunopositive terminal in asymmetric contact (filled arrow) with a dendrite. Bars (A-E) = 0.25μm.

striatum that these peptides are contained within medium-sized neurons where they could be colocalized with GABA (Bolam and Izzo, 1987).

Terminals containing SP primarily form symmetric junctions in nucleus accumbens; however, all axospinous contacts have asymmetric membrane specializations (Pickel et al., 1988a). According to Pickel and colleagues (1988a), 30% of all SP-immunoreactive terminals make contact with dendritic shafts, 8% with spines and 4% with perikarya. Some SP-containing terminals have large, dense granular or multivesicular inclusions. The perikarya contacted by SP endings are medium-sized and round with unindented nuclei and presumably represent the more common type of projection neuron. In addition to forming contacts with unlabelled elements in the nucleus, SP-immunoreactive terminals have axonic associations with other SP- or tyrosine hydroxylase (TH)-containing terminals (Pickel et al., 1988a).

Choline Acetyltransferase-immunoreactive Neurons

Presumptive cholinergic neurons are labelled when antisera directed against choline acetyltransferase (ChAT) are applied to nucleus accumbens (Phelps and Vaughn, 1986; Meredith et al., 1989). These neurons are most concentrated medially and ventrally in the nucleus (Meredith et al., 1989). Morphologically, they resemble the "macro" aspiny neuron described by Danner and Pfister (1981) i.e., large with long, aspiny dendrites (Phelps and Vaughn, 1986).

Each ChAT-immunoreactive perikaryon has a large nucleus with a single, central nucleolus and few, small condensations of chromatin close to the nuclear membrane; the nuclear envelope is deeply invaginated (Fig. 1B). The amount of cytoplasm is small, except at the poles, and abundant mitochondria and arrays of granular endoplasmic reticulum are present (Fig. 1B). Subsurface cisternae are commonly associated with the perikaryal membrane. The synaptic inputs to ChAT-immunoreactive perikarya are few in number. Most form typical symmetric specializations (Phelps and Vaughn, 1986). The asymmetric junctions, which are rare and usually possess sub-junctional dense bodies, are possibly of thalamic origin (see below). The ChAT-immunoreactive dendrites are smooth and the amount of synaptic input to them, as well as the proportion of asymmetric to symmetric synapses, is greater than on the somata (Phelps and Vaughn, 1986; Meredith and Wouterlood, 1990).

Choline acetyltransferase-immunoreactive terminals primarily form symmetric synaptic specializations (Fig. 2B) but a few establish asymmetric junctions (Phelps and Vaughn, 1986), generally with spines. According to Phelps and Vaughn (1986), 61% of ChAT-immunoreactive boutons form synapses with dendritic shafts, 13% with spines and 26% with perikarya. Further, most axosomatic junctions are with somata that resemble accumbens projection neurons (Phelps and Vaughn, 1986) indicating that the cholinergic neurons can directly affect output pathways. Although most synaptic contacts are with unlabelled structures, a few terminals contact ChAT-immunoreactive elements, usually dendrites. We also found occasional axo-axonic associations (Fig. 2B).

SYNAPTIC INPUT TO NEURONS

Corticostriatal Terminals

Neurons of the hippocampal formation in the allocortex project densely to nucleus accumbens (Groenewegen et al., 1990). Degenerating or labelled terminals exclusively form asymmetric synaptic contacts in the nucleus following lesions of the subiculum or the fimbria-fornix and after injections of the anterograde tracer, *Phaseolus vulgaris*-leucoagglutinin (PHA-L) into the subiculum. The major targets of these projections are spines (86% of the synaptic contacts; Fig. 2C). However, a few contact dendritic shafts (10%) and perikarya (2%). The axospinous synaptic contacts vary in size and shape but they are usually large and curved (type III of Hassler and Chung, 1976; Fig. 2C) and sometimes interrupted (type IV of Hassler and Chung, 1976). Axodendritic synaptic junctions are usually short and straight. Some boutons contact more than one element e.g., out of 274 terminals, 19 established synapses with two spines, two with three spines, and one with a spine and a dendrite (Meredith et al., 1990). Nonlabelled terminals often form symmetric junctions with the necks of spines that receive cortical inputs onto their heads (Totterdell and Smith, 1989).

The vast majority of corticostriatal targets in nucleus accumbens are GAD-immunoreactive spines. Moreover, the somata that are targeted are exclusively GAD-immunoreactive (Meredith et al., 1990). Only 2% of the corticostriatal input is directed onto ChAT-immunoreactive neurons and this onto distal parts of the neurons. These data suggest that cortical inputs have a strong influence on the GABAergic neurons, most of which are presumably projection neurons, but a minor effect on the cholinergic cells.

Thalamostriatal Terminals

Fibers originating from the midline thalamic nuclei project densely and in a highly compartmentalized manner to nucleus accumbens (Groenewegen et al., 1989). Our recent results (Meredith and Wouterlood, 1990) have shown that degenerating or labelled terminals are seen to establish typical asymmetric contacts in the nucleus with spines (33%), dendritic shafts (56%; Fig. 2D) and somata (4%) after electrolytic lesions or injections of PHA-L in midline thalamic nuclei. Terminals are rarely seen to contact more than one element. The axospinous synapses are generally large and curved (types III or IV of Hassler and Chung, '76) and occasionally interrupted (type IV of Hassler and Chung, '76). The axodendritic synapses (Fig. 2D) are straight (type VII of Hassler and Chung, '76). Occasionally, axodendritic and axosomatic synaptic contacts bear postsynaptic dense bodies; such terminals have been seen in contact with ChAT-immunoreactive elements (Meredith and Wouterlood, 1990).

Most thalamic terminals contact spines and distal dendritic shafts of cells resembling the medium-sized, spiny type of neuron. Fifteen percent of thalamic (paraventricular) fibers, however, terminate on ChAT-immunoreactive neurons and many of these contacts are proximally situated, on somata and large dendritic shafts (Meredith and Wouterlood, 1990). We know that the total synaptic input to proximal parts of cholinergic cells is very sparse and symmetric specializations that may be associated with inhibition, form the predominant synapses (Phelps and Vaughn, 1986; Meredith and Wouterlood, 1990). Therefore, the thalamic terminals, which primarily contact proximal cholinergic elements in nucleus accumbens and which are presumably excitatory (Chang and Wilson, 1990), could play an important role in the activation of the cholinergic neurons.

Dopamine- and Tyrosine Hydroxylase-immunoreactive Terminals

There is a dense DA- and TH-immunoreactive innervation of nucleus accumbens. The fibers are thin, delicate and highly varicose (Voorn et al., 1986; Totterdell and Smith, 1989). The ultrastructural appearances of DA- and TH-containing terminals are strikingly similar and most investigators presume TH-immunoreactive endings to be dopaminergic. The DA- and TH-immunoreactive boutons are small and contain both round and flattened vesicles; some contain cored vesicles (Voorn et al., 1986; Pickel et al., 1988b). Many labelled profiles appose structures without revealing any clear synaptic specialization; synaptic densities that can be recognized are clearly of the symmetric type (Pickel et al., 1988b; Totterdell and Smith, 1989). Less than 1% of the DA-positive terminals show asymmetric membrane specializations (Fig. 2E; Voorn et al., 1986). Sixteen percent of DA-immunoreactive terminals contact dendritic shafts, 11% spines and 3% perikarya (Voorn et al., 1986), whereas 22% of the TH-positive endings establish synapses with dendritic shafts, 24% with spines and 4% with perikarya (Pickel et al., 1988b). Neurons immunoreactive for GABA receive a monosynaptic input from TH-immunoreactive fibers. Indeed, 18% of TH-positive terminals contact GABA-immunoreactive perikarya and proximal dendrites (Pickel et al., 1988b). Moreover, TH-containing terminals are frequently found in axonic association with SP-containing boutons and often terminate on the same dendrite as the SP-immunoreactive ending (Pickel et al., 1988a).

There is recent evidence that TH-immunopositive (Totterdell and Smith, 1989) as well as DA-immunoreactive (Groenewegen et al., 1990) terminals converge with terminals of hippocampal fibers on the same neurons, even on the same structures, e.g., spines or dendrites. This convergence implies an interaction between dopaminergic and glutamatergic influences in nucleus accumbens.

CONCLUDING REMARKS

The interactions in nucleus accumbens between the GABAergic and cholinergic neuronal populations, on the one hand, and the afferent input, i.e. the excitatory (presumably amino acidergic) terminals of allocortical and thalamic origin and the terminals of dopaminergic fibers arising in the midbrain, on the other, have important implications for the regulation of the output of the nucleus. Little is known, however, of the anatomical substrate underlying these interactions. At the cellular level, the available data strongly indicate that the excitatory allocortical input is directed onto GABAergic neurons, while the thalamic excitatory input is directed to both cholinergic and spiny, presumably GABAergic, neurons (Meredith et al., 1990; Meredith and Wouterlood, 1990). Dopaminergic fibers also terminate on GABAergic cells (Pickel et al., 1988b) and even converge with hippocampal inputs onto the same neurons (Totterdell and Smith, 1989). Since DA can reduce the excitatory responses recorded from the output neurons following fornix stimulation (Yang and Mogenson, 1984), we expect that cortical and DA inputs interact through synapses on common target cells which contain GABA and project out of the nucleus.

The relationship between the afferent dopaminergic fibers and the intrinsic cholinergic cells in nucleus accumbens is unresolved, although this relationship has been investigated in the dorsal striatum (Chang, 1988). No anatomical data are available with respect to the possible GABAergic innervation of cholinergic cells in either the dorsal or ventral striatum. Pharmacological studies have shown that the turnover and release of ACh in *in vitro* slices of nucleus accumbens can be affected by glutamate and GABA (Henselmans and Stoof, personal communication). In contrast to the dorsal striatum, DA is much more limited in its ability to affect the release of ACh in the ventral striatum (Henselmans et al., 1988). Cholinergic perikarya and proximal dendrites form major targets of excitatory projections from the midline thalamus in accumbens. Moreover, stimulation of these thalamic fibers can increase the utilization of DA in nucleus accumbens (Jones et al., 1989) and cholinergic action can modulate the activity of DA in this nucleus, presumably through the presence of muscarinic receptors on dopaminergic terminals (de Belleroche et al., 1982). This would suggest that the thalamic input to cholinergic neurons is responsible for potentiating the release of DA in nucleus accumbens.

ACKNOWLEDGMENTS

We thank Drs. B. Wainer and W. Oertel for their generous gifts of antisera and Dr. P. Voorn for the photomicrograph in Figure 2E. We also thank A. Pattiselanno, P. Wentzel, S. Paniry, and D. de Jong for technical assistance. We appreciate the comments of Profs. H. Groenewegen and A. Lohman on the manuscript.

REFERENCES

Boeijinga, P.H., F.H. Lopes da Silva, and C.M.A. Pennartz, 1990, Long lasting paired-pulse facilitation in the nucleus accumbens following stimulation of subicular inputs in the rat, Neuroscience, in press.

Bolam, J.P. and P.N. Izzo, 1987, Possible sites of transmitter interaction in the neostratum: An anatomical approach, in: "Neurotransmitter Interactions in the Basal Ganglia," M. Sandler, C. Feuerstein, and B. Scatton, eds., Raven Press, New York.

Chang, H.T., 1988, Dopamine-acetylcholine interaction in the rat striatum: A dual-labeling immunocytochemical study, Brain Res. Bull., 21:295.

Chang, H.T. and S.T. Kitai, 1985, Projection neurons of the nucleus accumbens: an intracellular labeling study. Brain Res., 347:112.

Chang, H.T. and C. J. Wilson, 1990, Anatomical analysis of electro-physiologically characterized neurons in the rat strio-pallidal system, in: "Approaches to the Analysis of Neuronal Microcircuits and Synaptic Interactions. Handbook of Chemical Neuroanatomy, Vol. 8," F.G. Wouterlood and A.N. van den Pol, eds., Elsevier Science Publ., Amsterdam.

Costa, E., D.L. Cheney, C.C. Mao, and F. Moroni, 1978, Action of antischizophrenic drugs on the metabolism of γ-aminobutyric acid and acetylcholine in globus pallidus, striatum and n. accumbens, Fed. Proc., 37:2408.

Danner, V.H. and C. Pfister, 1981, Investigations on the cytoarchitecture of the nucleus accumbens septi of the rat, Anat. Anz. 150:264.

De Belleroche, J., I.C. Kilpatrick, N.J.M. Birdsall, and E.C. Hulme, 1982, Presynaptic muscarinic receptors on dopaminergic terminals in nucleus accumbens. Brain Res., 234:327.

Groenewegen H.J., G.E. Meredith, H.W. Berendse, P. Voorn, and J.G. Wolters, 1989, The compartmental organization of the ventral striatum in the rat, in: "Neural Mechanisms in Disorders of Movement," A.R. Crossman and M.A. Sambrook, eds., Libbey and Co. Ltd., London.

Groenewegen, H.J., H.W. Berendse, G.E. Meredith, S.N. Haber, P. Voorn, J.G. Wolters and A.H.M. Lohman, 1990, Functional anatomy of the ventral, limbic system-innervated striatum, in: "The Mesolimbic Dopamine System: From Motivation to Action," P. Willner and J. Scheel-Krüger, eds., John Wiley and Sons Ltd., Chichester, England.

Hamel, E. and A. Beaudet, 1987, Ultrastructural distribution of mu opioid receptors in rat neostriatum, in: "Neurotransmitter Interactions in the Basal Ganglia," M. Sandler, C. Feuerstein, and B. Scatton, eds., Raven Press, New York.

Hassler, R. and J.W. Chung, 1976, The discrimination of nine different types of synaptic boutons in the fundus striati (nucleus accumbens septi), Cell Tiss. Res., 168:489.

Henselmans, J.M.L., J.C. Stoof, and P.V.J.M. Hoogland, 1988, Regional difference in the D-2 receptor mediated inhibition of the release of acetylcholine in the nucleus accumbens but not in the neostriatum of the rat. Dutch Fed. Meetings, 29:165.

Jones, D.L., G.J. Mogenson, and M. Wu, 1981, Injections of dopaminergic, cholinergic, serotoninergic and GABAergic drugs into the nucleus accumbens: effects on locomotor activity in the rat, Neuropharm., 20:29.

Jones, M.W., I.C. Kilpatrick, and O.T. Phillipson, 1989, Regulation of dopamine function in the nucleus accumbens of the rat by the thalamic paraventricular nucleus and adjacent midline nuclei, Exp. Brain Res., 76:572.

Matthysse, S. 1981, Nucleus accumbens and schizophrenia, 1980, in: "The Neurobiology of Nucleus Accumbens," R.B. Chronister and J.F. DeFrance, eds., The Haer Institute for Electrophysiological Research, Brunswick, Maine.

Meredith, G.E. and F.G. Wouterlood, 1990, Hippocampal and midline thalamic fibers and terminals in relation to the choline acetyltransferase-immunoreactive neurons in nucleus accumbens of the rat: A light and electron microscopic study, J. Comp. Neurol., in press.

Meredith, G.E., B. Blank, and H.J. Groenewegen, 1989, The distribution and compartmental organization of the cholinergic neurons of nucleus accumbens in the rat, Neuroscience, 31:327.

Meredith, G.E., F.G. Wouterlood, and A. Pattiselanno, 1990, Hippocampal fibers make synaptic contacts with glutamate decarboxylase-immunoreactive neurons in the rat nucleus accumbens, Brain Res., in press.

Phelps, P.E. and J.E. Vaughn, 1986, Immunocytochemical localization of choline acetyltransferase in rat ventral striatum: a light and electron microscopic study. J. Neurocytol. 15:595.

Pickel, V.M., T. H. Joh, and J. Chan, 1988a, Substance P in the rat nucleus accumbens: ultrastructural localization in axon terminals and their relation to dopaminergic afferents. Brain Res., 444:247.

Pickel, V.M., A.C. Towle, T.H. Joh, and J. Chan, 1988b, Gamma-aminobutyric acid in the medial rat nucleus accumbens: Ultrastructural localization in neurons receiving monosynaptic input from catecholaminergic afferents, J. Comp. Neurol., 272:1.

Totterdell, S. and A.D. Smith, 1989, Convergence of hippocampal and dopaminergic input onto identified neurons in the nucleus accumbens of the rat, J. Chem. Neuroanat., 2:285.

Willner, P. and J. Scheel-Krüger, eds., 1990, "The Mesolimbic Dopamine System: From Motivation to Action," John Wiley and Sons Ltd., Chichester, England.

Voorn, P., B. Jorritsma-Byham, C. Van Dijk and R. M. Buijs, 1986, The dopaminergic innervation of the ventral striatum in the rat: A light- and electron-microscopical study with antibodies against dopamine, J. Comp. Neurol., 251:84.

Yang, C.R. and G.J. Mogenson, 1984, Electrophysiological responses of neurons in the nucleus accumbens to hippocampal stimulation and the attenuation of the excitatory responses by the mesolimbic dopaminergic system, Brain Res., 324:69.

Zahm, D.S. and L. Heimer, 1988, Ventral striatopallidal parts of the basal ganglia in the rat: I. Neurochemical compartmentation as reflected by the distributions of neurotensin and substance P immunoreactivity, J. Comp. Neurol., 272:516.

THE PRIMATE CENTRAL COMPLEX AS ONE OF THE BASAL GANGLIA

Percheron G., François C., Parent A.[*], Sadikot A.F.[*]
Fénelon G. and Yelnik J.

Laboratoire de Neuromorphologie et de Neurologie expérimen-
tale du mouvement, INSERM, Pav. Cl. Bernard, Hôpital de la
[*]Salpêtrière, 47 Boulevard de l'Hôpital, 75651 PARIS Cedex 13
Centre de Recherches en Neurobiologie, Hôpital de l'Enfant
Jésus, 1401, 18ème rue, Québec, Canada, G2J 1Z4

INTRODUCTION

It was in human brain, that Jules Bernard Luys (1865) first discovered what he called his "centre médian". Figure 1A shows that his "centre" was located lateral to Meynert's bundle and ventral to the medial
nucleus. For a long time it was considered as the ventral element of a so-
called medial part of the thalamus of which the nucleus medialis (hence
called medialis dorsalis) was the dorsal element. The description of the
nucleus medial to the bundle, the nucleus parafascicularis, was not done
until forty years later, in monkey brains, by Mrs Vogt (1909) and Friede-
man (1911). This was later included into the "hyperchromic circular formation" (Foix and Nicolesco, 1925) encircling the medial nucleus and also
comprising the intralaminar and periventricular elements. The centre
médian and the nucleus parafascicularis together are nowadays considered
to constitute a "complex" often called the centre médian-parafascicular
complex.

The role of this complex has been the subject of many successive
interpretations. The first, that of Luys, considered it as a sensory,
somesthetic relay. Experimental data obtained from non primate animals
seemed to reinforce this view. In the fifties, the complex became the
usual stereotactic target for pain relief. One particular sensory interpretation conceived the complex as a trigeminal relay. Spinothalamic
(Mehler, 1966) as well as cerebellar (Percheron, 1977) or nigral (François
et al., 1988) axons, indeed do cross through the complex on their way to
the more dorsal paralaminar region, but do not end in it. The description
of afferents from the brain stem reticular formation is linked to the most
common interpretation accepted nowadays, which follows Le Gros Clark
(1932) decision to include the complex within the more general system of
the so-called "intralaminar nuclei". There is now a considerable literature - extensively reviewed by Jones (1985) and Macchi and Bentivoglio
(1986) - and culminating with the elaboration of the "centrencephalon",
concerning the participation of the system in the "non specific" excitation of the cerebral cortex, the sleep-waking cycle and the polysensory
projection to the cortex. The inclusion of the complex within the intralaminar nuclei lato sensu has acted as a brake upon the search for its
distinctive role. There are also many arguments against such an inclusion.

The Basal Ganglia III, Edited by G. Bernardi *et al.*
Plenum Press, New York

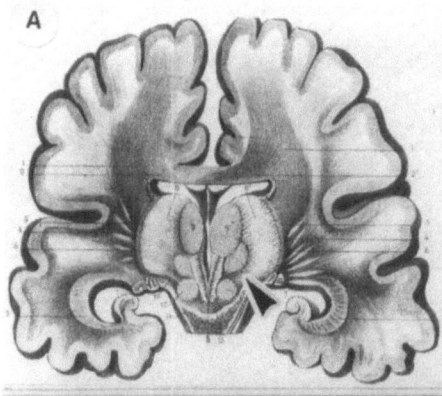

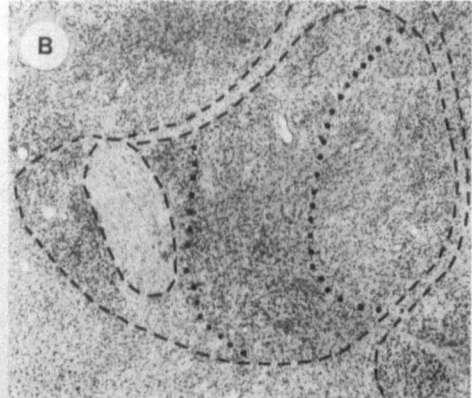

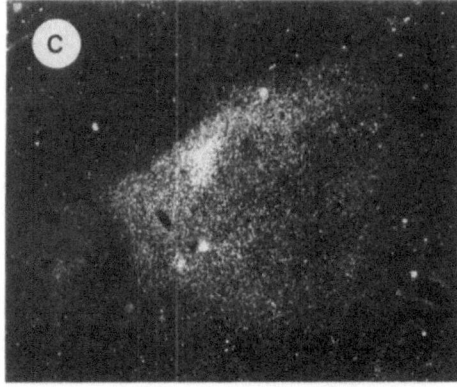

Fig. 1. A. Luys' (1865) drawing
showing his "center" (arrow).
B. Cytoarchitectonic picture
of the central complex of a
macaque showing the three
parts: parafascicularis,
media and paralateralis.
C. Dark field photography of
the intracentral distribution
of the axons from the medial
nucleus of the pallidum.

Phylogenetic studies, for instance, showed that the complex extensively
increases while intralaminar nuclei regress. Kuhlenbeck (1951) and Niimi
et al. (1960) among others, concluded that the centre médian was not a
derivative of the intralaminar nuclei but a "lateral differentiation of
the nucleus parafascicularis" which is itself a differentiation of the
periventricular grey.

This paper is an attempt to demonstrate that the so-called primate
centre médian-parafascicular complex is in fact a major element of the
basal ganglia system. The argumentation is hampered by major differences
observed in non primate and primate species. Our data have thus been
gathered only in human and simian brains.

ANATOMY OF THE CENTRAL COMPLEX

Figure 1B shows a cytoarchitectonic picture of the complex in a
macaque. Except medially, where it seems to emerge from the periventricular grey, its boundary can be easily traced. The complex is indeed
incompletely surrounded by a clearly visible capsule. The nature of this
capsule is an important element in determining if the complex is really
intralaminar or not. Two laminae, and not a single one, must be considered. Only the dorsal lamina separating the complex from the nucleus
medialis contains intralaminar neurons and is the caudo-ventral extension
of the internal medullary lamina (or lamina medialis). The lateral lamina
separating the complex from the lateral mass, the lamina centralis or
Vogts' (1941) lamella intermedia, could have a different embryologic

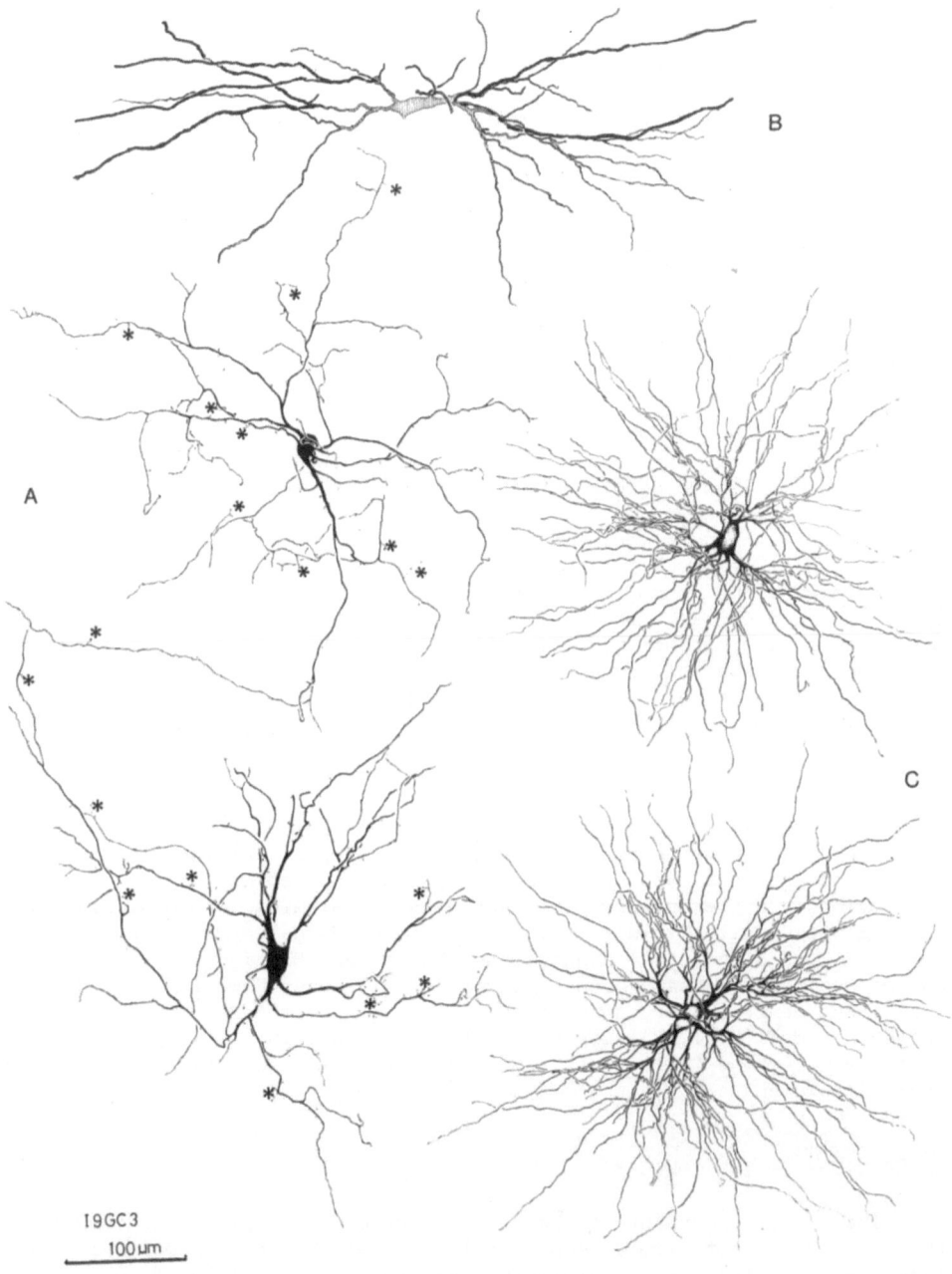

100 μm

Fig. 2. A. A Golgi picture of two neurons from the paralateral part of the central complex. The stars indicate thin axon-like processes. Central neurons are obviously different from the three other neurons.
B. One intralaminar neuron. C. Two tufted neurons from the lateral mass of the thalamus.

origin (Cooper, 1950) and be made up essentially of lemniscal axons (Yakovlev, 1969). Using this argument, Cooper (1950) stated that "the centre médian cannot be regarded as intralaminar". It is in fact infra-laminar.

Regarding the internal structure of the complex, cytoarchitectonic pictures show that there is a contrast between a rather dark and dense medial region and a rather pale and loose lateral region (Fig. 1B). The two are however separated by an intermediate region whose boundaries are irregular and fuzzy. In primates, it is now generally admitted that the complex is composed of three parts. This was first noted in man by the Vogts (1941), who divided the centre médian into dorso median, "magno-cellular" (CeMc) and ventrolateral, "parvocellular", parts (CePc). Niimi and his colleagues (1960) later also subdivided the complex into three parts but in a different way. The nucleus parafascicularis was subdivided into a lateral (Pfl) and a medial part (PfM). The centre médian was res-tricted to the most lateral, paucicellular, part. This conception was endorsed by Mehler (1966) and François et al. (1988). Subdivision into two nuclei is a pitfall in the case where three parts are recognized. In order to avoid an endless debate concerning whether the middle part belongs ei-ther to the nucleus parafascicularis or to the "centre médian", we chose to consider the whole complex as a single entity, the central complex (Formatio or Complexio centralis), with three parts: the pars parafasci-cularis (CPf) the pars media (CMe) and the pars paralateralis (CPl).

Our Golgi study of human and monkey brains showed that the dendrites of central neurons do not cross the capsule. Thus, except on its medial side, the complex, as a whole, is a closed nucleus. Our typological analy-sis based on the morphology of dendritic arborizations is only in its early days. Neurons of the pars parafascicularis belong to the leptoden-dritic family, i.e. of poorly branched neurons. Neurons of the pars paralateralis, are richly branched with numerous axon-like appendages indi-cated by the stars in Figure 2A. Neurons of the pars media have dendritic arborizations which, at first sight, seem intermediate between those of the pars parafascicularis and those of the pars paralateralis. They could however only have fewer axon-like processes and belong to the same neu-ronal species as the neurons of the paralateral part. At the moment, the main finding of our typological study is that the neurons of the complex are obviously different from the overlying intralaminar neurons (Fig. 2B). They do not usually stain with the same histochemical dyes, do not have the same kind of cortical connections (Macchi and Bentivoglio, 1986) and do not react in the same manner to cortical lesions (Powell and Cowan, 1967). But central neurons are also different from the usual thalamic tufted neurons (Fig. 2C). Given also the peculiar, infrathalamic, position of the complex one is led away from considering the complex as an ordinary thalamic element.

CONNECTIONS OF THE COMPLEX

The exact topography of the territories of cortical afferents within the complex has not firmly been established yet. The pars parafascicularis receives axons from the frontal eye field and from the frontal cortex (Akert and Hartmann-von Monakow, 1980). Data from Jürgens (1984) suggest a connection between the supplementary motor cortex and the pars media. Data from De Vito (1969), Künzle (1976), Jones et al. (1979), Akert and Hartmann-von Monakow (1980) and Leichnetz (1986), indicate that axonal endings from the motor cortex could be distributed according to a rough somatotopy in both pars media and paralateralis in the macaque. However, Kuypers (1966) showed that in the chimpanzee the projection from the motor cortex could be restricted to the pars paralateralis and that of the pre-motor cortex to the pars media. In the motor cortex, the pyramidal neurons which send their axons to the complex are located dorsal to the Betz neu-

rons (Castman-Berrevoets and Kuypers, 1978). At least some of the axons
from the motor cortex could be collaterals of axons of the pyramidal tract
(Blum, 1971). The organisation of the projection from the complex to the
cortex is not entirely known either. Data from Jones et al. (1979),
Ilinsky et al. (1985), Leichnetz (1986), Smith and Parent, (1986), Ghosh
et al. (1987) and Matelli et al. (1989) indicate that neurons of the pars
paralateralis could send their axons mainly to the motor cortex. Neurons
of the pars media could send their axons to the premotor and supplementary
motor cortex. The strong reciprocal link of the two lateral parts of the
complex with the motor, premotor and supplementary motor cortex is important evidence of their involvement in motor activities.

It has been known for a long time, that hemidecortication has no
effect on the neurons of the complex. The latter was said to have mainly
striatal connections. The centro-striatal connection was first demonstrated in human pathological material. The topographical organisation was
deduced from the study of vascular lesions by the Vogts (1941) and Simma
(1951). According to their findings the parafascicular nucleus (the pars
parafascicularis) projects to the fundus striati, the CeMc (the pars media) to the caudate nucleus and the CePc (the pars paralateralis) to the
putamen. This was disproved however by Martin (1970) who observed cases of
degenerative lesions predominating either in the putamen or in the caudate
nucleus. The literature dealing with experimental data in monkeys did not
appear to be topographically precise enough either.

We therefore decided to study anew the subcortical connections of the
complex according to very strict topographic conditions. This was done in
12 macaques by using either anterograde (autoradiography) or retrograde
(WGA-HRP) tracers and a cartographic system based on the CA-CP ventricular
coordinates (Percheron et al., 1986). A strong projection from the peri-
thalamic reticular nucleus was found to exist. However the main subcortical afference to the complex was confirmed to come from the medial nucleus of the pallidum. Medial pallidal neurons projecting to the complex
were found to lie essentially in the caudal and lateral part of the nucleus. This is the region which is crossed by the axons of the spiny
neurons from the striatal sensorimotor territory (Fénelon et al., 1989)
and where physiologists find neurons whose activity is related to movement
(DeLong et al., 1985). The cartography of the pallidal territory within
the complex made after the use of an anterograde tracer resulted in a sur-
prise. Not only are there no pallidal afferences to the pars parafascicularis but also only few to the pars paralateralis (Fig. 1C). The densest
part of the territory is crescent-shaped and its outline almost coincides
with that of the pars media (then named PfL, François et al., 1988). HRP
injections into the striatum were used in order to map retrogradely the
location of the central neurons projecting to the striatum. The pars parafascicularis projects profusely to the caudate nucleus. Injections into
the sensorimotor territory, delimited by the axons from the motor, premotor and somesthetic cortex in the caudo-dorsal part of the putamen
(Percheron et al., 1984a), always led to the retrograde labeling of neurons essentially located inside the pars media. Anterograde autoradiographic studies of the projection from the complex to the striatum (François
et al., 1989) confirmed the previous observation of Kalil (1978). The
intrastriatal distribution of central axons is patchy. Precise cartographic studies brought two other important data. First, the overall intrastriatal distribution of the axonal endings of central origin was found to
correspond strickly to the topography of the sensorimotor territory, which
is not thus a mere cortical territory. This intrastriatal distribution is
in contrast with that of the axonal endings from the amygdala (Russchen et
al., 85), which is restricted to the associative territory. The second
interesting topographic finding was that the patches of axonal endings of
central origin in fact have a complex 3-dimensional organisation. On
transverse sections, central patches can be observed to constitute 3 to 4
elongated oblique streaks parallel one to the other and to the direction

of the somatotopic stripes of the sensorimotor territory and are separated
by bands devoid of central axons. Central streaks make an angle of about
55° with the intercommissural plane. Figure 3 shows that a projection on a
plane having this inclination makes it possible to observe the intervals
between the streaks. When reconstructed in the anteroposterior dimension
and projected on such a plane the streaks appeared to form 3 to 4 long,
parasagittal, caudo-oral layers. Their three dimensional spatial pattern
is so peculiar, that it seems too early to be able to assert that the central intrastriatal layers in primates could be the sensorimotor equivalent
of either the matrix or the striosomes, mainly described within the striatal associative territory. More generally there are no conclusive data
concerning the existence of matrix-striosome compartments within the sensorimotor part of the putamen in primates.

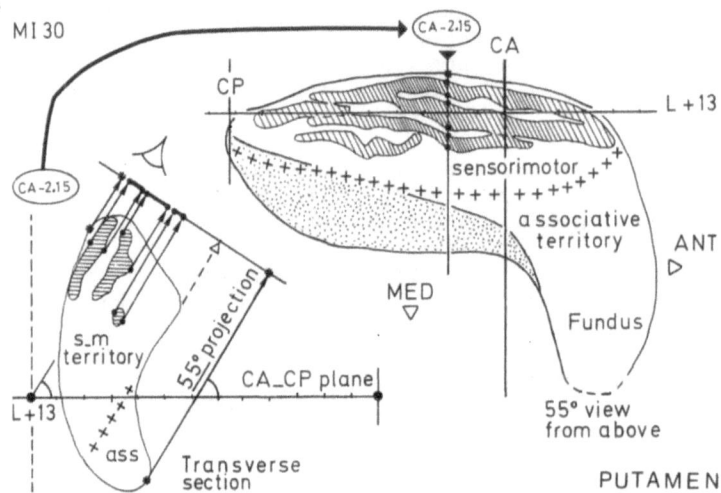

Fig. 3. Bottom left: transverse section of the putamen of a macaque showing three oblique streaks made by axonal endings of the central
complex neurons within the striatal sensorimotor (SM) territory.
Top right: a tridimensional reconstruction of the 3 to 4 longitudinal layers made by central axonal endings. Their overall topography coincides with that of the cortical sensorimotor territory
whose extent is indicated by crosses. ANT: anteriorwards; CA: anterior commissure; CP: posterior commissure; MED: medialwards.

POSITION OF THE COMPLEX WITHIN THE SYSTEM OF THE BASAL GANGLIA

In 1984 we attempted a rational definition of the system of the basal
ganglia (Percheron et al., 1984b). The set comprising the striatum and its
direct targets, the two nuclei of the pallidum and the pars lateralis and
reticulata of the substantia nigra, was said to constitute the "core of

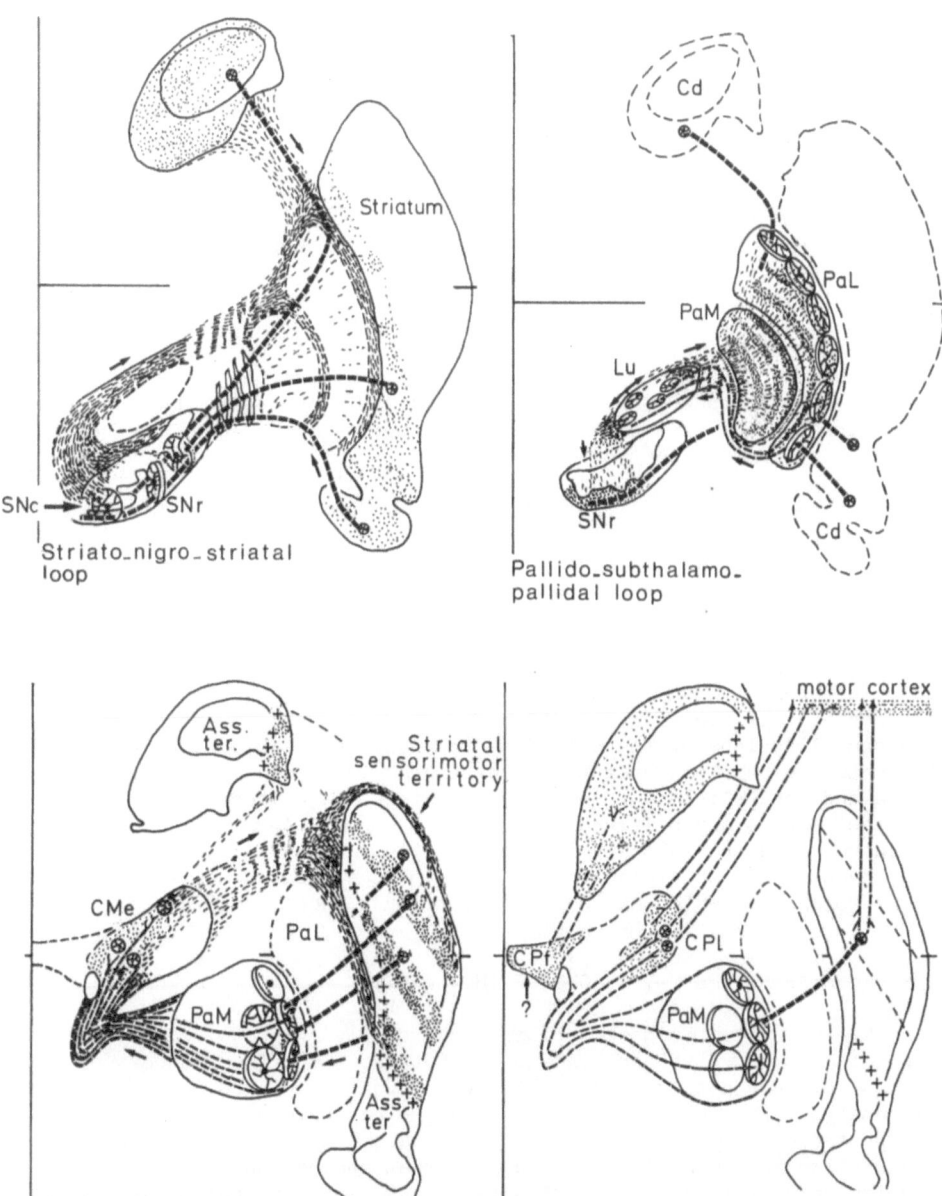

Fig. 4. Semi schematic drawings showing some subsystems of the basal
ganglia. The pars media of the central complex (bottom left) is
one of the three elements of the Nauta and Mehler' (1966) loop.
The extent of the sensorimotor territory of the striatum is indicated by crosses. Ass. ter.: associative territory; Cd: caudate
nucleus; CMe: pars media; CPf: pars parafascicularis; CPL: pars
paralateralis of the central complex; Lu: subthalamic nucleus;
PaL, PaM: lateral and medial nuclsi of the pallidum; SNc, SNr:
pars compacta and reticulata of the substantia nigra.

the basal ganglia". The striatum is first level and its targets are second
level basal ganglia. The core is primarily fed by cortical information
conveyed by axons from wide regions of the cortex. Several subsystems make

closed or key-ring loops with elements of the core. The best way to understand the position of the central complex within the system of the basal ganglia is to compare it to two other subsystems which also exert a direct influence on the core of the basal ganglia. This is done visually in Figure 4. The best known is the striato-nigro-striatal loop involving only two elements (Percheron et al., 1989). The spiny neurons of the striatum send their axons to the pars reticulata of the substantia nigra where they reach the ventral dendrites of the dopaminergic neurons of the pars compacta which in turn send back their axons to the striatum. The neurons of the pars reticulata are elements of a longer circuit involving the thalamus and the cortex. Another loop is made by the neurons of the subthalamic nucleus. Its most medial neurons send their axons to the pars reticulata of the substantia nigra but the greatest number of the subthalamic neurons send their axons to the two nuclei of the pallidum. The axons of the lateral pallidal neurons go back to the subthalamic nucleus and close a short loop. The neurons of the medial pallidal nucleus are elements of thalamo-striate and thalamo-cortical circuits.

The pars parafascicularis although projecting to the caudate nucleus does not receive axons either from the pallidum or from the substantia nigra (François et al., 1988) and is apparently not part of a loop of the basal ganglia system. The pars media and paralateralis, receive axons from the neurons of the sensorimotor part of the medial nucleus of the pallidum. The pars paralateralis, which sends only a few, if any, axons to the striatum is very likely essentially involved in centrocortical circuits which are still not entirely known. The pars media is the main source of a major afference to the sensorimotor territory of the striatum. The spiny neurons of this territory send their axons to the sensorimotor region of the medial nucleus of the pallidum, the neurons of which send back their axons to the pars media of the complex. The three consecutive elements of Nauta and Mehler's loop (1966) thus constitute a closed sensorimotor loop. The feedback is directly realised, in a strongly organized manner, on the striatum. Like the subthalamic nucleus, the pars media receives its afferent axons from one element of the core of the basal ganglia and sends its axons to another which may define third level basal ganglia.

QUESTIONS ABOUT THE MOTOR ROLE OF THE COMPLEX

The stereotactic destruction of the complex used to be used for the cure of movement disorders (Adams and Rutkin, 1965; Adams and Malamud, 1971). This did not lead to constant improvement and was thus abandoned. In monkeys also, lesions of the complex do not have clear motor effect (Carpenter et al., 1965). The study of cases of striatal degenerative lesions led Schulman (1957) to suggest that choreoathetoid movements could develop only if the centre médian is intact. The more recent work of Martin (1970) is less affirmative. Blum (1971) made the hypothesis that the complex could be an element of a feedforward system regulating motoneurons. In fact, we still do not know to this day the exact motor role of the central complex. However there remains strong anatomical evidence that it is a major part of the basal ganglia motor system.

REFERENCES

Adams, J.E., and Malamud, N., 1971, Severe chorea with degeneration of the nucleus centrum medianum, Arch. Neurol., 24:101-105.
Adams, J.E., and Rutkin, B.B., 1965, Lesions of the centrum medianum in the treatment of movement disorders, Confin. Neurol., 26:231-236.
Akert, K., and Hartmann-von Monkakow, K., 1979, Relationships of precentral, premotor and prefrontal cortex in the mediodorsal and intralaminar nuclei of the monkey thalamus, Act. Neurobiol., 40:7-25.

Blum, B., 1971, Microphysiological characterization of output channels and of impulse propagation from sensorimotor cortex. Including through pyramidal tract collaterals to the centre median nucleus of the cat and of the monkey, Int. J. Neurol., 8:178-189.

Carpenter, M.B., Strominger, N.L., and Weiss, A.H., 1965, Effects of lesions in the intralaminar nuclei upon subthalamic dyskinesia, Arch. Neurol. 13:113-125.

Castman-Berrevoets, C.E., and Kuypers, H.G.J.M., 1978, Differential laminar distribution of corticothalamic neurons projecting to the VL and the centre median: an HRP study in the cynomolgus monkey, Brain Res., 154:359-365.

Cooper, E.R.A., 1950, The development of the thalamus, Acta Anat., 9:201-226.

Delong, M.R., Crutcher, M.D., and Georgopoulos, A.P., 1985, Primate globus pallidus and subthalamic nucleus, functional organization, J. Neurophysiol., 53:530-543.

De Vito, J.L., 1969, Projections from the cerebral cortex to the intralaminar nuclei in monkey, J. Comp. Neurol., 136:193-202.

Fénelon, G., François, C., Percheron, G., and Yelnik, J., 1989, Topographical distribution of pallidal neurons projecting to the thalamus in macaque, accepted in Brain Res..

Foix, C., and Nicolesco, J., 1925, Les noyaux gris centraux et la région mésencéphalo-sous-optique, Masson, Paris.

François, C., Percheron, G., Yelnik, J., and Tandé, D., 1988, A topographic study of the course of nigral axons and of the distribution of pallidal axonal endings in the centre médian-parafascicular complex of macaques, Brain Res., 473:181-186.

François, C., Percheron, G., Parent, A., Sadikot, A.F., Fénelon, G., and Yelnik, J., 1989, Topography of the projection from the central complex of the thalamus to the sensorimotor striatal territory in monkeys, submitted.

Friedemann, M., 1911, Die Cytoarchitektonik des Zwishenhirns der Cercopitheken mit besonderer Berücksichtigung des Thalamus opticus, J. Psychol. Neurol. (Lpz), 18:309-378.

Ghosh, S., Brinkman, C., and Porter, R., 1987, A quantitative study of the distribution of neurons projecting to the precentral motor cortex in the monkey (M. fascicularis), J. Comp. Neurol., 259:424-444.

Ilinski, I.A., Jouandet, M.L., and Goldman-Rakic, P.S., 1985, Organization of the nigrothalamocortical system in the rhesus monkey, J. Comp. Neurol., 236:315-330.

Jones, E.G., 1985, "The thalamus," Plenum Press, New York.

Jones, E.G., Wise, S.P., and Coulter, J.D., 1979, Differential thalamic relationships of sensory-motor and parietal cortical fields in monkeys, J. Comp. Neurol., 183:833-882.

Jürgens, V., 1984, The efferent and afferent connections of the supplementary motor area, Brain Res., 300:63-81.

Kalil, K., 1978, Patch-like termination of thalamic fibers in the putamen of the rhesus monkey: an autoradiographic study, Brain Res., 140:333-339.

Kuhlenbech, H., 1951, The derivatives of thalamus dorsalis and epithalamus in the human brain: their relation to cortical and other centers, Mil. Surg., 108:205-256.

Künzle, H.,1976, Thalamic projections from the precentral motor cortex in Macaca fascicularis, Brain Res., 105:253-267.

Kuypers, H.G.J.M., 1966, Discussion, in: "The Thalamus," D.P. Purpura and M.D. Yahr, eds, Columbia Univ. Press, New York.

Le Gros Clark, W.E., 1932, The structure and connections of the thalamus, Brain, 55:406-470.

Leichnetz, G.R., 1986, Afferent and efferent connections of the dorsolateral precentral gyrus (area 4, hand/arm region) in the macaque monkey, with comparisons to area 8, J. Comp. Neurol., 254: 460-492.

Luys, J.B., 1865, "Recherches sur le système nerveux cérébro-spinal, sa structure, ses maladies, ses fonctions", Baillère, Paris.

Macchi, G., and Bentivoglio, M., 1986, The thalamic intralaminar nuclei and the cerebral cortex, in: "Cerebral cortex," E.G. Jones and A. Peters, eds, Plenum Publish. Corp., New York.

Martin, J.J., 1970, Contribution à l'étude de l'anatomie du thalamus et de sa pathologie au cours des maladies dégénératives dites abiotrophiques, Publ.Act. Med. Belg., Bruxelles.

Matelli, M., Luppino, G., Fogassi, L., and Rizzolatti, G., 1989, Thalamic input to inferior area 6 and area 4 in the macaque monkey, J. Comp. Neurol., 280:468-488.

Mehler, W.R., 1966, Further notes on the center median nucleus of Luys in: "The thalamus," D.P. Purpura and M.D. Yahr, eds, Columbia Univ. Press, New York.

Nauta, W.J.H., and Mehler, W.R., 1966, Projections of the lentiform nucleus in the monkey, Brain Res., 1:3-42.

Niimi, K., Katayama, K., Kanaseki, T., and Morimoto, K., 1960, Studies on the derivation of the centre median nucleus of Luys, Tokushima J. Exptl. Med., 6:261-268.

Percheron, G., 1977, The thalamic territory of cerebellar afferents and the lateral region of the thalamus of the macaque in stereotaxic ventricular coordinates, J. Hirnforsch., 18:375-400.

Percheron, G., Yelnik, J., and François, C., 1984a, A Golgi analysis of the primate globus pallidus III. Spatial organization of the striato-pallidal complex. J. Comp. Neurol., 227:214-227.

Percheron, G., Yelnik, J., and François, C., 1984b, The primate striato-pallido-nigral system. An integrative system for cortical information, in: "Basal ganglia: Structure and function," J.S. McKenzie, R.E. Kem and L.N. Wilcock, eds, Plenum Press, New York.

Percheron, G., Yelnik, J., and François, C., 1986, Systems of coordinates for stereotactic surgery and cerebral cartography: advantages of ventricular systems in monkeys, J. Neurosci. Methods, 17:69-88.

Percheron, G., François, C., Yelnik, J., and Fénelon, G., 1989, The primate nigro-striato-pallido-nigral system. Not a mere loop, in: "Neural mechanisms in disorders of movement," A. Crossman and M. Sambrook, eds, John Libbey, London.

Powell, T.P.S., and Cowan, W.M., 1967, The interpretation of the degenerative changes in the intralaminar nuclei of the thalamus, J. Neurol. Neurosurg. Psychiat., 30:140-153.

Russchen, F.T., Bakst, I., Amaral, D.G., and Price, J.L., 1985, The amygdalostriatal projection in the monkey. An anterograde tracing study, Brain Res., 329:241-257.

Schulman, S., 1957, Bilateral symmetrical degeneration of the thalamus. A clinico-pathological study, J. Neuropath. Exp. Neurol., 16:446-470.

Simma, K., 1951, Zur projektion des Centrum medianum and Nucleus parafascicularis thalami beim Menschen, Mnschr. Psychiat. Neurol., 122:32-46.

Smith, Y., and Parent A., 1986, Differential connections of caudate nucleus and putamen in the squirrel monkey (Saimiri sciureus), Neuroscience, 18:347-371.

Vogt, C., 1909, La myéloarchitectonique du thalamus du cercopithèque, J. Psychol. Neurol. (Lpz), 12:285-324.

Vogt, C., and Vogt, O., 1941, Thalamostudien III. 3 Das griseum centrale (Centrum medianum Luys), J. Psychol. Neurol. (Lpz), 50:75-154.

Yakovlev, P.E., 1969, Development of the nuclei of the dorsal thalamus and of the cerebral cortex, in: "Modern Neurology," S. Lokce, ed, Little Brown, Boston.

THE ULTRASTRUCTURAL CHEMOANATOMY OF THE BASAL GANGLIA: 1984-1989

I. THE NEOSTRIATUM

Pedro Pasik[1,2], Tauba Pasik[1] and Gay R. Holstein[1,2]

Departments of Neurology[1] and Cell Biology-Anatomy[2]
Mount Sinai School of Medicine, CUNY
New York, NY, USA

INTRODUCTION

Our last review on the ultrastructural chemoanatomy of the basal ganglia (Pasik et al., 1986) covered the articles published through part of 1984. In our concluding remarks, we emphasized that the significant gaps in our knowledge at the time, could be expected to be filled by the application of new techniques that were just beginning to show some success. They were: (1) the development of antibodies directed against small molecules such as the neuroactive substances, as opposed to their synthesizing enzymes; (2) the use of more subtle markers such as ferritin or gold particles, which do not obscure the subcellular details of labeled structures; and (3) the possibility of differentially visualizing more than one substance in the same preparation, thereby improving the chances for the discovery of the interactions which characterize the neuronal circuits operating in the basal ganglia.

A survey of the research literature from 1984 to 1989 indicates a similar pattern of concentration as in the previous 5 years, namely a strong emphasis on the neostriatum, with modest efforts in the pallidum and substantia nigra, little in the subthalamic nucleus, and none on the pedunculopontine nucleus. Since the majority of the studies were conducted in the rat, the following discussion refers to this animal, unless stated otherwise as in the few instances where the monkey (Old World or New World species), cat or ferret were the experimental subjects. The theme of the present article is the neostriatum. Other components of the basal ganglia system are covered in the next chapter.

ABBREVIATIONS

ACh	Acetylcholine
DBH	Dopamine beta hydroxylase
CBP	Calcium-binding proteins
CCK	Cholecystokinin
CGRP	Calcitonin gene-related peptide
ChAT	Choline acetyltransferase
CP	Caudate-putamen
DA	Dopamine
ENK	Enkephalins
5-HT	Serotonin

The Basal Ganglia III, Edited by G. Bernardi *et al.*
Plenum Press, New York

GABA	Gamma aminobutyric acid
GAD	Glutamate decarboxylase
GLU	Glutamate
GP	Globus pallidus
HRP	Horseradish peroxidase
NADPH-D	Nicotinamide adenine dinucleotide phosphate-diaphorase
NN	Neuromedin-N
NPY	Neuropeptide Y
NT	Neurotensin
PPN	Pedunculopontine nucleus
SN	Substantia nigra
SNc	Substantia nigra, pars compacta
SNl	Substantia nigra, pars lateralis
SNr	Substantia nigra, pars reticulata
SP	Substance P
SS	Somatostatin
SThN	Subthalamic nucleus
TH	Tyrosine hydroxylase
VIP	Vasoactive intestinal polypeptide
VTA	Ventral tegmental area
WGA	Wheat germ agglutinin

NEOSTRIATUM

Inputs

The organization of the striatal neuropil into circuits involving several chemically specified subsystems has advanced greatly during the period covered by this survey. Additional data have been obtained on the termination of dopaminergic fibers from the SNc following the original description by Pickel et al. (1981). Quantitative morphologic and biochemical studies offer much needed information on the number and proportion of dopaminergic terminals in the striatum. Thus, the density of boutons uptaking DA is $1.0-1.6 \times 10^8/mm^3$ (Doucet et al., 1986), and the fraction of vesicle-containing elements originating in the SN is 5-10%, as judged by the decrease in vesicle-associated proteins (synapsin I and synaptophysin) in the striatum after nigral destruction (Walaas et al., 1988). The TH+ fibers, representing to a great extent DA-containing elements, are unmyelinated with almost round, large vesicles, and few, usually non-synaptic varicosities (Freund et al., 1984). The thinner portions of the fibers form many *en passant* symmetric synapses mostly with the neck of spines, which also receive on their heads a TH- bouton in an asymmetric synaptic junction. The remaining synapses occur with the shafts of distal spiny dendrites, and less frequently with perikarya and axon initial segments. The target structures in this study are strionigral neurons of the medium spiny type as identified by simultaneous retrograde filling with HRP deposited in the SN, and by Golgi impregnation. As discussed earlier (Pasik et al., 1986), these authors also point out the difficulties in distinguishing symmetric from asymmetric contacts in standard immunostained material which always carries some degree of uncertainty. This reservation applies to their own report as well, and is particularly important when attempting to correlate structure and function. Thus, when TH is labeled by an immunoautoradiographic method which obscures less of the cytologic detail, 20% of the synapses appear to be asymmetric and on spines (Aoki and Pickel, 1988). Such a variation in synaptic morphology may be a reflection of different DA receptors, which may correlate with the diverse action of DA on the postsynaptic cell. In fact, there is evidence that D_1 and D_2 receptors coexist in the same striatal cells (Ohno et al., 1987). A series of studies using double labeling techniques examined the connectivity of TH+ terminals with somata and dendrites of chemically identified striatal cells. Some of these neurons are

188

immunoreactive for GAD and belong to 3 types: spiny, aspiny, and a large cell present only in the ventral region of the CP (Kubota et al., 1987a). Other spiny neurons are ENK+ (Kubota et al., 1986b) or SP+ (Kubota et al., 1986a), and some aspiny cells, postsynaptic to TH+ profiles, are either of medium size and NPY+ (Kubota et al., 1988), or large and ChAT-immunoreactive (Kubota et al., 1987b). The latter findings are at variance with a previous report that failed to demonstrate synaptic contacts between TH+ terminals and giant, presumably cholinergic, neurons (Freund et al., 1984).

The serotoninergic input was reexamined by both radioautography after intraventricular injection of tritiated 5-HT, and immunocytochemistry with an antiserotonin antibody (Soghomonian et al., 1989). In accord with the original studies in the monkey using the latter method (Pasik and Pasik, 1982; Pasik et al., 1984), labeled profiles are found restricted to axonal varicosities and terminals, as well as to some unmyelinated fibers. The most common pattern, encompassing about 90% of the observed elements, is the absence of morphologically defined synapses. The overwhelming majority of the remaining 5-HT+ varicosities form asymmetric synaptic junctions with either spines or dendritic shafts. In the case of spines, only occasionally is there a convergent input from a 5-HT- bouton in a symmetric junction. The presence of dense core vesicles in the labeled varicosities is also observed in this study, raising the possibility of colocalization of other neuroactive substances (ENK, SP, GABA) within those vesicles as demonstrated in other CNS structures. The variously reported actions of 5-HT on striatal neurons could be attributed to the differential effects of the synaptically released transmitter, which at least morphologically could be assumed to exert an excitatory effect given the marked asymmetry of the junctional complexes, versus the non-synaptic action of 5-HT from varicosities in direct apposition to other axons. The latter circuitry could explain the reported presynaptic action of 5-HT on DA and ACh terminals as suggested by some investigators (Soghomonian et al., 1989). These speculations may eventually find support through the localization of $5-HT_1$ and $5-HT_2$ receptors on striatal neurons and axon terminals, respectively.

Cortical terminals constitute 40-50% of all striatal boutons as revealed by the proportional decrease of vesicle-associated proteins in the striatum after cortical ablation (Walaas et al., 1988). Unfortunately, the nature of this input remains elusive, although biochemical studies strongly suggest it is glutamatergic. While antibodies against GLU have become available, immunocytochemical results are not definitive due to the difficulty in distinguishing between the neurotransmitter and the metabolic pool of this substance.

The intralaminar thalamic nuclei, including the centromedian-parafascicular complex, contain neurons immunoreactive for a variety of peptides. In the cat, some of these cells are SP+, and up to 20% of them can be retrogradely filled with HRP deposited in the striatum, indicating that at least some of the thalamostriatal afferents utilize this neuroactive substance (Sugimoto et al., 1984). Similar experiments demonstrate that other neurons within the same thalamic nuclei projecting to the striatum are CCK+ and/or VIP+, but are not immunoreactive for NT (Sugimoto et al., 1985). Although ENK+ cells are also found in the centromedian-parafascicular complex, no verification of their thalamostriatal nature is available (Covenas et al., 1986). A controversy persists as to the role of ACh in this pathway. The lack of evidence for decreased striatal ChAT activity after ibotenic acid induced destruction of thalamic neurons, negating the involvement of ACh in this circuit (Barrington-Ward et al., 1984), has been challenged by the demonstration of lower ChAT levels in the neostriatum after kainic acid lesions in the thalamus (Nieoullon et al., 1985). It should be noted, however, that no cholinergic neurons are recognized in the thalamus by ChAT immunocytochemistry (Wainer et al., 1984). An alternative explanation for the findings is that removal of the thalamostriatal input, not necessarily cholinergic, might result in a marked decrease of activity in cholinergic striatal interneurons either directly or through an increase in dopaminergic activation (Nieoullon, 1986).

A lesser known input to the neostriatum is represented by fibers immunostained for CCK and originating in the pyriform cortex and amygdala. Some ultrastructural features of these afferents, which predominate in the medial, ventral and caudal striatal regions, have recently been reported (Takagi et al., 1984). CCK+ boutons of probably extrinsic origin have rather large, sparsely distributed synaptic vesicles and form asymmetric synapses with dendritic spines and shafts. Whether CCK is colocalized with DA in some nigrostriatal fibers is not known, although there is evidence for this coexistence in mesolimbic dopaminergic pathways (Hökfelt et al., 1980). Similar in distribution to CCK+ axons, the ventral and caudal zones of the CP exhibit a rich plexus of fibers immunoreactive for CGRP. Stained boutons form asymmetric synapses with spines as well as the strictures of varicose dendrites (Okayama et al., 1989). The authors suggest that neurons in the "ventrolateral" thalamic nucleus, which project to the caudal CP, may be the origins of these fibers. CGRP, which is generated by alternate RNA processing of the calcitonin gene,.modulates DA metabolism in frontal cortex (Deutch and Roth, 1987), and is a potent inhibitor of SP degradation (Le Greves et al., 1985). These properties make it a good candidate to influence DA and SP actions in the ventrocaudal region of the CP.

Although only light microscopic observations have been reported, it is noteworthy that 20-25% of posterior hypothalamic neurons immunoreactive for histamine or histidine decarboxylase are retrogradely labeled with a tracer deposited into the CP. This is the first evidence for the existence of a histaminergic innervation of the neostriatum (Steinbusch et al., 1986). A similar experimental approach provides some indication for a cholinergic input to the striatum originating in the PPN (Woolf and Butcher, 1986). Finally, the possibly GABAergic input from the SNr and/or GP remains without positive identification.

Neurons

Spiny I. The best characterized neuronal type of the neostriatum is the medium size spiny, long-axoned efferent cell, exhibiting high density of spines beyond the secondary dendrites, and abundant axonal collateralization within the structure. This type was named the Spiny I neuron in monkey Golgi material (for equivalent nomenclature see Pasik et al., 1979). It was further characterized as having a centrally located unindented nucleus, and scanty cytoplasm, poor in organelles (DiFiglia et al., 1980). Hereafter, the term Spiny I neuron will be used to designate cells with these features, frequently referred to in the rat as the medium spiny neuron.

The use of a direct GABA antibody, which does not require colchicine pretreatment to interfere with axonal transport processes, allows the demonstration in the monkey of numerous immunolabeled Spiny I cells, but with weaker stain than that present in the fewer neurons of the Aspiny type (see below: Aspiny I) (Pasik et al., 1988). The immunoreactivity is present in the soma, dendritic shafts, bundles of myelinated axons, and less frequently in the spines. It is noteworthy that *in situ* hybridization experiments also show weaker expression of GAD mRNA in numerous cells and a stronger reaction in fewer neurons, all of medium size (Chesselet et al., 1987).

Spiny I neurons also exhibit immunoreactivity for several peptides, such as ENK, SP, and NT. All ENK+ cells are directly recognized as Spiny I neurons by concurrent Golgi impregnation in the cat and ferret (Izzo et al.,

1987). The simultaneous immunolabeling of GABA (with the PAP method) and ENK (with a ferritin marker) in the monkey confirms at the ultrastructural level the colocalization of both substances in the somata of some Spiny I neurons (Pasik et al., 1986). Additional support for these findings is given by the presence of GAD or ENK immunoreactivity in Spiny I neurons retrogradely-filled with HRP-WGA deposited into the SN (Aronin et al., 1986). Also, over 30% of GAD+ neurons are found to be ENK+ as well (Penny et al., 1986). New quantitative data on Spiny I neurons show that SP immunoreactivity is present in 30-35% of striatal cells (Kubota et al., 1986a; Penny et al., 1986)). All SP+ neurons are characterized as being of the Spiny I type by demonstration of this immunoreactivity in Golgi impregnated and gold toned material of the cat and ferret (Izzo et al., 1987). Lastly, retrogradely labeled neurons with Spiny I features projecting to the GP, SNc and VTA showed NT immunoreactivity. Half of these cells are also ENK+, but only 15% are GABA+. A few show colocalization of the three neuroactive substances, and none are SP+ (Sugimoto and Mizuno, 1987).

Although not directly related to synaptic transmission, it is of interest that neurons of the Spiny I type in rat and monkey exhibit immunoreactivity for CBP (Gerfen et al., 1985; DiFiglia et al., 1989). The latter study documents that the label is present in somata, dendritic shafts and spines, as well as in axonal boutons which form symmetric synapses with the shafts of labeled dendrites . The authors speculate that the presence of CBP specifically in Spiny I neurons, may regulate the high glutamatergic input to these cells, presumed to be of cortical origin.

Regarding the synaptic inputs to Spiny I neurons, several types of boutons form junctions with the soma, dendritic shafts and spines, as well as the axon initial segments. The extrinsic afferents of dopaminergic or serotoninergic nature, as well as those immunoreactive for CCK or CGRP, and others of cortical or thalamic origin have been discussed above (see: Inputs). In addition, the Spiny I neuron receives a host of inputs intrinsic to the striatum from axon collaterals of the same cell type, as well as from short-axoned cells of the Aspiny categories. The boutons may be immunoreactive for GABA, GAD, ChAT, and a variety of peptides such as ENK, SP, NPY, SS, CCK, and VIP.

GABA+ boutons usually contain rather large, ovoid vesicles, similar to those described in the pallidum, where they are considered to be of striatal origin (Pasik et al., 1988). It is highly probable that they derive from local axon collaterals of the same Spiny I neurons establishing symmetric synapses mostly with dendritic shafts of cells of similar type. These junctions are also made with the strictures of GABA+ varicose dendrites, i.e. of the Aspiny I type. Other, less frequent GABA+ terminals with different mitochondrial features, probably belonging to axons of Aspiny I neurons (see below: Aspiny I) form symmetric synapses with spines (Fig. 1).

A newly observed interaction between GABA+ elements is the establishment of gap junctions, as found in the monkey neostriatum. Fig. 2 illustrates such a finding with the characteristic heptameric arrangement (Sotelo and Korn, 1978) between two profiles, one apparently axonic, the other dendritic, both GABA+. This description correlates with recent reports of dye-coupling in the rat striatum, where a low molecular weight dye injected intracellularly into one cell results in the stain of other neurons (Cepeda et al., 1989). The fact that the distance between somata of coupled cells is about 50 μm indicates that the responsible contacts occur at dendritic and/or axonal regions. Gap junctions have been observed in the CNS in several combinations: soma-soma, soma-dendrite, and axon initial segment-dendrite (Sotelo and Korn, 1978). The example in Fig. 2 may be the first between an axonal bouton and a dendrite, with additional chemical specification of both partners. It should be noted that, contrary to the

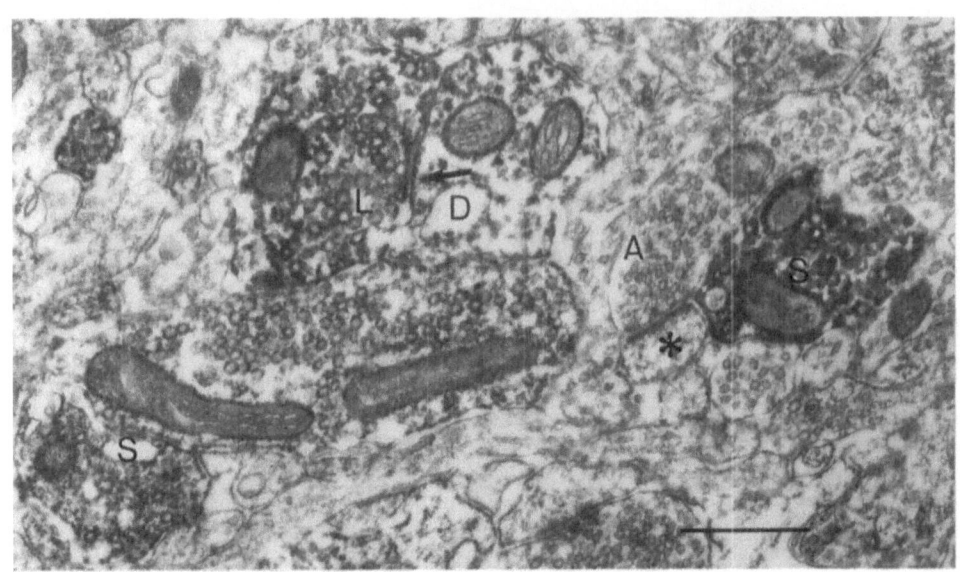

Fig. 1. GABA immunoreactivity in the monkey neostriatum.
Labeled boutons contain either large and ovoid (L)
or small and pleomorphic (S) vesicles. One of the
former is presynaptic to a GABA+ dendrite (D) at
arrow. A probable spine (*) receives a convergent
input from a GABA+ (S) and a GABA- (A) bouton,
forming symmetric and asymmetric synapses, respectively. Bar: 0.5 μm. (From Pasik et al., 1988).

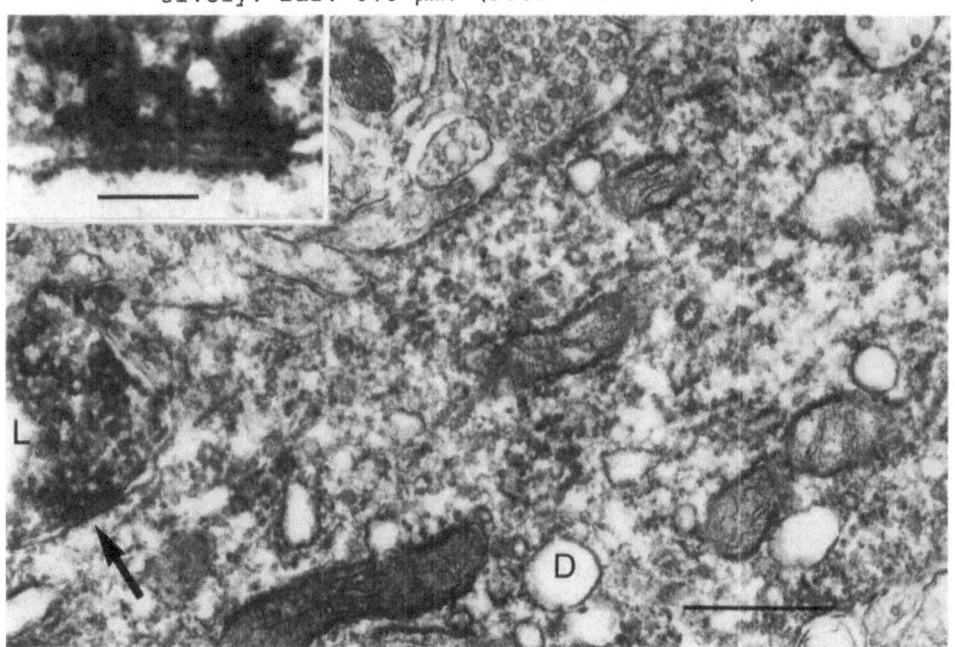

Fig. 2. Gap junction in monkey neostriatum. A GABA+ bouton
with large ovoid vesicles (L) in synaptic contact
with a strongly labeled dendrite (D) at arrow. Bar:
0.5 μm. Inset: Enlargement of the contact showing
the characteristic heptameric arrangement of a gap
junction. Bar: 0.1 μm.

findings in the inferior olivary complex (Bourrat and Sotelo, 1983), such electric synapses are infrequent in the adult striatum. It appears important to immunostain this structure at various developmental stages since the incidence of both gap junctions and dye-coupling decreases with age (Walsh et al., 1989). Such an inverse relationship has been attributed to the parallel age-related increase in DA which, at least in the retina, has an inhibiting role on the permeability of these junctions (Teranishi et al., 1983).

Identified Spiny I strionigral neurons are also postsynaptic to ENK+ boutons at the soma, proximal and distal dendrites and spines (Aronin et al., 1986; Kubota et al., 1986b). Most probably, these terminals derive from axon collaterals of Spiny I cells. Whether they contain both ENK and GABA has not yet been elucidated. SP+ boutons containing round, large vesicles form symmetric synapses most often with somata and proximal dendritic shafts of Spiny I neurons, and considerably less frequently with spines, some of which receive a convergent input from a SP- bouton in an asymmetric contact. In these studies, the Spiny I nature of the postsynaptic structures is verified by simultaneous Golgi impregnation and gold toning. Moreover, some of the cells are retrogradely filled with HRP-WGA injected into the SN, indicating their strionigral character. The origin of the SP+ terminals is attributed to collaterals of Spiny I axons and, to a lesser extent, to axonal boutons of aspiny interneurons (Bolam et al., 1986; Bolam and Izzo, 1988). Additional inputs to Spiny I neurons are represented by NPY+ boutons, which apparently establish symmetric synapses with dendritic shafts and the necks of spines. Some of these spines also have an asymmetric junction with an unlabeled terminal (Aoki and Pickel, 1988). Boutons of similar characteristics, but immunoreactive for SS, are observed in symmetric synaptic contacts with shafts of dendrites (DiFiglia and Aronin, 1984). It is possible that the same boutons contain both NPY and SS, as shown for their cell bodies of origin belonging to an aspiny interneuron type in the cat, monkey and human (Smith and Parent, 1986; Kowall et al., 1987). It has been suggested that Spiny I cells are postsynaptic to VIP+ boutons as well, and there is evidence that this peptide does not colocalize with SS (Theriault and Landis, 1987).

Finally, boutons immunostained for ChAT and containing pleomorphic vesicles form symmetric synapses with somata, proximal dendrites and axon initial segments of Spiny I neurons, as well as with the heads of spines, which also receive asymmetric junctions from ChAT- boutons (Phelps et al., 1985). Similar findings are reported in the monkey neostriatum, where this input is prevalent on spines, somewhat less so on shafts, and minimally on somata (DiFiglia, 1987). A different distribution is observed in the rat CP, with a greater proportion on dendritic shafts, followed by spines and perikarya (Izzo and Bolam, 1988). The latter study positively identifies the Spiny I nature of the postsynaptic partner by the electron microscopic examination of Golgi impregnated cells, as well as their strionigral category by retrogradely filling the cells with HRP deposited into the SN. It appears therefore that the cholinergic input to the Spiny I neuron is preponderant on distal segments in the monkey and more proximal regions in the rat.

Spiny II. The other long-axoned cell of the neostriatum, designated as Spiny II, is characterized in the monkey by the presence of spines less densely arranged than in the Spiny I, but appearing along the entire dendritic tree including proximal dendrites and occasionally even the soma. Another feature is the size spectrum, which extends from medium to large. The soma of the large Spiny II neuron is elongated and/or triangular, quite different than the globular, so-called, giant striatal cell or Aspiny II neuron (for review see Pasik et al., 1979). The neurochemical nature of the Spiny II cell is not clear. Some SP+ neurons may belong to this type as reviewed earlier (Pasik et al., 1986). An infrequent GABA+ cell is observed in the monkey neostriatum with some features of the large Spiny II type (Pasik et al., 1988), and similar GAD+ neurons are seen in the ventral region of the rat CP (Kubota et al., 1987a). Recent descriptions of ChAT+ cells in the rat striatum correlate better with the somatodendritic morphology of the large Spiny II cell than with the Aspiny II type (see below: Aspiny II) (Bolam et al., 1984; Phelps et al., 1985; Kubota et al., 1987b). This question, however, must remain open because of the long-axoned character of the Spiny II neuron that does not conform with the interneuronal nature of the striatal cholinergic cell.

Aspiny I. This neuronal type is characterized by its medium size, eccentric nucleus with deeply invaginated envelope, varicose dendrites, and locally arborizing beaded axons. Recently, a direct GABA antibody succeeded in labeling neurons of this sort in the monkey neostriatum, showing a considerably stronger immunoreactivity than that present in Spiny I neurons (Pasik et al., 1988). The results with GAD immunocytochemistry, which as a rule requires colchicine pretreatment to interfere with axonal transport processes, have varied. Some studies report only the stain of aspiny cells (Bolam et al., 1985), whereas others estimate at least 10-20% of the neurons, possibly aspiny, with strong reaction, and 30-40%, possibly spiny, with weak reaction (Kubota et al., 1987a). The Aspiny I neurons have been found to be immunoreactive to a variety of peptides: NPY in the rat, monkey and human (Smith and Parent, 1986; Kowall et al., 1987; Aoki and Pickel, 1988), as well as SS (DiFiglia and Aronin, 1984; Chesselet and Graybiel, 1986), CCK (Takagi et al., 1984), and SP (Bolam et al., 1986; Bolam and Izzo, 1988). The NPY+ cells are more abundant in the caudate nucleus than in the putamen, and the SS+ ones predominate in the "matrix" compartment. There is also evidence for colocalization of these two peptides (Smith and Parent, 1986; Kowall et al., 1987). CCK+ neurons prevail in the ventral half of the caudal CP. Although the possibility appears strong, whether any or all of these peptides are present within GABA+ Aspiny I neurons remains unresolved.

The GABA+ Aspiny I neurons are postsynaptic to a variety of elements, only two of which have been chemically specified, namely those being GAD+ (Bolam et al., 1985) or TH+ (Freund et al., 1984; Kubota et al., 1987a), both forming symmetric axosomatic and axodendritic synapses. The origin of the GABAergic terminals is not clear, since they may derive from Spiny I axon collaterals, other Aspiny I cells, and/or extrinsic axons from the SNr and GP (Bolam et al., 1985). The latter study did not report morphologic differences among the GAD+ terminals. GABA immunocytochemistry, however, allows the distinction of at least two kinds in the monkey neostriatum: one with rather large ovoid vesicles, which has been discussed above as belonging to collaterals of Spiny I axons, and the other with small pleomorphic vesicles interpreted as boutons of Aspiny I axons and/or of extrinsic origin (Pasik et al., 1988). In any event, the latter form symmetric synapses with spines having a convergent input from a GABA- terminal making an asymmetric junction (Fig. 1), as well as with GABA- somata and dendrites. Some of the non-GABAergic postsynaptic structures may belong to Spiny II and/or Aspiny II neurons. In fact, GAD+ boutons were observed in a presynaptic position to large GAD- neurons with ultrastructural features of ChAT+ cells (Bolam et al., 1985).

The earlier descriptions of SS immunoreactive Aspiny I neurons have been reviewed previously (Pasik et al., 1986). SS+ boutons derived from these cells represent 17% of the total number of boutons in the neostriatal neuropil, and make predominantly *en passant* symmetric and small axodendritic synapses (DiFiglia and Aronin, 1984). At least 26% of these contacts are found on spiny neurons.

194

The ultrastructural features of NPY+ neurons have recently been reported (Aoki and Pickel, 1988; Kubota et al., 1988; Vuillet et al., 1989a). Although these studies coincide in regard to the somatodendritic morphology of this cell type, they differ with respect to their catecholaminergic input. TH+ boutons are reported to form symmetric synapses with NPY+ somata and proximal dendrites by some authors (Kubota et al., 1988; Vuillet et al., 1989b), whereas only rare synapses are seen on dendrites and none on somata by others (Aoki and Pickel, 1988). Lastly, the latter study describes NPY+ unmyelinated varicose fibers and boutons, most probably belonging to the short axons of these aspiny neurons, forming symmetric contacts with dendritic shafts as well as with the neck and head of spines occasionally receiving a convergent asymmetric input from an unlabeled terminal. Very rarely, NPY+ boutons contact NPY+ dendrites. The staining intensity, as well as the number of NPY immunoreactive striatal neurons, increases significantly after 6-OH-DA lesions of the SN which deprive the striatum of DA input (Kerkerian et al., 1986), suggesting a tonic influence of DA on these neurons. Moreover, NPY induces an increase in DA turnover when injected into the striatum (Beal et al., 1986).

In addition to the presence of few CCK+ neurons with Aspiny I features, CCK+ boutons appear to be of two types: one with small, densely packed vesicles forming symmetric synapses with dendritic shafts, and the other with larger and less numerous vesicles establishing asymmetric synapses with shafts and spines (Takagi et al., 1984). The authors attribute the former to the CCK+ aspiny neurons, and the latter to an extrinsic source (see above: Inputs).

Aspiny II. Neurons with characteristics of the Aspiny II type (large, globular, with eccentric, deeply indented nucleus and abundant cytoplasm rich in organelles, with large Nissl bodies) exhibit ChAT immunoreactivity in rostral regions of the monkey caudate nucleus (DiFiglia, 1987). At variance with this study, the morphology of ChAT+ striatal cells in the rat resembles more the Spiny II than the Aspiny II neuron (Bolam et al., 1984; Phelps et al., 1985; Kubota et al., 1987b) (see above: Spiny II). If this discrepancy is confirmed further, particularly upon examination of other regions of the monkey striatum, it may indicate an interspecies variation in the cholinergic striatal interneuron.

Whatever their morphologic features, the only chemically specified inputs to the ChAT+ neurons, as revealed in double labeling studies, are that of TH+ and SP+ boutons. The former establish characteristically small, symmetric synapses with at least the soma and proximal dendrites (Kubota et al., 1987b). The latter, interpreted as collaterals of SP+ Spiny I neurons, also form numerous symmetric synapses of similar distribution (Bolam et al., 1986). However, many more classes of terminals synapse on the cholinergic cells, primarily on the dendrites (85%), both at varicosities and at strictures. There is indirect evidence that some of them, containing ovoid vesicles and establishing symmetric junctions, are GAD+ (Bolam et al., 1985). These properties make them likely to derive from collaterals of Spiny I axons. Other symmetric synapses are made by boutons with pleomorphic vesicles, some of which contact the axon initial segment in the rat but not the monkey. Still others with round vesicles, large or small, form asymmetric synapses with somata and dendrites, or just with dendrites, respectively (DiFiglia, 1987). It is noteworthy that some single unlabeled boutons are presynaptic to both ChAT+ dendrites and ChAT- dendritic shafts and spines (Phelps et al., 1985).

Some of the axons of ChAT+ neurons are myelinated in the rat (Phelps et al., 1985; Izzo and Bolam, 1988), a feature that is not apparent in the monkey (DiFiglia, 1987). It is possible that the myelinated axons may belong to ectopic cholinergic neurons of the nucleus basalis, or to axons from the latter structure passing through the striatum on their way to the cortex (Phelps et al., 1985). In fact, some of these neurons in the cat are filled retrogradely with HRP injected into various cortical fields (Dimova and Usunoff, 1989). All of these studies coincide in their descriptions of ChAT+ presynaptic varicosities as being small, packed with mostly pleomorphic vesicles and mitochondria, and forming short symmetric *en passant* synapses with Spiny I neurons. The location of these contacts has been reported as either the head of the spine (Phelps et al., 1985) or its neck (Izzo and Bolam, 1988). A frequently described arrangement includes a spine receiving a convergent input from a labeled terminal and an unlabeled bouton forming an asymmetric junction. Synaptic contacts of ChAT+ varicosities are also observed on proximal and distal dendrites, somata and axon initial segments, in this order of frequency. In addition, ChAT+ boutons synapse on somata of other neurons probably of the Aspiny I type (DiFiglia, 1987).

Aspiny III. This cell type has remained the most mysterious in the striatum, since its description in Golgi material as medium size, with rather straight, smooth dendrites and a very dense axonal arborization of fine beaded branches (for review see Pasik et al., 1979). There are indications, however, that this neuron could be VIP immunoreactive (Theriault and Landis, 1987). Such cells do not show immunoreactivity to SS, which is present in the Aspiny I type (see above: Aspiny I), or to NADPH-D which suggests that they may not contain NPY either (Vincent et al., 1983). Contrary to SS+ and SP+ neurons, VIP+ cells predominate in close association to traversing fiber bundles. The input to these neurons is mainly from unlabeled boutons forming symmetric axosomatic and axodendritic junctions, interpreted as deriving from collaterals of Spiny I axons. Postsynaptic targets include Spiny I dendritic shafts and spines, as well as VIP+ dendrites, some belonging to the same cell in an autaptic arrangement. It remains unknown whether the release of VIP in the striatum influences the cholinergic system, and/or contributes to the regulation of local energy metabolism as suggested in other CNS structures (Magistretti et al., 1981).

Undetermined Type. Neurons immunoreactive for TH are present in a rim of the neostriatum around all but the ventricular and pallidal borders (Dubach et al., 1987). These cells are found in monkeys but not in rodents. They have small somata (8x12 μm) with large indented nuclei, and receive axosomatic and axodendritic synapses, some of which are formed by TH+ boutons. No such cells are DBH immunoreactive, so that their dopaminergic nature is highly probable. Given their small size, they may belong to the neurogliform type, originally described by Ramón y Cajal (1911), and only seldom considered further (for review see Pasik et al., 1979).

SUMMARY

The preceding survey on the known chemically specified striatal neurons and their synaptic interactions strongly suggests that the functional unit of the neostriatum is the Spiny I cell, which receives multiple influences from within and outside the structure. As a rule, the extrinsic afferents of various sources contact more distal regions of the spine-laden dendritic tree and, with the exception of the majority of dopaminergic boutons, form asymmetric synapses, a morphology associated with excitatory function. The intrinsic afferents derive from collaterals of Spiny I axons and the axonal arborizations of the interneurons: Aspiny I (GABA and/or various peptides: NPY, SS, CCK), Aspiny II or a large version of a Spiny II (ACh), and Aspiny III (probably VIP). They all form symmetric synapses, suggesting inhibitory action, located in more proximal regions, namely dendritic shafts and somata, save for the cholinergic input which is also present in more distal zones. Particularly distinctive features are: (1) the convergence on individual spines of a symmetric and an asymmetric synapse, one of which is made by a bouton immunoreactive for TH, 5-HT, GABA, SP, NPY or ChAT; (2) the presence of gap junctions between processes of neighboring Spiny I cells; (3) the diffuse, non-synaptic influence of 5-HT varicosities. The Spiny I neuron itself is GABAergic, rich in Ca^{2+} binding proteins, and with the coexistence of one peptide (ENK, SP, NT) or even two (ENK and NT). The output of this cell provides the collaterals mentioned above, as well as the main striatal efferents innervating the pallidum and SN. The interneurons, in turn, are under the influence of at least some of the extrinsic inputs (dopaminergic) as well as the collaterals of Spiny I cells.

In the simplest terms, the excitatory inputs from the cortex, and perhaps also the thalamus, are remodeled at spatially restricted axospinous synapses by the modulatory action of DA, ACh, GABA and several peptides. The results of these interactions are then modified further by inhibitory synapses made on zones closer to the cell body.

ACKNOWLEDGEMENTS

The authors wish to thank Rosemary Lang and Victor Rodriguez for their expert assistance. The research and preparation of this survey were supported in part by Grants # R01 NS 22953, R01 NS 18657, R29 NS 24656, P50 NS 11631, and P30 EY 01867.

REFERENCES

A single list of references is given at the end of the following chapter for both articles.

THE ULTRASTRUCTURAL CHEMOANATOMY OF THE BASAL GANGLIA: 1984-1989

II. THE PALLIDUM, SUBSTANTIA NIGRA AND SUBTHALAMIC NUCLEUS

Pedro Pasik[1,2], Tauba Pasik[1] and Gay R. Holstein[1,2]

Departments of Neurology[1] and Cell Biology-Anatomy[2]
Mount Sinai School of Medicine, CUNY
New York, NY   USA

INTRODUCTION

The previous chapter covered our survey of the literature on the neostriatum. The present one completes the review of the other components of the basal ganglia system. A list of abbreviations can be found at the beginning of the preceding article.

PALLIDUM

Inputs

A few reports have provided additional information on the chemical nature of elements participating in pallidal circuits. The well-known inputs from the neostriatum, i.e. GABA+ and ENK+ to the lateral segment in the monkey, and the GP in non-primates, as well as GABA+ and SP+ to the medial segment and entopeduncular nucleus of the respective species, have been supplemented by the finding in the cat that some of the former but none of the latter fibers are also NT+ (Sugimoto and Mizuno, 1987). The results of combined retrograde labeling and postembedding GABA immunogold techniques, indicate that GABA+ boutons amount to 80% of the total input to those GP neurons projecting to the subthalamic nucleus in the rat. These terminals ensheathe pallidal dendrites, as is characteristic of the pallidal neuropil, and form symmetric synapses (Ingham et al., 1988). The boutons are considered to be primarily of striatal origin, although some may represent afferents from the SThN (Takada and Hattori, 1987). Other known inputs (5-HT, DA) have been reviewed earlier (Pasik et al., 1986). Novel afferents to pallidal neurons are fibers immunoreactive for CBP (DiFiglia et al., 1989) and CGRP (Chang and Kuo, 1989). The former are found in both segments of the monkey pallidum, where many labeled boutons establish symmetric axodendritic synapses, and are probably from the striatum. The CGRP axons predominate in caudal regions of the rat GP where they form a dense plexus, and their boutons make asymmetric synapses intermixed with many more symmetric contacts made on CGRP- profiles. It is noteworthy that retrograde tracers deposited into the GP and entopeduncular nucleus, label cells of the PPN, which are ChAT+, indicating that the latter structure provides cholinergic innervation to the pallidum (Woolf and Butcher, 1986). The input from the SThN appears to be glutamatergic (see below: Subthalamic Nucleus), and although there is indirect evidence that boutons of subthalamic origin establish asymmetric synapses mainly on dendritic shafts (Kita and Kitai,

The Basal Ganglia III, Edited by G. Bernardi *et al.*
Plenum Press, New York

1987), no ultrastructural information exists on their chemical specificity (Kitai and Kita, 1987; Smith and Parent, 1988).

Neurons

Earlier findings with the preembedding PAP technique in rat using GAD antibodies (Ribak et al., 1979; Mugnaini and Oertel, 1985), and monkey with GABA antibodies (Smith et al., 1987; Pasik et al., unpublished observations) show that most pallidal neurons are GABAergic. Additional support for this view derives from the labeling of many cells after the injection of ^{3}H-GABA into the SN (Takada and Hattori, 1987), a known target structure of the pallidal output. The latter study also reported numerous labeled neurons in the GP of the rat after depositing ^{3}H-GLY into the SThN, a finding which lends support to the earlier suggestion of the presence of GLY in the pallidosubthalamic pathway (for review see Pasik et al., 1986). GAD+ neurons, with all the characteristics of pallidal cells, have been observed in the ventral portion of the rat striatum (Bolam et al., 1985). They are rather few, and it is tempting to consider them as displaced pallidal neurons. In light of the results cited above, it is somewhat unsettling that GABA immunoreactivity is reportedly lacking in perikarya and dendrites of pallidal neurons labeled by the postembedding immunogold technique (Ingham et al., 1988).

Summary

A rather sketchy picture derives from the preceding review, allowing only a general formulation of the chemically specified systems in the pallidum, not far beyond our earlier treatment of the subject (Pasik et al. 1986). A potentially important, newly described, input to the pallidum is cholinergic, but no data on its synaptic articulations are available. The same applies to the presumably glutamatergic influence from the SThN. Evidence has accumulated on the GABAergic nature of most pallidal cells, although many of the pallidosubthalamic neurons also appear to be glycinergic. Whether or not GABA and GLY coexist in some of these cells is unknown. Similarly, the nature and connectivity of pallidal interneurons as well as the significance of the triadic synaptic arrangements present in the structure remain unresolved.

SUBSTANTIA NIGRA

Inputs

Two chemically identified inputs to the SN have received considerable attention during the period covered by this survey, namely those exhibiting GABA or ChAT immunoreactivity. Studies utilizing anterograde tracers and GABA immunogold techniques reveal that GABA+ terminals originating in the GP synapse on dendrites in the SNr and only few in the SNc (Smith and Bolam, 1989). The boutons are large, rich in mitochondria and pleomorphic vesicles. They form symmetric synapses with somata and large dendrites, and minimally with small dendrites. Moreover, ^{3}H-GABA deposited into the SN results in labeled pallidal neurons by retrograde transport of the specifically uptaken transmitter (Takata and Hattori, 1987). These investigations emphasize the role of GABA in the pallidonigral pathways. In the monkey, the most frequent type of GABA+ bouton in the SNr holds small and dark mitochondria, and is interpreted as of extrinsic, mostly striatal, origin (Holstein et al., 1986). The terminals establish symmetric synapses predominantly, but not exclusively, with GABA+ dendrites of the SNr of the class only partially covered with boutons. An example of this arrangement is given in Fig. 1. The simultaneous double labeling of GABA and ENK (Holstein and Pasik, 1987) shows that some of these boutons are immunoreactive for both substances (Fig. 2). Such a coexistence supports the view that they are of striatal

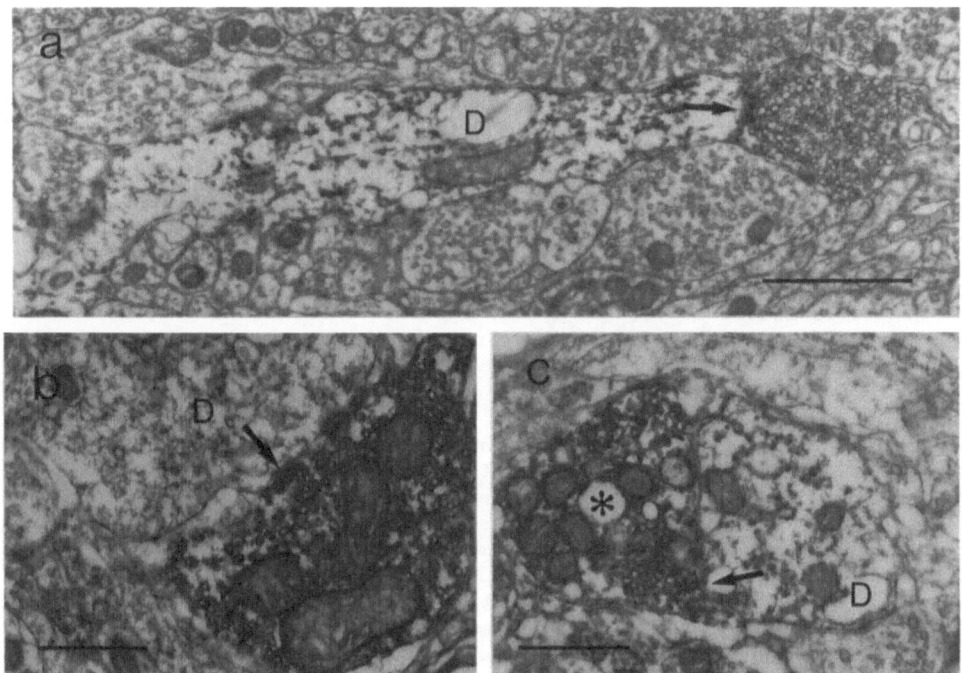

Fig. 1. GABA immunoreactivity in the monkey SNr. a: Labeled
        dendrite (D), only partially covered with boutons,
        one of which is also GABA+ and probably forms a
        synapse at arrow. Bar: 1.0 μm. b: Immunostained
        bouton synapses (arrow) with an unlabeled dendrite
        (D). c: Labeled bouton (*) containing darker and
        smaller mitochondria than those in b, possibly in
        synaptic contact (arrow) with a GABA+ dendrite (D).
        Bars: 0.5 μm. (From Holstein et al., 1986).

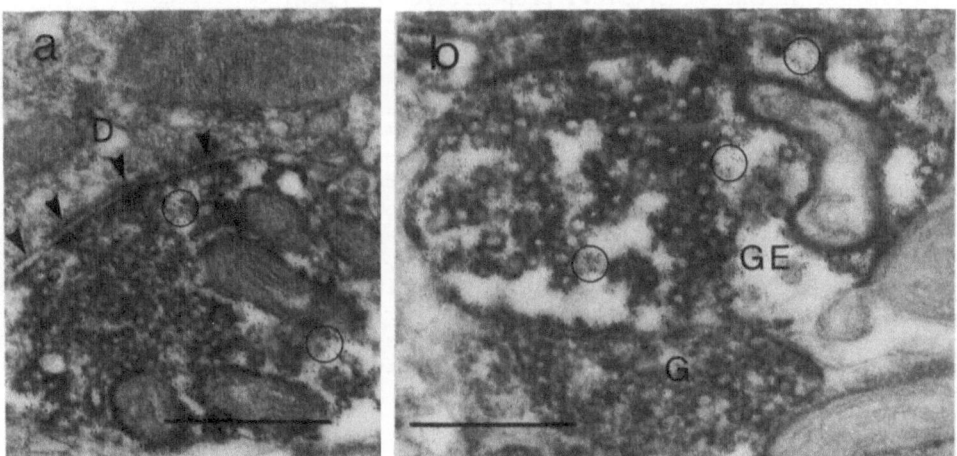

Fig. 2. GABA (coarse PAP label) and leu-enkephalin (7 nm
        ferritin particles) immunoreactivity in the monkey
        SNr. Some of the ferritin stipplings are encircled.
        a: A double-stained bouton forms a discontinuous
        synapse (arrow-heads) with unlabeled dendrite (D)
        bearing large mitochondria. b: Two immunoreactive
        boutons, either labeled for both GABA and ENK (GE)
        or only for GABA (G). Bars: 0.5 μm.

origin (see previous chapter). Quite characteristically, these terminals form discontinuous and extensive synaptic contacts with unlabeled dendrites. These earlier findings with GABA immunolabeling in the monkey have been confirmed more recently in the rat by GAD immunocytochemistry after lesions of the CP or interruption of the strionigral, and probably also pallidonigral, pathways (Nitsch and Riesenberg, 1988). Lastly, in a double labeling experiment, GAD+ boutons appear to synapse on TH+ somata and dendrites of the SNc, and on similar dendrites penetrating deeply into the SNr (Van den Pol et al., 1985). However, the extrinsic and/or intrinsic nature of the GABAergic elements is not specified.

There is little doubt that the SN receives a rich cholinergic innervation from the PPN (Woolf and Butcher, 1986; Benimato and Spencer, 1987). In the ferret, the plexus of ChAT+ fibers is denser in the SNc than in the SNr (Henderson and Greenfield, 1987), and a combined morphologic and pharmacologic experiment demonstrated that indeed ChAT+ neurons of the PPN directly influence cells of the SNc, and do so through nicotinic receptors (Clarke et al., 1987). ChAT+ boutons form asymmetric synapses with unlabeled, presumably dopaminergic, dendrites of the SNc (Martínez-Murillo et al., 1989a,b). Another chemically specified input to the SN in the cat is represented by fibers exhibiting NT immunoreactivity. These axons predominate in the ventrolateral region of the SNc, as well as in the VTA, and derive essentially from Spiny I striatal neurons (Sugimoto and Mizuno, 1987). A similar distribution is documented in a sequential double labeling study (Woulfe and Beaudet, 1989), which also shows that NT+ boutons form infrequent synapses with TH+ somata and dendrites as well as with TH-dendrites. The authors state that the NT antibody may also recognize NN, and note that both of these compounds activate dopaminergic neurons and DA metabolism.

Neurons

Most studies have focused on the SNr. In the monkey, numerous neurons of the SNr are GABA immunoreactive (Holstein et al., 1986; Smith et al., 1987). The latter investigation also reports a few small cells in the SNc with similar immunocytochemical properties. The GABA+ dendrites derived from SNr neurons bear large, pale mitochondria and are only partially covered by boutons, few of which are also GABA+ (Holstein et al., 1986). GABA-dendrites, presumably from SNc cells, are fully ensheathed with axon terminals, many of which are GABA+ and belong to at least two types, either with small and dark, or with large and pale mitochondria (Fig. 1). The latter are interpreted as collaterals of SNr neurons because they share mitochondrial features with the GABA+ dendrites and somata (Holstein et al., 1986), and are not ENK immunoreactive (Holstein and Pasik, 1987). The great majority of GABA+ boutons form symmetric synapses. Similar results are documented in the rat (Nitsch and Riesenberg, 1988), additionally showing that 60% of the boutons in the nigral neuropil are GAD+, and 60%-85% of them synapse on GAD+ dendrites. The latter study reports that GAD- dendrites, postsynaptic to GAD+ boutons may contain synaptic vesicles. This finding is reminiscent of the "presynaptic" dendrites present in the thalamus, and probably also in the pallidum.

Several ChAT immunocytochemical studies reveal the presence of some cholinergic neurons within or bordering the SN, although some negative findings have also been reported (Henderson and Greenfield, 1987). These cells predominate at caudal levels of the SNr and SNl (Woolf and Butcher, 1986; Gould and Butcher, 1986), but they are also observed in more rostral regions between these two subdivisions of the structure, and even intermingled with TH+ cells in the SNc (Martínez-Murillo et al., 1989b). The latter investigation describes the ChAT+ neurons as large (20-30μm), with a central nucleus showing few indentations. The soma and proximal dendrites are covered with glia except for a few unlabeled boutons with round vesicles forming asymmetric synapses, some with subjunctional dense bodies. Distal dendrites receive similar kinds of contacts, as well as some symmetric synapses made by boutons with flattened vesicles. The ChAT+ terminals cannot be ascribed with certainty to these cells, as opposed to afferents from the PPN (see above: Inputs). It should be noted that the ChAT+ neurons present in the SN are very similar to those in the PPN, and all the studies reviewed above suggest that they may represent ectopic elements of the pontomesencephalic tegmental system. Finally, it is of interest that some neurons of the SNc and VTA show coexistence of DA and NT (Hökfelt et al., 1984), and others exhibit immunoreactivity for CBP (Gerfen et al., 1985).

Summary

The present and earlier (Pasik et al., 1986) reviews allow some formulation, albeit quite incomplete, of the synaptic interactions among chemically specified components of the SN. Dopaminergic neurons, with somata in the SNc and dendrites in both SNc and SNr, are under the direct, probably excitatory, influence of the cortex (GLU?), striatum (SP), raphe nuclei (5-HT), and PPN (ACh acting on nicotinic receptors), as well as under inhibitory action from the striatum (GABA coexisting with ENK and/or NT) and GP (GABA). The latter inputs also impinge on the efferent GABAergic neurons of the SNr, the axon collaterals of which synapse on presumably dopaminergic dendrites. This circuitry helps to explain the apparently paradoxical finding of increased DA release upon striatal stimulation. Although such stimulation could initially inhibit DA neurons, it would also exert a similar action on SNr and pallidal cells, thereby removing their inhibitory influence on DA neurons which would thus be facilitated. The nature of the SNr interneurons and the significance of dendrodendritic synapses remains undefined.

SUBTHALAMIC NUCLEUS

Inputs

No noteworthy advances have been documented on the ultrastructure of chemically specified systems in the SThN. Yet, it is of interest that afferents from the GP uptake ^{3}H-GLY but not ^{3}H-GABA (Takata and Hattori, 1987), supporting the suggestion that GLY may play a role in the pallidosubthalamic pathway, as reviewed earlier (Pasik et al., 1986). Numerous GABA+ puncta are observed, however, surrounding GABA- cell bodies and dendrites in the monkey SThN (Smith et al., 1987). It is not clear, however, whether these terminals derive from the GP and/or other structures. Lastly, evidence has been provided for a cholinergic projection to the SThN from the PPN (Woolf and Butcher, 1986).

Neurons

The nature of SThN neurons projecting to the GP is probably glutamatergic, as suggested by their GLU immunoreactivity in both rat (Kitai and Kita, 1987) and monkey (Smith and Parent, 1988).

Summary

The SThN remains a poorly understood structure from the viewpoint of the synaptology of chemically specified elements. It is of interest, however, that contrary to the classic view, the majority if not all of its neurons appear to contain an excitatory amino acid. This finding, together with the asymmetric synapses made by terminals of subthalamopallidal fibers, help to explain why subthalamic lesions result in a hyperkinetic syndrome. Removal of the SThN excitatory action would disfacilitate the main target structures, the medial pallidal segment in primates and the entopeduncular nucleus in non-primate species, thereby decreasing their inhibitory influence upon the thalamocortical circuits.

CONCLUDING COMMENTS

The overall conclusion of this and the preceding chapter is that the past five years have witnessed an explosion of information that is leading toward a thorough understanding of the synaptic and, at times, non-synaptic transactions occurring in the basal ganglia, particularly regarding the neostriatum. Little is known, however, of the differential synaptology of the various histochemically recognized striatal compartments, and large gaps remain in our knowledge of several other components of the system, such as the pallidum, SThN and PPN. Although these issues stand unresolved, it is expected that during the next five years, most of the characters and their relationships will be recognized. This goal will be accomplished through the extension of the present techniques, and a new level of sophistication, presently in its infancy, which involves the ultrastructural visualization of specific receptor sites, second messenger molecules, and genetic determinants for the synthesis of both neuroactive substances and their receptors.

ACKNOWLEDGEMENTS

The authors wish to thank Rosemary Lang and Victor Rodriguez for their expert assistance. The research and preparation of this survey were supported in part by Grants # R01 NS 22953, R01 NS 18657, R29 NS 24656, P50 NS 11631, and P30 EY 01867.

REFERENCES

The following list applies to this and the preceding chapter.

Aoki, C., and Pickel, V.M. , 1988, Neuropeptide Y-containing neurons in the rat striatum: ultrastructure and cellular relations with tyrosine hydroxylase-containing terminals and with astrocytes, Brain Res., 459:205-225.

Aronin, N., Chase, K., and DiFiglia, M. , 1986, Glutamic acid decarboxylase and enkephalin immunoreactive axon terminals in the rat neostriatum synapse with striatonigral neurons, Brain Res., 365:151-158.

Barrington-Ward, S.J., Kilpatrick, I.C., Phillipson, O.T., and Pycock, C.J. , 1984, Evidence that thalamic efferent neurones are non-cholinergic: a study in the rat with special reference to the thalamostriatal pathway, Brain Res., 299:146-151.

Beal, M.F., Frank, R.C., Ellison, D.W., and Martin, J.B. , 1986, The effect of neuropeptide Y in striatal catecholamines, Neurosci. Lett., 71:118-125.

Beninato, M., and Spencer, R.F. , 1987, A cholinergic projection to the rat substantia nigra from the pedunculopontine tegmental nucleus, Brain Res., 412:169-174.

Bolam, J.P., Ingham, C.A., Izzo, P.N., Levey, A.I., Rye, D.B., Smith, A.D., and Wainer, B.H. , 1986, Substance P-containing terminals in synaptic contact with cholinergic neurons in the neostriatum and basal forebrain: a double immunocytochemical study in the rat, Brain Res., 397:279-289.

Bolam, J.P., and Izzo, P.N. , 1988, The postsynaptic targets of substance P-immunoreactive terminals in the rat neostriatum with particular reference to identified spiny striatonigral neurons, Exp. Brain Res., 70:361-377.

Bolam, J.P., Wainer, B.H., and Smith, A.D. , 1984, Characterization of cholinergic neurons in the rat neostriatum. A combination of choline acetyltransferase immunocytochemistry, Golgi impregnation and electron microscopy, Neuroscience, 12:711-718.

Bolam, J.P., Powell, J.F., Wu, J.-Y., and Smith, A.D. , 1985, Glutamate decarboxylase-immunoreactive structures in the rat neostriatum: A correlated light and electron microscopic study including a combination of Golgi impregnation with immunocytochemistry, J. Comp. Neurol., 237:1-20.

Bourrat, F., and Sotelo, C. , 1983, Postnatal development of the olivary complex in the rat. I. An electron microscopic study of the medial accessory olive, Develop. Brain Res., 8:291-310.

Cepeda, C., Walsh, J.P., Hull, C.D., Howard, S.G., Buchwald, N.A., and Levine, M.S., 1989, Dye-coupling in the neostriatum of the rat: I. Modulation by dopamine depleting lesions, Synapse, 4:229-237.

Chang, H.T., and Luo, H. , 1989, Calcitonin gene-related peptide (CGRP) in the rat substantia innominata and globus pallidus: a light and electron microscopic immuncytochemical study, Brain Res., 495:167-172.

Chesselet, M.-F., and Graybiel, A.M. , 1986, Striatal neurons expressing somatostatin-like immunoreactivity: Evidence for a peptidergic interneuronal system in the cat, Neuroscience, 17:547-571.

Chesselet, M.-F., Weiss, L., Wuenschell, C., Tobin, A.J., and Affolter, H.-U. , 1987, Comparative distribution of mRNAs for glutamic acid decarboxylase, tyrosine hydroxylase, and tachynins in the basal ganglia: an in situ hybridization study in the rodent brain, J. Comp. Neurol., 262:125-140.

Clarke, P.B.S., Homer, D.W., Pert, A., and Skirboll, L.R. , 1987, Innervation of substantia nigra neurons by cholinergic afferents from pedunculopontine nucleus in the rat: neuroanatomical and electrophysiological evidence, Neuroscience, 23:1011-1019.

Covenas, R., Romo, R., Cheramy, A., Cesselin, F., and Conrath, M. , 1986, Immunocytochemical study of enkephalin-like cell bodies in the thalamus of the cat, Brain Res., 377:355-361.

Deutch, A.Y., and Roth, R.H. , 1987, Calcitonin gene-related peptide in the ventral tegmental area: selective modulation of prefronatl cortical dopamine metabolism, Neurosci. Lett., 74:169-174.

DiFiglia, M., 1987, Synaptic organization of cholinergic neurons in the monkey neostriatum, J. Comp. Neurol., 255:245-258.

DiFiglia, M., and Aronin, N. , 1984, Quantitative electron microscopic study of immunoreactive somatostatin axons in the rat neostriatum, Neurosci. Lett., 50:325-331.

DiFiglia, M., Christakos, S., and Aronin, N. , 1989, Ultrastructural localization of immunoreactive calbindin-D$_{28k}$ in the rat and monkey basal ganglia, including subcellular distribution with colloidal gold labeling, J. Comp. Neurol., 279:653-665.

DiFiglia, M., Pasik, T., and Pasik, P. , 1980, Ultrastructure of Golgi impregnated and gold-toned spiny and aspiny neurons in the monkey neostriatum, J. Neurocytol., 9:471-492.

Dimova, R.V., and Usunoff, G. , 1989, Cortical projection of giant neostriatal neurons in the cat. Light and electron microscopic horseradish peroxidase study, Brain Res. Bull., 22:489-499.

Doucet, G., Descarries, L., and Garcia, S. , 1986, Quantification of the dopamine innervation in adult rat neostriatum, Neuroscience, 2:427-445.

Dubach, M., Schmidt, R., Kunkel, D., Bowden, D.M., Martin, R., and German, D.C. , 1987, Primate neostriatal neurons containing tyrosine hydroxylase: immunohistochemical evidence, Neurosci. Lett., 75:205-210.

Freund, T.F., Powell, J.F., and Smith, A.D. , 1984, Tyrosine hydroxylase-immunoreactive boutons in synaptic contact with identified striatonigral neurons, with particular reference to dendritic spines, Neuroscience, 13:1189-1215.

Gerfen, C.R., Baimbridge, K.G., and Miller, J.J. , 1985, The neostriatal mosaic: compartmental distribution of calcium-binding protein and parvalbumin in the basal ganglia of the rat and monkey, Proc. Natl. Acad. Sci. USA, 82:8780-8784.

Gould, E., and Butcher, L.L. , 1986, Cholinergic neurons in the rat substantia nigra, Neurosci. Lett., 63:315-319.

Henderson, Z., and Greenfield, S.A. ,.1987, Does the substantia nigra have a cholinergic innervation?, Neurosci. Lett., 23:109-113.

Hökfelt, T., Everitt, H.J., Theodorsson-Norheim, E., and Goldstein, M. , 1984, Occurrence of neurotensinlike immunoreactivity in subpopulations of hypothalamic, mesencephalic, and medullary catecholamine neurons, J. Comp. Neurol., 222:543-559.

Hökfelt, T., Skirboll, L., Rehfeld, M.F., Goldstein, M., Markey, K., and Dann, O. , 1980, A subpopulation of mesencephalic dopamine neurons projecting to limbic areas contains a cholecystokinin-like peptide. Evidence from immunohistochemistry combined with retrograde tracing, Neuroscience, 5:2093-2124.

Holstein, G.R., and Pasik, P. , 1987, Synaptology of immunocytochemically-identified GABAergic and enkephalinergic terminals in monkey substantia nigra, Soc. Neurosci. Abstr., 13:28.

Holstein, G.R., Pasik, P., and Hámori, J. , 1986, Synapses between GABA-immunoreactive axonal and dendritic elements in monkey substantia nigra, Neurosci. Lett., 66:316-322.

Ingham, C.A., Bolam, J.P., and Smith, A.D. , 1988, GABA-immunoreactive synaptic boutons in the rat basal forebrain: comparison of neurons that project to the neocortex with pallidosubthalamic neurons, J. Comp. Neurol., 273:263-282.

Izzo, P.N., and Bolam, J.P. , 1988, Cholinergic synaptic input to different parts of spiny striatonigral neurons in the rat, J. Comp. Neurol., 269:219-234.

Izzo, P.N., Graybiel, A.M., and Bolam, J.P. , 1987, Characterization of substance P- and [Met]enkephalin-immunoreactive neurons in the caudate nucleus of cat and ferret by a single section Golgi procedure, Neuroscience, 20:577-582.

Kita, H., and Kitai, S.T. , 1987, Efferent projections of the subthalamic nucleus in the rat: light and electron microscopic analysis with the PHA-L method, J. Comp. Neurol., 260:435-452.

Kitai, S.T., and Kita, H. , 1987, Anatomy and physiology of the subthalamic nucleus: a driving force of the basal ganglia, in: "The Basal Ganglia II. Structure and Function-Current Concepts," M.B. Carpenter and A. Jayaraman, eds., Adv. Behav. Biol., 32:357-373, Plenum Press, New York.

Kerkerian, L., Bosler, O., and Pelletier, G. , 1986, Striatal neuropeptide Y neurons are under the influence of the nigrostriatal dopaminergic pathway: immunohistochemical evidence, Neurosci. Lett., 66:106-112.

Kowall, N.W., Ferrante, R.J., Beal, M.F., Richardson Jr., E.P., Sofroniew, M.V., Cuello, A.C., and Martin, J.B. , 1987, Neuropeptide Y, somatostatin, and reduced nicotinamide adenine dinucleotide phospate diaphorase in the human striatum: a combined immunocytochemical and enzyme histochemical study, Neuroscience, 20:817-828.

Kubota, Y., Inagaki, S., and Kito, S. , 1986a, Innervation of substance P neurons by catecholaminergic terminals in the neostriatum, Brain Res., 375:163-167.

Kubota, Y., Inagaki, S., Kito, S., Takagi, H., and Smith, A.D. , 1986b, Ultrastructural evidence of dopaminergic input to enkephalinergic neurons in rat neostriatum, Brain Res., 367:374-378.

Kubota, Y., Inagaki, S., Kito, S., and Wu, J.-Y. , 1987a, Dopaminergic axons directly make synapses with GABAergic neurons in the rat neostriatum, Brain Res., 406:147-156.

Kubota, Y., Inagaki, S., Shimada, S.., Kito, S., Eckenstein, F., and
    Tohyama, M. , 1987b, Neostriatal cholinergic neurons receive direct
    synaptic inputs from dopaminergic axons, Brain Res., 413:179-184.
Kubota, Y., Inagaki, S., Kito, S., Shimada, S., Okayama, T., Hatanaka, H.,
    and Pelletier, G. , 1988, Neuropeptide Y-immunoreactive neurons
    receive synaptic inputs from dopaminergic axon terminals in the rat
    neostriatum, Brain Res., 458:389-393.
Le Greves, P., Nyberg, F., Terenius, L., and Hökfelt, T. , 1985, Calcitonin
    gene-related peptide is a potent inhibitor of substance P degradation,
    Eur. J. Pharmacol., 115:309-311.
Magistretti, P.J., Morrison, J.H., Shoemaker, W.J., Sapin, V., and Bloom,
    F.E. , 1981, Vasoactive intestinal polypeptide induces glycogenolysis
    in mouse cortical slices: A possible regulatory mechanism for the
    local control of energy metabolism, Proc. Ntl. Acad. Sci. USA,
    78:6535-6539.
Martínez-Murillo, R., Villalba, R., Montero-Caballero, M.I., and Rodrigo, J.
    , 1989a, Cholinergic somata and terminals in the rat substantia nigra:
    an immunocytochemical study with optical and electron microscopic
    techniques, J. Comp. Neurol., 281:397-415.
Martínez-Murillo, R., Villalba, R.M., and Rodrigo, J. , 1989b, Electron
    microscopic localization of cholinergic terminals in the rat
    substantia nigra: an immunocytochemical study, Neurosci. Lett.,
    96:121-126.
Mugnaini, E., and Oertel, W.H. , 1985, An atlas of the distribution of
    GABAergic neurons in the rat CNS as revealed by GAD
    immunohistochemistry, in: "Handbook of Chemical Neuroanatomy: GABA and
    Neuropeptides in the CNS," A. Björklund and T. Hökfelt, eds.,
    Elsevier, Amsterdam, pp. 436-608.
Nieoullon, A., 1986, Reply to the letter to the editor by Kilpatrick and
    Phillipson, Neurosci. Lett., 67:98-99.
Nieoullon, A., Scarfone, E., Kerkerian, L., Errami, M., and Dusticier, N. ,
    1985, Changes in choline acetyltransferase, glutamic acid
    decarboxylase, high-affinity glutamate uptake and dopaminergic
    activity induced by kainic acid lesion of the thalamostriatal neurons,
    Neurosci. Lett., 58:299-304.
Nitsch, C., and Riesenberg, R. , 1988, Immunocytochemical demonstration of
    GABAergic synaptic connections in rat substantia nigra after different
    lesions of the striatonigral projection, Brain Res., 461:127-142.
Ohno, Y., Sasa, M, and Takaori, S. , 1987, Coexistence of inhibitory
    dopamine D-1 and excitatory D-2 receptors on the same caudate nucleus
    neurons, Life Sci., 40:1937-1945.
Okayama, T., Kubota, Y., Kito, S., Funaki, H., Shimada, S., Takagi, H., and
    Inagaki, S. , 1989, A light and electron microscopic study of
    calcitonin gene-related peptide.in the rat caudate putamen, Brain Res.
    Bull., 22:657-663.
Pasik, P., Pasik, T., and DiFiglia, M. , 1979, The internal organization of
    the neostriatum in mammals, in: "The Neostriatum," I. Divac and R.G.E.
    Oberg, eds., Pergamon Press, Oxford, pp. 5-36.
Pasik, P., Pasik, T., Hámori, J., and Holstein, G.R. , 1986, Light and
    electron microscopic visualization of GABAergic elements in the monkey
    brain by means of a direct GABA antibody, in: "GABA in Endocrine
    Function," G. Racagni and A.O. Donoso, eds., Adv. Biochem.
    Psychopharmacol., 42:13-24, Raven Press, New York.
Pasik, P., Pasik, T., and Holstein, G.R. , 1986, Ultrastructural
    chemoanatomy of the basal ganglia: an overview, in: "Parkinson's
    Disease," M.D. Yahr and K.J. Bergmann, eds., Adv. Neurol., 45:59-66,
    Raven Press, New York.
Pasik, P., Pasik, T., Holstein, G.R., and Hámori, J. , 1988, GABAergic
    elements in the neuronal circuits of the monkey neostriatum: a light

and electron microscopic immunocytochemical study, <u>J. Comp. Neurol.</u>, 270:157-170.

Pasik, P., Pasik, T., Holstein, G.R., and Pecci Saavedra, J. , 1984, Serotoninergic innervation of the monkey basal ganglia: an immunocytochemical, light and electron microscopic study, <u>in</u>: "The Basal Ganglia. Structure and Function," J. McKenzie, R.E. Kemm and L.N. Wilcock, eds., Adv. Behav. Biol., 27:115-129, Plenum Press, New York.

Pasik, T., and Pasik, P. , 1982, Serotoninergic afferents in monkey neostriatum, <u>Acta Biol. Acad. Sci. Hung.</u>, 33:277-288.

Penny, G.R., Afsharpour, S., and Kitai, S.T. , 1986, The glutamate decarboxylase-, leucine enkephalin-, methionine enkephalin- and substance P-immunoreactive neurons in the neostriatum of the rat and cat: evidence for partial population overlap, <u>Neuroscience</u>, 17:1011-1045.

Phelps, P.E., Houser, C.R., and Vaughn, J.E. , 1985, Immunocytochemical localization of choline acetyltransferase within the rat neostriatum: a correlated light and electron microscopic study of cholinergic neurons and synapses, <u>J. Comp. Neurol.</u>, 238:286-307.

Pickel, V.M., Beckley, S.C., Joh, T.H., and Reis, D.J. , 1981, Ultrastructural immunocytochemical localization of tyrosine hydroxylase in neostriatum, <u>Brain Res.</u>, 225:373-385.

Ramón y Cajal, S., 1911, "Histologie du Système Nerveux de l'Homme et des Vertébrés, Vol. 2, Maloine, Paris, pp. 504-518.

Ribak, C.E., Vaughn, J.E., Saito, K., Barber, R., and Roberts, E. , 1976, Immunocytochemical localization of glutamate decarboxylase in rat substantia nigra, <u>Brain Res.</u>, 116:287-298.

Smith, Y., and Bolam, J.P. , 1989, Neurons of the substantia nigra reticulata receive a dense GABA-containing input from the globus pallidus in the rat, <u>Brain Res.</u>, 493:160-167.

Smith, Y., and Parent, A. , 1986, Neuropeptide Y-immunoreactive neurons in the striatum of cat and monkey: morphological characteristics, intrinsic organization and co-localization with somatostatin, <u>Brain Res.</u>, 372:241-252.

Smith, Y., and Parent, A. , 1988, Neurons of the subthalamic nucleus in primates display glutamate but not GABA immunoreactivity, <u>Brain Res.</u>, 453:353-356.

Smith, Y, Parent, A., Seguela, P., and Descarries, L. , 1987, Distribution of GABA-immunoreactive neurons in the basal ganglia of the squirrel monkey (Saimiri sciureus), <u>J. Comp. Neurol.</u>, 259:50-61.

Soghomonian, J.-J., Descarries, L., and Watkins, K.C. , 1989, Serotonin innervation in adult rat neostriatum. II. Ultrastructural features: a radioautographic and immunocytochemical study, <u>Brain Res.</u>, 481:67-86.

Sotelo, C., and Korn, H. , 1977, Morphological correlates of electrical and other interactions through low-resistance pathways between neurons of the vertebrate central nervous system, <u>Int. Rev. Cytol.</u>, 55:67-107.

Steinbusch, H.W.M., Sauren, Y., Groenewegen, H., Watanabe, T., and Mulder, A.H. , 1986, Histaminergic projections from the premammillary and posterior hypothalamic region to the caudate-putamen complex in the rat, <u>Brain Res.</u>, 368:389-393.

Sugimoto, T., Itoh, K., Yasui, Y., Kaneko, T., and Mizuno, N. , 1985, Coexistence of neuropeptides in projection neurons of the thalamus in the cat, <u>Brain Res.</u>, 347:381-384.

Sugimoto, T., and Mizuno, N. , 1987, Neurotensin in projection neurons of the striatum and nucleus accumbens, with reference to coexistence with enkephalin and GABA: an immunohistochemical study in the cat, <u>J. Comp.Neurol.</u>, 257:383-395.

Sugimoto, T., Takada, M., Kaneko, T., and Mizuno, N. , 1984, Substance P-positive thalamocaudate neurons in the center median-parafascicular complex in the cat, <u>Brain Res.</u>, 323:181-184.

Takada, M., and Hattori, T. , 1987, Glycine: an alternative transmitter candidate of the pallidosubthalamic projection neurons in the rat, J. Comp. Neurol., 262:165-172.

Takagi, H., Mizuta, H., Matsuda, T., Inagaki, S., Tateishi, K., and Hamaoka, T. , 1984, The occurrence of cholecystokinin-like immunoreactive neurons in the rat neostriatum: light and electron microscopic analysis, Brain Res., 309:346-349.

Teranishi, T., Negishi, K., and Kato, S. , 1983, Dopamine modulates S-potential amplitude and dye-coupling between external horizontal cells in carp retina, Nature, 301:243-246.

Theriault, E., and Landis, D.M.D. , 1987, Morphology of striatal neurons containing VIP-like immunoreactivity, J. Comp. Neurol., 256:1-13.

van den Pol, A.N., Smith, A.D.., and Powell, J.F , 1985, GABA axons in synaptic contact with dopamine neurons in the substantia nigra: double immunocytochemistry with biotin-peroxidase and protein A-colloidal gold, Brain Res., 348:146-15.

Vincent, S.R., and Johansson, O. , 1983, Striatal neurons containing both somatostatin and avian pancreatic polypeptide (APP)-like immunoreactivities and NADPH-diaphorase activity: A light and electron microscopic study, J. Comp. Neurol., 217:264-270.

Vuillet, J., Kerkerian, L., Kachidian, P., Bosler, O., and Nieoullon, A. , 1989a, Ultrastructural correlates of functional relationships between nigral dopaminergic or cortical afferent fibers and neuropeptide Y-containing neurons in the rat striatum, Neurosci. Lett., 100:99-104.

Vuillet, J., Kerkerian, L., Salin, P., and Nieoullon, A. , 1989b, Ultrastructural features of NPY-containing neurons in the rat striatum, Brain Res., 477:241-251.

Wainer, B.H., Levey, A.I., Mufson, E.J., and Mesulam, M.-M. , 1984, Cholinergic systems in mammalian brain identified with antibodies against choline acetyltransferase, Neurochem. Int., 6:163-182.

Walaas, S.I., Jahn, R., and Greengard, P. , 1988, Quantitation of nerve terminal populations: synaptic vesicle-associated proteins as markers for synaptic density in the rat neostriatum, Synapse, 2:516-520.

Walsh, J.P., Cepeda, C., Hull, C.D., Fisher, R.S., Levine, M.S., and Buchwald, N.A. , 1989, Dye-coupling in the neostriatum of the rat: II. Decreased coupling between neurons during development, Synapse, 4:238-247.

Woolf, N.J., and Butcher, L.L. , 1986, Cholinergic systems in the rat brain: III. Projections from the pontomesencephalic tegmentum to the thalamus, tectum, basal ganglia, and basal forebrain, Brain Res. Bull., 16:603-637.

Woulfe, J., and Beaudet, A. , 1989, Immunocytochemical evidence for direct connections between neurotensin-containing axons and dopaminergic neurons on the rat midbrain tegmentum, Brain Res., 479:402-406.

PHYSIOLOGY OF BASAL GANGLIA
COMPONENTS

DYE-COUPLING IN THE NEOSTRIATUM OF THE RAT

Carlos Cepeda, John P. Walsh, Chester D. Hull
Nathaniel A. Buchwald and Michael S. Levine

Mental Retardation Research Center
University of California
Los Angeles, CA 90024

INTRODUCTION

Morphologically, gap junctions are intermembranous channels between cells thought to mediate electrotonic transmission and to serve as a passage for small molecules in the CNS (for a review see Sotelo and Korn, 1978). After the discovery that the fluorescent dye Lucifer yellow crosses gap junctions to label adjacent cells (Stewart, 1978), dye-coupling was used as indirect evidence for the presence of gap junctions. Dye-coupling occurs in several neural areas, including the hippocampus (Knowles et el., 1982; MacVicar et al., 1982), the neocortex (Connors et al., 1983; Gutnick and Prince, 1981) and the hypothalamus (Andrew et al., 1981; Hatton et al., 1987). In parallel ultrastructural studies, gap junctions have been shown to be present in some of these structures (Sloper and Powell, 1978; Sotelo and Korn, 1978).

In the neostriatum, close appositions between cells have been observed (Adinolfi, 1977). However, to our knowledge there is no direct evidence for intercellular communication mediated by gap junctions. On the basis of experiments performed in the retina of fish and turtles, Rogawski (1987) suggested that one possible role for dopamine was the modulation of gap junction permeability. The neostriatum provides an ideal model system for analysis of dopaminergic modulation of gap junction permeability because dopamine is found in high concentrations and it is predominantly provided by unilateral projections from the substantia nigra.

In this paper we describe two experiments concerning the modulation of dye-coupling in the neostriatum. The first assessed the changes in the incidence of dye-coupling in the neostriatum of developing rats. Dopamine concentration increases over the first postnatal months (Tennyson et al., 1972; Santiago et al., 1987). Therefore, it is of interest to determine if decreases in the incidence of dye-coupling parallel increases in dopamine concentration. The second experiment assessed the effects of removing dopamine from the neostriatum in adult rats. In the retina, destruction of dopamine-containing horizontal cells increases dye-coupling as reflected by the diffusion of Lucifer yellow (Teranishi et al., 1983; 1984). Dopamine concentration in the neostriatum was reduced by damaging substantia nigra cells to determine if the incidence of coupling would increase.

The Basal Ganglia III, Edited by G. Bernardi *et al.*
Plenum Press, New York

213

MATERIAL AND METHODS

Experiment 1. Sprague-Dawley rats ranging in age from postnatal day (P)1 to 3 months were used in this developmental study. The animals were sacrificed by cervical dislocation, the brains removed and placed in ice-cold oxygenated Ringer's solution (concentration in mM; NaCl 124, KCl 5.0, $MgSO_4$ 2.0, $K_2H_2PO_4$ 1.25, $CaCl_2$ 2.0, $NaHCO_3$ 26.0 and glucose 10, pH 7.2-7.4). The neostriatum was sectioned into transverse or parasaggital slices (300-500 μm). Slices were placed upon filter paper in a recording chamber and superfused with the Ringer's (37°C). A warm and moist gas mixture (95% O_2-5% CO_2) flowed over the top surface of the slice. Intracellular microelectrodes (resistances ranged from 80-160 MΩ) were backfilled with 5% Lucifer yellow dissolved in distilled H_2O. The electrode above the tip was filled with 0.1 M LiCl. Lucifer yellow was injected by passing 0.5-1.0 nA of negative current for 1-5 min. After the experiment the slice was fixed with 4% paraformaldehyde in 0.1 M phosphate buffer, dehydrated with alcohols, cleared with methyl salicylate and mounted on glass slides. In some experiments, 3M K-acetate filled microelectrodes were used to assess for evidence of electrotonic transmission.

Experiment 2. Adult Sprague-Dawley rats were divided into 4 groups. The first group received unilateral electrolytic lesions of the substantia nigra and ventral tegmental area (2-4 mA anodal current for 10-30 sec) to deplete dopamine in the ipsilateral neostriatum. The second group received injections of 1.5-4 μl of the neurotoxin 6-hydroxydopamine (2 μg/μl with 0.5% ascorbic acid) to more selectively destroy dopamine-containing cells. A third group received aspirations of the anterior neocortex. These lesions do not reduce dopamine concentration but remove cortical afferents to the neostriatum. They serve as a control for nonspecific damage to a neostriatal input. The last group consisted of unoperated controls. Intracellular recordings and Lucifer yellow injections were performed in slices 3-5 weeks after surgery in lesioned or injected animals. Recordings were performed in the neostriata ipsilateral and contralateral to the lesion. The contralateral neostriatum served as an internal control since except for the lesion or 6-hydroxydopamine injection both sides of the brain received the same treatments. Preparation of the slices, recording and Lucifer yellow injection procedures are described above. The extent and placement of the lesions were verified histologically. In selected experiments both neostriata were dissected from one slice for biochemical analysis of dopamine concentration using high performance liquid chromatography.

RESULTS

Experiment 1. Dye-coupling in developing neostriatum

The incidence of dye-coupling was very high in the early postnatal days and decreased progressively with age (Table 1). Adult levels (about 10%) were reached by P25. The number of cells labelled after a single injection also declined with age (Table 1). In animals 1 or 2 days old clusters of as many as 6 cells were labelled (Fig. 1 - left panel). By P20, typically only two cells were labelled. All of the coupled cells appeared to be medium-sized spiny cells, although in the younger animals it was often difficult to determine the type of cell filled because spines were either absent or sparse.

Several recorded cells displayed evidence of electrotonic interactions (Baker and Llinás, 1971; Taylor and Dudek, 1982). Fast prepotentials in response to orthodromic stimulation were consistently observed in dye-coupled cells. These fast prepotentials did not collide

Table 1

Changes in Frequency of Dye-Coupling During Development

Postnatal Age (Days)	N	% Coupled	Average Number Cells Coupled
1-5	20	70	2.9
6-10	25	56	1.7
11-15	17	41	1.6
16-20	25	36	1.4
21-25	17	18	1.2
>25	20	10	1.1

with intracellularly evoked action potentials. In cells recorded with K-acetate filled electrodes and bathed in Ringer's containing Mn^{2+} to block synaptic transmission, extracellular stimulation evoked short-latency depolarizations. These potentials were also resistant to collision.

Experiment 2. Dye-coupling after dopamine-depleting lesions

Table 2 summarizes the results of this experiment. The incidence of dye-coupling is relatively low in the striatum of the adult rat (10%). In animals with unilateral substantia nigra lesions the incidence of coupling increased markedly (30% ipsilateral vs. 6% contralateral to the lesion). Injections of 6-hydroxydopamine also increased the frequency of dye-coupling. Overall, dye-coupling occurred more frequently in animals receiving electrolytic lesions than in animals receiving 6-hydroxydopamine treatment. There was also a greater decrease in dopamine concentration after electrolytic lesions than after 6-hydroxydopamine treatment (average depletions were 88±7% (SEM) vs. 68±13%, respectively). The frequency of dye-coupling in the control group was similar to that observed in the contralateral neostriatum. After cortical lesions dye-coupling did not occur in the ipsilateral neostriatum indicating that in order to increase the incidence of coupling the damage had to be specific to the substantia nigra.

Table 2

Frequency of Dye-Coupling

Side	Substantia Nigra Lesion		6-Hydroxydopamine		Combined*	
	Ipsi	Contra	Ipsi	Contra	Ipsi	Contra
Injected Cells	40	18	32	15	72	33
Dye-Coupled	15 (38%)	1 (6%)	6 (19%)	2 (13%)	21 (29%)	3 (9%)
Single	25 (62%)	17 (94%)	26 (81%)	13 (87%)	51 (71%)	30 (91%)

Side	Cortex Lesion		Control
	Ipsi	Contra	
Injected Cells	8	7	33
Dye-Coupled	0	1 (14%)	3 (9%)
Singles	8 (100%)	6 (86%)	30 (91%)

* Total of substantia nigra lesions and 6-hydroxydopamine injections.

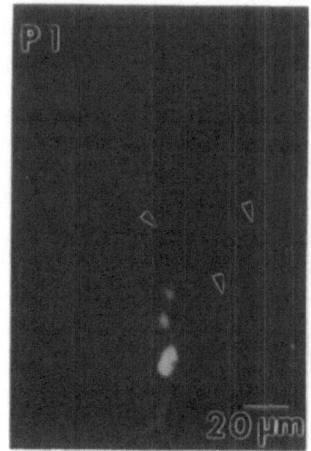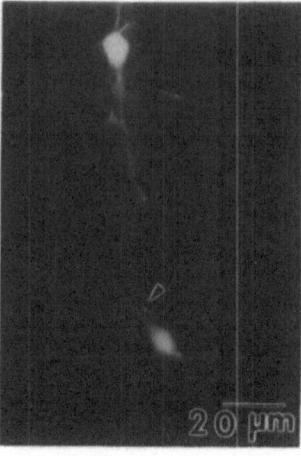

Fig. 1.   Left panel shows at least four cells labeled after a single
           injection of Lucifer yellow in the neostriatum of a rat pup
           at P1. Arrows indicate dendritic segments. Right panel shows
           dye-coupling after an electrolytic lesion of the substantia
           nigra.   The arrow indicates the tip of a dendrite from the
           upper cell very close to the dendritic shaft of the lower cell.

Typically, dye-coupling occurred between two cells. Figure 1 (right
panel) shows an example of dye-coupling after a substantia nigra lesion.
All injected cells were medium-sized spiny neurons. Most cases of dye-
coupling appeared to be dendro-dendritic or dendro-somatic since somata
of dye-coupled cells were separated (average intersomatic distance was
49±10 (SEM) μm). There were no apparent morphological differences between
single and dye-coupled cells.

Electrophysiological properties (action potential amplitude, input
resistance, evoked excitatory postsynaptic potential latency, duration
and amplitude) were compared between single and dye-coupled cells. In
general, measures were similar between the two groups. One consistent
electrophysiological difference emerged, however. Approximately 25% of
the cells recorded in the striatum ipsilateral to a substantia nigra
lesion (using K-acetate electrodes) displayed low amplitude spontaneous
depolarizations (1-4 mV). Similar potentials have been reported by
Galarraga et al. (1987). These potentials can be blocked by cholinergic
antagonists and therefore, probably do not reflect electrotonic
transmission.

CONCLUSIONS

These two experiments demonstrate that dye-coupling occurs in the
neostriatum and its frequency appears to be dependent upon striatal
dopamine concentration. The incidence of dye-coupling is very high during
the first postnatal week, when the concentration of striatal dopamine is
low. As the concentration of dopamine in the neostriatum increases during
postnatal maturation the frequency of dye-coupling in the neostriatum
decreases. The concentration of dopamine reaches adult levels between 3
and 6 months of age in the rat (Santiago et al., 1987). The frequency of
dye-coupling, although decreasing parallel to the increase in dopamine
concentration during the first postnatal month, appears to asymptote after
that time. The observation that the incidence of dye-coupling is high in

early postnatal periods in the neostriatum is consistent with other developmental studies (Bennett et al., 1981; Caveney, 1985; Connors et al., 1983; Das, 1977; Pannese et al., 1978; Ramoa et al., 1988). The high incidence of coupling during early development has led to the hypothesis that coupling may mediate intercellular communication responsible for growth and differentiation (Bennet, 1981; Caveney, 1985).

The results of the second experiment demonstrate that damage to substantia nigra dopaminergic neurons markedly increases the incidence of dye-coupling in the neostriatum. These findings are similar to results obtained in the retina of fish and turtles (Rogawski, 1987). If dye-coupling in the neostriatum is mediated by gap junctions then these findings imply that at least one role of dopamine may be to regulate gap junction permeability. Gap junctional permeability may be altered by dopamine in at least two ways. The lesion could produce a change in the number of functional gap junctions and/or the permeability of existing gap junctions may increase such that the Lucifer yellow is capable of moving from one cell to another. In the retina, the evidence implies that the permeability of gap junctions increases when dopamine concentration is decreased and decreases when dopamine is applied (Lasater, 1987; Lasater and Dowling, 1985; Piccolino et al., 1984; Teranishi et al., 1983; 1984). The latter effect occurs within minutes of application of dopamine indicating that it is unlikely that junctions are removed but more likely that their permeability is altered (Mangel and Dowling, 1985). Other studies have demonstrated that gap junction channels are not rigid structures (Bennett et al., 1988) and their permeability is sensitive to a number of extracellular and intracellular events (Baux et al., 1978; Rao et al., 1987; Spray et al., 1979). Unfortunately, our experiments cannot determine if dopamine alters the frequency of occurrence or the permeability of neostriatal gap junctions.

Although our studies provide evidence that dopamine regulates gap junctional permeability or expression, other factors may also be involved. For example, dendrotomy and/or axotomy may increase the incidence of coupling (Gutnick et al., 1985; Murphy et al., 1983). It is unlikely that the increase in dye-coupling was due to dendrotomy and/or axotomy as a consequence of the tissue slice procedure. Since both sides of the brain were treated in the same way, neurons on each side would be expected to have similar process truncations. The longer term effects of axotomy present a more difficult problem. Electrolytic lesions of the substantia nigra would be expected to damage the axons of neostriatal output neurons projecting to that structure (Bolam et al., 1981). It is possible that frequency of dye-coupling had been increased in these axotomized cells since studies in invertebrates have shown that axotomy can increase the incidence of dye-coupling (Murphy et al., 1983). In response to axotomy motoneurons also undergo a series of structural and metabolic changes (e.g., increase in RNA synthesis, chromatolysis, re-expression of neuropeptides) which are similar to events in early developmental periods (Haynes et al., 1982; Brown and Hardman, 1987) and may represent a return of the motoneuron to stages when it was growing, and making and eliminating synapses. The increased permeability and/or expression of gap junctions after substantia nigra electrolytic lesions could be viewed as part of a response that neostriatal neurons undergo after axotomy. However, since 6-hydroxydopamine lesions do not significantly damage neostriatal output neurons but still produce an increase in the frequency of dye-coupling, axotomy alone cannot explain the present findings. Further studies are necessary to clarify these issues.

The functional implications of dopamine's ability to modulate dye-coupling in the neostriatum remain speculative. In the retina, Dowling

(1986) proposed that dopamine, by uncoupling horizontal cells, reduces lateral inhibition and enhances the ability of receptor cells to pass information to ganglion cells. If a similar role is proposed for the neostriatum, the normally occurring high concentrations of dopamine would allow the neostriatal neurons to function more independently. As dopamine levels are reduced nearby cells become coupled and pairs of cells may respond more synchronously. One of the most accepted ideas is that coupling increases synchronous activity (Dudek, 1983; Llinás, 1985). Synchronous activity would allow incoming information to be transferred out of the neostriatum under conditions of minimal integration. In Parkinson's Disease, dopamine concentration is severely reduced because of degeneration of dopamine-containing cells in the substantia nigra (Hornykiewicz, 1982). The contribution of changes in gap junction permeability to the dysfunctions in this disorder remain to be elucidated. Previously, we have demonstrated that there is a lack of inhibition in the neostriatum of the developing cat during the first postnatal month (Levine et al., 1986; Morris et al.,1979). This lack of inhibition was related to immaturity of presynaptic terminals containing gamma aminobutyric acid (Fisher et al., 1987). The immaturity of inhibition and the increased incidence of dye-coupling may summate to prevent the neostriatum from integrating incoming information in this early developmental period.

ACKNOWLEDGEMENT

This research was supported by USPHS Grant HD05958.

REFERENCES

Adinolfi, A.M., 1977, The postnatal development of the caudate nucleus: a Golgi and electron microscopic study of kittens, Brain Res., 133:251-266.

Andrew, R.D., MacVicar, B.A., Dudek, F.E. and Hatton, G.I., 1981, Dye transfer through gap junctions between neuroendocrine cells of rat hypothalamus, Science, 211:1187-1189.

Baker, R. and Llinás, R., 1971, Electrotonic coupling between neurons in the rat mesencephalic nucleus, J. Physiol. (Lond)., 212:45-63.

Baux, G., Simonneau, M., Tauc, L. and Segundo, J.P., 1978, Uncoupling of electrotonic synapses by calcium. Proc. Natl. Acad. Sci. USA, 75:4577-4581.

Bennett, M.V.L., Spray, D.C. and Harris, A.L., 1981, Electrical coupling in development, Amer. Zool., 21:413-427.

Bennett, M.V.L., Verselis, V., White, R.L. and Spray, D.C., 1988, Gap Junctional Conductance: Gating. In: Modern Cell Biology. Vol. 7. Gap Junctions. E.L. Hertzberg and R.G. Johnson (eds.) A.R. Liss, Inc. New York, pp 287-304.

Bolam, J.P., Powell, J.F., Totterdell, S. and Smith, A.D., 1981, The proportion of neurons in the rat neostriatum that project to the substantia nigra demonstrated using horseradish peroxidase conjugated with wheat-germ agglutinin, Brain Res., 220:339-343.

Brown, M.C. and Hardman, V.J., 1987, Plasticity of vertebrate motoneurons. In: Growth and plasticity of neural connections. W. Winlow and C.R. McCrohan (eds.). Manchester University Press, pp. 36-55.

Caveney, S., 1985, The role of gap junctions in development, Ann. Rev. Physiol., 47:319-335.

Connors, B.W., Benardo, L.S. and Prince, D.A., 1983, Coupling between neurons of the developing rat neocortex, J. Neurosci., 3:773-782.

Das, G.D., 1977, Membrane-fusions and cytoplasmic bridges in the cells of the developing cerebellum, Cell Tiss. Res., 176:475-492.

Dowling, J.E., 1986, Dopamine: a retina neuromodulator? TINS, 9:236-240.

Dudek, F.E., Andrew, R.D., MacVicar, B.A., Snow, R.W. and Taylor, C.P. 1983, Recent evidence for and possible significance of gap junctions and electrotonic synapses in the mammalian brain. In H.H. Jasper and N.M. vanGelder (Eds.), Basic mechanisms of neuronal hyper-excitability, Alan R. Liss, New York, pp. 31-73.

Fisher, R.S., Levine, M.S., Adinolfi, A.M., Hull, C.D. and Buchwald, N.A., 1987, The morphogenesis of glutamic acid decarboxylase in the neostriatum of the cat: neuronal and ultrastructural localization, Brain Res., 33:215-234.

Galarraga, E., Bargas, J. Martinez-Fong, D. and Aceves, J., 1987, Spontaneous synaptic potentials in dopamine-denervated neostriatal neurons, Neurosci. Lett., 81:351-355.

Gutnick, M.J. and Prince, D.A., 1981, Dye coupling and possible electrotonic coupling in the guinea pig neocortical slice, Science, 211:67-70.

Gutnick, M.J., Lobel-Yaakov, R. and Rimon, G., 1985, Incidence of neuronal dye-coupling in neocortical slices depends on the plane of section, Neurosci., 15:659-666.

Hatton, G.I., Yang, Q.Z. and Cobbett, P., 1987, Dye coupling among immunocytochemically identified neurons in the supraoptic nucleus: Increased incidence in lactating rats, Neurosci., 21:923-930.

Haynes, L.W., Smyth, D.C. and Zakarian, S., 1982, Immunocytochemical localization of lipotropin C-fragment (B-endorphin) in the developing rat spinal cord, Brain Res., 232:115-128.

Hornykiewicz, O., 1982, Brain neurotransmitter changes in Parkinson's disease. In: Marsden, C.D and Fahn, S. (eds.), Movement disorders, Butterworth Scientific, London, pp. 41-58.

Knowles, W.D., Funch, P.G. and Schwartzkroin, P.A., 1982, Electrotonic and dye coupling in hippocampal CA1 pyramidal cells in vitro, Neurosci., 7:1713-1722.

Lasater, E.M., 1987, Retinal horizontal cell gap junctional conductance is modulated by dopamine through a cyclic AMP-dependent protein kinase, Proc. Natl. Acad. Sci. USA, 84:7319-7323.

Lasater, E.M. and Dowling, J.E., 1985, Dopamine decreases conductance of the electrical junctions between cultured retinal horizontal cells, Proc. Natl. Acad. Sci., 82:3025-3029.

Levine, M.S., Fisher, R.S., Hull, C.D. and Buchwald, N.A., 1986, Postnatal development of identified medium-sized caudate spiny neurons in the cat, Dev. Brain Res., 24:47-62.

Llinás, R.R., 1985, Electrotonic transmission in the mammalian central nervous system. In: Gap Junctions. Eds. M.V.L. Bennett and D.C. Spray, Cold Spring Harbor Laboratory, pp. 337-353.

MacVicar, B.A., Ropert, N. and Krnjevic, K., 1982, Dye-coupling between pyramidal cells of rat hippocampus in vivo, Brain. Res., 238:239-244.

Mangel, S.C., and Dowling, J.E., 1985, Responsiveness and receptive field size of carp horizontal cells are reduced by prolonged darkness and dopamine, Science, 29:1107-1109.

Morris, R., Levine, M.S., Cherubini, E., Buchwald, N.A. and Hull, C.D. 1979, Intracellular analysis of the development of responses of caudate neurons to stimulation of cortex, thalamus and substantia nigra in the kitten, Brain Res., 173:471-487.

Murphy, A.D., Hadley, R.D. and Kater, S.B., 1983, Axotomy-induced parallel increases in electrical and dye coupling between identified neurons of Helisoma, J. Neurosci., 3:1422-1429.

Pannese, E., Luciano, L. and Reale, E., 1978, Intercellular junctions in developing spinal ganglion, Zool., 6:129-138.

Piccolino, M., Neyton, J. and Gerschenfeld, H.M., 1984, Decrease of gap junction permeability induced by dopamine and cyclic adenosine 3', 5'-monophosphate in horizontal cells of turtle retina, J. Neurosci., 4:2477-2488.

Rao, G., Barnes, C.A. and McNaughton, B.L., 1987, Occlusion of hippocampal electrical junctions by intracellular calcium injection, Brain. Res., 408:267-270.

Ramoa, A.S., Campbell, G. and Shatz, C.J., 1988, Dendritic growth and remodeling of cat retinal ganglion cells during fetal and postnatal development, J. Neurosci., 8:4239-4261.

Rogawski, M.A., 1987, New directions in neurotransmitter action: Dopamine provides some important clues, TINS., 10:200-205.

Santiago, M., Cano, J., Machado, A. and Reinoso-Suarez, F., 1987, Postnatal development in the monoamine content in the striatum of the rat, Biogenic Amines, 4:381-389.

Sloper, J.J. and Powell, T.P.S., 1978, Gap junctions between dendrites and somata of neurons in the primate sensorimotor cortex, Proc. R. Soc. Lond. [Biol]., 203:39-47.

Sotelo, C. and Korn, H., 1978, Morphological correlates of electrical and other interactions through low-resistance pathways between neurons of the vertebrate central nervous system, Int. Rev. Cytol., 55:67-107.

Spray, D.C., Harris, A.L. and Bennett, M.V.L., 1979, Voltage dependence of junctional conductance in early amphibian embryos, Science, 204:432-434.

Stewart, W.W., 1978, Functional connections between cells as revealed by dye-coupling with a highly fluorescent naphthalimide tracer, Cell, 14:741-759.

Taylor, C.P. and Dudek, F.E., 1982, A physiological test for electrotonic coupling between CA1 pyramidal cells in rat hippocampal slices, Brain Res., 235:351-357.

Tennyson, V.M., Barrett, R.E., Cohen, G., Cote, L., Heikkila, R. and Mytilineou, C., 1972, The developing neostriatum of the rabbit: Correlation of fluorescence histochemistry, electron microscopy, endogenous dopamine levels, and [^{3}H] dopamine uptake, Brain Res., 46:251-285.

Teranishi, T., Negishi, K. and Kato, S., 1983, Dopamine modulates S-potential amplitude and dye-coupling between external horizontal cells in carp retina, Nature, 301:243-246.

Teranishi, T., Negishi, K. and Kato, S., 1984, Regulatory effect of dopamine on spatial properties of horizontal cells in carp retina, J. Neurosci., 4:1271-1280.

NEUROPHYSIOLOGICAL CHANGES IN AGED STRIATAL NEURONS IN RATS

Michael S. Levine and Carlos Cepeda

Mental Retardation Research Center
University of California at Los Angeles
Los Angeles, CA 90024

INTRODUCTION

Our laboratory has been studying some of the changes that occur in the basal ganglia of cats and rats as they age from young adults through middle age to senescence. In previous published research we identified a series of neurophysiological and morphological alterations in caudate nucleus neurons in aging cats (Levine et al., 1986a; Levine et al., 1987a; 1987b; Levine et al., 1988).

Extracellular recording experiments were performed in anesthetized and awake cats. Caudate neurons exhibited decreases in indices of excitation (Levine et al., 1987a; 1987b). Proportionately fewer initial excitatory responses to activation of monosynaptic inputs occurred and spontaneous firing rates were reduced. In addition, in aged cats thresholds for evoking excitatory responses were higher than thresholds for evoking inhibitory responses. The studies described in the present report are directed at elucidating the mechanisms underlying these neurophysiological alterations. As indicated above, the experiments in cats used extracellular recording techniques. The decrease in the proportion of initially excitatory responses when monosynaptic inputs were activated could be explained in more than one way. First, cells may have displayed excitatory postsynaptic potentials (EPSPs) that were subthreshold for action potential generation. Second, initially inhibitory postsynaptic potentials (IPSPs) may have predominated. Third, the biophysical characteristics inherent to caudate neurons may have been altered by the aging process (Boxer et al., 1988). These alternatives may not be mutually exclusive and could combine to affect decreases in excitation.

In order to determine how the characteristics of EPSPs and IPSPs were altered by the aging process intracellular recordings need to be performed. Rats were chosen instead of cats for this experiment for several reasons. Many of the characteristics of striatal electrophysiology are similar in cats and rats. For example, activation of monosynaptic inputs to striatal neurons produces initially excitatory responses in young adult animals of both species (Buchwald et al., 1973; Hull et al., 1973; Vandermaelen & Kitai., 1980). Using the rat as an alternative model permits generalization of results to another species. From a practical point of view the rat is also the species of choice for these experiments. The yield of successful impalements in individual intracellular recording experiments in vivo tends to be low. Thus, a large group of aged animals would need to be used to obtain a sufficient population of neurons. Since we do not have an unlimited supply of aged cats, using such animals for intracellular recording experiments would not be cost effective. In the present report we describe the effects of aging on striatal neuronal activity recorded intracellularly both in in vivo and in vitro preparations. The in vivo preparations provide more information about changes in afferent activation while the in vitro preparation, because there is probably less damage to neurons during impalement, can provide a more accurate assessment of changes in neuronal biophysical properties.

METHODS AND RESULTS

For in vivo recordings three age groups of male rats were used: young (3-5 months, n=16), middle-aged (10-12 months, n=11) and aged (>24 months, n=7). Procedural details for these experiments have been published (Cepeda et al., 1989). Briefly, all animals were anesthetized for the duration of the experiment with urethane. Stimulating electrodes were implanted in the frontal cortex just below the surface of the brain. Intracellular recordings were obtained using K-acetate-filled micropipettes with impedances between 30-60 MΩ. After impaling a neuron its response to stimulation of cortex was assessed. For each cell threshold currents for evoking both postsynaptic potentials and action potentials were determined by stimulating the cortex with a series of ascending intensities. The amplitudes and durations of the evoked postsynaptic responses (EPSPs and IPSPs) were measured from averaged records (3-5 trials).

The most important finding emerging from these in vivo intracellular experiments was that the threshold currents for evoking both EPSPs and synaptically driven action potentials in aged rats were significantly elevated compared to values obtained from young and middle-aged animals.

Intracellular recordings were obtained from 70 neurons localized in similar regions of the dorsal striatum (33 in young, 17 in middle-aged and 20 in aged rats). There were no particular difficulties in recording from neurons in the aged rats compared to the other groups. The average resting membrane potentials (RMPs) and action potential amplitudes were not significantly different in the three age groups (Table 1). Regardless of age, recorded cells displayed few spontaneous action potentials or were silent. In all groups rapid firing rates were an indication of cell damage.

The typical response evoked in striatal neurons when cortical inputs were activated consisted of an initial EPSP with an evoked action potential followed by a longer duration but smaller amplitude IPSP. In many cases a rebound excitation followed the IPSP. The basic components of these responses were consistent across the three age groups. However, in both middle-aged and aged groups proportionately fewer neurons displayed rebound excitations (53 and 55% in middle-aged and aged groups versus 64% in the young group). Initial IPSP responses were never observed.

Figure 1 compares responses of striatal neurons from a young and an aged rat to stimulation of the cortex at a series of ascending current intensities. Higher stimulus intensities were required to evoke excitatory responses in the neuron obtained from the aged rat. Quantitative analyses demonstrated that in middle-aged and aged groups the threshold currents necessary to evoke the EPSPs ($F=14.6$, $df=2/52$, $p<.01$) and action potentials ($F=11.2$, $df=2/42$, $p<.01$) were significantly higher (Table 1).

Table 1

Quantified Electrophysiological Parameters From In Vivo Experiments

	Age		
	3-5 Months	10-12 Months	24 Months
RESTING MEMBRANE POTENTIAL (mV)			
MEAN	-52.9	-47.6	-57.5
S.E.	2.1	1.9	3.1
ACTION POTENTIAL AMPLITUDE (mV)			
MEAN	39.8	38.3	38.6
S.E.	2.2	1.6	7.2
THRESHOLD INTENSITY TO EVOKE EPSPS (mA)			
MEAN	0.75	0.83 (11%)*	1.56 (108%)*
S.E.	0.09	0.16	0.13
THRESHOLD INTENSITY TO EVOKE ACTION POTENTIALS (mA)			
MEAN	1.30	1.45 (11%)*	2.30 (77%)*
S.E.	0.09	0.17	0.35
DURATION OF EVOKED EPSPS (msec)			
MEAN	40.3	39.5 (-2%)*	36.7 (-9%)*
S.E.	1.9	2.3	2.4
AMPLITUDE OF EVOKED EPSPS (mV)			
MEAN	7.4	6.6 (-11%)*	5.6 (-24%)*
S.E.	0.7	0.9	1.0

* Indicates percentage change from 3-5 month old group.

Not all neurons in aged rats had high threshold currents for evoking EPSPs and action potentials. In both middle-aged and aged groups a population of neurons had threshold values in the same range as those of young rats. Proportionately, the population of neurons that required higher threshold currents increased with age. In aged rats more than 45% of the neurons had thresholds greater than 1.0 mA. In both middle-aged and young animals, thresholds for evoking EPSPs were never greater than 1.0 mA. Similarly, in aged rats more than 40% of the neurons had elevated threshold currents (>2.0 mA) for orthodromic elicitation of action potentials. Only about 10% of the neurons in the young group had thresholds exceeding 2.0 mA.

EPSP amplitudes and durations were measured in all age groups in the same range of membrane potentials to account for voltage dependent changes. Although there was a decrease in average amplitude of the evoked EPSP in the aged group, the differences among these average values when assessed at twice threshold current were not statistically significant (Table 1). However, proportionately more neurons in aged rats had lower amplitude EPSPs. Almost 80% of the cells had EPSPs less than 4 mA in amplitude in aged rats. Only about 40% of the neurons in young rats had EPSP amplitudes less than 4 mA. The average duration of the EPSP decreased slightly with age (Table 1). This change was also not statistically significant when assessed at stimulation intensities of twice threshold. There was, however an increase in the proportion of neurons with short duration EPSPs in the aged group compared to the other two groups

No consistent differences were found in the amplitude or duration of the IPSP among the three groups.

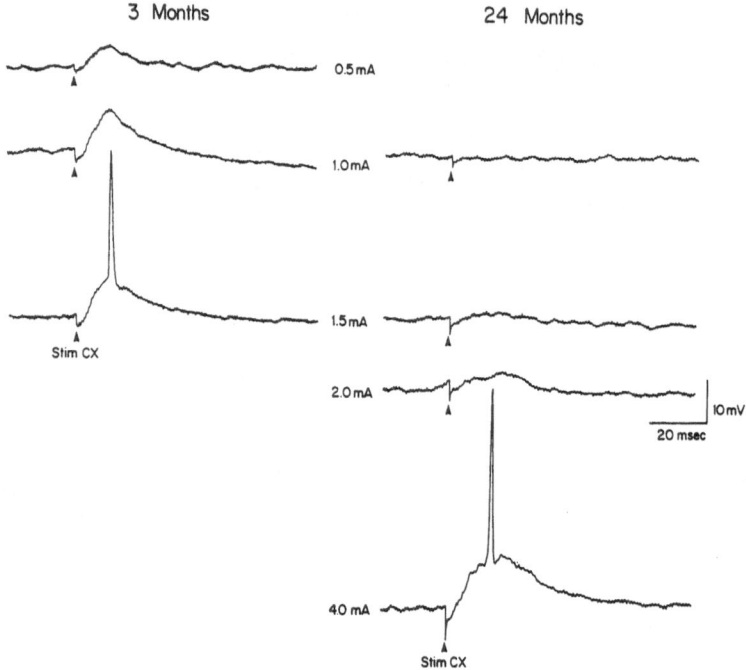

Fig. 1.   Comparison of threshold currents necessary to evoke EPSPs in
           striatal neurons in a young (3 month) and an aged animal (24
           months). Both cells had resting membrane potentials of -60 mV.
           Higher stimulus intensities were required to evoke responses
           in the neuron obtained from the aged rat.   In this neuron, a
           spike could only be evoked at an intensity of 4 mA. (Figure
           reprinted by permission from Cepeda et al., 1989)

     The ability of a neuron to generate a series of action potentials
when it is challenged with high frequency extracellular stimulation is
another measure of excitation. We tested this capacity by applying twin
pulses to the cortex at intensities 1.5X threshold current for action
potential elicitation in the striatum.   The delays between the two
stimulus pulses varied from 20-200 msec (in 10 msec steps). All the
cells tested in young rats (N= 4) could follow paired pulses at
interpulse intervals greater than 20 ms with little decrement in the
amplitude of the second spike. In contrast, in 80% (4/5) of the neurons
tested in aged rats a second action potential could not be generated at
interpulse intervals less than 80 ms.   The amplitude of the second
action potential was also markedly decreased in aged animals.

     In several experiments recorded neurons were identified by
intracellular injections of Lucifer Yellow (Walsh et al., 1988). Six
cells in the young and middle-aged groups were successfully filled with
Lucifer Yellow. All recovered cells were medium-sized spiny neurons. One
neuron was filled in the aged group. This neuron was also a medium-sized
spiny cell. There were fewer spines on distal dendritic segments and the
diameter of the dendrites appeared to be decreased compared to diameters
observed in neurons in young rats. These results were similar to our
observations in aged cats (Levine et al., 1986a). Finally, accumulations
of autofluorescent lipofuscin surrounded the soma.

     In order to provide a more detailed analysis of some of the changes
in passive and active membrane properties of aged striatal neurons, the

in vitro brain slice preparation was used. Procedures for preparation of striatal slices have been described (Walsh et al., 1988). Briefly, slices (coronal sections 400 μm thick) were perfused with oxygenated Ringer's solution and kept at 35-37°C. A bipolar stimulating electrode was positioned approximately 1 mm from the recording electrode to provide local extracellular stimulation.

Preliminary observations have been obtained from 44 cells in 17 young (3-5 months) and 41 cells in 17 aged (>24 months) rats. There were no group differences in average RMP, average action potential amplitude or average action potential duration. Average RMPs were -70 mV for both groups while average action potential amplitudes measured from the starting point of rapid rise of the action potential were 53 mV for both groups. Average action potential durations (measured at one-half amplitude) were 1.2 and 1.1 msec for young and aged groups, respectively. Similar to findings in anesthetized rats, in slices obtained from aged rats a population of cells had elevated thresholds for synaptically evoking both EPSPs and action potentials. Approximately 60% of the cells obtained from aged rats had EPSP thresholds greater than 300 μA. Only about 40% of the cells in young rats had thresholds exceeding 300 μA. Similarly, 60% of the cells in the aged group had thresholds for synaptically evoked action potentials that exceeded 600 μA. Fewer than 40% of the cells in the young group had thresholds for action potential elicitation exceeding this value. In addition, a small population of cells (15%) had higher input resistances (30-40 MΩ versus 10-25 MΩ) (Fig.

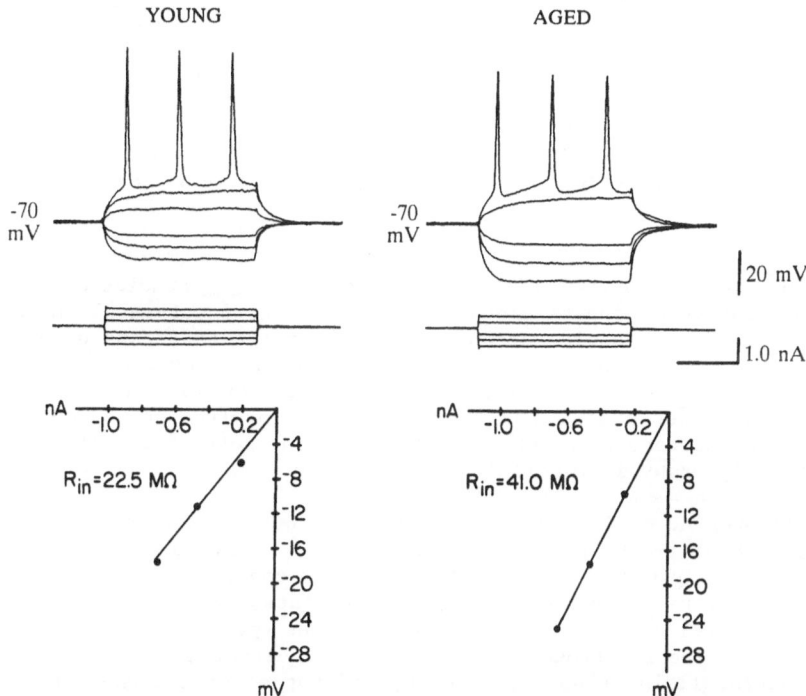

Fig. 2.   Examples of current-voltage plots in young and aged neurons in striatal slices. Top panels show membrane potential changes evoked by a series of intracellularly injected depolarizing and hyperpolarizing current pulses. RMPs were -70 mV for both neurons. Middle traces show amplitudes of current pulses. Bottom panels display current-voltage plots. The slope of the line is the input resistance (R_{in}).

2). Elevated thresholds for evoking action potentials by intracellular current injection also occurred (Fig. 3). In the young group, about 20% of the cells had thresholds of less than 0.50 nA for evoking action potentials by intracellular current injection. In aged rats this population comprised less than 5% of the cells. Finally, in the aged group over 30% of the cells displayed accommodation of repetitive spiking during intracellular injections of depolarizing current (Fig. 4). Accommodation was defined as the inability to produce repetitive action potentials in response to a depolarizing current pulse. This phenomenon did not occur frequently in the cells in the young group.

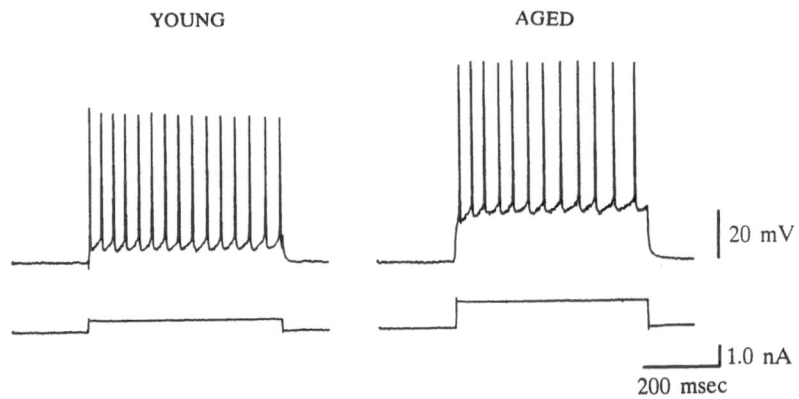

Fig. 3.   Examples of responses to intracellular injection of depolarizing current pulses in young and aged neurons obtained from striatal slices. Threshold for evoking a train of action potentials was higher in the neuron from the aged striatum. Bottom trace shows amplitude of the current pulse.

DISCUSSION

The combined results of the in vivo and in vitro studies demonstrate that activation of inputs produces initial EPSPs in striatal neurons in aged rats. The threshold currents for evoking the EPSPs and orthodromic action potentials in these neurons are elevated compared to values obtained in young rats.   A number of other age-related changes were observed in these experiments.   All of these involved decreased ability of striatal neurons to generate EPSPs or action potentials either evoked synaptically or by intracellular current injection.

The present findings in conjunction with our previous studies (Levine et al., 1987a; 1987b) provide evidence that striatal neurons in rats and cats display similar electrophysiological changes during aging.   In extracellular recordings obtained from aged cats, decreases in excitation were expressed mainly by decreases in the proportion of initially excitatory responses evoked by cortical and substantia nigra stimulation. These extracellular studies, however, did not elucidate whether changes in excitation were due to the inability of the cells to generate EPSPs, inability of the cells to generate action potentials or increased occurrence of initial IPSPs. The present findings using intracellular recordings in aged rats demonstrated that initial EPSPs occurred in every neuron. Increased proportions of initially inhibitory responses in extracellular recordings in aged cats are not related to the occurrence of initial IPSPs but rather to an increase in the threshold for EPSP and action potential generation in these animals.

In our previous studies in cats we demonstrated morphological changes in medium-sized spiny neurons in the striatum. When recorded striatal neurons are identified by intracellular injection of horseradish peroxidase or Lucifer Yellow, virtually all cells are medium-sized spiny neurons (Kitai et al., 1976; Levine et al., 1986b; Walsh et al., 1988).

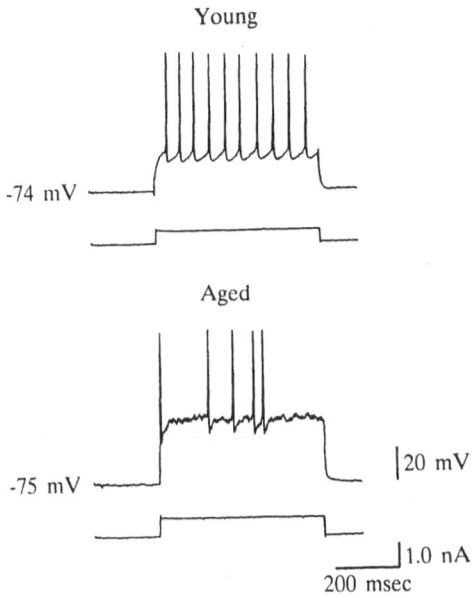

Fig. 4.   Example of accommodation in an aged neuron obtained from a striatal slice. Top trace shows that a depolarizing current pulse produces a repetitive train of action potentials in a young neuron. Bottom trace shows that during a depolarizing pulse the aged neuron was not capable of producing repetitive action potentials. Numbers to left of each trace are RMPs.

In the present experiment, all identified cells were medium-sized spiny neurons. The morphological changes we have observed in this type of cell using other techniques are probably valid indicators of morphological events occurring in the present experiments. There is a decrease in the density of synapses and of spines on distal dendritic segments as well as a loss of dendritic segments in older animals (Levine et al., 1986a; 1988). A majority of the inputs synapse on the spines of these neurons (Somogyi et al., 1981). Consequently, the morphological findings suggest that there should be decreases in the number of synaptic contacts underlying the generation of EPSPs.

It is now well established that not all neurons are equally affected by the aging process (Buell & Coleman, 1981; Flood et al., 1985; 1986; McNeill et al., 1985; Boxer et al., 1988; Chase et al., 1985; Morales et al., 1987; Scheibel et al., 1975). Our findings on the striatum in the aging rat and cat serve to underscore this idea. A continuum appears to exist in which certain neurons are more affected than are others. We recently proposed a model to account for the neurophysiological changes observed in aged striatal neurons that takes into account that not all neurons are equally affected by the aging process (Levine, 1988). In young adult animals most striatal neurons respond to activation of their monosynaptic inputs with an initial EPSP which is usually followed by an IPSP (Buchwald et al., 1973; Hull et al., 1970; 1973; Kitai et al., 1976). This IPSP is thought to be generated, in part, by mutual inhibition from neighboring cells (Hull et al., 1973; Levine et al., 1986b). In aged animals we hypothesized that a population of neurons received reduced excitatory input while another population received most of its excitatory input. When recorded from, neurons in the former population should display elevated thresholds for EPSP and action potential generation as well as reduced amplitude EPSPs as observed in the present experiments. We also hypothesized that IPSPs should be reduced in duration and amplitude because some of the surrounding neurons are not activated and cannot inhibit their neighbors. The present data do not appear to support this hypothesis. There was little consistent change in amplitude or duration of IPSPs. Age-related alterations in inhibitory events in the striatum will require further study to uncover their underlying mechanisms.

ACKNOWLEDGEMENTS

This research was supported by USPHS Grant AG 7462.

REFERENCES

Boxer, P.A., Morales, F.R. and Chase, M.H., 1988, Alteration of group 1A-motoneuron monosynaptic EPSPs in aged cats, Exp. Neurol., 100:583-595.

Buchwald, N.A., Price, D.D., Vernon, L. and Hull, C.D., 1973, Caudate intracellular responses to thalamic and cortical inputs, Exp. Neurol., 38:311-323.

Buell, S.J. and Coleman, P.D., 1981, Quantitative evidence for selective dendritic growth in normal human aging but not in senile dementia, Brain Res., 24:23-41.

Cepeda, C., Walsh, J.P., Hull, C.D., Buchwald, N.A. and Levine, M.S., 1989, Intracellular neurophysiological analysis reveals alterations in excitation in striatal neurons in aged rats, Brain Res., 494:215-226.

Chase, M.H., Morales, F.R., Boxer, P.A. and Fung, S.J., 1985, Aging of motoneurons and synaptic processes in the cat, Exp. Neurol., 90:471-478.

Flood, D.G., Buell, S.J., Defiore, C.H., Horowitz, G.J. and Coleman, P.D., 1985, Age-related dendritic growth in dentate gyrus of human brain is followed by regression in the 'oldest old', Brain Res., 345:366-368.

Flood, D.G., Guarnaccia, M., Coleman, P.D., 1986, Dendritic extent in human CA 2/3 hippocampal pyramidal neurons in normal aging and senile dementia, Soc. Neurosci. Abst., 12:272.

Hull, C.D., Bernardi, G. and Buchwald, N.A., 1970, Intracellular responses of caudate neurons to brain stem stimulation, Brain Res., 22:163-179.

Hull, C.D., Bernardi, G., Price, D.D. and Buchwald, N.A., 1973, Intracellular response of caudate neurons to temporally and spatially combined stimuli, Exp. Neurol., 38:324-336.

Kita, T., Kita, H. and Kitai, S.T., 1984, Passive electrical membrane properties of rat neostriatal neurons in an in vitro slice preparation, Brain Res., 300:129-139.

Kitai, S.T., Kocsis, J.D., Preston, R.J., and Sugimori, M., 1976, Mono-synaptic inputs to caudate neurons identified by intracellular injection of horseradish peroxidase, Brain Res., 109:601-606.

Levine, M.S., 1988, Neurophysiological and morphological alterations in caudate neurons in aged cats, Ann. New York Acad Sci., 515:314-328.

Levine, M.S., Adinolfi, A.M., Fisher, R.S., Hull, C.D., Buchwald N.A. and McAllister, J.P., 1986a, Quantitative morphology of medium-sized caudate spiny neurons in aged cats, Neurobiol. Aging, 7:277-286.

Levine, M.S., Adinolfi, A.M., Fisher, R.S., Hull, C.D., Guthrie, D. and Buchwald, N.A., 1988, Ultrastructural alteration in caudate nucleus in aged cats, Brain Res., 440:267-279.

Levine, M.S., Fisher, R.S., Hull, C.D. and Buchwald, N.A., 1986b, Postnatal development of identified medium-sized caudate spiny neurons in the cat, Dev. Brain Res., 24:47-62.

Levine, M.S., Lloyd R.L., Hull, C.D. Fisher, R.S. and Buchwald, N.A., 1987a, Neurophysiological alterations in caudate neurons in aged cats, Brain Res., 401:213-230.

Levine, M.S., Schneider, J.S., Lloyd, R.L., Hull, C.D. and Buchwald, N.A., 1987b, Aging reduces somatosensory responsiveness of caudate neurons in the awake cat, Brain Res., 405:389-394.

McNeill, T.H., Koek, L., Brown, S. and Rafols, J., 1985, Age-correlated dendritic changes in medium spiny striatal neurons of the C57BL/6NNIA mouse, Soc. Neurosc. Abst., 11:896.

McNeill, T.H., Brown, S.A., Rafols, J.A. and Shoulson, I., 1988, Atrophy of medium spiny striatal dendrites in advanced Parkinson's disease Brain Res., 455:148-152.

Morales, F.R., Boxer, P.A., Fung, S.J. and Chase, M.H., 1987, Basic electrophysiological properties of spinal cord motoneurons during old age in the cat, J. Neurophysiol., 58:180-194.

Scheibel, M.E., Lindsay, R.D., Tomiyasu, V. and Scheibel, A.B., 1975, Progressive dendritic changes in aging human cortex, Exp. Neurol., 47:392-403.

Somogyi, P., Bolam, J.P. and Smith, A.D., 1981, Monosynaptic cortical input and local axon collaterals of identified striatonigral neurons. A light and electron microscopic study using the Golgi-peroxidase transport-degeneration procedure, J. Comp. Neurol., 195:567-584.

Vandermaelen, C.P. and Kitai, S.T., 1980, Intracellular analysis of synaptic potentials in rat neostriatum following stimulation of cerebral cortex thalamus and substantia nigra, Brain Res. Bull., 5:725-733.

Walsh, J.P., Zhou, F.C., Hull, C.D., Fisher, R.S., Levine, M.S. and Buchwald, N.A., 1988, Physiological and morphological characterization of striatal neurons transplanted into the striatum of adult rats, Synapse, 2:37-44.

SHORT-TERM PLASTICITY IN THE VENTRAL STRIATUM

A COMPARISION BETWEEN IN VIVO AND IN VITRO MEASUREMENTS

P.H.Boeijinga, F.H.Lopes da Silva, and C.M.A.Pennartz

Dept. Exp.Zoology, University of Amsterdam
Kruislaan 320, 1098 SM
Amsterdam, The Netherlands

INTRODUCTION

In order to understand how information from the subiculum is mediated by way of the nucleus accumbens (Acb) to more caudally lying areas, it is of importance to know how the limbic inputs affect Acb output neurons. Here we report experiments using electrophysiological recording techniques and stimulation of the subiculum-accumbens pathways (i.e. the fornix) under various conditions to address this question. De France et al. (1985a) investigated a possible interaction between successive responses using paired pulse stimulation with intervals in the range of 10-40 ms. They showed that the response to the second, or test stimulus was suppressed at intervals below 30 ms. In a separate paper (De France et al. 1985b) these authors reported a suppressive effect of dopamine on the responses when the input fibers were stimulated once every 1-2 s. For stimulation with a rate of 6 pulses/s (intervals 167 ms) no such effect was found. The subiculum is a part of the hippocampal formation, that is known to have EEG activity predominantly in the theta range (4-12 Hz). Therefore, the transmission of the signals in the subiculum-accumbens pathways may be frequency-dependent.

These considerations led us to record in vivo both evoked potentials (EPs) and unit activity in the Acb following double pulse stimulation of the fornix fibers over a wide range of intervals. As regards the EPs paired-pulse facilitation (ppf) occurred at intervals between 80 and at least 200 ms. The same was found for unit firing; the probability of firing following the test stimulus of the pair with respect to that following the first, or conditioning pulse was found to be increased at intervals of more than 500 ms (Boeijinga et al. 1990). These results will be briefly summarized. This phenomenon of facilitation may depend on intrisic circuits of the Acb and/or on extrisic interconnections. Thus, in order to find out whether the Acb circuits can account for the ppf, experiments were also carried out in vitro in a slice preparation of the rat containing both the Acb and fornix input fibers. Preliminary results are presented.

The Basal Ganglia III, Edited by G. Bernardi *et al.*
Plenum Press, New York

METHODS

in vivo

The methods of the acute preparation and electrophysiological recording have been described in detail elsewhere (Boeijinga et al. 1990). In short, under halothane anaesthesia, stimulation electrodes were placed stereotaxically in the fornix, subiculum, and in the ventral pallidum. Glass micro-electrodes were placed in the Acb to record both field potentials and extracellular unit activity. The signals were amplified and digitized (CED interface), that was connected to an IBM-PC. Averaged evoked potentials (AEPs) and Post-Stimulus Time Histograms (PSTHs) were stored.
Off-line, peak amplitudes were determined for the AEPs, onset latencies and width of responses were extracted from the PSTHs.

in vitro

The methods for making slices was adapted from Pennartz et al. (1990). In short, male albino rats weighing about 100g, anaesthetized with ether and ketamine (150 mg/kg i.p.), were decapitated and the brains were quickly removed. The frontal pole was removed and a section was made in a direction perpendicular to the longitudinal axis of the brain under an angle of 40-45° with the horizontal plane. Slices of 400 µm were cut on a vibratome and transferred to a recording chamber, where they were continuously superfused with oxygenated standard Ringer containing both $CaCl_2$ and $MgSO_4$ (2.5 and 1.3 mM resp.). Bipolar stimulating electrodes and glass recording pipettes were placed à vue. Signals were sampled at a rate of 10 or 20 kHz and averaged using a Motorola Exorset and stored on diskette for off-line analysis.

RESULTS

spatial distribution

Examples of profiles of field potentials evoked by fornix stimulation, recorded in vivo are shown in Fig 1A. As can be seen, in the middle portion of the Acb the A(veraged)EPs consisted of two positive peaks (P1 and P2) at 10 and 25-30 ms, respectively, and a negative going deflection in between. This initial complex was followed by a long-lasting negative wave. Sometimes a positive going component was superimposed on this negativity. At the dorsal and medial margins of the Acb, a negative component clearly emerges between 10 and 20 ms. Both P1 and P2 were accompanied by increases in extracellularly measured unit activity (see below). The units that fired at a latency of around 10 ms, were interpreted as being monosynaptically driven single cells (Boeijinga et al. 1990).

In vitro, it was demonstrated that local electrical stimulation elicited two negative components at a latency of <2 ms and around 4 ms, the first of which could be identified as a compound action potential, the second as a synaptic component, since this was often accompanied by orthodromic cell firing (not shown, see however, Pennartz et al. 1990). Stimulating the fornix fibers we found only one negative component, between 5 and 9 ms (Fig 1B, N_s). In 5 slices tested so far, this component could only be recorded from the areas bordering the dorsal and medial margins of the Acb. This component is most likely equivalent to the synaptic component following local stimulation, although it occured at longer latency due to the larger distance between the sites of stimulation and recording. Deeper in the Acb, however, a positive going deflection (P_s) at about the same latency was found. This positive may be equivalent to the negative component at the dorsomedial border, but is not a mirror-image of the latter.

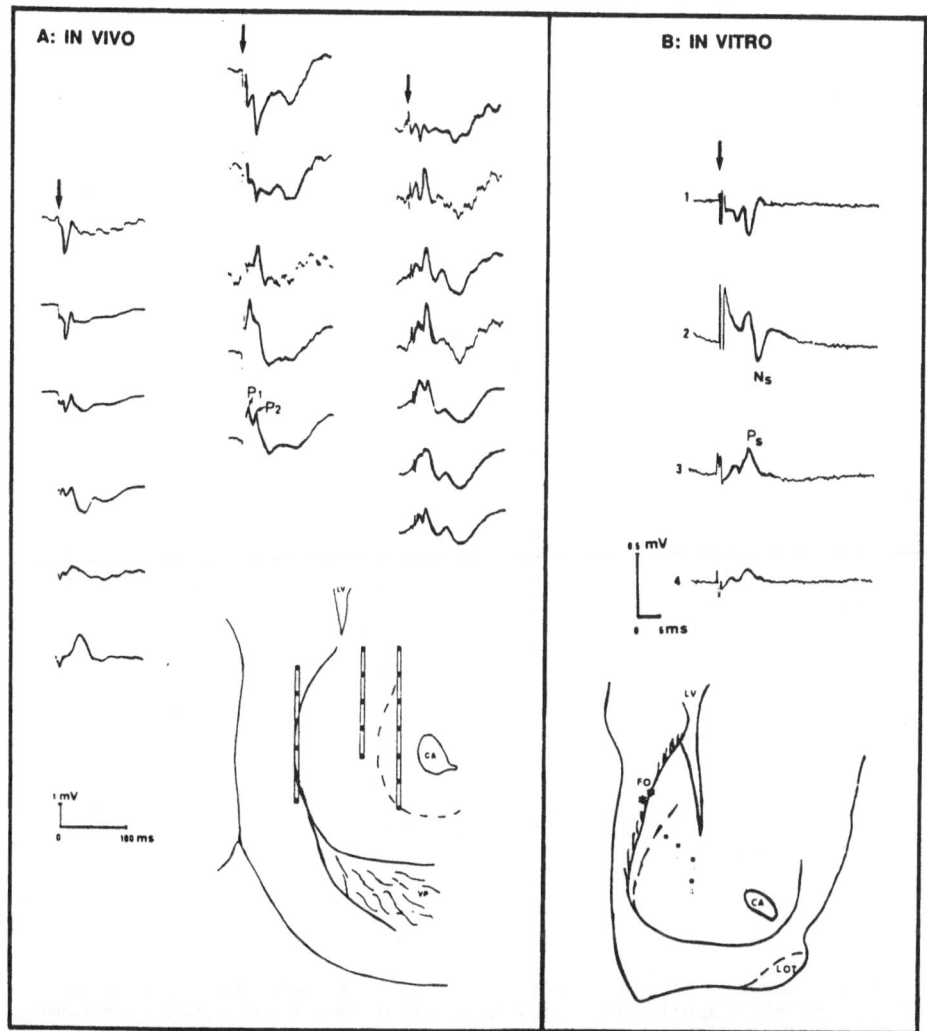

Fig 1 Topographic maps of field potentials:
A: EP profiles obtained in in vivo experiments. P1 and P2 are indicated; moment of stimulation given by arrows. In the lower right corner a reconstruction of the recording locations (black squares) in the frontal plane (A 9.0 before interaural according to Pellegrino et al. 1981) is shown for different experiments. Note the reversal in polarity around 10 ms at the dorsomedial part of the Acb.
B: Example of EP-profile recorded in vitro. Lower part shows recostruction, numbered locations correspond to those before each EP-trace. Abbreviations: CA = anterior commissure, FO = fornix, LOT = lateral olfactory tract, LV = lateral ventricle, VP = ventral pallidum.

In conclusion, in both the in vivo and vitro preparation fornix stimulation elicits similar synaptic events. It is likely that the P1 component of the AEPs in vivo, correspond to the positive going deflection in the evoked potentials recorded deep in the Acb slices; both correspond to an increase of firing of single cells in response to fornix stimulation.

PAIRED PULSE FACILITATION

These experiments allowed us to analyse ppf both in vivo and in vitro. Paired-pulse facilitation occurred, thus by definition the

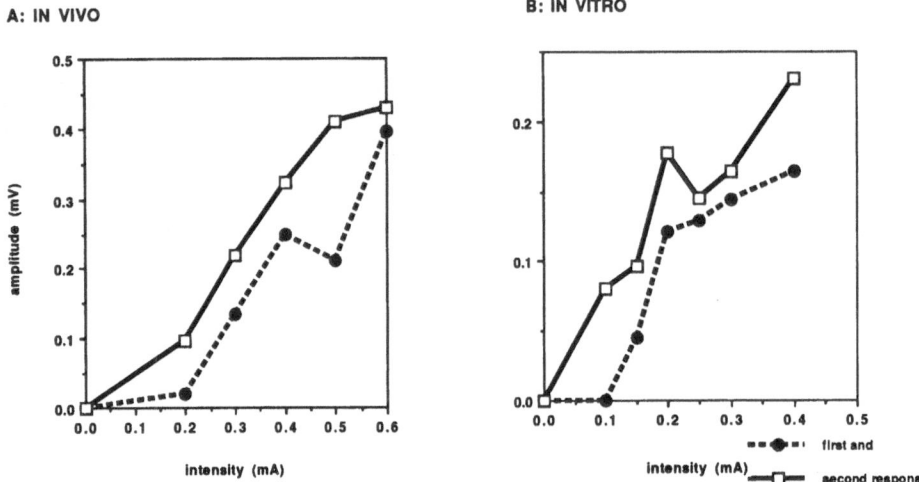

Fig 2   Amplitudes of responses as a function of stimulus intensity:
A: Example of parameters of first positive component for in vivo measurements with paired pulse interval of 100 ms.
B: N-values for in vitro measurements at interval of 50 ms.

amplitudes of the responses to the test stimulus were larger than those to the conditioning stimulus. First, we studied the responses as a function of stimulus intensity. In vivo, the amplitude differences between peak-value P1 and the subsequent negative trough were determined (Fig 2A), whereas in vitro the amplitude was defined by the difference between onset- and peak-value of the N_s-component (Fig 2B). It can be seen from these input/output curves that ppf was consistently present at each of the stimulus intensities that were used. Next, we quantificied the timespan of the ppf by way of extracting the amplitude parameters for the different intervals in vivo as shown in FIG 3A. At each interval, the value was tested against the control value obtained from single responses (sample t-test, 9 degrees of freedom). Significant ppf was present at intervals ranging from 60 to nearly 300 ms. Comparison with other experiments revealed that ppf ranged at least from 80 to 200 ms.

An example of the ppf for the amplitude parameter in vitro is shown in Fig 3B. As can be seen, maximal facilitation is present at intervals around 75 ms. Following local stimulation, similar results could be obtained, suggesting that, at least partially, the same mechanisms underlying ppf were in these cases in operation.

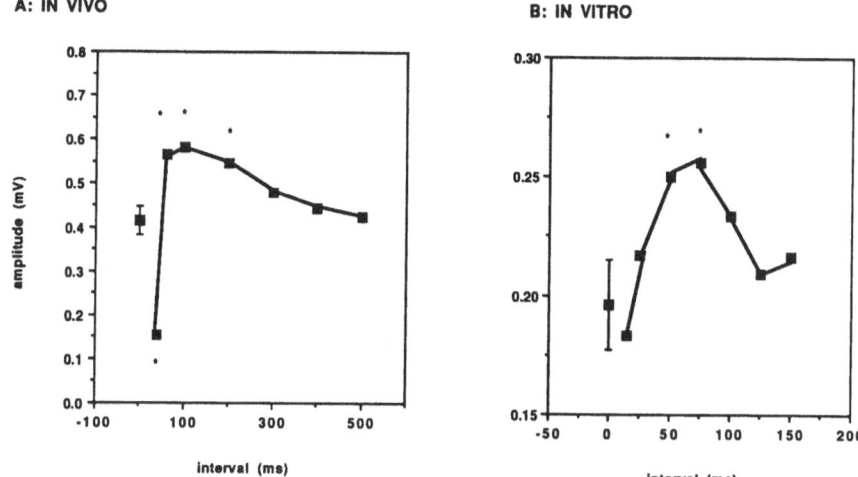

FIG 3 Amplitudes as a function of interval:
A: Example of first positive component obtained from EPs recorded in vivo for intervals from 40 to 500 ms.
B: Same as in A, for N_s-values of in vitro measurements. Note the smaller range of intervals. Asterisks indicate $p < 0.05$.

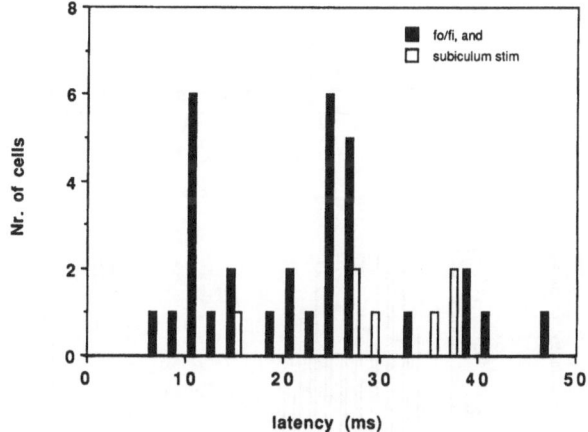

FIG 4 Distribution of onset latencies for extracellular recorded unit responses in vivo following stimulation of inputs to the Acb at the level of the fornix or subicular cortex.

UNIT ACTIVITY

Following fornix stimulation in vivo, the majority of responding cells showed excitatory activity. The distribution of latencies (FIG 4) showed two clusters, one around 10 ms, and the other around 24-26 ms. These two classes could also be distinguished on basis of the width of the PSTH peak. For the short latency units this was 3.5 +/- 1.5 ms, whereas for the longer latencies increased firing rates ranged from 7 to 48 ms after onset. These results are interpreted as indicating that the former group represents monosynaptically

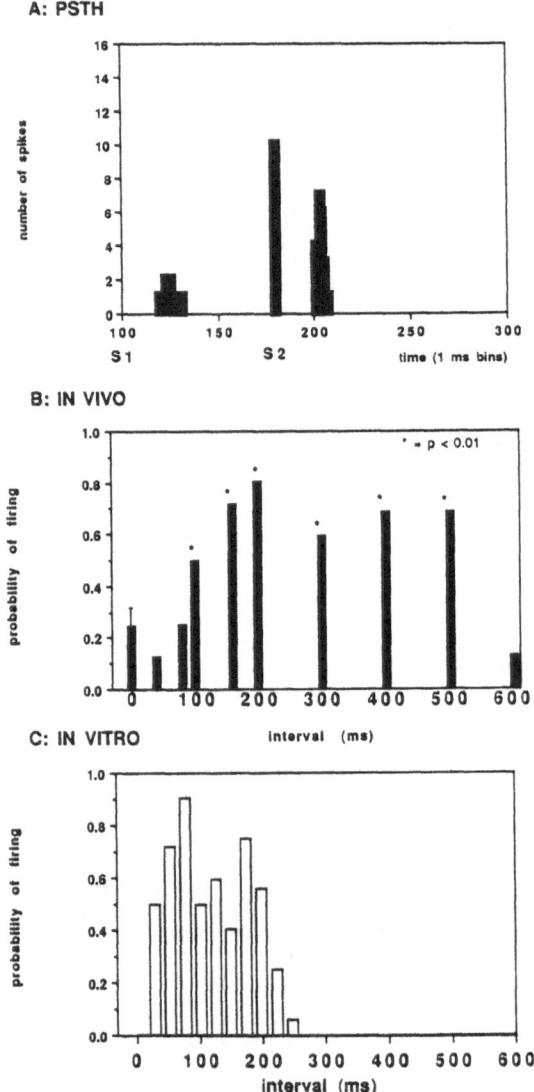

FIG 5 Paired-pulse facilitation of cellular discharges:
A: Example of a PSTH (32 sweeps, binwidth 1 ms) for an interval (S2-S1) of almost 200 ms. Note the increased response to the test stimulus(S2).
B: In vivo responding as a function of interval for the same cell as reported in A. Probability was determined by integrating spike activity in 7 consecutive bins and dividing by the number of sweeps.
C: for this in vitro example, stimulus intensity was adjusted that it was subthreshold for the conditiong response. Probability in this experiment was based on counts of 16 sweeps for each interval.

activated neurons in the Acb, whereas the second group possibly represents polysynaptic activation. Both types of excitatory responses showed strong paired-pulse facilitation (ppf), as revealed by the increased number of spikes following the test pulse with respect to those following the conditioning stimulus. An example of the enhanced number of spikes is shown in FIG 5A. We computed the probability of firing different stimulus intervals. It was found that ppf was present from 100 to 500 ms following the conditioning pulse (FIG 5B).

In vitro, local stimulation could give rise to increased firing probabilities at intervals up to 250 ms (FIG 5C).

CONCLUSIONS

Electrical stimulation of the subiculum inputs to the Acb both in vivo and in vitro consistently induces a characteristic field potential profile, and gives rise to paired-pulse facilitation. The intervals at which ppf is present in the field potentials in vivo, correspond to 5 to 12.5 Hz, that matches the theta range of the septohippocampal system of the mammalian brain (Lopes da Silva et al. (1990). Thus transmission in the subiculum-accumbens pathways may be facilitated while theta rhythm is present.

In vitro, the ppf seems to span a smaller range of intervals. The unit recordings indicate that the generators responsible for the field potentials and the form of plasticity described in this paper are located in the Acb. Regarding the mechanisms responsible for the facilitatory effects, it may be speculated that for the enhanced response an increase in transmitter release occurs, or that more synapses are activated (cf. Creager et al. 1980). We have indications that the activation of NMDA receptors may be involved in the ppf, since following local electrical stimulation the enhancement of the response to the test stimulus could, at least in part, be suppressed by application of AP-5 (Pennartz et al.1990).

Both in vivo and in vitro, the intervals at which increased probabilities of firing are seen, outlast those of the evoked potentials. The similarity justifies the use of the slice preparation to study cellular mechanisms of synaptic plasticity in the Acb.

ACKNOWLEDGEMENTS

This research was supported by grant 550-093 of the Foundation of Medical Research (MEDIGON) , which is subsidized by the Dutch Organization of Scientific Research (N.W.O.)

REFERENCES

Boeijinga, P.H., Pennartz, C.M.A. and Lopes da Silva, F.H., 1990, Paired-pulse facilitation in the nucleus accumbens following stimulation of subicular inputs in the rat. Neurosci., In press.

Creager, R., Dunwiddie,T. and Lynch,G., 1980, Paired-pulse and frequency facilitation in the CA1 region of the in vitro rat hippocampus, J.Phyiol., 299: 409

De France, J.F., Marchand, J.F., Sikes, R.W., Chronister,R.B. and Hubbard, J.I., 1985a, Characterization of fimbria input to nucleus accumbens, J.Neurophysiol., 54: 1553.

De France, J.F., Sikes, R.W. and Chronister,R.B., 1985b, Dopamine action in the nucleus accumbens, J.Neurophysiol., 54: 1568.

SPONTANEOUS DEPOLARIZING SYNAPTIC POTENTIALS IN THE NEOSTRIATUM

P. Calabresi, N.B. Mercuri, M. De Murtas, A. Stefani
and G. Bernardi

Clinica Neurologica - Dip. Sanita' Pubblica
II Universita' di Roma - Via O. Raimondo 8
00173 - Roma - Italy

INTRODUCTION

Although several "in vivo" and " in vitro" studies have previously
shown that neostriatal neurons have a very low frequency of spontaneous
firing activity, a common finding of some of these reports was the
presence of large spontaneous depolarizing potentials (SDPs) during
intracellular recordings (Hull et al. 1970; Buchwald et al. 1973;
Bernardi et al. 1976; Sugimori et al. 1978; Wilson and Groves 1981;
Bishop et al. 1982; Calabresi et al. 1987b). These SDPs were very
frequent "in vivo", but their amplitude was insufficient to trigger high
frequencies of firing activity. The finding that these potentials are
present also "in vitro", where they are smaller and less frequent,
seems to suggest that SDPs are synaptically mediated (Calabresi et al.
1990a; 1990b). In fact, the surgical manipulations carried out in the
slicing procedures could at least in part interrupt synaptic projections
originating from the afferent structures to the neostriatum.
In several neurons the SDPs cause brief bursts of action potentials
followed by relatively long pauses of firing activity (Wilson and Groves
1981; Calabresi et al. 1990a). The contribution of these potentials to
the regulation of the neuronal excitability of neostriatal cells may be
of major importance in controlling of the functions of the basal ganglia
during movements (De Long, 1973; Groves, 1981). In addition, the
impairment of the mechanisms from which these potentials originate may
significantly alter the complex balance between intrinsic neuronal
properties and afferent synaptic inputs within the neostriatum thus
generating some extrapyramidal syndromes (Bird and Iversen, 1974;
Groves, 1983; Calabresi et al. 1989). For these reasons in the present
paper we have studied some physiological and pharmacological
characteristics of the SDPs recorded both " in vivo " and " in vitro ".
In this report we also review previously published data concerning
these potentials (Calabresi 1987a; 1987b; 1988a; 1988b; 1990a; 1990b).

The Basal Ganglia III, Edited by G. Bernardi *et al.*
Plenum Press, New York

METHODS

"In vivo" recordings

Detailed descriptions of the methods may be found elsewhere (Mercuri et al., 1986 ; Calabresi et al., 1990a). In brief, male Wistar rats (200-350 g) were anesthetized with thiopental (50 mg/kg i.p.). The anesthesia was maintained by hourly injections of thiopental (8-10 mg/kg i.p.). Bipolar stimulating electrodes were inserted in the ipsilateral sensorimotor cortex and substantia nigra. After all the surgical manipulations were completed the animals were immobilized with succinylcholine and artificially ventilated. The iontophoretic electrodes were glued to the shank of the recording electrode and were used for drug ejection and automatic current balance.

"In vitro" recordings

Detailed descriptions of the methods may be found elsewhere (Calabresi et al., 1987; 1988; 1990). Briefly, slices of neostriatum (200-300 μM) were prepared from tissue blocks of neostriatum, using a glass guide. The standard saline for the pre-incubation of the slices (30 min) and perfusion of the recording chamber consisted of (mM): NaCl 124; KCl 5; $NaH_2 PO_4$ 1.22; $MgSO_4$ 1.3; $NaHCO_3$ 25.5; $CaCl_2$ 2.5; glucose 10; the pH was 7.4 and temperature 36 C. Intracellular recording electrodes were filled with 2M KCl or with 2M Cesium-Cl. For local stimulation, bipolar electrodes were positioned within the neostriatum near the recording electrode. Intracellular potentials were recorded using a M707-A electrometer, displayed on an oscilloscope and stored on magnetic tape.

RESULTS

"In vivo" recordings

As shown in fig. 1A, most of the neostriatal neurons (86%) intracellularly recorded "in vivo" showed SDPs. Their amplitude was 2-20 mV and their duration was 10-200 ms. The SDPs would either occur at regular frequencies or they would appear at irregular frequencies. In some cases the SDPs triggered action potentials. A characteristic finding was the increase of the SDPs amplitude during slight hyperpolarization of the membrane potential by DC negative current; this event was also coupled with the inhibition of the action potentials generated by the SDPs. On the contrary, when the hyperpolarization produced by negative current was larger than 20 mV and the membrane potential was more negative than -85 mV, the amplitude and the duration of the SDPs decreased. This observation was correlated with the strong inward rectification of these neurons at very negative membrane potentials (Calabresi 1987a; 1987b; 1988a; 1988b; 1990a; 1990b). The characteristic membrane rectification of the cells at hyperpolarized levels did not exclusively affect the SDPs, but also caused a voltage dependent behaviour of the synaptic potentials evoked either by cortical stimulation or by nigral activation. In fact, as shown in fig. 2A, the cortically evoked EPSP was increased by moderate hyperpolarization, but it was decreased during strong hyperpolarization of the membrane potential. The rectification of the membrane resistance at very negative levels also produced a clear shunt of the late excitation following the cortically-evoked synaptic potentials. As shown in fig. 2A, the activation of the ipsilateral cortex produced EPSP-IPSP sequences in most of the neurons (Bernardi et al. 1976; Herrling 1984)

240

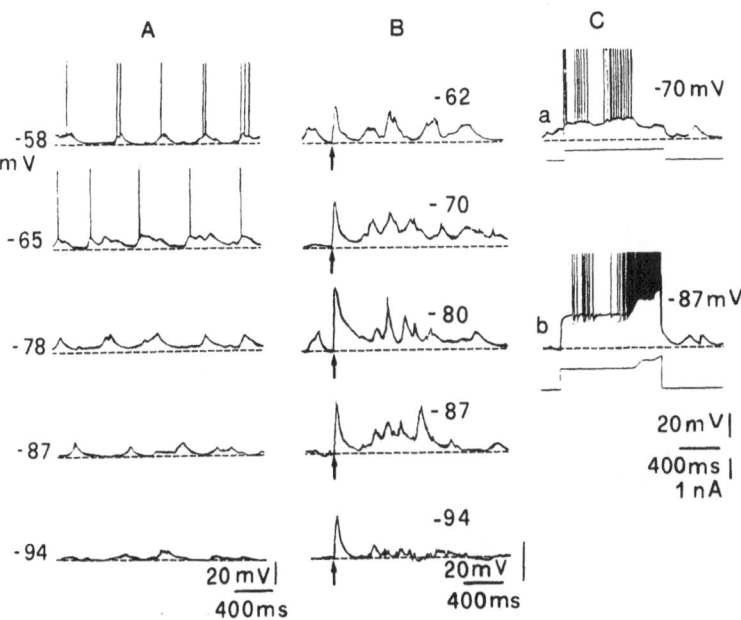

Fig. 1. Effect of membrane potential changes on the SDPs and on the
cortically-evoked EPSPs recorded "in vivo" from a neostriatal neuron.
(A) SDPs at resting level (-65mV) trigger single action
potentials. When the cell is depolarized by DC current (-58 mV),
the number of the action potentials increases and the SDPs amplitude
decreases. On the contrary, a slight membrane depolarization (-78mV)
blocks the action potentials, but it increases the amplitude of the SDPs.
When the neuron is hyperpolarized at very negative potentials (-87 and
-94 mV) the amplitude in the duration of the SDPs are greatly
reduced. (B) Synaptic stimulation of ipsilateral cortex produces EPSPs
followed by late excitation; the amplitude of the evoked potentials is
voltage dependent as the amplitude of the SDPs. (C) Direct
depolarization of the membrane produces bursts of spikes interrupted
by post-bursts firing pauses (Ca); the amplitude of the bursts is not
increased by membrane hyperpolarization and the increase of the
positive current injected during the depolarizing pulse abolishes the
firing pauses (Cb).

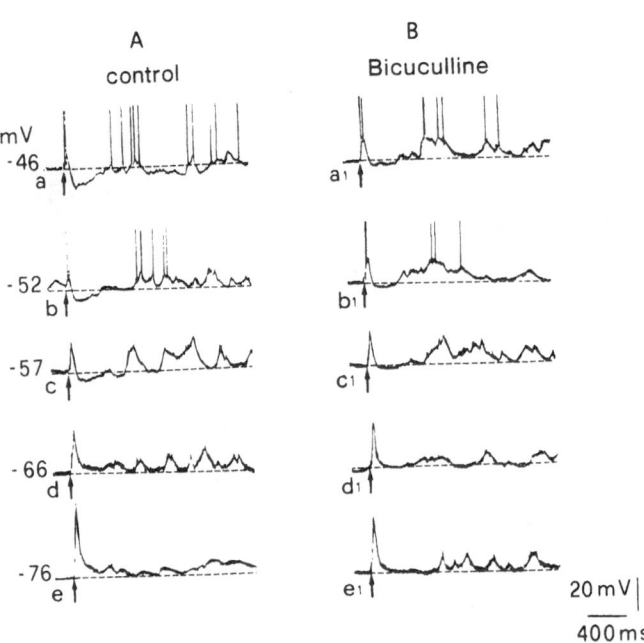

Fig. 2. Effects of iontophoretic applications of bicuculline on the cortically-evoked synaptic potentials at different membrane levels in a neostriatal neuron intracellularly recorded "in vivo".
(A) Under control conditions cortical stimulation produces an EPSP which is not followed by a hyperpolarizing IPSP at resting level (-66mV). Membrane depolarization by constant positive current (from -57 up to -46 mV) reveals a hyperpolarizing IPSP. (B) Iontophoretic application of bicuculline (100 nA) clearly reduces this hyperpolarizing potential.

The reversal potential for the IPSP was between -57 and -66 mV by using K-acetate electrodes. The iontophoretic application of bicuculline (fig. 2B) clearly decreased the amplitude the IPSP, but did not affect the frequency of most of the SDPs showing that these potentials, at least "in vivo", are not due to GABA-mediated events. Iontophoretic application of cadmium (10-100 nA) decreased or blocked both the EPSP-IPSP sequences and the SDPs.

"In vitro" recordings

SDPs were less frequent "in vitro" than "in vivo". In fact, we observed the presence of SDPs only in 52% of the neurons intracellularly recorded " in vitro " (in 5 mM potassium). The amplitude (2-15 mV) and the duration (4-100 ms) of these potentials was different from cell to cell. In some neurons we studied whether the SDPs produced summation or inhibition of intrastriatally-evoked EPSPs. As shown in fig. 3, we triggered evoked EPSPs whose amplitude was similar to the amplitude of the SDPs (Misgeld et al. 1982; Calabresi et al. 1987a). When the EPSP was generated in the absence of an SDP, the evoked potential was subthreshold (see b and d in fig. 3). On the contrary, when the EPSP was triggered during a SDP, the evoked potential reached the threshold for the action potential generation (see c in fig. 3). This finding indicates that SDPs produced summation rather than inhibition of the excitatory synaptic inputs within the neostriatum.

As shown in fig. 3e, membrane depolarizations produced either by synaptic inputs or by passive positive current increased the frequency of occurrence of SDPs. On the contrary, as described "in vivo", membrane hyperpolarization at very negative levels by constant negative current reduced the amplitude, the duration and frequency of the SDPs.

The possible role of potassium conductances in the generation of the membrane rectification affecting the amplitude of the SDPs at negative levels was examined by using internal and external potassium channels blockers. As shown in fig. 4 a-d, neurons recorded utilizing Cs-Cl electrodes (cesium blocks potassium channels from inside) showed SDPs of larger amplitude and duration. In addition, the shunt of the SDPs and of the EPSPs at very negative levels was decreased either by internal cesium or by external barium (0.1-1 mM) or TEA (10-20 mM).

The synaptic nature of the SDPs was studied perfusing the slices with low calcium plus cobalt (0.1-0.5 mM) containing solutions.

In this condition , the SDPs (see fig. 4 a1-d1) and the EPSPs (see fig. 4e1) were completely suppressed. TTX (1 μM) abolished most of the SDPs recorded " in vitro"; however, in some neurons we observed the presence of small TTX-resistant SDPs.

The neurotransmitters involved in the generation of the SDPs were studied utilizing either antagonists of the GABA A receptors or antagonists of the excitatory aminoacids receptors. Bicuculline (30-100 μM) blocked only part of these SDPs . The bicuculline-insensitive SDPs were reduced or suppressed by kynurenic acid (0.5-1 mM), an antagonist of the excitatory aminoacids.

DISCUSSION

Three main findings are the result of from this work. 1) Neostriatal neurons "in vivo" as well as "in vitro" can be phasically depolarized by SPDs. 2) These depolarizations are strongly influenced by the intrinsic membrane properties of the neostriatal neurons. 3) The SPDs are

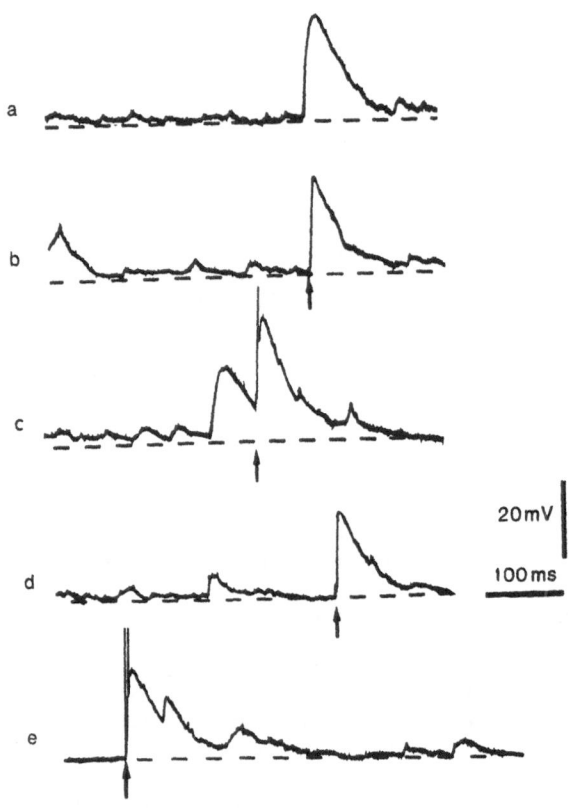

Fig. 3. Spontaneous depolarizing potentials recorded "in vitro" and their interaction with evoked EPSPs.(a): a subthreshold SDPs is recorded from a neostriatal cell (RMP = -72mV); (b) the intensity of intrastriatal stimulation is adjusted to produced a subthreshold EPSP;
(c): the same intensity of the stimulation delivered during the decay phase of the SDPs produces synaptic summation and the EPSP triggers an action potential ; (d): the amplitude of the evoked potential decreases again at a subthreshold level when the EPSP does not occur during the SDP;(e): in another neuron intrastriatal stimulation of high intensity produces a large EPSP which generates several SDPs during the decay phase. Calibrations in d apply also in a,b,c and e.

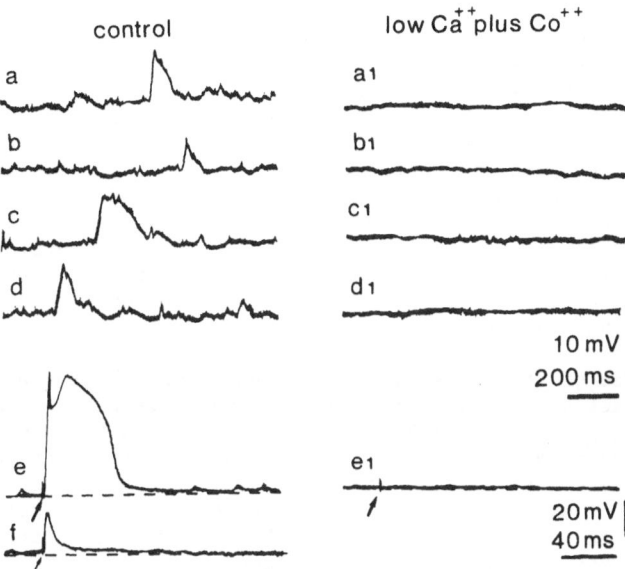

Cs Cl electrode

control   low Ca^{++} plus Co^{++}

a   a1

b   b1

c   c1

d   d1

10 mV
200 ms

e   e1

f

20mV
40ms

Fig. 4. Effects of internal cesium and of external cobalt on the SDPs and on the EPSPs recorded "in vitro" from a neostriatal neuron. (a, b, c, d): different traces recorded at resting level (-70 mV) from a cesium-loaded cell 20 minutes after the penetration; note the prolonged duration of the SDPs. (a1, b1, c1, d1): ten minutes after the onset of the bath application of a low calcium (0.5 mM) -cobalt (0.5 mM) containing solution the SDPs are abolished. (e,f): in the same cell, under control condition, the intrastriatal stimulation of two different intensities produces either an EPSPs which triggers a plateau spike (e: high intensity) or a subthreshold EPSP (f: low intensity). (e1):during the superfusion with the low calcium-cobalt containing solution the intrastriatally evoked EPSP is fully abolished. Calibrations in a1, b1, c1, d1 apply also in a, b, c, d. Calibrations in e1 apply also in e and f.

synaptically mediated and are probably generated by GABAergic and glutamatergic synapses.

1) The presence of SDPs "in vivo" suggests that these potentials may be relevant to produce bursts of neostriatal neurons activity. Since most of neostriatal cells are GABAergic projecting neurons , it is possible that these striatal SDPs represent phasic inhibitory influences on the structures which receive striatal projections (globus pallidus, thalamus and substantia nigra, pars reticolata) (Carpenter, 1976). These structures could, in turn, influence the neostriatum providing a possible feed-back control within the basal ganglia.

2) The finding that changes of the membrane potentials strongly modify the amplitude and the duration of the SDPs suggests that voltage dependent conductances operating at very negative potentials control the efficacy of the synaptic depolarization in these regions of the membrane potentials. The reduction of the shunt of the SDPs by internal cesium and/or by external barium or TEA suggests that voltage-dependent potassium channels are mainly involved in this rectifying behaviour. A possible explanation for the finding that membrane hyperpolarization not only affects the amplitude, but also affects the frequency of occurrence of the SDPs is that during membrane hyperpolarization the voltage signals produced by some synaptic inputs are greatly shunted and for this reason it is difficult to detect several spontaneous synaptic events.

3) The blockade of the SDPs by low calcium plus cobalt containing solutions suggests their synaptic nature. The finding that some SDPs persist in the presence of TTX indicates that in neostriatum, as well as previously described in the neuromuscolar junction (Katz and Miledi, 1963), a TTX-resistant release of neurotransmitters may occur.
Although we did not characterize the neurotransmitters involved in the generation of the SDPs observed "in vivo", the evidence that some of the SDPs recorded "in vitro" can be reduced by bicuculline while others are blocked by kynurenic acid, seems to indicate that both GABAergic and glutamatergic synapses are involved in the generation of these potentials.

ACKNOWLEDGMENTS

We thank Giuseppe Gattoni and Massimo Tolu for their excellent technical assistance.

REFERENCES

Bernardi, G., Marciani, M.G., Morocutti, C. & Giacomini, P. 1976. The action of picrotoxin and bicuculline on rat caudate neurons inhibited by GABA. Brain Res. 102: 379-384.

Bird, E.D. & Iversen, L.L. 1974. Huntington's Chorea. Brain 97: 457-472.

Bishop, G.A.; Chang, H.T. & Kitai S.T. 1982. Morphological and physiological properties of neostriatal neurons: an intracellular horseradish peroxidase study in the rat.. Neuroscience 7: 179-191.

Buchwald, N.A, Price, D.D., Vernon, L. & Hull, C.D. 1973. Caudate intracellular response to thalamic and cortical inputs. Exp. Neurol. 38: 311-323.

Calabresi, P., Benedetti, M., Mercuri, N.B. & Bernardi, G. 1988a. Depletion of catecholamine reveals inhibitory effects of bromocryptine and lysuride on neostriatal neurons recorded intracellularly in vitro. Neuropharmacology 27: 579-587.

Calabresi, P., Benedetti, M., Mercuri, N.B. & Bernardi, G. 1988b. Endogenous dopamine and dopaminergic agonist modulate synaptic excitation in neostriatum: intracellular studies from naive and cathecolamine-depleted rats. Neuroscience 27: 145-157.

Calabresi, P., Mercuri, N.B., Stanzione, P., Stefani, A. & Bernardi, G. 1987a. Intracellular studies on the dopamine-induced firing inhibition of neostriatal neurons in vitro: evidence for D1 receptor involvement. Neuroscience 20: 757-771.

Calabresi, P., Mercuri, N.B., Stefani, A. & Bernardi, G. 1990a. Synaptic and intrinsic control of the membrane excitability of neostriatal neurons. I. An "in vivo" analysis. J. Neurophysiol. in press.

Calabresi, P., Mercuri, N.B. & Bernardi G. 1990b. Synaptic and intrinsic control of the membrane excitability of neostriatal neurons. II. An "in vitro" analysis. J. Neurophysiol. in press.

Calabresi, P., Misgeld, U. & Dodt, H.B. 1987b. Intrinsic membrane properties of neostriatal neurons can account for their low level of spontaneous activity. Neuroscience 20: 293-303.

Calabresi, P., Stefani, A., Mercuri, N.B. & Bernardi, G. 1989. Acetylcholine-dopamine balance in the striatum: is it still a target for the antiparkinsonian therapy? in: Central cholinergic synaptic transmission, edited by M. Frotscher and U. Misgeld. Berlin: Springer Verlag. 315-321.

Carpenter, M.B. 1976. Anatomical organization of the corpus striatum and related nuclei. in: Basal Ganglia, edited by M.D. Yahr New York: Raven Press, 1-36.

De Long, M.R. 1973. Putamen: activity of single units during slow and rapid arm movements. Science 179: 1240-1242.

Groves, P.M. 1983. A theory of the functional organization of the neostriatum and the neostriatal control of voluntary movement. Brain Res. 5: 109-132.

Herrling, P.L. 1984. Evidence for early cortically evoked inhibition of caudate neurons. Exp. Brain Res. 55: 528-534.

Hull, C.D., Bernardi, G. & Buchwald, N.A. 1970. Intracellular responses of caudate neurons to brain stem stimulation. Brain Res. 22: 163-178.

Katz, B. & Miledi, R. 1963. A study of a spontaneous miniature potentials in spinal motoneurones. J. Physiol. 163: 389-422.

Mercuri, N.B., Bernardi, G., Calabresi, P., Cotugno, A., Levi, G. & Stanzione, P. 1985. Dopamine decreases cell excitability in rat striatal neurons by pre and postsynaptic mechanisms. Brain Res. 358: 119-121.

Misgeld, U., Wagner, A. & Ohno, T. 1982 Depolarizing IPSPs and depolarization by GABA on rat neostriatum cells in vitro. Exp. Brain Res. 45: 108-114

Sugimori, M., Preston, R.I. & Kitai, S.T. 1978. Response properties and electrical constant of caudate neurons in the cat. J. Neurophysiol. 41: 1662-1675.

Wilson, C.J. and Groves, P.M. 1981. Spontaneous firing patterns of identified spiny neurons in the rat neostriatum. Brain Res. 220: 67-80.

DOPAMINE D1 RECEPTORS AND TERMINAL EXCITABILITY

IN THE STRIATONIGRAL AND NIGROSTRIATAL SYSTEMS

Marco Diana*, Lawrence J. Ryan**
Stephen J.Young*** and Philip M. Groves***

*Dipartimento di Neuroscienze, Universita' di Cagliari
09100 Cagliari Italy
**Dept. of Psychology, Oregon State University
97330 Corvallis, Oregon
***Dept. of Psychiatry and Neuroscience, University
of California San Diego, 92093 La Jolla, California

INTRODUCTION

The terminal excitability method, pioneered by Wall (1958) to investigate the mechanisms of presynaptic inhibition in the spinal cord, is an electrophysiological procedure for observing the consequences of stimulation of receptors on axon terminals *in vivo*. In this technique, a stimulating electrode is placed in the vicinity of the axon terminal field in order to electrically elicit an antidromic response that is detected with a microelectrode located at the soma. The effects of terminal receptor stimulation or blockade are observed as a change in the threshold current just sufficient to elicit an antidromic response.

Over the past decade, studies from our laboratory and others have provided evidence that the electrical excitability of the axon terminal fields of dopaminergic neurons is modified by alterations in the stimulation of presynaptic receptors located on their axon terminals. The excitablity of dopamine axon terminals decreases following increased receptor stimulation due to local application of agonists such as apomorphine and amphetamine, or as the result of increased endogenous transmitter release due to increased impulse flow. Blockade of terminal receptors by antagonists such as haloperidol and sulpiride, increase excitabilty (Groves et al 1981, Tepper et. al.,1984 a & b, Mereu et. al., 1985, Garcia-Munoz et al 1986, Gariano et al 1989). Parallel effects on terminal excitability occur following changes in the activation of terminal receptors on adrenergic (Nakamura et. al., 1982; Ryan et. al., 1985) and serotoninergic neurons (Sawyer, 1986).

The availability of specific D1 and D2 dopamine antagonists and agonists has led to considerable behavioral, pharmacological, neurophysiological, and anatomical research concerning the specificity of function and localization of these dopamine receptors (Clark and White, 1989). We discuss here studies employing the terminal excitability technique which examine D1 effects on the nigral terminals of striatial output neurons, and on the terminal axons of nigrostriatal dopaminergic neurons.

Effects of D1 Agents on Striatonigral Terminal Excitability

Several lines of anatomical evidence suggest that dopamine D1 receptors appear as heteroreceptors on the axon terminals of non-dopaminergic striatal neurons that project to the substantia nigra. First, the dense autoradiographic labeling of D1 receptors in the substantia nigra declines greatly following chemical lesion of the neostriatum. In contrast there is a lesser reduction following selective lesion of nigral dopamine neurons suggesting that many D1 receptors are located on neostriatal axons (Altar and Hauser, 1987; Altar and Marien, 1987; Savasta et al., 1987). Second, D1 receptors increase in density on the neostriatal side of a transection of the mesencephalon between the substantia nigra and the neostriatum. This local accumulation probably indicates continued axonal transport of D1 receptors in neostriatal axons directed towards the substantia nigra (Aiso et al., 1987).

The physiological function of these D1 receptors is unclear. Like other presynaptic receptors, these heteroreceptors may modulate neurotransmitter release. The neostriatal neurons probably use GABA, substance K (or P), and dynorphin as neurotransmitters. In in vitro preparations made of mesencephalon containing the substantia nigra, exogeneously applied dopamine increases GABA release by activating D1 receptors (Reubi et al., 1977; Starr, 1987). Stimulation of these receptors may enhance GABA release from neostriatal axons, or, conversely, D1 stimulation may decrease release from these axon terminals leading to a disinhibition of GABAergic pars reticulata neurons and increased release from the axon collaterals of these neurons within pars reticulata (Di Chiara et al., 1979; Chevalier et al., 1985). This latter possibility appears more likely since the excitation of pars reticulata neurons by iontophoretically applied SKF38393 appears to be mediated indirectly by decreasing GABA release from neostriatal axon terminals (Waszczak and Martin, 1989).

Though stimulation of these presynaptic receptors by exgonously applied D1 agents may modulate striatonigral neurotransmission presynaptically, it is questionable whether endogenously released dopamine normally has access to these receptors. While dopamine is released from the dendrites of dopaminergic substantia nigra neurons, there is no anatomical evidence for presynaptic dopaminergic synapses in the substantia nigra. If dopamine were to reach the D1 heteroreceptors it would have to do so non-synaptically.

We have used the technique of terminal excitability to evaluate the function of dopamine D1 heteroreceptors on striatonigral axon terminals. Our results suggest that these receptors are functional, but that they are not normally activated by endogenously released dopamine.

Exogenous application of the D1 agonist R-SKF38393 directly into the substantia nigra axonal terminal fields of neostriatal projection neurons caused a dose-dependent decrease in axonal excitability. This effect is a specific receptor mediated response. as the decrease in excitability could be partially reversed by a subsequent infusion of the D1 antagonist SCH 23390 (Figure 1). Further, when SKF38393 was infused outside of the substantia nigra along the non-terminal portion of the striatonigral axons where D1 receptors are unlikely to reside, SKF38393 was without effect (Figure 1). Thus it appears that D1 receptors on the striatonigral axon terminals are functional. If endogenous dopamine has access to these heteroreceptors, they should be tonically activated and a D1 antagonist should have the opposite effect of the agonist, that is, it should increase axonal excitability. However, administration of the antagonist SCH 23390 or the non-specific dopamine antagonist, haloperidol, was without effect on terminal excitability even at a high doses (Figure 1). These results suggest that dopamine may not have access to these D1 receptors. This conclusion is also supported by the lack of effect on striatonigral terminal excitability of the dopamine releasing agent d-amphetamine sulfate. Amphetamine increases dopamine release from substantia nigra dopamine neuron dendrites (Paden et al., 1976; Cheramy et al., 1978; Kalivas et al 1989) and thus should enhance the possibility that endogenously released dopamine reaches and activates the D1 heteroreceptors. However, in both anesthetized (Figure 1) and freely-moving, behaving rats (Table 1) amphetamine did not alter terminal excitability at any tested dose. This lack of effect is all the more

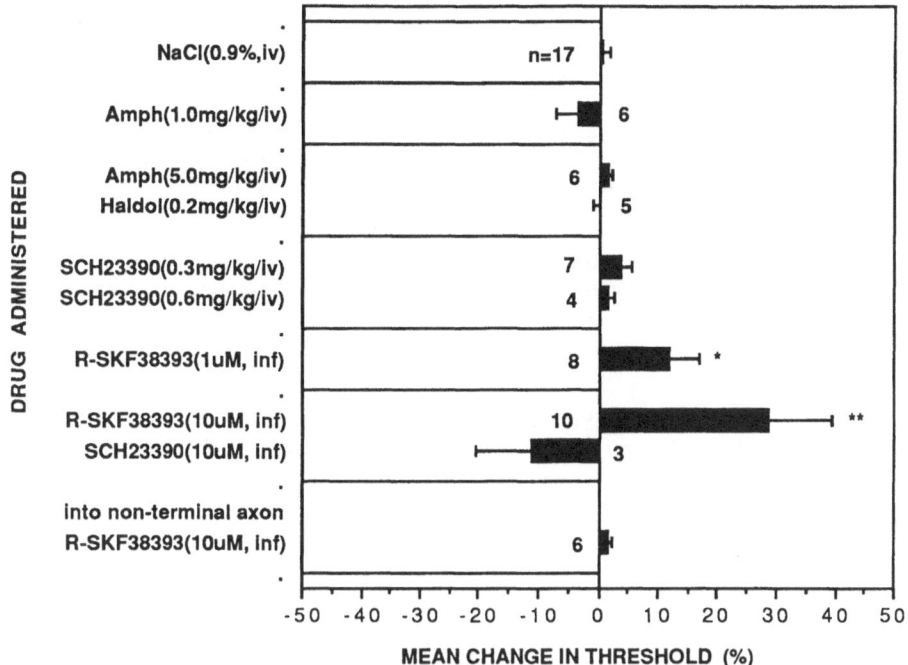

STRIATONIGRAL TERMINAL EXCITABILITY

DRUG ADMINISTERED

NaCl(0.9%,iv) — n=17
Amph(1.0mg/kg/iv) — 6
Amph(5.0mg/kg/iv) — 6
Haldol(0.2mg/kg/iv) — 5
SCH23390(0.3mg/kg/iv) — 7
SCH23390(0.6mg/kg/iv) — 4
R-SKF38393(1uM, inf) — 8 *
R-SKF38393(10uM, inf) — 10 **
SCH23390(10uM, inf) — 3
Into non-terminal axon
R-SKF38393(10uM, inf) — 6

MEAN CHANGE IN THRESHOLD (%)

FIG. 1 *Mean (and standard error) change in threshold terminal excitability induced by various dopaminergic agonists and antagonists. Paired drugs indicate subsequent administration. For example, SCH 23390 (10μM) was infused(inf) in three of the ten rats that had received R-SKF 38393 (10μM) within the previous ten minutes. Change in threshold is expressed relative to the threshold measured immediately before each drug administration. Infusions were administered in volumes of 300nL over 5 min. *p<0.1, **p<0.001. Redrawn from Ryan et al 1989.*

remarkable considering the intense behavioral activation and focussed stereotyped behavior elicited in behaving animals by 5.0 mg/kg amphetamine
Three lines of evidence suggest that these D1 receptors are functional in the sense that exogenously applied agents can activate the receptors and produce physiological responses. Thus D1 agents 1) modulate GABA release (Starr,1987), 2) indirectly excite substantia nigra pars reticulata neurons (Waszczak and Martin, 1989), and 3) decrease terminal excitability of striatonigral axons. Yet these receptors do not appear to be functioning under normal conditions. They do not appear to be tonically activated by dopamine in anesthetized animals (in which dopamine neurons fire at roughly the same rate and pattern as in awake animals) (Freeman and Bunney 1985; Diana et al 1989), nor in behaviorally activated animals under the influence of a pharmacological agent that releases endogenous dopamine.
Why, then, do non-dopamine striatonigral axons possess presynaptic D1 heteroreceptors if they do not function under normal physiological conditions? One speculation is that these D1 receptors play a role during development and are retained into adulthood. Such retention of juvenile traits is not unknown. Or, these receptors may be only functional within the

striatum. Another, and perhaps more likely possibility is that the striatonigral projection is heterogeneous. Although the projection appears to arise from a single cell type (or at least 99% of it (Bolam et al., 1981)), the neostriatum possesses at least two major interdigitated subregions, the striosomal and matrix, that have distinct biochemical identities (Graybiel and Ragsdale, 1983) and distinct targets in the substantia nigra (Gerfen, 1985). The more extensive projection deriving from the matrix region of the neostriatum is targeted at the substantia nigra pars reticulata whereas the striosomal projection is directed, at least in part, to the dopamine neurons in the lowest part of substantia nigra pars compacta and in the pars reticulata (Gerfen, 1985; 1987a). Perhaps those striosomal neurons that synapse directly on dopamine neurons have functioning D1 receptors whereas the matrix projection possesses functional but unused receptors. It is possible that we exclusively sampled terminal axons projecting from matrix since they are more numerous.

TABLE 1

Changes in Striatonigral Axonal Excitability Induced By
d-Amphetamine Sulfate in Freely-Moving, Behaving Rats

| | | Time Post-Drug | |
| | | 30 Minutes | 60 Minutes |
Dose	n	Mean(S.E.M.)	Mean(S.E.M.)
0.25 mg/kg	5	3.8% (6.3)	5.8(8.1)
1.00 mg/kg	7	-2.4% (2.4)	0.7(2.9)
2.50 mg/kg	5	-0.9% (5.7)	3.1(8.9)
5.0 mg/kg	4	-0.1% (3.2)	3.2(5.2)

Note: Amphetamine was administered s.c. Data after Ryan et al., 1988.

Effects of D1 agents on nigrostriatal terminal excitability

In contrast to the striatonigral system, the presence of D1 receptors on the nigrostriatal dopaminergic system is highly controversial. Most of the autoradiographic studies (Altar and Hauser 1987; Altar and Marien 1987; Morelli et al 1988) have indicated that dopamine autoreceptors are exclusively of the D2 type. In addition, in vitro release studies had previously shown that D1 selective agonists such as SKF 38393 and fenoldopam do not alter the electrically-induced overflow of $[^3H]$dopamine (Lehmann et al 1983). In spite of these findings, it has recently been shown that D1 selective agents, administered systemically, modify dopamine release as studied by brain dialysis (Zetterstrom et al 1986; Imperato et al 1987). Additionally, it appears, that SKF 38393 and SCH 23390 respectively, decrease and increase the release of dopamine when applied locally in the striatum through the dialysis probe (Imperato et al 1988). This evidence, considering the lack of terminal axo-axonic contacts in the rat striatum, raises questions concerning the absence of D1 autoreceptors in the nigrostriatal dopamine system. Furthermore, the presence of D1 receptors in the terminal region of nigrostriatal dopaminergic neurons would be consistent with studies (Savasta et al 1986; Porceddu et al 1987) which suggest that some of the D1 receptors in the substantia nigra are located on dopamine elements.

We have explored this issue using the terminal excitability paradigm. This procedure offers the unique possibility of studying the effect of presynaptic receptor stimulation at the single cell level. Our results raise the possibility that a terminal excitability-modulating autoreceptor of the D1 type indeed exists on the terminals of nigrostriatal dopaminergic neurons with pharmacological behavior indistinguishable from the better known D1 receptor of the striatonigral system.

Local striatal application of the D1 selective agonist R-SKF 38393 (1, 10, 100 µM) (Setler et al 1978) produced a dose-dependent decrease in dopaminergic terminal excitability (FIG 2). The R-SKF 38393-induced decrease (10 µM, 300 nl, 5 min) could be reversed by subsequent striatal administration of the D1 selective antagonist SCH 23390 (10µM, 300nl, 5 min) (Iorio et al 1983; Hyttel et al 1983) indicating a D1 receptor mediated effect. In contrast, local application of these compounds, to a non terminal region, along the medial forebrain bundle did not induce significant effects excluding nonspecific actions of the two molecules. In addition, since it has been proposed that SKF 38393 may be diminishing dopamine release indirectly (Imperato et al 1988; Galloway1988) by inhibiting the activity of the rate-limiting enzyme tyrosine hydroxylase, a group of rats was pretreated with the dopamine synthesis inhibitor alpha-methyl-para-tyrosine (Diana et al 1989c). In these dopamine-depleted rats the action of R-SKF 38393 was fully preserved, as was its reversal by SCH 23390 indicating a D1 mediated mechanism which does not depend on the presence endogenous dopamine. Striatal application of the neurotoxin kainic acid which destroys cell bodies and spares axons of passage was performed in a different group of rats in order to exclude the possibility that the D1-mediated effects could be mediated by D1 receptors located on striatal neurons (Altar and Marien 1987). Following this lesion, the action of R-SKF 38393 and its reversal by SCH 23390 did not differ from that observed previously, supporting the possibility that the effects induced by D1 selective agents are mediated by a D1 receptor located on the presynaptic dopaminergic terminal.

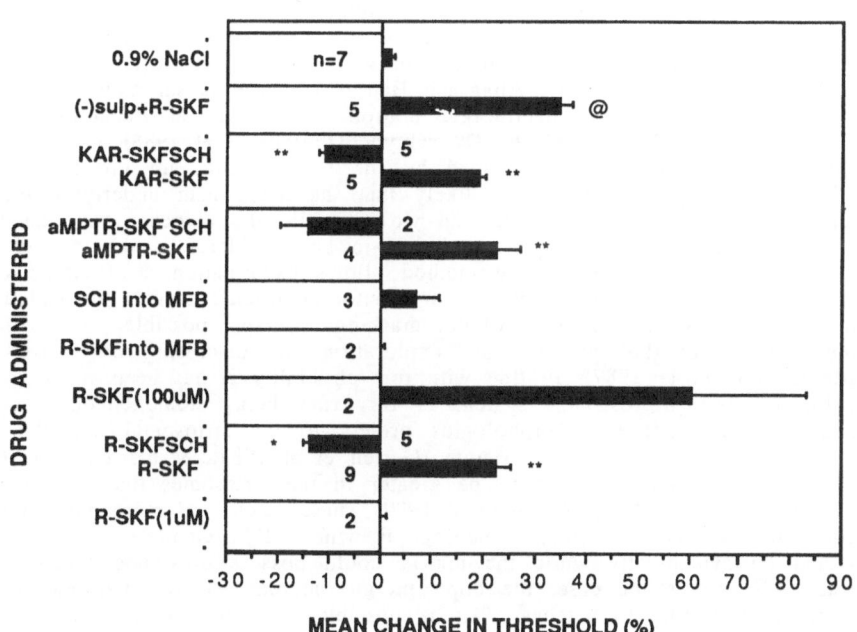

FIG. 2 *Mean change (and standard error) in threshold terminal excitability induced by various dopaminergic agonists and antagonists. Measurements are as described in Fig. 1. *p<0.05, **p<0.001. @ p< 0.05 as compared to R-SKF. Local infusions were delivered in 300nL volumes over 5 min. Drugs are administered at 10 µM unless otherwise indicated. Redrawn from Diana et al 1989c.*

Finally, rats were pretreated with the selective D2 receptor blocker (-)sulpiride 15 to 30 min. prior to striatal administration of R-SKF 38393 to investigate whether the D2 presynaptic receptor contributed in some manner to the decrease in terminal excitability induced by the D1 agonist R-SKF 38393. Surprisingly, the action of the D1 agonist was potentiated in this condition (Fig 2) This experiment is interpreted as indicating that while the D2 receptor does not mediate the D1-induced effects, it alters (magnifies) D1 responses because as we have previously shown, local administration of the selective D2 antagonist (-)sulpiride increases dopaminergic terminal excitability (Tepper et al, 1984b): therefore, the range over which subsequent administration of the D1 agonist decreases excitability is increased. It is interesting that under neuroleptic treatment the dopaminergic terminal is more sensitive to D1 agonist administration. If extrapolated to a clinical setting, this observation could lead to a new pharmacological strategy for schizoaffective disorders, namely concomitant administration of a D2 antagonist with a low, possibly presynaptic, dose of D1 agonist. This could possibly ameliorate schizoid symptomatology with lower total doses of therapeutic agents thereby reducing the risk, or delaying the onset, of extrapyramidal side effects.

Interestingly, the local infusion of the D1 antagonist SCH 23390 alone led to a paradoxical agonist-like decrease in excitability. This effect was prevented in animals pretreated with haloperidol and did not occur following dopamine depletion with alpha-methyl-para-tyrosine. These experiments, and the observation that the antagonist reversed the agonist induced decrease in excitability, suggest that SCH 23390 acts on a D1 receptor to increase the release of dopamine resulting in increased D2 stimulation and the paradoxical effect.

CONCLUSIONS

These experiments indicate that D1 selective agents are effective in modulating striatonigral and nigrostriatal terminal excitability by stimulation of a D1 receptor located on the synaptic terminal under study. While this contention is not surprising when referred to striatonigral projection neurons where there is strong evidence for the presence of dopaminergic receptors of the D1 type (Spano et al 1977; Altar and Hauser 1987), it is surprising in the case of the nigrostriatal dopaminergic neuron, where the autoreceptors are considered to be exclusively of the D2 subtype (Morelli et al 1988). Furthermore, since the effects induced by the same agents are virtually identical in the two systems, it is likely that the mechanism underlying these effects is also identical. How then can we reconcile the evidence reported here with the previous findings? One possibility to be considered is the method employed: the terminal excitability method allows examination at a single cell level in contrast to previous binding and autoradiographic studies in which a region is typically studied as a whole, masking, perhaps, possible inhomogeneities of the system under exploration and disclosing only gross details (Filloux et al 1987). In line with this possibility it has been reported that the dopaminergic nigrostriatal system is far from being homogeneous, components with different morphologies project to the striosomal and the matrix compartments of the rat striatum (Gerfen et al 1987a & b). The density of D1 receptors has been reported to be greater in the striosomes than in the matrix (Besson et al 1988; Murrin et al 1989) These patches of more dense D1 receptor binding could represent subregions where D1 receptors are present pre- and postsynaptically while the matrix could possess only post-synaptic D1 receptors. If this is the case, the dopaminergic neurons we sampled may have been those projecting to patches. This is possible since the neurons we studied were activated from a restricted region of the neostriatum (Diana et al 1989c) and thus possibly may represent a single population of nigrostriatal dopaminergic neurons.

These studies emphasize that understanding the physiological functions of D1 and D2 activation will require consideration of presynaptic as well as postsynaptic effects and interactions. In addition, the effects of D1 agents observed may be of pharmacological significance. Assuming that increases and decreases in excitability reflect corresponding changes in the facilitation of release, D1 agonists act to inhibit release. On the other hand, D1 antagonists may not affect release from striatonigral terminals, and, as a result of a paradoxical agonist-like action, may, after the initial increase, decrease release from dopamine terminals in the striatum.

REFERENCES

Aiso M, Potter WZ, Saavedra JM (1987) Axonal transport of dopamine D1 receptors in the rat brain. Brain Res 426: 392-396.

Altar CA and Hauser K (1987) Topography of substantia nigra innervation by D1 receptors in the rat brain. Brain Res 410: 1-11.

Altar CA and Marien MR (1987) Picomolar affinity of $[^{125}I]$-SCH23982 for D1 receptors in brain demonstrated with digital subtraction autoradiography. J Neurosci 7: 213-222.

Besson M.J., Graybiel A.M. and Nastuk M.A. (1988) $[^{3}H]$SCH 23390 binding to D1 dopamine receptors in the basal ganglia of the cat and primate: delineation of strisomal compartments and pallidal and nigral subdivision. Neuroscience 26(1): 101-119.

Bolam JP, Somogyi P, Totterdell S, Smith AD (1981) A second type of striatonigral neuron: a comparison between retrogradely labeled and Golgi-stained neurons at the light and electron microscopic levels. Neuroscience 6: 2141-2157.

Cheramy A, Nieouillon A, Glowinski J (1978) In vivo changes in dopamine release in cat caudate nucleus and substantia nigra induced by nigral application of various drugs including GABAergic agonists and antagonists. In: Interactions between putative neurotransmitters in the brain, S Garattini, JF Pujol, R Samanin (Eds.), Raven Press: New York, pp. 175-190.

Chevalier G, Vacher S, Deniau JM, Desban M (1985) Disinhibition as a basic process in the expression of striatal functions. 1. The striato-nigral influence on tecto-spinal/tecto-diencephalic neurons. Brain Res 334: 215-226.

Diana M., Young S.J. and Groves, P.M. (1989a) Modulation of dopaminergic terminal excitability by D1 selective agents. Neuropharmacology 28(1): 99-101.

Diana M., Garcia-Munoz M., Richards J. and Freed C.R. (1989b) Electrophysiological analysis of dopamine cells from the substantia nigra pars compacta of circling rats. Exptl. Brain Res. 74: 625-630.

Diana M. Young S.J. and Groves P.M. (1989c) Dopaminergic terminal excitability: neuropharmacological evidence for a D1 autoreceptor. Neuroscience (submitted).

Di Chiara G, Porceddu ML, Morelli M, Mulas ML, Gessa GL (1979) Evidence for a GABAergic projection from the substantia nigra to the ventromedial thalamus and to the superior colliculus of the rat. Brain Res 176: 273-284.

Filloux F.M. Wamsley J.K. and Dawson T.M. (1987) Dopamine D-2 auto- and postsynaptc receptors in the nigrostriatal system of the rat brain: localization by quantitative autoradiography with $[^{3}H]$sulpiride. Eur. J of Pharmacol. 138: 61-68.

Freeman A.S., Meltzer L.T. and Bunney B.S. (1985). Firing properties of substantia nigra dopaminergic neurons in freely moving rats. Life Sci 36: 1983-1994.

Garcia-Munoz, M. Patino, P., Chavez-Noriega, L. Arbuthnott, G. and Ryman, A. (1987) Dopamine control of excitability changes in nigrostriatal terminals. In *The Basal Ganglia II* (eds Carpenter, M.B. and Jayaraman, A.) pp. 149-155. Plenum Press, New York.

Gariano, R.F., Sawyer, S.F., Tepper, J.M., Young, S.J., and Groves, P.M. Mesocortical Dopaminergic Neurons: Electrophysiological consequences of terminal receptor activation. Brain Res Bul 22: 517-523.

Gerfen CR (1985) The neostriatal mosaic. I. Compartmental organization of projections from the striatum to the substantia nigra in the rat. J Comp Neurol 236: 454-476.

Gerfen C.R., Herkenham M. and Thibault J. (1987a) The neostriatal mosaic: II. Patch- and matrix-directed mesostriatal dopaminergic and non-dopaminergic systems. J. of Neurosci. 7: 3915-3934.

Gerfen C.R., Baimbridge K.G. and Thibault J. (1987b) The neostriatal mosaic: III. Biochemical and developmental dissociation of patch-matrix mesostriatal systems. J. of Neurosci. 7: 3935-3944.

Graybiel AM and Ragsdale CW (1983) Biochemical anatomy of the striatum. In: Chemical Neuroanatomy, PC Emson (Ed.), Raven Press: New York, pp. 427-503.

Groves, P.M., Fenster, G.A., Tepper, J.M., Nakamura, S. and Young, S.J. (1981) Changes in dopaminergic terminal excitability induced by amphetamine and haloperidol. Brain Res. 221: 425-431.

Hyttel J. (1983) SCH 23390- the first selective dopamine D-1 antagonist. Eur. J. Pharmacol. 91: 153-154.

Imperato A., Mulas A. and Di Chiara G. (1987) The D-1 antagonist SCH 23390 stimulates while the D-1 agonist SKF 38393 fails to affect dopamine release in the dorsal caudate of freely moving rats. Eur. J. of Pharmacol. 142: 177-181.

Imperato A. and Di Chiara G. (1988) Effects of locally applied D-1 and D-2 receptor agonists and antagonists studied with brain dialysis. Eur. J. of Pharmacol. 156: 385-393.

Iorio L.C., Barnett A., Leitz F.H., Houser V.P. and Korduba C.A. (1983) SCH 23390, a potential benzazepine antipsychotic with unique interactions on dopaminergic systems. J. Pharmacol. exp. Ther. 226: 462-468.

Kalivas P.W., Bourdelais A., Abhold R and Abbott L. (1989) Somatodendritic release of endogenous dopamine: in vivo dialysis in the A10 region. Neurosci. Lett. 100: 215-220.

Lehmann J., Briley M. and Langer S.Z. (1983) Characterization of dopamine autoreceptor and [^{3}H] spiperone binding sites in vitro with classical and novel dopamine receptor agonists. Eur. J. Pharmacol. 88: 11-26.

Mereu, G.P., Westfall, T.C. and Wang, R.Y. (1985) Modulation of terminal excitability of mesolimbic dopaminergic neurons by d_amphetamine and haloperidol. Brain Res. 359: 88-96.

Morelli M., Mennini T. and Di Chiara G. (1988) Nigral dopamine autoreceptors are exclusively of the D2 type: quantitative autoradiography of [^{125} I] iodosulpiride and [^{125}I] SCH 23390 in adjacent brain sections. Neuroscience 27: 865-870.

Murrin L.C. and Zeng W. (1989) Dopamine D1 receptor development in the rat striatum: early localization in striosomes. Brain Res. 480: 170-177.

Nakamura, S., Tepper, J.M., Young, S.J. and Groves, P.M. (1982) Changes in noradrenergic terminal excitability induced by amphetamine and their relation to impulse traffic. Neuroscience 7(9): 2217-2224.

Paden C, Wilson CJ, Groves PM (1976) Amphetamine-induced release of dopamine from the substantia nigra in vitro. Life Sci 19: 1499-1506.

Porceddu M.L., Giorgi O., De Montis G., Mele S., Cocco L., Ongini E. & Biggio G. (1987) 6-Hydroxydopamine-induced degeneration of nigral dopamine neurons: differential effect on nigral and striatal D-1 receptors. Life Sci. 41: 697-706.

Reubi J-C, Iversen LL, Jessell TM (1977) Dopamine selectively increases 3H-GABA release from slices of rat substantia nigra in vitro. Nature 268: 652-654.

Ryan, L.J. Tepper, S.F., Young, S.J., & Groves P.M. (1985) Amphetamine's effects on terminal excitability of noradrenergic locus coeruleus neurons are impulse dependent at low but not high doses. Brain Res 341: 155-163.

Ryan, L.J., Diana, M., Young, S.J. and Groves, P.M. (1989) Dopamine D1 heteroceptors on striatonigral axon terminals are not stimulated by endogenous dopamine either tonically or after amphetamine: evidence from terminal excitability. Exptl. Brain Res. 77: 161-165.

Ryan LJ, Young SJ, Segal DS, Groves PM (1989) Antidromically identified striatonigral projection neurons in the chronically implanted behaving rat: relations of cell firing to amphetamine-induced behaviors. Behavioral Neuroscience 103: 3-14.

Savasta, M., Dubois, A., Benavides, J. and Scatton, B. (1986) Different neuronal location of [^{3}H]SCH 23390 binding sites in pars reticulata and pars compacta of the substantia nigral in the rat. Neurosci Lett 72: 265-271.

Savasta M, Dubois A, Feuerstein C, Benavides J, Scatton B (1987) Localization of D1 dopamine receptors in the rat brain by quantitative autoradiography: effect of dopaminergic denervation. Biogenic Amines 4: 419-429.

Setler P., Sarau H., Zerkle C.L., Saunders H.L. (1978) The central effect of a novel dopamine agonist. Eur. J. of Pharmacol. 50: 413-440.

Spano P.F., Trabucchi M., Di Chiara G. (1977) Localization of nigral dopamine-sensitive adenylate-cyclase on neurons originating from corpus striatum. Science 196: 1343-1345.

Starr M (1987) Opposing roles of dopamine D1 and D2 receptors in nigral [3H] gamma-aminobutyric acid release? J Neurochem 49: 1042-1049.

Sawyer, S. F., Tepper, J.M., Young, S.J. and Groves, P.M. (1985) Activation of dorsal raphe neurons from neostriatum: physiological characterization and the effects of terminal autoreceptor activation. Brain Res: 332: 15-28.

Tepper, J.M. , Young, S.J. and Groves, P.M. (1984a) Autoreceptor mediated changes in dopaminergic terminal excitability: effects of increase in impulse flow. Brain Res.309: 309-316.

Tepper, J.M. , Young, S.J. and Groves, P.M. (1984b) Autoreceptor mediated changes in dopaminergic terminal excitability: effects of striatal drug infusions. Brain Res. 309: 317-333.

Wall, P.D. (1958) Excitability changes in afferent fibre terminations and their relation to slow potentials. J. Physiol. 142: 1-21.

Waszczak BL and Martin L (1989) Striatonigral lesions and intranigral injection of receptor inactivator EEDQ prevent D-1 agonist effects on substantia nigra pars reticulata (SNpr) neurons. Soc Neurosci Abs 15: 428.

Zettestrom T., Sharp T. and Ungersted U. (1986) Effect of dopamine D-1 and D-2 receptor selective drugs on dopamine release and metabolism in rat striatum in vivo. Naunyn-Schmiedeberg's Arch. Pharmacol. **334**: 117-124.

In Vivo Development of the Spontaneous Activity of Rat Nigrostriatal Dopaminergic Neurons

James M. Tepper[1], Francine Trent[2] and Shoji Nakamura[3]

[1]Center for Molecular and Behavioral Neuroscience, [2]Department of Biological Sciences, Rutgers, The State University of New Jersey, Newark, NJ 07102 and [3]Kanazawa University Faculty of Medicine, Kanazawa 920, Japan

INTRODUCTION

Dopaminergic neurons, along with other monoamine neurons, are known to be among the earliest neurons in the central nervous system to differentiate morphologically and neurochemically, and to send axons to their target regions (Olson and Seiger, 1972; Seiger and Olson, 1973; Voorn et al., 1988). Despite our knowledge of the morphological development of dopaminergic neurons, little is known about the time course of the development of the electrophysiological properties of these neurons. If dopaminergic neurons are physiologically functional early in ontogeny, they may play a role in the development of their target structures, analogous to that shown for noradrenergic neurons in a number of studies (Kasamatsu and Pettigrew, 1976; Pettigrew and Kasamatsu, 1976; Blue and Parnevelas, 1982). Information on the development of dopaminergic neurons *in situ* may also be relevant to our understanding of the physiological functioning of dopaminergic neurons grafted to the dopamine-denervated striatum. Thus the present experiments were carried out to characterize the developmental profile of the *in vivo* spontaneous activity of rat nigrostriatal dopaminergic neurons from birth through maturity.

METHODS

Subjects

Subjects consisted of 68 Sprague-Dawley pups derived from pregnant dams (obtained from Charles River or The Institute for Animal Behavior at Rutgers). Pregnant females were checked daily for the presence of new litters, and the day of birth was considered to be postnatal day 1 (PD1). The ages of the rat pups used ranged from PD1 to PD28, and their weights varied from 6.5 g to 74.5 g. Pups were anesthetized by intraperitoneal injection of urethane (1.3 mg/g body weight), and supplemented by inhalation of metofane (methoxyfluorane) if necessary, and installed into a modified stereotaxic device described by Nakamura and colleagues (Nakamura et al., 1987). Body temperature was maintained at 37 ± 1° C with a solid state heating pad.

For purposes of comparison with nigrostriatal dopaminergic neurons from adult rats, 12 male Sprague-Dawley rats, over 75 days of age (weights ranging from 240-460 g) were anesthetized with urethane (1.3 g/kg, i.p.) and installed in a stereotaxic frame according to standard procedures (e.g., Tepper et al., 1984).

Electrical Stimulation

For purposes of antidromic identification, a small burr hole was drilled overlying the anterior-lateral neostriatum (coordinates 0.5-0.7 mm anterior to bregma, 2.4-3.5 mm lateral to the midline) and an bipolar, stainless steel stimulating electrode (Tepper et al., 1984) was inserted to depths ranging from 2.2-3.4 mm below the cortical surface. In adult rats the stimulating electrode

The Basal Ganglia III, Edited by G. Bernardi *et al.*
Plenum Press, New York

was positioned in the anterior-lateral dorsal neostriatum at coordinates 1.0 mm anterior to bregma, 3.7 mm lateral and 4.0 mm below the cortical surface. Constant current stimuli, 0.1 to 5.0 mA at a pulse duration of 500 μsec were delivered at a rate of 0.67 Hz by a Winston A-65 timer/stimulator and Winston SC-100 stimulus isolation units.

Recording

A 1.5 mm diameter burr hole was drilled overlying the substantia nigra at coordinates 0.8-1.5 mm anterior to lambda and 0.7-1.5 mm lateral to the midline for neonates and 2.1 mm anterior to lambda and 2.0 mm lateral to the midline for adults. Single unit extracellular discharges were recorded with glass micropipettes filled with 2% Pontamine Sky Blue in 2 M NaCl, possessing *in vitro* impedances of 5-10 MOhms. Electrode signals were amplified with a Neurodata IR183 pre-amplifier and displayed on a Tektronix 5113 storage oscilloscope. Typical filter settings were 100 or 1 kHz low pass and 10 kHz or 30 kHz high pass. All data were recorded on magnetic tape for off-line analysis.

Data Analysis

Spike trains were played back from tape off-line and input to a Macintosh II microcomputer through a National Instruments MIO16L multifunction board. Spontaneous activity was analyzed for firing rate, and pattern of activity by means of first order interspike interval histograms and autocorrelograms and a statistical analysis of burst firing. Analyses of spike waveforms were obtained by digital averaging of 5-10 action potentials.

For purposes of statistical analysis, data from rats were pooled and assigned to one of the following groups: PD1-3, PD4-6, PD7-10, PD11-15, PD16-21, PD22-28 and ADULT. A one way analysis of variance was performed on a number of parameters including spontaneous firing rate, number of spikes per burst, burst length, mean interspike interval, duration of the extracellularly recorded spontaneous action potential, antidromic threshold, antidromic latency, and proportion of antidromic spikes consisting of full, initial segment-somadendritic (IS-SD) spikes. Where appropriate, differences between specific age groups were tested with Scheffé's F Test at the p<0.1 level of significance.

Histology

At the end of each experiment, the stimulating site was marked by a small DC lesion made with the stimulating electrode. The last recording site was marked by iontophoresis of Pontamine Sky Blue through the recording electrode. Animals were perfused with 10-20 ml normal saline followed by 50-70 ml of 10% formalin. The brains were removed, post-fixed in 10% formalin and sectioned on a vibratome. Sections were stained with neutral red and each stimulation site and the site of the last recording marked by a Pontamine Sky Blue dot were noted, photographed and/or drawn at 1X with a Nikon Labophot microscope equipped with a drawing tube.

RESULTS

Neuronal Identification

Extracellular recordings were obtained from 165 antidromically driven neurons in 68 rat pups and from 26 neurons from 12 adult rats presumed to be dopaminergic nigrostriatal neurons. Neurons recorded from neonates in this study were assumed to be dopaminergic nigrostriatal neurons provided that they could be antidromically activated from ipsilateral neostriatum with latencies greater than 9 ms, and provided that later histological analysis indicated that the neurons were located within the region of the substantia nigra, pars compacta. Evoked responses were considered antidromic provided that they collided with appropriately timed spontaneous spikes, or, for neurons that exhibited little or no spontaneous activity, could follow twin pulse stimulation with interstimulus intervals corresponding to a train of 200 Hz. Neurons encountered in the vicinity of the substantia nigra pars compacta that were not antidromically driven were excluded from study. All 26 neurons from adult rats fulfilled previously published electrophysiological criteria for nigrostriatal dopaminergic neurons (Guyenet and Aghajanian, 1978; Deniau et al., 1978; Grace and Bunney, 1983a).

Mean Firing Rates

Spontaneously active nigrostriatal dopaminergic neurons could be recorded as early as PD1, although a number of antidromically responsive neurons exhibited little or no spontaneous firing at this age. The mean firing rate of nigrostriatal dopaminergic neurons increased with age from PD1-3 through PD22-28 (F=30.18, df=6, 138, p<.001), with a concomitant decrease in the mean interspike interval (ISI) over this developmental span (F=11.99, df=6,132, p<.001). Although not subjected to a statistical analysis, the number of non-spontaneously active neurons appeared to decrease steadily through the first 3 postnatal weeks. By PD22-28, mean firing rates and mean ISIs no longer differed from those in adults. Developmental changes in mean firing rate and interspike interval are illustrated in Figure 1.

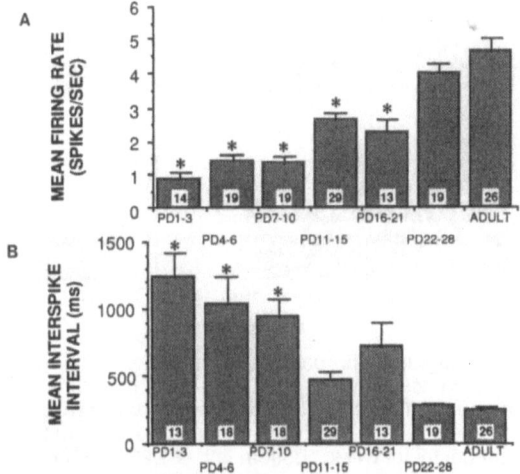

Figure 1. Changes in parameters of spontaneous activity of nigrostriatal neurons during development. A. Mean firing rate increases from PD1-3 through PD22-28. B. Mean interspike intervals decrease from PD1-3 through PD22-28. Numbers within the bars indicate number of neurons per group. Error bars represent SEM. Asterisks indicate significant difference from adult controls.

Spontaneous Firing Pattern

Firing pattern was analyzed by computing autocorrelograms from samples of spontaneous activity for each neuron and by computing the mean number of spikes per burst and the mean burst duration for neurons that exhibited at least one burst of 2 or more spikes. Burst onset was defined as an interspike interval of 80 ms or less, and the burst termination was defined as defined as the first interspike interval that exceeded 160 ms following the onset of a burst, according to criteria previously defined for nigral dopaminergic neurons in adult rats (Grace and Bunney, 1984).

The pattern of spike activity changed significantly over development. Dopaminergic neurons from the earliest postnatal group often displayed only sporadic spontaneous activity into which was imbedded long periods (up to several minutes but typically on the order of 5-45 seconds) of silence. PD1-3 neurons rarely fired in bursts. Over the next week, mean firing rate increased and some neurons began to exhibit short bursts consisting almost exclusively of two spikes with a very stereotyped interspike interval of 60.4±0.65 ms in an otherwise random firing pattern. The spontaneous firing rate continued to increase over the next week. During this stage (PD7-15) nigrostriatal neurons exhibited a transient phase of very regular, almost pacemaker-like rhythmic activity, occasionally interrupted by the two-spike bursts. This firing pattern was most apparent in autocorrelograms that displayed multiple initial peaks, resembling repetitive firing of nigral pars compacta neurons in adult rats (Wilson et al., 1977). Over the next week the incidence of rhythmic firing decreased, as the occurrence of doublet bursts as well as longer bursts increased. The was a significant developmental increase in both the duration of the burst (F=5.79, df=6, 78, p<.001) and the number of spikes per burst (F=4.88, df=6, 78, p<.001) but not in the interval between the first and second spikes in a burst. By PD16-21, neither the average burst duration nor the number of spikes per burst differed from adults. Sample spike trains and their associated autocorrelograms from different developmental phases are illustrated in Figure 2. Additional burst statistics are presented in Table 1.

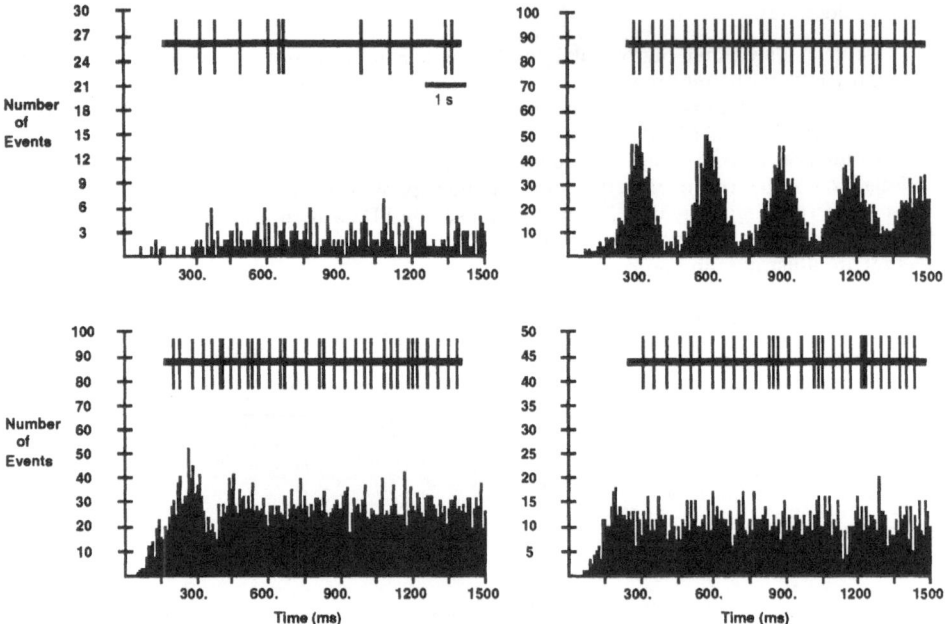

Figure 2. Autocorrelograms and portions of the spike trains from which they were constructed illustrating developmental changes in the pattern of spontaneous activity of nigrostriatal neurons. Top left. PD3 rat shows sporadic random firing. Top right. PD9 rat shows the very regular, almost pacemaker-like pattern transiently expressed during the second postnatal week. Bottom left. PD25 rat shows essentially mature firing pattern including some bursting. Bottom right. Adult control showing typical irregular pattern.

Action Potential Morphology

The morphology of the extracellularly recorded spontaneous action potential waveform also changed significantly during postnatal development, as illustrated in Figure 3. Neurons from the youngest animals exhibited action potentials of relatively low amplitude with a poor signal to noise ratio. Spike waveforms were of significantly greater duration in the neonates compared to adults (F=7.99, df=6, 125, p<.001), due in part to a broadening of the SD spike component in neurons from the youngest animals. In contrast, as soon as the IS component of the spike became reliably distinguishable (around PD 4-6), it did not change in either amplitude or duration through adulthood. There was a significantly greater delay between the IS and SD components of the action potential in the neonates than in adults. From birth through the third postnatal week, spike amplitudes increased continuously due to an increase in the amplitude of the SD component of the spike. A progressive shortening of the IS-SD delay was also apparent over this span. By the end of the third postnatal week, the morphology of the extracellularly recorded spike waveform was not significantly different from that of nigral neurons in adult rats.

Antidromic Response Properties

Antidromic responses could be reliably elicited by neostriatal stimulation in the earliest animals tested, pups as young as 6 hours post-partum, as illustrated for one PD1 rat in Figure 4. Similar to antidromic responses of nigrostriatal dopaminergic neurons from adults, striatal-evoked antidromic responses in neonatal rats most often consisted of the IS spike only, with the action potential failing to invade the SD portion of the neuron even at modest rates of stimulation (Guyenet and Aghajanian, 1978; Deniau et al., 1978, Tepper et al., 1984). However, the proportion of antidromic spikes consisting of a full IS-SD spike was significantly greater in neurons from neonates than from adults (F=4.5, df=6, 210, p<.001). Full spike (IS-SD) antidromic responses often exhibited a striking delay between the IS and SD components of the spike of up to a few milliseconds, particularly in neurons from the youngest animals. As previously reported for adult nigrostriatal neurons (Collingridge et al., 1980; Tepper et al., 1984), many neonatal nigrostriatal neurons

(53.2+7.7%) exhibited multiple, discrete antidromic latencies, even at constant stimulus currents. There were no significant developmental changes in the proportion of neurons exhibiting these multiple antidromic latencies. Neither the mean antidromic latency nor the mean antidromic

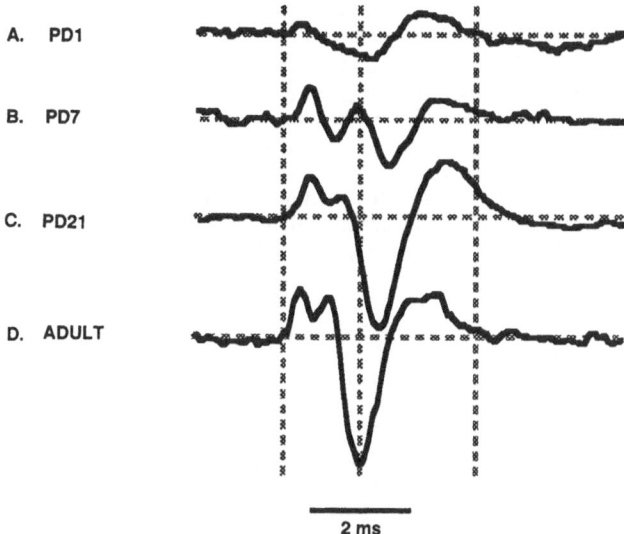

A. PD1

B. PD7

C. PD21

D. ADULT

2 ms

Figure 3. Representative developmental changes in the morphology and duration of extracellularly recorded spontaneous action potentials of nigrostriatal neurons. Dashed lines denote onset of IS spike, peak of SD spike, and return of SD spike to baseline for the adult control shown in D. Note progressive increase in the amplitude of the SD spike through PD21, and an overall reduction in spike duration due to a decrease in the IS-SD interval and a reduction in SD spike width. Each trace is the average of 5 single spikes from the same neuron. In this and in Figure 4, positivity is upwards.

threshold current (minimum current necessary to elicit 100% antidromic responding on non-collision trials) exhibited a significant developmental trend. Antidromic latencies and estimated conduction velocities are listed in Table 1 below.

Table 1. Developmental profile of additional neurophysiological parameters of nigral dopaminergic neurons. Cells bursting : Proportion of neurons firing at least one burst of greater than 2 spikes. Burst duration: Mean duration of all bursts of two or more spikes in ms. N Spikes/burst: Mean number of spikes per burst. AD latency: Antidromic latency in ms. Conduction velocity: Mean axonal conduction velocity in m/s estimated by dividing the straight line distance between recording and stimulating sites by the antidromic latency for each cell. Asterisks indicate significant difference from adult group.

	PD1-3	PD4-6	PD7-10	PD11-15	PD16-21	PD22-28	ADULT
CELLS BURSTING	3/13(23.1)	3/18(16.7)	1/18(5.6)	5/27(18.5)	4/13(30.8)	6/20 (30.0)	15/26(57.7)
BURST DURATION	60.1±2.2*	75.6±9.8*	61.0±2.8*	68.3±2.3*	115.8±20.4	108.4±17.1	167±28.7
N SPIKES/BURST	2.10±.10*	2.16±.09*	2.00±.01*	2.04±.01*	2.45±.18	2.33±.14	3.43±.44
AD LATENCY	15.5±.88	17.1±.43	17.08±.78	18.88±.69*	15.76±1.2	15.5±.93	14.95±.32
CONDUCTION VELOCITY	0.26±.027*	0.24±.011*	0.32±.027*	0.31±.022*	0.46±.032	0.46±.023	0.49±.011

DISCUSSION

Identification of Nigrostriatal Dopaminergic Neurons

Although the parameters of the spontaneous activity of nigral dopaminergic neurons in adult rats have been so well characterized as to rate, pattern and spike waveform that it is quite common to identify extracellular recordings as originating from dopaminergic neurons on the basis of these properties, since the electrophysiology of dopaminergic nigrostriatal neurons in neonates had not previously been characterized, this method was deemed unreliable for the present study. Instead, after it became clear that neonatal nigral neurons could be antidromically activated from neostriatum, and that the antidromic latencies and response properties were very similar to those in adults, neurons were tentatively identified as dopaminergic nigrostriatal neurons provided that they could be antidromically activated from ipsilateral neostriatum at appropriate latency (>9 ms), and provided that subsequent histological analysis indicated that the recording site was localized to the substantia nigra.

These criteria are sufficient to identify dopaminergic neurons since only two types of nigral neurons have been shown to project out of the substantia nigra: dopaminergic neurons located mainly in pars compacta but also to a lesser extent in pars reticulata which project to primarily to neostriatum, and non-dopaminergic neurons located mainly in pars reticulata which project primarily to tectum and thalamus, but occasionally to neostriatum (Guyenet and Aghajanian, 1978). In adults, discrimination between dopaminergic and non-dopaminergic nigrostriatal neurons is unequivocal since dopaminergic neurons exhibit antidromic latencies some 2-4 times greater than non-dopaminergic neurons, and because the antidromic response of dopaminergic neurons usually consists of IS spikes, even at low rates of stimulation (0.67 Hz) such as those employed in the present study (Guyenet and Aghajanian, 1978; Grace and Bunney, 1983a; Tepper et al., 1986). The nigrostriatal neurons recorded in the present study exhibited long-latency antidromic responses that usually consisted of the IS only spike, consistent with their identification as nigrostriatal dopaminergic neurons.

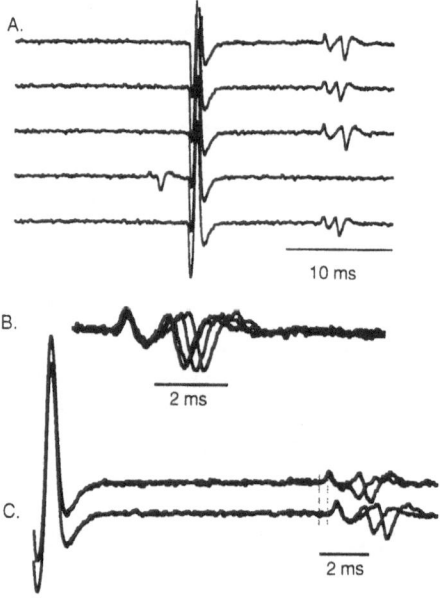

Figure 4. Antidromic responses of nigral dopaminergic neuron in a PD1 rat pup. A. Five consecutive sweeps illustrating antidromic responding from neostriatum. A collision with a spontaneous spike is shown in the fourth trace. Note variability of IS-SD delay. B. Four superimposed sweeps at higher time resolution showing constant latency to IS spike but long and variable IS-SD delay. C. Multiple discrete antidromic latencies to a constant stimulating current. Dashed lines indicate onset of IS spike for each pair of sweeps.

Firing Rate

The spontaneous firing rate of nigrostriatal neurons increased steadily from birth through the third postnatal week with a concomitant decrease in the mean interspike interval. This trend and its time course closely resembles that previously reported for locus coeruleus neurons in the rat (Nakamura et al., 1987), and confirm a report by Pitts et al., (1988) that the firing rates of nigrostriatal neurons from 4 week old rats do not differ from adults. Similar increases in spontaneous firing rates for feline basal ganglia neurons over postnatal development have been reported (Levine et al., 1982).

Firing Pattern

The pattern of spontaneous activity also changed markedly from birth through adulthood. In extracellular recordings obtained in adult rats *in situ*, nigral dopaminergic neurons fire principally in two different modes. The most commonly observed is an irregular pattern in which there is an initial prolonged trough in the autocorrelogram which then rises to an asymptotic value indicative of the mean firing rate. Less commonly observed is a pacemaker-like or repetitive mode, characterized by an initial trough in the autocorrelogram followed by a series of progressively diminishing peaks that eventually plateau out into the mean firing rate asymptote (Wilson et al., 1977). Bursting activity is seen imbedded in both of these modes, and is represented in the autocorrelogram by an early peak immediately following the initial trough.

Autocorrelograms from the youngest animals indicated that firing was almost totally random and irregular. There was very little bursting, and no repetitive activity. From the middle of the second week through the middle of the third week, neonatal nigral neurons exhibited a transient phase in which a large proportion of the neurons fired in a repetitive pacemaker pattern. Although not often seen in such high proportions in dopaminergic neurons from adult rats *in vivo*, this type of very regular firing pattern is typical of nigral dopaminergic neurons recorded in an *in vitro* slice preparation (Grace, 1987). From birth until PD15, when bursting occurred, it consisted of short bursts of only two spikes, with a very stereotyped interspike interval around 60 ms. By the fourth week, neonatal neurons exhibited autocorrelation histograms that displayed both types of adult firing modes, and proportion of spikes occurring in bursts and the average burst length did not differ from adults.

Spike Morphology

Recordings from the neonates tended to be relatively noisy, and the spikes were usually of lower amplitude and greater duration than that seen in the adult, except during periods of depolarization. During the early postnatal period, many neurons were encountered in which the IS spike appeared similar to that in the adult, but in which the SD spike followed at an unusually long delay, and was often equal to or smaller in amplitude than the IS spike. There was no apparent relationship between the firing rate of a neuron and the IS-SD delay. These phenomena are not typical of dopaminergic neurons from mature animals (Guyenet and Aghajanian, 1978; Tepper et al., 1984), except during spikes occurring late in bursts, when, as the soma becomes progressively more and more depolarized, the SD component of the action potential grows progressively smaller and increases in duration, presumably due to increasing depolarization-dependent inactivation of soma-dendritic sodium channels (Grace and Bunney, 1983a,b).

Antidromic Responses

The conduction times from substantia nigra to neostriatum, measured by the antidromic latency, remain constant from PD1 through adulthood, reflecting an increase in conduction velocity of approximately 2.5 times. A similar conservation of conduction time has also been noted for noradrenergic coeruleo-cortical axonal conduction in neonatal and adult rats (Nakamura et al., 1987). This suggests that whatever the roles played by the nigrostriatal dopaminergic system in neonatal neuronal development and adult motor functioning, the timing of dopaminergic nigrostriatal neurotransmission, presumably in relation to that of other afferents to neostriatum, would seem to be of critical importance.

As early as PD1, antidromic responses of many nigrostriatal neurons were observed to occur at discrete, multiple latencies. This phenomenon has been described previously for dopaminergic neurons in adult rats, and has been attributed to the highly branched nature of their terminal arborizations (Collingridge et al., 1980; Tepper et al., 1984). The fact that such multiple latencies were observed at the same frequency in neonates as in adults suggests that dopaminergic terminal fields are highly developed and capable of sustaining impulse traffic at least as early as the day of birth.

Antidromic responses consisted of full IS-SD spikes significantly more often in neonates than adults. When full IS-SD antidromic responses occurred, there was often a very long delay (up to several ms) between the IS and the SD components, similar to that seen in spontaneous spikes described above. The delay was sometimes so great that the waveforms of these neurons in some ways resembled the waveforms of two coupled cells, previously reported to occur in nigral dopaminergic neurons in adults (Grace and Bunney, 1983c; Freeman et al., 1985). That this phenomenon does not represent recordings from electrotonically coupled neurons in the present results is argued by the large temporal variability between the two spike components, the consistent obliteration of both components of the spike during collision with spontaneous spikes, the failure to observe single spike waveforms of "normal" morphology representing the firing of one member of the pair during spontaneous activity, and the absence of an IS-SD break on the initial component of the second spike component. This phenomenon did not seem to be an artifact of cell damage, as it was obtained in neurons that were recorded from for up to 45 minutes without showing any signs of deterioration. Rather, the high proportion of IS-SD antidromic spikes as well as the increased delay between the two components may reflect a state of relative somadendritic depolarization (Grace and Bunney, 1983a,b).

Conclusions

Many of the physiological characteristics of neonatal nigrostriatal neurons were remarkably similar to those obtained from fetal mesencephalic neurons grafted into the dopamine-denervated striatum by Fisher et al., (in press). These investigators reported that dopaminergic graft neurons display mean spontaneous firing rates below those of dopaminergic neurons *in situ* from adult rats, exhibit atypically long-duration action potentials, display a high frequency of pacemaker-type firing patterns, and burst activity that is largely constrained to the occurrence of two spikes with interspike intervals approximating 70 ms. These parameters appeared to change with the age of the graft, being most pronounced in grafts allowed to develop post-operatively for minimal times (~2 months), and more closely approximating values for mature nigral dopaminergic neurons *in situ* after post-operative development for 9 months. Interestingly, some properties i.e., spike duration and frequency and complexity of bursting activity did not change with time, and in the present study, measures of the complexity of bursting were among the last to mature. Thus, the electrophysiological properties of fetal dopaminergic neurons grafted to the striatum of 6-hydroxydopamine treated animals more closely resemble dopaminergic neurons *in situ* from neonatal animals than from adult animals, even after months of post-grafting development. These data suggest that grafted dopaminergic neurons may mature considerably more slowly than dopaminergic neurons *in situ*.

Several lines of evidence point towards the conclusion that dopaminergic nigral neurons in neonatal rats exist in a more depolarized state than in the adult. They are: (1) Decreased spike amplitude and increased spike duration. (2) Increased delay between the IS and SD components of both spontaneous and antidromically driven spikes. (3) Increased proportion of antidromic responses consisting of IS-SD spikes. The reason(s) for the relative depolarization of these neurons are not yet clear, but could be related to immature membrane properties of the neonatal neurons, as has been demonstrated with *in vitro* intracellular recordings in other systems (McCormick and Prince, 1987; Williams and Marshall, 1987), differences in tonic inhibitory GABAergic input from striatonigral and/or pallidonigral pathways (Graybiel and Ragsdale, 1983; Swann et al., 1989; Tepper et al., 1990), or an altered functioning or subsensitivity of somadendritic dopamine autoreceptors (Lacey et al., 1987; Pitts et al., 1988).

In summary, nigral dopaminergic neurons are active in neonatal rats at least as early as the day of birth, and show several signs of existing in a depolarized state relative to dopaminergic neurons of mature rats. At this time the dopaminergic axons are capable of conducting impulses to terminal zones in the neostriatum, and have already arborized to a considerable extent. Nigrostriatal conduction time is conserved from birth through maturity, over which time conduction velocity increases by a factor of 2.5. Between the third and fourth weeks of age, most of the properties of the spontaneous activity of these neurons have reached, or are close to those of dopaminergic nigral neurons in mature rats.

Acknowledgements

We thank Judith S. Rankin for excellent technical assistance and Dr. Anne Mayer and the Institute of Animal Behavior for supplying us with rat pups. This research was supported by MH 45286, a Rutgers University Research Council Grant and a Henry Rutgers Research Fellowship awarded to JMT.

REFERENCES

Blue, M.E., & Parnavelas, J.G. (1982) The effect of neonatal 6-hydroxydopamine treatment on synaptogenesis in the visual cortex of the rat. *J. Comp. Neurol. 205*:199-205.

Collingridge, G.L., James, T.A., & MacLeod, N.K. (1980) Antidromic latency variations in nigral compacta neurons. *Experientia 36*:970-971.

Deniau, J.M., Hammond, C., Riszk, A., & Feger, J. (1978) Electrophysiological properties of identified output neurons of the rat substantia nigra (pars compacta and pars reticulata): Evidence for the existence of branched pathways. *Exp. Brain Res. 32*:409-422.

Fisher, L.J., Young, S.J., Tepper, J.M., Groves, P.M. & Gage, F.H. (1990) Electrophysiological characteristics of cells within mesencephalon suspension grafts. *J. Neurosci.* (in press).

Freeman, A.S., Meltzer, L.T., & Bunney, B.S. (1985) Firing properties of substantia nigra dopaminergic neurons in freely moving rats. *Life Sci. 36*:1983-1994.

Grace, A.A. (1987) The regulation of dopamine neuron activity as determined by *in vivo* and *in vitro* intracellular recordings.In L.A. Chiodo and A.S. Freeman (Eds.) *Neurophysiology of Dopaminergic Systems - Current Status and Clinical Perspectives*, Lakeshore Publishing Co., Grosse Pt., pp.1-66.

Grace, A.A., & Bunney, B.S. (1983a) Intracellular and extracellular electrophysiology of nigral dopaminergic neurons-1. Identification and characterization. *Neuroscience 10*:301-315.

Grace, A.A., & Bunney, B.S. (1983b) Intracellular and extracellular electrophysiology of nigral dopaminergic neurons-2. Action potential generating mechanisms and morphological correlates. *Neuroscience 10*:317-331.

Grace, A.A., & Bunney, B.S. (1983c) Intracellular and extracellular electrophysiology of nigral dopaminergic neurons - 3. Evidence for electrotonic coupling. *Neuroscience 10*:333-348.

Grace, A.A., & Bunney, B.S. (1984) The control of firing pattern in nigral dopamine neurons: Burst firing. *J. Neurosci. 4*:2877-2890.

Graybiel, A.M., & Ragsdale, C.W., Jr. (1983) Biochemical anatomy of the striatum. In: P.C. Emson (Ed.)*Chemical Neuroanatomy*, Raven Press, New York, pp 427-504.

Guyenet, P.G., & Aghajanian, G.K. (1978) Antidromic identification of dopaminergic and other output neurons of the rat substantia nigra. *Brain Res. 150*:69-84.

Kasamatsu, T., & Pettigrew, J.D. (1976) Depletion of brain catecholamines: Failure of ocular dominance shift after monocular occlusion in kittens. *Science 194*:206-209.

Lacey, M.G., Mercuri, N.B., & North, R.A. (1987) Dopamine acts on D2 receptors to increase potassium conductance in neurons of the rat substantia nigra zona compacta. *J. Physiol. (Lond.) 392*:397-416.

Levine, M.S., Fisher, R.S., Hull, C.D., & Buchwald, N.A. (1982) Development of spontaneous neuronal activity in the caudate nucleus, globus pallidus-entopeduncular nucleus, and substantia nigra of the cat. *Dev. Brain Res. 3*:429-441.

McCormick, D.A., & Prince, D.A. (1987) Post-natal development of electrophysiological properties of rat cerebral cortical pyramidal neurones. *J. Physiol. (Lond.) 393*:743-762.

Nakamura, S., Kimura, F., & Sakaguchi, T. (1987) Postnatal development of electrical activity in the locus ceruleus. *J. Neurophysiol. 58*:510-524.

Olson, l., & Seiger, A. (1972) Early prenatal ontogeny of central monoamine neurons in the rat: Fluorescence histochemical observations. *Z. Anat. Entwickl.-Gesch. 137*:301-316.
Pettigrew, J.D., & Kasamatsu, T. (1978) Local perfusion of noradrenaline maintains visual cortical plasticity. *Nature 271*:761-763.

Pitts, D.K., Freeman, A.S., & Chiodo, L.A. (1988) Dopamine neuron ontogeny: Electrophysiological studies. *Soc. Neurosci. Abstr. 14*:408.

Seiger, A., & Olson, L. (1973) Late prenatal ontogeny of central monoamine neurons in the rat.Florescence histochemical observations. *Z. Anat. Entwickl.-Gesh. 140*:281-318.

Swann, J.W., Brady, R.J., & Martin, D.L. (1989) Postnatal development of GABA-mediated synaptic inhibition in rat hippocampus. *Neuroscience 28*:551-561.

Tepper, J.M., Nakamura, S., Young, S.J., & Groves, P.M. (1984) Autoreceptor-mediated changes in dopaminergic terminal excitability: Effects of striatal drug infusions. *Brain Res. 309*:317-333.

Tepper, J.M., Sawyer, S.F., Young, S.J., & Groves, P.M. (1986) Intracellular recording and HRP staining of rat nigral neurons. *Soc. Neurosci. Abstr. 12*:1542.

Tepper, J.M., Trent, F., & Nakamura, S. (1990) Postnatal development of the electrical activity of rat nigrostriatal dopaminergic neurons. *Dev. Brain. Res.*, in press.

Voorn, P., Kalsbeek, A., Jorritsma-Byham, B., & Groenewegen, H.J. (1988) The pre- and postnatal development of the dopaminergic cell groups in the ventral mesencephalon and the dopaminergic innervation of the striatum of the rat. *Neuroscience 25*:857-887.

Williams, J.T., & Marshall, K.C. (1987) Membrane properties of adrenergic responses in locus coeruleus neurons of young rats. *J. Neurosci. 7*:3687-3694.

Wilson. C.J., Young, S.J., & Groves, P.M.(1977) Statistical properties of neuronal spike trains in the substantia nigra: Cell types and their interactions. *Brain Res. 136*:243-260.

CHARACTERIZATION OF GABA AND GLYCINE ACTIONS ON SUBSTANTIA

NIGRA ZONA COMPACTA NEURONS

N. B. Mercuri, P. Calabresi, A. Stefani
M. De Murtas and G. Bernardi

Clinica Neurologica, Dipartimento di Sanita'
Pubblica - II Universita' di Roma, Italy

INTRODUCTION

GABA and glycine are regarded as inhibitory neurotransmitters in the mammalian central nervous system (Bernardi et al., 1975; Aprison and Nadi, 1978; Barker and Ransom, 1978; Nistri and Costanti, 1979; Alger and Nicoll, 1979; Levi et al., 1982; Bormann et al., 1987).
Although the effects of these aminoacids have been extensively studied in several brain structures, we have recently focussed most of our experimental work on the effects of both aminoacids on substantia nigra zona compacta cells (Mercuri et al., 1988; 1990).
Anatomical, biochemical, behavioral and electrophysiological works indicate γ-aminobutyric-acid (GABA) as the main inhibitory transmitter of the striato-nigral projection and of a population of neurons intrinsic to the substantia nigra (Fonnum and Storm-Mathisen, 1978; Dray, 1979; Lacey et al., 1988). Glycine, on the other hand, has been suggested to have a neurotransmitter role in the ventral mesencephalon (A9, A10 areas) (Straughan and James, 1978; Kerwin and Picock, 1979; Gundlach and Beart, 1981; McGeer et al., 1987).
In the present study, we used intracellular recordings (current and voltage clamp mode) from rat slices of the ventral mesencephalon to compare the electrophysiological effects of GABA and glycine on neurons located in the substantia nigra zona compacta.

METHODS

The methods we used have been published extensively elsewhere (Lacey et al., 1987). Briefly, slices of the rat ventral mesencephalon containing the substantia nigra were cut using a vibratome, placed in a recording chamber and superfused with a saline solution.
The standard solution contained (mM): NaCl 126, KCl 2.5, NaH$_2$PO$_4$ 1.2, MgCl$_2$ 1.2, CaCl$_2$ 2.4, Glucose 10, NaHCO$_3$ 26.
When CoCl$_2$ was added to the solution NaH$_2$PO$_4$ was omitted.
Low chloride solution was made by replacing NaCl with equimolar sodium isethionate. Intracellular potentials were obtained using KCL (2M) filled electrodes, 30-60 m Ω resistance.

The Basal Ganglia III, Edited by G. Bernardi *et al.*
Plenum Press, New York

The following drugs, GABA, muscimol, strychnine, bicuculline, picrotoxin, tetrodotoxin (TTX), 4,4-diisothiocyanatostilbene-2,2-disulfonic acid (DIDS), from Sigma and baclofen from Ciba-Geigy, were prepared as stock solutions in distilled water and then applied in the appropriate concentrations to the superfusion bath.

RESULTS

GABA and glycine, tested on the same substantia nigra zona compacta cells (n=25), produced a large decrease in membrane resistance and blocked the spontaneous and the evoked firing activity. The most common effect observed with both aminoacids, was a hyperpolarization of the membrane. This hyperpolarization was associated with an outward current in voltage clamp experiments (holding potential between - 50 and -60 mV).
GABA (1 mM) was more effective than glycine (1 mM) in hyperpolarizing the cells (- 7.6 +\- 1.2 versus - 3.7 +\- 0.41 mV n=17). There was no clear difference in the decrease of the input resistance produced by either GABA or glycine.
Furthermore, four cells were hyperpolarized by 1 mM GABA, but were depolarized by an equimolar concentration of glycine. Two cells were depolarized either by GABA or by glycine application.
The superfusion of GABA (1mM) and Glycine (1 mM) on two other cells produced a membrane hyperpolarization with GABA but a biphasic response with Glycine (hyperpolarization followed by a depolarization).
All the cellular responses to GABA and glycine were fast and had a rapid decay.
Figure 1 shows the actions of GABA, baclofen and glycine superfused on the same cell.

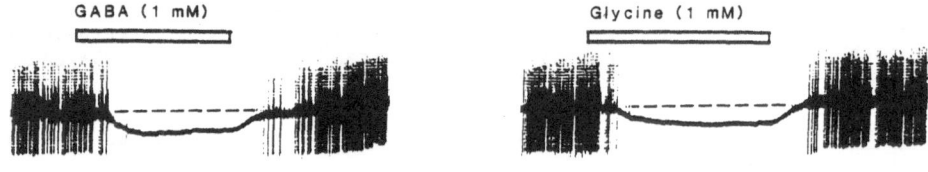

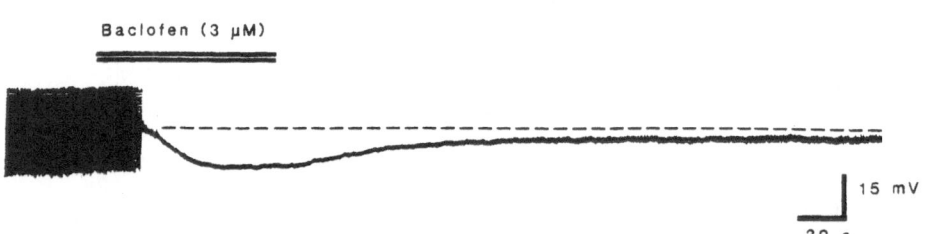

Fig. 1. Hyperpolarization of the membrane and block of the firing rate produced by the superfusion of GABA, glycine and baclofen on the same substantia nigra zona compacta neuron . The bars indicate the length of time in which the drugs were applied. Dotted line, indicates - 55 mV. The amplitude of the spontaneous action potentials is not fully reproduced.

A membrane hyperpolarization blocking spontaneous firing activity is observed with all the drugs. From this and initial results it is clear that GABA and glycine receptors (Lacey et al., 1988; Mercuri et al., 1988; 1990) coexist on the same substantia nigra zona compacta cells. The effects of GABA A receptors stimulation were antagonized by bicuculline and were possibly mediated by an increase in chloride conductance. GABA b effects were mimicked by baclofen and were bicuculline insensitive. It has been previously shown that these effects are mediated through an increase in potassium conductance (Lacey et al., 1988).

Glycine effects were antagonized by strychnine and were due to an increase in chloride conductance. The effects of glycine were also shown to be partially dependent on the concentration of potassium ions in the extracellular fluid (Mercuri et al., 1988; 1990).

Figure 2 shows a neuron hyperpolarized by GABA but depolarized by muscimol (a GABA A agonist) and by glycine.

This indicates that the effects of GABA b receptors, stimulated by the natural transmitter (GABA), prevail over those produced by GABA A receptor activation and that glycine causes, as GABA A agonists, a depolarizing response in this particular cell.

In previous experiments (Mercuri et al., 1990) we have shown that picrotoxin (a CL channel blocker) suppressed an inward current activated by GABA on substantia nigra zona compacta cells while did not affect the action of glycine.

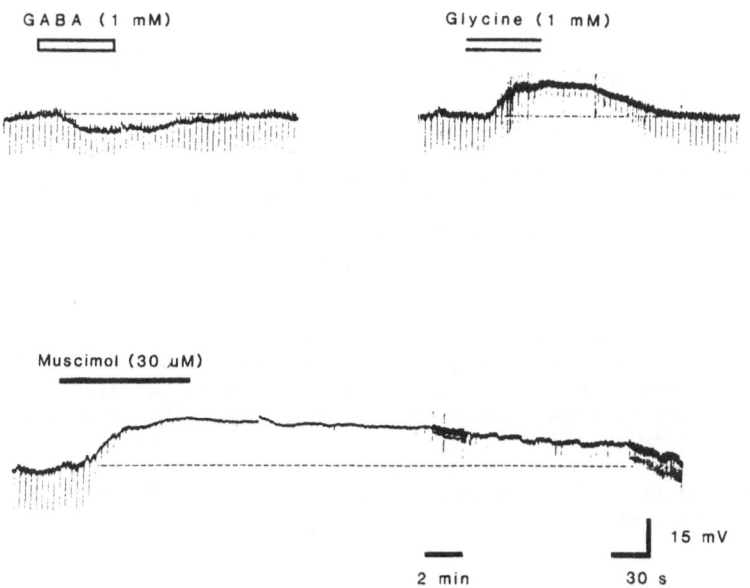

Fig. 2. Membrane responses of a substantia nigra zona compacta neuron to the superfusion of GABA, glycine and muscimol. GABA (1 mM) hyperpolarizes the membrane while glycine (1 mM) depolarizes this cell. A depolarization of the membrane is also observed with the application of muscimol (30 μM). Downward deflection are electrotonic potentials in response to hyperpolarizing current pulses (100 pA, 500 ms). Note the decrease in the input resistance produced by the three drugs. Few action potentials are observed during the glycine induced depolarization. Dotted line, - 59 mV.

In a new set of experiments we used DIDS, another CL channel blocker and, as observed with picrotoxin, DIDS (1-10 mM) did not prevent the actions of glycine on the dopaminergic neurons. When the superfusing medium contained tetrodotoxin (TTX) (1 μM) or calcium (0.5 mM) and cobalt (1-2 mM) the effects of GABA and glycine were still present.

DISCUSSION

This study has confirmed previous observations that GABA and glycine directly affect substantia nigra zona compacta cells (Lacey et al., 1988; Mercuri et al., 1988; 1990).
Bath application of both aminoacids (1 mM) evoked a membrane hyperpolarization in almost all the tested cells. The extent of this hyperpolarization was clearly smaller for the glycine than for the GABA applications.
The effects of GABA were attributed to the activation of two different GABA receptor sites that can be electrophysiologically and pharmacologically distinguished.
The GABA b receptor site, stimulated by GABA and baclofen, is insensitive to bicuculline and picrotoxin, while the GABA A receptor site, stimulated by GABA and muscimol, is antagonized by bicuculline and picrotoxin.
The GABA b effects are dependent on an increase in potassium conductance while the GABA A mediated events appear to be dependent on an increase in chloride conductance.
From the data presented, it is also clear that the GABA b effects prevail over the depolarizing GABA A response producing hyperpolarization of the membrane in CL loaded neurones. In addition, the results we have obtained with glycine show that this aminoacid also inhibits the firing discharge of dopaminergic cells.
The pharmacological features of the glycinergic responses were similar to those already described for a CL dependent and strychnine sensitive site (Nistri and Costanti, 1979; Hamill et al., 1983). However some electrophysiological events (eg. persistence of membrane hyperpolarization even in neurones impaled with KCL recording electrodes and the shift of the glycine reversal potential with increased concentrations of extracellular potassium) suggest that potassium ions may participate to the glycine induced inhibition (Mercuri et al., 1988).
The mean membrane hyperpolarization produced by GABA application was more pronounced in amplitude than that produced by glycine application. This could result from the prevalence of GABA b potassium-linked effects, when GABA is used as agonist, while the effects produced by the superfusion of glycine are mainly mediated by CL dependent processes.
In spite of the fact that a rapid and transient inhibition of substantia nigra zona compacta cells can be achieved with GABA and glycine, the respective role of both aminoacids in controlling the dopaminergic neurons activity is still unclear.
For this reason, further experiments must be performed in order to elucidate the physiological interplay of the aminoacids in the substantia nigra zona compacta.

ACKNOWLEDGEMENTS: The authors are grateful to R. Losacco for reviewing the English style.

REFERENCES

Alger, B.E., and Nicoll, R.A., 1979, GABA-mediated biphasic inhibitory responses in the hippocampus, Nature, 281: 315.

Aprison, M.H., and Nadi, N.S., 1978, Glycine: inhibition from the sacrum to the medulla, _in_ "Amino Acids as Chemical Transmitters", F. Fonnum, ed., Plenum Press, New York.

Barker, J.L., and Ransom B.R., 1978, Amino acid pharmacology of mammalian central neurones grown in tissue culture, _J. Physiol.,_, 280: 331.

Bernardi, G., Marciani, M.G., Morocutti, C., Giacomini, P., 1975, The action of GABA on rat caudate neurons recorded intracellularly, _Brain Res._, 92: 511.

Bormann, J., Hamill, O.P., and Sakmann, B., 1987, Mechanism of anion permeation through channels gated by glycine and γ-aminobutyric acid in mouse cultured spinal neurones, _J. Physiol._, 385: 243.

Dray, A., 1979, The striatum and substantia nigra: a commentary on their relationships, _Neuroscience._, 4: 1407.

Fonnum, F, and Storm-Mathisen, J., 1978, Location of GABA-ergic neurons in the CNS, _in_ "Handbook of Psychopharmacology", L.L., Iversen, S.D., Iversen and S. H., Snyder, eds., Plenum Press, New YorK.

Gundlach, A.L., and Beart, P.M., 1981, [H^3]Strychnine binding suggests glycine receptors in the ventral tegmental area of rat brain, _Neurosci. Lett._, 22: 289.

Hamil, O.P., Bormann, J., and Sakmann, B., 1983, Activation of multiple-conductance state chloride channels in spinal neurones by glycine and GABA, _Nature_, 305: 805.

Kerwin, R.W., and Piçock C.J., 1979, Specific stimulating effect of glycine on [^{3}H]-dopamine efflux from substantia slices of the rat, _Eur. J. Pharmacol._, 54: 93.

Lacey, M.G., Mercuri N.B., and North R.A., 1987, Dopamine acts on D2 receptors to increase potassium conductance in neurones of the rat substantia nigra zona compacta, _J. Physiol._, 392: 397.

Lacey, M.G., Mercuri N.B., and North R.A., 1988, On the potassium conductance increase activated by GABA$_b$ and dopamine D$_2$ receptors in rat substantia nigra neurones, _J. Physiol._, 401: 437.

Levi, G., Bernardi, G., Cherubini, E., Gallo, V., Marciani, M.G. and Stanzione, P., 1982, Evidence in favor of a neurotransmitter role of glycine in the rat cerebral cortex, _Brain Res_., 236: 121.

McGeer, P.L., Eccles, J. and McGeer, E.G., 1987, Anatomical distribution of glycine, _in_: "Molecular neurobiology of the mammalian brain", P.L. McGeer, J. Eccles, J. and E.G. McGeer, eds., Plenum Press, New York.

Mercuri, N.B., Calabresi, P. and Bernardi G., 1988, Potassium ions play a role in the glycine induced inhibition of rat substantia nigra zona compacta neurones, _Brain Res._, 462: 199.

Mercuri, N.B., Calabresi, P. and Bernardi G., 1990, The effects of glycine on neurons in the rat substantia nigra zona compacta: an in vitro electrophysiological study, _Synapse_, in press.

Nistri, A. and Costanti, A., 1979, Pharmacological characterization of different types of GABA and glutamate receptors in vertebrate and invertebrate, _Progr. in Neurobiol._, 13: 117.

Straughan, D.W. and James, T.A., 1978, Microphysiological and pharmacological studies on transmitters in the substantia nigra, _in_: "Advances in Pharmacology and Therapeutics", P. Simon, ed., Pergamon Press, Oxford.

CHOLINERGIC EXCITATION OF A9 AND A10 DOPAMINERGIC NEURONES IN VITRO THROUGH BOTH NICOTINIC AND MUSCARINIC RECEPTORS

M.G Lacey[*], P.Calabresi[+] and R.A.North

Vollum Institute, Oregon Health Sciences University
Portland, Oregon 97201, USA

INTRODUCTION

There are no cholinergic neurones within either substantia nigra (A9) or ventral tegmental (A10) area (Mesulam et al., 1983). However choline acetyltransferase-immunoreactive terminals within substantia nigra pars compacta have been observed (Beninato & Spencer, 1988). Such innervation may arise from the cholinergic neurones of the pedunculopontine nucleus (PPN) (Woolfe & Butcher, 1986; Clarke et al., 1987; but see Lee et al., 1988).

[^{3}H]-nicotine binding, apparently on dopaminergic neurones, has been demonstrated (Clarke & Pert, 1985), but muscarinic binding sites in A9 and A10 regions are both sparse (Nonoka & Moroji, 1984) and largely unaffected by 6-hydroxydopamine (Cross & Waddington, 1980). Excitation of dopaminergic neurones in both A9 and A10 regions in the rat has been reported following ionophoresis or systemic injection of nicotine (Lichtensteiger et al, 1982; Clarke et al., 1985; Grenhoff et al., 1986). In contrast, in only one case has acetylcholine (ACh; applied by ionophoresis) caused an excitation attributed to muscarinic receptor activation (Scarnati et al., 1986). Nonetheless, there is behavioural evidence for a role of muscarinic receptors in A9 and/or A10 in enhancing dopamine-dependent behaviours (Winn et al., 1983; Yeomans et al., 1984; Niijima & Yoshida, 1988).

Dopamine neuronal systems, particularly the A10 - limbic/cortical pathway, are of importance in locomotion, motivation and reward (Fibiger & Phillips, 1986). It has been proposed that the excitation by nicotine of dopamine neurones (particularly those in A10) plays a role in the rewarding effects of nicotine consumption and in tobacco dependence (Grenhof et al., 1986). A muscarinic component of rewarding electrical stimulation of the medial forebrain bundle (which excites A10 neurones) additionally implicates cholinergic innervation of dopamine neurones in reward processes (Yeomans et al., 1984).

The receptor and ionic mechanisms of action of ACh were investigated in the present study using intracellular recording from A9 and A10 neurones in slice preparations of rat ventral midbrain. Both nicotinic (Calabresi et al., 1989) and muscarinic (Lacey et al., 1990) effects were identified and separately examined.

Present addresses: [*]Department of Pharmacology, Smith Kline & French Research Ltd., Welwyn Herts, AL6 9AR, U.K. [+]II Universita degli Studi di Roma, Via Orazio Raimondo 8, 00173 Rome Italy

METHODS

Detailed descriptions of the methods may be found elsewhere (Lacey et al., 1987; 1989). In essence, coronal slices (300 μm) of rat midbrain were maintained at 36 °C in a chamber continuously superfused in an oxygenated solution containing (mM): NaCl, 126; KCl, 2.5; NaH$_2$PO$_4$, 1.2; MgCl$_2$, 1.3; CaCl$_2$, 2.4; glucose, 10; NaHCO$_3$, 26. Intracellular recordings were made from neurones in either A9 or A10 regions, which could be discriminated visually, using glass microelectrodes containing 2M KCl or 2M CsCl. Signals were amplified with an Axoclamp 2A amplifier, which also permitted single electrode voltage clamp recording. Drugs were applied by their addition to the superfusate in known concentrations. ACh (100 mM) was applied by ejection by pressure pulse of a few nanolitres from the tip of a pipette positioned close to the recording site.

RESULTS

Recordings were made from a total of 108 A10 and 35 A9 neurones. All cells had electrophysiological properties ascribed to dopaminergic neurones and those tested were hyperpolarised by dopamine (Lacey et al., 1989). No difference between A9 and A10 neurones was apparent in these properties or in those described below.

Effect of Acetylcholine

Pressure ejection of ACh caused a depolarisation and increase in firing rate or, in voltage clamp, an inward current in 75% of cells tested. In the presence of the acetylcholinesterase inhibitor neostigmine (0.1 - 10 μM) (which was itself without effect) the response to ACh was both enhanced and present in 100% of cells tested. The response comprised a fast early transient component followed by a slower late component (Fig. 1). The late component was blocked by the muscarinic receptor antagonist scopolamine (1 -10 μM; Fig. 1).

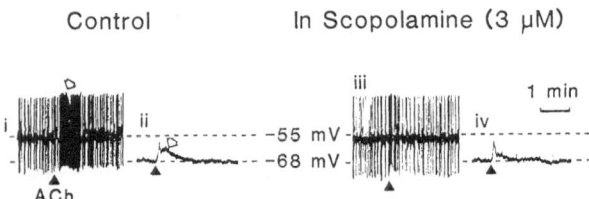

Fig. 1. Membrane potential records from a single neurone showing excitatory nicotinic and muscarinic actions of ACh. ACh applied from a pipette (triangles) during spontaneous firing at rest (i and iii) and with membrane held at -68 mV with constant hyperpolarising current injection (ii and iv). Biphasic firing rate increase at rest (i) was mirrored by a two-component depolarisation at -68 mV. Late component (open arrows) was blocked by scopolamine (3 μM; iii and iv). From Calabresi et al. (1989). Reproduced with permission.

The Nicotinic Response

The fast response to ACh was examined in the presence of scopolamine (1 - 10 μM) and neostigmine (0.1 - 10 μM).

<u>Reversal potential</u>. The depolarisation or inward current caused by ACh (Fig. 2) or carbachol (also in scopolamine) increased in amplitude with membrane hyperpolarisation. When membrane rectification at potentials less

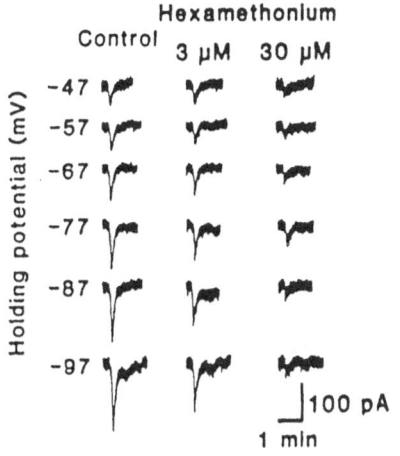

Fig. 2. ACh currents increase with membrane hyperpolarisation, as does the degree of block by hexamethonium. Inward membrane currents evoked by ACh in a single cell at the potentials indicated before (control, left column) and in the presence of hexamethonium (3 and 10 µM; centre and right column). Scopolamine (3 µM), neostigmine (0.1 µM), TTX (1 µM) present. From Calabresi et al. (1989). Reproduced with permission.

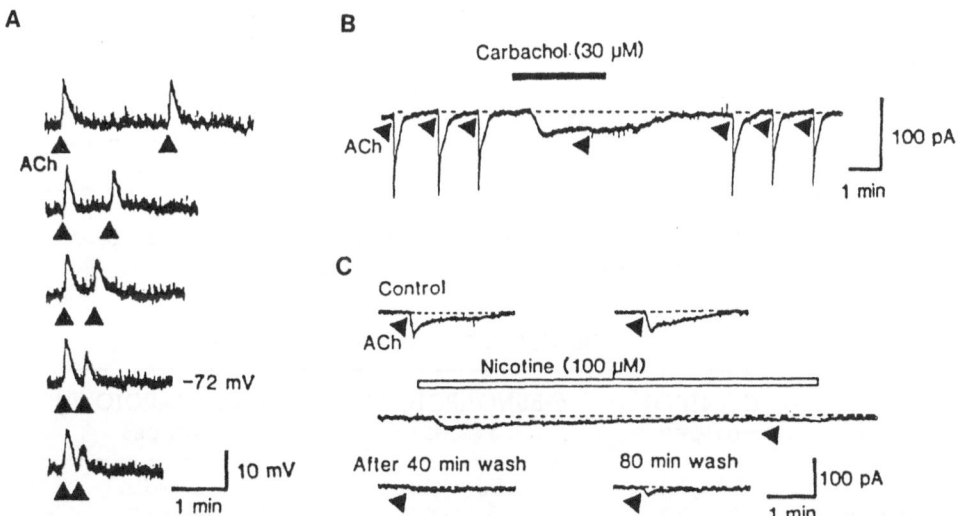

Fig. 3. Desensitisation and cross-desensitisation of nicotinic responses. <u>A</u>. ACh depolarisations became smaller when interval between applications (triangles) was decreased. Continuous records of membrane potential of cell held at -72 mV with constant current. <u>B</u>. Carbachol (30 µM, solid bar) caused an inward current (membrane potential clamped at -60 mV), during which time ACh response (triangles) were reversibly abolished. <u>C</u>. Nicotine (100 µM, open bar) causes an inward current (at -65 mV) that declined during application. Nicotine additionally blocked the ACh response, which recovered only partially after 80 min wash of nicotine. Records from 3 different cells, each in scopolamine (3 µM) and neostigmine. From Calabresi et al. (1989). Reproduced with permission.

negative to rest was reduced with tetrodotoxin (TTX, 1 µM), cobalt (2 - 5 mM) in combination with either barium (1 mM) plus tetraethylammonium (TEA, 30 mM) or 2 M Cs-filled microelectrodes, the response to ACh or carbachol reversed polarity at 3.7 $\pm$ 2.4 mV (S.E.M.; n=4).

Cobalt and TTX resistance. The persistence of the response in TTX and cobalt makes it likely that the site of action was on the impaled neurone, rather than due to an indirect action of an unknown transmitter.

Voltage-dependent block by hexamethonium. Hexamethonium (1 - 100 µM) reversibly reduced the response to ACh; the block was greater at more negative potentials (6 cells; Fig 2). ACh responses were also reduced by around 50% by d-tubocurarine (10 µM).

Desensitisation. Responses to ACh declined in amplitude when the interval between applications was reduced (Fig. 3A). The minimum inter-response interval required for reproducible responses was increased by neostigmine. Depolarisations or inward currents were also produced by superfusion of carbachol (30 µM) and nicotine (10 - 100 µM). Response amplitudes declined while these agonists were still present (Figs. 3B, C). Decline of concomitant responses to ACh in the presence of these two agonists also occurred (Figs. 3B, C). These observations may be attributed to receptor desensitisation and cross-desensitisation between different agonists acting at the same site; its duration appeared agonist-specific with nicotine > carbachol > ACh.

Block by k-bungarotoxin. The relative ability of α- and k-bungarotoxin to block the nicotinic response was assessed as follows: slices were incubated for 2h at 36°C in superfusion medium containing added α- (1 µM) or k- (0.5 µM) bungarotoxin (gift of Dr. V.A. Chiappinelli) or controls. Recordings were then made from a total of 37 A10 neurones and, with membrane potential held at -70 $\pm$ 0.7 mV, responsiveness to ACh and carbachol was assessed in the presence of scopolamine (3 µM) and neostigmine (0.1 µM). k-Bungarotoxin reduced the nicotinic depolarisations in response to both ACh and carbachol by more than 50% (p > 0.01, Dunnett's t-test modification), whereas α-bungarotoxin was without effect (Table 1).

Table 1. Depolarisations obtained from neurones in slices that had been pretreated with α-or k-bungarotoxin (or controls) in response to ACh and carbachol. Values are means $\pm$ SEM.

	DEPOLARIZATION AMPLITUDE (mV)		
	CONTROL 6 slices	α-BUNGAROTOXIN 3 slices	κ-BUNGAROTOXIN 4 slices
ACH (max. response)	6.7 $\pm$ 1.3 (9 cells)	5.3 $\pm$ 1.1 (8 cells)	2.5 $\pm$ 0.6 (14 cells)
CARBACHOL (30 µM)	3.8 $\pm$ 0.5 (12 cells)	3.8 $\pm$ 0.7 (9 cells)	1.4 $\pm$ 0.3 (14 cells)

The Muscarinic Response

All but 5% of over 100 cells tested were excited by muscarine (3 - 100 µM): muscarine caused an increase in firing rate or, at potentials negative to rest, a depolarisation or, in voltage clamp, an inward current (Fig. 4). The response to muscarine was variable from cell to cell, attributable, at least

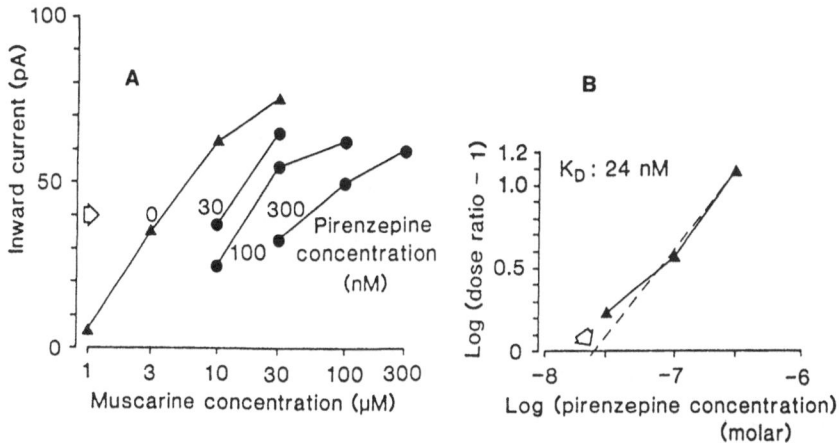

Fig. 4. Muscarine (30 µM, solid bar) depolarised the membrane, inducing firing (upper records), or caused an inward current under voltage clamp (lower records). Full amplitude of action potentials and their afterhyperpolarisations are not reproduced. Records from same cell held at -65 mV with -105 pA constant direct current (upper pair of records) or voltage clamped at -65 mV (lower records). Muscarinic depolarisation was accompanied by apparent input resistance increase. From Lacey et al. (1990). Reproduced with permission.

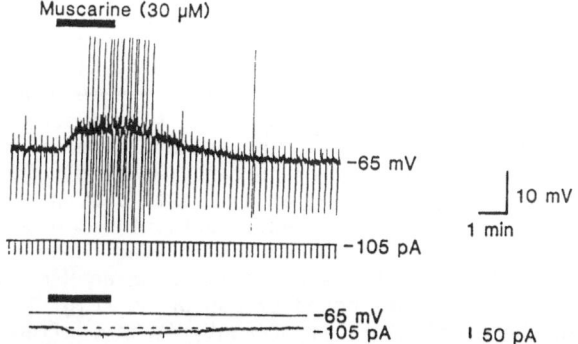

Fig. 5. Pirenzepine K_D of 24 nM indicates involvement of M_1 receptor subtype in muscarinic response. A, inward currents measured at -55 mV from a single cell plotted against muscarine concentration in presence of 0, 30, 100 and 300 nM pirenzepine, showing rightward shift. B, Schild transformation of data in A with dose ratio calculated at response level of 40 pA. The dashed line passing close to all three points had slope of 1, indicating competitive antagonism, and its X-intercept gives the K_D for pirenzepine.

in part, to the pronounced voltage dependence of the response around resting potential (see below and Fig. 6) and differences from cell to cell in resting or holding potential at which muscarine was tested.

Muscarine acts through the M_1 receptor. Concentration-effect curves were constructed from inward currents in response to muscarine on individual cells. Muscarinic responses were reduced by pirenzepine (30 - 1000 nM), but this antagonism was surmountable with higher concentrations of muscarine. Thus the concentration-effect curves were shifted to the right (Fig. 5A). Schild analysis from two cells indicated an equilibrium dissociation constant (K_D) for pirenzepine of around 20 nM (Fig. 5B). The mean value of K_D estimates from four cells was 14 $\pm$ 3.6 nM, consistent with involvement of an M_1 receptor subtype (Hammer et al., 1980).

Dependence upon extracellular calcium ions. Muscarinic depolarisations or inward currents were reversibly reduced or abolished a low calcium (0.25 mM) / high magnesium (10 mM) solution, but not by TTX (1 μM). These solutions also blocked spontaneous action potential firing. Two interpretations of this result are possible: either calcium entry into the cell is required for the muscarinic response and/or the effect of muscarine is indirect and results from calcium-dependent (but TTX-resistant) release of other transmitters within the preparation.

Ionic mechanism of muscarinic response. The muscarinic inward current decreased in amplitude at membrane potentials negative to rest (Fig. 6). As suggested by the increase in input resistance accompanying the muscarinic depolarisation (Fig. 4), the inward current was accompanied by a decrease in membrane conductance at potentials around rest (Fig. 6).
Excitatory actions of muscarine on neurones are often due to a decrease in resting membrane potassium conductance (North et al., 1985; Madison et al., 1987; see also Christie & North, 1988). However, muscarinic responses failed to reverse polarity close to the expected potassium equilibrium potential (Fig. 6), irrespective of whether the potassium concentration in the superfusate was 2.5, 4.5, 10.5 or 20.5 mM. Indeed, at potentials negative to around -75 mV (in 2.5 mM potassium) the muscarinic current appeared not be associated with a change in membrane conductance (Fig. 6)
The action potential afterhyperpolarisation in some neurones has also been reported to be inhibited by muscarine (North & Tokimasa, 1983; Madison et al., 1985), but this was not observed in the present experiments (see Fig. 4).
The M-current is a third potassium current reportedly inhibited by muscarine (Brown & Adams, 1980; Madison et al., 1987). Step hyperpolarisations from around -40 mV (the range for M-current inactivation) did cause a slowly-developing current, but the lack of a clear associated conductance decrease (Fig. 7) argues for caution in considering this to be an M-current relaxation. Nonetheless, this current was unaffected by muscarine (Fig. 7). Thus either M-current is absent in these cells or, if present, it is unaffected by muscarine.
Taken together, these experiments provide no evidence for the muscarinic current being due to a change in membrane potassium conductance. It remains possible that such a change is occurring, but possibly in a voltage-dependent manner (but not involving M-current) and/or in conjunction with a change in conductance to one or more other ions.

DISCUSSION

Excitatory responses mediated by both nicotinic and muscarinic receptors, were a near-ubiquitous phenomenon in rat midbrain dopamine neurones, both in A9 and A10 regions. Functional postsynaptic targets for cholinergic innervation of dopamine neurones are therefore clearly present.

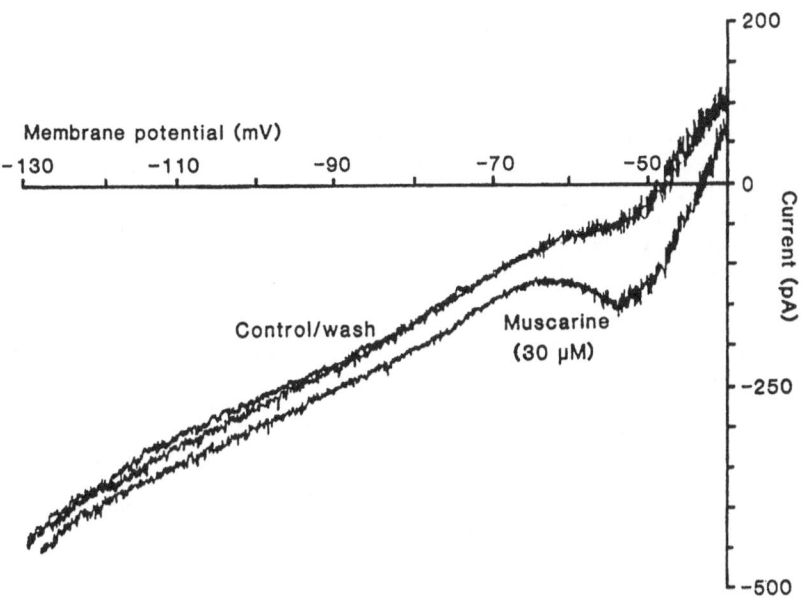

Fig. 6. Voltage-dependence of current caused by muscarine. Steady-state
current/voltage relations resulting from slow ramp depolarisations (2 mV/s)
from −130 to −40 mV under voltage clamp. Muscarine (30 μM) caused an inward
current, reversible on washout, throughout this voltage range. From Lacey et
al. (1990). Reproduced with permission.

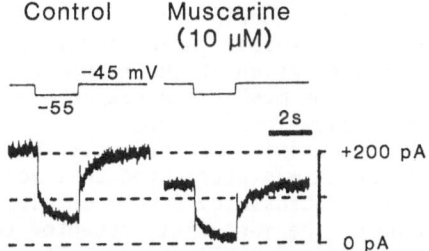

Fig. 7. Muscarine does not inhibit M-current. Pair of records of both
membrane voltage (upper traces) and current (lower traces) showing membrane
currents resulting from 2s hyperpolarising steps from −45 mV to −55 mV under
voltage clamp in absence (control) and presence of muscarine (10 μM).
Muscarine caused an inward current at −45 mV and at −55 mV, accompanied by a
conductance decrease. However, the time-dependent current developing during
the voltage step, which may be an outward relaxation of M-current, was
unaffected by muscarine. From Lacey et al. (1990). Reproduced with permission.

The nicotinic response appears to be a direct effect on these neurones, whereas this is uncertain for the muscarinic response due to its requirement for extracellular calcium (but see below).

The Nicotinic Response

The fast, scopolamine-resistant response to ACh resembled the nicotinic response in ganglia and at the neuromuscular junction in that it was an inward current with reversal potential around -4 mV, showed desensitisation and cross-desensitisation to carbachol and nicotine and inhibition by d-tubocurarine. However its block by hexamethonium (Rang, 1982) and by k-bungarotoxin, but insensitivity to α-bungarotoxin (Chiappinelli, 1983; Loring & Zigmond, 1984), places the receptor type in the 'ganglionic' or 'neuronal' class, as opposed to those of the neuromuscular junction. Voltage-dependent block by hexamethonium has been shown in neurones in other brain regions by Egan & North (1986) and McCormick & Prince (1987).

Sensitivity to k-bungarotoxin is seen in some, but not all of the functional nicotinic receptors expressed in Xenopus oocytes following their injection with RNA encoding various combinations of the different α and β receptor subunits (Duvoisin et al., 1989). The incomplete suppression of the nicotinic response by k-bungarotoxin pretreatment might indicate a mixed population of nicotinic receptor subtypes on dopamine neurones, some of which were insensitive to k-bungarotoxin. Alternatively, slow dissociation of the toxin from the receptors while recordings were made following incubation might be responsible for a less than total block of the response.

The Muscarinic Response

Muscarinic responses were antagonised by pirenzepine (0.03 - 1 μM), with an apparent K_D of 14 nM. This places the muscarinic receptor in the M_1 muscarinic receptor class as defined by high affinity (10 - 25 nM) pirenzepine binding (Hammer et al., 1980), comparable with other functional studies (eg. Gil & Wolfe, 1985; North et al., 1985).

M_1 receptors frequently couple to the breakdown of phosphatidyl inositol via phospholipase C (eg. Gil & Wolfe, 1985; Peralta et al., 1988), a process which is calcium-dependent to a large extent (Berridge, 1987). This may explain the sensitivity of the muscarinic response to extracellular calcium concentration, rather than this result indicating a requirement for transmitter release for manifestation of the response.

The ionic mechanism of the muscarinic response does not appear principally to involve potassium ions. Insofar as it was examined here, the voltage dependence of the muscarinic current resembles that described in intestinal smooth muscle, which was attributed to a voltage dependent, non-selective cation conductance increase by Benham et al., (1985). It was proposed that the coupling of the muscarinic receptor to the channels requires calcium, but is independent of phospholipase C (Bolton & Lim, 1989). Further experiments will be required to see if such a mechanism pertains in dopaminergic neurones.

Implications for Dopamine System Function

The mechanisms described here for nicotinic and muscarinic cholinergic excitation of dopamine neurones at the level of their cell bodies in A9 and A10 regions may also function at their terminals. Thus nicotinic receptors sensitive to k-bungarotoxin mediate the transient release of dopamine from slices of striatum caused by nicotine (Schulz & Zigmond, 1989). Also, spontaneous release of dopamine from striatal slices by muscarinic agonists is TTX-insensitive, but calcium-dependent (Giorguieff et al., 1977) and mediated by M_1 receptors (De Belleroche & Gardner, 1985).

The present results provide further support for both nicotinic and muscarinic receptors as mediators of excitation of dopaminergic neurones, which are important neuronal substrates of behavioural reward. It is hoped that their pharmacological characterisation will permit more precise targeting of drugs to modify and dissect the circuitry of reward and motivation. One consequence of such an approach might be in the development of therapeutic agents to reduce the gratification of tobacco consumption.

Supported by USDHHS grants MH 40416 and DA 03161.

REFERENCES

Beninato, M. & Spencer, R.F. 1988. The cholinergic innervation of the rat substantia nigra: a light and electron microscopic immunohistochemical study. Exp. Brain Res. 72: 178-184.

Benham, C.D., Bolton, T.B. & Lang, R.J. 1985. Acetylcholine activates an inward current in single mammalian smooth muscle cells. Nature 316: 345-347.

Berridge, M.J. 1987. Inositol triphosphate and diacylglycerol: two interacting second messengers. Ann. Rev. Biochem. 56: 159-193.

Bolton, T.B. & Lim, S.P. 1989. Properties of calcium stores and transient outward currents in single smooth muscle cells of rabbit intestine. J. Physiol. 409: 385-402.

Brown, D.A. & Adams, P.R. 1980. Muscarinic suppression of a novel voltage-sensitive K^+-current in a vertebrate neurone. Nature 283: 673-676

Calabresi, P., Lacey, M.G. & North, R.A. 1989. Nicotinic excitation of rat ventral tegmental neurones in vitro studied by intracellular recording. Brit. J. Pharmacol. 98: 135-140.

Chiappinelli, V.A. 1983. Kappa-bungarotoxin: a probe for the neuronal nicotinic receptor in the avian ciliary ganglion. Brain Res. 277: 9-21.

Christie, M.J. & North, R.A. 1988. Control of ion conductances by muscarinic receptors. Trends Pharmacol. Sci. Subtypes of Muscarinic Receptors III: 30-34.

Clarke, P.B.S. & Pert, A. 1985. Autoradiographic evidence for nicotinic receptors on nigrostriatal and mesolimbic dopaminergic neurons. Brain Res. 348: 355-358.

Clarke, P.B.S., Pert, A., Hommer, D.W. & Skirboll, L.R. 1985. Electrophysiological actions of nicotine on substantia nigra single units. Brit. J. Pharmacol. 85: 827-835.

Clarke, P.B.S., Hommer, D.W., Pert, A. & Skirboll, L.R. 1987. Innervation of substantia nigra neurons by cholinergic afferents from pedunculopontine nucleus in the rat: neuroanatomical and electrophysiological evidence. Neurosci. 23: 1011-1019.

Cross, A.J. & Waddington, J.L. 1980. [^{3}H]quinuclidinyl benzylate and [^{3}H]GABA receptor binding in rat substantia nigra after 6-hydroxydopamine lesions. Neurosci. Lett. 17: 271-275.

De Belleroche, J. & Gardner, I.M. 1985. Muscarinic receptors discriminated by pirenzepine are involved in the regulation of neurotransmitter release in rat nucleus accumbens. Br. J Pharmac. 86: 505-508.

Duvoisin, R.M., Deneris, E.S., Patrick, J. & Heinemann, S. 1989. The functional diversity of the neuronal nicotinic acetylcholine receptors is increased by a novel subunit: β4. Neuron 3:487-496.

Egan, T.M. & North, R.A. 1986. Actions of acetylcholine and nicotine on rat locus coeruleus neurones in vitro. Neuroscience 19: 565-571.

Fibiger, H.C. & Phillips, A.G. 1986. Reward, motivation, cognition: psychobiology of mesotelencephalic dopamine systems. In: Handbook of Physiology section 1, Vol. IV, pp. 647-676. American Physiological Society, Bethesda, USA.

Gil, D.W. & Wolfe, B.B. 1985. Pirenzepine distinguishes between muscarinic receptor-mediated phosphoinositide breakdown and inhibition of adenylate cyclase. J. Pharmac. Exp. Ther. 232: 608-616.

Giorguieff, M.F., Le Floc'h, M.L., Glowinski, J. & Besson, M.J. 1977. Involvement of cholinergic presynaptic receptors of nicotinic and muscarinic types in the control of the spontaneous release of dopamine from striatal dopaminergic terminals in the rat. J. Pharmac. Exp. Ther. 200, 535-544.

Grenhoff, J., Aston-Jones, G. & Svensson, T.H. 1986. Nicotinic effects on the firing pattern of midbrain dopamine neurons. Acta Physiol. Scand. 128: 351-358.

Hammer, R., Berrie, C.P., Birdsall, N.M.J., Burgen, A.S.V. & Hulme, E.C. 1980. Pirenzepine distinguishes between different subclasses of muscarinic receptors. _Nature_ 283: 90-92.

Lacey, M.G., Mercuri, N.B. & North, R.A. 1987. Dopamine acts on D_2 receptors to increase potassium conductance in neurones of the rat substantia nigra zona compacta. _J. Physiol._ 392: 397-416.

Lacey, M.G., Mercuri, N.B. & North, R.A. 1989. Two cell types in the rat substantia nigra zona compacta distinguished by membrane properties and the actions of dopamine and opioids. _J. Neurosci._ 9: 1233-1241.

Lacey, M.G., Calabresi, P. & North, R.A. 1990. Muscarine depolarizes rat substantia nigra zona compacta and ventral tegmental neurons _in vitro_ through M_1 receptors. _J. Pharmacol. Exp. Ther._ (in press).

Lee, H.L., Rye, D.B., Hallanger, A., Levey, A.I. & Wainer, B.H. 1988. Cholinergic vs. noncholinergic efferents from the mesopontine tegmentum to the extrapyramidal motor system. _J.Comp. Neurol._ 275: 469-492.

Lichtensteiger, W., Hefti, F., Felix, D., Huwyler, T., Melamed, E. & Schlumpf, M. 1982. Stimulation of nigrostriatal dopamine neurones by nicotine. _Neuropharmacol._ 21: 963-968.

Loring, R.H. & Zigmond, R.E. 1988. Characterization of neuronal nicotinic receptors by snake venom neurotoxins _Trends Neurosci._ 11: 73-78.

Madison, D.V., Lancaster, B. & Nicoll, R.A. 1987. Voltage clamp analysis of cholinergic action in the hippocampus. _J. Neurosci._ 7: 733-741.

McCormick, D.A. & Prince, D.A. 1987. Acetylcholine causes rapid excitation in the medial habenular nucleus of guinea pig. _in vitro_. _J. Neurosci._ 7: 742-752.

Mesulam, M.-M., Mufson, E.J., Wainer, B.H. & Level, A.I. 1983. Central cholinergic pathways in the rat: an overview based on an alternative nomenclature (Ch1-Ch6). _Neurosci._ 10: 1185-1201.

Niijima, K. & Yoshida, M. 1988. Activation of mesencephalic dopamine neurons by chemical stimulation of the nucleus tegmenti pedunculopontinus pars compacta. _Brain Res._ 451, 163-171.

Nonoka, R. & Moroji, T. 1984. Quantitative autoradiography of muscarinic cholinergic receptors in the rat brain. _Brain Res._ 296: 295-303.

North, R.A. & Tokimasa, T. 1983. Depression of calcium-dependent potassium conductance of guinea-pig myenteric neurones by muscarinic agonists. _J. Physiol._ 342: 253-266.

North, R.A., Slack, B.E. & Surprenant, A. 1985. Muscarinic M_1 and M_2 receptors mediate depolarization and presynaptic inhibition in guinea pig enteric nervous system. _J. Physiol._ 368: 435-452.

Peralta, E.G., Ashkenazi, A., Winslow, J.W. Ramachandran, J. & Capon, D.J. 1988. Differential regulation of PI hydrolysis and adenyl cyclase by muscarinic receptor subtypes. _Nature_ 334: 434-437.

Rang, H.P. 1982. The action of ganglionic blocking drugs on the synaptic responses of rat submandibular ganglion cells. _Br. J. Pharmacol._ 75, 151-168.

Scarnati, E., Proia, A., Campana, E. & Pacitti, C. 1986. A microiontophoretic study on the nature of the putative synaptic neurotransmitter involved in the pedunculopontine-substantia nigra pars compacta excitatory pathway of the rat. _Exp. Brain Res._ 62: 470-478.

Schulz, D.W. & Zigmond, R.E. 1989. Neuronal bungarotoxin blocks the nicotinic stimulation of dopamine release from rat striatum. _Neurosci. Letts._ 98: 310-316.

Winn, P., Farrell, A., Maconick, A. & Robbins, T.W. 1983. Behavioural and pharmacological specificity of the feeding elicited by cholinergic stimulation of the substantia nigra in the rat. _Behav. Neurosci._ 97: 794-809.

Woolfe, N.J. & Butcher, L.L. 1986. Cholinergic systems in the rat brain. III. Projections from the pontomesencephalic tegmentum to the thalamus, tectum, basal ganglia, and basal forebrain. _Brain. Res. Bull._ 16: 603-637.

Yeomans, J.S., Kofman, O. & McFarlane, V. 1984. Cholinergic involvement in lateral hypothalamic rewarding brain stimulation. _Brain Res._ 329: 19-26.

INWARD RECTIFYING PROPERTIES OF NUCLEUS ACCUMBENS NEURONES IN VITRO

N. Uchimura[1], E. Cherubini[2] and R.A. North

Vollum Institute, Oregon Health Sciences University, Portland, Oregon 97201, USA

INTRODUCTION

Neurones from the nucleus accumbens septi when recorded in vivo (White and Wang, 1984; Yang and Mogenson, 1984) or in vitro (Uchimura et al., 1989a) have the particular characteristic that they do not fire spontaneously. In fact these cells have a resting membrane potential in the range of -70, -80 mV which is below the threshold for spike activation (-60 mV). Cell excitability is regulated by a number of extrinsic and intrinsic factors, such as neurotransmitters acting on specific membrane receptors and voltage-gated conductances. Among the voltage-dependent conductances, the anomalous rectifier play a crucial role. This membrane current was first studied by Katz (1949) who noticed that in frog skeletal muscle fibres hyperpolarization is associated with an increase in membrane potassium conductance. It was called anomalous rectification because it operates in the opposite direction to that predicted by the electrodiffusion equation (Goldman, 1943; Hodgkin and Katz, 1949); therefore this current is inwardly directed. Subsequent studies have shown the existence of two classes of inward rectifiers: one with fast kinetics, carried by potassium ions, which is selectively blocked by barium and another, with slow kinetics, selective to sodium and potassium ions, which is insensitive to barium but it is blocked by cesium. We now report that both these currents are present in nucleus accumbens neurones, where they contribute to the resting membrane conductance.

METHODS

The technique for preparing nucleus accumbens slices has been already reported (Uchimura et al., 1986; Uchimura et al., 1989a). Briefly a slice (300 µm thick) containing a section through the nucleus accumbens was placed in a recording chamber and superfused at 3 ml/min with artificial cerebrospinal fluid (ACSF) of the following composition (mM): NaCl, 126; KCl, 2.5; NaH_2PO_4, 1.2; $MgCl_2$, 1.3; $CaCl_2$, 2.4; glucose 10; $NaHCO_3$, 26; gassed with 95% O_2 and 5% CO_2 (37° C). In some experiments the solution contained added salts (barium, cesium, strontium or potassium chloride).

---

[1] present address: Departments of Physiology and Neuropsychiatry, Kurume University Medical School, 67 Asahi-machi, Kurume, 830 Japan.
[2] present address: INSERM U. 029, 123 Bd. de Port Royal, 75014 Paris, France.

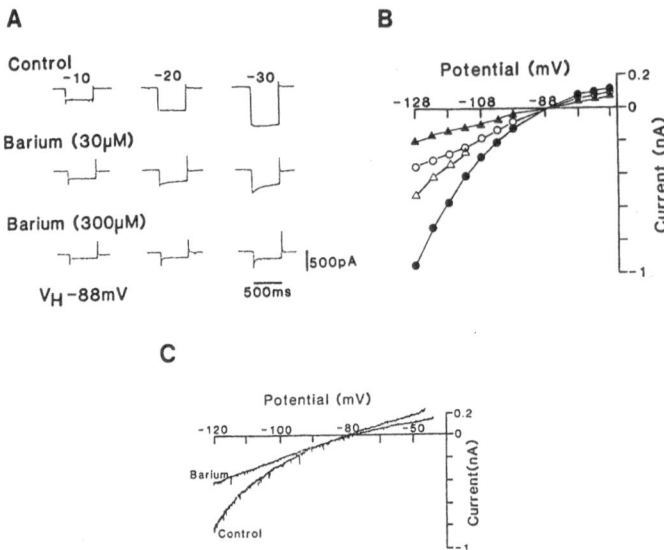

Fig. 1 The fast inward rectifier is blocked by barium.

A. Membrane currents evoked by hyperpolarizing voltage steps of 10, 20 and 30 mV from a holding potential of - 88 mV in control conditions and during superfusion of 30 and 300 μM of barium. Note the time dependent block of the current in the presence of low concentrations of barium.

B. The amplitude of the instantaneous and steady state currents shown in A are plotted against the membrane potential. (●) instantaneous and steady state values in control conditions; instantaneous (Δ) and steady state (o) values during superfusion of barium (30 μM); instantaneous and steady state values (▲) during superfusion of barium (300 μM).

C. Steady state current/voltage plots from slow depolarizing ramp before and during barium (10 μM) superfusion.

The nucleus accumbens was identified with respect to the anterior commissure. Intracellular recordings were obtained from neurones in the dorsomedial part of the nucleus, within 300 μm medial or ventral to the anterior commissure. Recording electrodes containing 2M potassium chloride (resistance 40-80 MΩ) were used. Current was passed through the recording electrode using the active bridge circuit. Bridge balance was checked repeatedly during the course of the impalement and capacitance was fully compensated. Membrane currents were recorded with a single electrode voltage clamp amplifier (Axoclamp 2A). The sampling frequency was 2.5-3.5 kHz, 30% duty cycle. To ensure correct operation of the clamp, the voltage at the head stage amplifier was continously monitored on a separate oscilloscope. Membrane potentials and currents were recorded on Gould 2400 chart recorder. Current/voltage (I/V) plots were constructed by stepping the voltage to various potentials and measuring the current (istantaneous or steady values). In some experiments current was plotted directly as a function of voltage on an X/Y plotter by using a slow depolarizing ramp (1mV/s from -130 to -55 mV).

RESULTS

Intracellular recordings were performed from 52 neurones having resting membrane potential of -78.9 ± 1.4 mV (mean ± S.E.M.) and membrane conductance ranging from 4 to 36 nS. The input resistance measured with large hyperpolarizing pulses (70-200 ms, > 0.5 nA) delivered at rest

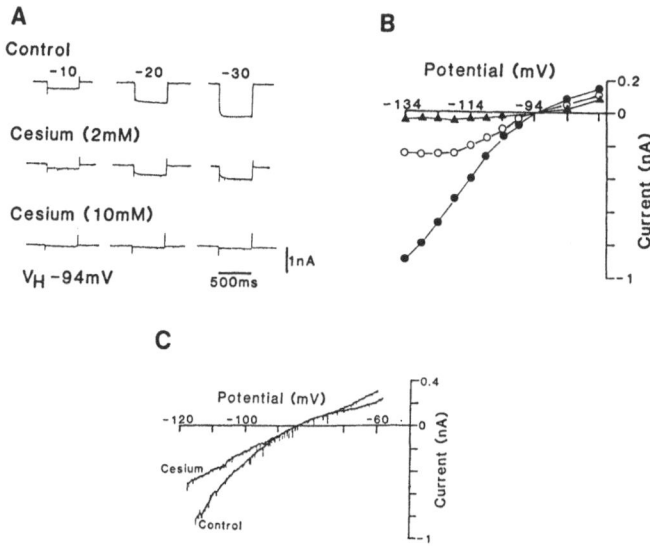

Fig. 2   Blockade of the fast inward rectifier by cesium.
A. Membrane currents elicited by hyperpolarizing voltage steps of 10, 20 and
30 mV from a holding potential of -94 mV in control conditions and during
bath applications of two different concentrations of cesium.
B. I/V plot of the amplitude of the steady state currents shown in A in control
conditions (●), during superfusion of cesium 2 mM (o) and cesium 10 mM (▲).
C. Steady state current/voltage plots from slow depolarizing ramp before and
during cesium (10 mM).

decreased as the cell was hyperpolarized. At -90 mV the input resistance was 50% of the control
value (Uchimura et al., 1989a). In the majority of cells (47=90.3%), the apparent fall in input
resistance developed rapidly (within a few ms) and was thereafter unchanged for up to 2s. In a
minority of neurones (5=9.7%), the fall in input resistance was time dependent. The rapid membrane hyperpolarization decayed to a plateau level 100-200 ms after the onset of application of the
hyperpolarizing current pulse and was followed at the termination of the pulse by a rebound depolarization. Although no differences were found in membrane input resistance and time constant
between these two population of neurones, we noticed that the resting membrane potential of the
cells having a time dependent anomalous rectification was less negative (-67.6 ± 2.2 mV against -
79.9 ± 1.4 mV).

    In the following voltage-clamp experiments we tryed to characterize the ionic mechanisms
and pharmacological properties of both fast and time dependent anomalous rectification.

Fast anomalous rectification

    As already reported (Uchimura et al., 1989b) the steady state current/voltage relationship
showed a portion of relatively linearity between -60 and -90 mV. The conductance greatly increased with hyperpolarization (Figure 1). In 2.5 mM concentration of potassium the conductance between -60 and -70 mV was 8.6 ± 0.6 nS and it was 44.5 ± 1.4 nS at potentials between -120 and -130
mV. The membrane currents activated by hyperpolarizing voltage steps were not time dependent
within the resolution afforded by the limited settling time of the single electrode voltage clamp
system (10-50 ms). When potasssium concentration was progressively increased from 2.5 to 4.5, 6.5.
8.5 and 10.5 mM the conductance increased throughout the voltage range tested and shifted to a

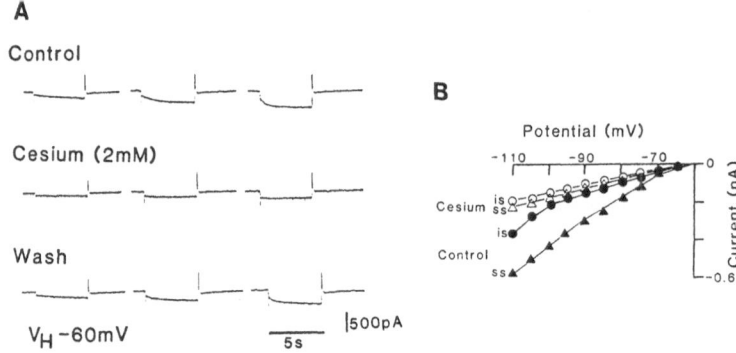

Fig. 3   Cesium blocks the time dependent inward rectifier.
A. Membrane currents evoked by hyperpolarizing voltage steps of 20, 30 and
40 mV from a holding potential of -60 mV before and during superfusion of
cesium (2 mM). Note the different time scale in comparison to Fig. 1 and 2.
B. I/V plot of the instantaneous (● o) and steady state values (▲ Δ) of the
membrane currents shown in A, before and during bath application of cesium.

less negative value the potential at which the inward rectification occurred. Figure 1 shows also
that low concentrations of barium (30 μM) reduced the inward current in a time and voltage dependent way, the instantaneous current being less reduced than the steady state current. The time
constant of the decline of the inward current in the presence of 1 μM of barium was 1.5 s at -120 mV
and 600 ms at -140 mV. The block occurred more quickly when higher concentrations of barium
(100 μM) were used. At resting membrane potential, barium induced also an inward current which
was dose dependent (40 pA with 1 μM and 130 pA with 100 μM).

Tetraethylammonium (TEA, 10 and 20 mM) also reduced the inward rectification of the
membrane. The effects of 10 mM and 20 mM of TEA were approximately the same as those of 1 and
3 μM of barium, except that the effects were not time dependent.

Figure 2 shows that cesium chloride (2 mM) reduced the inward rectification of the membrane. This effect was more pronounced at more hyperpolarized potentials, beyond the potassium
equilibrium potential, indicating that the blockade of the inward current was voltage dependent.
Cesium (2 mM) produced also an inward current (40-160 pA) at resting membrane potential.
However, the effect of cesium was time independent. Rubidium (2 mM) caused a time independent
depression of the inward current, but unlike barium and cesium, it increased the membrane
conductance between -55 and 105 mV. The effects of 2 mM of rubidium were in this respect similar to
that of increasing potassium concentration from 2.5 to 4.5 mM. Divalent cations such as strontium
(10 mM), calcium (10 mM) and magnesium (10 mM), all reduced the inward rectification; however
the effect was weak. Whereas the effect of strontium was time dependent, the effects of calcium
and magnesium were time independent.

Time dependent anomalous rectification

As for the fast inward rectifier, the time dependent inward rectifier was characterized by a
conductance increase with hyperpolarization. In the example of Figure 3 hyperpolarizing voltage
steps of 10, 20 and 30 mV from a holding potential of -60 mV caused an instantaneous inward current
followed by a slow inward relaxation (due to a slow time dependent conductance increase). An
inward tail current was observed on repolarization at the end of the hyperpolarizing pulse. Both
the conductance increase during the inward relaxation and the amplitude of the tail current were
voltage dependent, increasing at more negative potentials. The time course of the inward relaxa-

tion conformed reasonably to a single exponential with a time constant of 1.1s at -80 mV and 0.8s at -100 mV.

In agreement with other studies (Halliwell and Adams, 1982; Mayer and Westbrook, 1983; Spain et al., 1987), this current was mediated by potassium and sodium ions, since it was increased by raising the potassium concentrations to 4.5 and 6.5 mM and decreased in low sodium solutions (in which 126 mM of Na^+ were substituted by choline in the presence of scopolamine, 50 μM). In low sodium solutions, the tail currents completely disappeared. Unlike the fast inward rectifier, barium (100-300 μM) was ineffective in reducing both the instantaneous and steady state current, whereas cesium (2 mM) blocked it. The plot of Figure 3B shows the instantaneous and steady state current voltage relationship of a neurone in control conditions and during bath application of cesium. Cesium (2 mM) was very efficient in reducing both the instantaneous and steady state conductances particularly at more negative potentials. Rubidium (2 mM) increased the conductance in the same way as increasing potassium concentration from 2.5 to 4.5 mM.

DISCUSSION

We have shown that two classes of inward rectifiers (fast and time dependent) are present in nucleus accumbens neurones. Both are activated by hyperpolarization from resting membrane potential and are characterized by a non linearity in the I/V relationship. The time dependent inward rectifier (which has been observed in the minority of the neurones) occurs at potentials less negative than E_K. The fast inward rectifier bears many similarities with the 'classical' inward rectifier which has been described in skeletal muscle (Katz, 1949; Leech and Stanfield, 1981), in tunicate and starfish egg cells (Hagiwara et al., 1976; Ohmori, 1978), in submucous plexus (Mihara et al., 1987), and in central neurones (Constanti and Galvan, 1983; Williams et al., 1988a and 1988b; Yoshimura and Jessell, 1989). The conductance increase which occurs almost instantaneously is relatively selective for potasium ions and is activated at potentials near E_K. Like skeletal muscle (Standen and Stanfield, 1978) tunicate and starfish egg cells (Hagiwara et al., 1976; Ohmori, 1978) and other rat brain neurones (Constanti and Galvan, 1983; Williams et al. 1988a and 1988b; Yoshimura and Jessell, 1989), this conductance is higly sensitive to barium ions, which induce a voltage and time dependent block. Barium at low concentrations induces at the resting potential an inward current probably due to the blockade of the fast inward rectifier. This suggests that this current participates in the resting membrane conductance. This is in contrast with locus coeruleus and olfactory cortex neurones where the inward rectifier is activated below the resting membrane potential and therefore does not contribute to the resting conductance of the cells.

The time dependent inward rectifier in many features is similar to I_H or I_Q which has been described in smooth muscle (Benham et al., 1987), rod photoreceptors (Hestrin, 1987) and central neurons (Halliwell and Adams, 1982; Mayer and Westbrook, 1983; Crepel and Penit-Soria, 1986; Spain et al., 1987; Williams et al., 1988a; Yoshimura and Jessell, 1989) or i_f which has been described in the heart (Di Francesco et al., 1986). In agreement with other reports (Halliwell and Adams, 1982; Mayer and Westbrook, 1983), this conductance is permeable to both sodium and potassium ions. It is insensitive to barium but it is blocked by cesium.

What could be the physiological role of the fast and time dependent inward rectifier in nucleus accumbens neurones? First, as already mentioned, the fast inward rectifier contributes to the resting membrane conductance as suggested by the negative resting potential and relatively low input resistance (Uchimura et al., 1989a and 1989b). Second, this conductance is the target of several neurotransmitters which act either increasing or decreasing membrane potassium conductance (Uchimura et al., 1986; Mihara et al., 1987; Williams et al., 1988a and 1988b; Higashi et al., 1989; North and Uchimura, 1989; Uchimura and North, 1990). This will ensure the effectiveness of transmitters at the resting potential but also will amplify their effects on membrane potential since any hyperpolarization will result in a further increment in conductance. Nucleus accumbens neurones receive an important dopaminergic projection from the ventral tegmental area and project back to the ventral tegmental area either directly or to the ventral pallidum. This pathway is involved in the control of movements and some rewarding effects of brain stimulation (Wise and Bozarth, 1984; Nakajima and Mckenzie, 1986). Dopamine released from dopaminergic nerve terminals acts on D1 and D2 receptors to increase and decrease potassium conductance respectively (Uchimura et al., 1986; Higashi et al., 1989). Interestingly, it has been found recently that substances like 5-hydroxytryptamine (North and Uchimura, 1989) and muscarine (Uchimura and North, 1990) depolarize and excite nucleus accumbens neurones by blocking the inward rectifier conductance.

It has been suggested in other systems that the time dependent inward rectifier acts in a regenerative way maintaining the pacemaker activity in spontaneous firing neurones (Spain et al., 1987; see also Di Francesco et al., 1986). It will turn on slowly during the hyperpolarization which follows the action potential and will shorten the quiescent periods between the action potentials. Nucleus accumbens neurones do not fire spontaneously, but as in cerebellum (Llinas and Sugimori, 1980) the firing rises from subthreshold non synaptic membrane oscillations due to the activation of a non-inactivating sodium conductance (Uchimura et al., 1989a). Activation of a non-inactivating sodium conductance and inactivation of a time dependent inward rectifier would contribute to the oscillatory behaviour of nucleus accumbens neurones. Such a mechanism has been recently proposed to explain subthreshold, sodium dependent oscillations in stellate cells of the enthorinal cortex (Alonso and Llinas, 1989).

REFERENCES

Alonso, A. and Llinas, R., 1989, Subthreshold Na^+-dependent theta like rhythmicity in stellate cells of entorhinal cortex layer II, **Nature**, 342: 175.
Benham, C.D., Bolton, T.B., Denbigh, J.S. and Lang, R.J., 1987, Inward rectification in freshly isolated single smooth muscle cells of the rabbit jejunum, **J. Physiol.**, 383: 461.
Constanti, A. and Galvan, M., 1983, Fast inward-rectifying current accounts for anomalous rectification in olfactory cortex neurones, **J. Physiol.**,335: 153.
Crepel, F. and Penit-Soria, J., 1986, Inward rectification and low threshold calcium conductance in rat cerebellar purkinje cells: An in vitro study, **J. Physiol.**, 372: 1.
Di Francesco, D., Ferroni, A., Mazzanti, M. and Tromba, C., 1986, Properties of the hyperpolarizing-activated current (i_f) in cells isolated from the rat sino atrial node, **J. Physiol.**, 377: 61.
Goldman, D.E., 1943, Potential, impedance and rectification in membranes, **J. Gen Physiol.**, 27: 37.
Hagiwara, S., Miyazaki, S. and Rosenthal, N.P., 1976, Potassium current and the effect of cesium on this current during anomalous rectification of the egg cell membrane of a starfish, **J. Gen. Physiol.**, 67: 621.
Hagiwara, S., Miyazaki, S., Moody, W. and Patlak, J., 1978, Blocking effects of barium and hydogen ions of the potassium current during anomalous rectification in the starfish egg, **J. Physiol.**, 279: 167.
Halliwell, J.V. and Adams, P.R., 1982, Voltage clamp analysis of muscarinic excitation in hippocampal neurons, **Brain Res.**, 250: 71.
Hestrin, S., 1987, The properties and function of inward rectification in rod photoreceptors of the tiger salamander, **J. Physiol.**, 390: 319.
Higashi, H., Inanaga, K., Nishi, S. and Uchimura, N., 1989, Enhancement of dopamine actions on rat nucleus accumbens neurones in vitro after methamphetamine pre-treatment, **J.Physiol.**,408: 587.
Hodgkin, A.L. and Katz, B., 1949, The effect of sodium ions on the electrical activity of the giant axon of the squid, **J.Physiol.**, 108: 37.
Katz, B., 1949, Les constantes electriques de la membrane du muscle, **Archs. Sci. physiol.**, 3: 285.
Leech, C.A. and Stanfield, P.R., 1981, Inward rectification in frog skeletal muscle fibres and its dependance on membrane potential and external potassium, **J. Physiol.**, 319: 295.
Llinas, R. and Sugimori, M., 1980, Electrophysiological properties of in vitro Purkinje cell somata in mammalian cerebellar slices, **J. Physiol.**, 305: 171.
Mayer, M.L. and Westbrook, G.L., 1983, A voltage clamp analysis of inward (anomalous) rectification in mouse spinal sensory ganglion neurones, **J. Physiol.**, 340: 19.

Mihara, S., North, R.A. and Surprenant, A., 1987, Somatostatin increases an inwardly rectifying potassium conductance in guinea pig submucous plexus neurones, J. Physiol., 390: 335.

Nakajima, S. and Mckenzie, G.M., 1986, Reduction of the rewarding effect of brain stimulation by a blockade of dpopamine D1 receptors with SCH 23390, Pharmacol. Biochem. Behav., 24: 919.

North, R.A. and Uchimura, N., 1989. 5-hydroxytryptamine acts at 5-HT$_2$ receptors to decrease potassium conductance in rat nucleus accumbens neurones, J.Physiol., 417: 1.

Ohmori, H., 1978, Inactivation kinetics and steady state current noise in the anomalous rectifier of tunicate egg cell membranes, J. Physiol., 281: 77.

Spain, W.J., Schwindt, P.C. and Crill, W.E., 1987, Anomalous rectification in neurons from cat sensorimotor cortex in vitro, J. Neurophysiol., 57: 1555.

Standen, N.B. and Stanfield, P.R., 1978, A potential and time dependent blockade of inward rectification in frog skeletal muscle fibres by barium and strontium ions, J. Physiol., 280: 169.

Uchimura, N., Higashi, H. and Nishi, S., 1986, Hyperpolarizing and depolarizing actions of dopamine via D-1 and D-2 receptors on nucleus accumbens neurons, Brain Res., 375: 368.

Uchimura, N., Higashi, H. and Nishi, S., 1989a, Membrane properties and synaptic responses of the guinea pig nucleus accumbens neurons in vitro, J. Neurophysiol., 61: 769.

Uchimura, N., Cherubini, E. and North, R.A., 1989b, Inward rectification in rat nucleus accumbens neurons, J. Neurophysiol., in press.

Uchimura, N. and North, R.A., 1990, Muscarine reduces inwardly rectifying potassium conductance in rat nucleus accumbens neurones, J. Physiol., in press.

White, F.J. and Wang, R.Y., 1984, Interactions of cholecystokinin and dopamine on nucleus accumbens neurons, Brain Res., 300: 161.

Williams, J.T., Colmers, W.F. and Pan, Z.Z., 1988a, Voltage and ligand activated inwardly rectifying currents in dorsal raphe neurons in vitro, J. Neurosci., 8: 3499.

Williams, J.T., North, R.A. and Tokimasa, T., 1988b, Inward rectification of resting and opiate-activated potassium currents in rat locus coeruleus neurons, J. Neurosci., 8: 4299.

Wise, R.A. and Bozarth, M.A., 1984, Brain reward circuits: four circuit elements 'wired' in apparent series, Brain Res. Bull., 12: 202.

Yang, C.R. and Mogenson, G.T., 1984, Electrophysiological responses of neurons in the nucleus accumbens to hippocampal stimulation and the attenuation of excitatory responses by the mesolimbic dopaminergic system, Brain Res., 324: 69.

Yoshimura, M. and Jessell, T.M., 1989, Membrane properties of rat substantia gelatinosa neurons in vitro, J. Neurophysiol., 62: 109.

THE GLUTAMATERGIC AND GABAERGIC TRANSMITTER SYSTEM IN THE RAT NUCLEUS

ACCUMBENS: AN ELECTROPHYSIOLOGICAL IN VITRO STUDY

Cyriel M.A. Pennartz, Peter H. Boeijinga and Fernando H. Lopes da Silva

Department of Experimental Zoology, University of Amsterdam Kruislaan 320, 1098 SM, Amsterdam, Netherlands

SUMMARY

In this study, postsynaptic potentials elicited by local stimulation or stimulation of the corpus callosum were investigated by extra- and intracellular recordings in slices of the rat Nucleus Accumbens (Acb). In extracellular recordings, the postsynaptic response consisted of a population spike and a mixed EPSP-IPSP component. This response was strongly and reversibly antagonized by quisqualate/kainate (Q/K) receptor antagonists but not by acetylcholine, dopamine and enkephalin antagonists. In intracellular recordings the predominant role of Q/K receptors in mediating the synaptic response was confirmed. In addition, a small NMDA receptor mediated component was found which contributed to the late phase of the EPSP. Application of a GABA(A) antagonist particularly enhanced and prolonged the NMDA receptor mediated component of the synaptic response. The relevance of these findings for studying the role of the Acb in regulating locomotor activity is discussed.

INTRODUCTION

The Nucleus Accumbens is considered to be a basal forebrain region integrating limbic inputs and converting them into motor behaviour (Mogenson et al. 1980; Koob and Swerdlow 1988). The dopaminergic transmitter system of this structure, which originates in the ventral tegmental area, has been implicated in various neurological and mental disorders, most notably Parkinson's disease, schizophrenia and drug addiction (Price et al. 1978; Seeman 1987; Koob and Bloom 1988). One of the most widely used behavioral models to study the function of the mesolimbic dopamine system is induction of locomotor hyperactivity by bilateral injection of dopamine agonists into the Acb (Pijnenburg and Van Rossum 1973; Costall and Naylor 1975). In recent years, the behavioral significance of two other important transmitter systems of the Acb has been outlined. Bilateral injection of glutamate receptor agonists into the Acb has been repeatedly shown to enhance locomotor activity (Arnt 1981; Hamilton et al. 1986; Shreve and Uretsky 1988). Locomotion has also been shown to be stimulated following injection of the GABA(A) antagonist picrotoxin (Wachtel and Anden 1978; Morgenstern et al. 1984).

Glutamate, or a related excitatory amino acid, is probably utilized by afferent fibers originating in the subiculum, amygdala, prefrontal cortex and midline thalamic nuclei (Christie et al. 1987; Walaas 1981; Fuller et al. 1987). Many fibers originating in these structures are likely to terminate on medium-sized spiny neurons in the Acb (DeFrance et al. 1985; Pennartz and Kitai, unpublished observations). A substantial part of these cells are projection neurons and are probably GABAergic and peptidergic (Walaas and Fonnum

1980; Chang and Kitai 1985; Penny et al. 1986). Major efferent pathways of the Acb project to the ventral pallidum, substantia nigra and ventral tegmental area (Nauta et al. 1978).

These biochemical and anatomical findings prompted us to examine the electrophysiology of the glutamatergic and GABAergic transmitter system in the rat Acb. In characterizing glutamate-mediated synaptic transmissionwe investigated the contribution of Q/K and NMDA receptors (Watkins and Evans 1981) to postsynaptic potentials. Furthermore, we investigated whether a GABA-mediated IPSP contributes to the postsynaptic response and how it interacts with the NMDA receptor mediated component. The electrophysiological findings presented here are consistent with a number of behavioral studies using glutamate and GABA receptor agonists and antagonists. The results may be of importance in elucidating striatal mechanisms mediating locomotor activity.

METHODS

The methods used for preparation, stimulation and recording have been described elsewhere (Pennartz et al. 1989). Briefly, slices were prepared from male Wistar albino rats (100 - 180 g) that had been anaesthesized with ketamine (150 mg/kg, i.m.). After cooling the brain in Ringer solution at 3 - 7 C for 1 - 3 min., slices of 400 μm thickness were cut frontally by a vibroslice (Campden, UK) and transferred to the recording chamber. They were submerged and continuously superfused (2 - 3 ml/min.) with oxygenated (95 % O2, 5 % CO2) Ringer solution (34-36 C, pH 7.3) of the following composition (in mM): NaCl 132, KCl 3.5, MgSO4 1.3, CaCl2 2.5, NaH2PO4 1.0, NaHCO3 26.2, D-glucose 10.0. Postsynaptic responses were elicited by applying bipolar, biphasic rectangular current pulses (50 - 500 μA, 0.2 msec. duration, 0.1- 0.2 Hz) through two 60 μm thick stainless steel electrodes, insulated except at the tip and separated by 100 - 200 μm. The stimulation electrodes were placed onto the slice surface, either within the Acb or in the rostral corpus callosum overlying the Acb (Paxinos and Watson 1986). The recording electrodes were glass micropipettes filled with 4 M KAc (80-120 MOhm) for intracellular recordings and 3 M NaCl for extracellular recordings (3 - 8 MOhm). Intracellular and extracellular potentials were recorded using an Axoclamp 2A amplifier (Axon Instruments) and displayed on a digital Nicolet 3091 oscilloscope. Postsynaptic responses were usually averaged (N=4) on line using a Motorola Exorset microcomputer and stored on disk for further analysis. The DC membrane potential was continuously monitored on a chart recorder.

Stock solutions of kynurenic acid (Sigma), D(-)-2-amino-5-phosphonopentanoic acid (D-AP5; Tocris Neuramin), glycine (Sigma), and picrotoxin (Sigma) were made up in distilled water. 6-Cyano-7-nitroquinoxaline-2,3-dione (CNQX; Tocris Neuramin) was dissolved in dimethylsulfoxide and diluted 10,000 times for preparing test solutions. Dimethylsulfoxide itself had no effects on postsynaptic potentials (4 slices tested). Statistical evaluation of pharmacological effects was done using Wilcoxon's matched pairs signed rank test. Numerical values are expressed as mean ± SEM.

RESULTS

Field potentials were usually elicited by local electrical stimulation within the Acb. Stimulation of the rostral part of the corpus callosum elicited responses very similar to those evoked by local stimulation; therefore the results obtained using these different methods of stimulation were pooled.

Extracellular Recordings

The evoked field potential consisted of an early negative peak (N1; mean peak latency: 1.45 ± 0.04 msec.; N= 60) and a late negative peak (N2; mean peak latency 4.0 ± 0.1 msec.; N= 65). N1 was usually followed by a small positive component, P1, and N2 by a relatively long-lasting positive wave, P2 (fig.1A). This field potential could be recorded in all subregions of the Acb, although P2 was most clearly seen in the region surrounding the anterior commissure. The position of the stimulation electrode was not critical in

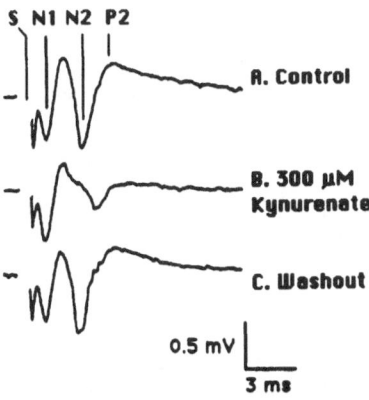

Fig.1. The NMDA- and non-NMDA receptor antagonist kynurenic acid (300 µM) reversibly reduces the N2 and P2 component of the locally evoked field potential but does not affect the N1 component. (A), in this control response the stimulus artefact (S) and the different components of the locally evoked potential are indicated. A P1-component is ocassionally observed between N1 and N2 and is non-synaptic (see text for further explanation) The baseline is indicated by the trace preceding the stimulus artifact. Positive is upward in all figures. (B), response during application of kynurenic acid; (C), washout.

evoking the response: most subregions of the Acb gave rise to N1 and N2 components when electrically stimulated. In order to identify the different components of the locally evoked potential, unit activity was studied. Antidromically activated units were distinguished from orthodromically activated units by lack of jitter, high frequency following (3-6 pulses, 100 or 150 Hz) and absence of changes in discharge latency with varying stimulation intensity (cf. Lemon and Prochazka 1984). The latency of antidromically activated unitary discharges generally corresponded to the latency of the accompanying N1 component, and likewise, orthodromically activated units coincided with the N2 component (results not shown). This finding suggests that N1 is a (non-synaptic) compound action potential and N2 a monosynaptic population spike (cf. Pennartz et al. 1989). As pointed out in the discussion, the P2 component can be identified as a mixed EPSP-IPSP sequence.

To investigate which transmitter system is primarily involved in generating the postsynaptic response, a number of receptor antagonists were applied to the slice. Before each application, a control period of 10 min. was taken into account. The amplitude of the population spike was not significantly attenuated by D-Tubocurarine (14 µM; N=9), atropine (100 µM; N=6), haloperidol (5 µM; N=6) or naloxone (1 µM; N=6), suggesting that the response is not mediated by acetylcholine, dopamine or enkephalins. In contrast, the broadly acting glutamate receptor antagonist kynurenic acid (300 µM) reduced the postsynaptic response throughout all subregions of the Acb, while leaving the compound action potential unaffected (fig.1). Specifically, the amplitude of the population spike was reduced to 29 ± 6 % of control values (P < 0.01; N= 10) and the amplitude of the PSP-component to 16 ± 7 % (P < 0.02; N = 7). After washout, the population spike recovered to 100 ± 4 % and the PSP-component to 91 ± 6 % of control (not significant).The potent Q/K receptor antagonist CNQX (2 μM; Honore et al. 1988) had a similar effect: during application the population spike was reduced to 8 ± 3 % of control values (P < 0.01; N= 9) and the PSP-component to 3 ± 3 % (N=5). Again, the effects were reversible: the population spike recovered to 94 ± 4 % and the PSP-component to 93 ± 6 % of control (not significant).The involvement of glutamate receptors was further confirmed by assessment of NMDA receptor activity (cf. Watkins and Evans 1981; Herron et al. 1986). When magnesium-free medium was applied to the slice, the duration of the P2-component was found to increase significantly up to 155 ± 17 % of control values (N=6; P < 0.05). This effect was reversed by addition of the NMDA receptor antagonist D-AP5 (50 μM; results not shown).

Intracellular Recordings

Stable impalements were obtained from 46 neurons, having a mean resting membrane potential of -80 ± 2 mV. Action potentials generated in response to injection of depolarizing current were 1.5 ± 0.1 msec in duration and 65 ± 1 mV in amplitude (N=34); the spike threshold amounted to -41 ± 2 mV. In most cells, responses to hyperpolarizing current pulses were smaller than to depolarizing current pulses of equal strength. The input resistance as estimated from recordings using small (0.2 - 0.6 nA) depolarizing currents was 37 ± 3 MOhm (N=40).

Postsynaptic potentials recorded in a total number of 34 neurons were evoked by local or corpus callosum stimulation. In 32 neurons, monophasic, monosynaptic depolarizing postsynaptic potentials (DPSPs) were elicited (fig.2A), while 2 cells generated spikes due to antidromic activation. When the stimulation current was adjusted just below the intensity needed to elicit a spike (this intensity will be termed 'spike threshold intensity' - STI), the onset and peak latency of the DPSP were 1.7 ± 0.1 and 5.9 ± 0.3 msec. respectively; the peak amplitude of the DPSP was 30.4 ± 1.5 mV (N=32). In all cells tested, stimulus currents exceeding the STI evoked only 1 spike in the postsynaptic response. The stimulus current corresponding to the STI was 310 ± 30 μA. At an intensity of 1.3 to 1.7 times STI, the spike latency amounted to 3.7 ± 0.2 msec. (N=27).

The postsynaptic response was investigated at depolarized levels of membrane potential using injecting of DC current. In the membrane potential range of -65 to -45 mV revealed a biphasic response pattern in 5 of 12 neurons tested, which consisted of a DPSP followed by a HPSP that lasted 100 to 250 msec. This HPSP was abolished in the presence of 50 μM picrotoxin. In the remaining 7 neurons, a prolonged DPSP was found instead of a DPSP followed by a HPSP (results not shown).

The contribution of different glutamate receptor subtypes in mediating the postsynaptic potential was tested using CNQX and D-AP5. The stimulation intensity was adjusted below the STI. During CNQX (4 μM) superfusion the peak amplitude of the DPSP was reduced to 14 ± 2 % of control values (fig.2A-B; P < 0.01; N=9). A nearly complete washout of this effect was established in 4 cells (DPSP amplitude 81 ± 18 % of control; fig.2D). To exclude antagonistic actions of CNQX at the glycine-binding site of the NMDA receptor-complex (Harris and Miller 1989; Yamada et al. 1989), 30 μM glycine was added to the superfusate already containing CNQX, and no recovery of the DPSP was found (fig.2B-C; 4 cells tested). The membrane potential remained constant during glycine application. Furthermore, the small DPSP remaining during CNQX application was found not to be abolished or reduced by application of 50 μM D-AP5 (4 cells tested).

To assess the contribution of NMDA receptors to DPSPs in the presence of functionally active Q/K receptors, effects of D-AP5 (50 μM) on postsynaptic potentials were tested. In 16 neurons, a DPSP peak amplitude of 32 ± 2 mV was measured in the control period. In the presence of D-AP5, the DPSP amplitude was not significantly changed (99 ± 2 % of control, N=16; washout value: 98 ± 4 % of control, N = 10). However, a relatively small reduction was noted in the late phase of the EPSP. This change was quantified by measuring the peak-to-half decay time of the postsynaptic response. Thus, the half-decay time in the control situation was 12 ± 1 msec. and during D-AP5 it was significantly

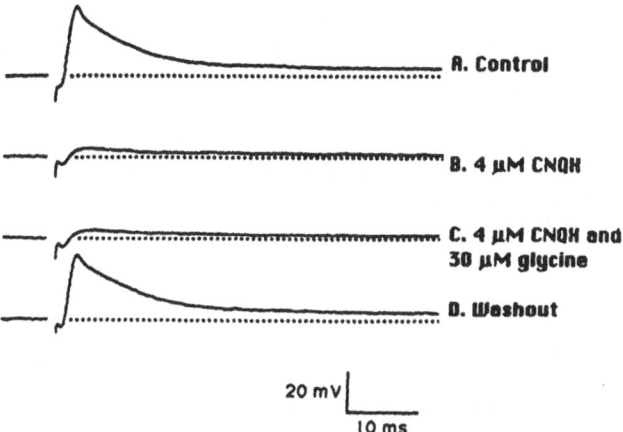

20 mV
10 ms

Fig.2. The quisqualate/kainate receptor antagonist CNQX reversibly reduces
the intracellularly recorded DPSP evoked by local or corpus callosum
stimulation. (A), control response; (B), 4 μM CNQX almost completely abolishes
the DPSP; (C), addition of 30 μM glycine to the medium already containing 4 μM
CNQX does not restore the DPSP; (D), washout of CNQX and glycine. In this cell
the resting membrane potential was -84 mV.

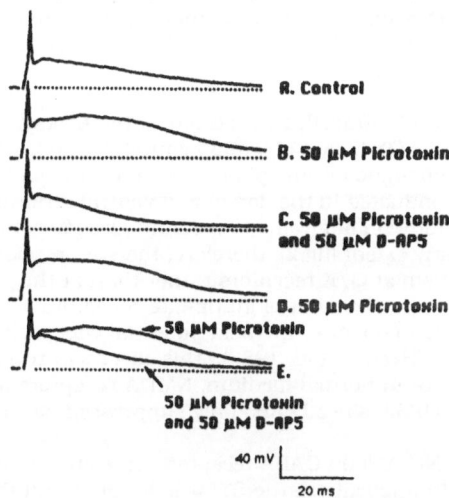

40 mV
20 ms

Fig.3. Effects of a NMDA receptor antagonist on DPSPs that were first modified
by a GABA(A) antagonist. (A-B), 50 μM picrotoxin enhanced and prolonged the
late phase of the DPSP; (C-D), 50 μM D-AP5 reversibly abolished the effect
induced by picrotoxin. (E), trace (B) and (C) are superimposed to show the
contribution of NMDA receptors to the postsynaptic response more clearly.
The resting membrane potential of this cell was -85 mV.

reduced to 85 ± 2 % of control (P < 0.001; N = 16; washout value: 97 ± 3 % of control; N = 11; not significant). Neither D-AP5 nor CNQX had effects on the resting membrane potential or input resistance (5 cells tested).

To examine whether the picrotoxin-sensitive component of the response, which can be identified as a GABAergic IPSP, suppresses the NMDA receptor mediated component of the EPSP, experiments using both picrotoxin (50 µM) and D-AP5 (50 µM) were carried out. In 4 of 5 cells tested, picrotoxin was observed to lower the STI and to prolong the EPSP. This prolongation was particularly clear at stimulation intensities exceeding the STI (fig.3A-B). Changes in resting membrane potential and input resistance were absent during picrotoxin application. D-AP5 was shown to reversibly abolish the prolongation of the EPSP in all of 4 cells tested (fig.3B-E). The interaction between GABA(A) and NMDA receptors was investigated in only a small number of cells since they are in line with previous extracellular recordings (Pennartz et al. 1989).

DISCUSSION

In both intracellular and extracellular recordings, local stimulation or stimulation of the rostral corpus callosum elicited a monosynaptic response (Pennartz et al. 1989; Chang and Kitai 1986). By correlating unit activity to field potentials it was found that the extracellularly recorded N1 and N2 components of this response corresponded to a (non-synaptic) compound action potential and a population spike. This finding was further corroborated by intracellular recordings, since the latency of action potentials generated on top of the DPSP fell in the same range as the latency of the N2 component. In comparing extra- and intracellular recordings, we observed that the P2 field component had a similar time course as the intracellularly recorded DPSP. Furthermore, picrotoxin prolonged both the P2-component and the intracellularly recorded DPSP to about the same extent (cf. Pennartz et al. 1989), supporting the notion that P2 corresponds to a DPSP. By injecting depolarizing currents and applying picrotoxin, it was found that in about half of the neurons the DPSP was composed of an EPSP followed by a GABA(A) receptor mediated IPSP.

In both extracellular and intracellular recordings, the postsynaptic potential was reversibly suppressed by CNQX (2 or 4 µM) and kynurenic acid (300 µM) but not by nicotinic, muscarinic, dopamine or enkephalin receptor antagonists. These results are in line with other studies conducted in the dorsal and ventral striatum (Cherubini et al. 1988; Uchimura et al. 1989). Addition of glycine to medium already containing CNQX did not restore the response to any extent (fig.2); therefore, the suppression of the DPSP can be explained by an antagonism at Q/K receptors rather than at the glycine-binding site of the NMDA receptor-complex. Involvement of glutamate receptors was further demonstrated by showing that the P2 component of the field potential was prolonged by application of magnesium-free medium (Herron et al. 1986); this effect was reversed by the NMDA receptor antagonist D-AP5. In normal medium, NMDA receptors appear to generate only a minor component of the DPSP, since D-AP5 only suppressed part of its decay phase.

Interactions between NMDA and GABA receptors were studied using picrotoxin and D-AP5. Following picrotoxin application, the STI was lowered and the late part of the EPSP was enhanced and prolonged, in agreement with previous extracellular observations (Pennartz et al. 1989). Since this prolonged component was effectively antagonized by D-AP5, it is suggested that NMDA receptor mediated activity is enhanced under conditions of reduced GABAergic inhibition. Whereas the glutamate-receptor mediated EPSP most probably arises from excited cortical, allocortical and possibly thalamic afferents (Christie et al. 1987; Fuller et al. 1987; Pennartz and Kitai, unpublished observations), the origin of the GABA-mediated IPSP is less clear. Feed-forward inhibition may exist in certain afferent pathways (Yim and Mogenson 1988), but lateral inhibition arising from firing activity of intrinsic Acb neurons may contribute as well. In agreement with an earlier study (Chang and Kitai 1985), we have recently found that at least a large part of Acb cells generating the postsynaptic response described above are medium-sized spiny neurons and that they possess axon collaterals and varicosities located within the Acb. This suggests that lateral synaptic interactions between intrinsic Acb neurons may

298

regulate their activity. In the dorsal striatum, the existence of recurrent inhibition has been supported by findings of Park et al. (1980). Thus, it is suggested that the synaptic excitability of Acb medium-sized spiny neurons is attenuated by GABAergic IPSPs arising from excited recurrent collaterals of adjacent neurons.

These findings can be related to a number of behavioral studies investigating the role of the Acb in regulating locomotor activity of the rat. Recently, Yang and Mogenson (1989) have proposed a mechanism by which physiological dopamine effects in the Acb may be linked to dopamine-induced locomotor hyperactivity. The ventral pallidum, which receives a GABAergic input from the Acb, has been shown to be critically involved in relaying locomotor-hyperactivity inducing signals evoked in the Acb (Jones and Mogenson 1981; Koob and Swerdlow 1988). Yim and Mogenson (1988) have suggested that dopamine attenuates EPSPs in the Acb in vivo. Dopamine injection into the Acb in vivo is thought to result in inhibition of firing activity of projection neurons and in decreased GABA release at their terminals in the ventral pallidum. Indeed, injection of GABA agonists into the ventral pallidum has been found to attenuate locomotor hyperactivity induced by dopamine injection into the Acb. Furthermore, injections of dopamine into the Acb result in increased firing rates of ventral pallidal neurons in vivo. In summary, Yang and Mogenson (1989) suggest that enhanced levels of locomotor activity correspond to *decreased* firing activity in ventral striatopallidal projections neurons.

This hypothesis is challenged by the electrophysiological findings presented here in conjunction with a number of behavioral studies. It is suggested here that locomotor hyperactivity corresponds to *increased* firing acitivity in Acb medium-sized spiny neurons, many of which are known to project to areas outside the Acb (Chang and Kitai 1985; H. Kita, pers. comm.). First, it has been shown that bilateral injection of glutamate receptor agonists into the Acb enhances locomotor activity and that this action can be reduced by glutamate receptor antagonists (Arnt 1981; Hamilton et al. 1986; Shreve and Uretsky 1988). Very high doses of glutamate agonists elicit less hyperactivity than moderate doses, a finding that may be explained by excessive depolarization leading to spike inactivation. Secondly, bilateral injection of kynurenic acid into the Acb reduces exploratory locomotion and increases the latency to initiate movements (Schacter et al. 1989). Administration of D-AP5 does not suppress locomotor activity (Hamilton et al. 1986). Thirdly, bilateral injection of GABA agonists into the Acb reduces spontaneous locomotion (Wachtel and Anden 1978;). In contrast, bilateral injection of picrotoxin elicits locomotor hyperactivity (Morgenstern et al. 1984; Wachtel and Anden 1978).

These behavioral findings can be explained with the help of the results outlined above. Injection of moderate amounts of glutamate agonists is likely to enhance the firing rate of Acb neurons (cf. Uchimura et al. 1989). Similarly, firing levels may be increased by injection of picrotoxin, since it will enhance afferent excitatory activity that is fed into the Acb during on-going activity of the animal. Suppression of this activity by Q/K receptor antagonists will result in locomotor hypoactivity. Our finding that application of D-AP5 affects postsynaptic responses of Acb neurons only to a minor extent is markedly parallelled by the lack of effect of D-AP5 injection on locomotor activity. Finally, the reduction of spontaneous locomotion by bilateral injection of GABA agonists may be explained by assuming that these substances suppress excitatory inputs by increasing GABAergic inhibition. A major part of the cells responding with DPSPs to local or corpus callosum stimulation are medium-sized spiny neurons (Pennartz and Kitai, unpublished observations). Altogether, these findings suggest that the firing rate of medium-sized spiny neurons may be positively correlated to the level of locomotor activity by an as yet undefined mechanism.

At the present time it is difficult to choose between these contrasting hypotheses, since a detailed knowledge of the ventral striatopallidal region is lacking. First, the precise functions of GABAergic interneurons and other types of striatal neurons are unknown. Secondly, subregions having differential functions in regulating locomotor activity may exist in both the Acb and ventral pallidum. Thirdly, while Yim and Mogenson (1988) recorded effects of dopamine on a short-lasting time scale, its long-lasting effects are unknown. Thus, identification of the striatal mechanisms underlying regulation of locomotor activity must await further investigations.

ACKNOWLEDGEMENTS

We wish to thank H. Groenewegen, S. Kitai, A. Stefani, M. Takada , N. de Vries and J. Whittaker for their advice and comments on the manuscript. This project was supported by the Netherlands Organization for Scientific Research and the Amsterdam Neuroscience Program.

REFERENCES

Arnt, J., 1981, Hyperactivity following injection of a glutamate agonist and 6,7-ADTN into rat nucleus accumbens and its inhibition by THIP, Life Sci., 28:1597.

Chang, H. T., and Kitai, S. T., 1985, Projection neurons of the nucleus accumbens: an intracellular labeling study, Brain Res.,347:112.

Chang, H. T., and Kitai, S. T., 1986, Intracellular recordings from rat nucleus accumbens in vitro, Brain Res., 366:392.

Cherubini, E., Herrling, P. L., Lanfumey, L., and Stanzione, P., 1988, Excitatory amino acids in synaptic excitation of rat striatal neurones in vitro, J. Physiol., 400:677.

Christie, M. J., Summers, R. J., Stephenson, J. A., Cook, C. J., and Beart, P. M., 1987, Excitatory amino acid projections to the nucleus accumbens septi in the rat: a retrograde transport study utilizing D[3H]aspartate and [3H]GABA, Neuroscience, 22:425.

Costall, B., and Naylor, R. J., 1975, The behavioral effects of dopamine applied intracerebrally to areas of the mesolimbic system, Eur. J. Pharmacol., 32:87.

DeFrance, J. F., Marchand, J. F., Sikes, R. W., Chronister, R. B., and Hubbard, J. I., 1985, Characterization of fimbria input to nucleus accumbens, J. Neurophysiol., 54:1553.

Fuller, T. A., Russchen, F. T., and Price, J. L., 1987, Sources of presumptive glutamergic/aspartergic afferents to the rat ventral striatopallidal region, J. Comp. Neurol., 258:317.

Hamilton, M. H., De Belleroche, J. S., Gardiner, I. M., and Herberg, L. J., 1986, Stimulatory effect of N-methyl-D-aspartate on locomotor activity and transmitter release from rat nucleus accumbens, Pharmacol. Biochem. Behav., 25:943.

Harris, K. M., and Miller, R. J., 1989, CNQX (6-cyano-7-nitroquinoxaline-2,3-dione) antagonizes NMDA-evoked [3H] GABA release from cultured cortical neurons via an inhibitory action at the strychnine-insensitive glycine site, Brain Res., 489:185.

Herron, C. E., Lester, R. A. J., Coan, E. J., and Collingridge, G. L., 1986, Frequency-dependent involvement of NMDA receptors in the hippocampus: a novel synaptic mechanism, Nature, 322: 265.

Honore, T., Davies, S. N., Drejer, J., Fletcher, E.J., Jacobsen, P., Lodge, D., and Nielsen, F. E., 1988, Quinoxalinediones: potent competitive non-NMDA glutamate receptor antagonists, Science, 241:701.

Jones, D. L., and Mogenson, G. J., 1980, Nucleus accumbens to globus pallidus GABA projection subserving ambulatory activity, Am. J. Physiol., 238:R63.

Koob, G. F. and Bloom, F. E., 1988, Cellular and molecular mechanisms of drug dependence, Science, 242:715.

Koob, G. F., and Swerdlow, N. R., 1988, The functional output of the mesolimbic dopamine system, Ann. N. Y. Acad. Sci., 537:216.

Lemon, R., and Prochazka, A., 1984, "Methods for neuronal recording in conscious animals", Wiley, New York.

Mogenson, G. J., Jones, D. L., and Yim, C. Y., 1980, From motivation to action: functional interface between the limbic system and the motor system, Progr. Neurobiol.,14:69.

Morgenstern, R., Mende, T., Gold, R., Lemme, P., and Oelssner, W., 1984, Drug-induced modulation of locomotor hyperactivity induced by picrotoxin in nucleus accumbens, Pharmacol. Biochem. Behav., 21:501.

Nauta, W. J. H., Smith, G. P., Faull, R. L. M., and Domesick, V. B., 1978, Efferent connections and nigral afferents of the nucleus accumbens septi in the rat, Neuroscience, 3:385.

Park, M. R., Lighthall, J. W., and Kitai, S. T., 1980, Recurrent inhibition in the rat neostriatum, Brain Res.,194:359.

Paxinos, G., and Watson, C., 1986, "The rat brain in stereotaxic coordinates", Academic Press, New York.

Pennartz, C. M. A., Boeijinga, P. H., and Lopes da Silva, F. H., 1989, Locally evoked potentials in slices of the rat nucleus accumbens: NMDA and non-NMDA receptor mediated components and modulation by GABA, Brain Res., in press.

Penny, G. R., Afsharpour, S., and Kitai, S. T., 1986, The glutamate decarboxylase-, leucine enkephalin-, methionine enkephalin- and substance P-immunoreactive neurons in the neostriatum of the rat and cat: evidence for partial population overlap, Neuroscience, 17:1011.

Price, K. S., Farley, I. J., and Hornykiewicz, O., 1978, Neurochemistry of Parkinson's disease: relation between striatal and limbic dopamine, Adv. Biochem. Psychopharmacol.,19:293.

Pijnenburg, A. J. J., and Van Rossum, J. M., 1973, Stimulation of locomotor activity following injection of dopamine into the nucleus accumbens, J. Pharm. Pharmacol., 25:1003.

Schacter, G. B., Yang, C. R., Innis, N. K., and Mogenson, G. J., 1989, The role of the hippocampal-nucleus accumbens pathway in radial-arm maze performance, Brain Res., 494: 339.

Seeman, P., 1987, Dopamine receptors and the dopamine hypothesis of schizophrenia, Synapse, 1:133.

Shreve, P. E., and Uretsky, N. J., 1988, Role of quisqualic acid receptors in the hypermotility response produced by the injection of AMPA into the nucleus accumbens, Pharmacol. Biochem. Behav., 30:379.

Uchimura, N., Higashi, H., and Nishi, S., 1989, Membrane properties and synaptic responses of the guinea pig nucleus accumbens neurons in vitro, J. Neurophysiol., 61:769.

Wachtel, H., and Anden, N., 1978, Motor activity of rats following intracerebral injections of drugs influencing GABA mechanisms, Naunyn-Schmiedeberg's Arch. Pharmacol., 302:133.

Walaas, I., 1981, Biochemical evidence for overlapping neocortical and allocortical glutamate projections to the Nucleus Accumbens and rostral caudatoputamen in the rat brain, Neuroscience, 6:399.

Walaas, I., and Fonnum, F., 1980, Biochemical evidence for gamma-aminobutyrate containing fibres from the nucleus accumbens to the substantia nigra and ventral tegmental area in the rat, Neuroscience, 5:63.

Watkins, J. C., and Evans, R. H., 1981, Excitatory amino acid transmitters, Ann. Rev. Pharmacol. Toxicol., 21:165.

Yamada, K. A., Dubinsky, J. M., and Rothman, S. M., Quantitative physiological characterization of a quinoxalinedione non-NMDA receptor antagonist, J. Neurosci., 9:3230.

Yang, C. R., and Mogenson, G. J., 1989, Ventral pallidal neuronal responses to dopamine receptor stimulation in the nucleus accumbens, Brain Res., 489:237.

Yim, C. Y., and Mogenson, G. J., 1988, Neuromodulatory action of dopamine in the nucleus accumbens: an in vivo intracellular study, Neuroscience, 26:403.

A COMPARISON OF SINGLE UNIT ACTIVITY IN PRIMATE CAUDATE

NUCLEUS AND PUTAMEN IN A SENSORY CUED MOTOR TASK

J.W. Aldridge, D. Jaeger, and S. Gilman

Department of Neurology
University of Michigan
Ann Arbor, MI 48104

ABSTRACT

The behavioral correlates of single unit activity in the putamen and
caudate nucleus were compared in primates. Single unit neuronal activity
was recorded during the performance of a sensory cued reaching task. Units
in the putamen were more often modulated with motor events than units in
the caudate nucleus. The latter had many more units with anticipatory
responses to sensory events of the task and more responses to sensory cues.
It is suggested that the putamen plays a more important role in movement
regulation than the caudate nucleus and that the caudate nucleus has a role
in switching behavioral states.

INTRODUCTION

The putamen and caudate nucleus, together the striatum, have an
integrative role in motor behavior. Both physiological [9] and clinical
evidence [20] support this idea. Additionally, the striatum may integrate
information related to behavioral state [5,21], that is, information about
the learned significance of sensory stimuli and/or motor events. The
differences and similarities between the two striatal components in these
integrative mechanisms are not well understood.

The cerebral cortical connections to the putamen and caudate nucleus
are distinctive and different. The putamen receives a large number of motor
cortical afferents [19,23], suggesting a role in regulating voluntary
movement. The caudate nucleus receives a smaller quantity of motor cortical
afferents [19,23], but it receives a substantial set of inputs from
association and prefrontal cerebral cortical areas [11], suggesting an
integrative role for sensory or cognitive information. Even with these
anatomical differences, strict functional subdivisions may be blurred by
the overlap and convergence of cerebral cortical and subcortical striatal
afferents. Additionally, interneuronal communication may distribute
information processing functions throughout the striatum. The similarity of
neuronal architecture and biochemical characteristics [12] suggests that
similar internal mechanisms of information processing might exist. From a
clinical perspective, the striatum has a conspicuous role in motor control
and it is speculated that regional differences in processing may be
important [1]. The appearance of cognitive deficits in basal ganglia

The Basal Ganglia III, Edited by G. Bernardi et al.
Plenum Press, New York

disorders indicates the importance of non-motor afferents to this brain area [10,24] and suggests a role in integrating non-motor behavior.

Single unit recording studies in behaving animals have emphasized the relationships of the putamen to movement regulation. Unit activity in the putamen of primates has a somatotopic functional organization [7] and it is modulated by movement direction [4,8]. Putaminal activity related to movement also has sensory set-dependent properties [17,18]. In the caudate nucleus, the set-dependent properties of single unit activity are observed frequently. Set-dependence is defined here as an increased probability for a change in discharge that depends on the behavioral state or context. Sensory cues that trigger learned movements [2] and the presentation of visual stimuli that provide precue information for subsequent movements [16] are correlated to changes in unit activity in the caudate nucleus. Other diverse environmental stimuli and movements activate these units [6,22]. Saccadic eye movements are also correlated to discharge in the caudate nucleus [13]. This relationship is complex and depends on the anticipatory properties of learned behavior [14,15]. A comparable dependence on the behavioral context has been observed for limb movements [3]. The caudate nucleus may have a role related to the sequential organization of behavior [25]. The modulatory influence of the behavioral state on caudate nucleus unit activity supports this idea.

In this study we have directly compared the caudate nucleus and the putamen, examining the behavioral correlates of single unit activity in a sensory-cued reaching task. It was anticipated that units in the putamen would be closely linked to arm movements and that units in the caudate nucleus would be coupled to the presentation of sensory cues and task sequences. In part, these hypotheses were confirmed. This paper presents preliminary analyses that distinguish the properties of these two structures. Detailed reports on other aspects of the data are in preparation.

METHODS

Task: The motor task required the animal to reach out and touch buttons in two vertical arrays, one for each limb (Figure 1A). The 12 buttons were touch-sensitive metal knobs 2.5 cm in diameter. Each button had a light emitting diode (LED) in its center. A computer controlled task timing and sensory cue presentation with LEDs and auditory go-cues and monitored button contact. A randomly varied period of time from 0.5 to 1.5 seconds preceded all sensory cues. The task had 4 phases (Figure 1B). The first phase (initial position phase, Figure 1B) began with a go-cue and the illumination of one of the two initial position knobs near the middle of the left vertical array. The animal had 1500 ms to touch and hold the button with its left hand. Left button contact triggered the presentation of similar cues for a reaching movement to one of the right limb initial position buttons. There was a 1000 ms time limit. Maintained contact with both initial position buttons triggered the precue phase (Figure 1B). During the precue phase a subset of button LEDs was flashed for 2 seconds and provided partial to complete information about the target location in the touch target phase. Steady contact with the initial position knobs was required during and after the precue. In the touch target phase, one button was illuminated and the go-cue was sounded. The animal had 800 ms to touch and hold that button. Successful completion resulted in the delivery of fruit juice in the reward phase.

Experimental procedure: Monkeys were trained for approximately 8 months. After training, a stainless steel recording chamber was surgically

attached to the skull with dental acrylic and bone screws under
pentobarbital anesthesia. The animals were removed from their cages and
brought to the laboratory for each recording session, which lasted 3-4
hours. Multiwire electrodes recording 1 to 6 units simultaneously were
manipulated to striatal recording sites with a hydraulic microdrive secured
to the chamber. Extracellular potentials were amplified (x10000), filtered
(300-10000 Hz), and recorded with a parallel processor computer system. For
each spike, peak amplitudes and peak to peak durations were recorded and
stored on disk. Spikes were discriminated off-line. During some sessions,
eye movements were recorded with surface EOG electrodes and in others, EMG
activity from muscles in the forearm, upper arm, shoulder and paraspinal
regions were recorded. To evaluate neuronal responses, peri-event time
histograms were constructed for sensory cues and movements onsets and
offsets. Changes in discharge activity were assessed statistically and
categorized by the temporal structure of the response.

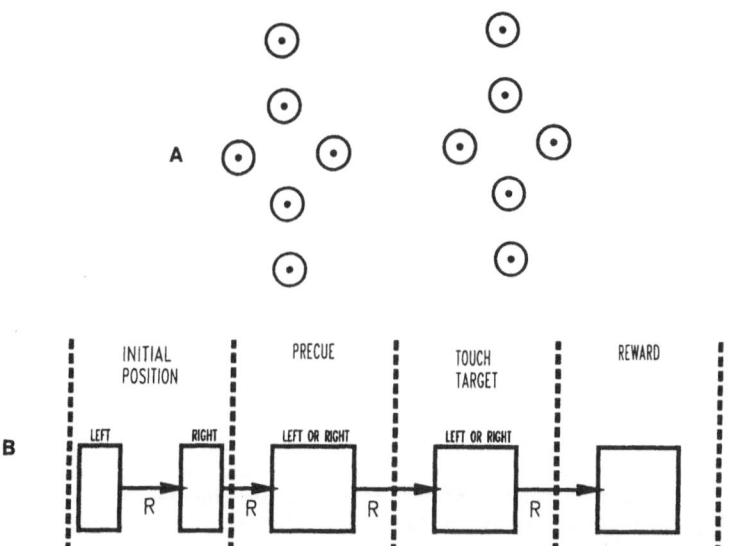

Figure 1. Behavioral task. The touch-sensitive metal knobs for the reaching
task were configured as illustrated in A. In the center of each vertical
array, two horizontally placed buttons served as the initial position
locations. In B, the four task phases are schematized. Each cue
presentation is represented by a box. A random delay period (R) separated
different task events.

RESULTS

A total of 421 units in the putamen and 229 units in the caudate
nucleus from 2 monkeys are included in this analysis. Most units were
responsive in at least one task phase. The proportions of responsive units
were similar in the two structures (Figure 2A). Spontaneous firing rates
measured prior to the initial position phase had wide distributions (Figure
2B). These distributions, which were dominated by slowly firing units,
included some rapidly active units. Since there was a substantial overlap
between the distributions of responsive and non-responsive units (Figure
2B), it was not possible to predict whether a unit would be correlated with
the task on the basis of firing rate alone. This was the case for both the
caudate nucleus and the putamen.

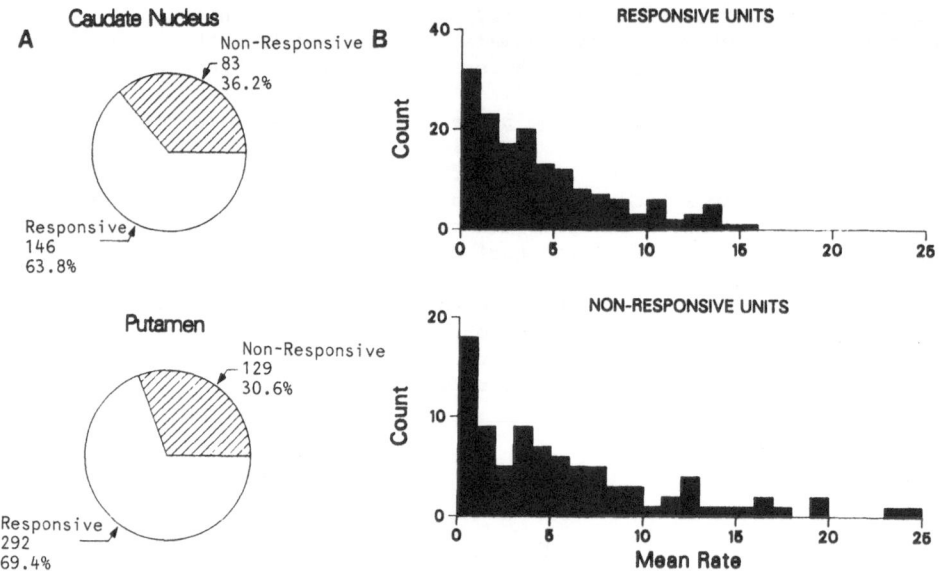

Figure 2. A. The putamen and caudate nucleus had similar proportions of responsive units. A unit was responsive if in at least one phase of the task it had a significant change in discharge rate correlated with a sensory or motor event. In B, the distributions of mean discharge rates for responsive units (top) and non-responsive units (bottom) are shown for one monkey.

Two types of movement-related responses were observed. Most had a temporal structure similar to the EMG of arm and shoulder muscles. These units (Figure 3A) were activated at about the same time as or just after EMG onset. This type of movement-related response was detected in 27% of units in the putamen compared to 7% of units in the caudate nucleus. Responses related to movement of the contralateral limbs dominated, but in 6% of putaminal units and 3% of caudate nucleus units, ipsilateral limb movements were correlated to unit discharge. A second movement-related response suggested a preparatory function. These units were activated prior to a cue and the activity change persisted until the movement was completed (Figure 3B). These responses were observed in 4% of the units in the putamen and 2% of units in the caudate nucleus. Responses to both limbs were found in movement-related units and many had responses to both the initial position and the touch target phase. Most responses were excitatory, but in 5% of all units in putamen and 1% of caudate nucleus units, an inhibitory response was detected. The arm area of the putamen had the greatest proportion of responses and also had more units with directional responses and discharge activity that preceded EMG onset.

A frequently observed response consisted of discharge activity that "anticipated" (preceded) the presentation of a sensory cue. Anticipatory responses were defined by a discharge pattern that began prior to a sensory cue and stopped before movement. In the caudate nucleus, anticipatory responses were evident in the precue phase (Figure 4B). This phase required a stable limb posture while a visual stimulus was presented. Although a random delay preceded the precue stimulus, its sequential position in the task was consistent and predictable. The exact time at which a stimulus would occur, however, was unpredictable to the animal. In units with an anticipatory response to the precue, peri-movement histograms for contact

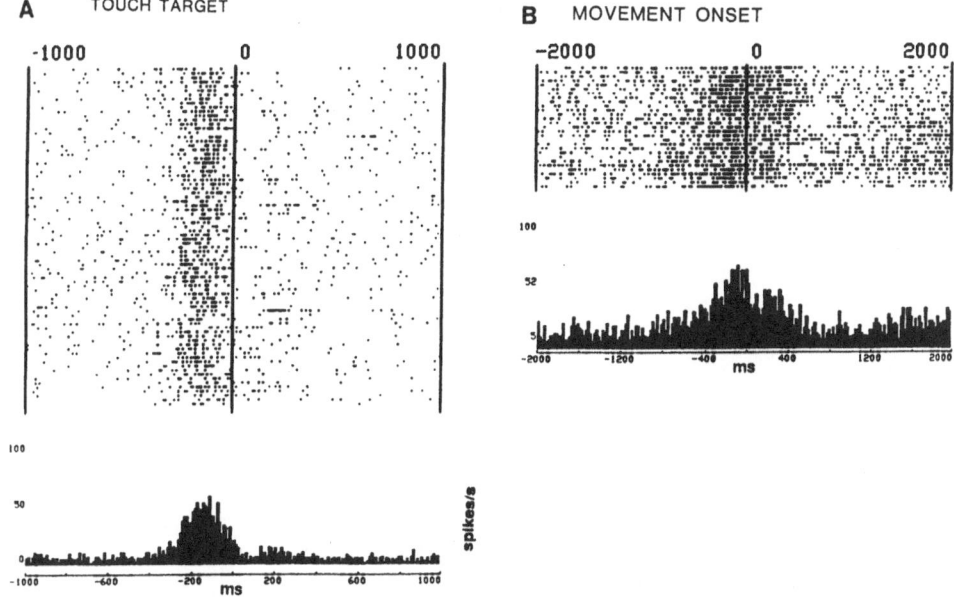

Figure 3. A. The pattern of movement-related discharge activity illustrated here was observed commonly. In this raster and perimovement time histogram diagram from a unit in the putamen, spike trains were aligned at time zero to the point in time at which the movement ceased, i.e., touch target. This type of response began at approximately the same time or just after EMG discharge, 150-250 ms prior to movement, and ended with the movement completion. In B, a preparatory type of response in a unit from the putamen had discharge that begins prior to the go-cue and more than 500 ms prior to movement onset at the time zero alignment point.

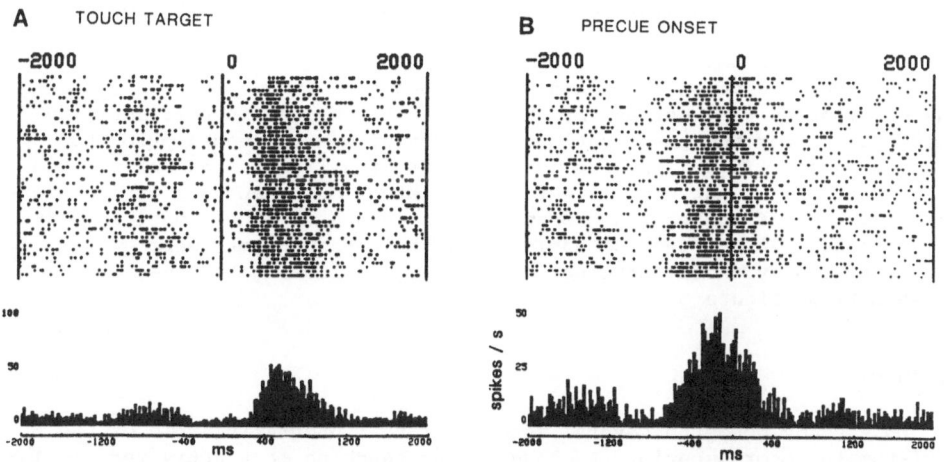

Figure 4. Two perievent time histograms of sequential events from a unit in the caudate nucleus illustrate anticipatory responses. In A, the histogram was aligned to the end of the right limb movement of the initial position phase (Figure 1). In B, the same spike trains were aligned to the onset of the precue, which followed the movement. An increase in discharge activity preceded the precue by more than 600 ms. The histogram was more sharply defined about the movement offset because there was a randomized delay between movement completion and precue. The discharge lasted throughout the variable delay precue and ended 300 ms after the precue onset.

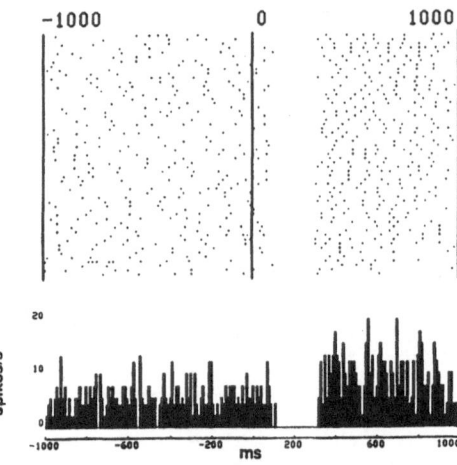

Figure 5. Cue-related response in the caudate nucleus. The perievent histogram constructed about the presentation of an initial position cue demonstrates a short duration (200 ms) decrease in discharge activity.

with the right initial position button, the event immediately prior to the precue, were well-synchronized with a latency of around 300 ms (Figure 4A). Given the random delay, this latency was about 200 ms prior to the earliest possible cue presentation. Other predicable cues (movement go-cues and rewards) also elicited anticipatory responses. In the caudate nucleus, 18% of all units had anticipatory responses, whereas, in the putamen only 6% had such responses. Excitatory changes in unit activity dominated this category of responses. Only one unit with an inhibitory response was found.

Both the caudate nucleus and putamen had sensory cue-related responses. These responses were defined by short duration changes in unit activity that began after the presentation of a sensory cue and ended prior to movement onset. In the caudate nucleus, 18% of all units had cue-related responses. In the putamen, 9% were cue-related. Cue responses were observed in all phases of the task and preceding movements of both limbs. Some 7% of the responses in the caudate nucleus and 5% in the putamen were characterized by short decreases in discharge following the cue presentation (Figure 5).

DISCUSSION

This study concerns several aspects of the integrative role of the striatum in motor behavior. A sensory-cued reaching task activated similar proportions of units in both the putamen and caudate nucleus, however, the relationships of unit activity with various behavioral events were different in the two nuclei. Because of the overlap of firing rate distributions in responsive and non-responsive units, it was not possible to predict a unit's response probability based on spontaneous mean firing rate. An analysis of the task-related discharge characteristics demonstrated that movement-related responses were more prevalent in the putamen than the caudate nucleus. The caudate nucleus had more set-dependent responses that anticipated the next event in a behavioral sequence. Cue-related discharge was common in both structures, but more common in the caudate nucleus. Many units were inhibited by the cue presentation. Responses associated with contralateral and ipsilateral limbs were present in both nuclei.

No response pattern was unique to either the caudate nucleus or the putamen. Whether a unit was located in a particular structure could not be determined simply from its set of peri-event histograms. Examples of anticipatory, movement-related and cue-related responses were found in both structures. There was, however, a dissimilarity in the proportions of different responses in the two nuclei. These differences were marked. The putamen, as expected, had many more movement-related responses than the caudate nucleus. This is consistent with the massive afferent projection from motor cortex. Nevertheless, the putamen also had cue-related and anticipatory responses. The caudate nucleus receives only a small number of motor cortical afferents and this may explain the smaller number of movement-related responses. Divergence of cortical afferents within the striatum and information transfer through interneurons may also contribute to the distributed processing observed. The higher proportion of cue-related and anticipatory responses in the caudate nucleus presumably reflects the differential effect of its particular set of cortical afferents. It is possible that some cue-related responses may not be functionally involved in stimulus evaluation per se, but could be involved in neuronal circuitry that is assigning the behavioral significance to a stimulus. Also, task-related eye or head movements may contribute to what appears to be a stimulus response based on timing.

These findings suggest that unit activity in caudate nucleus is strongly related to switching of behavioral states in a well-learned task whereas unit activity in the putamen is more directly linked to motor control. Both of these processes are important to behavior and the combination of this information in the globus pallidus may be the ultimate information processing objective of the basal ganglia.

REFERENCES

1. Albin, R.L., Young, A.B. and Penney, J.B., The functional anatomy of basal ganglia disorders, Trends Neurosci., 12 (1989) pp. 366-375.
2. Aldridge, J.W., Anderson, R.J. and Murphy, J.T., The role of the basal ganglia in controlling a movement initiated by a visually presented cue, Brain Res., 192 (1980a) pp. 3-16.
3. Aldridge, J.W., Anderson, R.J. and Murphy, J.T., Sensory motor processing in the caudate nucleus and globus pallidus: A single unit study in behaving primates, Can. J. Physiol. Pharmacol., 58 (1980b) pp. 1192-1201.
4. Alexander, G.E., Selective neuronal discharge in monkey putamen reflects intended direction of planned limb movements, Exp. Brain Res., 67 (1987) pp. 623-634.
5. Buchwald, N.A., Hull, C.D., Levine, M.S. and Villablanca, J., The basal ganglia and the regulation of response and cognitive sets. In M.A.B. Brazier (ed.), Growth and Development of the Brain, Raven Press, New York, 1975, pp. 171-189.
6. Buser, P., Pouderoux, G. and Mereaux, J., Single unit recording in the caudate nucleus during sessions with elaborate movements in the awake monkey, Brain Res., 71 (1974) pp. 337-344.
7. Crutcher, M.D. and DeLong, M.R., Single cell studies of the primate putamen. I. Functional organization, Exp. Brain Res., 53 (1984a) pp. 233-243.
8. Crutcher, M.D. and DeLong, M.R., Single cell studies of the primate putamen. II. Relations to direction of movement and pattern of

muscular activity, Exp. Brain Res., 53 (1984b) pp. 244-258.

9. DeLong, M.R. and Georgopoulos, A.P., Motor functions of the basal ganglia. In V. Brooks (ed.), Handbook of physiology. Sec. 1. Nervous system. Volume 2. Motor control, Williams and Wilkins, Baltimore, 1981, pp. 1017-1061.

10. Girotti, F., Marano, R., Soliveri, P., Geminiani, G. and Scigliano, G., Relationship between motor and cognitive disorders in Huntington's disease, J. Neurology, 235 (1988) pp. 454-457.

11. Goldman, P.S. and Nauta, W.J.H., An intricately patterned prefonto-caudate projection in the rhesus monkey, J. Comp. Neurol., 171 (1977) pp. 369-386.

12. Graybiel, A.M. and Ragsdale Jr., C.W., Biochemical anatomy of the striatum. In P.C. Emson (ed.), Chemical Neuroanatomy, Raven Press, New York, 1983, pp. 427-504.

13. Hikosaka, O., Sakamoto, M. and Usui, S., Functional properties of monkey caudate neurons. 1. Activities related to saccadic eye movements, J. Neurophysiol., 61 (1989a) pp. 780-798.

14. Hikosaka, O., Sakamoto, M. and Usui, S., Functional properties of monkey caudate neurons. 2. Visual and auditory responses, J. Neurophysiol., 61 (1989b) pp. 799-813.

15. Hikosaka, O., Sakamoto, M. and Usui, S., Functional properties of monkey caudate neurons. 3. Activities related to expectation of target and reward, J. Neurophysiol., 61 (1989c) pp. 814-832.

16. Jaeger, D., Gilman, S. and Aldridge, J.W., Single unit activity of primate caudate nucleus in a precue task, Soc. Neurosci. Abst., 14 (1988) p. 719.

17. Kimura, M., The role of primate putamen neurons in the association of sensory stimuli with movement, Neurosci. Res., 3 (1986) pp. 436-443.

18. Kimura, M., Rajkowski, J.R. and Evarts, E., Tonically discharging putamen neurons exhibit set-dependent responses, Proc. Natl. Acad. Sci. USA, 81 (1984) pp. 4998-5001.

19. Kunzle, H., Bilateral projections from precentral motor cortex to the putamen and other parts of the basal ganglia. An autoradiographic study in Macaca fascicularis, Brain Res., 88 (1975) pp. 195-209.

20. Marsden, C.D., The mysterious motor function of the basal ganglia: the Robert Wartenberg lecture, Neurology, 32 (1982) pp. 514-539.

21. Oberg, R.G.E. and Divac, I., "Cognitive" functions of the neostriatum. In I. Divac and R.G.E. Oberg (eds.), The Neostriatum, Pergamon Press, Oxford, 1979, pp. 291-313.

22. Rolls, E.T., Thorpe, S.J. and Maddison, S.P., Responses of striatal neurons in the behaving monkey. 1. Head of the caudate nucleus, Behav. Brain Res., 7 (1983) pp. 179-210.

23. Selemon, L.D. and Goldman-Rakic, P.S., Longitudinal topography and interdigitation of corticostriatal projections in the Rhesus monkey, J. Neurosci., 5 (1985) pp. 776-794.

24. Starkstein, S.E., Preziosi, T.J., Berthier, M.L., Bolduc, P.L., Mayberg, H.S. and Robinson, R.G., Depression and cognitive impairment in Parkinson's disease, Brain, 112 (1989) pp. 1141-1153.

25. van den Bercken, J.H.L. and Cools, A.R., Evidence for a role of the caudate nucleus in the sequential organization of behaviour, Behav. Brain Res., 4 (1982) pp. 319-337.

PHARMACOLOGY OF DOPAMINE AND
RELATIONSHIPS WITH OTHER
NEUROTRANSMITTERS

COMPARATIVE DISTRIBUTION OF D1 RECEPTORS AND μ OPIATE RECEPTORS IN THE NEONATAL RAT STRIATUM : EFFECTS OF AN EARLY LESION OF DOPAMINERGIC FIBERS

Monique Rogard, Jocelyne Caboche
and Marie-Jo Besson

Laboratoire de Neurochimie-Anatomie
IDN-CNRS, 9 quai Saint Bernard
75005 Paris, France

INTRODUCTION

The compartmental organization of the striatum in striosomes and matrix (Graybiel and Ragsdale, 1978) is followed by a large variety of neuronal markers. Peptide-containing fibers and peptide-containing perikarya (Graybiel and Ragsdale, 1983 ; Graybiel, 1984a), afferent fibers (Gerfen et al., 1987a ; Gerfen, 1989; Jimenez-Castellanos and Graybiel, 1987) efferent neurons (Jimenez-Castellanos and Graybiel, 1989), are mostly distributed according to this two compartments. Various specific receptor ligands have also been shown to label binding sites organized according to a striosomal pattern, as for instance μ opiate receptors (Herkenham and Pert, 1981), muscarinic M1 receptors (Nastuk and Graybiel, 1985), benzodiazepine type II sites (Faull and Villiger, 1988). With regard to dopamine (DA) receptors, a compartmental distribution has been observed in the striatum of cat and monkey, with heightened density of D1 receptors (Besson et al., 1988) in striosomes and heightened density of D2 receptors in the matrix (Beckstead et al., 1988). No such heterogeneity has been described for D1 or D2 receptors in the striatum of the adult rat (Boyson et al., 1986 ; Savasta et al., 1986). However, it has been shown in feline (Richfield at al., 1987) and rodents (Murrin and Zeng, 1989 ; Ohta and al., 1989) that both striatal DA fibers and striatal D1 receptors are clustered during the neonatal period. The purpose of our study was to reexamine, in the rat striatum, the postnatal evolution of the distribution of D1 receptors, and to analyze the possible role of the incoming DA fibers in the disappearance of their patchy organization. The distribution of D1 receptors was compared to that of μ opiate receptors during the first three postnatal weeks in intact rats and in rats treated at postnatal day 2 (P2) by an intrastriatal 6-OHDA injection.

The Basal Ganglia III, Edited by G. Bernardi *et al.*
Plenum Press, New York

RESULTS AND DISCUSSION

Striosomal organization of D1 receptors in the neonatal rat striatum

D1 receptors were visualized autoradiographically on rat brain sections by the binding of 3H-SCH 23390. At postnatal day 0 (P0) D1 receptors are present in the striatum and there, are more abundant than in any other region but the choroïd plexus. The binding of 3H-SCH 23390 appears along a dense line at the external border of the striatum and within patches mainly distributed in the lateral two-thirds of this structure. This organization which is similar to that previously described by Murrin et al., (1989), clearly shows a striosomal ordering of D1 binding sites. Patches with heightened density of D1 receptors are in register with tyrosine hydroxylase (TH) immunoreactive-rich islands and densely acetylcholinesterase-(AChE) stained zones, which both define striosomes in young mammals (Graybiel, 1984b).

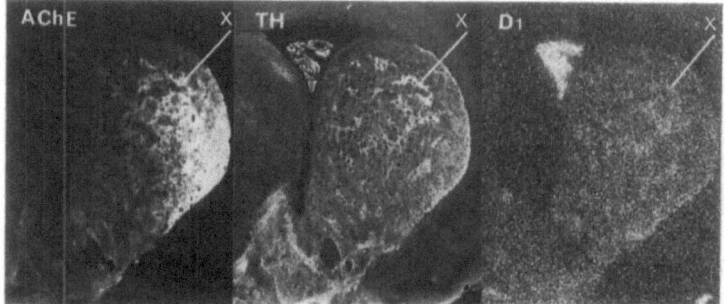

Fig. 1. Dark-field photographs of serial coronal sections of neonatal rat at postnatal day 4 (P4). Patches rich in acetylcholinesterase staining (AChE), tyrosine hydroxylase-immunoreactive fibers (TH) and D1 receptors (D1), appear lighter than the surrounding area. x indicates a patch of dense D1 binding sites in register with an AChE- and a TH- rich island. Note the dense TH-immunoreactivity and the heightened D1 binding sites at the level of choroïd plexus.

Quantitative measurements show a 50% difference in D1 binding site density between striosomes and the surrounding matrix. In contrast, at P0 binding sites labelled by 3H-DAGO, a selective ligand for μ receptors, have a relatively homogeneous distribution, an observation which is in agreement with previously reported data (Lança et al., 1986 ; Moon Edley and Herkenham, 1984 ; Van Der Kooy, 1984). Later on, at P3-P4, D1 binding patches are distributed in the whole striatum. They are still found in register with TH-immunoreactive fibers and AChE staining (Fig. 1). They are also in correspondence with μ opiate receptors which show at this age a clear patchy organization. At P3-P4, 3H-SCH 23390 binding sites are significantly increased in patches and matrix when compared to the values obtained in these compartments at P0. A 40% difference in the density between these two compartments is still detected.

Progressive disappearance of the striosomal organization of D1 receptors

The period around P9 appears to be crucial for the dissociation of the correspondence between patches of TH- immunoreactive fibers and D1 receptors. Islandic TH-immunoreactive fibers well observed in the dorsal striatum become barely visible in the ventral striatum, whereas the reverse pattern is found for D1 receptors. In fact, the patchy distribution of D1 binding sites disappears in the dorsal part of the striatum, but is preserved in the ventral part. Moreover, the density of D1 binding sites measured in the whole dorsal striatum is significantly higher than in striatal patches of P4 rats. This indicates that the disappearance of D1 receptor patches in the dorsal striatum results from a more rapid increase of D1 binding sites in matrix than in striosomes. In the ventral striatum, the density in patches remains significantly higher (+35%) than in the matrix, and the values in both compartments increased comparatively to the P4 values. In the whole striatum μ opiate receptors become clearly and almost exclusively distributed in patches. In the ventral region, they are in alignment with patches of D1 receptors.

At P20, both D1 binding sites and TH-immunoreactive fibers are homogeneously distributed in the striatum with no striosomal pattern, either in the dorsal or in the ventral part. There is however in the whole striatum a further increase in D1 binding sites as indicated by the higher density values at P20 than at P9. Contrasting with the homogeneous distribution of D1 receptors, μ opiate receptors are very patchy and the density of binding sites is 2.5 to 3 times higher in patches than in surrounding matrix.

Thus, during the postnatal period, D1 and μ opiate receptors show an almost inverse pattern of evolution and their organization in relation to DA innervation seems complexe. The similar distribution of TH-immunoreactive fiber islands and D1 receptor patches, observed at P0 could infer a close relationship between these respective pre- and postsynaptic markers of DA innervation. However, prenatal analyses have revealed that the clustering of striatal cells containing DARPP-32, a DA- and cAMP-regulated phosphoprotein linked to D1 receptors appears at embryonic day 19 (E19) and precedes by at least 2 days the islandic organization of TH-immunoreactive fibers clearly seen at P0 (Foster et al., 1987). Therefore the striosomal pattern of D1 receptors does not seem to be governed by islandic arrangement of their corresponding presynaptic elements. By contrast, μ opiate receptors appear in a striosomal organization one or two days after the clustering of TH-immunoreactive fibers. This delay suggests that the striosomal ordering of μ opiate receptors is partly dependent on the incoming DA fibers.

Effect of an early lesion of DA fibers on the organization of D1 and μ opiate receptors

The clustering of both TH-immunoreactive fibers and D1 receptors, in correspondence between P0 and P4, progressively disappears and is replaced by an homogeneous distribution of these two markers. We hypothesized that DA fibers which progressively innervate the matrix could influence the occurence of D1 receptors in that compartment, leading to the disappearance of their patchy distribution. To examine this possibility, DA fibers were lesioned by an intrastriatal injection of 6-OHDA, performed in 2-day old rats. Seven (P9) and 20 (P22) days later, the distribution of D1 binding sites was analyzed on the injected and on the intact side and compared to that of μ opiate binding sites. Locally, in the dorsal striatum,

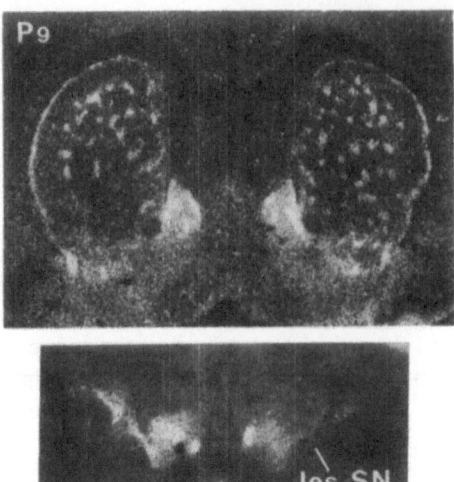

Fig. 2. Dark-field photographs of an autoradiogram illustrating the distribution of 3H-DAGO binding in a coronal section of the striatum (upper part) and of the AChE staining in a coronal section of the substantia nigra (lower part) of a P9 rat which received at P2 a 6-OHDA injection in the right striatum. Note the similar distribution of 3H-DAGO binding in the striatum ipsilateral and contralateral to the 6-OHDA injection and the almost complete disappearance of AChE staining in the substantia nigra ipsilateral to the 6-OHDA injection indicating the degenerescence of DA neurons.

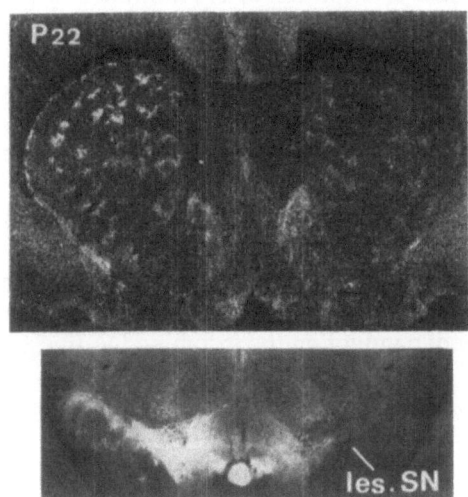

Fig. 3. Dark-field photographs of an autoradiogram illustrating the distribution of 3H-DAGO binding in a coronal section of the striatum (upper part) and of the AChE staining in a coronal section of the substantia nigra (lower part) of a P22 rat which received at P2 a 6-OHDA injection in the right striatum. Note the disorganization of patches of 3H-DAGO binding sites in the striatum ipsilateral to the 6-OHDA injection as compared to the contralateral side, and the almost complete disappearance of AChE staining in ipsilateral substantia nigra.

the 6-OHDA injection produced a complete disappearance of TH-immunoreactive fibers. Distally, in the substantia nigra, a decrease of around 80% of AChE staining (Fig. 2 and 3) and of TH-immunoreactive cell bodies was observed, indicating a retrograde degenerescence of DA neurons.

In P9 rats, no difference in the macroscopic distribution of D1 receptors could be detected between the injected and the intact side, either in the ventral striatum, where the patchy ordering remained unaffected, or in the dorsal striatum where the distribution was homogeneous. Furthermore there was no difference in the density of D1 binding sites between the two sides. The spatial distribution of μ opiate receptors patches and the density of these sites within patches were also not altered 7 days after the 6-OHDA injection (Fig. 2).

Twenty days after the 6-OHDA injection (P22 rats) the distribution and the density of D1 receptors were still unaffected. Indeed, D1 receptors were not any more organized in patches in the lesioned side as in the intact side. Moreover, no difference in the receptor density between these two sides could be detected either in the dorsal or in the ventral parts of the striatum. By contrast, μ opiate receptor patches were not as sharply delineated in the lesioned side as in the intact side. Quantitative measurements show that the density of μ binding sites within patches was significantly lower (-25%) in the injected side than in the intact side (Fig. 3).

The disorganization of the striosomal pattern of D1 binding sites during the second postnatal week, seems to occur independently of the growth of extrastriosomal DA fibers. Furthermore, the supersensitivity of DA receptors usually observed in adult rats after a 6-OHDA lesion of DA neurons (Buonamici et al., 1986) was not apparent suggesting that following lesion of DA neurons in neonatal rats, the reactivity of D1 receptors could be different from that observed in adult rats. Interestingly, and in agreement with our findings, it is noteworthy that the mutant mouse Weaver, which exhibits a failure in the development of the nigrostriatal innervation, possesses a pattern of distribution of D1 binding sites similar to control mice, with a striosomal organization at P7 and an homogeneous distribution at the adult age (Ohta et al., 1989). On this model no supersensitivity of D1 receptors was detected, and even a decrease of D1 binding sites was observed.

Concerning the distribution of μ opiate receptors, a more complex interaction with DA fibers must be stressed. Previously reported data (Pollard et al. 1978) have shown that following a 6-OHDA lesion, there was a large disappearance of μ binding sites in the striatum, and it has been suggested that these receptors were presynaptically localized on DA fibers. These findings were supported by results obtained from autoradiograms indicating a disparition of m opiate receptor patches after a lesion of DA neurons (Gerfen et al., 1987b). However in both types of studies, the effects of 6-OHDA lesion on μ receptor density and distribution were examined one month after the injection. From our data it appears that a disorganization of μ receptor patches associated with a decrease of binding sites is observed when the delay after the injection is prolonged. Early after the DA fiber lesion (one week), while the TH-immunoreactivity has disappeared in the striatum and in the substantia nigra, the distribution of μ receptors is not disorganized and the density is unchanged. This observation suggests that μ opiate receptors are not localized on DA fibers but most likely are postsynaptic to these aminergic fibers. In agreement with this hypothesis is the observation of a dissociation between the patchy appearance of TH-immunoreactive fibers and of μ receptors during ontogenesis. In favour also with a postsynaptic localization of μ receptors is the reduction of these binding sites observed after chronic treatment with haloperidol at prenatal stages (Moon, 1984), or in adult rats after an injection of palmitate ester of pipothiazine (a long lasting neuroleptic) (Trovero et al., submitted). These results taken altogether, indicate however that DA has a crucial role in the expression of μ receptors. This amine could preferentially activate the expression of these receptors within striosomes, leading to their patchy organization.

CONCLUSION

During ontogenesis, incoming DA fibers do not appear to control the macroscopic distribution of D1 receptors, neither during the prenatal period (Foster et al., 1987) nor during the postnatal development. One could suggest that striosomal neurons, expressing first D1 receptors are responsible for the attraction and/or the stabilisation of DA fibers coming from the substantia nigra. However it has been reported that the chronic blockade of DA receptors with haloperidol during the prenatal period does not modify the pattern of distribution of DA fibers (Moon Edley, 1983). Thus functional DA receptors are likely not responsible for the postnatal organization of DA fibers but other factors expressed by striosomal cells might intervene.

On the contrary, the early clusters of DA fibers could induce regulatory processes promoting a differential expression of μ receptors in striosomal neurons leading to the patchy appearance of these receptors. In that case, one might suppose that DA fibers innervating matrix and striosomes and originating from distinct neuronal populations of the substantia nigra (Gerfen et al., 1987a ; Jimenez-Castellanos and Graybiel, 1987) exert different controls on the expression of μ receptors.

ACKNOWLEDGEMENTS

This work was supported by UPMC, CNRS and INSERM n°87-6002.

REFERENCES

Beckstead, R.M., Wooten, G.F., and Trugman, J.M., 1988, Distribution of D1 and D2 dopamine receptors in the basal ganglia of the cat determined by quantitative autoradiography, J. Comp. Neurol., 268 : 131-145.

Besson, M.J., Graybiel, A.M., and Nastuk, M.A., 1988, [3H] SCH 23390 binding to D1 dopamine receptors in the basal ganglia of the cat and primate: delineation of striosomal compartments and pallidal and nigral subdivisions, Neuroscience, 26 : 101-119.

Boyson, S.J., McGonigle, P. and Molinoff, P.B., 1986, Quantitative autoradiographic localization of D1 and D2 subtypes of dopamine receptors in rat brain, J. Neurosci. 6 : 3177-3188.

Buonamici, M., Caccia, C., Carpentieri, M., Pegrassi, L., Ross, A.C. and DiChiara, G., 1986, D1 receptors supersensitivity in the rat striatum after unilateral 6-hydroxydopamine lesion, Eur. J. Pharmacol., 126 : 347-348.

Faull, R.L.M. and Villiger, J.W., 1988, Multiple benzodiazepine receptors in the human basal ganglia : a detailed pharmocological and anatomical study, Neuroscience, 24 : 433-451.

Foster, G.A., Schultzberger, M., Hökfelt, T., Goldstein, M., Hemmings, H.C., Ouimet, Jr C.C., Walaas, S.I. and Greengard, P., 1987, Developpement of a dopamine-

and cyclic adenosine 3', 5'-monophosphate-regulated phosphoprotein (DARPP-32) in the prenatal rat central nervous system, and its relationship to the arrival of presumptive dopaminergic innervation, J. Neurosci., 7 : 1994-2018.

Gerfen, C.R., Herkenham, M., and Thibault, J., 1987a, The neostriatal mosaic : II. Patch and matrix-directed mesostriatal dopaminergic and non-dopaminergic systems, J. Neurosci., 7 : 3915-3934.

Gerfen, C.R., Braimbridge, K.G. and Thibault, J., 1987b, The neostriatal mosaic : III. Biochemical and developmental dissociation of patch-matrix mesostriatal systems, J. Neurosci., 7 : 3935-3944.

Gerfen, C.R., 1989, The neostriatal mosaic: striatal patch-matrix organization is related to cortical lamination, Science, 246 : 385-387.

Graybiel, A.M. and Ragsdale, C.W., 1978, Histochemically distinct compartments in the striatum of human, monkey and cat demonstrated by acetylthiocholinesterase staining, Proc. Natl. Acad. Sci. USA, 75 : 5723-5726.

Graybiel, A.M., and Ragsdale, C.W., 1983, Biochemical anatomy of the striatum. In: Chemical Neuroanatomy, Emson P.L., Ed., Raven Press, New York, 427-504.

Graybiel, A.M., 1984a, Neurochemically specified subsystems in the basal ganglia, In: Functions of basal ganglia, Ciba fundation symposium, Pitman, Eds., London, 104 : 114-149.

Graybiel, A.M., 1984b, Correspondence between the dopamine islands and striosomes of the mammalian striatum, Neuroscience, 13 : 1157-1187.

Herkenham, M. and Pert, C.B. 1981, Mosaic distribution of opiate receptors, parafasicular projections and acetylcholinesterase in rat striatum, Nature, 291 : 415-418.

Jiménez-Castellanos, J., and Graybiel, A.M., 1987, Subdivisions of the dopamine-containing A8-A9-A10 complex identified by their differential mesostriatal innervation of striosomes and extrastriosomal matrix, Neuroscience., 23 : 223-242.

Jiménez-Castellanos, J., and Graybiel, A.M., 1989, Compartmental origins of striatal efferent projections in the cat, Neuroscience, 32 : 297-321.

Lança, A.J., Boyd, S., Kolb, B. and Van Der Kooy, D., 1986, The development of a patchy organization of the rat striatum, Dev. Brain Res., 27 : 1-10.

Moon Edley, S., 1983, Effects of prenatal haloperidol on receptors in the developing rat striatum: opposite changes in naloxone and spiperone binding, Soc. Neurosci. Abstr., 9 : 874.

Moon Edley, S. and Herkenham, M., 1984, Comparative development of striatal opiate receptors and dopamine revealed by autoradiography and histofluorescence, Brain Res., 305 : 27-42.

Moon, S.L., 1984, Prenatal haloperidol alters striatal dopamine and opiate receptors, Brain Res., 323 : 109-113.

Murrin, L.C., and Zeng, W., 1989, Dopamine D_1 receptor development in the rat striatum: early localization in striosomes, Brain Res., 480 : 170-177.

Nastuk, M.A., and Graybiel, A.M., 1985, Patterns of muscarinic cholinergic binding in the striatum and their relation to dopamine islands and striosomes, J. Comp. Neurol., 237 : 176-194.

Ohta, K., Graybiel, A.M., and Roffler-Tarlov, S., 1989, Dopamine D1 binding sites in the striatum of the mutant mouse weaver, Neuroscience, 28 : 69-82.

Pollard, H., Llorens, C., Schwartz, J.C., Gros, C., and Dray, F., 1978, Localization of opiate receptors and enkephalins in the rat striatum in relationship with the nigrostriatal dopaminergic system: lesion studies, Brain Res., 151 : 392-398.

Richfield, E.K., Debowey, D.L., Penney, J.B. and Young, A.B., 1987, Basal ganglia and cerebral cortical distribution of dopamine D1 and D2 receptors in neonatal and adult cat brain, Neurosci. Lett., 73 : 203-208.

Savasta, M., Dubois, A., and Scatton, B., 1986, Autoradiographic localization of D1 dopamine receptors in the rat brain with 3H SCH 23390, Brain Res., 375 : 291-301.

Trovero, F., Hervé, D., Desban, M., Glowinski, J., and Tassin, J.P., Striatal opiate μ receptors are not located on dopamine nerve endings in the rat, (submitted to Neuroscience).

Van Der Kooy, D., 1984, Developmental relationships between opiate receptors and dopamine in the formation of caudate-putamen patches, Develop. Brain Res., 14 : 300-303.

DOPAMINERGIC AND MUSCAREINC CHOLINERGIC RECEPTOR SUBTYPE: LOCALIZATION TO NEUROTRANSMITTERS SPECIFIC COMPONENTS OF THE STRIATUM

James K. Wamsley and Mary A. Hunt

Neuropsychiatric Research Institute

700 First Avenue South, Fargo, North Dakota, USA

INTRODUCTION

Disorders of the striatum (for review see Albin et al., 1989) usually surface as disturbances in motor function (see Delong and Georgopoulos, 1982), although other functions are most certainly associated with the caudate as well. These disorders result in alterations of receptor populations within the caudate-putamen (Seeman et al., 1987). Knowledge of the neurotransmitter specific connections and their accompanying receptors within the striatum would provide a means of predicting the effects of a pharmacological agent, acting at dopaminergic or cholinergic receptor subtypes, on the overall output of the system. Such information would be invaluable in drug development and would add tc the general body of knowledge concerning the striatum's role in movement. One way investigators have sought to understand the receptor specific connections within the striatum associated with dopaminergic and cholinergic receptor subtypes is by combining the techniques of autoradiography (Kuhar et al., 1986) and neurotoxin induced lesions of specific cell populations contributing to the striatal system.

RECEPTOR LABELING

Localization of dopamine receptor subtypes (D1 and D2) within the Central Nervous System (CNS) has been described (see Dawson et al., 1986b; Wamsley et al., 1989). Utilizing the methodology for obtaining specific labeling of each subtype has allowed examination of these receptors following various lesions within the striatum. D1 receptors have been localized using [^{3}H]SCH23390 (Dawson et al., 1985; 1986a) and D2 receptors have been localized using [^{3}H]sulpiride (Gehlert and Wamsley, 1984; 1985) after performing various neurochemical or neurosurgical lesions in the brain (Filloux et al., 1988a; 1988b; Dawson et al., 1988; 1990). The conditions for labeling of dopamine receptors are presented in Table I. D1 receptor activation is associated with an increase in adenylate cyclase activity (see Creese et al., 1983) and changes in [^{3}H]forskolin binding to adenylate cyclase (Gehlert et al., 1984; 1985) has been shown to follow the alterations in D1 receptors seen with the lesions (Gehlert et al., 1987; Filloux et al., 1988a).

Other lesion studies involve the localization of muscarinic receptor subtypes (Dawson et al., 1990; Filloux et al., 1990) which historically were broken down into two major subtypes, M1 and M2 (see Potter et al., 1984; Watson et al., 1986). Localization studies, however, indicate discrepancies in the labeling patterns of ligands thought to be specific for either subtype (Wang et al., 1989). For instance, pirenzepine (an M1 receptor ligand) displacement of tritiated quinuclidinyl benzilate, [^{3}H]QNB (a ligand which should label all subtypes), leaves the radioligand bound to more sites than are displaceable with AFDX116 or AFDX384 (M2 receptor ligands) even after allowing for nonspecific binding (Hunt and Wamsley, unpublished observations). To date, five distinct muscarinic receptor subtypes have been identified by molecular biological techniques (Kubo et al., 1986; Bonner et al., 1987). The messenger RNA for four of these muscarinic receptors (m_1-m_4) have been identified in the brain by in situ hybridization (Buckley et al., 1988). Binding studies show pirenzepine, at low concentrations, preferentially binds to the m_1 sites (Doods et al., 1987). Therefore, localization studies using [^{3}H]pirenzepine identify the m_1 subtypes (Yamamura et al., 1983; Wamsley et al., 1984), while pirenzepine displacement of [^{3}H]QNB binding shows the localization of m_2-m_4 receptors (again, after allowing for nonspecific binding). The m_2 sites can be directly labeled with [^{3}H]AFDX116 (Regenold et al., 1987; Wang et al., 1989). These conditions are shown in Table I.

The presence of dopamine vs. acetylcholine containing cell populations were verified by labeling the terminals of cells for receptor autoradiographic localization of the respective uptake sites (dopamine vs. choline). To localize dopamine terminals and to monitor the effectiveness of lesion of dopamine-containing neurons with 6-hydroxydopamine (6-OHDA), dopamine uptake sites were localized (Filloux et al., 1989) with tritiated [1-(2-benzo(B)thiophenyl) cyclohexyl] piperidine ([^{3}H]BTCP). With regard to the cholinergic lesions (Dawson et al., 1990), the sites associated with sodium dependent-high affinity-choline uptake were localized (Vickroy et al., 1985) with [^{3}H]hemicholinium-3 ([^{3}H]HC-3). Again, the conditions for radiolabeling these various receptor populations can be found in Table I.

LESIONS

To remove the cell bodies of the dopamine neurons (which project from the substantia nigra-zona compacta to the caudate-putamen), 6-OHDA was injected stereotaxically (Filloux et al., 1987a; 1988a) into the medial forebrain bundle (MFB) or substantia nigra. Overlap of this neurotoxin onto the noradrenergic catecholamine system was reduced by pretreating the animals with desipramine. The effectiveness of these lesions was ascertained by examining the animals' response to subcutaneous injection of amphetamines. If an effective lesion were present, the animals would vigorously circle to the ipsilateral side (i.e. injection on the left side of the brain caused the animal to circle to the left). Only animals which vigorously circled in response to the injection of amphetamine were assumed to have an effective lesion and included in the study. Controls consisted of animals injected under the same conditions with vehicle only

Table I

Conditions for Obtaining Labeling of Specific Receptor Sites

Receptor	Tritiated Ligand [ ]	Buffer	Incubation time temp Rinse time temp	Displacer [ ]	Reference
D1	SCH23390 1.0 nM	Tris-HCl (50mM) pH = 7.4 NaCl (120 mM) KCl (5 mM) CaCl$_2$ (2 mM) MgCl$_2$ (1 mM)	30 min 22°C 2x5 min 0-4°C	piflutixol 10 μM	Dawson et al., 1985; 1986
D2	sulpiride 20 nM	Tris-HCl (0.17 M) pH = 7.6 NaCl (120 mM) KCl (5 mM) CaCl$_2$ (2 mM) ascorbate (0.001%)	20 min 22°C 4x1 min 0-4°C	haloperidol 1 μM	Gehlert and Wamsley 1984; 1985
DA uptake	BTCP 10 nM	Tris-HCl (50 mM) pH = 7.0 NaCl (120 mM)	60 min 22°C 4x5 min 0-4°C	GBR12909 1 μM	Filloux et al., 1989
adenylate cyclase	forskolin 20 nM	Tris-HCl (50 mM) pH = 7.5 MgCl$_2$ (5 mM) sucrose (180 mM)	60 min 22°C 2x1 min 0-4°C	forskolin 20 μM	Gehlert et al., 1984; 1985
m$_1$	PZ 20 nM	Kreb's phosphate pH = 7.4	60 min 22°C 3x4 min 0-4°C	atropine 1 μM	Yamamura et al., 1983; Wamsley et al., 1984
m$_2$	AFDX 116 10 nM	Kreb's phosphate pH = 7.4	60 min 22°C 3x4 min 0-4°C	atropine 1 μM	Regenold et al., 1987; Wang et al., 1989

(continues)

Table I (cont.)

Conditions for Obtaining Labeling of Specific Receptor Sites

Receptor	Tritiated Ligand [ ]	Buffer	Incubation time temp Rinse time temp	Displacer [ ]	Reference
m_2-m_4	QNB 1 nM	Kreb's phosphate pH = 7.4	60 min 22°C 2x5 min 0-4°C	atropine 1 μM	Wamsley et al., 1981
choline uptake	HC-3 2.5 nM	phosphate-buffered saline pH = 7.4 NaCl (120 mM) KCl (4.8 mM) MgSO$_4$ (1.2 mM) CaCl$_2$ (1.3 mM) NaH$_2$PO$_4$ (20.3 mM) D-glucose (10 mM)	30 min 22°C 2x1 min 0-4°C	HC-3 1 μM	Vickroy et al., 1985

(minus the neurotoxin). The presence of dopaminergic neurons, within the caudate-putamen, was determined in similarly lesioned animals by labeling sections with [^{3}H]BTCP (see Wamsley et al., 1990).

To remove the cholinergic interneurons of the caudate-putamen, the cholinergic neurotoxin AF64A was utilized (Dawson et al., 1988; 1990). Under appropriate conditions, this cholinotoxin is relatively specific when used very sparingly (Sandberg et al., 1984; McGurk et al., 1987; Dawson et al., 1990). The effectiveness of the lesion was monitored by examining [^{3}H]HC-3 binding in the brain of lesioned and control animals. The results of AF64A lesion of the cholinergic interneurons were then differentiated from the results obtained with a general neurotoxin which would eliminate all of the cell bodies within the caudate-putamen, irrespective of their neurotransmitter. The latter lesion was accomplished by multiple injections of ibotenic acid into parts of the striatum to indiscriminantly remove cell bodies (Filloux et al., 1988a).

Compromising blood flow through the middle cerebral artery produced infarction of the cerebral cortex while sparing subcortical structures including the caudate-putamen (Chen et al., 1986). The lesion reduced the corticostriate neuronal population (as demonstrated by receptor autoradiography) and provided a means for determining associated receptor subtypes (Filloux et al., 1988b; 1990).

CONSEQUENCES OF THE LESIONS

Lesion of the dopaminergic input to the caudate-putamen of the rat resulted in a decrement in D2 receptor binding within the substantia nigra pars compacta (Filloux et al., 1987a; 1988a). However, the D2 receptor population of the caudate-putamen actually increased following 6-OHDA lesion of the MFB. The D1 receptor population did not appear to change initially, but longer time intervals following the lesion show up-regulation in this receptor subtype as well (see Wamsley et al., 1990). Lesion of the cell bodies within the caudate-putamen (with ibotenic acid), after lesion of the dopaminergic input (with 6-OHDA), showed a slight increase in D2 receptor binding in comparison to the effects of the ibotenic acid lesion alone. The 6-OHDA lesion of the MFB resulted in a decrease in muscarinic receptor binding in the ipsilateral caudate-putamen (see Wamsley et al., 1990).

Results of cholinotoxin (AF64A) lesion of cholinergic interneurons within the caudate-putamen, indicated that D2 receptors diminished following this lesion along with non-m_1 muscarinic (presumably m_2) receptors (Dawson et al., 1990). D1 receptors did not change and m_1 receptors increased following the loss of these cholinergic interneurons. Ibotenic acid lesion of the caudate-putamen, on the other hand, resulted in virtually a total loss of D1 receptor binding and forskolin binding to adenylate cyclase in regions of the lesion. A reduction in m_1 receptors and other muscarinic receptor subtypes occurred as well. These same lesions, restricted to the caudate-putamen, also resulted in a dramatic and topographically related loss of D1 receptors and adenylate cyclase within the substantia nigra pars reticulata (Gehlert et al., 1987; Filloux et al., 1988a).

Compromising the corticostriatal input by infarction resulted in the loss of D2 receptors within the caudate-putamen and showed a surprising increase in muscarinic receptor binding (Filloux et al., 1988b; 1990) as well as an increase in glutamate receptor binding (Filloux and Wamsley, unpublished observations).

SURVEY OF THE RESULTS

Receptor autoradiographic techniques provide a means of analyzing potential sites of action of various pharmacological compounds and allow visualization and quantitation of receptor sites for these agents. Combining of localization techniques with lesion of receptor populations within the CNS, provides an opportunity to examine receptor specific connections within the brain. One very important region where this methodology has been applied is the caudate-putamen. Compromising the nigrostriatal and corticostriatal inputs and lesioning various cell groups within the caudate itself, indicate the presence of both pre- and postsynaptic receptors within the striatum (Figure 1).

DOPAMINE RECEPTOR SUBTYPES

Lesion of the dopaminergic projections to the caudate-putamen causes a dramatic reduction in [^{3}H]BTCP binding (see Wamsley et al., 1990) reflecting the loss of dopamine terminals in this structure. This lesion should result in the loss of presynaptic D2 autoreceptors on dopaminergic projections from the substantia nigra pars compacta (see Filloux et al., 1988a). The results of autoradiographic experiments indicate, however, that the D2 receptor population does not diminish, but rather is increased in response to this lesion. Some of the D2 receptors are postsynaptic (actually the majority of the D2 receptor populations are thought to be postsynaptic) and it is possible that, following the lesion, a D2 dopamine receptor denervation supersensitivity occurs and the resulting increase in postsynaptic D2 receptor binding overwhelms the slight decrease in D2 receptors which accompanies the loss of dopamine terminals. In an attempt to address this possibility, double-lesion experiments have been performed (Filloux et al., 1988a). The results of these experiments represent a situation where postsynaptic receptor up-regulation in response to denervation should be reduced. In these animals, the ibotenic acid lesion causes a loss of D2 receptor binding in the caudate-putamen and the subsequent 6-OHDA lesion still results in a slight increase in D2 receptor binding. These data do not support the proposed existence of D2 presynaptic autoreceptors on the terminals of dopamine neurons in the caudate-putamen. A small (perhaps very small) population of these receptors could have been overlooked, however, since other populations of presynaptic D2 receptors would not be affected by the lesion. These other D2 receptors would be expected to up-regulate and could still negate any receptor loss caused by the combined lesions. It is also feasible that the presence of the intact dopamine terminals somehow protects the postsynaptic cells from the effects of the ibotenic acid lesion.

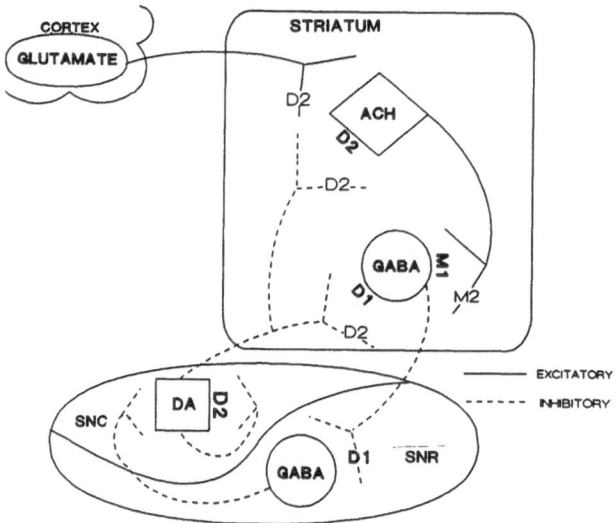

Figure 1. Schematic representation of neuroreceptor relationships in the striatum.

Another D2 receptor population is thought to be present presynaptically on corticostriate fibers and their terminals within the caudate-putamen (see Filloux et al., 1988b). By compromising the blood flow through the middle cerebral artery, it is possible to create a relatively specific cortical lesion which results in minimal subcortical damage (Chen et al., 1986). Utilization of D2 receptor autoradiography following this infarction shows that there is a slight decrement in D2 receptor binding (predominantly in the lateral quadrant of the contralateral caudate-putamen, with a slight decrease in the ipsilateral nucleus as well) verifying the presence of a small population of presynaptic D2 receptors on corticostriate projections (Filloux et al., 1988b). This localization is still controversial, however, since not all autoradiographic studies, employing various techniques to lesion the cortex, result in similar findings (Joyce and Marshall, 1987; Trugman et al., 1986). Lesion of the corticostriate neurons could conceivably cause them to dump their glutamate and result in damage, due to overstimulation, of a small population of cells within the caudate-putamen (thus, reducing postsynaptic D2 receptors). However, the discovery of D2 receptor coding messenger RNA within cells of the cortex (which could project to the striatum) appear to verify the association of D2 receptors with these cells (Weiner and Brann, 1989).

Most of the D2 receptors within the caudate-putamen exist postsynaptically. Lesion of the cholinergic interneurons within the caudate-putamen results in a dramatic loss of D2 receptor binding with no apparent change in D1's (Dawson et al., 1990). Thus, a majority of the D2 receptors are associated with the cholinergic interneurons (Joyce and Marshall, 1985; Fage and Scatton, 1986; Dawson et al., 1988) where they exist postsynaptic to the dopamine terminals of the nigrostriatal system (Scatton 1982; Lehmann and Langer, 1983; Kubota et al., 1987b). A 6-OHDA lesion of the dopaminergic input to the striatum causes an increase in D1 receptor binding and dopamine-stimulated adenylate cyclase activity in the caudate-putamen (Mishra et al., 1974; Bounamici et al., 1986; Wamsley et al., 1990). Likewise, lesion of the caudate-putamen with ibotenic acid, results in virtually a complete loss of D1 receptor binding within the confines of the lesion boundary (Filloux et al., 1987b; 1988a). This lesion-induced receptor loss is mimicked by a reduction in [^{3}H]forskolin binding (Worley et al., 1986; Gehlert et al., 1987; Filloux et al., 1988a) and the D1 receptor phosphoprotein DARPP-32 (Walaas and Greengard, 1984). Thus most, if not all, D1 receptors exist postsynaptically on noncholinergic cell bodies within the caudate-putamen which receive inputs from dopaminergic neurons (Scheel-Krüger, 1986; Kubota et al., 1987a). This same lesion results in an almost total loss (depending directly on the magnitude of the lesion in the caudate-putamen) of D1 receptor binding and forskolin binding in the ipsilateral substantia nigra pars reticulata (Porceddu et al., 1986; Savasta et al., 1986; Gehlert et al., 1987; Filloux et al., 1988a; Wamsley et al., 1990). Thus, the cells that have D1 receptors associated with their somata in the caudate-putamen, also have D1 receptors associated with their terminals in the pars reticulata (striatonigral fibers). There appears to be a topographical relationship between D1 receptors in the reticulata and caudate-putamen. D1 receptors remaining in areas outside of the lesion site resemble those found in the substantia nigra such that it is as if the caudate-putamen was turned on its side with the base of the caudate extending laterally and somewhat dorsally to represent the distribution of receptor sites remaining in the pars reticulata. Similar ibotenic acid lesions of the caudate-putamen result in a decrease of D2 receptor binding of >60% within the structure, without a concomitant (at least in magnitude) reduction in the pars compacta (Filloux et al., 1988a). Thus, most of the receptors of the D2 variety are postsynaptic in the caudate-putamen.

MUSCARINIC RECEPTOR SUBTYPES

Lesion of the dopamine system causes a down-regulation of muscarinic receptor populations within the caudate-putamen (see Wamsley et al., 1990). Presumably this is due to the decrease in inhibition of the cholinergic interneurons caused by the reduction in dopamine release. The lesion results in overactivity of the cholinergic neuron (disinhibition) and the increased release of acetylcholine causes a subsequent down-regulation of the muscarinic receptor populations. Elimination of the corticostriate input, which is presumably glutamatergic (see Albin et al., 1989), causes an up-regulation of glutamate receptor binding in the caudate (presumably due to denervation supersensitivity) and results in an up-regulation of muscarinic cholinergic binding, due to the lack of stimulatory release of acetylcholine (Filloux et al., 1990). Lesion of the cholinergic system with the cholinotoxin AF64A causes a loss (slight) in non-m_1 receptor binding while an increase in m_1 receptor binding results (Dawson et al., 1990). Thus, both pre- and postsynaptic muscarinic receptors exist, but the major population of m_1 receptors appears to be postsynaptic and there appears to be a significant population of presynaptic non-m_1 receptors which exist as autoreceptors on the cholinergic interneuron populations. Ibotenic acid lesion of the caudate-putamen causes a profound decrease in muscarinic receptor binding (see Wamsley et al., 1990), supporting the conclusion that most of the muscarinic receptors, in general, are postsynaptic.

It would appear then that at least a portion of the receptor subtypes associated with the dopaminergic and cholinergic neuronal systems exists on specific cell types within the caudate-putamen. Some of these sites are presynaptic and some are postsynaptic indicating differential effects would result from stimulation or blockade of these sites when appropriate drug compounds are on board. A schematic representation of these putative relationships is presented in Figure 1. The predicted nature of this, when coupled with the wiring diagram being developed for some of the more simplified connections of the caudate-putamen, may have some value in determining how an agent will affect the caudate-putamen and become manifest as a behavioral alteration. These localizations could also be important in determining what receptor change would be appropriate to look for in postmortem tissues where a certain neuronal population is known to be involved in a disease. These changes could be monitored in living brain tissue, utilizing positron emission tomography (PET) scanning techniques, to examine the receptor sites affected by the course of a neuropsychiatric disorder (Wong et al., 1986).

FUTURE TRENDS

It is possible to transplant fetal nigral cells into the denervated striatum of an adult animal. Lesioned animals, which would normally circle vigorously in response to an injection of amphetamine, do not appear to have their motor function disturbed by the drug after receiving a transplant of these fetal mesencephalic cells (see Wamsley et al., 1990). Studies indicate that these cells are growing, extending processes, and establishing functional synaptic connections where they release dopamine within the confines of the denervated striatum (see Bjorklund et al., 1987). The receptor alterations, which have been seen following ibotenic acid lesions of the caudate-putamen (a decrease in both D1 and D2 receptor populations), are restored to normal several months after receiving these nigral grafts. Thus, not only are the mechanics and structural components being established in these denervated tissues by the transplant, but the receptor effects resulting from the lesion also appear to be reversed. This has some very interesting and provocative aspects when one considers that it is possible to culture and cryopreserve fetal mesencephalic cells from various species (Redmond et al., 1988). Release of dopamine from these implants could reverse the receptor changes caused by the disease and hopefully improve the progressive movement disturbances seen with Parkinson's disease.

D1 and D2 receptor alterations may be involved in other neuropsychiatric disorders as well (Seeman et al., 1987). Efforts to address the effectiveness of the blockade of individual dopamine receptor subtypes would benefit from the development of specific antagonists. D2 drugs are available for clinical applications, but adequate D1 antagonists are being developed more slowly (see Barnett et al., 1986; Waddington and O'Boyle, 1989). Schering Corporation has developed a D1 receptor antagonist which appears to be more metabolically stable in primates than were its predecessors, and is very specific for the D1 receptor subtype (Chipkin et al., 1988). This new compound SCH39166 will be available for analysis in patients and, however the results turn out, will be very informative concerning the theories which involve D1 vs. D2 receptors in neuropsychiatric medicine.

CONCLUSIONS

Evidence has been provided for the existence of presynaptic D1 receptors on the terminals of GABAergic striatonigral fibers, D2 receptors on the cell bodies of the dopaminergic nigrostriatal cells, D2 receptors on cholinergic interneurons intrinsic to the striatum, D2 receptors on the terminals of glutamatergic corticostriate fibers within the interstices of the caudate-putamen, and D1 receptors associated with cell bodies (presumably GABAergic) within the striatal system. Muscarinic cholinergic receptor subtypes have been shown to exist presynaptically as cholinergic autoreceptors (non-m_1 receptor subtypes) and postsynaptically on cells intrinsic to the striatum (m_1 and perhaps other subtypes). Knowledge of the relationship of dopaminergic and cholinergic receptors to their respective neurons could indicate a way of predicting a drug effect and ascertaining which drug would be useful to obtain appropriate responses in the system with regard to movement disorders in neuropsychiatric medicine.

ACKNOWLEDGMENTS

The authors wish to thank Karen Meidinger for her excellent secretarial assistance and Diane Nordeng for her help in the library. The authors were supported by grants from the Public Health Service (NS22033, HD22702, and DA05167) during the period when the research reviewed in this article was undertaken.

REFERENCES

Albin, R.L., Young, A.B., and Penny, J.B., 1989, The functional anatomy of basal ganglia disorders, Trends Neurosci., 12:366-375.

Barnett, A., Iorio, L.C., McQuade, R.D., and Chipkin, R.E., 1986, Pharmacological and behavioral effects of D1 dopamine antagonists, in "Central D1 Dopamine Receptors," M. Goldstein, K. Fuxe and I. Tabachnick, ed., Plenum Press, New York, pp. 137-144.

Bjorklund, A., Lindvall, O., Isacson, O., Brundin, P., Wictorin, K., Strecker, R.E., Clarke, D.J., and Dunnett, S.B., 1987, Mechanisms of action of intracerebral neural implants: studies on nigral and striatal grafts to the lesioned striatum, Trends Neurosci., 10:509-516.

Bonner, T.I., Buckley, N.J., Yound, A.C., and Brann, M.R., 1987, Identification of a family of muscarinic acetylcholine receptor genes, Science, 237:527-532.

Bounamici, M., Caccia, D., Carpentieri, M., Pegrassi, L., Rossi, A.C., and Di Chiara, G., 1986, D-1 Receptor supersensitivity in the rat striatum after unilateral 6-hydroxydopamine lesions, Eur. J. Pharmacol., 126:347-348.

Buckley, N.J., Bonner, T.I., and Brann, M.R., 1988, Localization of a family of muscarinic receptor mRNAs in rat brain, J. Neurosci., 8:4624-4652.

Chen, S.T., Hsu, C.Y., Hogan, E.L., Maricq, H., and Balentine, J.D., 1986, A model of focal ischemic stroke in the rat: Reproducible extensive cortical infarction, Stroke, 17:738-743.

Chipkin, R.E., Iorio, L.C., Coffin, V.L., McQuade, R.D., Berger, J.G., and Barnett, A., 1988, Pharmacological profile of SCH39166: A dopamine D1 selective benzonaphthazepine with potential antipsychotic activity, J. Pharmacol. Exp. Ther., 247(3):1093-1102.

Creese, I., Sibley, D.R., Hamblin, M.W., and Leff, S.E., 1983, The classification of dopamine receptors: Relationship to radioligand binding, Ann. Rev. Neurosci., 6:43-71.

Dawson, T.M., Gehlert, D.R., McCabe, R.T., Barnett, A., and Wamsley, J.K., 1986a, D-1 dopamine receptors in the rat brain: A quantitative autoradiographic analysis, J. Neurosci., 8:2352-2365.

Dawson, T.M., Gehlert, D.R., and Wamsley, J.K., 1986b, Quantitative autoradiographic localization of central dopamine D-1 and D-2 receptors, in: "Neurobiology and Central D1-Dopamine Receptors," G.R. Breese and I. Creese, eds., Plenum Press, New York, pp. 93-118.

Dawson, T.M., Gehlert, D.R., Yamamura, H.I., Barnett, A., and Wamsley, J.K., 1985, D-1 dopamine receptors in the rat brain: Autoradiographic localization using [^{3}H]SCH23390, Eur. J. Pharmacol., 180:323-325.

Dawson, V.L., Dawson, T.M., Filloux, F.M., and Wamsley, J.K., 1988, Evidence for dopamine D-2 receptors on cholinergic interneurons in the rat caudate-putamen, Life Sci., 42:1933-1939.

Dawson, V.L., Dawson, T.M., and Wamsley, J.K., 1990, Muscarinic M-2 autoreceptors and postsynaptic D-2 receptors on striatal cholinergic interneurons in the rat brain: An autoradiographic study after intrastriatal injection of the cholinotoxin AF64A, Synapse, [in press].

Delong, M.R. and Georgopoulos, A.P., 1982, Motor functions of the basal ganglia, in: Handbook of Physiology: The Nervous System II, Am. Physiol. Soc., V.B. Brooks Ed., Washington, D.C.

Doods, H.N., Mathy, M.J., Davidesko, D., van Charllorp, K.J., De Jonge, A. and van Zwieten, P.A., 1987, Selectivity of muscarinic antagonists in radioligand and in vivo experiments for the putative M_1, M_2 and M_3 receptors. J. Pharmacol. Exp. Ther., 242:257-262.

Fage, D., and Scatton, B., 1986, Opposing effects of D-1 and D-2 receptor antagonists on acetylcholine levels in rat striatum, Eur. J. Pharmacol., 129:359-362.

Filloux, F., Dawson, T.M., and Wamsley, J.K., 1988a, Localization of nigrostriatal dopamine receptor subtypes and adenylate cyclase, Brain Res. Bull., 20:447-459.

Filloux F., Hunt, M.A., and Wamsley, J.K., 1989, Localization of the dopamine uptake complex using [^{3}H]-[1-(2-benzo(B)thiophenyl) cyclohexyl] piperidine ([^{3}H]BTCP) in rat brain, Neurosci. Lett., 100:105-110.

Filloux, F., Liu, T.H., Hsu, C.Y., Hunt, M.A., and Wamsley, J.K., 1988b, Selective cortical infarction reduces [^{3}H]sulpiride binding in rat caudate-putamen: Autoradiographic evidence for presynaptic D2 receptors on corticostriate terminals, Synapse, 2:521-531.

Filloux, F., Wamsley, J.K., and Dawson, T.M., 1987a, Dopamine D-2 auto- and postsynaptic receptors in the nigrostriatal system of the rat brain: localization by quantitative autoradiography with [^{3}H]sulpiride, Eur. J. Pharmacol., 138:61-68.

Filloux, F., Wamsley, J.K., and Dawson, T.M., 1987b, Presynaptic and postsynaptic D-1 dopamine receptors in the nigrostriatal system of the rat brain: A quantitative autoradiographic study using the selective D-1 antagonist [^{3}H]SCH 23390, Brain Res., 408:205-209.

Filloux, F., Hsu, C.Y., Liu, T.H., Hunt, M.A. and Wamsley, J.K., Selective, unilateral cortical infarction increases striatal muscarinic receptor binding: Potential evidence for cortical modulation of intrastriatal cholinergic transmission, J. Chem. Neuroanat., in press.

Gehlert, D.R., Dawson, T.M., Filloux, F.M., Sanna, E., Hanbauer, I., and Wamsley, J.K., 1987, Evidence that [^{3}H]forskolin binding in the substantia nigra is intrinsic to a striatal-nigral projection: An autoradiographic study of rat brain, Neurosci. Lett., 73:114-118.

Gehlert, D.R., Dawson, T.M., Yamamura, H.I., and Wamsley, J.K., 1985, Quantitative autoradiography of [^{3}H]-forskolin binding sites in the rat brain, Brain Res., 361:351-360.

Gehlert, D.R., Dawson, T.M., Yamamura, H.I., and Wamsley, J.K., 1984, Localization of [^{3}H]-forskolin binding sites in the rat brain using quantitative autoradiography. Eur. J. Pharmacol., 106:223-225.

Gehlert, D.R., and Wamsley, J.K., 1984, Autoradiographic localization of [^{3}H]sulpiride binding sites in the rat brain, Eur. J. Pharmacol., 98:311-312.

Gehlert, D.R., and Wamsley, J.K., 1985, Dopamine receptors in the rat brain: Quantitative autoradiographic localization using [^{3}H]sulpiride, Neurochem. Int., 7:717-723.

Herman, J.P., Choulli, K., Le Moal, M., 1985, Hyper-reactivity to amphetamine in rats with dopaminergic grafts, Exp. Brain Res., 60:521-526.

Joyce, J.N., and Marshall, J.F., 1987, Quantitative autoradiography of dopamine D_2 sites in rat caudate-putamen: Localization to intrinsic neurons and not to neocortical afferents, Neurosci., 20:773-795.

Joyce, J.N., and Marshall, J.F., 1985, Striatal topography of D-2 receptors correlates with indexes of cholinergic neuron localization, Neurosci. Lett., 53:127-131.

Kubo, T., Maeda, A., Sugimoto, K., Akiba, I., Mikami, A., Takahashi, H., Mishina, H., Haga, T., Haga, K., Ichiyama, A., Kangawa, K., Kojima, M., Matuso, M., Hirose, T., and Numa, S., 1986, Cloning, sequencing and expression of complementary DNA encoding the muscarinic acetylcholine receptor, Nature, 323:411-416.

Kubota, Y., Inagaki, S., Kito, S., and Wu, J.-Y., 1987a, Dopaminergic axons directly make synapses with GABAergic neurons in the rat neostriatum, Brain Res., 406:147-156.

Kubota, Y., Inagaki, S., Shimada, S., Kito, S., Eckenstein, F., and Tohyama, M., 1987b, Neostriatal cholinergic neurons receive direct synaptic inputs from dopaminergic axons, Brain Res., 413:179-184.

Kuhar, M.J., DeSouza, E.B., and Unnerstall, J.R., 1986, Neurotransmitter receptor mapping by autoradiography and other methods, Ann Rev. Neurosci., 9:27-59.

Lehmann, J., and Langer, S.Z., 1983, The striatal cholinergic interneuron: Synaptic target of dopaminergic terminals?, Neurosci., 10:1105-1120.

McGurk, S.R., Hartgraves, S.L., Kelly, P.H., Gordon, P.H., and Butcher, L.L., 1987, Is ethylcholine mustard aziridinium ion a specific cholinergic neurotoxin, Neurosci., 222:215-224.

Mishra, R.K., Gardner, E.L., Katzman, R., and Makman, M.H., 1974, Enhancement of dopamine-stimulated adenylate cyclase activity in rat caudate after lesions in the substantia nigra: Evidence for denervation supersensitivity, Proc. Natl. Acad. Sci. U.S.A., 71:3883-3887.

Porceddu, M.L., Giorgi, O., Ongini, E., Mele, S., and Biggio, G., 1986, ^{3}H-SCH23390 binding sites in the rat substantia nigra: Evidence for a presynaptic localization and innervation by dopamine, Life Sci., 39:321-328.

Potter, L.T., Flynn, D.D., Hanchet, H.E., Kalinoski, D.L., Luber-Narod, J., and Mash, D.C., 1984, Independent M_1 and M_2 receptors, ligands, autoradiography and function, Trends Pharmacol. Sci., Suppl 22-31.

Redmond, Jr., D.E., Naftolin, F., Collier, T.J., Leranth, C., Robbins, R.J., Sladek, C.D., Roth, R.H., and Sladek, Jr., J.R., 1988, Cryopreservation, Culture, and Transplantation of Human Fetal Mesencephalic Tissue into Monkeys, Science, 242:768-770.

Regenold, W., Araujo, D. and Quirion, R., 1987, Direct visualization of brain M_2 muscarinic receptors using the selective antagonist [^{3}H]AF-DX116. Eur. J. Pharmacol., 144:417-419.

Sandberg, K., Hanin, I., Fisher, A., and Coyle, J.T., 1984, Selective cholinergic neurotoxin AF64A's effects in rat striatum, Brain Res., 293:49-55.

Savasta, M., Dubois, A., Benavides, J., and Scatton, B., 1986, Different neuronal location of [^{3}H]SCH23390 binding sites in pars reticulata and pars compacta of the substantia nigra in the rat, Neurosci. Lett., 72:265-271.

Scatton, B., 1982, Further evidence for the involvement of D_2, but not D_1 dopamine receptors in dopaminergic control of striatal cholinergic transmission, Life Sci., 31:2883-2890.

Scheel-Krüger, J., 1986, Dopamine-GABA Interactions: Evidence that GABA transmits, modulates, and mediates dopaminergic functions in the basal ganglia and limbic system, Acta. Neurol. Scand., 73:S107.

Seeman, P., Bzowej, N.H., Guan, H.C., Bergeron, C., Reynolds, G.P., Bird, E.D., Riederer, P., Jellinger, K., and Tourtellotte, W.W., 1987, Human brain D_1 and D_2 dopamine receptors in schizophrenia, Alzheimer's, Parkinson's, and Huntington's diseases, Neuropsychopharmacol., 1:5-15.

Trugman, J.M., Geary, II, W.A., and Wooten, G.F., 1986, Localization of D-2 dopamine receptors to intrinsic striatal neurones by quantitative autoradiography, Nature, 323:267-269.

Vickroy, T.W., Roeske, W.R., Gehlert, D.R., Wamsley, J.K., and Yamamura, H.I., 1985, Quantitative light microscopic autoradiographic of [^{3}H]-hemicholinium-3 binding sites in the rat central nervous system: A novel biochemical marker for mapping the distribution of cholinergic nerve terminals, Brain Res., 329:368-378.

Waddington, J.L., and O'Boyle, K.M., 1989, Drugs acting on brain dopamine receptors: A conceptual reevaluation five years after the first selective D-1 antagonist, Pharmacol. Ther., 43(1):1-52.

Walaas, S.I. and Greengard, P., 1984, DARPP-32, A dopamine- and adenosine 3':5'-Monophosphate-regulated phosphoprotein enriched in dopamine-innervated brain regions: I. Regional and cellular distribution in the rat brain, J. Neurosci., 4:84-98.

Wamsley, J.K., Gehlert, D.R., Filloux, F.M., and Dawson, T.M., 1989, Comparison of the density and distribution of D-1 and D-2 dopamine receptors in the rat brain, J. Chem. Neuroanat., 2:119-137.

Wamsley, J.K., Gehlert, D.R., Roeske, W.R., and Yamamura, H.I., 1984, Muscarinic antagonist binding site heterogeneity as evidenced by autoradiography after direct labeling with [^{3}H]-QNB and [^{3}H]-pirenzepine, Life Sci., 34:1395-1402.

Wamsley, J.K., Lewis, M.S., Young III, W.S., and Kuhar, M.J., 1981, Autoradiogrpahic localization of muscarinic cholinergic receptors in rat brainstem, J. Neurosci., 1:176-191.

Wang, J., Roeske, W.R., Hawkins, K.N., Gehlert, D.R., and Yamamura, H.I., 1989, Quantitative autoradiography of M_2 muscarinic receptors in the rat brain identified by using a selective radioligand [^{3}H]AF-DX 116, Brain Res., 477:322-326.

Watson, M., Roeske, W.R., Vickroy, T.W., Smith, T.L., Akiyama, K., Guyla, K., Duckles, S.P., Serra, M., Adem, A., Nordberg, A., Gehlert, D.R., Wamsley, J.K., and Yamamura, H.I., 1986, Biochemical and functional basis of putative muscarinic receptor subtypes and its implications, Trends Pharmacol. Sci., 7:(Supp. II) 46-55.

Weiner, D.M., and Brann, M.R., 1989, The distribution of a dopamine D2 receptor mRNA in rat brain, FEBS Letters, 253:207-213.

Wong, D.F., Wagner, H.N., Tune, L.E., Dannals, R.F., Pearlson, G.D., Links, J.M., Tamminga, C.A., Broussolle, E.P., Ravert, H.T., Wilson, A.A., Toung, J.K.T., Malat, J., Williams, J.A., O'Tuama, L.A., Snyder, S.H., Kuhar, M.J., and Gjedde, A., 1986, Positron emission tomography reveals elevated D_2 dopamine receptors in drug-naive schizophrenics, Science, 234:1558-1563.

Worley, P.F., Baraban, J.M., DeSouza, E.B., and Snyder, S.H., 1986, Mapping second messenger systems in the brain: Differential localizations of adenylate cyclase and protein kinase, C. Proc. Natl. Acad. Sci., 83:4053-4057.

Yamamura, H.I., Wamsley, J.K., Deshmukh, P., and Roeske, W.R., 1983, Differential light microscopic autoradiographic localization of muscarinic cholinergic receptors in the brainstem and spinal cord of the rat using [^{3}H]-pirenzepine, Eur. J. Pharmacol., 91:147-149.

PHARMACOLOGICAL AND MORPHOLOGICAL PROPERTIES OF TWO
FUNCTIONALLY DISTINCT SUBPOPULATIONS OF STRIATAL NEURONS

Theodore W. Berger, Eric S. Nisenbaum, Shao-pii Onn and
Anthony A. Grace

Departments of Behavioral Neuroscience and Psychiatry
Center for Neuroscience, University of Pittsburgh
Pittsburgh, PA 15260 U.S.A.

INTRODUCTION

We have shown recently that two subpopulations of striatal neurons in the rat, Type I and Type II, can be distinguished on the basis of their electrophysiological responses to paired impulse stimulation of cortical afferents (Berger et al., 1987; Nisenbaum et al., 1988b). The experimental paradigm involves recording from single striatal neurons while stimulating the corticostriatal pathway with pairs of electrical impulses; intervals between the impulses of each pair vary systematically, e.g., from 10-1000 ms. The excitatory cortical volley evoked by the first impulse elicits a spike discharge from Type I and Type II neurons. As a consequence of this initial response, however, voltage-dependent conductances of target striatal neurons are activated, and activity is generated within local circuitry (e.g., axon collateral systems and interneurons) and within efferent projections to other brain systems with which the striatum is reciprocally connected (e.g., substantia nigra). These additional events provide the basis for feedforward/feedback modulation of striatal cell response to subsequent cortical input. The second impulse of each pair of impulses reveals the net effect of this modulation. Our analyses have shown that, for a wide range of interstimulus intervals (ISIs), the probabilities of Type I and Type II spike discharge to the second impulse are inversely related, indicating that the two populations are modulated by different feedforward and/or feedback mechanisms.

Two possible sources of feedback to the striatum are the extensive collateral system of intrinsic GABAergic neurons (Fisher et al., 1986) and the dopamine (DA)-containing afferents of the substantia nigra (activated through the striatonigral projection; see Grace and Bunney, 1985). In the experiments reviewed here, we have used both *in vivo* and *in vitro* preparations to investigate the possibility that feedback from GABAergic and DAergic systems is different for Type I and II neurons. If so, manipulations of either neurotransmitter should have a differential effect on the paired impulse functions of the two cell classes. In addition, because the striatum has been shown to consist of several morphologically distinct classes of neurons (Chang et al., 1981), we have used intracellular recording and staining of striatal cells *in vivo* to determine if the functional identity of Type I or Type II neurons is associated with a particular morphology.

METHODS

Preparations

All experiments were conducted using male, Sprague-Dawley rats. For experiments involving *in vivo* preparations, animals were anesthetized with chloral hydrate (400 mg/kg) throughout electrophysiological recording. For experiments involving *in vitro* preparations, animals first were anesthetized, then decapitated, and their brains were removed rapidly and immersed in a 2° C ringer solution (concentrations in mM: NaCl 125.0, KCL 5.0, MgCl 1.5, KH$_2$PO$_4$ 1.25, CaCl$_2$ 2.0, NaHCO$_3$ 26.0, glucose 10.0, pH 7.3-7.4). Using a vibratome, horizontal sections 350-700 μm in thickness were cut through the dorsoventral extent of the brain, maintaining intact the striatum and the adjacent cortex and its white matter.

Extracellular Recording

The extracellular responses of single Type I and Type II neurons were recorded from the anterodorsal portion of the striatum using either single- or three-barrel glass micropipettes; recording barrels were filled with 2M NaCl, and had input impedances of 1-10 MΩ. For iontophoretic experiments, each of the remaining two barrels was filled with either 5 mM bicuculline methiodide or 4 M NaCl for neutralization of tip currents. For both *in vivo* and *in vitro* preparations, bipolar stimulation electrodes were placed in the white matter containing the corticostriatal pathway. All cells were isolated by their orthodromic response to single stimulation impulses (100-200 μs duration, biphasic) applied to the corticostriatal pathway.

Intracellular Recording and Staining

Experiments involving intracellular recording and dye-filling were conducted using *in vivo* preparations only. Recording electrodes were filled with 3M KAc or 10% Lucifer yellow in distilled water, and typically had tip impedances of 100-250 MΩ. After neurons were identified as Type I or Type II, Lucifer yellow was injected into the neurons for 5-15 min using 1 nA hyperpolarizing constant current impulses interrupted by depolarizing impulses delivered at 7 Hz for 5-15 min. Animals then were perfused with 500 ml of 4% paraformaldehyde or 10% formalin. Serial sections (25-75 μm in thickness) were cut in a frontal or sagittal plane on a freezing microtome and cleared in dimethylsulfoxide (DMSO) (Grace and Llinás, 1985). Sections were examined with fluorescence microscopy, using a Leitz I3 filter cube (excitation-emission wavelength of 438-535 nm).

Paired Impulse Stimulation

For extracellular recording experiments, a series of impulse pairs with variable inter-impulse intervals were delivered by computer control to bipolar stimulation electrodes implanted in the corticostriatal pathway. Three ranges of inter-impulse intervals were used depending on the experiment: 10-250 ms, 10-500 ms, and 10-1000 ms. Pairs of impulses were delivered once every 5 s, and alternated between short and long intervals. Stimulation intensity was adjusted so that the probability of discharge to the first impulse of each pair was approximately 0.5 (threshold). In response to each stimulation impulse, responses were digitized continuously (5-20k Hz) and collected and stored for later analysis. The following subsequently was calculated and plotted for each cell: 1) the probability of discharge in response to the first impulse of all pairs of impulses, which was defined as the baseline response, and 2) the probabilities of discharge to the second impulse of each impulse pair as a function of ISI.

Paired impulse stimulation of corticostriatal afferents also was used to identify the functional subtype of intracellularly recorded neurons, though the range of inter-impulse intervals tested was more limited. For most experiments, ISIs were 10-20 ms and 100 ms; for a number of experiments, intervals of 10-100 ms were used. Only spike responses were used to determine the paired impulse profile, so as to remain comparable with the extracellular recording studies. Cells showing an inhibition or facilitation to 10-20 ms and a facilitation to 100 ms were classified as Type II neurons. Cells exhibiting facilitation to 10-20 ms and inhibition to 100 ms were classified as Type I neurons.

Striatal DA Depletions

DA-depleting lesions were produced by administering 20 μl (10 μl in each hemisphere) containing 6-hydroxydopamine hydrobromide (6-HDA) (100-250 μg dissolved in 0.9% NaCl and 0.1% ascorbic acid), into the lateral ventricles. All animals were pretreated with atropine methylnitrate (10 mg/kg, i.p.), pargyline (40 mg/kg, i.p.), and desmethylimipramine (25 mg/kg, i.p.). After recording was completed, each animal was sacrificed and its brain was removed rapidly, covered with crushed dry ice, and stored at -70° C for subsequent analysis of striatal DA content using HPLC.

RESULTS

Defining Characteristics of Type I and Type II Striatal Cell Responses to Paired Impulse Stimulation of the Corticostriatal Pathway *in vivo*

Paired impulse stimulation of the corticostriatal pathway elicits markedly different responses from Type I (N = 35) and Type II (N = 292) striatal cells over a wide range of ISIs. Type I striatal cells (Figure 1) display an increased probability of spike discharge to the second impulse in response to ISIs of approximately 10-30 ms (average maximum = 0.84 ± 0.04; mean ± S.E.M.). In response to ISIs of 50-240 ms, Type I neurons display a robust inhibitory response (average maximum = 0.09 ± 0.04. Following this inhibition, a second facilitative response is seen for ISIs of approximately 270-750 ms. Thus, the paired impulse curves generated by Type I cells are characterized by a facilitation-inhibition-facilitation response.

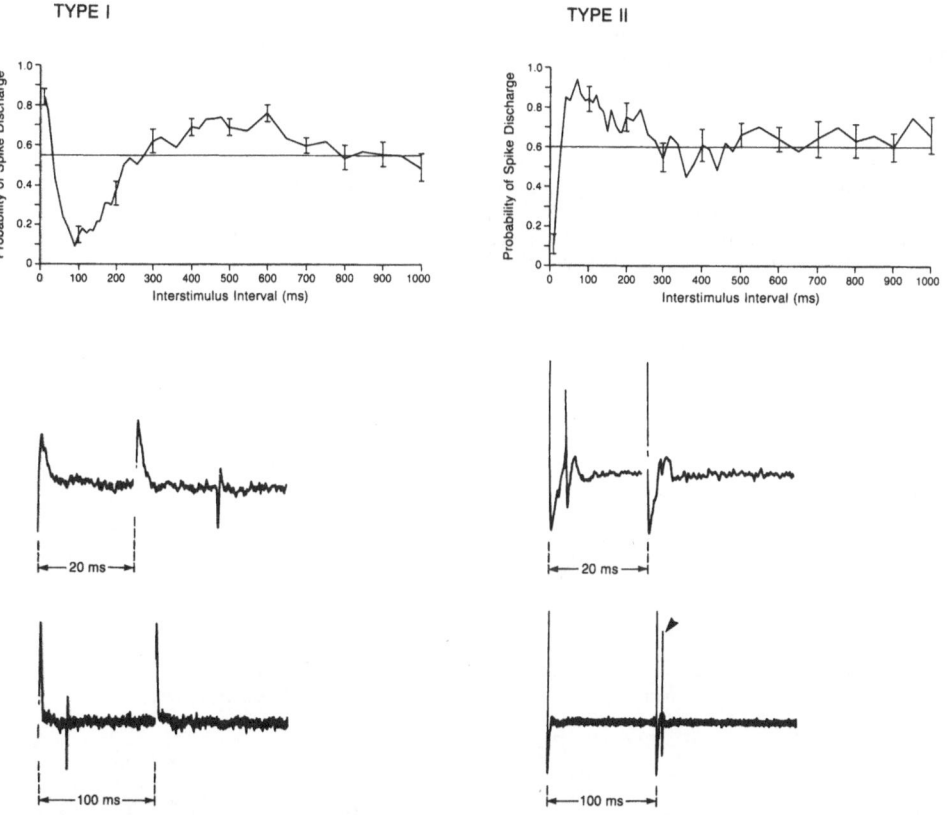

Figure 1. Paired impulse responses of Type I and Type II striatal neurons to stimulation of the corticostriatal pathway. The average paired impulse responses for Type I and Type II neurons are presented in the upper panels. In this and all similar figures, the horizontal lines are equal to the probability of action potential discharge to the first impulse of all pairs of impulses, and the probabilities of spike discharge to the second impulse of each pair are presented as a function of ISI. The error bars indicate standard errors of the means. Note that the range of ISIs was 10-1000 ms. Representative examples of Type I and Type II responses to select ISIs 20 ms and 100 ms at threshold stimulation intensities are presented in the lower panels. As has been previously reported Type I neurons consistently exhibit negative-positive waveforms and Type II neurons display positive-negative waveforms (Nisenbaum et al., 1988b; Skirboll and Bunney, 1979). Calibration bars represent 100 µV and either 5 or 20 ms (Type I) and 200 µV and either 5 or 20 ms (Type II). Arrow denotes Type II spike discharge.

337

The paired impulse profile of Type II neurons is qualitatively different and inversely related to that of Type I cells (Figure 1). Approximately 50% of Type II neurons show an initial decrease in the probability of spike discharge in response to ISIs of 10-30 ms (average maximum = 0.11 ± 0.13). All Type II neurons display a pronounced facilitatory response to ISIs of 30-250 ms such that the average probability of spike discharge approaches 1.0 (0.94 ± 0.04). Following this facilitation, and again in contrast to Type I cells, Type II neurons exhibit a decrease in probability of firing to intervals of approximately 280-500 ms. Thus, Type II cells respond to paired impulse stimulation with an inhibition-facilitation-inhibition pattern. Finally, striatal cells exhibiting the Type II paired impulse profile are encountered more frequently (approximately 90% of all neurons sampled) than neurons displaying the profile characteristic of Type I cells.

Effects of Bicuculline on the Paired Impulse Responses of Type I and Type II Neurons *in vivo*

The relative contribution of GABA$_A$ receptors to the functional differences between Type I (N = 19) and Type II (N = 14) neurons was investigated by examining the effects of iontophoretically applied bicuculline. The spontaneous firing rates of Type I cells were markedly affected by 5 nA of bicuculline, increasing by an average 360% (± 131). In contrast, iontophoresis of bicuculline did not alter the spontaneous firing rate of Type II cells, even with ejection currents of 25 nA, though threshold for spike discharge was decreased.

Despite the robust effects on spontaneous activity, bicuculline did not alter the paired impulse responses of Type I neurons (Figure 2). Type I cells continued to display a facilitation of spike discharge in response to ISIs of 10-30 ms and an inhibition to intervals of 50-240 ms even when drug ejection currents were increased from 5 to 25 nA. In some experiments, bicuculline was applied continuously to Type I cells for approximately 25 min, with no change in the paired impulse responses.

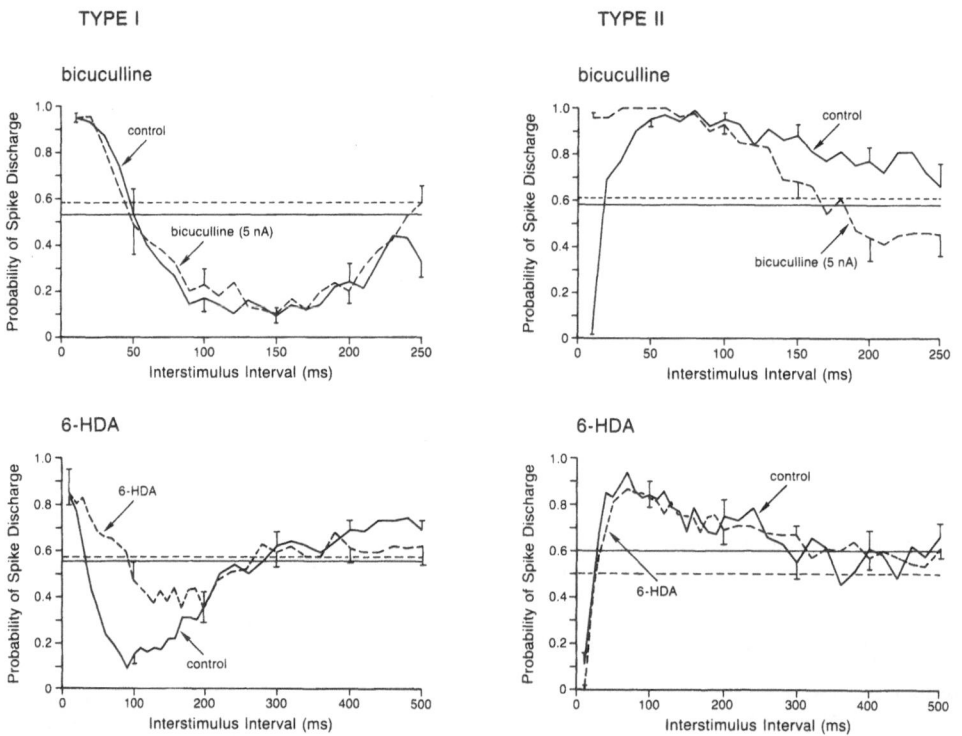

Figure 2. Effects of iontophoretically applied bicuculline and injection of 6-HDA on the paired impulse responses of Type I and Type II neurons. Note that the range of ISIs for bicuculline and 6-HDA experiments was 10-250 ms and 10-500 ms, respectively.

In contrast, the paired impulse responses of Type II neurons clearly were affected by iontophoretic application of bicuculline (Figure 2). The inhibition of Type II neurons in response to short ISIs (10-20 ms) was completely antagonized by 5 nA of bicuculline. Blockade occurred within 10 s after drug ejection began, and this disinhibitory response persisted until the application of bicuculline was terminated. The effects of bicuculline were relatively selective for the short-interval inhibition of Type II neurons, although the range of facilitation (10-160 ms) also was reduced when compared with control values (20-240 ms).

The results of these studies indicate that GABAergic elements within the striatum make no detectable contribution to the paired impulse responses of Type I neurons. In contrast, GABAergic afferents have a relatively selective and robust effect on the paired impulse responses of Type II neurons, such that interruption of GABAergic input completely blocks the short-interval inhibition of these cells. The bicuculline-induced increase in the spontaneous firing of Type I cells and the decrease in spike threshold for Type II cells suggests that both cell classes receive a tonic GABAergic input.

Effects of DAergic Denervation on the Paired Impulse Responses of Type I and Type II Neurons *in vivo*

The contribution of DAergic afferents to the paired impulse profiles of Type I (N = 10) and Type II (N = 20) neurons was examined in animals with near-total striatal DA depletions (95-99%) produced with the neurotoxin, 6-HDA. Electrophysiological recordings were completed within 1 week after 6-HDA administration. For Type I cells, DA depletion was associated with both an increase in spontaneous firing rate (Orr et al., 1987) and an alteration in paired impulse response characteristics. Although facilitation in response to short ISIs was similar to that observed in control animals, the inhibition of spike discharge to longer ISIs was reduced from an average of 0.09 ± 0.04 in control animals to 0.35 ± 0.05 in 6-HDA-treated animals (Figure 2). The interval associated with maximum inhibition was 170 ms in 6-HDA-treated animals, compared to 90 ms in control animals.

Type II cells also exhibited a significant increase in their spontaneous discharge rates after DA depletion (Orr et al., 1986; Nisenbaum et al., 1986). In contrast to Type I cells, however, the paired impulse responses of Type II neurons were altered only with respect to ISIs of 10-20 ms (Figure 2). The average maximum probability in response to short ISIs decreased from 0.35 ± 0.07 in control animals to 0.08 ± 0.05 in 6-HDA treated animals. Moreover, the percentage of Type II cells exhibiting short-interval (10-20 ms) inhibition increased from approximately 50% in control animals to approximately 90% in animals treated with 6-HDA. For Type II neurons displaying such inhibition, the magnitude of this inhibitory response was increased from 0.11 ± 0.05 in control animals to 0.005 ± 0.01 in 6-HDA-treated animals (data not shown). In contrast to these changes, Type II cells still displayed a peak facilitatory response at a 70 ms ISI, however, striatal DA depletions increased the range of this facilitation from 30-250 ms in control animals to 30-460 ms in 6-HDA-treated animals.

These results demonstrate that both Type I and Type II subpopulations of striatal neurons are influenced by DAergic input from the substantia nigra. The predominant effects were a decrease in the magnitude of the long-interval inhibition of Type I cells and an increase in the percentage of Type II neurons exhibiting short-interval inhibition.

Paired Impulse Responses of Striatal Neurons in an *in vitro* Corticostriatal Slice Preparation

In the corticostriatal slice preparation that we developed recently (Nisenbaum et al., 1988a), the adjacent cortex and white matter are kept intact in order to preserve cortical afferents to the striatum, so that these afferents can be selectively stimulated in a manner similar to that used *in vivo*. Using this preparation, the paired impulse profiles of *in vitro* neurons then could be compared with the responses of Type I and Type II cells in order to determine how reduction of the intrinsic GABAergic collateral network and DAergic feedback from the substantia nigra altered the responses of *in vitro* neurons. Results from the *in vivo* experiments described above would predict that the short-interval inhibition of Type II neurons would be reduced, and that the inhibition of Type I cells in response to long ISIs also would be attenuated.

The responses of striatal neurons *in vitro* (N = 25) to paired impulse stimulation of the corticostriatal pathway were remarkably homogeneous. In control slices, striatal cells displayed a marked increase in probability of spike discharge to the second impulse in response to short ISIs (10-30 ms) (Figure 3). Moreover, this facilitation of spike discharge persisted in response to longer ISIs of 40-350 ms, and for some neurons the probability of spike discharge never completely returned to baseline values (within the range tested, 10-500 ms). In contrast to the responses of Type I and Type II neurons *in vivo*, paired impulse stimulation of the corticostriatal pathway rarely elicited inhibitory responses from striatal neurons *in vitro*. Only 2 of 25 cells *in vitro* displayed inhibition in response to short ISIs similar to that seen for Type II neurons. In addition, none of the neurons *in vitro* exhibited long-interval inhibition similar to that characteristic of Type I neurons.

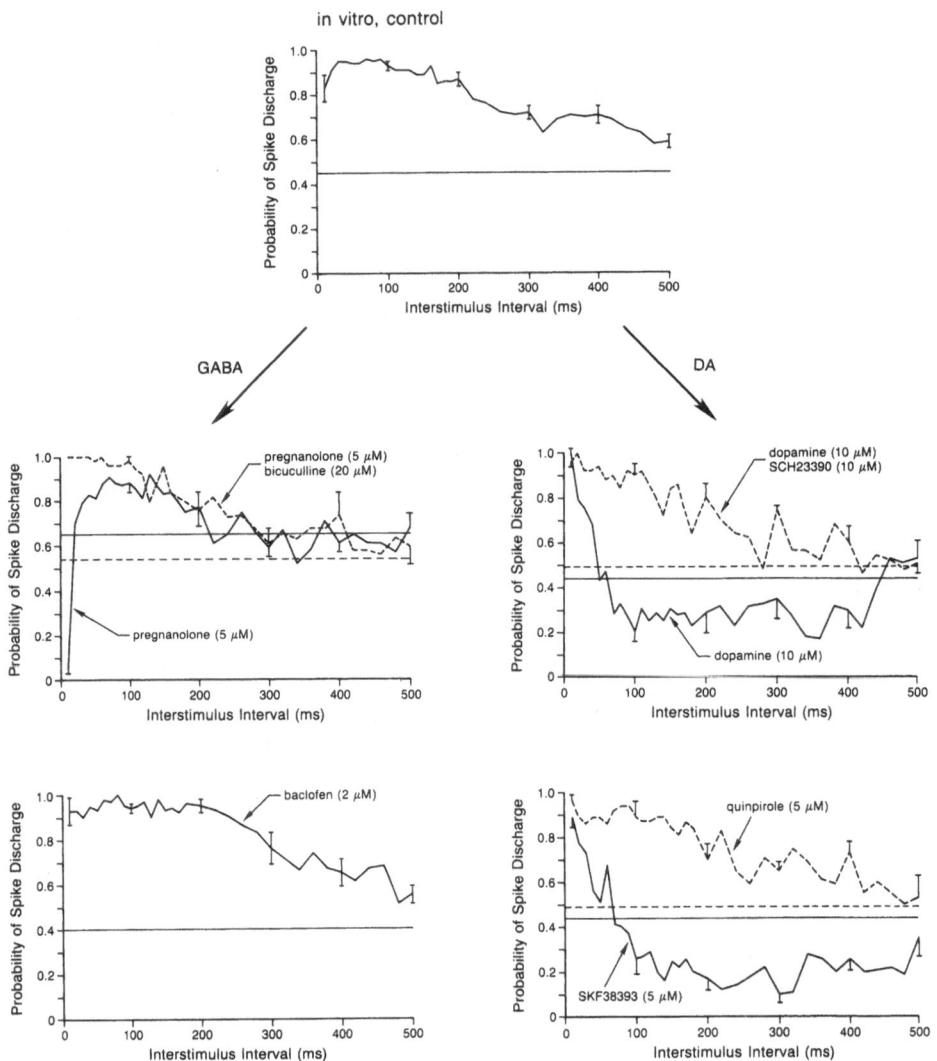

Figure 3. Paired impulse responses of *in vitro* striatal neurons to stimulation of the corticostriatal pathway. The responses of *in vitro* cells in control media and in the presence of specific GABAergic and DAergic receptor agonists and antagonists are presented. Note that the range of ISIs was 10-500 ms.

The absence of any inhibition in response to the wide range of ISIs intervals tested is consistent with the hypothesis that much of the GABAergic and DAergic input has been severed in this slice preparation. Subsequent experiments were designed to examine the extent to which enhancing GABAergic and DAergic activity could reinstate the distinguishing characteristics of Type I and Type II neurons *in vitro*.

Effects of GABAergic Agonists on the Paired Impulse Responses of Striatal Neurons *in vitro*: Identification of Type II-like Neurons

Contribution of GABA$_A$ Receptors. The initial effect of the allosteric GABA$_A$ agonists, pentobarbital (50 µM) and pregnanolone (5 µM), was an increase in the threshold for spike discharge (N = 31). When stimulation intensity was increased to a level sufficient for threshold activation to the first impulse, approximately 50% of striatal neurons *in vitro* exhibited a pronounced inhibition in response to short ISIs (10-20 ms), similar to that of Type II neurons *in vivo* (Figure 3). Also similar to the responses of Type II cells, spike discharge remained facilitated for ISIs of 30-200 ms. Higher concentrations of pentobarbital (75-100 µM) or pregnanolone (10 µM) increased the range of intervals (extending to 200 ms) associated with inhibition, though the magnitude of inhibition remained inversely proportional to the ISI. Bicuculline blocked the effects of both pentobarbital and pregnanolone, (N = 18), suggesting that the drug-induced inhibitory response was mediated by activation of GABA$_A$ receptors. Thus, in the presence of an appropriate dose of GABAergic allosteric agonist, the majority of striatal neurons *in vitro* displayed paired impulse responses similar to the profiles of Type II striatal neurons *in vivo* (Figure 4).

Possible consequences of the higher stimulation intensities required in the presence of pentobarbital and pregnanolone was evaluated in control slices by using intensities sufficient to evoke a spike discharge with a probability >0.80. No increase in the magnitude of short-interval inhibition was observed. Thus, short-interval inhibition in slices treated with pentobarbital or pregnanolone was not due to effects associated with increasing stimulation current (e.g., recruitment of additional fibers).

Contribution of GABA$_B$ Receptors. To further determine the specificity of the pregnanolone and pentobarbital effects, the paired impulse responses of neurons were examined in slices treated with baclofen (1-5 µM), a GABA$_B$ receptor agonist. Similar to the effects of the GABA$_A$ receptor agonists, baclofen produced an increase in the threshold for spike discharge. In contrast, baclofen had no effect on paired impulse response profiles (N = 10) (Figure 3). These results indicate not only that the effects of pregnanolone and pentobarbital are mediated by GABA$_A$ receptors, but also provide additional evidence that the enhancement of short-interval inhibition by GABA$_A$ receptor agonists is not due solely their effects on spike threshold.

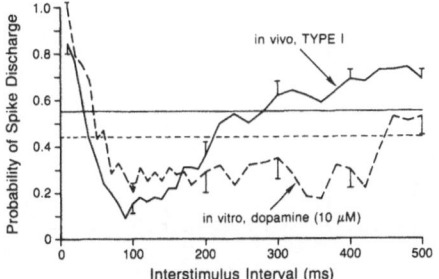 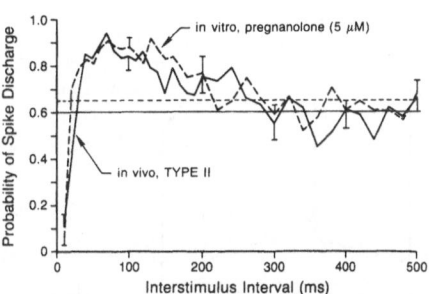

Figure 4. Identification of putative Type I and Type II neurons *in vitro*. Enhancement of DAergic activity *in vitro* revealed in a small population of neurons, paired impulse responses similar to those of Type I neurons *in vivo*. In addition, following potentiation of GABAergic input, *in vitro* neurons displaying profiles similar to Type II neurons *in vivo* were observed.

Effects of DAergic Agonists on the Paired Impulse Responses of Striatal Neurons *in vitro*: Identification of Type I-like Neurons

The effects of enhancing DAergic activity on the paired impulse responses of neurons *in vitro* was investigated by adding exogenous DA to the superfusion media. Similar to the responses of neurons in control media, all cells displayed an increase in the probability of spike discharge in response to short ISIs of 10-40 ms in the presence of 10 µM DA (Figure 3). However, in contrast to cells in control slices, a small percentage (approximately 10%) of cells *in vitro* exhibited a pronounced inhibition in response to intervals of 70-420 ms. The probability of spike discharge returned to baseline only in response to the longest intervals tested (440-500 ms). Thus, in the presence of DA, a small population of striatal neurons *in vitro* displayed paired impulse responses similar to the profiles of Type I neurons *in vivo* (Figure 4).

The effect of 10 µM DA was blocked by the addition of 10 µM of SCH23390, a D_1 receptor antagonist, to the superfusion media (Figure 3). Facilitation to shorter intervals was unaffected. In contrast, 10 µM of sulpiride, a D_2 receptor antagonist, had no effect on DA-induced inhibition. The D_1 agonist, SKF38393 (5 µM), but not the D_2 agonist quinpirole (5-10 µM), produced a similar long-interval inhibitory response, also for a small (10%) population of neurons *in vitro* (Figure 3). These results indicate that the DA-induced long-interval inhibition of neurons *in vitro* is mediated by stimulation of D_1, and not D_2 receptors.

Morphological Characteristics of Type I and Type II Neurons

In an attempt to identify morphological characteristics that may distinguish Type I and Type II striatal cells, we have recorded intracellularly from striatal neurons *in vivo* (N = 48) and injected cells from both classes with Lucifer yellow. Both Type I and Type II neurons have been identified, with the majority of cells (39/48) exhibiting a Type II paired impulse profile and approximately 20% (9/48) exhibiting a Type I profile (Figure 5).

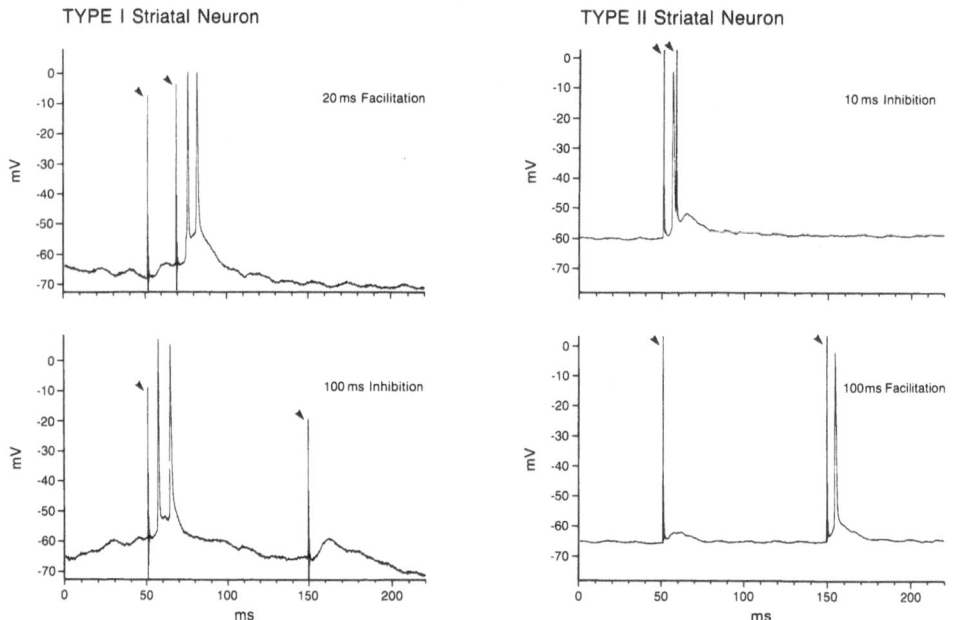

Figure 5. Intracellular responses of Type I and Type II neurons to paired impulse stimulation of the corticostriatal pathway. Representative examples of Type I and Type II responses to select ISIs of 20 ms and 100 ms (Type I) and 10 and 100 ms (Type II) are presented. Arrows denote stimulus artifacts.

All Type II neurons successfully recovered to date have had a morphology similar to that of Spiny I neurons (Chang et al., 1981): medium-sized somata of oval or bipolar shape which give rise to 6-8 primary dendrites. The secondary and tertiary processes are densely spinous and are angular or recursive in nature. Type I neurons stained to date have exhibited a more variable morphology, i.e., they seemed to consist of a variety of different cell types. Five neurons classified as Type I neurons shared some characteristics of Spiny I cell morphology: medium-sized soma giving rise to 3-4 primary dendritic branches with spines on distal dendritic processes. The remaining four identified Type I cells had a morphology unlike that of Spiny I neurons: medium to large-sized soma of multipolar shape with aspinous or sparsely spined dendritic branches.

DISCUSSION

The results of these experiments reveal that both GABA and DA are major determinants of the paired impulse response profiles that distinguish Type I and Type II striatal neurons. Studies investigating the role of GABA found that the inhibition of Type II cells to short ISIs is mediated by $GABA_A$ and not $GABA_B$ receptors. Neither $GABA_A$ nor $GABA_B$ receptors influence the facilitation of Type II neurons to longer ISIs. These results indicate that GABA is released phasically onto Type II neurons within the first 20 ms following cortical stimulation, i.e., a potent $GABA_A$-mediated inhibitory input to Type II cells occurs via a feedforward and/or feedback pathway. Initial results from intracellular recording and dye-filling experiments reveal that the vast majority of cells exhibiting the Type II paired impulse profile have morphological characteristics of Spiny I neurons, which also have been shown to be GABA-containing (Fisher et al., 1986; Kita and Kitai, 1988) and are the source for dense collateral plexus that ramifies among neighboring striatal neurons (Preston et al., 1980). Thus, Type II paired impulse profiles may represent functional characteristics of GABAergic Spiny I neurons, with the inhibition to short ISIs due to input from collaterals of neighboring Spiny I cells.

In contrast to Type II neurons, the paired impulse response of Type I cells is completely unaffected by $GABA_A$ receptor blockade. However, Type I neurons receive a potent $GABA_A$-mediated input, because the spontaneous firing rate of this cell class is increased by several fold in the presence of bicuculline. Thus, $GABA_A$-mediated inhibition of Type I cells also is substantial, but is provided by neuronal elements outside feedforward or feedback pathways activated by cortical input. As a consequence, only tonic firing rate is affected by bicuculline.

The fact that bicuculline markedly altered the evoked responses of Type II neurons without changing the evoked responses of Type I cells may be relevant to the issue of functional interactions between the two striatal cell classes. If it is assumed that the iontophoretic application of bicuculline affected both Type I and neighboring Type II cells (studies in progress using micro-injection of bicuculline have replicated the effects reported here for Type I cells), our results suggest that Type II cells do not project to Type I cells. It remains possible that Type I neurons provide synaptic input to Type II cells. If the hypothesis is correct that Type II neurons are GABA-containing, our findings also suggest that a second class of GABAergic neuron (non-Type II) provides inhibitory input to Type I cells.

DA also influenced the paired impulse profiles of Type I and Type II neurons, but manipulations of DAergic input affected different ranges of ISIs for the two cell classes. Depletion of striatal DA attenuated the inhibition of Type I neurons in response to long ISIs (50-240 ms), suggesting that phasically active DAergic feedback from substantia nigra contributes to this inhibitory response. However, in the presence of either DA or a D_1 receptor agonist, long-interval inhibition characteristic of Type I cells *in vivo* could be induced for a subpopulation of striatal neurons *in vitro*, a condition in which the cell bodies of DA-containing terminals remain severed from their terminals. These results suggest that effects of DA on Type I neurons are mediated by a presynaptic action or by modification of a DA-sensitive property of Type I cell membranes (e.g., voltage-dependent conductance). Surprisingly, D_2 receptor-mediated activity appeared to be without effect on the paired impulse profiles of Type I or Type II cells.

DA depletion also selectively altered the inhibition of Type II cells in response to short ISIs. However, the results of experiments using bicuculline showed that this inhibition is due to GABA$_A$ receptor activation. A concomitant effect of DA depletion was an increase in the excitability of Type II cells, as evidenced by the increase in spontaneous firing rate. If the inhibition to short ISIs is mediated by collateral projections among Type II neurons, then this inhibitory response may increase after DA depletion due to an increase in the number of Type II neurons activated per impulse.

In summary, these findings demonstrate that both GABA and DA play critical roles in determining the dynamics of striatal cell responses to cortical input. Procedures such as those outlined here, as well as more theoretically based analytical approaches (Berger et al., 1988), should allow a systematic evaluation of the individual contribution of each of the neurotransmitters identified for striatal neurons and their afferents, as well as a determination of the functional consequences of their interactions.

ACKNOWLEDGEMENTS

This research was supported by grants from NINDS (NS19608), NIMH (MH00343), (MH09717), (NS08288), (MH42217).

REFERENCES

Berger, T.W., Nisenbaum, E.S., Stricker, E.M., and Zigmond, M.J. Evidence for two functionally distinct subpopulations of neurons within striatum and their differential sensitivity to dopamine. In: Neurophysiology of Dopaminergic Systems: Current Status and Clinical Perspectives, L.A. Chiodo and A.S. Freeman (Eds.), Detroit: Lake Shore Publishers, 253-284 (1987).

Berger, T.W., Sidney, J.P., Nisenbaum, E.S., and Sclabassi, R.J. Nonlinear response characteristics of striatal neurons to random impulse train stimulation of cortical afferents. Soc. Neurosci. Abstr., 14: 75 (1988).

Chang, H.T., Wilson, C.J., and Kitai, S.T. A Golgi study of rat neostriatal neurons: Light microscopic analysis. J. Comparative Neurology, 208: 107-126 (1982).

Fisher, R.S., Buchwald, N.A., Hull, C.D., and Levine, M.S. The GABAergic striatonigral neurons of the cat: demonstration by double peroxidase labeling. Brain Res., 398: 148-156 (1986).

Grace, A.A. and Bunney, B.S. Opposing effects of striatonigral feedback pathways on midbrain dopamine cell activity. Brain Res., 333: 271-284 (1985).

Grace, A.A. and Llinás, R. Dehydration-induced morphological artifacts in intracellularly stained neurons: circumvention using rapid DMSO clearing. Neurosci., 16: 461-475 (1985).

Kita, H. and Kitai, S.T. Glutamate decarboxylase immunoreactive neurons in rat neostriatum: their morphological types and populations. Brain Res., 447: 346-352 (1988).

Nisenbaum, E.S., Stricker, E.M., Zigmond, M.J. and Berger, T.W. Long-term effects of dopamine-depleting brain lesions on spontaneous activity of Type II striatal neurons: relation to behavioral recovery. Brain Research, 398: 221-230 (1986).

Nisenbaum, E.S., Grace, A.A., and Berger, T.W. Responses of striatal neurons to paired impulse stimulation in a corticostriatal slice preparation. Soc. Neurosci. Abstr., 14: 75 (1988a).

Nisenbaum, E.S., Orr, W.B. and Berger, T.W. Evidence for two functionally distinct subpopulations of neurons within the rat striatum. J. Neurosci., 8: 4138-4150 (1988b).

Orr, W.B., Gardiner, T.W., Stricker, E.M., Zigmond, M.J., and Berger, T.W. Short-term effects of dopamine-depleting brain lesions on spontaneous activity of Type II striatal neurons: relation to local striatal dopamine levels and behavior. Brain Res., 376: 20-28 (1986).

Orr, W.B., Stricker, E.M., Zigmond, M.J., and Berger, T.W. Short-term effects of dopamine-depleting brain lesions on spontaneous activity of Type I striatal neurons: relation to local dopamine levels and behavior. Synapse, 1: 461-469 (1987).

Preston, R.J., Bishop, G.A., and Kitai, S.T. Medium spiny neuron projection from the rat striatum: an intracellular horseradish peroxidase study. Brain Res., 183: 253-263 (1980).

Skirboll, L.R., and Bunney, B.S. The effects of acute and chronic haloperidol treatment on the spontaneously firing neurons in caudate nucleus of the rat. Life Sci., 25: 1419-1434 (1979).

INFLUENCE OF CORTICO-STRIATAL GLUTAMATERGIC NEURONS ON

DOPAMINERGIC TRANSMISSION IN THE STRIATUM

J. GLOWINSKI, L. BARBEITO and A. CHERAMY

COLLEGE DE FRANCE - INSERM U 114
11, place Marcelin Berthelot
75231 PARIS cedex 05 - France

INTRODUCTION

Several studies performed on rat striatal slices and in some cases on synaptosomes have indicated that dopamine (DA) and most neurotransmitters found in the striatum regulate presynaptically the release of DA from nerve terminals of the nigro-striatal DA neurones (chesselet, 1984). These neurotransmitters include excitatory or inhibitory amino-acids, amines and several neuropeptides found either in afferent fibers, interneurones or in collaterals of efferent projections from the striatum. These regulations are either direct or indirect, ie mediated through receptors located on DA nerve terminals or on neurones in contact with these nerve terminals. This has been mainly shown by examining the persistance or disappearance of these presynaptic regulations in the presence of tetrodotoxin (TTX), a neurotoxine which interrupts nerve impulse flow. There is some evidence that some neurotransmitters may diffuse at some distance from their release sites before their inactivation. Nevertheless, the physiological significance of some presynaptic regulations of DA release has been often challenged by neuroanatomists since up to now few axo-axonic contacts have been observed between DA neurones and other neurones innervating the striatum (Bouyer et al., 1984). In vivo studies, in which identified pathways projecting to the striatum can be activated are of great value for determining the physiological relevance of these presynaptic regulations. Therefore, we will first summurize results of in vivo experiments made in halothane anesthetized cats implanted with push-pull cannulae which have allowed to demonstrate the role of the cortico-striatal glutamatergic projection in the presynaptic control of DA release. We will also describe more recent experiments made either in vivo or in vitro in which attempts were made to identify the types of glutamatergic receptors involved in the presynaptic control of DA release and to distinguish direct and indirect effects mediated by local circuits.

The Basal Ganglia III, Edited by G. Bernardi *et al.*
Plenum Press, New York

In vivo evidence for a role of the cortico-striatal glutamatergic projection in the presynaptic control of dopamine release in the cat caudate nucleus

As expected, the application of a high concentration of GABA (10^{-3}M) into thalamic motor nuclei induces an inhibition of the firing rate of a majority of cells in these nuclei (Romo et al., 1986a). However, some of these cells are activated when a small concentration of GABA (10^{-5}M) is used (Romo et al., 1986a). This is likely due to the higher sensitivity to GABA of presynaptic receptors which control the release of GABA from afferent fibers. In fact, we have shown that the unilateral application of this small concentration of GABA (10^{-5}M) into motor thalamic nuclei (fig 1) produces an activation of a thalamo-cortico striatal neuronal loop which involves the bilateral cortico-striatal glutamatergic projection. Indeed, this treatment induces a marked and prolonged increase in the release of glutamate in both caudate nuclei and the contralateral effect is prevented by a section of the corpus callosum (fig 2) (Romo et al., 1984). Interestingly enough, parallel changes in the release of DA can be seen in these conditions. The GABA local treatment enhanced the release of ^{3}H-DA continuously formed from ^{3}H-tyrosine in both caudate nuclei and an ipsilateral response was seen only in cats with an acute section of the corpus callosum as observed previously following the unilateral electrical stimulation of the peri-cruciate cortex (Nieoullon, et al., 1978). This indicates that the experimental paradigm used is suitable for investigation on the role of striatal glutamatergic neurones in the presynaptic control of DA release.

Other experiments revealed that the increased release of DA evoked in both caudate nuclei by the unilateral application of GABA (10^{-5}M) into motor thalamic nuclei is abolished following the ablation of the peri-cruciate motor cortex and that the enhanced release of DA in the ipsilateral caudate nucleus is completely suppressed by the local application of riluzole (PK 26124), a compound which prevents glutamatergic transmission (Romo et al., 1986b). It should be added that the changes in DA seen in both caudate nuclei under the unilateral application of GABA (10^{-5}M) into motor thalamic nuclei were seen in spite of the inhibition of the firing rate of DA cells. This was shown by measuring simultaneously the activity of nigral DA neurons by extracellular recording (Romo et al., 1986a). Finally, the evoked release of DA in the ipsilateral caudate nucleus was still observed following the acute transection of the nigro-striatal DA pathway, a procedure, which as expected, decreased markedly the spontaneous release of DA by interrupting completely the propagation of nerve impulse flow (Romo et al., 1986a). All together these observations demonstrate that the activation of cortico-striatal glutamatergic neurones facilitates presynaptically the release of DA from nerve terminals of the nigro-striatal DA neurones.

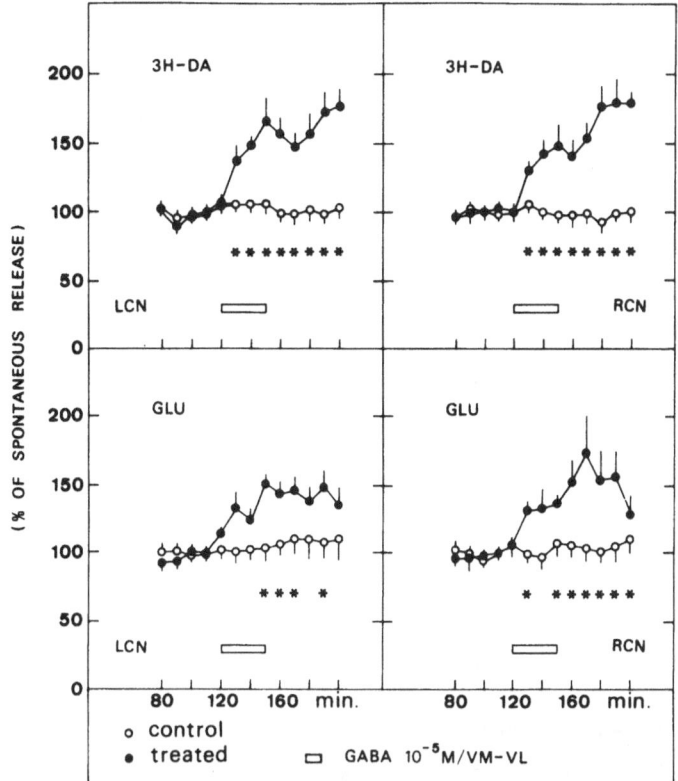

Fig 1 . Effects of application of GABA into the left thalamic
motor nuclei on the release of 3H-dopamine and
glutamate in the two caudate nuclei.

In unlesioned halothane-anaesthetized cats, an artificial CSF
containing 3H-tyrosine (50 µCi/ml) was delivered continuously
(25 µl/min) to push-pull cannulae implanted into the left
(LCN) and right (RCN) caudate nuclei. Another push-pull
cannula was inserted either into the left ventralis medialis
or ventralis lateralis (VM-VL) thalamic nuclei and GABA was
added for 30 min into the superfusion medium (open box), 2
hours after the onset of superfusion. In each animal 3H-
dopamine (3H-DA, upper panels) and glutamate (GLU, lower
panels) were measured in successive 10 min superfusate
fractions from LCN and RCN. Results are expressed as a
percentage of the average spontaneous release calculated from
5 fractions collected before GABA application. Data are the
mean ± SEM of results obtained in 7 animals (black circles)
* p<0.05 when compared with corresponding control values
obtained in 7 untreated animals (open circles).

Experiments were carried out and the results expressed as in Fig 1 except that an anterior sagittal section (corpus callosum, septum pellucidum, fornix and commissura anterior) was made just before the implantation of push-pull cannulae. In each case, the mean ± SEM of data obtained from six

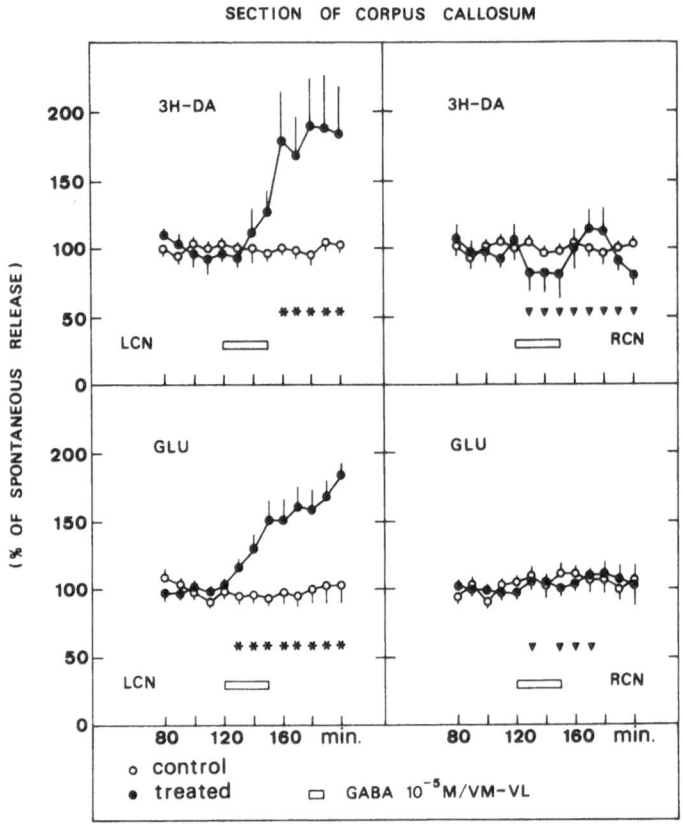

SECTION OF CORPUS CALLOSUM

Fig 2 . Effect of an anterior sagittal section on the release of ³H-dopamine and glutamate induced by GABA application into the left thalamic motor nuclei.

control (sectioned but not GABA-treated) or six treated (sectioned and GABA treated) cats was calculated. * p<0.05 when compared to corresponding control values. Black triangles: p<0.05 when data from sectioned and GABA-treated, animals are compared with those obtained from GABA-treated animals without section (Fig 1).

Types of glutamate receptors involved in the direct presynaptic control of dopamine release by cortico-striatal glutamatergic fibers

When applied locally into the cat caudate nucleus, glutamate stimulates the release of newly synthesized ^{3}H-DA through calcium-dependent and TTX-resistant processes (Cheramy et al., 1986). These in vivo observations are similar to those previously seen on rat striatal slices (Giorguieff et al., 1977; Roberts and Sharif, 1978; Roberts and Anderson, 1979) and favor a direct presynaptic control of DA release by cortico-striatal glutamatergic neurones. In fact, there is anatomical evidence for the existence of axo-axonic contacts between cortico-striatal neurones and nigro-striatal DA fibers (Bouyer et al., 1984).

In the presence of TTX, the in vivo stimulatory effect of glutamate on DA release is concentrations-dependent and can be seen with concentrations as low as 10^{-6}M. A TTX-resistant-evoked release of DA was also induced by kainate (10^{-5}M) but not by quisqualate (10^{-5}M) or NMDA (10^{-5}M, in the presence of Mg^{++}) (Barbeito et al., 1990). The stimulatory effects of glutamate and kainate were antagonized not only by riluzole (10^{-5}M) but also by glutamate diethyl ester (10^{-5}M) and gamma-D-glutamyl-glycine (10^{-5}M) (Barbeito et al., 1990). The lack of effect of quisqualate seems to be due to a rapid desensitization of the receptors as indicated by experiments made with concanavalin A, a lectin which has been shown to prevent the quisqualate-induced desensitization of glutamate receptors on striatal neurones in primary culture. Indeed, the in vivo application of concanavalin A into the caudate nucleus allowed the detection of the excitatory effect of quisqualate on DA release (fig 3) (Barbeito et al., 1990). Furthemore, in the absence of concanavalin A, the prior application of quisqualate (10^{-4}M) prevented the stimulatory effects of both glutamate (10^{-4}M) and kainate (10^{-5}M) on DA release (fig 3). These results suggest that receptors of a quisqualate/kainate subtype are present on DA nerve terminals. More definitive evidence for a localization of these receptors on DA nerve terminals has been obtained recently by using purified synaptosomes of the rat striatum. Kainate stimulates in a calcium-dependent manner the release of newly synthesized ^{3}H-DA in this preparation. This effect is concentration-dependent and can be antagonized by both riluzole (10^{-5}M) and the prior application of quisqualate (10^{-5}M) which alone is without effect (Desce et al., in preparation).

As indicated previously, in normal conditions (presence of Mg^{++} in the superfusion fluid), NMDA does not stimulate the release of DA through a TTX-resistant process in the cat caudate nucleus when used at a concentration of 10^{-5}M. Very likely, such an effect should be seen with higher concentrations of the agonist. In fact, other experiments indicate that besides quisqualate/kainate receptors, NMDA receptors are also involved in the presynaptic control of DA release.

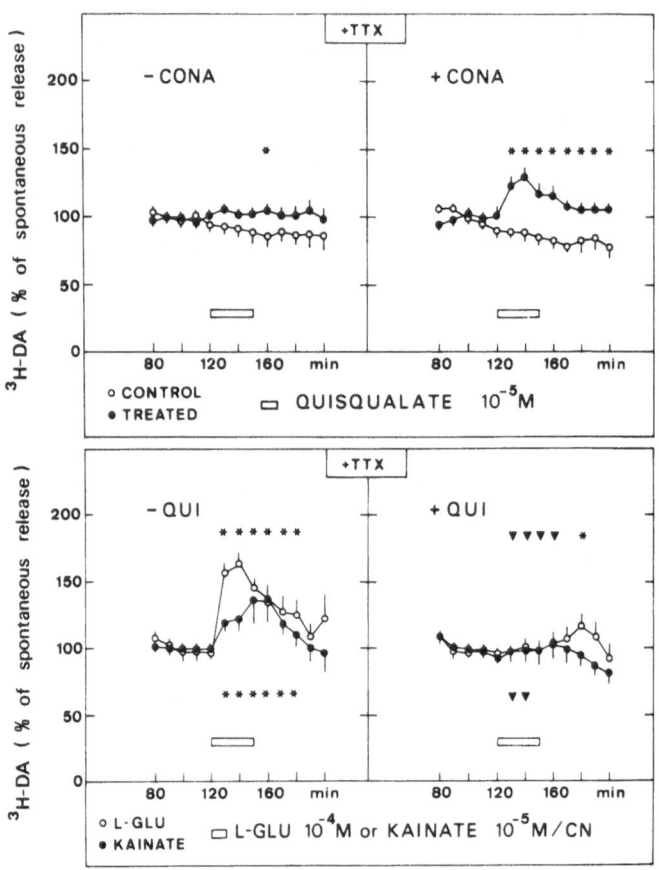

Fig 3. A single quisqualate/Kainate receptor substype is
involved in the direct (tetrodotoxine resistant)
glutamate-evoked release of ^{3}H-dopamine.

An artificial CSF containing ^{3}H-tyrosine (50 µCi/ml) and TTX
(10^{-6}M) was delivered through a push-pull cannula implanted
into the left caudate nucleus (CN) of halothane-anaesthetized
cats. ^{3}H-DA was estimated in 10 min fractions of superfu-
sates.
Upper panels : The CSF contained (+ CONA) or not (-CONA)
concanavalin A(10^{-7}M), quisqualate (QUI, 10^{-5}M) being added
for 30 min (open box). Open circles: controls; black circles:
QUI-treated.
Lower panels : Kainate (KAI, 10^{-5}M, black circles) or
glutamate (GLU, 10^{-4}M, open circles) were added (open box) in
the presence (+QUI) or absence (-QUI) of quisqualate (10^{-5}M),
QUI being applied for 45 min, 15 min before KAI or GLU.
Results are expressed as a percentage of the average
spontaneous release and are the mean ± SEM of data obtained
from groups of 6 to 16 animals. * p<0.05 when compared to
corresponding controls. Black triangles: p<0.05 when compared
to data obtained with KAI or GLU in animals not pretreated
with QUI (left lower panel).

Indeed, using a new superfusion procedure allowing to study the release of transmitters from discrete areas of striatal slices, we have recently observed that NMDA (5.10^{-5} M) stimulates the release of newly synthesized ^{3}H-DA from rat striatal slices and that this response is potentiated by glycine (10^{-6} M) through strychnine-resistant sites (Krebs et al., 1989). The NMDA-evoked release of ^{3}H-DA was blocked by MK 801 and by the addition of Mg^{++} and the potentiating effect of glycine could be reversed by kynurenate (Krebs et al., 1989). The NMDA-evoked response was observed both in the presence or in the absence of TTX in agreement with results of other workers. Recently, we have obtained results similar to those found in striatal slices in the presence of TTX using a purified preparation of striatal synaptosomes (Desce et al., in preparation). This demonstrates further that a population of NMDA receptors located on DA nerve terminals contribute also to the presynaptic control of DA release.

Evidence for indirect presynaptic regulations of dopamine release by glutamate and glutamate agonists

Elegant anatomical studies have demonstrated that the cortico-striatal glutamatergic neurones innervate the medium sized spiny neurones which represent the largest population of efferent cells in the striatum (Bolam, 1984). These neurones contain GABA and in addition peptides which may act as co-transmitter. Depending on the population of GABAergic efferent neurones involved, tachykinins (substance P, substance K), enkephaline and/or dynorphine may be co-localized with GABA (Gerfen, 1984; Graybiel, 1986). Since these efferent neurones have numerous collaterals within the striatum, they could regulate either directly or indirectly the release of DA by presynaptic processes. Indeed GABA and the above mentioned peptides have been shown to influence in vitro the spontaneous or the evoked release of DA (Chesselet, 1984). Similarly, several data indicate that the cortico-striatal neurones innervate the striatal cholinergic neurones and it is well established that ACh controls presynaptically both directly and indirectly the release of DA (Bolam, 1984, Chesselet, 1984). Therefore besides its direct effect on DA nerve terminals, glutamate released from cortico-striatal glutamatergic neurones could also indirectly modulate the release of DA.

Three groups of observations indicate indeed that glutamate may also intervene indirectly in the presynaptic control of DA release. 1) In vivo, in the absence of TTX, glutamate used in high concentration (10^{-4} M) decreased the spontaneous release of ^{3}H-DA in the cat caudate nucleus. This effect was prevented by riluzole but also by bicuculine indicating that GABAergic containing neurones contribute to this inhibitory response (Cheramy et al., 1986). 2) In contrast to that observed in the presence of TTX and in a way similar to that described for glutamate, kainate did not stimulate the release of ^{3}H-DA but exerted a slight inhibitory effect in the cat caudate nucleus (Barbeito et al., 1990). 3) A shown in vitro on striatal slices, in some areas enriched in matrix, NMDA stimulates much more the release of DA in the absence than in the presence of TTX. In addition, this response was amplified markedly in the presence of cholinergic antagonists (Krebs et al., in preparation). Although further experiments are required in order to elucidate precisely the nature of the local circuits involved in the indirect presynaptic control of DA release, these data suggest already that glutamate may act through different subtypes of receptors.

REFERENCES

Barbeito, L., Cheramy, A., Godeheu, G., Desce, J.M. and Glowinski, J. 1990, Glutamate receptors of the quisqualate-kainate subtype are involved in the presynaptic regulation of dopamine release in the cat caudate nucleus in vivo, Eur. J. Neuroscience, (in press).

Barbeito, L., Girault, J.A., Godeheu, G., Pittaluga, A., Glowinski, J., and Cheramy, A., 1989, Activation of the bilateral cortico-striatal glutamatergic projection by infusion of GABA into thalamic motor nuclei in the cat: an in vivo release study. Neuroscience, 28:365-374.

Bolam, J.P. 1984, Synapses of identified neurons in the neostriatum. in: Functions of the Basal Ganglia (eds Evered D. and O'Connors M.) pp 30-47, Pitman, London.

Bouyer, J.J., Park, D.H., Joh, T.H., and Pickel, V.M., 1984, Chemical and structural analysis of the relation between cortical inputs and tyrosine hydroxylase-containing terminals in rat neostriatum. Brain Res., 302:267-275.

Cheramy, A., Romo, R., Godeheu, G., Baruch, P., and Glowinski, J., 1986, In vivo presynaptic control of dopamine release in the cat caudate nucleus; II. Facilitatory or inhibitory influence of L-glutamate. Neuroscience, 12:1081-1090.

Chesselet, M.F., 1984, Presynaptic regulation of neurotransmitter release in the brain: Facts and hypothesis. Neuroscience, 12:347-275.

Gerfen, C.R., 1984, The neostriatal mosaic: Compartmentalisation of corticostriatal input and striatal output systems. Nature, 311:461-463.

Giorguieff, M.F., Kemel, M.L., and Glowinski, J., 1977, Presynaptic effect of L-glutamic acid on the release of dopamine in rat striatal slices. Neurosci. Lett. 6:73-77.

Graybiel, A.M., 1986, Neuropeptides in the basal ganglia. in: Neuropeptides in Neurologic and Psychiatric Diseases (J.B. Martin and J.D. Barchas, Eds.), Raven Press, New York, pp 135-161.

Krebs, M.O., Kemel, M.L., Gauchy, C., Desban, M., and Glowinski, J., 1986, Glycine potentiates the NMDA-induced release of dopamine through a strychnine-insensitive site in the rat striatum. Europ. J. Pharmacol., 166:567-570.

Nieoullon, A., Cheramy, A., and Glowinski, J., 1978, Release of dopamine evoked by electrical stimulation of the motor and visual areas of the cerebral cortex in both caudate nuclei and in the substantia nigra in the cat. Brain Res., 145:69-83.

Roberts, P.J., and Anderson, S.D., 1979, Stimulatory effect of L-glutamate and related amino acids on ³H-dopamine release from rat striatum: an in vitro model for glutamate actions. J. Neurochem. 32:1539-1545.

Roberts, P.J., and Sharif, N.A., 1978, Effects of L-glutamate and related amino acids upon the release of ³H-dopamine from rat striatal slices. Brain Res., 157:391-395.

Romo, R., Cheramy, A., Godeheu, G., and Glowinski, J., 1984, Distinct commissural pathways are involved in the enhanced release of dopamine induced in the contralateral caudate nucleus and substantia nigra by the unilateral application of GABA in the cat thalamic motor nuclei. Brain Res., 308:43-52.

Romo, R., Cheramy, A., Godeheu, G., and Glowinski, J., 1986a, In vivo presynaptic control of dopamine release in the cat caudate nucleus: I. Opposite changes in neuronal activity and release evoked from thalamic motor nuclei. Neuroscience, 19:1091-1099.

Romo, R., Cheramy, A., Godeheu, G., and Glowinski, J., 1986b, In vivo presynaptic control of dopamine release in the cat caudate nucleus. III. Further evidence for the implication of cortico-striatal glutamatergic neurons. Neuroscience, 19:1091-1099.

MODULATION OF STRIATAL DOPAMINE AND ACETYLCHOLINE
RELEASE BY DIFFERENT GLUTAMATE RECEPTORS: STUDIES
WITH IN VIVO MICRODIALYSIS

Mario Herrera-Marschitz

Department of Pharmacology, Karolinska institutet
Stockholm, Sweden

INTRODUCTION

There is evidence that striatal dopamine (DA) release is
presynaptically modulated by a glutamatergic cortical input (see[1,2,3]).
In the caudate of the push-pull cannulated cats[4], as well as in striatal
slice preparations in rats[5], glutamate stimulates DA release. It was
proposed[4], that the DA stimulation produced by glutamate reflected
direct axonal interactions between cortical glutamatergic and
mesencephalic dopaminergic afferents, since the stimulating effect of L-
glutamate was still observed in the presence of tetrodotoxin. This
hypothesis has received some support from biochemical and histochemical
studies showing direct intrastriatal axonal interactions between
cortico-striatal projections and nigro-striatal DA terminals labeled with
antibodies against the enzyme tyrosine hydroxylase[5,6]. However, the
majority of the striatal afferents from the cortex and the substantia
nigra make axodendritic synaptic contacts with striatal neurons[7], giving a
basis for polysynaptic loops including gamma-aminobutyric acid (GABA)
and/or acetylcholine (ACh) neurons, by which cortical glutamate neurons
could also modulate striatal dopamine release. There is now evidence that
the excitatory action of glutamate is conveyed by multiple receptors,
which have been pharmacologically differentiated and characterized, i.e.
N-methyl-D-aspartate (NMDA)-, quisqualate- and kainate-receptors. All
these receptors are found in the striatum (see[8]), but they appear located
on different cells or associated with different synapses, thereby related
to different functions. Indeed, it has been shown, that, in the cat
caudate, cortically evoked monosynaptic excitatory postsynaptic
potentials (EPSPs) are mediated by kainate or quisqualate, but not by NMDA
receptors[9]. The present studies with **in vivo** microdialysis[10] show that, in
the striatum, in situ kainic acid (KA) stimulation increases DA, but
decreases ACh. However, while NMDA is less potent that KA on DA release,
it induces a dose-dependent increase in striatal AC release. Furthermore,
it was found that cortical stimulation with either KA or NMDA produced a

dose-dependent increase in striatal ACh release, but only a minor increase in striatal DA release.

EXPERIMENTAL PROCEDURES

Microdialysis- Male Sprague-Dawley rats weighing around 500 g were anaesthetized with halothane and placed in a David Kopf stereotaxic frame. The anaesthesia was maintained by free breathing into a mask fitted over the nose of the rat (1.0-1.5% of halothane in an air flow of 1.5 l/min). Two microdialysis probes (CMA 10, Carnegie Medicin AB, Stockholm, Sweden) (dialysing length= 4 mm; diameter= 0.5 mm) were implanted: one diagonally into the left fronto-parietal cortex (coordinates, according to Zilles atlas[11], B 1.7, L -1.5, V 5.0; inserted with a 40° angle from vertical in the coronal plane); the other vertically into the lateral portion of the corpus of the left striatum (coordinates: B 0.7, L -3.5, V 7.5). The microdialysis probes were perfused with Ringer (147 mmol Na^+, 2.3 mmol Ca^{2+}, 4 mmol K^+ and 155.6 mmol Cl^-, pH adjusted to approximately 7 by degassing with helium), Ringer containing 10 μM Neostigmine (Sigma, St. Lous, Mo, USA) and Ringer containing neostigmine and KA (Sigma) or NMDA (Sigma). A constant flow of 2 μl/min was maintained with a microdialysis pump (CMA 100, Carnegie Medicin AB). All probes had an **in vitro** recovery of 15-25% for both ACh and DA. Changes in the perfusion medium were performed by using a liquid switch (CMA 110, Carnegie Medicin AB), which made possible to instantaneously shift syringes containing different perfusion media without introducing air into the system. On completion of the microdialysis experiment, the brain was rapidly dissected out and stored in 10% formaldehyde for confirmation of the location of the dialysis probes.

Biochemistry -HPLC assays- DA and metabolites were assayed in 20 min fractions on a reverse phase ion-pair High Performance Liquid Chromatography (HPLC) system (BAS, West Lafayette, IN, USA) with electrochemical detection (for details see[12]). ACh and Ch were determinated in 10 μl aliquots using a column-reactor system (BAS). ACh and Ch were first separated on a polymeric column using a sodium phosphate buffer containing 1 mM sodium octanesulfonate as ion-pairing reagent. An enzymatic post column reactor with immobilized acetylcholinesterase and Ch oxidase converted ACh and choline to hydrogen peroxide and betaine. Hydrogen peroxide was electrochemically detected at a platinium electrode poised at 500 mV (vs Ag/AgCl) (for details see[13]).

Statistics. Means and standard errors of the means (SEM) have been calculated. Changes in DA and ACh release are expressed as the percentage over the values obtained during the immediately precedent control period. Differences between the means were tested by Student's t-test for replicated data. A level of $P < 0.05$ for the one tail test was considered critical for assigning significant statistical differences.

RESULTS

In situ stimulation: Table 1 shows the changes in striatal ACh and DA release produced by KA or NMDA included in the microdialysis perfusion medium for a 20 min period. A vehicle treated group is also shown, indicating in parenthesis the ACh and DA concentrations found at the 220-240 min control period.

Table 1. Changes (%)[a] in striatal dopamine (DA) and acetylcholine (ACh) release during **in situ**[b] 20 min stimulation with either kainic acid (KA) or N-methyl-D-aspartate (NMDA).

Treatment	N	ACh	DA
Vehicle	6	103 ± 3	90 ± 9%
		(231 ± 31 nM)[c]	(10 ± 1 nM)[c]
KA 10^{-5}M	5	165 ± 29% *	146 ± 23% *
KA 10^{-4}M	5	69 ± 8% *	418 ± 99% **
KA 10^{-3}M	5	62 ± 14% *	628 ± 115%**
NMDA 10^{-5}M	5	205 ± 44% *	111 ± 7%
NMDA 10^{-4}M	4	388 ± 37% *	138 ± 15% *

[a] Changes in DA and ACh release are expressed as the percentage over the values obtained during the immediately precedent control period.
[b] 240 min after the microdialysis implantation, KA or NMDA was included for 20 min in the perfusion medium of a microdialysis probe into the left striatum.
[c] ACh or DA release (perfusion medium= Ringer including 10 μM neostigmine; halothane anaesthesia) measured at the 220-240 min control perfusion period.
* P < 0.05; ** P < 0.01.

After **in situ** KA stimulation striatal DA release was dose-dependently increased, while ACh release was decreased. At the same concentrations, NMDA produced only a minor increase in striatal DA release, while ACh was significantly increased.

Cortical stimulation: Since the rats were simultaneously implanted with two microdialysis probes, one into the frontoparietal cortex and the other into the ipsilateral striatum, substances could be independently administered to these brain structures and the effects on each structure be simultaneously monitored. Table 2 shows the effect of a 20 min cortical KA or NMDA stimulation on striatal ACh and DA release. A vehicle treated group is included, indicating in parenthesis the ACh and DA concentrations found at the 140-160 min control period.

Striatal ACh release was increased, in a dose-dependent manner, by cortical KA or NMDA stimulation. Cortical stimulation with 10^{-4}M of KA produced also a significant increase in striatal DA release. This effect was, however, less remarkable than that on ACh release. Cortical NMDA stimulation did not produce any significant change on striatal DA release.

Table 2. Changes (%)[a] in striatal dopamine (DA) and acetylcholine (ACh) release during 20 min cortical stimulation with either kainic acid (KA) or N-methyl-D-aspartate (NMDA).

Treatment	N	ACh	DA
Vehicle	6	112 ± 23%	92 ± 3%
		(381 ± 13 nM)[c]	(12 ± 2 nM)[c]
KA 10^{-5}M	5	74 ± 13%	122 ± 9%
KA 10^{-4}M	6	405 ± 74% **	178 ± 45%*
NMDA 10^{-5}M	5	171 ± 28% *	118 ± 16%
NMDA 10^{-4}M	4	414 ± 62% **	115 ± 22%

[a] Changes in DA and ACh release are expressed as the percentage over the values obtained during the immediately precedent control period.
[b] 160 min after the microdialysis implantation, KA or NMDA was included for 20 min in the perfusion medium of a microdialysis probe into the left frontoparietal cortex.
[c] ACh or DA release (perfusion medium= Ringer including 10 μM neostigmine; halothane anaesthesia) measured at the 140-160 min control perfusion period.
* P < 0.05; ** P < 0.01.

DISCUSSION

The present **in vivo** microdialysis experiments give evidence that glutamate can modulate striatal DA and ACh transmission via different receptor types. KA, a selective agonist for kainate type of glutamate receptors, produced, following **in situ** administration, a dose-dependent increase in striatal DA. The glutamate receptor agonist, NMDA was less effective in increasing striatal DA release, but induced instead a strong increase in ACh release. In contrast, striatal ACh release was decreased by KA. In order to further investigate whether these effects were mimicking the modulatory actions of a cortico-striatal pathway on DA or ACh release, KA or NMDA were administered into the ipsilateral cortex, by using a microdialysis probe implanted into the deepest layers of the frontoparietal cortex. A 20 min cortical stimulation pulse with KA or NMDA produced a dose-dependent increase in striatal ACh release, but only a minor effect on striatal DA release.

The present results are in agreement with the idea of an excitatory regulation of striatal neurotransmission by glutamatergic pathways[4,5]. The stimulation produced by striatal KA on striatal DA could support the hypothesis of direct axonal interactions between cortical glutamate and striatal DA terminals[6,7]. The parallel decrease observed in striatal ACh could be secondary to an increase in DA release. However, it is probable that glutamate terminals have a main stimulatory action on striatal ACh neurons, since ACh release was increased by both striatal and cortical NMDA stimulation. Indeed, cortical KA stimulation produced also a strong increase in striatal ACh release, but only a minor increase in striatal DA release (c.f. Table 1 and 2). It is probable that the striatal ACh increase produced by cortical stimulation exerts in turn a stimulatory effect on striatal DA terminals as well. In agreement, Carter et al.[14] have presented results of a trans-striatal dialysis study, suggesting that while both NMDA and kainate receptor stimulations are responsible for a glutamate-evoked liberation of striatal DA, the stimulation produced by NMDA receptor may be conveyed by a polysynaptic loop involving a striatal ACh neuron, since the effect on striatal DA release is fully blocked by atropine and tetrodotoxin. Indeed, there is evidence showing that the majority of the striatal afferents from the cortex and the substantia nigra make axodendritic synaptic contacts with striatal neurons[7], supporting the idea of a monosynaptic input on acetylcholine (ACh) neuron, by which cortical glutamate neurons could also modulate striatal dopamine release.

In conclusion, the present results using **in vivo** microdialysis give evidence for the existence of glutamate cortico-striatal pathways modulating both ACh and DA release. It is probable, however, that the major part of the cortico-striatal modulation of dopamine is mainly exerted via polysynaptic loops involving ACh neurons, although a monosynaptic input should also be considered. Indeed, as previously suggested by Herrling[9], the excitatory action of glutamate can be conveyed by multiple receptors, i.e. NMDA-receptors on ACh neurons modulating DA release and KA-receptors exerting a direct modulation on DA terminals.

Acknowledgments. This study was supported by grants from the Medical Research Council (K88-21P-08154-02B; B89-14X-08669), Karolinska institutet fonder, The Swedish Medical Society, Loo och Hans Ostermans fond and Carnegie Medicin AB.

REFERENCES

1. A. Cheramy, R Romo, G. Godeheu, P. Baruch and J. Glowinski, In vivo presynaptic control of dopamine release in the cat caudate nucleus - II. Facilitatory or inhibitory influence of l-glutamate, Neuroscience 19: 1081 (1986).
2. R. Romo, A. Cheramy, G. Godeheu, and J. Glowinski, In vivo presynaptic control of dopamine release in the cat caudate nucleus -I. Opposite changes in neuronal activity and release evoked from thalamic motor nuclei, Neuroscience 19: 1079 (1986).
3. R. Romo, A. Cheramy, G. Godeheu and J. Glowinski, In vivo presynaptic control of dopamine release in the cat caudate nucleus -III. Further evidence for the implication of corticostriatal glutamatergic neurons, Neuroscience 19: 1091 (1986).
4. M.F. Giorguieff, M.L. Kemel and J. Glowinski, Presynaptic effect of L-glutamic acid on the release of dopamine in rat striatal slices, Neuroscience Lett. 6: 73 (1977).
5. P.J. Roberts and N.A. Sharif, Striatal glutamatergic function: modifications following specific lesions, Brain Res. 235: 83 (1982).

structural analysis of the relation between cortico inputs and tyrosine hydroxylase-containing terminals in rat neostriatum, Brain Res. 302: 267 (1984).

7. P. Somogyi, J.P. Bolam and A.D. Smith, Monosynaptic cortical input and local axon collaterals of identified striatonigral neurons. A light and electron microscopy study using the Golgi-peroxidase transport-degeneration procedure, J. Comp. Neurol. 195: 567 (1981).

8. C.W. Cotman, D.T. Monaghan, O.P. Ottersen and J. Storm-Mathisen, Anatomical organization of excitatory amino acid receptors and their pathways, TINS 10: 273 (1987).

9. P.L. Herrling, Pharmacology of the corticocaudate excitatory postsynaptic potential in the cat: evidence for its mediation by quisqualate- or kainate-receptors, Neuroscience 14: 417 (1985).

10. U. Ungerstedt, M. Herrera-Marschitz, U. Jungnelius, L. Ståhle, L. Tossman and T. Zetterström, Dopamine synaptic mechanisms reflected in studies combining behavioural recordings and brain dialysis, in: "Advances in dopamine research, Advances in the Biosciences 37", M. Kohsaka, T. Shomori, Y. Tsukada and G.N. Woodruff, eds, Pergamon Press, Oxford (1982).

11. K. Zilles, The cortex of the rat. A stereotaxic atlas, Springer, Berlin (1985).

12. M. Reid, M. Herrera-Marschitz, T. Hökfelt, L. Terenius and U. Ungerstedt, Differential modulation of striatal dopamine release by intranigral injection of gamma-aminobutyric acid (GABA), dynorphin A and substance P, Eur. J. Pharmacology 147: 411 (1988)

13. D. Maysinger, M. Herrera-Marschitz, A. Carlsson, L. Garofalo, A.C. Cuello and U. Ungerstedt, Striatal and cortical acetylcholine release in vivo in rats with unilateral decortication: effects of treatments with monoganglioside GM1, Brain Res. 461: 355 (1988).

14. C.J. Carter, R. L'Heureux and B. Scatton, Differential control of N-methyl-D-aspartate and kainate of striatal dopamine release in vivo: a trans-striatal dialysis study, J. Neurochemistry 51: 462 (1988).

DISTINCT CHOLINERGIC CONTROL OF DOPAMINE RELEASE IN STRIOSOMAL AND MATRIX AREAS OF THE CAT CAUDATE NUCLEUS

M.L. Kemel, M. Desban, J. Glowinski, and C. Gauchy

INSERM U114, chaire de neuropharmacologie, Collège de France 11 pl. Marcellin Berthelot 75231 Paris France

INTRODUCTION

Two anatomical compartments can be distinguished in the striatum of the adult cat and other mammalian species including the man. These compartments were revealed first by using acetylcholinesterase (AChE) as a marker; areas poor and rich in AChE being denominated respectively striosomes and matrix[18]. Then several other studies have shown that, these two main compartments can be distinguished further with other biochemical markers but also by their afferent and efferent projections. However, few investigations have been made yet on the functional properties of the striosomes and the matrix and on their relationships.

Besides their poor content in AChE, the striosomes are particularly characterized by their high levels in opiate[20] and M1 cholinergic[28] receptors while the matrix contains neurones which express the D28KD calbindin protein[13] and the cell bodies of the cholinergic[17] and somatostatin[6] interneurones. A higher proportion of D1[1] than D2[23] dopamine (DA) receptors is found in the striosomes while the reverse is seen in the matrix. Similarly, neurones containing mRNAs for substance P, dynorphin or Met-enkephalin are distributed in both compartments but they seem to be present in slightly higher proportion in the striosomes than in the matrix[15]. Demonstrating further the complexity of the anatomical organization of the striatum, binding studies have indicated that neurotensin receptors are located in an annular region superimposed with the AChE negative border zone lying between the striosomal and matrix compartments[10].

Several studies have been made already to look for differences in the cortical, thalamic or mesencephalic innervation of the striosomes and the matrix. For instance, the prefrontal cortex sends inputs to the striosomal compartment whereas most other cortical areas including the motor and visual cortices project to the matrix[11;25]. More recently, Gerfen[12] has reported interesting informations on the laminar cortical

The Basal Ganglia III, Edited by G. Bernardi *et al.*
Plenum Press, New York

organization of the cortico-striatal projections innervating the striosomes and the matrix. In each cortical area (prefrontal, cingulate and motor), deep and superficial layers project respectively to the striosomes and the matrix but the prelimbic cortex sends a larger number of inputs to the striosomes than to the matrix while the reverse is found for the motor cortex. The striatal compartments differ also by their mesencephalic DA innervation since DA cells located in the so-called dense cellular zone of the substantia nigra pars compacta innervate exclusively the striosomes while those located in the A8 group project to the matrix (other DA cells innervating the striatum project to both compartments)[14;21]. Interestingly enough, the substantia nigra pars compacta is innervated by neurones originating in the striosomes while the substantia nigra pars reticulata and the globus pallidus receive projections from the matrix[9;11;22].

As shown by electrophysiological recordings[8] or the 2-deoxyglucose method[26], responses restricted to limited areas of the striatum can be seen following either electrical stimulation of the motor cortex or application of VIP within the striatum. However, in these studies, no correlation was made between the sites of the responses and the striosomes and matrix delineations. On the contrary, using combined electrophysiological and anatomical methods , Malach and Graybiel[25] have observed that cells of the primary somatic sensory cortex, which respond to peripheral sensory stimulation, project to a small well defined sector of the dorsolateral corner of the caudate nucleus within the matrix. Using the in situ hybridization method, Chesselet and Robbins[7] have also described differences in the properties of populations of cells in the striatal compartments since tachykinin-containing neurones in the striosomes were found to have a higher rate of tachykinin synthesis than those located in the surrounding matrix.

Both from in vivo and in vitro studies, it is well established that the release of DA from nerve terminals of the nigro-striatal DA neurones is regulated presynaptically by different transmitters contained either in striatal interneurones or afferent fibers[5]. These regulations are either direct (tetrodotoxin (TTX)-resistant) or indirect (TTX-sensitive), ie mediated by receptors located either on the DA fibers or on other cells or fibers in contact with DA fibers. Due to differences seen between the striosomes and the matrix in the identity or concentration of some receptors and in afferent or efferent neurones, some direct or indirect presynaptic regulations of DA release should differ in both compartments. In order to investigate such a possibility, we have developped an in vitro method allowing the study of DA release in well defined striosomal and matrix areas[24]. This was achieved partly on the basis of a three-dimensional analysis of the organization of the striosomal compartment in the cat caudate nucleus[9]. In the present review, we will describe the main results of this anatomical study and then summarize data indicating that the cholinergic presynaptic regulation of DA release is much more complex in the matrix than in the striosomes.

Spatial organization of the striosomal compartment

AChE staining on successive frontal or saggital sections was used to determine the three-dimensional organization of the striosomal and matrix compartments in the adult cat caudate nucleus[9]. Reconstruction drawings of the AChE poor zones (striosomes) indicated that the striosomal compartment is a labyrinthine network organized in the rostrocaudal and mediolateral axis which is reproducible from one animal to another. Four main anteroposterior channels converging in the mediolateral pole of the caudate nucleus were distinguished. These channels were crossed by seven to eight diagonally oriented channels running in the mediolateral axis in the central core of the caudate nucleus. The pattern of organization of the numerous and tortuous striosomal channels was more complicated medially, while the lateral part of the caudate nucleus was represented mainly by the matrix compartment. It could be argued that the striosomes cannot be easily visualized in this lateral area of the caudate nucleus using AChE as a marker. However, similar conclusions on the general organization of the striosomal network were made by Groves et al.[19] who used Met-enkephalin immunoreactivity as a striosomal marker.

As already mentioned, most striosomes were found to have a reproducible localization from one animal to another. Particularly, a prominent striosomal area located in the core of the caudate nucleus (A16 to A17,5) was seen in each animal. At the same anteriority, the ventral part of the lateral corner of the caudate nucleus was always devoid of AChE poor zones and considered as a matrix area. These striosomal and matrix areas were selected in our study on the cholinergic presynaptic control of DA release in both compartments.

Cholinergic control of dopamine release in striosomal and matrix compartments

Several reasons led us to investigate the cholinergic control of DA release in the striosomal and matrix areas of the cat caudate nucleus: 1) In the rat, as shown on slices and synaptosomes, muscarinic and nicotinic receptors are both involved in this regulation[5;16]. 2) Unpublished observations made in our laboratory, have suggested the occurrence of regional variations in the stimulatory effect of ACh on DA release. 3) As shown in the cat, a higher concentration of muscarinic binding sites has been observed in striosomes when compared to the matrix[28]. 4) According to several anatomical[3] and biochemical[29] studies the cortico-striatal glutamatergic neurones innervate and regulate (through NMDA receptors particularly) the cholinergic interneurones and it has been suggested that, NMDA used in a high concentration, stimulates indirectly the release of DA through a cholinergic link[4]. 5) Finally, there is also some evidence for an excitatory control of medium size spiny GABAergic neurones by cholinergic neurones suggesting, that ACh may indirectly influence the release of DA through GABA which is released from the collaterals of efferent GABAergic neurones[2].

Two adjacent frontal slices (1mm thick) located between the anterior planes 17,9-16,9 and 16,9-15,9 (according to the atlas of Snider and Niemer[30]) were cut using a vibratome, and then placed in a specially designed superfusion chamber. They were continuously superfused with an oxygenated CSF kept at 33°C. Microsuperfusion devices were then vertically applied on the selected matrix and striosomal areas using a micromanipulator and a dissecting microscope. An oxygenated CSF enriched in L 3,5 ^{3}H-tyrosine (50 Ci mmole^{-1}, 60 μci ml^{-1}) was continuously delivered at a rate of 50 μl min^{-1} to each microsuperfusion device. The release of ^{3}H-DA (newly synthesized from ^{3}H-tyrosine) was estimated in successive 5 min fractions, ^{3}H-DA being separated from ^{3}H-tyrosine and ^{3}H-metabolites by ion-exchange chromatography and alumina adsorption[24].

When applied for 25 min, into either the striosomal or the matrix areas, ACh (50 μM) stimulated the release of ^{3}H-DA. This effect lasted up to the end of the ACh application in the striosomal area but was of shorter duration in the matrix . In the presence of tetrodotoxine (TTX, 1 μM), ACh induced a long lasting stimulation of ^{3}H-DA release in both compartments. These results indicate that cholinergic receptors are involved in a direct presynaptic control of DA release in the striosomal and matrix areas, but they also suggest that the cholinergic regulation of DA release is more complex in the matrix area.

In the striosome, the ACh-evoked release of ^{3}H-DA was blocked by atropine (1μM, a muscarinic antagonist) but it remained unaffected by pempidine (10μM, a nicotinic antagonist) both in the absence or presence of TTX. Therefore, muscarinic receptors located on DA nerve terminals are implicated in the presynaptic ACh facilitatory control of DA release.

In the matrix, the short lasting ACh-evoked response observed only in the absence but not in the presence of TTX, suggests that ACh exerts also a secondary indirect inhibitory effect. Nevertheless, the ACh-evoked release of ^{3}H-DA seen in the presence of TTX was antagonized by atropine but not by pempidine, indicating that, muscarinic receptors located on DA nerve terminals are involved in a direct presynaptic control of DA release similarly to that observed in the striosome. In the absence of TTX however, although the initial stimulatory effect of ACh on 3H-DA release was decreased (50%) by atropine (1μM) , a long lasting effect was observed thereafter (Fig.1). This suggests that: 1) atropine prevents also an inhibitory action of ACh, and that a population of muscarinic receptors is involved in the indirect ACh-mediated inhibitory response, 2) by stimulating nicotinic receptors ACh facilitates also the release of ^{3}H-DA through a TTX-sensitive process. Nicotinic receptors are indeed involved in the latter effect since, pempidine not only reduced (50%) the initial stimulatory effect of ACh seen in the absence of atropine but also abolished completely the long lasting ACh-evoked release of ^{3}H-DA observed in the presence of atropine. Therefore, by acting on muscarinic or nicotinic cholinergic receptors, ACh inhibits or stimulates respectively the release of ^{3}H-DA through two distinct TTX-sensitive processes.

Dynorphin-containing neurones contribute to the acetylcholine mediated inhibition of dopamine release in the matrix

Further experiments were performed in order to identify the neurones responsible for the indirect ACh-mediated inhibitory response on 3H-DA release.

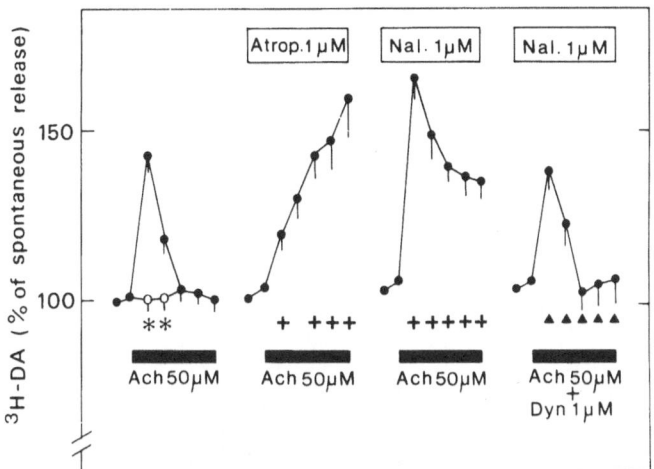

Fig.1. Indirect acetylcholine-evoked regulation of ^{3}H-dopamine release in the matrix

Superfusion experiments were performed as previously described. When applied, atropine (Atrop.) or naloxone (Nal.) were added into the superfusion medium during all the superfusion. Acetylcholine (ACh) or ACh with dynorphin (DYN) were applied during the last 25 min of the experiments, 65 min after the onset of the superfusion. In each experiment, the amount of ^{3}H-DA recovered in successive 5 min fractions was expressed as a percentage of the mean spontaneous release of ^{3}H-DA determined on the basis of estimations made in the four fractions which preceded the onset of ACh application. Results are the mean $^+$/- SEM of data obtained in 8 to 12 experiments.
* $p<.01$ when compared to data obtained in 8 to 10 control experiments (o)
+ $p<.01$ (ACh + atropine) or (ACh + naloxone) when compared to data obtained in experiments in which ACh was applied alone.
 $p<.01$ (ACh + DYN + naloxone) when compared to data obtained in experiments in which ACh was applied in the presence of naloxone.

Immunohistochemical, in situ hybridization and binding studies have revealed the presence of both dynorphin-containing neurones and K opiate receptors in the striatum. In addition, dynorphin has been shown to inhibit both the spontaneous and the evoked release of ^{3}H-DA from rat striatal slices, these effects being blocked by naloxone, a non selective opiate antagonist[27]. These observations led us to determine whether or not naloxone could prevent the indirect inhibitory effect of ACh on ^{3}H-DA release in the matrix area. An ACh-evoked release of 3H-DA of long duration was observed in the presence of naloxone (1μM) (Fig.1). Furthermore, the effect of naloxone was reversed when dynorphin (1μM) was applied simultaneously with ACh (Fig.1), restoring thus the short lasting ACh-induced stimulatory response. These results favor of the involvement of dynorphin-containing neurones in the indirect ACh-mediated inhibitory response on ^{3}H-DA release.

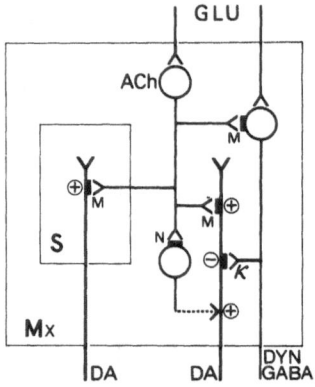

Fig.2. Schematic representation of the complex role of the cholinergic neurones in the regulation of dopamine release in both striatal compartments.

The cells bodies of cholinergic neurones are located in the matrix but they send sparse and dense fibers in the striosomes (S) and the matrix (Mx) respectively. Glutamatergic (GLU) corticostriatal inputs innervate both cholinergic interneurones and dynorphin (DYN) and/or GABA efferent neurones of the striatum. The present study shows that cholinergic neurones may directly stimulate (+) DA release through muscarinic (M) receptors located on DA nerve terminals both in the matrix and striosomal areas. In addition, in the matrix, ACh indirectly stimulates (+) and/or inhibits (-) DA release by acting respectively on nicotinic (N) and muscarinic receptors, nicotinic receptors being located on yet unidentified neurones. The indirect inhibition of DA release involves the dynorphin-containing neurones, released dynorphin acting on K receptors located on DA nerve terminals.

Additional experiments were made in the matrix in the presence of TTX. The long lasting TTX-resistant ACh-evoked release of ^{3}H-DA was prevented by the additional application of dynorphin (1μM) or U50488 (1μM, a selective K agonist). These results indicate that, the indirect inhibitory effect of ACh on ^{3}H-DA release is mediated by dynorphin released from dynorphin-containing neurones, dynorphin acting on K receptors located on DA nerve terminals.

Concluding remarks

The striatal cholinergic neurones, whose cell bodies are located in the matrix[17] and which are innervated by cortico-striatal fibers,[3;29] could play a role in the transfert of information between the matrix and striosomal compartments. They contribute indeed to a presynaptic control of DA release in both compartments. In spite of the presence of a higher density of muscarinic receptors in the striosomes[28], through muscarinic receptors located on DA nerve terminals, ACh exerts a direct presynaptic facilitation of DA release in both compartments. In addition, in the matrix, ACh induced two indirect (TTX-sensitive) and opposing effects: a stimulation of DA release mediated by nicotinic receptors and an inhibition of DA release mediated by muscarinic receptors (Fig.2). The latter effect results from the activation of the inhibitory dynorphin-containing neurones, dynorphin released acting directly on K receptors located on DA nerve terminals. The identity of the neurones implicated in the indirect ACh facilitatory response mediated by nicotinic receptors has still to be determined.

Since dynorphin seems to be colocalized with GABA in some of the inhibitory neurones which innervate both the caudate nucleus (collaterals) and the substantia nigra (axon terminals)[15], the complete suppression by naloxone of the ACh-evoked inhibitory response seen in the matrix area, is in someways surprising. Indeed, GABA could have been involved as well in this indirect inhibitory process. In fact, preliminary experiments made in the selected matrix area have indicated that bicuculline is without effect on the ACh-mediated inhibitory response. However, experiments are in progress to understand better the complexity of the local circuits implicated in the indirect control of DA release in the matrix. In fact by examining the sites of origin of the striato-nigral projections and comparing their localization with that of AChE poor zones (striosomes), we have recently shown that the matrix has a patchy organization, cells of the efferents projections being distributed in these patches (mainly in the lateral part of the caudate nucleus)[9]. Undoubtedly, this complex anatomical organization of the matrix will have to be also taken into account for the elucidation of the mecanisms involved in the local control of DA release.

References

1. Besson M.J., Graybiel A.M. and Nastuk M.A. (1988) Neurosci. 26, 101-119
2. Besson M.J., Kemel M.L., Gauchy C. and Glowinski J. (1982) Brain Res. 241, 241-248
3. Bolam J.P. (1984) Ciba Found. Symp.107, 30-47
4. Carter C.J., L'heureux R. and Scatton B. (1988) J.Neurochem. 51, 462-468
5. Chesselet M.F. (1984) Neurosci. 12, 347-375
6. Chesselet M.F. and Graybiel A.M. (1986) Neurosci. 17, 547-571
7. Chesselet M.F. and Robbins E. (1989) Neurosc. Lett. 96, 47-53
8. Crutcher M.D. and Delong M.R. (1983) Exp. Brain. Res. 130, 1-11
9. Desban M., Gauchy C., Kemel M.L., Besson M.J. and Glowinski J. (1989) Neurosci.29, 551-566

10. Faull R.L.M., Dragunow M. and Villiger J.W. (1988) Brain Res. 488, 381-386

11. Gerfen C.R. (1984) Nature 311, 461-464

12. Gerfen C.R. (1989) Science 246, 385-388

13. Gerfen C.R., Baimbridge K.G. and Miller J.J. (1985) Proc. Natl. Acad. Sci. USA 82, 8780-8784.

14. Gerfen C.R., Herkenham M. and Thibault J. (1987) J. Neurosci. 7, 3915-3934.

15. Gerfen C.R. and Scott Young W.S. (1988) Brain Res. 460, 161-167

16. Giorguieff M.F., Le Floc'h M.L., Glowinski J. and Besson M.J. (1977) J.P.E.T. 200, 535-544

17. Graybiel A.M., Baughman R.W. and Eckenstein F. (1986) Nature 323, 625-627.

18. Graybiel A.M. and Ragsdale C.W. (1978) Proc. Natl. Acad. Sci. USA 75, 5723-5726.

19. Groves P.M., Martone M., Young S.J. and Armstrong D.M. (1988) J. Neurosci. 8, 892-900.

20. Herkenham M. and Pert C.B. (1981) Nature 291, 415-417.

21. Jimenez-Castellanos J. and Graybiel A.M. (1987) Neurosci. 23, 223-243

22. Jimenez-Castellanos J. and Graybiel A.M. (1989) Neurosci. 32,297-321

23. Joyce J.N., Sapp D.W. and Marshall J.F. (1986) Proc. Natl. Acad. Sci. USA 83, 8002-8006

24. Kemel M.L., Desban M., Glowinski J. and Gauchy C. (1989) Proc. Natl. Acad. Sci. USA 86, 9006-9010.

25. Malach R. and Graybiel A.M. (1986) J. Neurosci. 6, 3436-3558

26. Mc Culloch J., Kelly P.A.T., Uddman R. and Edvinsson L. (1983) Proc. Natl. Acad.Sci. USA 80, 1472-1476.

27. Mulder A.H., Wardeh G., Hogenboom F. and Frankhuyzen A.L. (1984) Nature 308, 278-280.

28. Nastuk M.A. and Graybiel A.M. (1988) J. Neurosci. 8, 1052-1062

29. Scatton B. and Lehmann J. Nature . (1982) 297, 422-424

30. Snider R.S. and Niemer W.T. A stereotaxic atlas of the cat brain. The university of Chicago Press Chicago IL (1961).

EFFECT OF NEUROTENSIN ON DOPAMINE AND MUSCARINIC ACETYLCHOLINE RECEPTORS

IN THE RAT STRIATUM

Rie Miyoshi, Shozo Kito and Akiko Tanaka

Third Department of Internal Medicine, Hiroshima University
School of Medicine, 1-2-3 Kasumi, Minamiku, Hiroshima 734
Japan

INTRODUCTION

Neurotensin (NT) is a tridecapeptide first isolated and sequenced from bovine hypothalamus in 1973 (1). This peptide has a wide distribution within central and peripheral tissues where it is present in both neuronal and endocrine structures (2). In the mammalian central nervous system, NT has been established as a candidate of neurotransmitter/neuromodulator. NT-containing fibers (3,4) and NT receptors (5,6) have been observed in high density within the striatum. An interaction between dopamine and NT has been well demonstrated in the nigro-striatal neuronal pathway. For instance, it has been reported that NT receptors are not only located on dopaminergic perikarya and dendrites (7) but also on dopaminergic nerve terminals (8). In addition, NT facilitates endogenous or K^+-evoked dopamine release (9,10,11). From receptor binding assay, it has been observed that NT decreases the affinity of D2 receptor agonist binding (12). On the other hand, NT dose not modulate the binding characteristics of ^{3}H-spiperone, which is a D2 receptor antagonist (13).

Dopamine receptors have been classified into two major subtypes on the basis of their pharmacological and biochemical characteristics (14,15). Functional significance of D2 receptor in the central nervous system has been well discussed from viewpoints of pharmacotherapy of psychosis and movement disorders. Recently, availability of D1 selective ligands has made it possible to study physiological roles of D1 receptors in the central nervous system (16,17). First, in the present paper, the effect of NT on D1 receptor binding was investigated in the rat striatum.

In the striatum, acetylcholine (ACh)-containing neurons are also found in very high density, which are considered to be mostly interneurons. An interaction between ACh and NT has not been well described in the central nervous system. There are a few papers showing the ACh-NT interaction in the nucleus basalis (18) and parietal cortex (19). The authors examined the modulatory effect of NT on muscarinic acetylcholine receptors (mAChR) in the striatum.

The Basal Ganglia III, Edited by G. Bernardi *et al.*
Plenum Press, New York

MATERIALS AND METHODS

Effect of NT on Dopamine Dl Receptors

Wistar strain male rats weighing 200-250 g were used in these experiments. After decapitation, the striatum was rapidly removed and P_2 fractions of the tissue were prepared. ^{3}H-SCH 23390 was used as Dl selective antagonist. For observation of Dl antagonist binding sites, saturation experiments with ^{3}H-SCH 23390 were performed. For Dl agonist binding sites, dopamine/^{3}H-SCH 23390 inhibition experiments were done. As assay medium, 50 mM Tris HCl buffer (pH7.4) containing 5 mM KCl, 1 mM $MgCl_2$, 2 mM $CaCl_2$, 30 mg/l bacitracin and 2 g/l bovine serum albumin was used. Non-specific binding was defined as the binding in presence of 10^{-5} M SCH 23390. To study the effect of NT, 1 μM NT or its analogs were added into the incubation medium. Aliquots of the tissue preparation were incubated at 22°C for 30 min. Membrane bound ^{3}H-ligands were trapped by the rapid vacuum filtration method.

Effect of NT on mAChR

Striatal P_2 fractions were used. Saturation experiments of ^{3}H-QNB and ^{3}H-oxotremorine-M (oxo-M) were done in order to observe mAChR antagonist and agonist binding sites, respectively. As assay buffer, 0.1 M phosphate buffer (pH7.4) containing 30 mg/l bacitracin and 2 g/l bovine serum albumin was used. Non-specific binding was determined using 1 μM atropine for ^{3}H-QNB binding and 10^{-4} M tremorine for ^{3}H-oxo-M binding. As for ^{3}H-QNB binding, the aliquots were incubated at 25°C for 60 min. For ^{3}H-oxo-M, they were incubated at 30°C for 8 min.

It has been demonstrated that activations of both mAChR and NT receptors stimulate phosphatidylinositol (PI) turnover in various tissues. In this study, the synergic effect of carbachol, a mAChR agonist, and NT on PI turnover was investigated. Striatal slices with 400 μm thickness were used. Slices were incubated in Krebs-Ringer bicarbonate medium containing 0.45 μM ^{3}H-myo-inositol at 37°C for 90 min. Slices were then incubated with carbachol and/or NT in presence of 10 mM LiCl. The incubation was terminated 30 min later by the addition of chloroform/methanol. After addition of chloroform and water, the samples were centrifuged to facilitate phase separation and an aliquot of the aqueous phase was chromatographed with a Dowex-1 resin column. ^{3}H-inositol-1-phosphate (IP_1) was eluted with 0.2 M ammonium formate/0.1 M formic acid.

RESULTS

Effect of NT on Dopamine Dl Receptors

Fig. 1 shows Scatchard plots of saturation experiments of ^{3}H-SCH 23390 in the rat striatum. ^{3}H-SCH 23390 binding sites had a single high affinity binding site whose Kd and Bmax values were 1.57 nM and 907.6 fmol/mg protein, respectively. NT had no effect on ^{3}H-SCH 23390 binding.

Dopamine/^{3}H-SCH 23390 inhibition curves are shown in Fig. 2. Curves without and with NT exhibited heterogeneous characteristics with Hill coefficients much less than 1. In the absence of NT, the binding data were fitted best by a two-site model, characterized by two dissociation constants: K_H value of 7.76×10^{-7} M and K_L value of 2.45×10^{-5} M. Percentages of high (R_H) and low (R_L) affinity binding sites to the total binding capacity were 74.6 % and 25.4 %, respectively. After

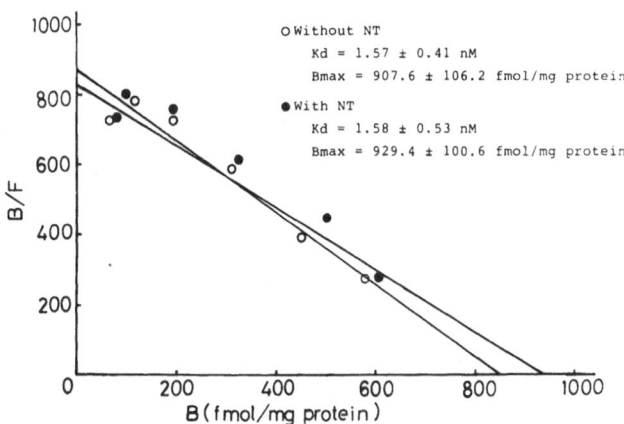

Fig. 1 Scatchard plots of saturation data of ^{3}H–SCH 23390 with and without 1 μM NT in the rat striatum. These values were mean ± S.D. in 6 repeated experiments.

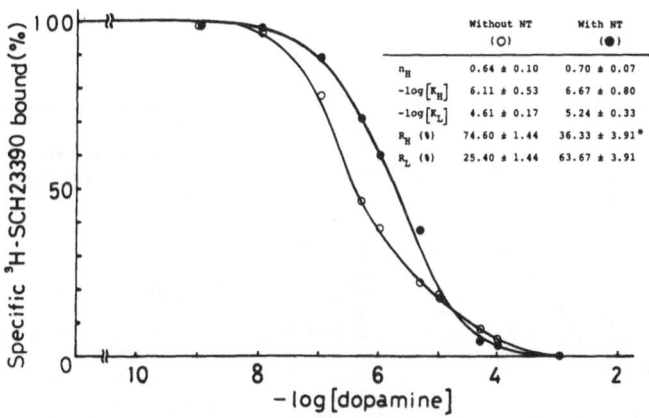

	Without NT (O)	With NT (●)
n_H	0.64 ± 0.10	0.70 ± 0.07
$-\log[K_H]$	6.11 ± 0.53	6.67 ± 0.80
$-\log[K_L]$	4.61 ± 0.17	5.24 ± 0.33
R_H (%)	74.60 ± 1.44	36.33 ± 3.91*
R_L (%)	25.40 ± 1.44	63.67 ± 3.91

Fig. 2 Effect of 1 μM NT on dopamine/^{3}H–SCH 23390 inhibition experiments in the rat striatum. n_H: Hill coefficient. The concentration of ^{3}H–SCH 23390 was 1 nM.
*: Significant difference from the R_H value of experiments without NT, when examined by Student's t-test, $p < 0.05$. n=3.

Table 1 Effect of NT, its analogs and GTP on dopamine binding to the rat striatum measured by competition with ^{3}H-SCH 23390.

	n_H	$-\log[K_H]$	$-\log[K_L]$	R_H (%)	R_L (%)
	0.64 ± 0.10	6.11 ± 0.53	4.61 ± 0.17	74.60 ± 1.44	25.40 ± 1.44
NT					
0.01 µM	0.70 ± 0.04	6.79 ± 0.20	5.51 ± 0.14	60.87 ± 3.72	39.13 ± 3.72
0.1 µM	0.77 ± 0.08	6.79 ± 0.16	5.74 ± 0.14	45.27 ± 8.76*	54.73 ± 8.76
1 µM	0.70 ± 0.07	6.67 ± 0.80	5.24 ± 0.33	36.33 ± 3.91*	63.67 ± 3.91
10 µM	0.71 ± 0.10	7.89 ± 0.38	5.91 ± 0.07	20.19 ± 3.45*	79.81 ± 3.45
NT_{1-8} µM	0.66 ± 0.13	6.42 ± 0.35	4.73 ± 0.29	60.82 ± 7.71	39.18 ± 7.71
NT_{8-13} µM	0.72 ± 0.17	6.71 ± 0.68	5.24 ± 0.33	36.39 ± 5.57*	63.61 ± 5.57
GTP					
1 µM	0.79 ± 0.05	6.42 ± 0.17	5.38 ± 0.07	64.86 ± 4.09	35.14 ± 4.09
10 µM	0.76 ± 0.03	6.51 ± 0.10	5.47 ± 0.02	55.35 ± 4.64*	44.65 ± 4.64
50 µM	0.71 ± 0.20	6.99 ± 1.47	4.93 ± 0.49	48.13 ± 7.53*	51.87 ± 7.53
100 µM	0.79 ± 0.09	6.97 ± 0.52	5.74 ± 0.08	28.49 ± 6.71*	71.51 ± 6.71
50 µM GTP + 1 µM NT	0.58 ± 0.26	6.43 ± 0.01	5.68 ± 0.01	48.39 ± 17.8*	51.61 ± 17.8

*: Significant difference from the R_H value of control experiments, when examined by Student's t-test, p < 0.05. n=3-8.

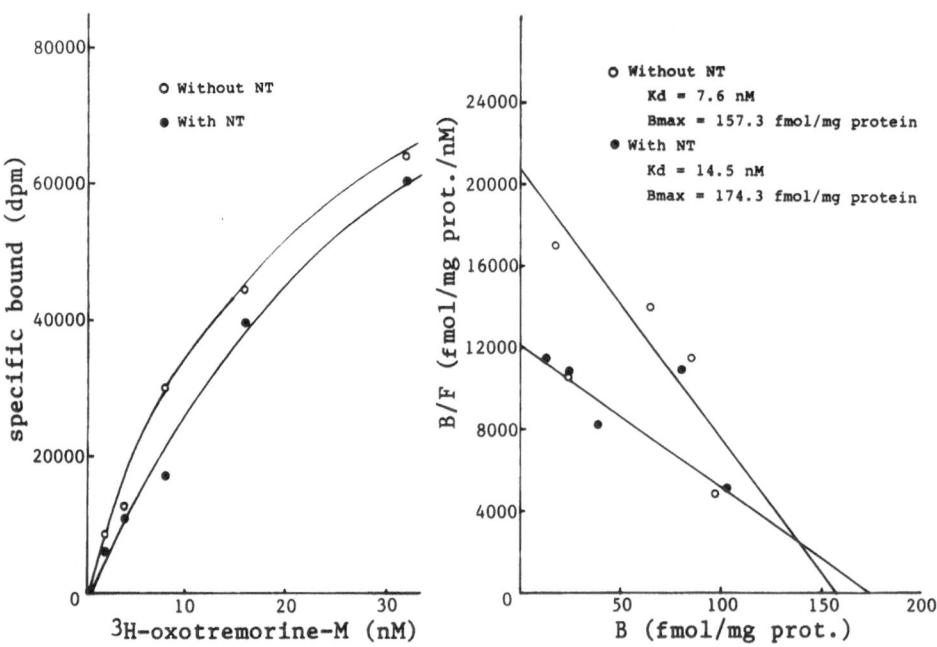

Fig. 3 Effect of NT on mAChR agonist binding in the rat striatum. The left figure shows saturation curves of specific binding of ^{3}H-oxo-M with and without 1 µM NT. The right shows Scatchard plots. Values were mean ± S.D. in 3 repeated experiments.

adding NT, R_H decreased to 36.3 % and R_L increased to 63.7 %, with their affinity constants unchanged. As shown in Table 1, the effect of NT was found to be dose-dependent when the concentrations from 10^{-8} M to 10^{-5} M were used. NT_{8-13} was equipotent with NT itself in this assay, whereas NT_{1-8} was less potent to modulate dopamine binding. Previous reports have demonstrated that guanine nucleotide converts a part of agonist high affinity binding sites into low affinity ones in the striatal D1 receptors (20). In our experimental system, GTP converted dopamine high affinity binding sites to low affinity ones in a dose-dependent manner. When both 1 μM NT and 50 μM GTP were added, NT did not cause any further conversion.

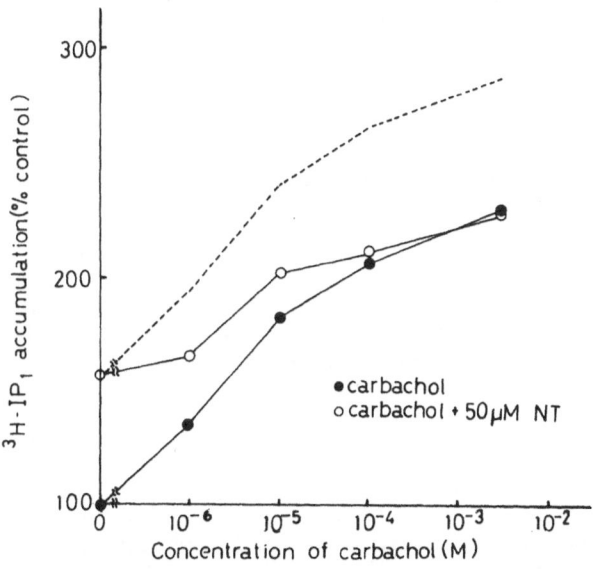

Fig. 4 Synergic effect of carbachol and NT on PI turnover in the rat striatum. Values of the ordinate represent % of those in experiments in which 10 mM LiCl alone was added to the incubation medium. The dotted line is a theoretical curve in case of an additive effect of carbachol and NT on PI turnover.

Effect of NT on mAChR

NT did not affect ^{3}H-QNB binding in the rat striatum (data not shown). Fig. 3 shows the result of saturation experiments of ^{3}H-oxo-M. ^{3}H-oxo-M had a single high affinity binding site whose Kd and Bmax values were 7.6 nM and 157.3 fmol/mg protein, respectively. By adding NT, the Kd value was increased from 7.6 nM to 14.5 nM with the Bmax value unchanged.

Both carbachol and NT caused basal accumulation of ^{3}H-IP$_1$ in rat striatal slices. The simultaneous addition of carbachol and NT elicited a non-additive convergent stimulation of PI turnover (Fig. 4).

DISCUSSION

There has been strong evidence to indicate an interaction between NT and D2 receptors in the brain (21,22,23). It has been reported that NT dose not affect dopamine D2 receptor antagonist binding (13), while it decreases the affinity of agonist binding (12). In our experiments, the authors investigated effects of NT on D1 receptors in the rat striatum. As the result, it was noticed that NT modulated D1 receptors in the same manner as D2 receptors. D1 and D2 receptors cause an opposing effect on adenylate cyclase activity, that is, D1 and D2 receptors are linked to cyclic AMP formation in a stimulatory and inhibitory way, respectively (24). However, these two subtype receptors have been demonstrated to cause a synergic effect on electrophysiological and behavioral responses (25,26,27). It is noteworthy that NT reduces the affinity of agonist binding for both D1 and D2 receptors. It is considered that activation of NT receptors affects the functional state of dopamine receptor proteins through the intramembrane transduction signalling system, since the peptide had no effect on dopamine antagonist binding.

It has been demonstrated that NT_{8-13} is the shortest fragment with full biological activity of NT analogs and NT_{1-8} shows no or less activity (28,29). Our data likewise showed a full activity of NT_{8-13} and a negligible activity of NT_{1-8} in modulating D1 agonist receptor binding.

GTP has been found to convert a part of D1 high affinity binding sites into a low affinity state (20). This effect of GTP on agonist binding has been accepted to be the result of the dissociation of a GTP binding protein from a receptor protein. In our experiments, GTP and NT showed the same effect on D1 agonist binding independently and NT did not affect D1 agonist binding sites in presence of GTP. It appears that NT modulates dopamine agonist binding sites in relation with a G protein.

NT also decreased the affinity of mAChR agonist binding, while it did not the antagonist binding. A similar mechanism may occur in the modulation of mAChR by NT to the case of that of dopamine receptors. As previously described (30), both carbachol and NT caused stimulation of PI turnover in the rat striatum. The simultaneous addition of carbachol and NT elicited a non-additive convergent stimulation of PI turnover. It is possible that mAChR and NT receptors couple to a common second messenger system in the rat striatum.

ACKNOWLEDGEMENTS

This study was supported in part by a Grant-in-Aid for Science Research from the Ministry of Education, Science and Culture, Japan.

REFERENCES

1. R. Carraway and S. E. Leeman, The isolation of a new hypotensive peptide, neurotensin, from bovine hypothalami, J. Biol. Chem., 248:6854 (1973)

2. C. B. Nemeroff, D. Luttinger, and A. J. Prange, Neurotensin and bombesin, in: "Handbook of Psychopharmacology", L. L. Iversen, S. D. Iversen, and S. H. Snyder, eds., Plenum Press, New York, p.363 (1983)

3. G. R. Uhl, M. J. Kuhar, and S. H. Snyder, Neurotensin: immuno-histochemical localization in rat central nervous system, _Proc. Natl. Acad. Sci._, 74:4059 (1977)

4. M. Goedert, P. W. Mantyh, P. C. Emson, and S. P. Hunt, Inverse relationship between neurotensin receptors and neurotensin-like immunoreactivity in cat striatum, _Nature_, 307:543 (1984)

5. W. S. III Young and M. J. Kuhar, Neurotensin receptor localization by light microscopic autoradiography in rat brain, _Brain Res._, 206:273 (1981)

6. M. Goedert, K. Pittaway, B. J. Williams, and P. C. Emson, Specific binding of tritiated neurotensin to rat brain membranes: characterization and regional distribution, _Brain Res._, 304:71 (1984)

7. J. M. Palacios and M. J. Kuhar, Neurotensin receptors are located on dopamine-containing neurones in rat midbrain, _Nature_, 294:587 (1981)

8. M. Goedert, K. Pittaway, and P. C. Emson, Neurotensin receptors in the rat striatum: lesion studies, _Brain Res._, 299:164 (1984)

9. Y. Okuma, Y. Fukuda, and Y. Osumi, Neurotensin potentiates the potassium-induced release of endogenous dopamine from rat striatal slices, _Eur. J. Pharmacol._, 93:27 (1983)

10. M. E. Quidt and P. C. Emson, Neurotensin facilitates dopamine release in vitro from rat striatal slices, _Brain Res._, 274:376 (1983)

11. D. Jiang, D. Feng, L. Zhu, and Y. Cheng, Neurotensin potentiates the endogenous dopamine release from striatal synaptosomes of rat, _Life Sci._, 43:27 (1988)

12. G. Euler and K. Fuxe, Neurotensin reduces the affinity of D-2 dopamine receptors in rat striatal membranes, _Acta Physiol. Scand._, 131:625 (1987)

13. C. B. Nemeroff, D. Luttinger, D. E. Hernandez, R. B. Mailman, G. A. Mason, S. D. Davis, E. Widerlov, G. D. Frye, C. A. Kilts, K. Beaumont, G. R. Breese, and A. J. Prange, Interactions of neurotensin with brain dopamine systems: biochemical and behavioral studies, _J. Pharmacol. Exp. Ther._, 225:337 (1983)

14. S. E. Leff and I. Creese, Dopamine receptors re-explained, _Trends Pharmacol. Sci._, 463 (1983)

15. J. C. Stoof and J. W. Kebabian, Two dopamine receptors: biochemistry, physiology and pharmacology, _Life Sci._, 35:2281 (1984)

16. L. C. Iorio, A. Barnett, F. H. Leits, V. P. Houser, and C. A. Korduba, SCH 23390, a potential benzazepine antipsychotic with unique interactions on dopaminergic systems, _J. Pharmacol. Exp. Ther._, 226:462 (1983)

17. K. M. O'Boyle and J. L. Waddington, Selective and stereospecific interactions of R-SK & F 38393 with (^{3}H)piflutixol but not (^{3}H)spiperone binding to striatal D_1 and D_2 dopamine receptors: comparisons with SCH 23390, _Eur. J. Pharmacol._, 98:433 (1984)

18. E. Szigethy, G. L. Wenk, and A. Beaudet, Anatomical substrate for neurotensin-acetylcholine interactions in the rat basal forebrain, Peptides, 9:1227 (1989)

19. D. Malthe-Sorenssen, P. L. Wood, D. L. Cheney, and E. Costa, Modulation of the turnover rate of acetylcholine in rat brain by intraventricular injections of thyrotropin-releasing hormone, somatostatin, neurotensin and angiotensin II, J. Neurochem., 31:685 (1978)

20. P. Seeman, C. Ulpian, D. Grigoriadis, I. Pri-Bar, and O. Buchman, Conversion of dopamine D_1 receptors from high to low affinity for dopamine, Biochem. Pharmacol., 34:151(1985)

21. C. B. Nemeroff, D. E. Hernandez, D. Luttinger, P. W. Kalivas, and A. J. Prange, Interactions of neurotensin with brain dopamine systems, Ann. N. Y. Acad. Sci., 400:330 (1982)

22. G. R. Uhl and M. J. Kuhar, Chronic neuroleptic treatment enhances neurotensin receptor binding in human and rat substantia nigra, Nature, 309:350 (1984)

23. S. M. Simasko and G. A. Weiland, Effect of neurotensin, substance P and TRH on the regulation of dopamine receptors in rat brain, Eur. J. Pharmacol., 106:653 (1985)

24. J. C. Stoof and J. W. Kebabian, Opposing roles for D-1 and D-2 dopamine receptors in efflux of cyclic AMP from rat neostriatum, Nature, 294:366 (1981)

25. J. R. Walters, D. A. Bergstrom, J. H. Carlson, T. N. Chase, and A. R. Braun, D_1 dopamine receptor activation required for postsynaptic expression of D_2 agonist effects, Science, 236:719 (1987)

26. O. Gershanik, R. E. Heikkila, and R. C. Duvoisin, Behavioral correlations of dopamine receptor activation, Neurology, 33:1489 (1983)

27. D. M. Jackson and M. Hashizume, Bromocriptine induces marked locomotor stimulation in dopamine-depleted mice when D-1 dopamine receptors are stimulated with SK&F 38393, Psychopharmacol., 90:147 (1986)

28. M. Goedert, K. Pittaway, B. J. Williams, and P. C. Emson, Specific binding of tritiated neurotensin to rat brain membranes: characterization and regional distribution, Brain Res., 304:71 (1984)

29. K. S. Kanba and E. Richelson, Comparison of the stimulation of inositol phospholipid hydrolysis and of cyclic GMP formation by neurotensin, some of its analogs, and neuromedin N in neuroblastoma clone N1E-115, Biochem. Pharmacol., 36:869 (1987)

30. M. Goedert, R. D. Pinnock, C. P. Downes, P. W. Mantyh, and P. C. Emson, Neurotensin stimulates inositol phospholipid hydrolysis in rat brain slices, Brain Res., 323:193 (1984)

REGULATION OF NT RECEPTORS AFTER CHRONIC TREATMENT WITH

TYPICAL AND ATYPICAL NEUROLEPTIC DRUGS

Laura Calzà, Luciana Giardino, *Pier Vincenzo Piazza and *Giuseppe Amato

Inst. Human Physiology, Medical School University of Cagliari, Italy; *Inst. Human Physiology, Medical School, University of Palermo Italy

INTRODUCTION

Neurotensin is a tridecapeptide, which produces central effects such as hypotension, hypothermia, muscle relaxation, analgesia, antinociception, and reduces locomotor activity (Nemeroff et al., 1977). In the recent years, several lines of evidences indicated the great importance of the dopamine (DA)- neurotensin (NT) interaction taking place mostly in the mesolimbocortical areas, both at cell bodies (A10 DA-ergic group located in the ventral tegmental area -VTA-) and nerve terminal level (cerebral cortex, n. accumbens, olfactory tubercle, ventral n. caudato-putamen) (Quirion, 1983; Nemeroff, et al., 1984; Nemeroff, 1986). Immunocytochemical data indicated the presence of NT-like immunoreactivity in DA cell bodies in the VTA, including midline structures, with only single example of coexistence in the substantia nigra (Hokfelt et al., 1984). Anatomical data indicate that NT-like immunoreactive fibres, nerve terminals and also cell bodies are present in the basal ganglia (Kalivas, 1984; Sugimoto and Mizuno, 1987; Uhl et al, 1979). NT receptors are localized on the DA-ergic cell bodies in the ventral mesencephalon (Young and Kuhar, 1981). NT receptors are also reported to exist in the basal ganglia of the rat, on DA nerve terminals, intrinsic neurons and corticostriatal fibres (FIG. 1) (Quirion et al, 1985; Goedert et al., 1984).

Conventional antipsychotic drugs possess as their most common side-effect the ability to produce extrapyramidal symptoms. Certain of these agents produce a low incidence of extrapyramidal side-effects and have been termed atypical neuroleptics. A different action of typical and atypical neuroleptics on the activity of A9 and A10 midbrain dopaminergic neurons has been indicated as an argument for site specificity in order to partially explain the different clinical and behavioral properties of these two classes of drugs (Chiodo and Bunney, 1983; White and Wang, 1983; Boris and Diamond, 1983).

The Basal Ganglia III, Edited by G. Bernardi et al.
Plenum Press, New York

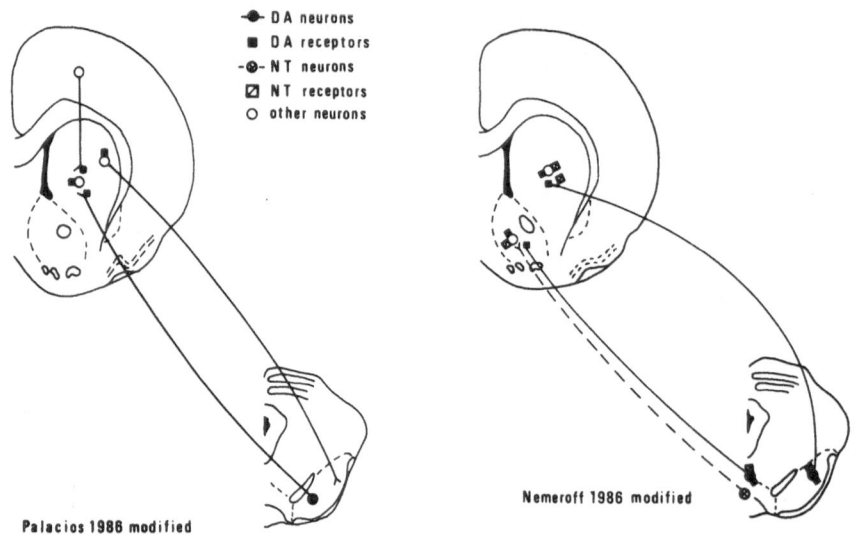

- ● D A neurons
- ■ D A receptors
- -⊙- N T neurons
- ▨ N T receptors
- ○ other neurons

Palacios 1986 modified

Nemeroff 1986 modified

Fig. 1. Schematic representation of the DA and NT receptor
localization on cell bodies and nerve terminals in
mesencephalic ascending pathways and related
structures.

In view of the different NT/DA interaction in the
mesostriatal and mesolimbocortical systems, we investigated
by means of quantitative receptor autoradiography the
modifications induced by the long-term treatment with
haloperidol low dosage, thioridazine, clozapine and
chlorpromazine on the DA2 and NT receptors in the substantia
nigra and ventral tegmental area, and the effect produced by
chronic clozapine on the DA2/NT receptor balance in the
target areas of the mesolimbocortical and mesostriatal
systems.

MATERIALS AND METHODS

Male Sprague Dawley rats, 200-250 gr body weight, were
used. The animals were housed in standard light and dark
conditions, with food pellets and water ad libitum. Groups of
10 rats were treated with the following drugs at the
indicated doses: haloperidol 0.5 mg/kg i.p., thioridazine 5
mg/kg in 0.2 ml saline orally, chlorpromazine 20 mg/Kg i.p.,
clozapine 20 mg/Kg i.p.. Saline 0.2 ml orally was used as
control of thioridazine treated rats, saline 0.2 ml i.p. as
control of other treatments. All the groups were treated for
21 days once a day. At the end of the treatments and after 3
days washout, the rats were sacrificed. Briefly, under
ketamine anaesthesia (10 mg/kg), the rats were perfused
through the ascending aorta (100 ml saline solution followed
by 100 ml paraformaldehyde 0.1% in PBS pH 7.4), then the
brains were quickly removed, frozen and sectioned in a
cryostat (Leitz Kriostat 1720, -25°C, 20 μm thickness). 5
series of 4 consecutive sections (2 sections for each isotope, total and unspecific binding) were collected on gelatine
coated slides at the rostro-caudal level 1.2 anterior to the
bregma according to the Paxinos and Watson stereotaxic atlas

(1986) and at mesencephalic level. 3H-spiperone was used for the DA2 receptor labelling. 2 sections (total and unspecific binding) were incubated with 3H-spiperone (4nM; DU PONT NEN Products Division, Firenze, Italy; spec. act. 23.2 Ci/mmol), in 170 mM Tris-HCl buffer, pH 7.4, containing 120 mM NaCl, 5 mM KCl, 1 mM MgCl2, 2 mM CaCl2 for 60 min at room temperature. Binding of 3H-spiperone to 5HT-2 receptors was prevented by adding 50 nM ketanserin to the assay buffer. The 3H-spiperone binding to DA receptors was assessed in the presence of 1 μM cold (+)butaclamol. Both series of slides were rinsed with two washes in the same buffer at 4°C for 10 min each, followed by one wash in bidistilled water, buffered with Tris-HCl, pH 7.5 (4°C, 1 min). Two sections (total and unspecific binding) were incubated with 3H-neurotensin (5 nM; DU PONT NEN Products Division, Firenze, Italy; spec. act. 85 Ci/mmol), in 170 mM Tris-HCl buffer, pH 7.5, containing 0.05% BSA and 10 μM bacitracin for 60 min at 4°C. The unspecific binding was assessed in the presence of 1 μM cold neurotensin. Both series of slides were washed in the same buffer 2 x 5 min each. After 6 weeks (3H-spiperone) or 3 months (3H-neurotensin) exposure, the films (Amersham, 3H-Hyperfilm) were developed using Kodak D-19 developer. The quantification of the autoradiograms was performed by means of computerized microdensitometry (Image Analyzer Tesak VDC 501, computer Digital PDP 11), using the standard 3H-microscale by Amersham (England) for the convertion of the optical density values into fmol/mg protein (Fabbri et al., 1985; Giardino et al., 1989; McEachron 1986). In each animal, the 5 series of sections prepared for each isotope at each rostrocaudal level were measured. Analysis of the variance (ANOVA) and the Dunnet test were used for the statistical analysis.

RESULTS

The autoradiographic visualization of NT receptors in a coronal section of the rat brain at mesencephalic level is shown in FIG. 2A. In this figure the distribution of NT binding sites in the substantia nigra pars compacta and in the ventral tegmental area is well evident. These two areas contain most of the DA-ergic cell bodies localized at mesencephalic level, as indicated by the radioimmunochemical visualization of the DA-marker enzyme tyrosine hydroxilase reported in FIG. 2B. The distribution of 3H-spiperone binding sites is also shown (FIG. 2C).

The modifications of the NT and DA2 receptor density in the substantia nigra (SN) (pars compacta and pars lateralis) and in the ventral tegmental area (VTA) induced by the chronic treatment with haloperidol low dosage, chlorpromazine, thioridazine and clozapine are reported in TABLE I. We found an increase of DA2 receptor density induced by chronic thioridazine and haloperidol both in the SN and in the VTA, whereas chlorpromazine is uneffective and clozapine induces a DA2 upregulation in the SN. The NT receptor density is decreased in the VTA of thioridazine and clozapine treated animals, increased in the SN after chronic chlorpromazine treatment, unchanged by chronic haloperidol.

In FIG. 3, we reported the modification of the DA2/NT receptor balance induced by the chronic treatment with clozapine in the target areas of the mesolimbocortical and

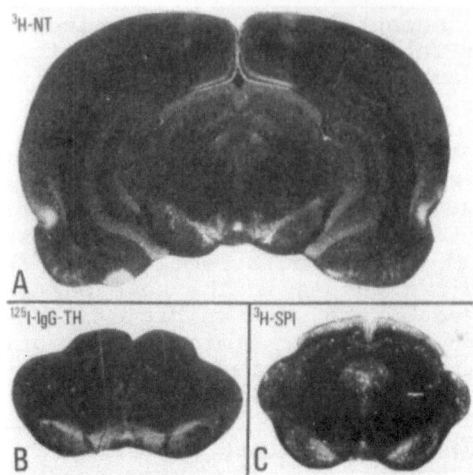

Fig. 2. Autoradiographic visualization of 3H-NT- (A), 3H-SPI-
(C) binding sites and tyrosine hydroxilase immunoreactivity (B) on a coronal section of the rat
mesencephalon.

Table 1. Modifications of NT and DA2 receptor density in the
ventral mesencephalon after chronic treatment with
neuroleptic drugs. The results are expressed in
percentage of the treated groups respect to the
saline. Abbreviations: VTA ventral tegmental area;
SNC substantia nigra pars compacta; SNR substantia
nigra pars reticolare.

		haloperidol 0.5mg/kg	thioridazine 5mg/kg	chlorpromazine 20mg/kg	clozapine 20mg/kg
VTA	3H-SPI	+ 50.2**	+ 56.7**	+ 21.0	+ 6.5
	3H-NT	- 3.3	- 21.9*	+ 7.8	- 19.3*
SNC	3H-SPI	+ 40.8**	+ 34.8*	- 6.4	+ 24.5*
	3H-NT	+ 13.3*	+ 5.7	+ 60.4**	+ 3.7
SNR	3H-SPI	+ 65.4**	+ 6.9	- 7.8	+ 13.3
	3H-NT	+ 12.9	- 18.4	+ 14.4	- 1.6

mesostriatal systems. According to the most recent
classification of the DA-ergic mesencephalic ascending
pathways (Bjorklund and Lindvall, 1984), we divided the
analyzed areas at basal ganglia level in "pure
mesolimbocortical" including the cerebral cortex (medial,

dorsal and lateral, which correspond to the anterior cingulate, frontoparietal motor and frontoparietal somatosensory), the nucleus accumbens and the olfactory tubercle, "pure mesostriatal" including the dorsal part of the n. caudato-putamen (dorso-medial, dorso-central and dorso-lateral) and "mixed mesostriatal-mesolimbocortical" including the ventral part of the n. caudato-putamen (ventro-medial, ventro-central, ventro-lateral). The results are expressed in percentage of the treated groups respect to the saline. The chronic clozapine treatment induces a decrease of the DA2 receptor density in the cerebral cortex and in the ventral striatum, i.e. in the areas receiving a pure A10 or a mixed A10-A9 input. In some of these areas, i.e. the anterior cingulate cortex and in the ventral part of the n. caudato-putamen, NT receptor density also decreases.

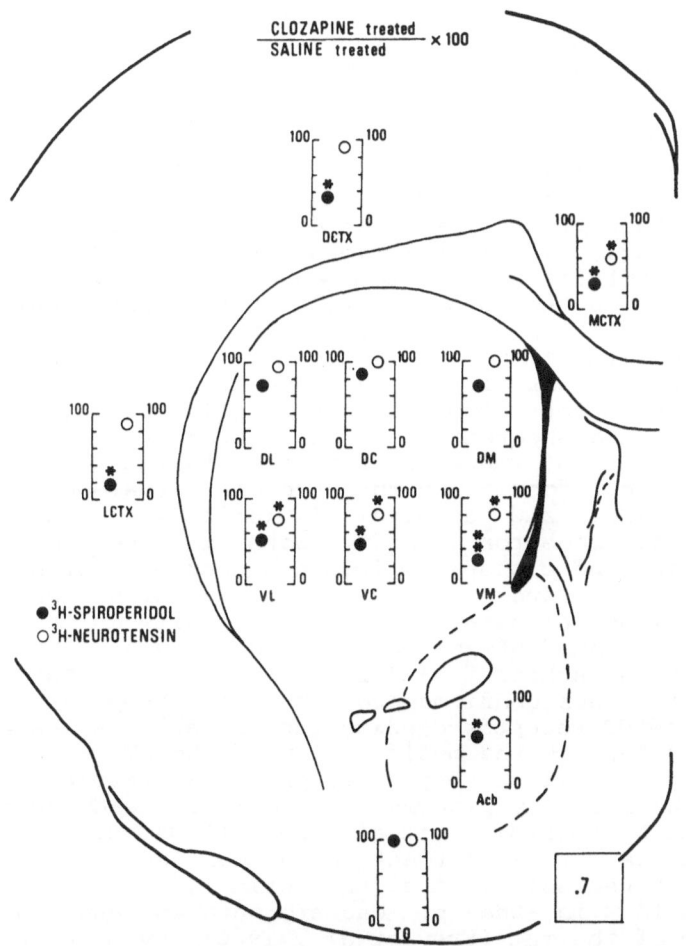

Fig. 3. Modifications of the DA2/NT receptor balance induced by the chronic treatment with clozapine in the target areas of the mesolimbocortical and mesostriatal systems. The results are expressed in percentage of the treated group respect to the control.

DISCUSSION

Because of the extremely complex anatomical and neurochemical organization of the central nervous system (see Graybiel and Ragsdale, 1983 for a revision of the biochemical anatomy of the striatum), the preservation of the anatomical integrity of the tissue should be considered an indispensable condition in the study of the relationship between the behavioral and neurochemical action of all drugs (Piazza et al., 1989; Giardino et al., 1990; Piazza et al., 1990). The use of receptor binding studies performed on tissue homogenate is probably a source of confusion and one of the reasons of the contrasting results present in the literature on the receptor action of neuroleptic drugs on the mesolimbocortical and mesostriatal DA-ergic systems. In fact, by means of this approach is almost impossible to distinguish the portions of striatum receiving DA-ergic inputs from the substantia nigra and from the ventral tegmental area, respectively. The DA-ergic projections from the mesencephalon to the basal ganglia are in fact topographically organized in medio-lateral and antero-posterior fashion (Bjorklund and Lindvall, 1984). The n. caudato-putamen receives projection from the substantia nigra, but also from the ventral tegmental area, in the medial part at all rostro-caudal levels, and also in the ventral part of its head. Moreover, the DA2 receptor distribution in the n. caudato-putamen, such as other transmitters, peptides and receptors, is not uniform (Joyce et al., 1985; Richfield et al., 1987). In this contest, quantitative receptor autoradiography appears as a crucial tool in the identification of the anatomical site of action of a certain pharmacological treatment.

The modifications we found of DA2 and NT receptor density in the ventral mesencephalon after chronic treatment with typical and atypical neuroleptic drugs, could be helpful in the understanding the different behavioral and clinical action of these drugs respect to the DA/NT receptor balance. In fact, both typical (haloperidol and chlorpromazine) and atypical (thioridazine) neuroleptics increase DA2 receptor density both in the SN and VTA. Thioridazine and clozapine induce a decrease of NT binding sites in VTA and are uneffective in the SN. Haloperidol and chlorpromazine increase NT receptor density in the SNC. The increase we found of NT receptor density on the DA-ergic neurons of the substantia nigra after chronic treatment with neuroleptic drugs confirms previous results by Uhl and Kuhar (1984). The NT receptor density can probably be changed in a direct way through the modification of the NT concentration, or in an indirect way through the regulation of a transmitter-identify pathway anatomically and functionally related to NT (Giardino et al., submitted). Several reports indicated that the treatment with neuroleptics interfering with DA transmission, i.e. haloperidol, increases the NT concentration in the n. caudato-putamen and in the n. accumbens of the rat (Frey et al., 1986; Govoni et al., 1980; Nemeroff, 1986; Radke et al., 1989). It has also been suggested that DA inputs regulate the density of postsynaptic NT binding sites through cortical and subcortical DA receptors (Herve et al., 1986).

The decrease of DA2 receptor density in the target areas of the mesolimbocortical system produced by clozapine is not surprising considering that clozapine has characteristics that differ from other atypical neuroleptics (Clasghorn et al., 1987). It is an effective antpsychotic drug with minimal extrapyramidal side-effects. The preferential effect on the mesolimbocortical system is also confirmed by our results.

Our data further support the regulation of NT transmission by drug acting on the DA-ergic systems and indicate the receptor as a possible, indirect synaptic level of action. The study of the NT/DA balance is obviously not enough to investigate the mechanism of action of the antipsychotic drugs, in view of the close anatomical and functional relationship of these systems with other transmitter-identified pathways. The knowledge of the subcellular localization of the different receptor types (Fig. 4) is also a crucial point in order to convert the hypothesis concerning the "synapsis regulation by drugs" into functionally reliable theory.

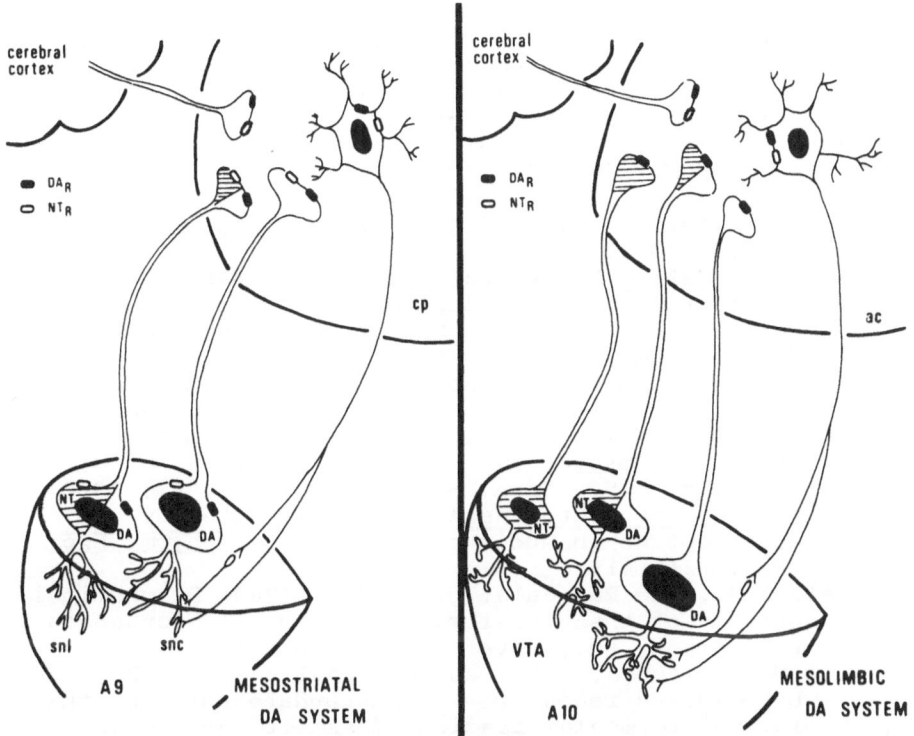

Fig. 4. Localization of NT and DA receptors in the neurons of the mesostriatal and mesolimbocortical DA-ergic systems. DA, NT and DA/NT neurons are reported. Abbreviations: ac n. accumbens; cp n.caudato-putamen, DAR dopamine receptors; NTR neurotensin receptors; snc substantia nigra, pars compacta; snr substantia nigra, pars reticolata; VTA ventral tegmental area.

REFERENCES

Bjorklund, A., and Lindvall, O., 1984, Dopamine containing systems in the CNS, in: "Handbook of Chemical Neuroanatomy. Vol.2: Classical Transmitters in the CNS, Part I," A. Bjorklund and T. Hokfelt, eds., Elsevier, Amsterdam.

Borison, R.L., and Diamond, B.I., 1983, Regional selectivity
     of neuroleptic drugs: an argument for site
     specificity, Brain Res. Bull., 11:215.
Chiodo, L.A., and Bunney, B.S., 1983, Typical and atypical
     neuroleptics: differential effects of chronic
     administration on the activity of A9 and A10 midbrain
     dopaminergic neurons, J. Neurosci., 3:1607.
Claghorn, J., Honigfeld, G., Abuzahab, F.S., Wang, R.,
     Steinbook, R., Tuason, V., and Klerman, G., 1987, The
     risks and benefits of clozapine versus
     chlorpromazine, J. Clin. Psychopharmacol., 7:377.
Fabbri, P.L., Agnati, L.F., Fuxe, K., Battistini, N., Zini,
     I., and Zoli, M., 1985, Principles for the
     construction of the software for image analysis of
     transmitter-identified neurons, in: "Quantitative
     Neuroanatomy in Transmitter Research," L.F. Agnati
     and K. Fuxe, eds., Macmillan Press, London.
Frey, P., Fuxe, K., Eneroth, P., and Agnati, L.F., 1986,
     Effects of acute and long-term treatment with
     neuroleptics on regional neurotensin levels in the
     male rat, Neurochem. Int., 12:33.
Giardino, L., Calzà, L., Zanni, M., Velardo, A., Pantaleoni,
     M., and Marrama, P., 1989, Daily modifications of
     3H-naloxone binding sites in the rat brain: a
     quantitative autoradiographic study, Chronobiol.
     Int., 6:203.
Giardino, L., Calzà, L., Piazza, P.V., and Amato, G., 1990,
     Opiate receptor modifications in the rat brain after
     chronic treatment with haloperidol and sulpiride, J.
     Psychopharm., 4, in press.
Goedert, M., Mantyh, P.W., Emson, P.C., and Hunt, S.P., 1984,
     Inverse relationship between neurotensin receptors
     and neurotensin-like immunoreactivity in cat
     striatum, Nature, 307:543.
Govoni, S., Hong, J.S., Yang, H.-Y., and Costa, E., 1980,
     Increase of neurotensin content elicited by
     neuroleptics in nucleus accumbens, J. Pharm. Exp.
     Ther., 215:413.
Graybiel, A.M., and Ragsdale, C.W., Jr., 1983, Biochemical
     anatomy of the striatum, in: "Chemical Neuroanatomy,"
     P.C. Emson, ed., Raven Press, New York.
Joyce, J.N., Loeschen, S.K., and Marshall, J.F., 1985,
     Dopamine D2 receptors in rat caudate-putamen: the
     lateral to medial gradient does not correspond to
     dopaminergic innervation, Brain Res., 338:209.
Herve, D., Tassin, J.P., Studler, J.M., Dana, C., Kitabgi,
     P., Vincent, J.P., and Glowinsky, J., 1986,
     Dopaminergic control of 125I-labeled neurotensin
     binding sites density in corticolimbic structures of
     the rat brain, Proc. Natl. Acad. Sci. USA., 83:6203.
Hokfelt, T., Everitt, D.J., Theodorsson-Norheim, E., and
     Goldstein, M., 1984, Occurrence of neurotensin like
     immunoreactivity in subpopulations of hypothalamic,
     mesencephalic and medullary catecholamine neurons, J.
     Comp. Neurol., 222:543.
Kalivas, P.W., 1984, Neurotensin in the ventromedial
     mesencephalon of the rat: anatomical and functional
     considerations, J. Comp. Neurol., 226:495.
McEachron, D.L., 1986, Functional mapping in biology and
     medicine: computer assisted autoradiography, Karger,
     Basel.

Nemeroff, C.B., 1986, The interaction of neurotensin with dopaminergic pathways in the central nervous system: basic neurobiology and implications for the pathogenesis and treatment of schizofrenia, Psychoneuroendocrinol., 11:15.

Nemeroff, C.B., Bissette, G., Prange, A.J., Jr., Loosen, P.T. Barlow, T.S., and Lipton, M.A., 1977, Neurotensin: central nervous system effects of a hypothalamic peptide, Brain Res., 128:485.

Nemeroff, C.B., Kalivas, P.W., and Prange, A.J., Jr., 1984, Interaction of neurotensin and dopamine in limbic structures, in: "Catecholamines: Neuropharmacology and Central Nervous System - Theoretical aspects," E. Usdin, A. Carlsson, A. Dahlstrom, eds., Alan R. Liss, New York (1984).

Palacios, J.M., 1986, Dopamine receptor disputes, Nature, 323:205.

Paxinos, G., and Watson, C., 1986, The rat brain in stereotaxic coordinates, Academic Press, New York.

Piazza, P.V., Giardino, L., Calzà, L., and Amato, G., 1989, Behavioral and neurochemical modifications induced by apomorphine treatment, Neurosci. Lett., 100:265.

Piazza, P.V., Calzà, L., Giardino, L., and Amato, G., 1990, Chronic thioridazine treatment differently affects DA receptors in striatum and in mesolimbo-cortical systems, Pharm. Biochem. Behav., 35, in press.

Quirion, R., 1983, Interactions between neurotensin and dopamine in the brain: An overview, Peptides, 4:609.

Quirion, R., Chiueh, C.C., Everist, H.D., and Pert, A., 1985, Comparative localization of neurotensin receptors on nigrostriatal and mesolimbic dopaminergic terminals, Brain Res., 327:385.

Radke, J.M., MacLennan, A.J., Beinfeld, M.C., Bissette, G., Nemeroff, C.B., Vincent, S.R., and Fibiger, H.C., 1989, Effects of short- and long-term haloperidol administration and withdrawal on regional brain cholecystokinin and neurotensin concentrations in the rat, Brain Res., 480:176.

Richfield, E.K., Young, A.B., and Penney, J.B., 1987, Comparative distribution of dopamine D-1 and D-2 receptors in the basal ganglia of turtles, pigeons, rats, cats, and monkeys, J. Comp. Neurol., 262:446.

Sugimoto, T., and Mizuno, N., 1987, Neurotensin in projection neurons of the striatum and nucleus accumbens, with reference to coexistence with enkephalin and GABA: an immunohistochemical study in the cat, J. Comp. Neurol., 257:383.

Uhl, G.R., Goodman, R.R., and Snyder, S.H., 1979, Neurotensin containing cell bodies, fibers and nerve terminals in the brain stem of the rat: immunohistochemical mapping, Brain Res., 167:77.

Uhl, G.R., and Kuhar, M.J., 1984, Chronic neuroleptic treatment enhances neurotensin receptor binding in human and rat substantia nigra, Nature, 309:350.

Young, W.S.III, and Kuhar, M.J., 1981, Neurotensin receptor localization by light microscopic autoradiography in rat brain, Brain Res., 206:273.

White, F.J, and Wang, R.Y., 1983, Differential effects of classical and atypical antipsychotic drugs on A9 and A10 dopamine neurons, Science, 221:1054.

EVIDENCE FOR D2-DOPAMINE RECEPTORS MODULATING

NORADRENALINE RELEASE IN THE RAT FRONTAL CORTEX

Z.L. Rossetti, L. Pani, C. Portas, S. Carboni
and G.L. Gessa

"B.B. Brodie" Department of Neuroscience
University of Cagliari
Via Porcell 4, 09124 Cagliari, Italy

INTRODUCTION

The frontal cortex of the rat receives dopaminergic and noradrenergic nerve terminals from the ventral tegmental area and the locus coeruleus, respectively. Different categories of catecholamine receptors are present in this area, including alpha2-adrenoceptors and D1 and D2 dopamine (DA) receptors. However the location and the role of such receptors are not clear. Peripheral noradrenergic nerve terminals have both alpha2 and D2 presynaptic receptors, the stimulation of either ones resulting in the inhibition of noradrenaline (NA) release (see 1). In the central noradrenergic nerve terminals the presence of presynaptic alpha2-adrenoceptors has been shown in vitro by using brain slices and synaptosomes (2,3) and, more recently, by the use of the technique of brain dialysis in anesthetized rats (4). In the latter study the cortical outflow of NA was shown to be inhibited by clonidine, an alpha2-adrenoceptor agonist, and to be increased by idazoxan, a blocker of alpha2-adrenoceptors.

While considerable evidence in vivo and in vitro indicates that presynaptic D2 receptors modulate the release and/or synthesis of DA in the frontal cortex (5-8), little is known of the presence of presynaptic D2 receptors on noradrenergic nerve terminals in the frontal cortex. Galzin et al. (2) found in rabbit hypothalamic slices that two dopaminergic agonists, pergolide and apomorphine, inhibited the electrically evoked [^{3}H] NA release and that the inhibition was antagonized by (-)sulpiride, a specific D2 antagonist.

The Basal Ganglia III, Edited by G. Bernardi *et al.*
Plenum Press, New York

Misu et al. (3), using rat hypothalamic slices, found that DA and apomorphine had a biphasic action on the electrically evoked release of [^{3}H] NA, i.e. an inhibitory effect at low concentrations, which seemed to be mediated by D2 receptors on noradrenergic nerve terminals, and a stimulant one, at higher concentrations, possibly mediated by presynaptic-beta receptors.

We have used the technique of intracerebral dialysis in conscious freely moving rats in order to investigate if D2 receptors, other than alpha2-adrenoceptors, might control the release of NA from nerve terminals in the frontal cortex.

MATERIALS AND METHODS

Brain microdialysis was performed in male Sprague-Dawley CD 350-400g (Charles River, Italy) freely moving rats essentially as described by Imperato and Di Chiara (9), using bicarbonate Krebs-Ringer buffer pH 7.4 as perfusing solution. Dialysis membranes (acrylonitrile-methallyl sulphonate fibers, 220 μm i.d., Hospal, France) were implanted transversely through the frontal cortex (A: 2.2; V: 2.0 from bregma) (10) 24 hr before the experiment. The correct location of the dialysis membrane in the brain was verified by anatomical examination after each experiment. The perfusates (50 μl/30 min) were directly applied to the HPLC-EC equipment, using a computer-controlled, fully automated system which avoided the manipulation of the animals. The mobile phase in the HPLC was citrate/acetate buffer 50 mM, pH 4.8, containing sodium octyl sulphate 70 mg/l, EDTA 1 mM and methanol 3.5% v/v. The column was a 25 cm octadecyl silica, 5 μm particle size (Supelco), the flow rate 0.9 ml/min (Waters 510 pump) and the applied potential 0.7 Volts vs. an Ag/AgCl reference electrode (Waters M460 electrochemical detector) with a glassy carbon electrode). An integrator (Waters 710 Data Module) was used to record the signal.

After obtaining a stable baseline for the release of the amine, the animals received drugs or saline. NA identity in the perfusates was assessed by collecting the peak obtained from 15-20 samples, adsorbing the catecholamine fraction into alumina and injecting the acid extract from alumina into the HPLC at various chromatographic conditions.

RESULTS

A stable baseline of NA release was observed within 120 min after the beginning of the perfusion and was main-tained for at least 6 hrs. Baseline NA output measured in 30 min fractions in the perfusate corresponded to 21.3 ± 2 fmol/min (n=45). within 90 min after treatment, remained elevated for more than 3 hrs (Fig. 1).

As expected from previous results (4), the adminis-tration of yohimbine (5 mg/kg i.p.) resulted in a sharp increase of cortical NA release, which reached its maximum

(150% above baseline). Vice versa, the administration of clonidine (0.3 mg/kg i.p.) reduced NA release by approximately 50% within 30 min after treatment, the reduction persisting for about 2 hrs (Fig. 2).

As shown in Fig. 3, the subcutaneous administration of LY 171555, a selective D2 receptor agonist, caused a dose-dependent decrease in NA release. A significant inhibition of NA outflow was produced by the dose of 50 ug/kg. The dose of 150 ug/kg produced a maximal inhibition of about 50% of baseline. No further increase was observed with a higher dosage.

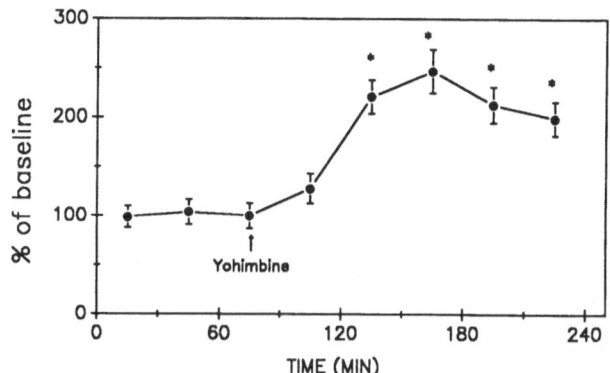

Figure 1. Effect of yohimbine (5 mg/kg i.p.) on cortical NA release in the conscious freely moving rat, perfused transcortically with a dialysis fiber, at a perfusion rate of 1.6 ul/min. NA was assayed by HPLC-EC in 30 min samples. Results are expressed as per cent of baseline and are means ± S.E.M. of data obtained from 4 rats. P < 0.01 versus basal values. Baseline NA release was 9.6 ± 3.3 fmol/min.

As Fig. 4 shows, the intraperitoneal administration of (-)sulpiride, a selective D2 receptor antagonist, at the dose of 100 mg/kg, produced a 150% increase in NA release. NA output remained elevated for over 2 hrs.

The inhibitory effect of clonidine (0.3 mg/kg) was prevented by yohimbine (5 mg/kg), administered i.p. 30 min beforehand, which, however, failed to prevent the response to LY 171555 (50 μg/kg s.c.).

Vice versa, while the response to LY 171555 (50 μg/kg

s.c.) was prevented by (-)sulpiride (150 mg/kg i.p., 30 min beforehand), the latter failed to antagonize the inhibitory effect of clonidine (0.3 mg/kg i.p.) (Rossetti et al., submitted).

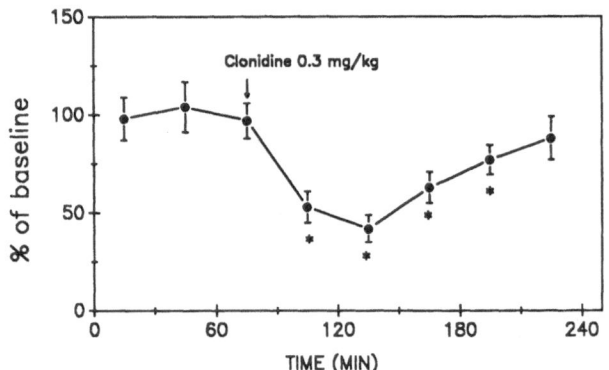

Figure 2. Effect of clonidine (0.3 mg/kg i.p.) on cortical NA release in rats under the same conditions described in the legend of Fig. 1. Each value is the mean ± S.E.M. of data obtained from 4-6 rats. *P < 0.01 versus basal values.

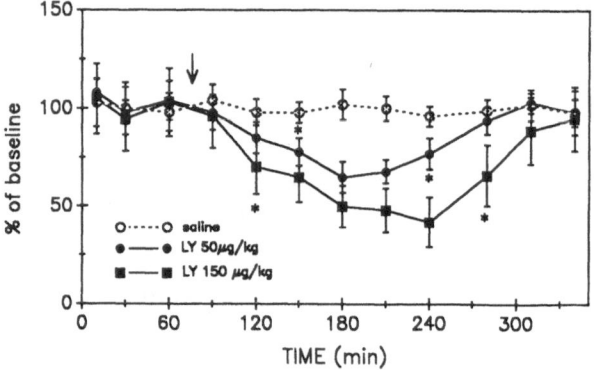

Figure 3. Effect of LY 171555 on cortical NA release in rats under conditions described in the legend of Fig.1. Each value is the mean ± S.E.M. of data obtained from 4-6 animals. P < 0.01 versus basal values.

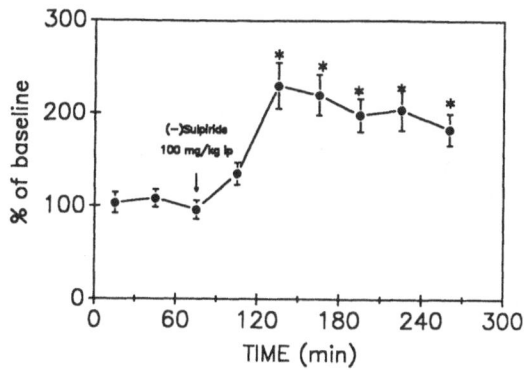

Figure 4. Effect of (-)sulpiride on cortical NA release in rats, under same conditions in Fig. 1. Each value is the mean ± S.E.M. of data obtained from 4-5 animals. *P < 0.01 versus basal value.

DISCUSSION

We have confirmed in freely moving rats that clonidine and yohimbine modify NA outflow in the frontal cortex in an opposite manner: clonidine inhibiting and yohimbine increasing NA output (4).

Moreover, we found that NA outflow in the frontal cortex was inhibited by LY 171555, a selective D2 agonist, while it was increased by (-)sulpiride, a selective D2 receptor antagonist. The effect of clonidine and that of LY 171555 appear to result from actions at alpha2 and D2 receptors, respectively, as they were differentially blocked by the alpha2-adrenergic receptor antagonist, yohimbine, and by the D2 antagonist (-)sulpiride, respectively.

Thus, our results indicate the existence of D2 receptors modulating the release of NA in the frontal cortex. Moreover, the finding that NA output in this area was increased after (-)sulpiride administration, suggests the presence of a tonic D2 receptor-mediated inhibitory control of NA release by DA.

Our experimental model does not allow any indication on the exact location of the D2 receptors modulating NA release. However, since DA agonists are known to be ineffective in inhibiting the firing rate of noradrenergic neurons (11) it is likely that D2 receptors inhibiting NA release might be located in the innervated areas. From previous observations in brain slices and studies on peripheral adrenergic nerve terminals one might suggest that the inhibitory D2 receptors are located on noradrenergic nerve terminals in the frontal area.

Besides the problem of the exact location of such receptors our findings raise the possibility that D2-mediated modulation of NA release might be involved in the depressant effect of minute doses of apomorphine and other DA agonists and that, vice versa, a blockade of the above receptors might mediate the motor stimulant and, perhaps, the antidepressant effect of (-)sulpiride (12).

ACKNOWLEDGMENTS

We thank Mr. Antonio Boi for his skillful technical assistance. This work was partially supported by CNR (Z.L.R.), contract no. 88.00626.04. Financial support from Ravizza S.p.A., Muggiò, Italy, is also acknowledged.

REFERENCES

1. S. Z. Langer, Presynaptic regulation of the release of catecholamines. Pharmacol. Rev. 32:337 (1980)
2. A. M. Galzin, M. L. Dubocovich, and S. Z. Langer, Presynaptic inhibition by dopamine receptor agonists of noradrenergic neurotransmission in the rabbit hypothalamus. J. Pharmacol. Exp. Therap. 221:461 (1982).
3. Y. Misu, Y. Goshima, H. Ueda, and T. Kubo, Presynaptic inhibitory dopamine receptors on noradrenergic nerve terminals: analysis of biphasic actions of dopamine and apomorphine on the release of endogenous norepinephrine in rat hypothalamic slices. J. Pharmacol. Exp. Therap. 235:771 (1985).
4. R. L'Heureux, T. Dennis, O. Curet, and B. Scatton, Measurement of endogenous noradrenaline release in the rat cerebral cortex in vivo by transcortical dialysis: effects of drugs affecting noradrenergic transmission. J.Neurochem. 46:1794 (1986).
5. F. Fadda, G.L. Gessa, M. Marcou, E. Mosca, and Z.L. Rossetti, Evidence for dopamine autoreceptors in mesocortical dopamine neurons. Brain Res. 293:67 (1985).
6. R. K. Talmaciu, I.S. Hoffmann, and L.X. Cubeddu, Dopamine autoreceptors modulate dopamine release from the prefrontal cortex. J. Neurochem. 47:865 (1986).
7. C. A. Altar, W. C. Boyar, E. Oei, and P. L. Wood, Dopamine autoreceptors modulate the in vivo release of dopamine in the frontal, cingulate and entorhinal cortices. J.Pharmacol. Exp. Therap. 242:115 (1987).
8. M. P. Galloway, M. E. Wolf, and R. H. Roth, Regulation of dopamine synthesis in the medial prefrontal cortex is mediated by release modulating autoreceptors: studies in vivo. J. Pharmacol. Exp. Therap., 36: 689 (1986).

9. A. Imperato, and G. Di Chiara, Trans-striatal dialysis coupled to reverse phase high performance liquid chromatography with electrochemical detection: a new method for the study of the in vivo release of endogenous dopamine and metabolites. J. Neurosci., 4:966 (1984).

10. J. F. Konig, and R. A. Klippel, The rat brain, Williams & Wilkins, Baltimore (1960).

11. J. M. Cedarbaum, and G. K. Aghajanian, Catecholamine receptors on locus coeruleus neurons: pharmacological caracterization. Eur. J. Pharmacol. 44:375 (1977).

12. G. U. Corsini, M. Del Zompo, G. B. Melis, A. Mangoni, and G. L. Gessa, (-)Sulpiride as a specific antagonist of "low-dose" effects of apomorphine in man, in "Sulpiride and Other Benzamides, Experimental and Clinical Pharmacology", P.F. Spano, M. Trabucchi, G.U. Corsini, and G.L. Gessa eds., Italian Brain Research Foundation Press, Milano (1979).

COMPARATIVE INVESTIGATIONS OF TERGURIDE ISOMERS ON

CENTRAL DOPAMINE AND NORADRENALINE FUNCTIONS

H. Wachtel, K.-J. Rettig, P.-A. Löschmann and
G. Sauer

Research Laboratories of Schering AG, Berlin
and Bergkamen, F.R.G.

INTRODUCTION

Terguride is an 8α-ergoline derived from the dopamine
(DA) receptor agonist lisuride. In previous neuropharmacological (Wachtel and Dorow, 1983) and neurobiochemical
(Kehr, 1984) experiments terguride has been shown to
exhibit a profile of action which characterises the compound as a DA partial agonist. Thus, at normosensitive central DA receptors the DA antagonistic component of action
prevails (Wachtel and Dorow, 1983) whereas the DA agonistic
component of action becomes evident at non-synaptic DA
receptors like those located on pituitary lactotrophs
(Wachtel and Dorow, 1983; Wachtel et al., 1984a) or is
unmasked at supersensitive central DA receptors as judged
from the release of contralateral rotations in rats bearing
unilateral nigrostriatal lesions (Dlabac and Krejci, 1980),
the production of stereotypies and hyperlocomotion in chronically reserpinised rats (Wachtel et al., 1984b) or the
release of contralateral rotations in hemiparkinsonian 1-
methy-4-phenyl-1,2,3,6-tetrahydropyridine   (MPTP)-treated
monkeys (Brücke et al., 1988). The DA partial agonism of
terguride might also contribute to reducing the incidence
of unwanted side effects like nausea and emesis being associated with the use of DA agonists in man (Wachtel and
Dorow, 1983).
This unique profile of action of terguride suggested the
usefulness of the compound for the treatment of diseases
which are thought to arise from disturbed central DA activity. Thus, terguride has been shown to be effective in
patients with Parkinson's disease (Corsini et al., 1985;
Brücke et al., 1986; Suchy et al., 1986; Critchley and
Parkes, 1987; Filipova et al., 1988) where one might expect
denervation supersensitivity of striatal DA receptors due
to the degeneration of DA neurones of the substantia nigra.
Furthermore, terguride effectively lowered prolactin secretion in hyperprolactinaemic patients (Ciccarelli et al.,
1988). Recently, terguride has been successfully employed
for the treatment of negative symptoms in patients

Structure of terguride isomers

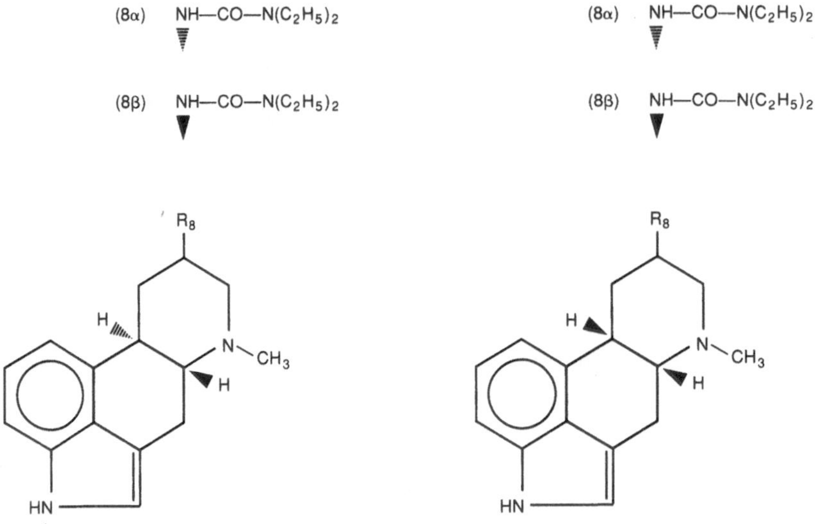

(8α) NH—CO—N(C$_2$H$_5$)$_2$

(8β) NH—CO—N(C$_2$H$_5$)$_2$

(8α) NH—CO—N(C$_2$H$_5$)$_2$

(8β) NH—CO—N(C$_2$H$_5$)$_2$

(5,10)-Trans-Dihydrolisuride (TDHL)

(5,10)-Cis-Dihydrolisuride (CDHL)

Fig. 1

suffering from chronic schizophrenia (Olbrich and Schanz, 1988).
Terguride can be synthesised stereoselectively via reduction of the 8α-ergoline lisuride with lithium in ammonia (Sauer et al., 1986). On principle, hydrogenation of the 9,10-double bond of lisuride can result in 5,10-trans configurated dihydrolisuride (8α-TDHL = terguride) or the 5,10-cis configurated dihydrolisuride (8α-CDHL) (Fig. 1). To get an impression on steric influences on the neurotropic profile of terguride, 8α-CDHL and the corresponding 8β-analogues (8β-TDHL; 8β-CDHL) (Fig. 1) were synthesised and tested for their effect on central DA- and α$_2$-receptor function in comparison with terguride.

MATERIAL AND METHODS

Drugs and Solutions

The following drugs were used: Apomorphine hydrochloride (Sandoz AG, Basel, Switzerland); clonidine hydrochloride (Boehringer, Ingelheim, F.R.G.). The 8α- and 8β-isomers of 5,10-trans configurated dihydrolisuride (8α-TDHL = terguride; 8β-TDHL) and of 5,10-cis configurated dihydrolisuride (8α-CDHL; 8β-CDHL) were synthesised by Dr. G. Sauer (Dept. of Pharmaceutical Chemistry, Schering AG,

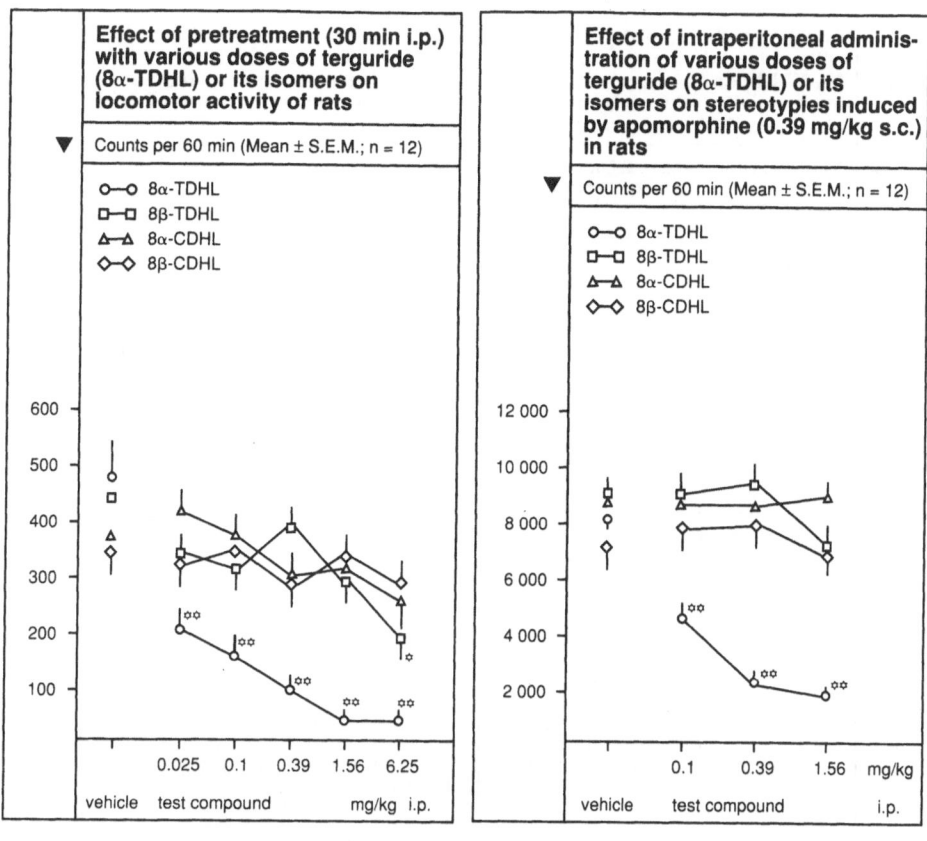

Fig. 2   Fig. 3

Berlin/Bergkamen, F.R.G.) Apomorphine and clonidine were
dissolved in isotonic saline solution whereas the ergolines
were suspended in isotonic saline solution containing 10 %
w/v Cremophor EL^R (polyethoxylated castor oil, BASF,
Ludwigshafen, F.R.G.). For experiments in dogs suspensions
of the ergolines in isotonic saline solution containing
0.085 % w/v Myrj 53 (Atlas Chemie, Essen, F.R.G.) were
used. Ascorbic acid (0.1 % w/v) was added to the apomorphine solutions.

Animals and Treatment Schedules

Male NMRI mice weighing 20-25 g and male Wistar rats
weighing 80-120 g (for behavioural studies) or 180-220 g
(for prolactin studies) were used (Dept. of Animal Breeding
and Housing, Schering AG). Adult male beagle dogs (9-18 kg)
were used for investigations on emetic activity. The drugs
were administered i.p. except for apomorphine which was
given s.c. or i.p. For studies on prolactin secretion
(rats) and emetic action (dogs) the ergolines were administered s.c. All doses were calculated as base and administered in a volume of 0.1 ml per 10 g body weight (mice),
0.5 ml per 100 g body weight (rats) and 0.5 ml per 1000 g
body weight (dogs) with randomised allocation of treatments. Control animals received the corresponding volume of
the vehicle.

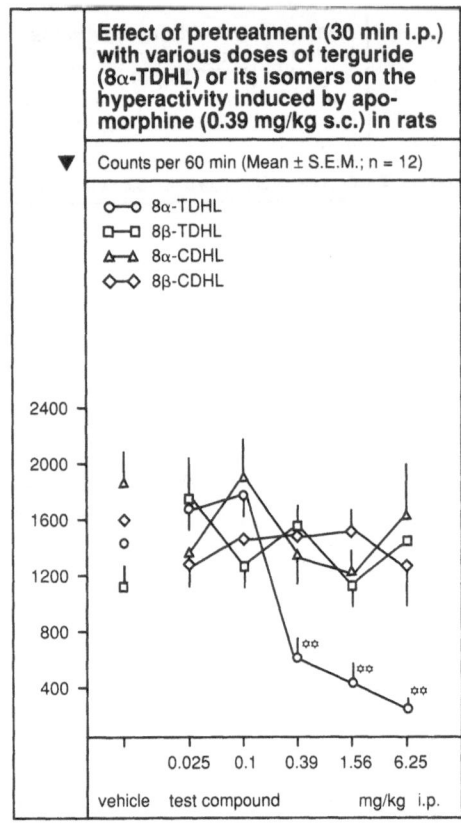

Fig. 4

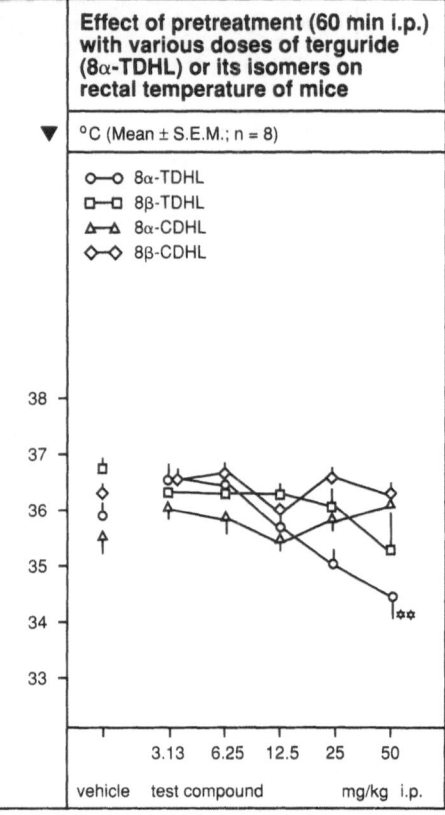

Fig. 5

Test Methods

Catalepsy: The cataleptogenic effect of terguride and its isomers was tested in mice according to a modification of a previously described method (Timsit, 1967). One hour after i.p. pretreatment with various drug dosages or with the vehicle mice were gripped by the scruff and placed in the center of a test area (7 x 8.5 cm) marked on a vertical plane wire grid; the duration of time the animals spent within the test area was then measured. Animals remaining within the test area for > 15 seconds were considered cataleptic.

Antagonism of stereotyped behaviour: Antagonism of behavioural stereotypies was determined either by observation of the animals or by automatic registration.
(A) Assessment by observation: One hour before experimentation individual mice were placed into acrylic glass cages (9 x 10 x 8.5 cm). Animals were given vehicle or various doses of the ergolines i.p. 30 minutes prior to apomorphine (1.56 mg/kg s.c.). Thirty minutes after apomorphine the mice were observed for 2 minutes for the presence of stereotyped behaviour (continuous sniffing, licking or gnawing movements for more than 30 seconds within this time) by an experienced observer unaware of the previous treatment.

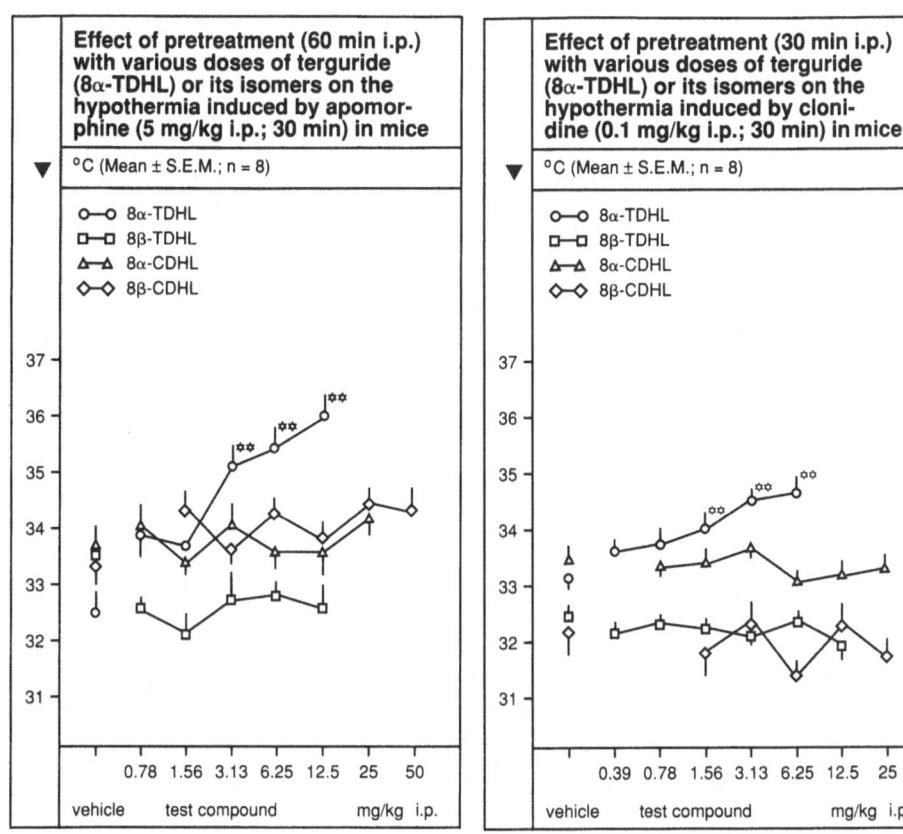

Fig. 6   Fig. 7

(B) Automatic registration (Horowski, 1978): Various doses of the ergolines or the vehicle were administered i.p. together with apomorphine (0.39 mg/kg s.c.). Immediately following the injections individual rats were placed in tubular restraining cages of acrylic glass. Inside the tubular cage in front of the animal's head a stainless steel gnawing rod fixed in the center of a stainless steel beaker was installed; at the opposite end a stainless sheet steel lining was attached inside the cage. The measurement is based on registration of the change in resistance between the rear sheet steel lining and the gnawing rod with beaker fixed at the front, the change being caused by the animal's licking, chewing and gnawing movements. Signals were transformed to digital pulses and counted separately for each cage. The number of counts accumulated during 10 minutes intervals was recorded for 60 minutes.

Locomotor activity: Locomotor activity of rats was measured using circular photocell activity cages. Thirty minutes following the i.p. pretreatment with various doses of the ergolines or with the vehicle, individual animals were placed into the motility cages and the number of light beam interruptions accumulated at 10 minutes intervals were recorded. To examine the influence of terguride or its isomers on the hyperactivity induced by apomorphine (0.39 mg/kg s.c.), the rats were given various doses of the test compounds or the vehicle i.p. 30 minutes prior to apomorphine. Immediately following the injection of apomorphine the locomotor activity was recorded as described above.

Body temperature: Mice were individually placed into acrylic glass cages (9 x 10 x 8.5 cm) one hour prior to experimentation. Body temperature was measured with an electric thermometer for 20 seconds after the introduction of a rectal probe. Dose-response relationships were determined at 60 minutes after the i.p. administration of terguride or its isomers. For interaction studies mice were given vehicle or various doses of the ergolines i.p. 30 minutes prior to a hypothermic dose of apomorphine (5 mg/kg i.p.) or of clonidine (0.1 mg/kg i.p.) and rectal temperature was measured 30 minutes after the injection of apomorphine and clonidine, respectively.

Prolactin: Rats were given vehicle or various doses of the ergolines s.c. 2 hours prior to decapitation. Blood from the trunk was collected in centrifuge tubes, allowed to clot at 4°C and centrifuged at 3000 x g for 10 minutes. The serum was removed and stored at -20°C until assayed. Prolactin was determined using the radioimmunoassay kit for rat prolactin kindly supplied by NIAMDD.

Emesis: After fasting for 24 hours dogs were given a standard diet one hour prior to s.c. administration of vehicle or of various doses of the ergolines. The incidence and frequency of vomiting were assessed during an observation period of 2 hours. A washout period of 14 days elapsed between the trials.

Statistics

The mean effective doses (ED_{50}) and 95 % confidence limits were determined by probit analysis. Means $\pm$ S.E.M. were calculated and the statistical significances of the differences between the means of the various drug doses and the corresponding control were determined by one way analysis of variance in conjunction with the Dunnett test (x : p < 0.05; xx : p < 0.01)

RESULTS

Catalepsy: Terguride (8α-TDHL) in high doses caused catalepsy in mice. The ED_{50} was 50.3 mg/kg i.p. (95 % confidence range: 23.4 to 110.0 mg/kg i.p.). The terguride isomers 8β-TDHL, 8α-CDHL and 8β-CDHL were inactive up to doses of 100 mg/kg i.p.

Antagonism of stereotyped behaviour: Terguride dose-dependently prevented apomorphine-induced stereotypies in mice. The ED_{50} was 1.96 mg/kg i.p. (95 % confidence range: 1.10 to 3.38 mg/kg i.p.). The terguride isomers 8β-TDHL, 8α-CDHL and 8β-CDHL were ineffective up to doses 12.5 mg/kg i.p.
At doses ranging from 0.1 to 1.56 mg/kg i.p. terguride also in rats dose-dependently prevented apomorphine-induced stereotypies whereas its isomers were inactive in this respect (Fig. 2).

Locomotor activity: Terguride, at doses ranging from 0.025 to 6.25 mg/kg i.p., dose-dependently inhibited locomotor activity of rats; this effect was statistically significant at the dose of 0.025 mg/kg. 8β-TDHL significantly reduced locomotor activityonly at the highest dose (6.25 mg/kg) whereas 8α-CDHL and 8β-CDHL were inactive (Fig. 3).
At doses ranging from 0.025 to 6.25 mg/kg i.p., only pretreatment with terguride prevented the apomorphine-induced hypermotility in rats whereas its isomers were ineffective in this respect (Fig. 4).

Body temperature: Terguride at high doses lowered rectal temperature of mice whereas its isomers did not influence thermoregulation up to doses of 50 mg/kg i.p. (Fig. 5).
Pretreatment with terguride at dosages, which per se did not influence thermoregulation, dose-dependently reversed the hypothermia caused by apomorphine (Fig. 6) or clonidine (Fig. 7) in mice. 8β-TDHL, 8α-CDHL and 8β-CDHL were inactive in both interaction tests.

Prolactin: Two hours after pretreatment with terguride or its isomers at dosages ranging from 0.025 to 6.25 mg/kg s.c., serum prolactin levels were lowered only in terguride-treated rats; this effect was statistically significant at the lowest dose tested (21.5 $\pm$ 2.5 % of controls; n = 8). 8β-TDHL, 8α-CDHL and 8β-CDHL had no influence on prolactin secretion.

Emesis: At dosages ranging from 0.03 to 0.3 mg/kg s.c., terguride caused vomiting in one of 5 dogs at the highest dose tested (0.3 mg/kg s.c.) whereas its isomers 8β-TDHL, 8α-CDHL and 8β-CDHL were ineffective in this respect.

DISCUSSION

The present data confirm and extend the previously published evidence on the DA partial agonistic activity of terguride (Wachtel and Dorow, 1983). The agonistic activity is revealed at the DA receptors of the pituitary lactotrophs and of the chemoreceptor trigger zone. The potent prolactin lowering effect in rats contrasts with the rather weak emetic activity in dogs, a findig also confirmed in man (Ciccarelli et al., 1988). On the other hand, the antagonistic activity of terguride is manifest at the receptors of the mesohypothalamic, nigrostriatal and mesolimbic DA systems as judged from the reversal of the apomorphine-induced hypothermia, the antagonism of the apomorphine-induced stereotypies, the inhibition of locomotor activity and the suppression of the apomorphine-induced hyperactivity. This could indicate that even under conditions of physiological receptor occupancy with endogenous DA there exist gradual differences in the sensitivity of effector systems towards the physiological transmitter within a given brain structure which can be disclosed by drugs possessing mixed DA agonist-antagonist properties like terguride. Therefore, the functional state of the DA receptors

(Carlsson, 1983) will determine the response towards terguride.

The present findings also indicate that terguride, in addition to its DA partial agonistic activity, possesses central α_2-receptor blocking properties as judged from the reversal of the clonidine-induced hypothermia in mice. Blockade of central α_2-receptors, e.g. by mianserin or idazoxan, is thought to be linked to the clinical antidepressant effect of these compounds (Baumann and Maitre, 1977; Crossley, 1984). As there is a high incidence of depression in Parkinsonian patients (Santamaria et al., 1986), the additional central α_2-receptor blockade might be useful for the management of mental depression accompanying Parkinson's disease.

From our data obtained in the various test models it can be concluded that, in contrast to terguride, its isomers 8β-TDHL, 8α-CDHL and 8β-CDHL exhibited no DA partial agonistic or central α_2-receptor blocking activity. These findings suggest that the steric configuration of the urea residue in position 8 and of the hydrogen atoms in the positions 5 and 10 strongly determines the unique neurotropic profile of terguride.

ACKNOWLEDGEMENTS

We thank Margrit Kunow, Petra Pietzuch, Verena Schulze, Astrid Seltz and Peter Böttcher for technical assistance and Manuela Stappenbeck for typing the manuscript.

REFERENCES

Baumann, P.A. and Maitre, L., 1977, Blockade of presynaptic receptors and of amine uptake in the rat brain by the antidepressant mianserin, Naunyn-Schmiedeberg's Arch. Pharmacol., 300: 31-37.

Brücke, T., Danielczyk, W., Simányi, M., Sofic, E. and Riederer, P., 1986, Terguride: Partial dopamine agonist in the treatment of Parkinson's disease, Adv. Neurol., 45: 573-576.

Brücke, T., Bankiewicz, K., Harvey-White, J. and Kopin, I., 1988, The partial dopamine agonist terguride in the MPTP-induced hemiparkinsonian monkey model, Eur. J. Pharmacol., 148: 445-448.

Carlsson, A., 1983, Dopamine receptor agonists: Intrinsic activity vs. state of receptor, J. Neural Transm., 57: 309-315.

Ciccarelli, E., Touzel, R., Besser, M. and Grossman, A., 1988, Terguride - a new dopamine agonist drug: A comparison of its neuroendocrine and side effect profile with bromocriptine, Fertil. Steril., 49: 589-594.

Corsini, G.U., Bonuccelli, U. Rainer, E. and del Zompo, M., 1985, Therapeutic efficacy of a partial dopamine agonist in drug-free Parkinsonian patients, J. Neural Transm., 64: 105-111.

Critchley, P. and Parkes, D., 1987, Transdihydrolisuride in parkinsonism, Clin. Neuropharmacol., 10: 57-64.

Crossley, D.I., 1984, The effects of idazoxan, an α_2-adrenoceptor antagonist in depression - a preliminary investigation, Abstr. 9th IUPHAR Congress, London 1724 P.

Dallabonzana, D., Liuzzi, A., Oppizzi, G., Cozzi, R., Verde, G., Chiodini, P., Rainer, E., Dorow, R. and Horowski, R., 1986, Chronic treatment of pathological hyperprolactinemia and acromegaly with the new ergot derivative terguride, J. Clin. Endocrinol. Metab., 63: 1002-1007.

Dlabac, A. and Krejci, I., 1980, Central dopaminergic effects of ergoline derivatives, Activ. Nerv. Sup., 22: 208-209.

Filipová, M., Filip, V., Mazek, Z., Müllerová, J., Kás, S., Zizková, B., Krivka, J., Votavá, M. and Krejcová, H., 1988, Terguride in parkinsonism - a multicenter trial, Eur. Arch. Psychiat. Neurol. Sci., 237: 298-303.

Gräf, K.-J., Köhler, D., Horowski, R. and Dorow, R, 1986, Rapid regression of macroprolactinomas by the new dopamine partial agonist terguride, Acta Endocrinol., 111: 460-466.

Horowski, R., 1978, A new method for automatic registration of stereotyped behaviour in the rat, Arzneimittelforsch., 12: 2281-2286.

Kehr, W., 1984, Transdihydrolisuride, a partial dopamine receptor antagonist: Effects on monoamine metabolism, Eur. J. Pharmacol., 97: 111-119.

Olbrich, R. and Schanz, H., 1988, The effect of the partial dopamine agonist terguride on negative symptoms in schizophrenics, Pharmacopsychiat., 21: 389-390.

Santamaria, J., Tolosa, E.S., Vallés, A., Bayes, A., Blesa, R. and Masana, J., 1986, Mental depression in untreated Parkinson's disease of recent onset, Adv. Neurol., 45: 443-446.

Sauer, G., Haffer, G. and Wachtel, H., 1986, Reduction of 8α-substituted 9,10-didehydroergolines, Synthesis, 12: 1007-1012.

Suchy, I., Rinne, U.K. and Wachtel, H., 1986, Evaluation of terguride in patients with Parkinson's disease, Adv. Neurol., 45: 577-581.

Timsit, J., 1967, Activité cataleptigène de quelques neuroleptiques et quelques parasympathomiméthiques chez la souris, Therapie, 22: 885-893.

Venturini, P.L., Horowski, R., Fasce, V., Valenzano, M., Ferreri, C., Badino, G., Rainer, E., Scholz, A. and de Cecco, L., 1988, Suppression of puerperal lactaton by terguride: A double-blind study, Gynecol. Obstet. Invest., 26: 33-38.

Wachtel, H. and Dorow, R., 1983, Dual action on central dopamine function of transdihydrolisuride, a 9,10-dihydrogenated analogue of the ergot dopamine agonist lisuride, Life Sci., 32: 421-432.

Wachtel, H., Dorow, R. and Sauer, G., 1984a, Novel 8α-ergolines with inhibitory and stimulatory effects on prolactin secretion in rats, Life Sci., 35: 1859-1867.

Wachtel, H., Rettig, K.-J. and Seltz, A., 1984b, The central dopamine agonistic action of transdihydrolisuride is unmasked at supersensitive receptors, Naunyn-Schmiedeberg's Arch. Pharmacol., 325 (Suppl.): R 80.

SPECIFICALLY EVOKED RELEASE OF NEWLY SYNTHESIZED

OR STORED DOPAMINE BY DIFFERENT TREATMENTS

Vincent Leviel, Alain Gobert, Bernard Guibert

Laboratoire de physiologie nerveuse

CNRS 91190 Gif sur Yvette, France

INTRODUCTION

It is now well admitted that intraterminal dopamine (DA) is distributed between various compartments . Numerous studies have been devoted to the problem of the DA compartmentation (Javoy and Glowinski, 1971; Groppetti et al., 1977, Mc Millen et al., 1980) and we have recently proposed a new functional model of the intraterminal organization of the metabolic pathways leading to the synthesis and the release of DA (Leviel et al., 1989). In brief, the greatest part of the amine appears to be stored; besides, the newly synthesized molecules constitute a second pool under close dependence of the synthesis and being the first to be released under basal conditions (Besson et al.,1969; Leviel and Guibert, 1987). In consequence, an evoked overflow of DA could result from the involvement of either the stored or the newly synthesized amine.

In addition, various mechanisms regulating the release of DA from the terminals of nigrostriatal cells were also described. It results from several studies that the release of DA is not only determined by the firing of the dopaminergic cells (Gonon and Buda, 1985) and that presynaptic influences can be exerted to the terminal level (Besson et al.,1969; Langer, 1981; Glowinski et al., 1988). The corticostriatal glutamatergic pathway was, for instance, proposed to mediate such a preterminal control (Cheramy et al., 1986). It has also been demonstrated that DA can modulate its own release through an activation of autoreceptors located on the DA terminals (Farnebo and Hamberger, 1971; Tepper et al. 1984). However it is not known whether the various mechanisms regulating the release act on the same or on different metabolic steps in the DA terminals. Thus, the present study was undertaken to investigate if treatments inducing an increased overflow of DA through 1) a glutamatergic stimulation, 2) a terminal depolarisation or 3) an increased firing, affect similarly the storage and the synthesis of the amine.

The preferential involvement of the newly synthesized amine to the release provides a simple method to differentiate treatments affecting mainly the storage from those affecting the synthesis. Indeed, a continuous superfusion of the striatum of anesthetised rats with [³H]tyrosine induces the release of tritiated DA ([³H]DA). At a steady state level, it can be expected that an evoked increase of [³H]DA output should be the mark of an evoked synthesis. Conversely, the involvement of the stored amine (poorly tritiated) should induce a decrease of the ratio between tritiated and unlabelled amine. Thus, this ratio was continuously calculated during local applications of potassium (8 and 16 mM), glutamate (10^{-8} M) and electrical stimulation of afferent fibres to caudate nucleus. Expecting that glutamate and potassium applications enhance the DA synthesis, the tyrosine hydroxylase (TH) activity was assayed on rat striatal slices preincubated with each subtance.

MATERIAL AND METHODS

Rats (300 g) were anesthetized by ventilation with pure oxygen containing 1% halothane. They were implanted with a push pull cannula in the head of the caudate nucleus. Cannula was supplied with an artificial CSF. After a resting period of 1 hour [³H]tyrosine was added in the superfusing fluid and the successive 20 min superfusate fractions were analysed for their content in DA by high performance liquid chromatography and electrochemical detection (LCED). Radioactivity corresponding to the amine was also measured and the ratio between total DA and [³H]DA was calculated. The method of superfusion and evaluation of DA ratio has already been described in details (Leviel et al. 1989).

The activity of the tyrosine hydroxylase enzyme was assayed on the soluble extract from rat striatal slices. The caudate nucleus of rats was sliced (400 um) and preincubated at 37°C in artificial CSF added or not with glutamate (10^{-4} and 10^{-8} M, 30 min) or potassium (56 mM, 5 min). At the end of the preincubation the tissue was homogeneized in phosphate buffer (5 mM, triton 2%, pH 6.0) and centrifugated (10,000 g, 10 min). The enzymatic activity of TH was evaluated by the measurement of DOPA synthesized during the incubation of the supernatant with tyrosine (100 uM) and 6-MPH4 (1mM) as TH cofactor. The DOPA was assayed by LCED. Incubation and DOPA assay were described elsewhere (Leviel et al. 1990).

RESULTS

Effect of a local application of glutamate (10^{-8} M) in the caudate nucleus

When superfusing fluid was added during 20 minutes with glutamate (10^{-8} M), an increase of the [^{3}H]DA release (70 %) could

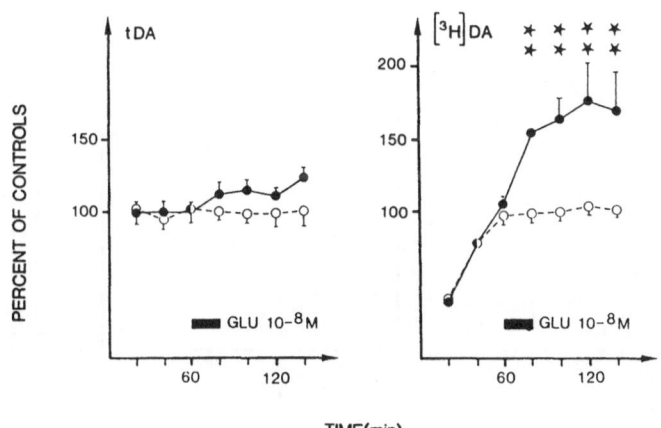

Figure 1. Total (tDA) and tritiated ([^{3}H]DA) dopamine released (0-0) in the caudate nucleus of anesthetised rats, during superfusion with 10^{-8} M glutamate. Comparison is presented with corresponding fractions of control rats (0-0). n=6, **, P<0.05

be observed (Fig 1). This effect did not concern the release of the total DA which was only slightly enhanced (10 %). This effect was long lasting and 80 minutes after the beginning of the application (black bar), the [³H]DA release remained elevated. The consequence of the difference between the release of total DA and [³H]DA was an increase of the ratio between the two forms of the amine (Fig 2).

Effect of local applications of potassium (8,16 mM) in the striatum

Potassium was added in superfusing fluid supplying the push pull cannula during a period of 10 min. A dose of 8 mM nearly remained without effect on the DA release. A moderately increased release of the amine (230%) was obtained after a 16 mM application. During this treatment, as presented on the fig. 2, the ratio between the two forms of the DA was unmodified from control values. This was the sign that tritiated or unlabelled DA were similarly affected by the presence of potassium.

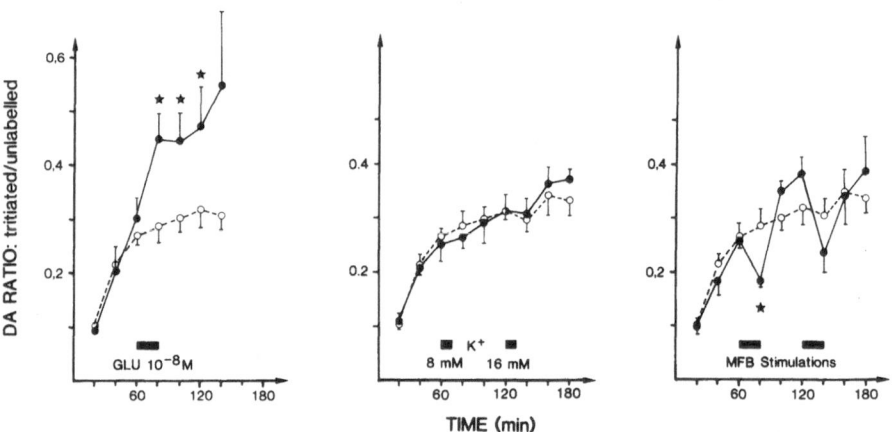

Figure 2. Ratio between total (tDA) and tritiated ([³H]DA) dopamine during treatments increasing the amine release (●-●). Treatments were chosen because they enhanced about twofold the spontaneous release. Comparison is presented with corresponding fractions of control rats (O-O). n=6, **:P<0.05.

Effect of an electrical stimulation of the afferent fibres to the caudate nucleus

The DA fibres were stimulated in the region of the lateral hypothalamus (Ant:4, lat:1.2 ht:1.5) with a concentric bipolar electrode simultaneously implanted with the cannula. The stimulation was constituted of 24 square pulses of 0.3 ms with a frequency of 20 Hz repeated every 1.6 second during 10 minutes. The consequence of the stimulation was an increase of the total DA release (220%) lasting the time of the stimulation. The DA ratio was largely decreased during this stimulation (Fig.2) due to the output of unlabelled DA predominantly to the tritiated DA output. The control values were rapidly reached after the end of the stimulation. The effects of the two successive stimulations are presented on the fig2.

Effects of glutamate and potassium on the enzymatic activity of tyrosine hydroxylase assayed on striatal slices

The effect of glutamate suggesting an increased synthesis of the amine, the activity of TH was assayed on rat striatal slices in presence or not of two doses of glutamate (10^{-8} M, 10^{-4} M). The

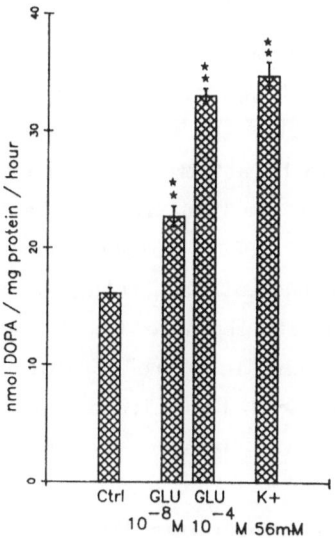

Figure 3. Activity of tyrosine hydroxylase in rat striatal slices preincubated with glutamate (10^{-8} M, 10^{-4} M, 30 min) or potassium (56 mM, 5 min). **, P 0.05. n=6.

effects were compared with that of a potassium application (56 mM) well known to enhance TH activity in such conditions.

The results, presented on the Fig. 3, show that preincubation of the slices with increasing doses of glutamate produced an activation of the enzymatic activity. This effects can be compared with the activation obtained after treatment with potassium.

DISCUSSION

The present study evidenced metabolic differences between three processes increasing DA release in the striatum and involving respectively the firing of DA cells (stimulation of the lateral hypothalamus), the depolarization of DA terminals (local potassium) and the presynaptic receptors (local glutamate). The two main results were the following: first, glutamate mainly affects the DA synthesis, producing in turn an increase of the output of the newly synthesized amine. Second, an electrical stimulation mimicking an increased firing activity leads to the release of the stored amine. Thus it can be concluded that DA is released in the striatum by two different mechanisms triggered by specific stimuli at least.

The conclusion, that glutamate induces the DA synthesis is supported by two observations: the addition of the amino acid in the superfusing fluid induced an increase of the extracellular content of tritiated DA. Surprisingly, the total DA was only slightly affected and this increase could be attributed to the output of the tritiated amine which is a part of the total released amine. As we have already reported (Leviel et al. 1989), higher doses of glutamate did not induce a further enhancement but rather a decrease of the amine output. The only role attributable to glutamate is an enhanced TH activity as confirmed on "in vitro" preparation. This increase is comparable in amplitude with that already obtained in the presence of potassium. Thus glutamate appears to affect exclusively the DA synthesis.

The effect of potassium differs from that of glutamate, since in its presence, the ratio between total and tritiated amine remained unmodified in spite of an increased release of the total DA. This effect is likely to be the result of the simultaneous release of the stored amine (with a low specific activity) and an increased synthesis.

The DA release produced by an electrical stimulation of afferent fibres was responsible for a decrease of the ratio showing a preferential release of the stored amine. This confirms the proposal of Suaud-Chagny (1989) that electrically-evoked release of DA depends more on DA stores. In this case the release of the total DA appears not immediately compensated by a neosynthesis. Nevertheless, an increase of the DA synthesis in response to an increased firing activity along the DA axon have been evidenced (Murrin and Roth, 1976). Our results are not really in conflict with this observation. Indeed, an enhancement of the synthesis likely occurs with some delay and this is attested by the fact that the ratio does not return to the control value between the two stimulations but remains elevated. The same delayed increase of the synthesis was observed by Herdon (1985) after application of potassium on a comparable model "in vitro". Another hypothesis can be proposed to explain the absence of a rapid increase of the synthesis in response to the electrically induced release. As a matter of fact, the total quantities of amine released in the three experimental situations, could be quite different and in the case of the electrical stimulation insufficient to induce an increased [^{3}H]DA overflow in response to each stimulation.

In addition to their effects on the storage and the synthesis, a major difference between the effects of the three treatments come from the duration of the response. In fact an increased synthesis induces a long lasting effect (more than 80 min) able to produce an important output of the amine. On the contrary, an increase of the firing appears to produce the release of the stored amine with a fast return to control values; the synthesis likely occuring secondarily. These observations should be considered in the light of the possible role of the DA in the striatum. Indeed, two modes of action were sometimes attributed to the biogenic amine in the forebrain: first, the direct transmission of a signal toward post synaptic cells and second, a more general "neuro-humoral" action on various distant cells (Beaudet and Descarries, 1978, Fuxe et al., 1989). The first type of action could be the fact of a rapid mechanism of release as that evidenced by stimulation of the caudate-innervating DA fibres.

In opposite, a striatal presynaptic mechanism could be responsible for a large and long lasting release interfering with distant receptors or receptors located on DA terminals. It should be kept in mind that glutamate is a putative transmitter of the corticostriatal pathway known to modulate DA relase in the caudate nucleus. We have recently proposed that its dual modulation is exerted through both an increase and a decrease of the DA synthesis (Leviel et. al., in press). Thus it can be proposed that glutamate could mainly act on the "neuro-humoral" action of the DA in the striatum.

To conclude, whatever the involved trigger is, a double mechanism appears responsible for the release of DA in the striatum. Experiments are needed to better define their relationships and the physiological significance of this duality.

REFERENCES

Beaudet, A. and Descarries, L., 1978, The monoamine innervation of the rat cerebral cortex synaptic and non-synaptic axon terminals, Neuroscience, 3:851-860.

Besson , M.J., Cheramy, A., Feltz, P. and Glowinski J., 1969, Release of newly synthesized dopamine from dopamine containing terminals in the striatum of the rat, Proc.Natl.Acad.Sci.USA, 62:741.

Cheramy, A., Romo, R., Godeheu, G., Baruch, P. and Glowinski, J., 1986, In vivo presynaptic control of dopamine release in the cat caudate nucleus II Facilitatory or inhibitory influence of L-Glutamate, Neuroscience, 19:1081-1090.

Farnebo, L.O. and Hamberger, B., 1971, Drug-induced changes in the release of ^{3}H-monoamines from field stimulated rat brain slices, Acta Physiol.Scand.Suppl., 371:35-44.

Fuxe, K, Agnati, L.F., Zoli, M., Bjelke, B. and Zini, I., 1989, Some aspects of the communicational and computational organization of the brain, Acta Physiol.Scand., 135:203-216.

Glowinski, J., Cheramy, A., Romo, R. and Barbeito, L., 1988 Presynaptic regulation of dopaminergic transmission in the striatum, Cell.Mol.Neurobiol., 8:7-17.

Gonon, F.G. and Buda, M.J., 1985, Regulation of dopamine release by impulse flow and by auto receptors as studied by in vivo voltammetry in the rat striatum, <u>Neuroscience</u>, 3:765-774.

Groppetti, A., Algeri,S., Cattabeni, F., DiGiulio, A.M., Galli A.M., Ponzio, F. and Spano P.F., 1977, Changes in specific activity of dopamine metabolites as evidence of a multiple compartimentation of dopamine in striatal neurons, <u>J.Neurochem.</u>, 28:193-197.

Herdon, H., Strupish, J. and Nahorski, S.R., 1985, Differences between the release of radiolebelled and endogenous dopamine from superfused rat brain slices: Effects of depolarizing stimuli, amphetamine and synthesis inhibition, <u>Brain Res.</u>, 348:309-320.

Javoy, F. and Glowinski, J., 1971, Dynamic characteristic of the functional compartment of dopamine in dopaminergic terminals of the rat striatum, <u>J.Neurochem.</u>,18:1305-1311.

Langer, S.Z., 1981, Presynaptic regulation of the release of catecholamines, <u>Pharmac.Rev.</u>, 32:337-362.

Leviel, V. and Guibert, B., 1987, Involvement of intraterminal dopamine compartments in the amine release in the cat striatum, <u>Neurosci.Letters</u>, 76:197-202.

Leviel, V., Gobert, A. and Guibert, B., 1989, Direct observation of dopamine compartmentation in striatal nerve terminal by 'in vivo' measurement of the specific activity of released dopamine, <u>Brain Res.</u>, 499:205-213.

Leviel, V., Gobert , A. and Guibert, B., 1989, Intraterminal metabolism of dopamine in rat sriatum under different conditions of release, 3°IBAGS Meeting, Cagliary,Italy,June.

Leviel, V., Fayada, C., Guibert, B., Chaminade, M., Machek, G., Mallet, J. and Faucon Biguet, N., 1990, Short and long term alterations of gene expression in limbic structures by repeated electroconvulsive induced seizures, <u>J.Neurochem.</u>, "in press".

McMillen, B.A., German, D.C. and Shore, P.A., 1980, Functional and pharmacoligical significance of brain dopamine and norpinephrine storage pools, <u>Biochem.Pharmacol.</u>, 29:3045-3050.

Murrin, L.C. and Roth, R., 1976, Dopaminergic neurons: Effects of electrical stimulation on dopamine biosynthesis, Mol.Pharmac., 12:463-475.

Suaud-Chagny, M.F., Buda, M. and Gonon, F.G., 1989, Pharmacology of electrically evoked dopamine release studied in rat olfactory tubercule by in vivo electrochemistry. Eur.J.Pharmac., 164:273-283.

Tepper, J.M., Young, S.J. and Groves, P., 1984, Autoreceptor mediated changes in dopaminergic terminal excitability: Effects of increases in impulse flow, Brain Res., 309:309-316.

INTERACTIONS BETWEEN D1 and D2 DOPAMINE RECEPTOR MECHANISMS: THE IMMEDIATE

EARLY RESPONSE GENE C-FOS IN LONG-TERM CHANGES IN THE STRIATUM

H.A. Robertson, M.L. Paul and G.S. Robertson

Department of Pharmacology
Dalhousie University
Halifax, Nova Scotia, Canada B3H 4H7

INTRODUCTION

As recently as 1984, it was possible to refer to the D2 dopamine site as the only functional, behaviourally relevant dopamine receptor; the significance of the D1 site, beyond adenylate cyclase stimulation, was unknown (Grigoriadis and Seeman, 1984). It is now clear that D1 and D2 dopamine receptor mechanisms interact in such a way as to produce additive or synergistic effects on many behavioural and movement-related functions in rats, mice, monkeys and, most likely, in man. The exact mechanisms by which this occurs remains obscure. Here we review the evidence for additive or synergistic effects of D1 and D2 receptor mechanisms and discuss the anatomical implications of these findings. Interaction between D1 and D2 dopamine receptor mechanism may also have important long-term consequences. These long-term consequences include important clinical phenomena such as the tardive dyskinesias and dystonias. In addition to these movement disorders, there is at least reason to suspect that these mechanisms (D1/D2 receptor mechanism interaction, immediate early gene activation) might be involved in major psychiatric illnesses such as the schizophrenias. In animal studies, the phenomenon of dopaminergic "priming" may be a model for the sorts of long-term changes that occur clinically in these psychiatric illnesses and motor disturbances.

Interactions between D1 and D2 dopamine receptor agonists

It is a curious fact that while D1 and D2 dopamine receptor activation often leads to opposite effects on many biochemical systems, simultaneous activation of D1 and D2 dopamine receptors usually produces effects that are either additive, or in some instances, are clearly synergistic. While there are some clear-cut instances in which D1 and D2 dopamine receptor agonists have antagonistic effects on behaviour when administered simultaneously (Rosengarten et al, 1983), for the most part low doses of D1 and D2 agonists, when given together, produce either additive or synergistic effects. The evidence for this has been reviewed on a number of occasions recently (Robertson and Robertson, 1987a,b; Clark and White, 1987; Waddington and O'Boyle, 1987). Recent evidence has confirmed earlier findings (Rouillard and Bedard, 1988; Anden and Grabowska-Anden, 1988; Jackson et al, 1988; Rubinstein et al, 1988; Sonsalla et al, 1988).

The Basal Ganglia III, Edited by G. Bernardi *et al.*
Plenum Press, New York

A neuroanatomical basis for D1/D2 system interactions

It is unlikely that the interactions between D1 and D2 agonists will be understood without taking into account the anatomical localization of D1 and D2 receptors. Both the location of these receptors on neurons (presynaptic, postsynaptic) and the location within the brain (caudate-putamen, olfactory tubercle, nucleus accumbens, substantia nigra pars compacta, substantia nigra pars reticulata) must be considered (see Robertson and Robertson, 1987a;b;1988;1989). The synergistic interaction of D1 and D2 agonists is paradoxical in that all earlier biochemical evidence suggests that D1 and D2 agonists have opposite effects. We suggested that this paradox might be explained if D1 and D2 dopamine receptor agonists have separate sites of action and further proposed that this might involve D2 sites in the striatum and D1 sites in the substantia nigra pars reticulata (Robertson and Robertson, 1987a). There is now good evidence to suggests that this model is probably correct (Robertson and Robertson, 1989) but does not account for all of the observations. The most important omission in this hypothesis is that we know that the striatum and nucleus accumbens both have a very high level of both D1 and D2 dopamine receptor (see Beckstead et al, 1988, for example) and there is behavioural evidence for synergistic activation of locomotion after direct injection of the D1 agonist SKF 38393 and D2 agonist LY 171555 into the nucleus accumbens (Dreher and Jackson, 1989). While the biochemical evidence suggest that there is no antagonistic interaction between D1 and D2 dopamine receptor mechanisms in the nucleus accumbens (Stoof and Verheijden, 1986), neither is there any evidence for additive biochemical effects.

Studies of the interaction between D1 and D2 receptor agonists have been facilitated by recognition that D1 dopamine receptors in the substantia nigra are located exclusively on presynaptic terminals of striatonigral neurons (Savasta et al, 1986, Filloux et al, 1988a), many of which are GABAergic. D1 dopamine agonists probably act at this site to increase GABA release (Geffen et al, 1976). Similarly, most of the D2 receptors in the substantia nigra are located on the cell bodies of the dopaminergic neurons (Morelli et al, 1988). In the striatum, the dopaminergic terminals of the substantia nigra neurons have presynaptic receptors of the D2 type (Starke et al, 1983). There are also postsynaptic D1 and D2 dopamine receptors in the striatum. The exact location of some of these receptors is controversial. For example, there is evidence both for and against the existence of D2 dopamine receptors on the terminals of the glutamatergic corticostriatal afferents (see Trugman et al, 1987; Filloux et al, 1988b; Maura et al, 1988). However, we can make some tentative conclusions that will be useful in resolving the situation. For example, we know that the dopamine neurons in the substantia nigra pars compacta have D2 receptors on the cell body and dendrites and as presynaptic receptors in striatum. Clearly, dopamine neurons express the gene for the D2 receptor and the product appears in dendrites, cell body and terminals (Meador-Woodruff et al, 1989; Mengod et al, 1989); dopaminergic neurons appear not to express D1 receptor (Savasta et al, 1986; Filloux et al, 1988a). Similarly, we know that striatonigral neurons have D1 and not D2 receptor on the terminals (Savasta et al, 1986; Filloux et al, 1988a). If we accept the principal that if a neuron has receptors on its terminals (presynaptic), it must possess similar receptors on the perikarya, it follows that striatonigral neurons express the message for the D1 receptor and possess D1 but not D2 receptors on the perikarya. There is now a good deal of evidence to suggest that the synergistic effects of D1 and D2 dopamine receptor agonists can involve activation of D2 dopamine receptors in striatum and D1 receptors in the substantia nigra (Robertson and Robertson, 1987;1988a,b;1989). However, recent results from our laboratory on D1-mediated activation of c-fos (Robertson et al, in preparation) and the results of Jackson and colleagues

418

indicate that D1/D2 additivity is also possible within the basal ganglia (Dreher and Jackson, 1989). Thus the simple situation that we presented as a hypothesis in 1987 (Robertson and Robertson, 1987a,b) appears to be only partially true; clearly, D1 agonists have effects in the caudate-putamen and other forebrain areas as well. The idea, first proposed by Herrera-Marschitz and Ungerstedt (1984), that dopamine agonist rotation might depend on different output pathways is another possible resolution. Herrera-Marschitz and Ungerstedt (1984) showed that apomorphine-induced rotation (where D1 effects are greater than D2 effect) is mainly dependent on striatonigral pathways while pergolide-induced rotation (where D2 effects predominate) is mainly dependent on other striatal efferents (presumably those to globus pallidus). Thus, the synergistic effects of D1 and D2 agonists may be the result of D1 activation of presynaptic D1 dopamine receptors in nigra but also the simultaneous activation of both efferent pathways from the caudate-putamen. It may be that the dopaminergic system has a good deal of redundancy built in and dopamine will regulate nigrostriatal output in at least 3 different sites using 2 different receptor mechanisms.

Activation of c-fos by D1 dopaminergic mechanisms

In addition to stimulation of rotation, dopaminergic agonists also have profound long-term effects on striatum. For example, either D1- or D2-dopamine agonist "priming" promotes D1-agonist supersensitivity (Morelli and Di Chiara, 1987; Morelli et al, 1987; Criswell et al, 1989). Although D2-dopamine receptor activation can contribute to priming of D1-dopamine receptors, it appears that D2-dopamine receptor stimulation is not necessary for induction or expression of behavioural supersensitivity to D1 agonists (Criswell et al, 1989). It is reported that the priming effect lasts for at least six months (Criswell et al, 1989) and this raises the question of the mechanism by which such changes occur. Recent speculation has centered on the possibility that long-term changes in the central nervous system may rely on an interaction between neuronal activity and the genome (Berridge, 1986). For example, we now know that a kindling stimulation in the hippocampus will activate the c-fos proto-oncogene in dentate granule cells (Dragunow and Robertson, 1987); these same cells later undergo a number of important and long-lasting biochemical and morphological changes (Morris et al, 1988; Sutula et al, 1988). These findings and the suggestion that c-fos might be useful in mapping functional pathways (Sagar et al, 1988) prompted us to examine the effects of dopaminergic agonists on c-fos. In these studies, we demonstrated that L-Dopa and selective D1-dopamine receptor agonists activate c-fos in the striatum ipsilateral to a 6-OHDA lesion of the substantia nigra (Robertson et al, 1989a,b).

Animals with unilateral 6-OHDA lesions rotated away from the side of the lesion when treated with either the D1-selective agonists SKF 38393 or CY 208-243 or with the D2-selective agonist LY 171555. Immunocytochemical examination revealed that the D1-selective agonists but not the D2-selective agonist produced a dramatic activation of c-fos in the denervated striatum; this induction of c-fos seemed to reflect the degree of supersensitivity of dopamine receptors (Robertson et al, 1989a,b). D-amphetamine, which acts by releasing dopamine from the intact dopamine-containing terminals on the unlesioned side, induces ipsilateral rotation and a selective activation of c-fos in the striatum on the intact side (see fig. 1). Thus activation of c-fos appears to reflect the activation of dopamine receptors. Induction of c-fos protein was always confined to the caudate-putamen; no c-fos activation was observed in globus pallidus. The D1-selective agonist SCH 23390 given 10 min prior to the potent D1 agonist CY 208-243 prevents both rotational behaviour and c-fos induction.

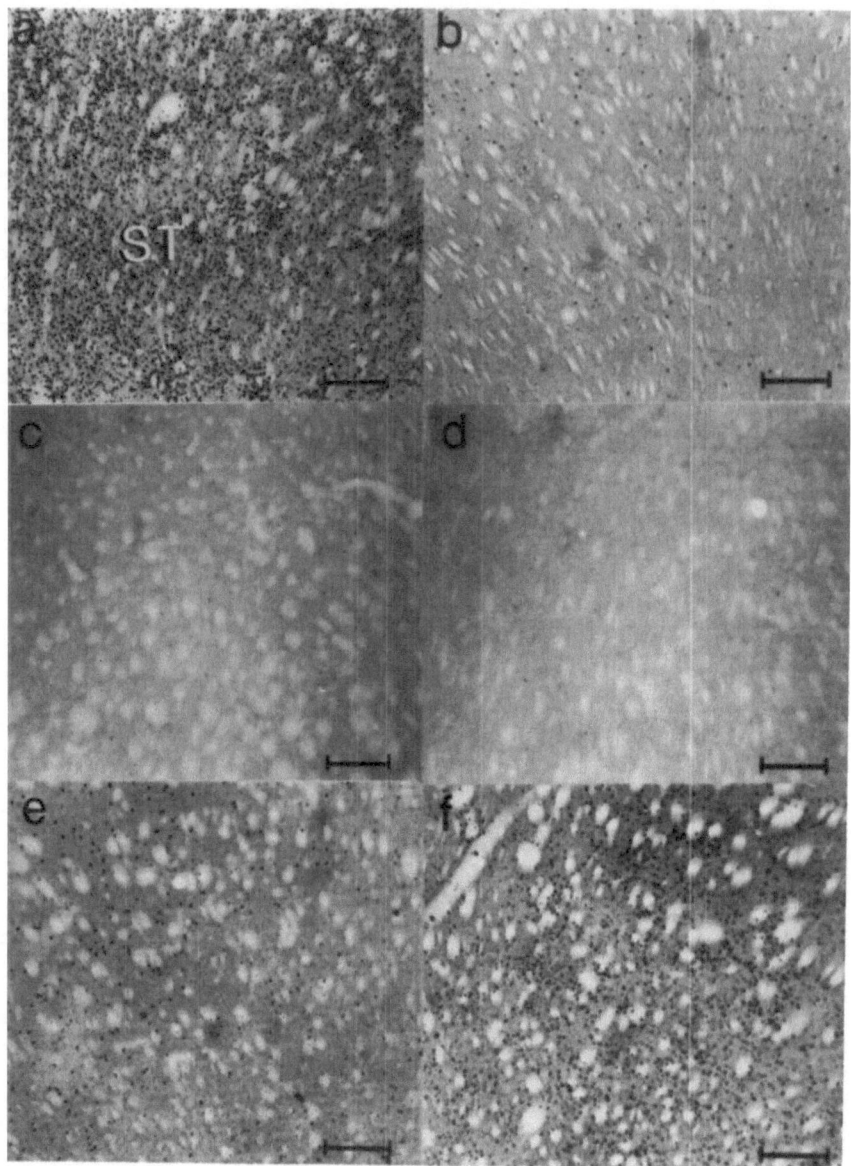

Fig. 1. Effect of dopamine agonists on c-fos-like immunoreactivity in rat
striatum after a 6-hydroxydopamine lesion. a) denervated and b)
intact striatum 2 hours after the D1 agonist SKF 38393 (4 mg/kg);
c) denervated and d) intact striatum 2 hours after the D2 agonist
LY 171555 (0.2 mg/kg); e) and f) denervated and intact striatum 2
hours after D-amphetamine (2.5 mg/kg). bar = 1 mm.

Since the D2-agonist LY 171555 induces rotation but no increase in c-fos is observed in the striatum, it seems unlikely that rotation per se produces the increase in the level of the proto-oncogene. This is confirmed by other two observations. First, injections of SKF 38393 into substantia nigra produce rotation but no c-fos induction in the striatum (or substantia nigra). Second, animals given the D1 agonist CY 208-243 to produce rotation, but maintained on anaesthetic (pentobarbital, 50 mg/kg, i.p.) to prevent rotation for the 2 hour period after the D1 agonist, still show induction of the c-fos protein in striatum. It therefore seems likely that fos activation was not the result of sensory input resulting from the turning process. This conclusion was also supported by the finding that c-fos induction was only observed in rats which were "good turners" (Robertson et al, 1989a).

It is known that activation of dopamine receptors in the striatum has long-term consequences (Morelli and Di Chiara, 1987; Morelli et al, 1987; Criswell et al, 1989). Our results demonstrate that activation of D1-dopamine receptors selectively leads to a dramatic increase in the levels of the protein associated with the c-fos proto-oncogene. The results with the D2 agonist LY 171555 suggest a dissociation between rotation, per se, and c-fos activation. This is confirmed in experiments where L-Dopa-induced rotation but not fos activation was prevented with anaesthetic doses of pentobarbital. It is interesting that L-Dopa and the selective D1 receptor agonist SKF 38393 have similar effects in this system and that the selective D2 receptor agonist LY 17 1555 does not produce these effects. This result reinforces the idea that L-Dopa exerts many of its effects via activation of D1-dopamine receptors (Trugman and Wooten, 1987; Robertson and Robertson, 1988; 1989a).

The demonstration that D1-dopamine receptor stimulation can activate a proto-oncogene mechanism implicated in long-term changes in brain may have important consequences for our understanding of the role of dopamine. The use of dopaminergic drugs is often associated with irreversible pathological changes such as tardive dyskinesia following chronic dopamine receptor blockade. Activation of c-fos may also provide another useful method (in addition to 2-deoxyglucose studies (Trugman and Wooten, 1987) for studying dopamine-activated pathways in caudate-putamen.

REFERENCES

Anden, N.-E. and M. Grabowska-Anden, 1988, Stimulation of D1 dopamine receptors reveals direct effects of the preferential dopamine autoreceptor agonist B-HT 920 on postsynaptic dopamine receptors, Acta Physiol. Scand., 134:285.
Beckstead, R.M., Wooten, G.F. and Trugman, J.M., 1988, Distribution of D1 and D2 dopamine receptors in the basal ganglia of the cat determined by quantitative autoradiography, J. Comp. Neurol. 268:131.
Berridge, M.J., 1986, Second messenger dualism in neuromodulation and memory. Nature 323:294.
Clark, D. and White, F.J., 1987, D1 dopamine receptor-the search for a function, Synapse 1:347.
Criswell, H., Mueller, R.A. and Breese, G.R., 1989, Priming of D1-dopamine receptor responses: long-lasting behavioral supersensitivity to a D1-dopamine agonist following repeated administration to neonatal 6-OHDA-lesioned rats, J. Neurosci. 9:125.
Dragunow, M. and Robertson, H.A., 1987, Kindling stimulation induces c-fos protein(s) in granule cells of the rat dentate gyrus, Nature 329:441.
Dreher, J.K. and Jackson, D.M., 1989, Role of D1 and D2 dopamine receptors in mediating locomotor activity elicited from the nucleus accumbens of rats, Brain Res. 487:267.

Filloux, F., Dawson, T.M. and Wamsley, J.K., 1988a, Localization of nigrostriatal dopamine receptor subtypes and adenylate cyclase, Brain Res. Bull. 20:447.

Filloux, F., Liu, T.H., Hsu, C.Y., Hunt, M.A. and Wamsley, J.K., 1988b, Selective cortical infarction reduces 3H-sulpiride binding in rat caudate-putamen: autoradiographic evidence for presynaptic D2 receptors on corticostriate terminals, Synapse 2:521.

Grigoriadis, D. and Seeman, P., 1984, The dopamine/neuroleptic receptor, Can. J. Neurol. Sci. 4:108.

Jackson, D.M., Ross, S.B. and Hashizume, M., 1988, Further studies on the interaction between bromocriptine and SKF 38393 in reserpine and alpha methyl-para-tyrosine-treated mice, Psychopharmacol. 94:321.

Liebman, J.M., Gerber, R., Hall, N.R. and Altar, C.A., 1988, Heterologous rotational responsiveness in 6-hydroxydopamine-denervated rats: pharmacological and neurochemical characterization, Psychopharmacol. 96:477.

Maura, G., Giardi, A. and Raiteri, M., 1988, Release-regulating D2 dopamine receptors are located on striatal glutamatergic nerve terminals, J. Pharmacol. Exp. Therap. 247:680.

Meador-Woodruff, J.H., Mansour, A., Bunzow, J.R., Van Tol, H.H.M., Watson Jr., S.J. and Civelli, O., 1989, Distribution of D2 dopamine receptor mRNA in rat brain, Proc. Nat. Acad. Sci. U.S.A. 86: 7625 (1989).

Mengod, G., Martinez-Mir, M.I., Vilaro, M.T. and Palacios, J.M., 1989, Localization of the mRNA for the dopamine D2 receptor in the rat brain by in situ hybridization histochemistry, Proc. Nat. Acad. Sci. U.S.A. 86:8560.

Morelli, M. and Di Chiara, G., 1987, Agonist-induced homologous and heterologous sensitization to D-1 and D-dependent contraversive turning, Eur. J. Pharmacol. 141:101.

Morelli, M., Fenu, S. and Di Chiara, G., 1987, Behavioural expression of D1-receptor supersensitivity depends on previous stimulation of D-2 receptors, Life Sci. 40:245.

Morelli, M., Mennini, T. and Di Chiara, G., 1988, Nigral dopamine autoreceptors are exclusively of the D2 type: quantitative autoradiography of [^{125}I]iodosulpride and [^{125}I]SCH 23982 in adjacent brain sections, Neuroscience 27:865.

B.Morris, B.J., Feasey, K.J., ten Bruggencate, G., Herz, A. and Höllt, V., 1988, Electrical stimulation in vivo increases the expression of proenkephalin mRNA and decreases the expression of prodynorphin mRNA in rat hippocampal granule cells, Proc. Natl. Acad. Sci. U.S.A. 85:3226.

Robertson, H.A. and Robertson, G.S., 1987a, Combined L-dopa and bromocriptine therapy for Parkinson's disease: a proposed mechanism of action, Clinical Neuropharmacol. 10:384.

Robertson, G.S. and Robertson, H.A., 1987b, D1 and D2 agonist synergism: separate sites of action? Trends Pharmacol. Sci. 8:295.

Robertson, G.S. and Robertson, H.A., 1988, Evidence that the substantia nigra is a site of action for L-dopa, Neurosci. Lett. 89:204.

Robertson, G.S. and Robertson, H.A., 1989, Evidence that L-Dopa-induced Rotational Behaviour is Dependent on Both Striatal and Nigral Mechanisms, J. Neurosci. 9:3326.

Robertson, G.S., Herrera, D.G., Dragunow, M. and Robertson, H.A., 1989a, L-DOPA activates c-fos in the striatum ipsilateral to a 6-hydroxydopamine lesion of the substantia nigra, Eur. J. Pharmacol. 159:99.

Robertson, H.A., Peterson, M.R., Murphy K. and Robertson, G.S., 1989b, D1 dopamine receptor agonists selectively activate striatal c-fos independent of rotational behaviour, Brain Research 503:346.

Rosengarten, H., Schweitzer, J.W. and Friedhoff, A.J., 1983, Induction of oral dyskinesias in naive rats by D1 stimulation, Life Sci. 33:2479.

Rouillard, C. and Bedard, P.J., 1988, Specific D1 and D2 dopamine agonists have synergistic effects in the 6-hydroxydopamine circling model in the rat, Neuropharmacol. 27:1257.

Rubinstein, M., Gershanik, O. and Stefano, F.J.E., 1988, Different roles of D1 and D2 receptors involved in locomotor activity in supersensitive mice. Eur. J. Pharmacol. 148:419.

Sagar, S.M., Sharp, F.R. and Curran, T., 1988, Expression of c-fos protein in brain: Metabolic mapping at the cellular level, Science 240:1328.

Savasta, M., Dubois, A. and Scatton, B., 1986, Autoradiographic localization of D1 dopamine receptors in the rat brain with [3H]-SCH 23390, Brain Res. 375:291.

Sonsalla, P.K., Manzino, L. and Heikkila, R.E., 1988, Interactions of D1 and D2 dopamine receptors on the ipsilateral vs. contralateral side in rats with unilateral lesions of the dopaminergic nigrostriatal pathway, J. Pharmacol. Exp. Therap. 247:180.

Starke, K., Spath, L., Lang, J.D. and Adelung, C., 1983, Further functional in vitro comparison of pre- and postsynaptic dopamine receptors in the rabbit caudate nucleus, Naunyn-Schmiedeberg's Arch. Pharmacol. 323:298.

Stoof, J.C. and Verheijden, P.F.H.M., 1986, D2 receptor stimulation inhibits cyclic AMP formation brought about by D1 receptor stimulation in the rat neostriatum but not nucleus accumbens, Eur. J. Pharmacol. 129:205.

Sutula, T., Xiao-Xian, H., Cavazos, J. and Scott, G., 1988, Synaptic reorganization in the hippocampus induced by abnormal functional activity, Science 239:1147.

Trugman, J.M., Geary, II, W.A. and Wooten, G.F., 1986, Localization of D2 dopamine receptors to intrinsic striatal neurons by quantitative autoradiography, Nature 292:463.

Trugman, J.M. and G.F. Wooten, G.F., 1987, Selective D-1 and D-2 dopamine agonists differentially alter basal ganglia glucose utilization in rats with unilateral 6-Hydroxydopamine substantia nigra lesions. J. Neurosci. 7:2927.

Ungerstedt, U., 1971, Postsynaptic supersensitivity after 6-hydroxydopamine induced degeneration of the nigrostriatal dopamine system, Acta Physiol. Scand. (suppl.) 367:69.

Waddington, J.L. and O'Boyle, K.M., 1987 The D1 dopamine receptor and the search for its functional role: from neurochemistry to behaviour, Reviews in the Neurosciences 1:157.

INHIBITORY ROLE OF DOPAMINERGIC D2 RECEPTORS IN THE EXPRESSION OF GLUTAMIC ACID DECARBOXYLASE AND PREPROENKEPHALIN mRNA IN THE RAT STRIATUM

*Jocelyne Caboche, **Philippe Vernier, **Jean-François Julien
*Monique Rogard, **Jacques Mallet, and *Marie-Jo Besson

*Laboratoire de Neurochimie-Anatomie, Institut des Neurosciences
Université P. et M. Curie, 75005 Paris
and **Departement de Génétique Moléculaire
Laboratoire de Neurobiologie Cellulaire et Moléculaire
C.N.R.S., 91198 Gif-sur-Yvette, France

INTRODUCTION

Dopamine (DA) contained in the dense network of DA fibers innervating the striatum has been shown to play a major role in the regulation of striatal neuron activity. This has been approached by analyzing either changes in electrophysiological properties of neurons (Bernardi et al., 1978; Calabresi et al., 1988) or changes in biochemical parameters such as the levels and/or the synthesis (Hong et al., 1978a and b; Hanson et al., 1988; Li et al., 1987) or the release (Lehman and Langer, 1983; Girault et al., 1986a) of specific neurotransmitters localized in striatal neurons. A useful tool in understanding the regulation of neuronal activity is now offered by molecular biology which allows the identification of specific mRNAs, reflecting earlier events than neurotransmitter contents. In this study, we tried to elucidate the DA receptor subtype which mediates dopaminergic modulation of the expression of messengers encoding glutamic acid decarboxylase (GAD) and preproenkephalin (PPE), two major markers contained in projection neurons of the striatum (Fonnum et al., 1974; Cuello et al. 1978). This analysis has been initiated by previous findings showing that a 6-hydroxydopamine (6-OHDA) lesion of nigro-striatal DA neurons produced increased levels of mRNAs encoding GAD (Vernier et al., 1988) and PPE (Angulo et al. 1986; Young et al., 1986; Normand et al. 1988), and that these effects could be reproduced by chronic treatments with haloperidol (Sabol et al., 1983; Tang et al., 1983; Sivam et al., 1986; Normand et al., 1987; Morris et al., 1988; Vernier unpublished observations). These previous studies suggested an inhibitory control of DA on the expression of both GAD- and PPE-mRNA. The action of DA being mediated by an interaction with at least two receptor subtypes, coupled to different second messenger systems (Stoof and Kebabian, 1981), it was of interest to elucidate the receptor subtype involved in these effects.

In a first step of the study, chronic D1 receptor blockade was produced by repeated injection of SCH 23390, a compound which selectively blocks DA-stimulated adenylate cyclase activity (Seeman, 1980) and possesses part of the pharmacological profile of neuroleptics (Christensen et al., 1984). Then in view to determine, whether the observed effects were due to the blockade of D1 receptors or to the activation by DA of the unaffected D2 receptors, we associated to the SCH 23390 treatment, a D2 agonist administration. The levels of GAD- and PPE-mRNA were measured by differentiating the dorsal and the ventral striatum, since we had previously shown that these two striatal regions differed in their reactivities to selective DA agents (Girault et al., 1986a).

METHODS

Rats were subcutaneously injected (twice a day, during 15 days) with the D1 antagonist SCH 23390 (0.2 mg/kg) alone or associated with the D2 agonist Ru 24926 (0.3 mg/kg). In parallel, control animals were injected with the vehicle. Twenty four hours after the last injection, rats were sacrified. Brains were frozen, 500 mm thick coronal sections were prepared and the dorsal striatum (dorsal caudate-putamen) and the ventral striatum (including the ventral caudate-putamen, the nucleus accumbens and the olfactory tubercles) were dissected. Total RNAs were extracted and similar amount of RNA per sample was submitted to electrophoresis and blotted according to a method previously described (Faucon-Biguet et al., 1986). Specific mRNAs were detected by successive hybridization with appropriate ^{32}P-cDNA probes to identify PPE, GAD and α-tubulin. The hybridized α-tubulin-mRNA was used as an internal standard for the quantification of GAD- and PPE-mRNA. After radioautography specific mRNAs were measured by densitometry with an image analyzer (Imstar, France).

RESULTS AND DISCUSSION

Effects of a chronic blockade of D1 receptors on GAD- and PPE- mRNA levels in the dorsal and the ventral striatum.

The long term blockade of D1 receptors by SCH 23390 treatment produced a significant decrease of both GAD- and PPE- mRNA content in the dorsal striatum (-30%) (upper panel of figure) when compared to control rats. By contrast, no significant modification of these mRNAs was found in the ventral striatum following SCH 23390 treatment (lower panel of figure).

The density of D1 receptors in these two regions is relatively homogeneous (Boyson et al., 1986; Dawson et al., 1986; Savasta et al., 1986) and cannot account likely for the differential modulation of GAD- and PPE-mRNA in the dorsal and the ventral striatum. It can be noticed that during the blockade of D1 receptors, endogenous DA can still act on the unaffected D2 receptors. Here again, the relatively homogeneous distribution of these receptors in the striatal regions investigated can hardly explain the results obtained (Boyson et al., 1986). However, SCH 23390 treatment could affect differently the release of DA in the ventral and the dorsal striatum leading to a preferential activation of D2 receptors in one of these two regions. In fact, it has been shown that SCH 23390 treatment produced a stimulation of DA release in the dorsal striatum (Imperato et al., 1987). This effect was attributed to an indirect desinhibition of nigral neurons by SCH 23390 acting on

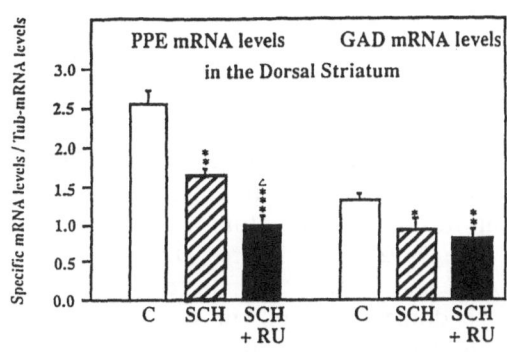

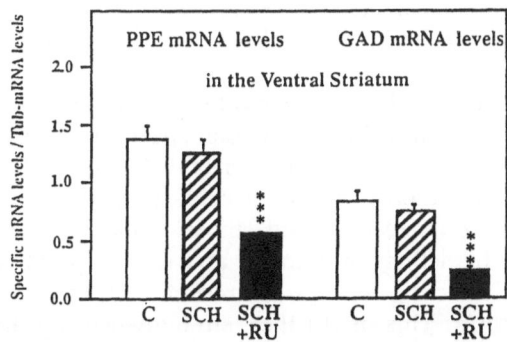

FIGURE 1 Effects of a chronic treatment with SCH 23390 alone and in association with Ru 24926 on PPE- and GAD-mRNA levels in the dorsal (upper panel) and in the ventral striatum (lower panel).

Levels of PPE-, GAD- and α-tubulin-mRNA were determined by densitometric measurements of autoradiograms obtained after northern-blot analysis. Columns correspond to the ratio of PPE-mRNA over α-tubulin-(Tub) mRNA (left part of the figure) or to the ratio of GAD-mRNA over α-tubulin-(Tub) mRNA (right part of the figure). Student's t test was utilized for statistical analysis.
* p < 0.05 ; ** p < 0.005 ; *** p <0.001 for comparison between treated and control animals. p < 0.001 for comparison between SCH 23390 (SCH) treated rats and SCH 23390 associated to Ru 24926 (SCH + Ru) treated rats.

D1 receptors present in high concentration in the substantia nigra and mainly located on afferent GABA terminals (Gale et al., 1977). Such a SCH 23390 mediated DA release is likely to be weak in the ventral striatum. Only a very low density of D1 receptors is observed in the ventral tegmental area where are located the DA neurons innervating the ventral striatum (Björklund and Lindvall, 1984).

Thus, the decrease of GAD- and PPE-mRNA expression obtained in the dorsal striatum following a SCH 23390 treatment could result from a preferential stimulation by DA of the unaffected D2 receptors.

Effects of the combined chronic blockade of D1 receptors and activation of D2 receptors.

To test the hypothesis that the decrease in GAD- and PPE-mRNA levels observed after a chronic blockade of D1 receptors was mainly due to an inhibition mediated by DA acting through D2 receptors we coadministrated SCH 23390 and a D2 receptor agonist : Ru 24926, to reinforce D2 receptors stimulation. After this chronic treatment GAD- and PPE-mRNA decreased in both the dorsal and the ventral striatum. In the dorsal striatum, the decrease of GAD-mRNA levels was slightly accentuated (-38%) and the down regulation of PPE-mRNA was significantly amplified (-57%) when compared to the effects of SCH 23390 alone. In the ventral striatum, a marked decrease in mRNA levels was observed. Interestingly, the decrease was similar for both GAD-mRNA (-67%) and PPE-mRNA (-62%) (see figure).

These results indicate that the simultaneous D1 receptor blockade and D2 receptor stimulation produce in the whole striatum opposite effects to those described after a 6-OHDA lesion of nigro-striatal DA neurons (Vernier et al., 1988) or after repeated administrations of Haloperidol (Vernier et al., submitted). The comparison between our results and those obtained following Haloperidol treatment, a neuroleptic acting on both D1 and D2 receptors, strongly suggests an inhibitory action of DA acting through D2 receptors in the control of expression of GAD- and PPE-mRNA. This is further strengthened by the observation that Sulpiride, a specific D2 receptor antagonist, increases GAD- and PPE-mRNA in the striatum (Caboche et al., submitted).

In the striatum, D2 receptors have a presynaptic localization on DA fibers and at this level negatively regulate the release of DA. They are also distributed postsynaptically to DA fibers on cortical afferences where they have been implicated in the inhibition of the release of glutamate (Rowlands et al., 1980) and on striatal neurons where they modulate the activity (Calabresi et al., 1988).

Actually, two D2 receptors have been cloned differing by the presence or absence of one exon in the third putative cytoplasme loop, (Dal Toso et al., 1989) but their transduction system is not yet completely elucidated. From pharmacological and electrophysiological studies, several coupling mechanisms have been determined. In the striatum the negative coupling of D2 receptors to adenylate cyclase is a well characterized transduction mechanism. A negative coupling to the phosphoinositide transduction system has been considered but its existence in the striatum is more controversial. A link to a G protein modulating K^+ chanels is also possible (Freedman and Weight, 1988).

It is likely that activation of D2 receptors located on striatal neurons decreases cAMP levels, an effect which can be further amplified by the concommittant blockade of D1 receptors. This decrease might trigger an inhibition of GAD- and PPE gene expression. The interaction of DA and DA agonist Ru 24926 on D2 receptors localized on cortical efferences could also lead to an inhibition of GAD- and PPE-mRNA expression. The overall effect produced by such an interaction, resulting in a decrease of released glutamate , is a reduction in neuronal activity and in intracellular Ca^{++} concentration. Both parameters could be involved in the regulation of expression of GAD- and PPE-mRNA. Interestingly, it has been shown that a cortical lesion, which is known to reduce the release of glutamate in the striatum (Girault et al., 1986b), decreases the expression of PPE-mRNA in this structure (Uhl et al., 1988). Thus, the marked reduction in GAD- and PPE-mRNA observed following a D2 agonist treatment associated with a D1 antagonist implies likely various regulating factors as suggested by the multiple interactions of DA in the striatum. Interestingly, PPE gene possesses 5' upstream regulatory elements confering its inducibility by cAMP-dependent protein kinase and Ca^{++}-dependant protein kinase (Comb et al., 1988). Whether GAD gene possesses such regulatory elements is not yet known, but can be suggested.

A question which is raised also is whether inhibition of GAD- and PPE-mRNA occurs in neurons where these specific neurotransmitter markers are colocalized. Particularly, GAD and Enkephalin coexist in striato-pallidal neurons (Aronin et al., 1984). These neurons possess a D2/D1 receptor ratio higher than the other efferent striatal neurons (Boyson et al., 1986; Beckstead et al., 1988). The reduction of GAD- and PPE-mRNA observed in the dorsal striatum after a chronic D1 receptor blockade and which appears to result from a stimulation of the unaffected D2 receptors could predominantly take place in these neurons. The generalized decrease of GAD-mRNA obtained by the co-administration of SCH 23390 and Ru 24926 might also involve other neuronal populations. Many striatal neurons contain GAD and, particularly, striatoentopeduncular and striatonigral neurons account for two major neuronal populations of the dorsal striatum (Fonnum et al., 1974). It would be interesting to determine whether in these neurons where the D2/D1 receptors ratio is much lower than in striatopallidal neurons, GAD-mRNA expression is similarly regulated by D2 receptors. It can be noted that in these projection neurons GAD is colocalized with peptides as tachykinins and dynorphin and inverse effects of D2 agonists on the expression of their prepropeptides mRNAs has been described (Young et al., 1986). Many striatal interneurons, also contained GAD and GAD-mRNA in high concentration and the regulation of GAD-mRNA expression (Chesselet and Robbins, 1989) at their level would be interesting to consider.

CONCLUSION

Our data strongly suggest that GAD-mRNA and PPE-mRNA expression is negatively regulated by DA acting mainly on D2 receptors. Due to their localization on various postsynaptic elements and to their linkage to different transduction systems, their inhibitory control on GAD- and PPE-mRNA seems to be exerted through several regulating factors. It is likely that D2 receptor control involves cAMP and Ca^{++} dependent mechanisms.

ACKNOWLEDGMENTS

This work was supported by grants from INSERM (87-6002), CNRS and University P. and M. Curie (for the laboratory of Neurochimie Anatomie) and from CNRS, INSERM, DRET and Rhône Poulenc Santé (for the laboratory of Genetique Moléculaire).

REFERENCES

Angulo, J.A., Davis, L.G., Burkhart, B.A., Christoph, G.R., 1986, Reduction of striatal dopaminergic neurotransmission elevates striatal proenkephalin mRNA, Eur. J. Pharmacol., 130: 343-344.

Aronin, N., DiFiglia, M., Graveland, G.A., Schwartz, W.J. and Wu, J.-Y., 1984, Localization of immunoreactive enkephalins in GABA synthesizing neurons of the rat neostriatum, Brain Res., 300: 376-380.

Beckstead, R.M., Wooten, G.F., and Trugman, J.M., 1988, Distribution of D1 and D2 dopamine receptors in the basal ganglia of the cat determined by quantitative autoradiography, J. Comp. Neurol., 268 :131-145.

Bernardi, G., Marciani, M.G., Morocutti, C., Pavone F., and Stanzione, P., 1978, The action of dopamine on rat caudate neurons intracellularly recorded, Neurosci. Letts, 8: 235-240.

Björklund, A., and Lindvall, O., Dopamine-containing systems in the CNS., 1984, "Handbook of Chemical Neuroanatomy, Vol. 2, Classical Neurotransmitters in the CNS", Björklund A. and Hökfelt T. eds, Part I, pp. 55-112. Elsevier, Amsterdam.

Boyson, S.J., Mc Gonigle, P., and Molinoff, P.B., 1986, Quantitative autoradiographic localization of the D1 and D2 subtypes of dopamine receptors in rat brain, J. Neurosci., 6: 3177-3188.

Calabresi, P, Mercuri, N., Stanzione, P., Stefeni, A., and Bernardi, G., 1986, Intracellular studies on the dopamine-induced firing inhibition of neostriatal neurons in vitro: evidence for D1 receptor involvement, Neurosci., 20: 757-771.

Calabresi, P., Benedetti, M., Mercuri, N.B., and Bernardi, G., 1988, Endogenous dopamine and dopaminergic agonists modulate synaptic excitation in neostriatum: intracellular studies from naive and catecholamine-depleted rats, Neurosci., 27: 145-157.

Chesselet, M.F. and Robbins, E., 1989, Characterization of striatal neurons expressing high levels of glutamic acid decarboxylase messenger RNA, Brain Res., 492: 237-244.

Christensen, A.V., Arnt, J., Hyttel, J., Larsen, J.-J., and Svendsen, O., 1984, Pharmacological effects of a specific dopamine D-1 antagonist SCH 23390 in comparison with neuroleptics, Life Sci., 34: 1529-1540.

Comb, M., Mermod, N., Hyman, S.E., Pearlberg, J., Ross, M.E., and Goodman, H.M., 1988, Proteins bound at adjacent DNA elements act synergistically to regulate human proenkephalin cAMP inducible transcription, EMBO J. 7: 3793-3805.

Cuello, A.C., and Paxinos, G., 1978, Evidence for a long Leu-enkephalin striatopallidal pathway in rat brain, Nature, 271: 178-180.

Dal Toso, R., Sommer, B., Ewert, M., Herb, A., Pritchett, D.B., Bach, A., Shivers, B.D., and Seeburg, P., 1989, The dopamine D2 receptor: two molecular forms generated by alternative splicing, EMBO J., 8: 4025-4034.

Dawson, T.M., Gehlert, D.R., McCabe, R.T., Barnett, A., and Wamsley, J.K., 1986, D-1 dopamine receptors in the rat brain : a quantitative autoradiographic analysis, J. Neurosci., 6: 2353-2365.

Faucon Biguet, N., Buda, M., Lamouroux, A., Samolyk D., and Mallet, J., 1986, Time course of the changes of TH mRNA in rat brain and adrenal medulla after a single injection of reserpine, EMBO J., 5: 287-291.

Freedman, J.E. and Weight, F.F., 1988, Single K+ channels activaed by D2 copmaine receptors in acutely dissociated neurons from rat corpus striatum, Proc. Natl. Acad. Sci. USA, 85: 3618-3622.

Fonnum, F., Gottesfeld, Z., and Grofova, I., 1974, Distribution of glutamate decarboxylase, choline acetyltransferase and aromatic amino-acid decarboxylase in the basal ganglia of normal and operated rats : evidence for striatopallidal, striatoentopeduncular and striatonigral GABAergic fibers, Brain Res. 143: 125-138.

Gale, K., Guidotti, A., and Costa, E., 1977, Dopamine-sensitive adenylate cyclase: location in substantia nigra, Science, 195: 503-507.

Girault, J.A., Spampinato, U., Savaki, H.E., Glowinski, J., and Besson, M.J., 1986a, In vivo release of (^{3}H)o-aminobutyric acid in the rat neostriatum - I.Characterization and topographical heterogeneity of the effects of dopaminergic and cholinergic agents, Neuroscience, 19: 1101-1108.

Girault, J.A., Barbeito, L., Spampinato, U., Gozlan, H., Glowinski, J., and Besson, M.-J., 1986b, In vivo release of endogenous amino acids from the rat striatum: further evidence for a role of glutamate and aspartate in corticostriatal neurotransmission, J. Neurochem., 47: 98-106.

Hanson, G.R., Merchant, K.M., Letter, A.A., Bush, L., and Gibb, J.W., 1988, Characterization of methamphetamine effects on the striatal-nigral dynorphin system, Eur. J. Pharmacol., 155: 11-18.

Hong, J.S., Yang, H.-Y. T., Fratta, W., and Costa E., 1978a, Rat striatal methionine-enkephalin content after chronic treatment with cataleptogenic and non-cataleptogenic antischizophrenic drugs, J. Pharmacol. exp. Ther., 205: 141-147.

Hong, J.S., Yang, H.Y.T., and Costa, E., 1978b, Substance P content of substantia nigra after chronic treatment with antischizophrenic drugs, Neuropharmacol., 17: 83-85.

Imperato, A., Mulas, A., and Di Chiara, G., 1987, The D-1 antagonist SCH 23390 stimulates while the D-1 agonist SKF 23393 fails to affect dopamine release in the dorsal caudate of freely moving rats, Eur. J. Pharmacol., 142: 177-181.

Lehmann, J., and Langer, S.Z., 1983, The striatal cholinergic interneurons: synaptic target of dopaminergic terminals?, Neuroscience, 10: 1105-1120.

Li, S., Sivam, S.P., McGinty, J.F., Huang Y.S., and Hong, J.S., 1987, Dopaminergic regulation of tachykinin metabolism in the striatonigral pathway, J. Pharmacol. Exp. Ther., 243: 792-798.

Morris, B.J., Höllt, V., and Herz, A., 1988, Dopaminergic regulation of striatal proenkephalin mRNA and prodynorphin mRNA: contrasting effects of D1 and D2 antagonists, Neurosci., 25: 525-532.

Normand, E., Popovici, T., Fellmann, D., and Bloch, B., 1987, Anatomical study of enkephalin gene expression in the rat forebrain following haloperidol treatment, Neurosci. Lett., 83: 232-236.

Normand, E., Popovici, T., Onteniente, B., Fellmann, D., Piatier-Tonneau, D., Auffray, C., and Bloch, B., 1988, Dopaminergic neurons of the substantia nigra modulate preproenkephalin A gene expression in rat striatal neurons, Brain Res., 439: 39-46.

Pizzi, M., Da Prada, M., Valerio, A., Memo, M., Spano, P.F., and Haefely, W.E. ,1988, Dopamine D2 receptor stimulation inhibits inositol phosphate generating system in rat striatal slices, Br. Res., 456: 235-240.

Rowlands, G.J., and Roberts, P.J., 1980, Activation of dopamine receptors inhibits calcium-dependant glutamate release from cortico-striatal terminals in vitro, Eur. J. Pharmacol. 62: 241-242.

Sabol, S.L., Yoshikawa, K., and Hong, J.-S., 1983, Regulation of methionine-enkephalin precursof messenger RNA in rat striatum by haloperidol and lithium, Biochem. and Byophys. Res. Communic., 113: 391-399.

Savasta, M., Dubois, A., and Scatton, B., 1986, Autoradiographic localization of D1 dopamine receptors in the rat brain with (3H)SCH 23390, Brain Res., 375: 291-301.

Seeman, P., 1980, Brain dopamine receptors, Pharmacol. Rev., 32: 229-313.

Sivam, S.P., Strunk, C., Smith, D.R., and Hong, J.S., 1986, Proenkephalin-A gene regulation in the rat striatum : influence of lithium and haloperidol, Molec. Pharmacol., 30: 186-191.

Stoof, J.C., and Kebabian J.W., 1981, Opposing roles for D_1 and D_2 dopamine receptors in efflux of cyclic AMP from rat neostriatum, Nature, 294: 366- 368.

Tang, F., Costa, E., and Schwartz, J.P., 1983, Increase of proenkephalin mRNA and enkephalin content of rat striatum after daily injection of haloperidol for 2 to 3 weeks, Proc. Natl. Acad. Sci. USA, 80: 3841-3844.

Uhl, G.R., Navia, B., and Douglas, J., 1988, Differential expression of preproenkephalin and preprodynorphin mRNAs in striatal neurons: high levels of preproenkephalin expression depend on cerebral cortical afferents, J. Neurosci., 8: 4755-4564.

Vernier, P., Julien, J.F., Rataboul, P., Fourrier, O., Feuerstein, C., and Mallet, J., 1988, Similar time course changes in striatal levels of glutamic acid decarboxylase and proenkephalin mRNA following dopaminergic deafferentation in the rat, J. Neurochem., 51: 1375-1380.

Young, W.S., Bonner, T.I., and Brann, M.R., 1986, Mesencephalic dopamine neurons regulate the expression of neuropeptide mRNAs in the rat forebrain, Proc. Natl. Acad. Sci. USA, 83: 9827-9831.

MODULATION OF GENE EXPRESSION BY

NEURONAL LESIONS IN THE RAT STRIATUM

Philippe Vernier, Elisabeth Brault, Jean-François Julien
Pierre Rataboul, Sylvie Berrard and Jacques Mallet

Laboratoire de Neurobiologie Cellulaire et Moléculaire
C.N.R.S., 91198 Gif-sur-Yvette Cedex France

THE ROLE OF GENE REGULATION IN THE NEURONAL ADAPTATION TO LESIONS: THE EXAMPLE OF STRIATAL DOPAMINERGIC DEAFFERENTATION

Neurons have the ability not only to modify their physiological characteristics but also their morphology and their synaptic contacts, in response to lesions of the cells with which they are connected. These functional and anatomical modifications are termed "adaptative", without preconceiving their contribution to a process of restoration of the neural function (Zigmond and Bowers 1981). Adaptative changes are clearly observed during the course of neurological disorders like neurodegenerative diseases, where they contribute to the genesis of clinical symptoms.

Such a long-lasting remodeling of the anatomy and function of the affected cells depends upon an integrated modulation of genetic expression (Changeux et al., 1987). Some of these genes, which determine the cellular identity, encode proteins that are readily accessible to biochemical and pharmacological analysis (for example, membrane receptors or rate-limiting enzymes in neurotransmitter metabolism). Gene expression may be regulated at different points of the protein synthesis pathway (fig. 1), modifying either the functional activity or the amount of a given protein. This regulation operates in a spatially defined group of cells (subset of neurons, microglia or glial cells) and exhibits variable temporal patterns, some short-lasting events triggering durable changes in cell activity. This "adaptative program" is likely to be specified and acquired during cellular determination and differentiation of the nervous system.

In models of post-lesional plasticity in the peripheral nervous system, the neuromuscular junction (Changeux et al. 1987) or the superior cervical ganglion (Black et al. 1987), significant advances have already been made in understanding events linking changes in the cellular environment to the modulation of gene expression. However, there has been at present little attempt to systematically analyze, in the CNS, the molecular changes elicited by lesions of a particular pathway. The adaptative changes promoted in the striatum by a 6-hydroxydopamine lesion of the nigrostriatal dopaminergic pathway in the rat (Ungerstedt, 1971), provides a convenient model to study the underlying modulation of gene expression. This experimental lesion also represents an animal model of Parkinson's disease in which the degeneration of substantia nigra is the main characteristic. The nigral lesion elicits compensatory changes in the afferent neurons when the system is partially damaged as well as

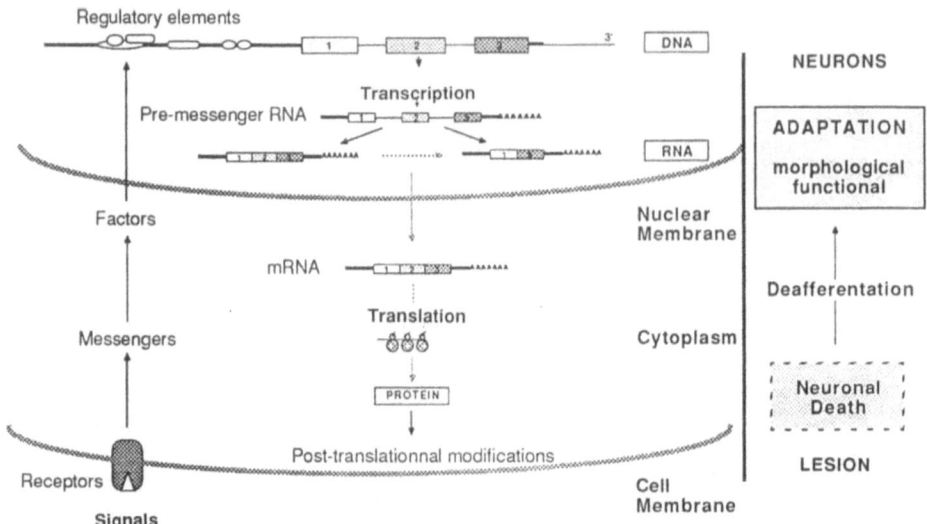

Fig.1. Schematic drawing of the different steps of gene expression (left part of the figure) which could be regulated by extracellular signals as a consequence of neuronal deafferentation (right part).

adaptative modifications of the deafferented neurons, phenomenons which have been extensively studied (Marshall, 1985, Zigmond and Stricker, 1984). When considering only the plasticity of the deafferented striatal systems, important features have been revealed, including: (i) enhancement of the energetic metabolism and the electrical activity of the striatal cells (Schultz and Ungerstedt, 1978, Wooten and Collins 1981), (ii) prominent modifications in the turnover of various neurotransmitters like GABA (Scheel-Krüger, 1986), acetylcholine (Lehman and Langer, 1984), enkephalin, substance P and dynorphin (Young et al., 1986, Morris et al., 1989), (iii) supersensitivity of striatal cells to dopaminergic agonists, which is, at least in part, mediated by an increase in the number of D2 receptors (Creese and Snyder, 1979, Staunton et al., 1981).

The rapid development of the molecular biology techniques and the recent cloning of genes relevant to this experimental model, allowed us to approach the underlying genetic regulation.

MODULATION OF GAD, PPE AND ChAT mRNA

We have first considered the modulation in the expression of genes encoding some pivotal proteins involved in the biosynthesis of striatal neurotransmitters, i.e. proenkephalin (PPE), the precursor of enkephalin, glutamate decarboxylase (GAD) and choline acetyltransferase (ChAT), the enzymes synthesizing GABA and acetylcholine respectively. GABA- and enkephalin-containing neurons are primary targets of the dopaminergic nigrostriatal pathway (Kubota et al., 1986, Kubota et al., 1987) and these two transmitters are colocalized in a subset of striatal neurons (Penny et al., 1986). In addition, dopamine has been shown to affect GABAergic (Vincent et al., 1978, Girault et al., 1986) and enkephalinergic (Morris et al., 1989) activity in the striatum.

The time course of changes in the levels of GAD and PPE mRNA in the rat striatum, after unilateral lesion of the substantia nigra, was studied using a quantitative northern blot technique described by Faucon Biguet et al. (1986), and specific radioactive cDNA probes (Yoshikawa et al. 1982, Julien et al., 1987, 1990). Amounts of GAD and PPE mRNA (3.7 kb and 1.4 kb in size respectively) rose shortly after the lesion, with a twofold increase at ten days post lesion (fig. 2). This level remained

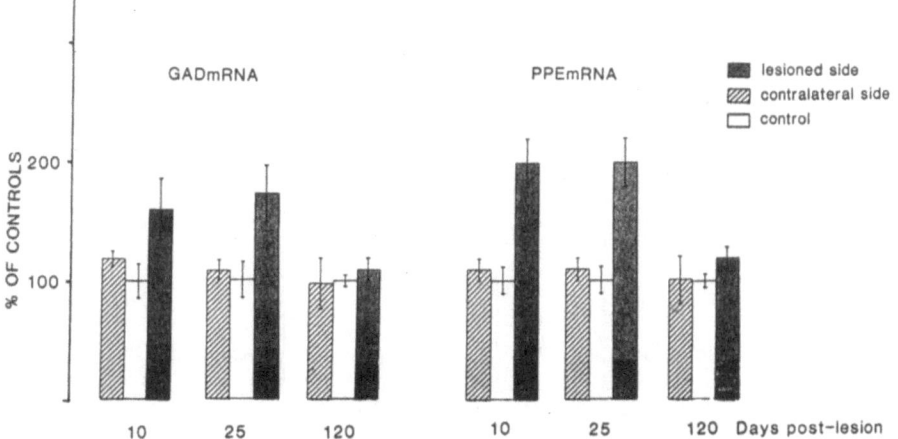

Fig. 2. Time course changes of the GAD (left pannel) and PPE mRNA levels (right pannel) in the contralateral and deafferented striatum. Total RNA, extracted from both striata at 10, 25 and 120 days after the lesion, were loaded on the same gel. Amount of specific mRNA in deafferented and contralateral striatum is expressed as a percent (± SEM) of mRNA levels measured in the control striatum. (p<0.005, n= 5-10).

unchanged up to 25 days post-lesion, and then declined slowly. However, while the GAD mRNA reached the control value at 3 months, PPE mRNA remained slightly but significantly elevated (Vernier et al.1988). This time course of mRNA modulations is very similar to that described for the GAD activity in the same experimental conditions (Segovia and Garcia-Munoz, 1987). These data indicate that the modifications in neurotransmitter metabolism, following the changes in cellular activity, are accompanied by parallel changes of specific mRNA levels.

Northern blot hybridization with the the GAD cDNA probe also revealed another transcript of higher molecular weight, exhibiting the same tissue distribution. Interestingly, the increase of this RNA levels was more pronounced than that of the shorter GAD mRNA in the denervated striatum, and followed the same time course. This observation suggests that the large transcript represents a premessenger RNA and is indicative of an enhancement in the transcription rate of the GAD gene. However, it cannot be excluded that it corresponds to an alternative transcript of the same gene (Vernier et al., 1988). Recently, Bond et al. (1988) have shown that expression of the high molecular weight RNA dominated shortly after birth, when the GABA synthesis is maximum.

It should also be noted that a similar increase of GAD and PPE mRNA occurred, when dopamine receptors are chronically blocked by haloperidol (Vernier et al., submitted). Therefore, the lack of dopamine receptor stimulation alone is able to trigger the adaptative modulation of gene expression. The respective role of D1 and D2 receptors in this regulation is described, in collaboration with J. Caboche and M.J. Besson, in the present volume.

Beside its influence on the GABAergic output system, dopamine has long been proposed to interact with the cholinergic interneurons in the striatum (Lehman and Langer, 1983). This interaction which is thought to be mutually inhibitory (Guyenet et al. 1975), may be based at present on anatomical ground (Kubota et al. 1987, Chang, 1988). Moreover, the control of acetylcholine release has been reported to be modulated as a function of dopamine depletion induced by the nigral lesion (Fage et al., 1984, MacKenzie et al., 1989).

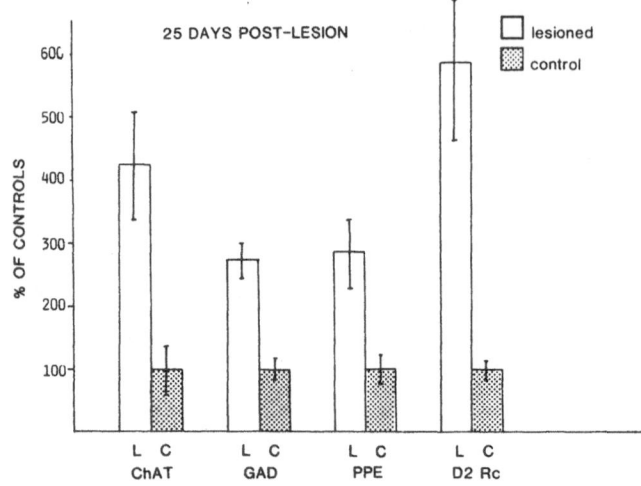

Fig.3. Modulation of mRNA encoding α-tubulin (Tub), GAD, PPE, ChAT, and D2 receptor (D2Rc), 25 days after a 6-OHDA nigral lesion. Poly(A)+ RNA, extracted from two pools of 5 lesioned and control striata, were analysed on the same blot. Amount of specific mRNA in deafferented striatum is expressed as a percent of mRNA levels measured in the control striatum. The mean of 2 independent experiments is presented.

The recent isolation of cDNA encoding ChAT (Berrard et al., 1988, Brice et al. 1989) allowed us to study the modulation of the corresponding mRNA in the deafferented striatum. We have compared the ChAT mRNA levels, 25 days after a nigral lesion to the levels of GAD, PPE, and α-tubulin mRNAs on the same blot of striatal poly(A)+ RNA. ChAT mRNA was found to be elevated 4 fold in the deafferented striatum as compared to controls, an increase which is more prominent than that of GAD and PPE mRNA (fig 3). However, in the same animals, the ChAT enzymatic activity was not modified by the deafferentation. Although it is of course attractive to propose that an increase in the protein turnover could account for this observation, a more careful analysis is needed before firm conclusions can be drawn.

Taken together, these results suggest that dopamine exerts an overall tonic inhibition on the activity of GABA-, enkephalin- and acetylcholine-containing neurons. The neurons rapidly react to the suppression of this influence, and then develop long-term adaptation underlied by a modulation of gene expression. The signals controlling this regulation, and their precise role in the adaptive processes need to be elucidated.

THE LONG-TERM INCREASE OF D2 RECEPTORS

The possibility for a cell to modulate the number of receptors expressed at its surface, as well as the transduction mechanisms which are associated with them, play a crucial role in the cellular adaptation, allowing to modify the sensitivity to neurotransmitters. An increase in the D2 receptor number is a hallmark of dopaminergic deafferentation in the striatum (Creese and Snyder, 1975). The cloning of a D2 receptor cDNA (Bunzow et al., 1988) provided the tools to analyze whether the rise in receptor number corresponds to an increased synthesis or to a prolonged half-life of the protein.

We have used the complete coding sequence of the rat D2 receptor, synthetized by the polymerase-chain-reaction technique, to probe the same RNA blots as those utilized to quantify the Chat mRNA (fig. 3). 25 days after the nigral injury, the level of D2 receptor mRNA is increased 6 fold in the deafferented striatum as compared to controls. This elevation differed markedly from the increase in receptor number, which was shown to be much less pronounced (between 30% and 50%). Although the modulations of mRNA levels are very often ampler than that of the corresponding protein, reflecting differences in the turnover of the molecules, several other considerations may account for this disparity. First, the mRNA could be not translated as efficiently as in the normally afferented striatum, as suggested for the β2-adrenergic receptor in another model of cellular adaptation (Allen et al., 1990). Second, mRNA is contained within the cellular bodies present in the striatum, whereas D2 receptors are borne by afferent nerve terminals as well (Filloux et al.,1988). These receptors could therefore not participate to the increase of D2 binding sites as intrinsic striatal receptors do. Third, D2 autoreceptors, harbored by the dopaminergic terminals might have disappeared as a consequence of the lesion, but this fact has never been clearly demonstrated. The augmented amount of D2 receptor mRNA provides a strong evidence for an increased synthesis of receptor protein, in response to dopaminergic denervation.

Nevertheless, the contribution of the increase in receptor number to the dopaminergic supersensitivity of striatal cells is still controversial. Indeed, behavioral and biochemical supersensitivity develops much earlier than the measurable elevation in binding sites, and requires a less severe depletion of striatal dopamine (Staunton et al, 1981, Fage et al., 1984, MacKenzie et al., 1989). It could be suggested that changes in the ability of D2 receptors to activate intracellular signals also participate to the supersensitivity phenomenon. This assumption is strengthened by the recent finding of two molecular forms of the D2 receptor (Dal Toso et al.,1989). They are generated by alternative splicing of the premessenger RNA, and differ by the presence or absence of a 29 amino-acid sequence. The sequences diverge in a region which is required to interact with the intracellular signal proteins (G proteins). Therefore, it would be of interest to determine whether a regulation of the the D2 mRNA processing, leading to different forms of receptor molecules, can contribute to the post-lesional plasticity in the striatum. In contrast, the long-term adaptation which develops in striatal cells involves an increased expression of the D2 receptor protein.

LACK OF A PROMINENT LONG LASTING TROPHIC RESPONSE IN THE STRIATUM FOLLOWING DOPAMINERGIC DEAFFERENTATION

In addition to the biochemical modulations exhibited by the deafferented neurons, it has been stressed in various lesional models, that these cells also modify their shape and synaptic connectivity for a long period of time. To address the question of the long-term modulation of genes encoding proteins which play a role in cellular growth, we have first measured the levels of α- and β-tubulin mRNA in the striatum, 10, 25 and 120 days after the nigral lesion (fig. 4). No significant change in the level of tubulin mRNA has been observed, (Vernier et al. 1988 and unpublished results), although an increase of these mRNA species was described at shorter post-lesion intervals in a similar model (Yavin et al., 1987). However, the type of striatal cells in which this reaction occurs is still unknown. They could be astrocytes, since these cells reacts very early to the injury of nerve terminals (see below), and since a significant sprouting or other anatomical changes have never been observed in deafferented striatal neurons (Ingham et al. 1990).

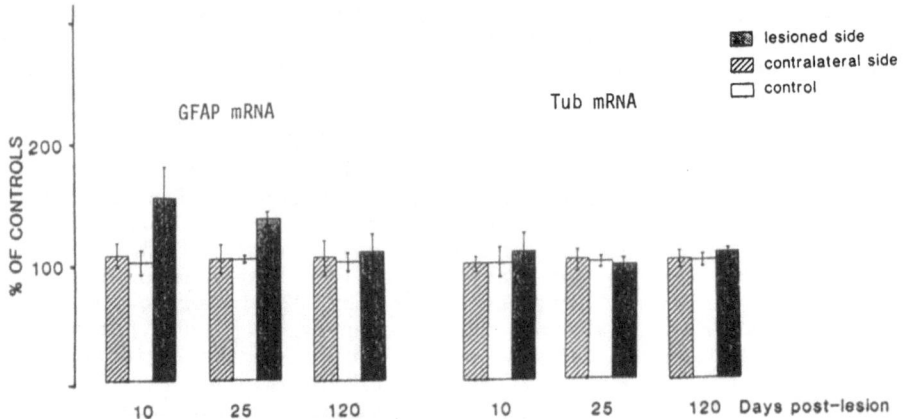

Fig.4. Time course changes of the α-tubulin (Tub) (left pannel) and GFAP mRNA levels (right pannel) in both striata from 6-OHDA lesioned rats. Total RNA were analysed on the same gel for each post-lesion time (10, 25 and 120 days). Amount of specific mRNA in deafferented and contralateral striatum is expressed as a percent (± SEM) of mRNA levels measured in the control striatum. ($p < 0.005$, n= 5-10).

Moreover, we have also used cDNA probes belonging to the ras gene superfamily, rab 1 and rab 2, whose expression has been demonstrated to be associated with cell proliferation and neurite outgrowth, respectively (Ayala et al., 1989). These mRNA exhibited also no modulation, at the post-lesion times we have studied. This clearly indicates that dopaminergic deafferentation does not elicit noticeable long-term changes in the expression of genes encoding proteins involved in neuronal growth mechanisms. It contrasts sharply with other lesional models, where cellular remodeling plays an important role in the plasticity of the system.

THE STRIATAL ASTROCYTIC RESPONSE TO DOPAMINERGIC DEAFFERENTATION

Beside neurons, astrocytes also participate to the adaptative response of the striatum to the lesion of the dopaminergic pathway. In our experiments, a GFAP cDNA probe was used to monitor this astrocytic reaction (Rataboul et al. 1987). GFAP mRNA was the most elevated at the shorter post-lesion interval we have studied (10 days), an then declined continuously with time (fig 4).

In fact, the astrocytic reaction appears very shortly after the lesion, as it has been demonstrated in another experimental condition, the ibotenic acid lesion of the striatum. In this model, the increase of GFAP mRNA level was maximal 2 days after the injury at the lesion site, and was slightly delayed in the substantia nigra, one of the projection areas of the damaged neurons (Rataboul et al. 1988). The modulation of GFAP mRNA levels also preceded that of the corresponding immunoreactivity. Moreover, since the turnover of the polymerized GFAP is low, the immunoreactivity persisted for a long time after the mRNA has return to lower levels (Stromberg et al. 1986). It is therefore difficult to use GFAP immunolabeling to infer how long the reactive astrocytosis persists, and to what extend it participates to the pathogenesis of scarring in the CNS. Conversely, GFAP mRNA allows a more accurate analysis of the astrocytic reaction. It also highlights the important implication of astrocytes in the healing response in cooperation with microglia (Akiyama and McGeer, 1989), both at the lesion site and in the degeneration areas of nerve terminals.

CONCLUSION

In the present study, we have described the modulation of different classes of mRNA, expressed within different cell types in the striatum, as a response to severe dopaminergic deafferentation. The corresponding genes stand as targets of the regulation pathways which operate on the postsynaptic side of the cellular adaptation. How these postsynaptic modifications are integrated into the general adaptative phenomenons occuring in other afferent and efferent systems, and which is the nature of the intracellular signals triggering the modulation of genetic expression, represent the future issues to be addressed. In addition, we have undertaken, without any a priori, to isolate striatal genes which are modulated by dopaminergic denervation by the mean of differential screening of cDNA libraries (Rhyner et al., 1986). This approach should allow a more detailed analysis of the genetic regulation of the post-lesional adaptation in the CNS.

ACKNOWLEDGMENTS

We wish to thank E. Jean-Gilles, J. Clot-Humbert and C. Helin for skillful technical assistance, and Ph. Samama, J. Stinnakre and S. Vyas for careful readings of the manuscript. This work is supported by grants from CNRS, INSERM, DRET and Rhône-Poulenc Santé.

REFERENCES

Akiyama, H. and McGeer, P.L. (1989) Microglial response to 6-hydroxy-dopamine-induced substantia nigra lesions. Brain Res., 489: 247-253.

Allen, J.M., Abrass I.B. and Palmiter, R.D. (1989) β2-adrenergic receptor regulation after transfection into a cell line deficient in the cAMP-dependent proteine kinase. Mol. Pharmacol., 36: 248-255.

Ayala, J., Olofsson, B., Touchot, N., Zahraoui, A., Tavitian, A. and Prochiantz, A. (1989) Developmental and regional expression of three new members of the ras-gene family in the mouse brain. J. Neurosci. Res. 22: 384-389.

Berrard, S., Brice, A., Lottspeich, F., Braun, A., Barde, Y.A. and Mallet, J. (1987) cDNA cloning and complete sequence of porcine choline acetyltransferase: In vitro translation of the corresponding mRNA yields an active protein. Proc. Natl. Acad. Sci. USA, 84: 9280-9284.

Black, I.B., Adler, J.E., Dreyfus, C.F., Friedman, W.F., LaGamma, E.F. & Roach, A.H. (1987) Biochemistry in information storage in the nervous system. Science, 236, 1263-1268.

Brice, A., Berrard, S., Raynaud, B., Ansieau, S., Coppola, T., Weber, M. and Mallet, J. (1989) Complete sequence of a cDNA encoding an active rat choline acetyltransferase: A tool to investigate the plasticity of cholinergic phenotype expression. J. Neurosci.Res. 23: 266-273.

Bunzow, J.R., VanTol, H.H.M., Grandy, D.K., Albert, P., Salon, J., Christie, M., Machida, C.A., Neve, K.A. and Civelli, O. (1988) Cloning and expression of a rat D2 dopamine receptor. Nature 336: 783-787.

Chang, H.T. (1988) Dopamine-acetylcholine interaction in the rat striatum: a dual-labeling immunocytochemical study. Brain Res. Bull. 21: 295-304.

Changeux, J.P., Klarsfeld, A. & Heidmann T. (1987) The acetylcholine receptor and molecular models for short-term and long-term memory in: The Neural and Molecular Bases of Learning, eds Changeux, J.P. & Konishi, M. J.Wiley & Sons Ld, Chichester, pp. 31-84.

Creese, I. and Snyder, S.H. (1979) Nigrostriatal lesions enhance striatal [3H]-apomorphine and [3H]-spiroperidol binding. Eur. J. Pharmacol., 56: 277-281.

Dal Toso, R., Sommer, B., Ewert, M., Herb, A., Pritchett, D.B., Bach, A., Shivers, B. and Seeburg, P.H., (1989) The dopamine D2 receptor: Two molecular forms generated by alternative splicing. EMBO J. 8: 4025-4034.

Fage, D., Guerin, B., Feuerstein, C., Demenge, P. & Scatton, B. (1984) Time course of the changes in striatal acetylcholine levels induced by pergolide and haloperidol after lesion of the nigro-striatal dopaminergic pathway in the rat. Brain Res., 310, 379-383.

Faucon Biguet, N., Buda, M., Lamouroux, A., Samolyk, D. & Mallet, J. (1986) Time course of the changes of TH mRNA in rat brain and adrenal medulla after a single injection of reserpine. EMBO J. 5, 287-291.

Filloux, F., Dawson, T.M. and Wamsley, J.K. (1988) Localisation of nigrostriatal dopamine receptor subtypes and adenylate cyclase. Brain Res. Bull., 20: 447-459.

Girault, J.A., Spampinato, U., Glowinski, J. & Besson, M.J. (1986) In vivo release of (3H) -aminobutyric acid in the rat neostriatum-I. Characterization and topographical heterogeneity of the effects of dopaminergic and cholinergic agents. Neuroscience, 19, 1109-1117.

Ingham, C. and Arbuthnott, G.W. (1990) Spine density on neostriatal neurones changes with 6-hydroxydopamine lesions and with age. Brain Res. 503: 334-338.

Julien, J.F., Legay, F., Dumas, S., Tappaz, M. & Mallet, J. (1987) Molecular cloning, expression and in situ hybridization of rat brain glutamic acid decarboxylase mRNA. Neurosci. Lett., 173- 180.

Julien, J.F., Samama, Ph. and Mallet J. (1990) Rat brain glutamic acid decarboxylase sequence deduced from a cloned cDNA. J. Neurochem. 54: 703- 705.

Kubota, Y., Inagaki, S., Kito, S., Takagi, H. & Smith, A.D. (1986) Ultrastructural evidence of dopaminergic input to enkephalinergic neurons in rat neostriatum. Brain Res. 367, 374-378.

Kubota, Y., Inagaki, S., Kito, S. & Wu, J.Y. (1987) Dopaminergic axons directly make synapses with GABAergic neurons in the rat striatum. Brain Res., 406, 147-156.

Kubota, Y., Inagaki, S., Shimada, S., Kito, S., Eckenstein, F.and Tohyama, M. (1987) Neostriatal cholinergic neurons receive direct synaptic inputs from dopaminergic axons. Brain Res. 413: 179-184.

Lehman, J. and Langer, S.Z. (1983) The striatal cholinergic interneuron: Synaptic target of dopaminergic terminals? Neuroscience 10: 1105-1120.

MacKenzie, R.G., Stachowiak, M.K. and Zigmond, M.J. (1989) Dopaminergic inhibition of striatal acetylcholine release after 6-hydroxydopamine. Eur. J. Pharmacol. 168: 43-52.

Marshall, J.F. (1985) Neural plasticity and recovery of function after brain injury. in International Review of Neurobiology, eds. Smythies, J.R. & Bradley, R.J., vol. 26 (Academic Press, Inc. London), pp. 201-247.

Morris, B.J., Herz, A. and HÖllt, V. (1989) Localization of striatal opioid gene expresion, and its modulation by the mesostriatal dopamine pathway: an in situ hybridization study. J. Mol. Neurosci. 1: 9-18.

Penny, G.R., Afsharpour, S. & Kitai, S.T. (1986) The glutamate decarboxylase, leucine enkephalin-methionine enkephalin and substance P immunoreactive neurons in the neostriatum of the rat and cat : evidence for partial population overlap. Neuroscience, 17, 1011-1045.

Rataboul, P., Faucon Biguet, N., Vernier, P., De Vitry, F., Boularand, S., Privat, A. & Mallet, J. (1987) Identification of human GFAP cDNA : a tool for the molecular analysis of reactive gliosis in the mammalian central nervous system. J. Neurosci. Res., 20: 165-175.

Rataboul, P., Vernier, P., Faucon Biguet, N., Mallet, J., Poulat, P. and Privat, A. (1988) Modulation of GFAP mRNA levels following toxic lesions in the basal ganglia of the rat. Brain Res. Bull., 22: 155-161.

Rhyner, T.A., Faucon Biguet, N., Berrard, S., Borbély, A.A. and Mallet, J. (1986) An efficient approach for the selective isolation of specific transcripts from complex brain mRNA populations. J. Neurosci. Res., 16: 167-181.

Scheel-Krüger, J. (1986) Dopamine-GABA interactions : evidence that GABA
tranmits, modulates and mediates dopaminergic functions in the basal
ganglia and the system. Acta Neurol. Scand. Suppl. 107, 1-47.

Schultz, W. and Ungerstedt, U. (1978) Short-term increase and long term
reversion of striatal cell activity after degeneration of the nigrostriatal dopamine system. Exp. Brain Res., 33: 159-171.

Segovia, J. and Garcia-Munoz, M. (1987) Changes in the activity of GAD in
the basal ganglia of the rat after striatal dopaminergic denervation.
Neuropharmacology, 26: 1449-1451.

Staunton, D.A., Wolfe, B.A., Groves, P.M. and P.B. Molinoff. (1981)
Dopamine receptor changes following destruction of the nigrostriatal
pathway: lack of relationship to rotational behavior. Brain Res. 211:
315-

Stromberg,I., Bjorklund, H., Dahl, D., Jonsson, G., Sundstrom, E. &
Olson, L. (1986) Astrocyte responses to dopaminergic denervations by
6-hydroxydopamine and 1-methyl-4-phenyl- 1,2,3,6-tetrahydropyridine as
evidenced by glial fibrillary acidic protein immunohistochemistry.
Brain Res. Bull., 17, 225-236.

Ungerstedt, U. (1971) Postsynaptic supersensitivity after 6- hydroxydopamine induced degeneration of nigrostriaral dopaminergic system.
Acta Physiol. Scand., Suppl. 367: 95-122.

Vernier, P., Julien, J.F., Rataboul, P., Fourrier, O., Feuerstein, C. and
Mallet, J. (1988) Similar time course changes in striatal levels of
glutamic acid decarboxylase and proenkephalin mRNA following
dopaminergic deafferentation in the rat. J.Neurochem. 51: 1375-1380.

Vincent, S.R., Nagy, J.I. & Fibiger, H.C (1978) Increased striatal
glutamate decarboxylase after lesions of the nigrostriatal pathway.
Brain Res., 143, 168-173.

Yavin, E., Gil, S., Consolazione, A., dal Toso, R. and Leon, A. (1987)
Selective enhancement of tubulin gene expression and increase in
oligo(dT)-bound RNA in the rat brain after nigrostriatal pathway
unilateral lesion and treatment with ganglioside. J. Neurosci. Res.,
18: 615-620.

Yoshikawa, K., Williams, C. & Sabol, S.L. (1984) Rat brain preproenkephalin mRNA. J. Biol. Chem., 259, 14301-14308.

Young, S.W., Bonner, T.I. & Brann, M.R. (1986) Mesencephalic dopamine
neurons regulate the expression of neuropeptides mRNAs in the rat
forebrain. Proc. Nat. Acad. Sci. USA., 83, 9827-9831.

Zigmond, R.E. and Bowers, C.W. (1981) Influence of nerve activity on the
macromolecular content of neurons and their effector organs. Ann. Rev.
Physiol. 43: 673-687.

Zigmond, M.J. and Stricker, E.M. (1984) Parkinson's disease: Studies with
an animal model. Life Sci. 35: 5-18.

PRIMING OF THE BEHAVIORAL EXPRESSION OF DOPAMINE-RECEPTOR SUPERSENSITIVITY

IN THE BASAL GANGLIA: PHARMACOLOGICAL AND BIOCHEMICAL STUDIES

Micaela Morelli, Graziella De Montis[*], Sandro Fenu, Alberto Cozzolino and Gaetano Di Chiara

Institute of Experimental Pharmacology and Toxicology and [*]Institute of Pharmacology and Biochemical Phatology, University of Cagliari, Italy

In drug-naive rats unilaterally lesioned with 6-hydroxydopamine (6-OHDA) from 17 days, stimulation of D-1 receptors by SKF 38393 (2 mg/kg s.c.) failed to induce contralateral turning. Administration, three days before, of a single dose of a dopaminergic agonist which elicited contralateral turning, made SKF 38393 very active in producing contralateral turning (priming). The effectiveness of the D-1/D-2 agonist apomorphine as a primer of SKF 38393-induced turning was critically dependent on the interval between the administration of the two agonists. Effectiveness was minimal after 3 h, increased after 6-12 h, peaked at 72 h and was reduced after 10 days. In drug-naive 6-OHDA lesioned rats, administration of the selective antagonists SCH 23390 (D-1) or raclopride (D-2) abolished apomorphine induced contralateral turning, while in primed rats both antagonists only modified the pattern of apomorphine turning but failed to abolish it. Analysis of D-1 receptor binding in striata of drug-naive and primed rats, showed no change in the B_{max} and K_d, while dopamine stimulated adenylate cyclase showed a decreased K_m for dopamine in the lesioned side after priming. Finally, administration of the N-Methyl-D-Aspartate (NMDA) receptor antagonist (+) MK 801 in conjunction with apomorphine, prevented the ability of apomorphine to act as a primer, indicating that the NMDA receptor exerts a permissive role on priming.

INTRODUCTION

The behavioral actions of dopaminergic (DA) agonists are mediated by two different receptors: D-1, liked to the stimulation of adenylate cyclase activity and D-2 which either fail to influence or inhibit the activity of DA-sensitive cyclase (Garau et al., 1978; Kebabian and Calne, 1979; Onali et al., 1984).

In normal intact rats a complex interaction between the two receptors exists and concurrent D-1 and D-2 receptor stimulation is necessary for the full expression of DA-mediated effects in the basal ganglia (Gershanik et al., 1983; Barone et al., 1986; Longoni et al., 1987 a,b). After 6-hydroxydopamine (6-OHDA) denervation, however full behavioral stimulation can be mediated by the two receptors independently. In particular, in the unilateral 6-OHDA model (Ungerstedt, 1971), systemic administration of both D-1 or D-2 receptor agonists elicits turning behavior contralateral to the lesion side as a result of DA-receptor denervation supersensitivity (Herrera-Marshitz and Ungerstedt, 1984; Arnt and Hyttel, 1985).

We have recently shown that in rats unilaterally lesioned with 6-OHDA, previous stimulation of DA receptors provides a priming of the ability of the D-1 receptor agonist SKF 38393 to induce contralateral turning (Morelli et al., 1987 a,b; Morelli et al., 1989). In this chapter we examine the pharmacological and biochemical characteristics of this phenomenon.

EXPERIMENTAL PROCEDURES

Male Sprague-Dawley rats of 275-300 g of weight were injected in the left medial forebrain bundle (MFB) with 8 μg/4 μl of 6-OHDA in order to lesion the DA nigro-striatal bundle. Spontaneous ipsilateral turning behavior was measured every day for five days after 6-OHDA lesion; only rats positive to the test for the five days were used for the subsequent experiments.

For recording of behavior, rats were placed in plastic hemispherical bowls connected with automated rotameters 30 min before administration of drugs or vehicle. The number of contralateral turns (360°) made in 3 min every 10 min or the number of total turns was measured.

For biochemical studies, two weeks after lesioning rats were divided in two groups, placed in rotameters and injected with vehicle or primed with 0.2 mg/kg s.c. of the D-2 agonist LY 171555. Three days later, rats were sacrificed and the striata of both sides were dissected out and used for the biochemical experiments. The binding of the D-1 antagonist [^{3}H] SCH 23390 was performed in striatal homogenates according to the method of Billard et al. (1984) A concentration of 50 nM mianserine was added to the incubation buffer to occlude binding to 5-HT receptors. Non specific binding was determined in the presence of 1 μM SCH 23390.

DA stimulated adenylate cyclase activity was assayed by a modification of the method of Solomon et al. 1974 (Olianas et al. 1983) in the presence of 1 μM (-) sulpiride in order to block D-2 receptors. K_m and V_{max} were calculated by linear regression analysis of Eadie-Holfstee plots obtained from concentration-response curves.

RESULTS

Drug naive rats

Administration of the D-1 receptor agonist SKF 38393 (2 mg/kg s.c.) to drug-naive 6-OHDA lesioned rats, failed to induce contralateral turning 17 days after lesion. Contralateral turning (mean 14 turns/3 min) was observed 60 days after lesion, only after 90 days turning was intense (Fig. 1A). In drug-naive rats, lesioned 17 days earlier, no or slight contralateral turning was observed after 2 and 4 mg/Kg s.c. of SKF 38393 and a low intensity but dose-dependent contralateral turning was obtained after 6 and 10 mg/kg s.c. of SKF 38393 (Fig 1B).

Priming by DA-agonists

Administration of the D-1/D-2 agonist apomorphine (0.1 mg/kg s.c.), to rats unilaterally lesioned with 6-OHDA from 14 days, made the otherwise ineffective dose of 2.0 mg/kg s.c. of SKF 38393, given 3 days later, capable of producing strong and long-lasting contralateral turning (priming). Sensitization to SKF 38393 induced turning was also obtained by pre-administration of the D-2 agonist LY 171555 (0.2 mg/kg s.c.) or by SKF 38393 (10 mg/kg s.c.) itself (Fig 2).

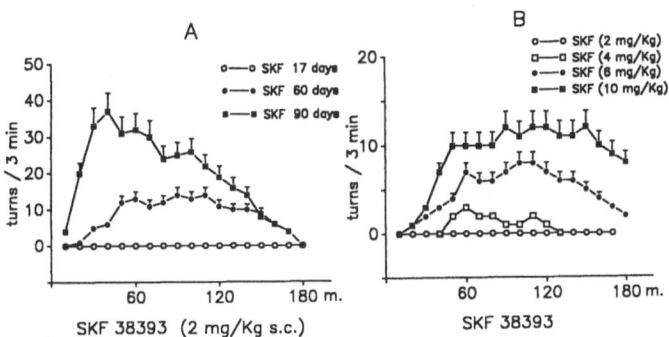

Fig 1. A: contralateral turning after administration of SKF 38393 (2 mg/kg s.c.) to drug-naive rats, lesioned with 6-OHDA at different times. F (2.51) = 40.85, P < 0.005 (two-way ANOVA).
B: contralateral turning after administration of different doses of SKF 38393 (2-4-6-10 mg/kg s.c.) to drug-naive rats, lesioned with 6-OHDA 17 days earlier. F (3.68) = 56.61, P < 0.005 (two-way ANOVA). The abscissa indicates the time after SKF 38393 administration, the ordinate the mean ± SEM of the number of contralateral turns made in 3 min.

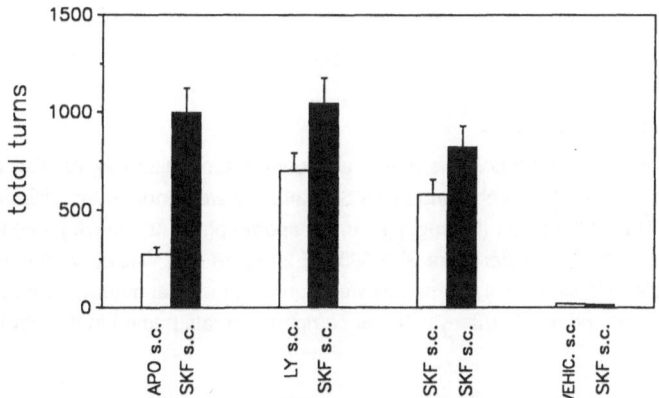

Fig 2. Contralateral turning in unilaterally 6-OHDA lesioned rats. Fourteen days after lesion, rats received apomorphine (0.1 mg/kg s.c.), LY 171555 (0.2 mg/kg s.c.) or SKF 38393 (10 mg/kg s.c.) or vehicle (empty columns). Three days later, rats received 2 mg/kg s.c. of SKF 38393 (filled columns).

Priming: time relationship

Priming with apomorphine (0.1 mg/kg s.c.) differentially affected contralateral turning by SKF 38393 depending on the time interval between priming and SKF 38393 challenge. Thus, as shown in tab. 1, while SKF 38393 (2 mg/kg s.c.) elicited sporadic contralateral turning 3 h after apomorphine administration, (total turns 30 ± 9) it was able to induce consistent contralateral turning 6 h (total turns 300 ± 58) or 12 h (total turns 392 ± 49) after apomorphine. Intense and long lasting contralateral turning was instead obtained 3 days after apomorphine priming (total turns 1216 ± 150). Further delay of the SKF 38393 challenge (10 days) resulted in a reduction of turning (total turns 250 ± 51) as compared with the results obtained at 3 days.

TABLE 1

DRUG	TIME	DRUG	TOTAL TURNS (after SKF38393)
APO	3 h	SKF	30 ± 9
APO	6 h	SKF	300 ± 58
APO	12 h	SKF	392 ± 49
APO	3 d	SKF	1216 ± 150
APO	10 d	SKF	250 ± 51

Number of total contralateral turns after SKF 38393 (2 mg/kg s.c.) in rats primed with apomorphine (0.1 mg/kg s.c.) 14 days after 6-OHDA lesion. SKF 38393 was administered to 5 different groups of rats 3, 6, 12 hours or 3, 10 days after apomorphine. $F_{(4,85)}$ = 31.73, $P < 0.005$ (two-way ANOVA). h= hours, d=days.

On the role of place conditioning

In order to investigate the possible role of environmental conditioning, priming with apomorphine or LY 17155 and challenge with SKF 38393 were performed in different environments. Thus, LY 171555 (0.2 mg/kg s.c.) or apomorphine (0.1 mg/kg s.c.) were administered in a plastic cylinder while SKF 38393 (2 mg/kg s.c.) was administered in a rotameter bowl, 3 days later. No significant differences were found in the total number of turns in response to SKF 38393 in rats primed in the cylinder as compared to rats primed in the bowl (data not shown).

Interaction between D-1 and D-2 receptors

Administration of 0.1 mg/kg s.c. of the D-1/ D-2 agonist apomorphine to drug-naive rats unilaterally lesioned with 6-OHDA from seventeen days, elicited contralateral turning behavior with a monophasic time course (Fig. 3A and 4A). Pretreatment with the D-1 antagonist SCH 23390 (0,1 mg/kg s.c.), 30 min before, or with the D-2 antagonist raclopride, (3 mg/kg s.c.), 45 min before, completely antagonized the contralateral turning induced by apomorphine in drug-naive rats (Fig. 3A and 4A).

Challenge with 0.1 mg/kg s.c. of apomorphine to 6-OHDA lesioned rats which had been primed, 3 days before, with the same dose of apomorphine, resulted in a two peak pattern of contralateral turning (total turns 245 and 252) Fig. 3B and 4B. Pretreatment with SCH 23390 (0.1 mg/kg s.c.), 30 min before, changed the time-course of contralateral turning from a two peaks into a one peak and significantly reduced the number of total turns (148 ± 16; p<0.05) Fig. 3B. Pretreatment with raclopride (3 mg/kg s.c.), 45 min before, also abolished the two peak pattern of contralateral turning in response to apomorphine but did not significantly reduced the total number of contralateral turns (233 ± 36). Summing up the number of total turns obtained under D-1 receptor blockade with that obtained under D-2 receptor blockade gave a value of 381 which was higher than that of the turning after apomorphine alone (total turns 245 and 252).

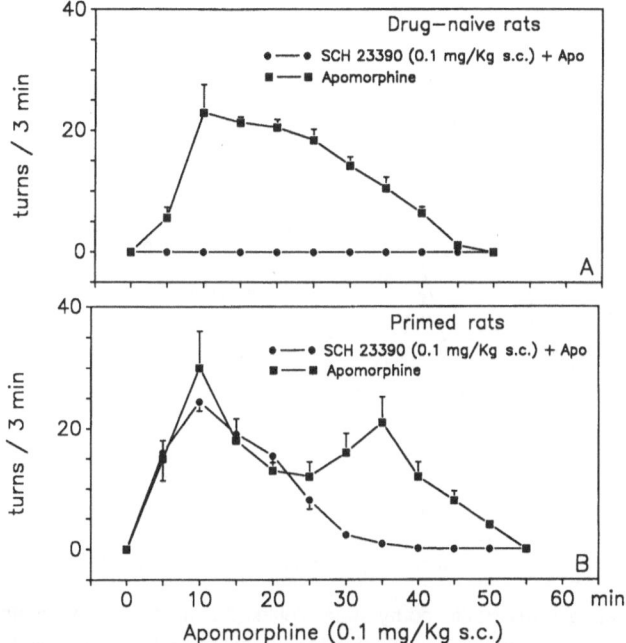

Fig 3. Contralateral turning induced by apomorphine 0.1 mg/kg s.c. in rats unilaterally lesioned, 17 days before, with 6-OHDA. SCH 23390 (0.1 mg/kg s.c.) was administered 30 min before apomorphine. Three days before the experiment the rats received H$_2$O (drug-naive rats) or 0.1 mg/kg s.c. of apomorphine (primed rats). The abscissa indicates the time after apomorphine administration, the ordinate the mean ± SEM of contralateral turns made in 3 min by 8 rats.

D-1 receptor-adenylate-cyclase complex

Table 2 shows the B$_{max}$ and the K$_d$ of the specific [^{3}H] SCH 23390 binding in homogenates of striata correspondent to the intact or the 6-OHDA injected side. No significant differences were found between the striata of the two sides and between the drug-naive and primed group of rats.

Similarly the V$_{max}$ of the DA stimulated adenylate cyclase did not show any difference between the intact and the lesioned side and between drug-naive and primed rats, while in the lesioned striatum of primed rats the K$_m$ value was lower (-41%) than that of drug-naive rats (table 2). Basal activity of the enzyme was not modified in the two experimental groups.

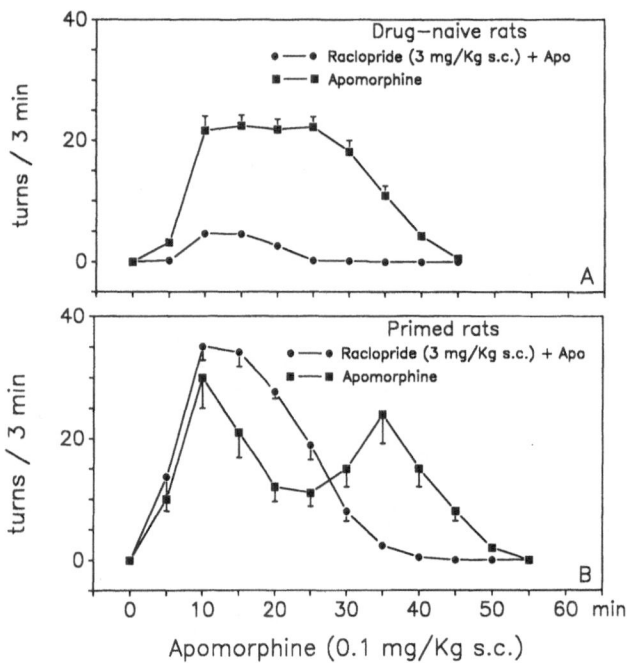

Fig 4. Contralateral turning induced by apomorphine 0.1 mg/kg s.c. in rats unilaterally lesioned, 17 days before, with 6-OHDA. Raclopride (3 mg/kg s.c.) was administered 45 min before apomorphine. Three days before the experiment the rats received H_2O s.c. (drug-naive rats) or apomorphine 0.1 mg/kg s.c. (primed rats). The abscissa indicates the time after apomorphine administration, the ordinate the mean number ± SEM of contralateral turns made in 3 min by 12 rats.

TABLE 2

[^{3}H] SCH 23390 binding and adenylate cyclase activity in striata of drug-naive and primed rats.

		[^{3}H] SCH 23390 binding		cAMP	
		B_{max}(fmol/mg prot)	K_d(nM)	V_{max}	K_m
Drug-naive	L	1198±89	0.79±0.07	112±12	4.51±0.41
	R	1110±90	0.8±0.07	102±10	4.97±0.5
Primed	L	1150±99	0.82±0.09	111±11	2.66±0.34*
	R	1080±79	0.8±0.08	110±10	4.85±0.42

Rats were lesioned with 6-OHDA in the left MFB and injected on day 14th with vehicle s.c. (drug-naive), or with 0.2 mg/kg s.c. of LY 171555 (primed) and then sacrificed 3 days later. cAMP formation was expressed as pmol/mg prot/min. The values are in units of μM. Results are the mean ± SEM of at least 5 experiments. L = left, R = right; *P<0.01

Effect of MK 801

In order to study the role of glutamate receptors of the NMDA-type, rats unilaterally lesioned with 6-OHDA from 14 days, were treated with the non competitive antagonist MK 801, (Wong et al. 1988) (0.1 mg/kg i.p.) or with vehicle and 15 min later were administered with 0.1 mg/kg s.c. of apomorphine. Pretreatment with (+) MK 801, the active isomer of MK 801, significantly increased the number of contralateral turns elicited by apomorphine (total turns = 172.4 ± 18 p<0.05) as compared with vehicle + apomorphine treated rats (total turns = 110.5 ± 12), while (-) MK 801 failed to influence the acute effect of apomorphine (total turns = 121 ± 15) (Fig. 5). Three days later administration of 3 mg/kg s.c. of SKF 38393 elicited vigorous contralateral turning in rats pretreated with vehicle + apomorphine (total turns = 966 ± 98), while produced a very weak contralateral turning in rats pretreated with (+) MK 801 + apomorphine (total turns = 154 ± 28, p<0.005). This effect was stereospecific as (-) MK 801 failed to influence the ability of SKF 38393 to induce turning. (Fig. 5)

DISCUSSION

The present results show that D-1 receptor stimulation by SKF 38393, is poorly effective in eliciting contralateral turning in drug-naive rats unilaterally lesioned with 6-OHDA from 14 days. Only after high doses of SKF 38393 (6-10 mg/Kg) or longer post-lesion interval, (60-90 days), contralateral turning was obtained. Therefore SKF 38393 can produce contralateral turning in drug-naive 6-OHDA lesioned rats, provided that a sufficient high dose or a long post-lesion interval is given. However, administration of a single dose of a DA agonist which produce contralateral turning by itself, sensitizes 6-OHDA lesioned rats to the expression of contralateral turning in response to the subsequent D-1 receptor stimulation by SKF 38393 (priming).

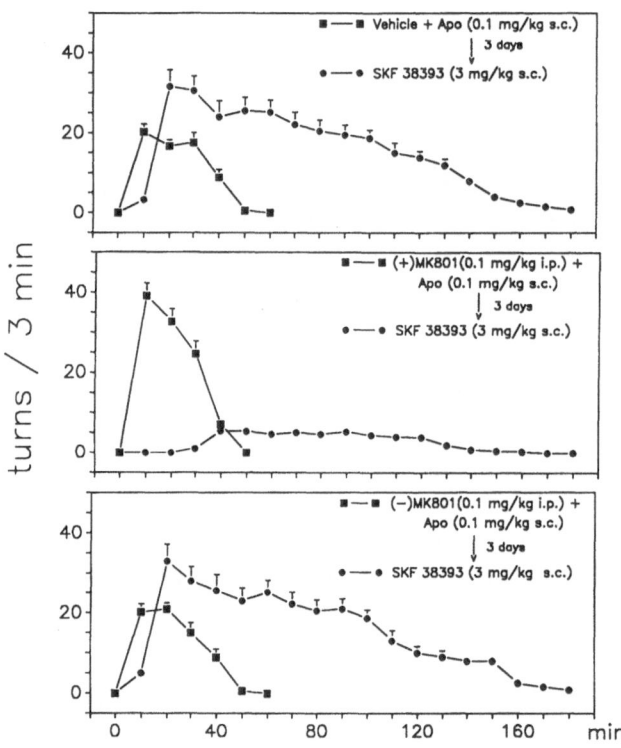

Fig 5. Contralateral turning in unilaterally 6-OHDA lesioned rats. Fourteen days after lesion, 3 different groups of rats received vehicle + apomorphine (0.1 mg/kg s.c.), (+) or (-) MK 801 (0.1 mg/kg i.p.) + apomorphine (0.1 mg/kg s.c.). Three days later rats received 3 mg/kg s.c. of SKF 38393. The abscissa indicates the time after administration of apomorphine or SKF 38393, the ordinate the means ± SEM of contralateral turns made in 3 min.

Priming therefore is not an absolute requirement for the behavioral expression of D-1 receptor supersensitivity but it strongly facilitates it.

Priming was unrelated to environmental conditioning, in fact, priming with apomorphine or LY 171555 was equally effective when performed in an environment different from that where challenge with SKF 38393 took place; therefore priming appears quite different from "paradoxical" conditioned turning (Silverman and Ho, 1981).

The priming effect of apomorphine was strictly dependent on the interval between priming and challenge with SKF 38393; thus the effectiveness of priming increased progressively after intervals from 3 to 72 h, but decreased after a 10 days interval. Such a temporal relationship might be due to the time necessary for the synthesis of a molecular substrate of the priming effect, and is consistent with some form of mnemonic process. In this regard it is interesting that blockade of the NMDA receptor, which has been shown to play an important role in various forms of adaptive mechanisms in the CNS (Collindridge and Bliss, 1987, Morris et al., 1986), is able to abolish the priming effect of apomorphine. Therefore glutamatergic transmission through NMDA receptors might exert a permissive role on the ability of DA-receptor stimulation to promote priming.

The modification of D-1 receptor mediated behavior after priming is associated with a change in D-1/D-2 receptor interaction. In drug-naive lesioned rats, in fact, apomorphine induced a turning with a monophasic time-course which was blocked by both D-1 or D-2 receptor antagonist, indicating that in drug-naive rats, concurrent stimulation of D-1 and D-2 receptors is necessary for the full expression of turning behavior.

After priming, apomorphine induced the typical two peak turning and the selective blockade of D-1 or D-2 receptor eventually decreased but did not abolish the contralateral turning induced by the drug, suggesting that after priming a different interaction between D-1 and D-2 receptors takes place and each receptor becomes able to evoke turning behavior independently from the other. Indeed, combined stimulation of D-1 and D-2 receptor in primed rats (e.g. apomorphine alone) resulted in a lower amount of turning than the sum of the turning obtained by selective stimulation of each receptor type (e.g. apomorphine plus raclopride or SCH 23390).

Biochemical analysis of the D-1 receptor-adenylate-cyclase complex shows that priming increases the affinity of adenylate cyclase for DA in the lesioned striatum (decrease of K_m) while the number (B_{max}) and the affinity (K_d) of D-1 receptor sites are not changed. These results suggest that the behavioral expression of D-1 receptor supersensitivity might be associated with an alteration in the transduction mechanism of the D-1 receptor, rather than a modification of the recognition site and can explain why priming increases the potency of SKF 38393 in eliciting contralateral turning in 6-OHDA lesioned rats.

In conclusion, priming can be considered a process of behavioral sensitization promoted by previous stimulation of DA-receptors and can provide a simplified model to study the role of Basal Ganglia in motor behavior.

REFERENCES

Arnt, J. and Hyttel, J., 1985, Differential involvement of dopamine D-1 and D-2 receptors in the circling behaviour induced by apomorphine, SKF 38393, pergolide and LY 171555 in 6-hydroxydopamine-lesioned rats, Psychopharmacology 85, 346.

Barone, P., Davis, T.A., Braun, A.R. and Chase, T.N., 1986, Dopaminergic mechanisms and motor function: characterization of D-1 and D-2 dopamine receptor interactions, European J. Pharmacol., 123:109.

Billard, W., Ruperto, V., Crosby, G., Iorio, L.C. and Barnett, A., 1984, Characterization of the binding of ^{3}H-SCH 23390, a selective D-1 receptor antagonist ligand, in rat striatum, Life Sci. 35, 1985.

Collingridge, G.L. and Bliss, T.V.P., 1987, NMDA receptors their role in long-term potentiation, Trends Neurosci., 10: 288-293.

Garau, L., Govoni, S., Stefanini, E., Trabucchi M. and Spano, P.F., 1978, Dopamine receptors: pharmacological and anatomical evidence indicate that two distinct dopamine receptors populations are present in rat striatum, Life Sci. 23, 1745.

Gershanik, O., Heikkila, R.E. and Duvoisin, R.C., 1983, Behavioural correlations of dopamine receptor activation, Neurology, 33:1489.

Herrera-Marschitz, M. and Ungerstedt, U., 1984, Evidence that apomorphine and pergolide induce rotation in rats by different actions on D-1 and D-2 receptor sites, European J. Pharmacol., 98:165.

Kebabian, J.W. and Calne, D.B., 1979, Multiple receptors for dopamine, Nature 277, 93.

Longoni, R., Spina, L. and Di Chiara, G., 1987 a, Permissive role of D-1 receptor stimulation by endogenous dopamine for the expression of postsynaptic D-2-mediated behavioural responses. Yawning in rats, European J. Pharmacol., 134:163.

Longoni, R., Spina, L. and Di Chiara, G., 1987 b, Permissive role of D-1 receptor stimulation for the expression of D-2 mediated behavioural responses: a quantitative phenomenological study in rats, Life Sci., 41:2135.

Morelli, M. and Di Chiara, G., 1987 a, Agonist-induced homologous and heterologous sensitization to D-1 and D-2 dependent contraversive turning, European J. Pharmacol., 141:101

Morelli, M., Fenu, S. and Di Chiara, G., 1987 b, Behavioural expression of D-1 receptor supersensitivity depends on previous stimulation of D-2 receptors, Life Sci., 40: 245

Morelli, M., Fenu,S., Garau, L., and Di Chiara, G., 1989, Time and dose dependence of the priming of the expression of dopamine receptor supersensitivity, European J. Pharmacol., 162:329.

Morris, R.G.M., Anderson, E., Lynch, G.S. and Baudry, M., 1986 Selective impairment of learning and blockade of long-term potentiation by an N-methyl-D-aspartate-receptor antagonist, AP5. Nature 319: 774-776.

Olianas, M.C., Onali, P., Neff, N.H. and Costa, E., 1983, Adenylate cyclase activity of synaptic membranes from rat striatum, Mol. Pharmacol. 23, 393.

Onali, P., Olianas, M.C. and Gessa, G.L., 1984, Selective blockade of dopamine D1 receptors by SCH 23390 discloses striatal dopamine D2 receptors mediating inhibition of adenylate cyclase in rats. European J. Pharmacol., 99:127.

Silverman, P.B. and Ho, B.T., 1981, Persistent behavioural effect of apomorphine in 6-hydroxydopamine-lesioned rats. Nature 294: 475

Solomon, Y., Londos, C. and Rodbell, M., 1974, A highly sensitive adenylate cyclase assay Analyt. Biochem., 58, 541.

Ungerstedt, U., 1971, Postsynaptic supersensitivity after 6-hydroxydopamine induced degeneration of the nigro-striatal dopamine system, Acta Physiol. Scand., 376,1.

Wong, E.H.F., Knight, A.R. and Woodruff, G.N., 1988, [^{3}H] MK 801 labels a site on the N-Methyl D-Aspartate receptor channel complex in rat brain membranes, J. Neurochem., 50:274.

PHARMACOLOGICAL EFFECTS UPON
BEHAVIOUR

BEHAVIORAL EFFECTS OF SINGLE AND REPEATED TREATMENTS WITH THE COMBINATION

OF D1 AND D2 ANTAGONISTS

Ottavio Gandolfi and Rossella Dall'Olio

Institute of Pharmacology, University of Bologna, Bologna
Italy

It is generally accepted that cerebral D2 dopamine receptors mediate the antidopaminergic activity of neuroleptics (Creese et al. 1976). This view is supported by the observations that all clinically effective antipsychotics are either mixed D1/D2 blockers or selective D2 dopamine antagonists (Christensen et al. 1984) and that a high correlation exists between the ability of neuroleptics to interact with the D2 receptors "in vitro" and their clinical efficacy (Seeman 1980). Nevertheless, the substituted benzamide sulpiride, which selectively blocks D2 dopamine receptors, is much weaker than classic neuroleptics in suppressing dopamine mediated behaviours (Jenner et al. 1978) and fails to antagonize apomorphine-induced stereotyped behaviour and to induce catalepsy (Costall and Naylor 1975; Jenner et al. 1978; Serra et al. 1983). In contrast, the selective D1 antagonist SCH 23390 does induce catalepsy and inhibits agonist-elicited stereotypy and hyperactivity (Iorio et al. 1983; Christensen et al. 1984), thus behaving as a classic neuroleptic agent. Therefore some antidopaminergic effects could be mediated by the concomitant blockade of D1 receptors which could contribute to the full activity of the typical antipsychotic drugs.

Therefore in this study we wondered whether the concomitant blockade of D1 receptors would allow the D2 antagonist (-)-sulpiride to exhibit a wider spectrum of neuroleptic activity in rats.

Rats, treated with a combination of subthreshold doses of (-)-sulpiride and SCH 23390, were evaluated for exploratory activity, apomorphine-induced stereotyped behaviour and hypermotility elicited by the D2 selective agonist LY 171555.

It is known that in rats, chronic neuroleptic blockade of central dopamine receptors is associated with increased striatal dopamine receptor binding (Burt et al. 1977) and with behavioral supersensitivity measured by enhanced stereotyped response to apomorphine (Tarsy and Baldessarini, 1974). Therefore the increased density of striatal dopaminergic receptors has been proposed as a contributing factor in the development of tardive dyskinesia, the neurological disorder associated with long-term administration of neuroleptics (Klawans and Rubowitz, 1972).

Thus, the blockade of striatal dopaminergic transmission appears to be connected to the extrapyramidal side effects of the neuroleptics, whereas the antipsychotic action of these drugs seems to be related to impairment of limbic dopaminergic transmission (Bartholini, 1976). In contrast, repeated administrations with the atypical antipsychotic sulpiride do not increase striatal domaminergic receptor density (Trabucchi et al., 1980)

The Basal Ganglia III, Edited by G. Bernardi *et al.*
Plenum Press, New York

or are slightly effective only at very high doses (Rupniak et al. 1984) and do not induce extrapyramidal side effects either in man (Mielke et al. 1977) or in rat (Montanaro et al. 1982).

Therefore it was intriguing enough to study whether repeated treatments with subthreshold doses of SCH 23390 given alone or in combination with (-)-sulpiride, besides potentiating the antidopaminergic effect of the benzamide, could induce dopaminergic behavioral supersensitivity to a threshold dose of apomorphine.

MATERIAL AND METHODS

Animals

Male Sprague Dawley rats (200-250 g) from Charles River (Como, Italy) were used. The animals were housed four per cage under controlled conditions of light (from 7:00 a.m. to 7:00 p.m.), temperature (22 + 2°C) and humidity (60%) and were allowed free access to standard laboratory diet and tap water. They were tested behaviourally only once between 9:00 a.m. and 3:00 p.m..

Drugs

LY 171555 (Lilly, Indianapolis, IN), SCH 23390 (Schering-Plough, Bloomfield, NJ) and apomorphine (Sigma, St.Louis, MO) were dissolved in saline. (-)-Sulpiride, kindly gifted by Ravizza (Muggiò, Italy) and haloperidol (Serenase, kindly gifted by Lusofarmaco, Milano, Italy), were used as vial solution. (-)-Sulpiride and haloperidol were always injected IP 60 min before the behavioural trials; SCH 23390 was administered SC 30 min before the trials.

Apparatus

Exploratory activity and LY 171555-induced hypermotility were evaluated in 38 x 30 x 25 cm actometric cages. A DC current (65 V, 25 μA) was continuously delivered to the stainless steel grid floor of the cage and every closure of the circuit performed by the rat feet was recorded as one motility count by an electronic counter: in this way, only horizontal displacements of the animal across the cage were recorded, whereas rearing, forepaw treading or tramples did not by themselves activate the circuit unless associated with animal locomotion. The cages had the front panel and the top cover made of transparent Plexiglas, in order to observe the animal's behaviour, and were located into a sound proofed and dimly illuminated room.

Circular transparent Plexiglas cages (18 cm diameter) with corrugated paper covering the floor were used to observe apomorphine-induced stereotyped behaviour.

Procedure

Exploratory activity, apomorphine-induced stereotyped behaviour and LY 171555-induced hypermotility were studied in rats treated with different doses of (-)-sulpiride or SCH 23390. Thereafter, each behavioural response was assayed following acute treatments with a combination of ineffective dose of both antagonists.

Exploratory activity. Seven groups of eight naive rats were placed in the actometric cages after the injection of (-)-sulpiride (10, 20 and 40 mg/kg IP, 60 min before), SCH 23390 (5, 10 and 20 μg/kg, 30 min before) or the combined administration of (-)-sulpiride (10 and 20 mg/kg IP) plus SCH 23390 (5 μg/kg SC, 30 min later). Controls were treated with saline. Motility counts were recorded every 5 min for 1 hour.

<u>Apomorphine-induced stereotyped behavior</u>. Apomorphine (0.5 mg/kg SC) was injected in rats (8 animals per group) previously treated with saline, (-)-sulpiride (10, 20 and 40 mg/kg), SCH 23390 (5, 10 and 20 µg/kg), or with (-)-sulpiride (10 and 20 mg/kg) in combination with SCH 23390 (5 µg/kg). Immediately after apomorphine injection, the rats were placed in the transparent circular cages and their stereotyped behaviour was evaluated every 10 min for 30 min by an observer unaware of the treatments according to the following rating scale: 0= no change in normal behaviour; 1= intermittent sniffing; 2= continuous sniffing; 3= intermittent licking or biting; 4= continuous licking or biting. The individual data were the mean scores of the three 10-min spaced observations.

<u>LY 171555-induced hypermotility</u>. Seven groups of eight rats, accustomed to the actometric cages for 1 hour, were injected with LY 171555 (0.5 mg/kg, IP) following the administration of saline, (-)-sulpiride (2.5, 5 and 10 mg/kg, 60 min before), SCH 23390 (2.5 and 5 µg/kg, 30 min before) or (-)-sulpiride (2.5 mg/kg) plus SCH 23390 (2.5 µg/kg). Locomotor activity was recorded every 10 min for 120 min.

Different groups of rats received repeated administrations (21 consecutive days) of (-)-sulpiride (20 mg/kg IP), SCH 23390 (20 or 5 µg/kg SC), (-)-sulpiride + SCH 23390 (20 mg/kg IP and 5 µg/kg SC), haloperidol (0.2 mg/kg IP) or saline.

At the 15th day of treatment, the rats were administered subcutaneously with apomorphine (0.5 mg/kg SC, 60 min after the daily (-)-sulpiride or haloperidol injection; 30 min after the daily SCH 23390 administration) and individually placed in the circular cages for the observation of stereotyped behavior. After the 21 days repeated treatment, following a washout period of 5 days, rat's stereotyped response to a subthreshold dose of apomorphine (0.25 mg/kg SC) was evaluated.

Motility counts were analyzed by ANOVA followed by single comparisons of the means (Dunnet t test). Stereotyped behaviour was statistically evaluated by the Mann-Whitney U test.

Table 1. Exploratory motility exhibited by rats after the injection of saline, (-)-sulpiride (60 min before), SCH 23390 (30 min before) or after the combined administration of (-)-sulpiride and SCH 23390.

PRETREATMENT	MOTILITY COUNTS
SALINE	543 + 52
(-)-Sulpiride 10 mg/kg	588 + 61
(-)-Sulpiride 20 mg/kg	552 + 65
(-)-Sulpiride 40 mg/kg	443 + 56
SCH 23390 5 µg/kg	618 + 68
SCH 23390 10 µg/kg	364 + 68
SCH 23390 20 µg/kg	212 + 20*
(-)-Sulpiride 10 mg/kg + SCH 23390 5 µg/kg	383 + 43*
(-)-Sulpiride 20 mg/kg + SCH 23390 5 µg/kg	352 + 36*

Mean value + SEM of 60 min motility counts (N = 8 per group).
* $P < 0.05$ in comparison to saline pretreated group (Dunnet T Test after ANOVA.

Table 2. Stereotyped behavior induced by apomorphine (0.5 mg/kg) in rats pretreated with saline, (-)-sulpiride (60 min before), SCH 23390 (30 min before) or with the combined administration of (-)-sulpiride and SCH 23390.

PRETREATMENT	STEREOTYPED SCORES
SALINE	3.1 + 0.4
(-)-Sulpiride 10 mg/kg	3.0 + 0.3
(-)-Sulpiride 20 mg/kg	3.0 + 0.4
(-)-Sulpiride 40 mg/kg	2.9 + 0.3
SCH 23390 5 µg/kg	3.2 + 0.3
SCH 23390 10 µg/kg	2.3 + 0.2
SCH 23390 20 µg/kg	1.3 + 0.3*
(-)-Sulpiride 10 mg/kg + SCH 23390 5 µg/kg	1.5 + 0.2*
(-)-Sulpiride 20 mg/kg + SCH 23390 5 µg/kg	1.4 + 0.2*

Mean values + SEM of individual mean scores obtained from 3 observations taken at intervals of 10 min for 30 min immediately after the dopaminomimetic injection.

* $P < 0.05$ in comparison to saline pretreated group (Mann-Whitney U Test).

RESULTS

Table 1 shows rat exploratory activity following various doses of SCH 23390 (5, 10 and 20 µg/kg) or (-)-sulpiride (10, 20 and 40 mg/kg). The administration of the highest dose of SCH 23390 significantly reduced locomotor activity while the selective D2 receptor blocker (-)-sulpiride did not affect this behaviour. When the rats were treated with the combination of the two antagonists at doses ineffective by themselves, a significant inhibition of the exploratory activity was observed.

Apomorphine (0.5 mg/kg)-induced stereotyped behaviour was antagonized only by the highest dose of SCH 23390 (20 µg/kg) whereas it was unaffected by pretreatment with (-)-sulpiride. When rats were treated with (-)-sulpiride (10 or 20 mg/kg) followed by a dose of SCH 23390 devoid of direct effects (5 µg/kg), a significant inhibition of the stereotyped behaviour was observed (table 2).

Table 3 shows that the lower doses of (-)-sulpiride (2.5 mg/kg) or SCH 23390 (2.5 µg/kg) did not modify LY 171555-induced hypermotility; the combination of the same low doses of the two antagonists reduced the locomotor response to the D2 receptor stimulation in an amount comparable to higher (5 or 10 mg/kg) doses of (-)-sulpiride.

The effects of repeated administrations (21 days) of the different treatments on the stereotyped response to apomorphine are shown in table 4. At the 15th day, apomorphine (0.5 mg/kg)-induced stereotyped behavior was dramatically inhibited in the animals receiving haloperidol (0.2 mg/kg), SCH 23390 (20 µg/kg) or the combination of (-)-sulpiride (20 mg/kg) plus the lower dose of SCH 23390 (5 µg/kg). In contrast, treatments with (-)-sulpiride or with the lower dose of SCH 23390 (5 µg/kg) alone exhibited stereotypy of the same magnitude as the controls. Following a washout pe-

Table 3. Locomotor activity induced by 0.5 mg/kg IP LY 171555 in rats pretreated with (-)-sulpiride (60 min before), and SCH 23390 (30 min before) given separately or in combination.

PRETREATMENT	MOTILITY COUNTS
SALINE	719 + 131
(-)-Sulpiride 2.5 mg/kg	709 + 146
(-)-Sulpiride 5.0 mg/kg	344 + 66*
(-)-Sulpiride 10,0 mg/kg	254 + 87*
SCH 23390 2.5 µg/kg	614 + 90
SCH 23390 5.0 µg/kg	211 + 39*
(-)-Sulpiride (2.5 mg/kg) + SCH 23390 (2.5 µg/kg)	352 + 51*

Mean values + SEM of motility counts recorded from 20 to 120 min after drug administration. N = 8 animals per group.
* P < 0.05 in comparison to saline-pretreated group (Dunnet t test after ANOVA).

riod of 4 days, the animals previously treated with haloperidol or the full dose of SCH 23390 showed an increased response to the threshold dose of apomorphine (0.25 mg/kg); in contrast, the rats treated with (-)-sulpiride, the lower dose of SCH 23390 or with (-)-sulpiride + SCH 23390 (5 µg/kg) failed to exhibit any sign of behavioural supersensitivity.

DISCUSSION

In the first part of this study, the D1 receptor blocker SCH 23390 not only antagonized in a dose-dependent manner exploratory motility and apomorphine-induced stereotyped behaviour, but was able to prevent the appearance of typical D2 mediated responses (LY 171555-induced hyperactivity). Therefore, our results, in agreement with others (Iorio et al. 1983; Pugh et al. 1985; Arnt, 1985; Breese and Mueller, 1985), support the view that a strict correlation between D1 and D2 receptors in the modulation of several dopamine-mediated behaviours exists in rat brain.

The selective D2 blocker (-)-sulpiride, when given alone, potently antagonized the hyperactivity induced by the D2 agonist LY 171555, whereas it failed to change both spontaneous exploratory activity or apomorphine-induced stereotyped behaviour (these results and Costall and Naylor, 1975; Jenner et al. 1978; Serra et al. 1983). Nevertheless, when the benzamide was administered in combination with doses of the D1 blocker which by themselves were without effect, all these behaviours were greatly inhibited and a full neuroleptic effect was expressed.

Our results therefore suggest that D1 receptor blockade plays an important role in the antidopaminergic activity of neuroleptics. Moreover, besides the full neuroleptic activity, repeated administrations of the combination of low doses of D1 and D2 receptor antagonists appear free from the risk of inducing dopaminergic supersensitivity, while haloperidol or higher doses of SCH 23390 enhanced the stereotyped response to apomorphine. Since tardive dyskinesia is thought to be the reflection of an increased striatal dopamine receptor number which brings to enhanced striatal behavioural response, the last observation indicates that the increase of striatal D1 receptors subtype could be responsible for the dyskinetic symptoms following chronic treatment with neuroleptics. In contrast repeated treatments with (-)-sulpiride, failed to inhibit apomorphine-induced stereotypy, a response related to stimulation of dopaminergic receptors in the striatum (Costall

Table 4. Effect of repeated treatment with neuroleptics on stereotyped response to apomorphine.

PRETREATMENTS (21 days)	APO 0.5 mg/kg (during daily treatment)	APO 0.25 mg/kg (following washout period)
SALINE	3.0 + 0.3	1.6 + 0.2
(-)-Sulpiride 20 mg/kg	3.1 + 0.3	1.8 + 0.2
SCH 23390 5 µg/kg	3.0 + 0.3	1.8 + 0.2
SCH 23390 20 µg/kg	0*	2.9 + 0.3*
Haloperidol 0.2 mg/kg	0*	3.2 + 0.3*
(-)-Sulpiride 20 mg/kg + SCH 23390 5 µg/kg	0.5 + 0.1*	1.8 + 0.2

The dopaminomimetic was administered at a full dose (0.5 mg/kg) during daily antipsychotic injections (evaluation of neuroleptic activity) or at a sub-threshold dose (0.25 mg/kg) after the withdrawal (5 days) from repeated treatment (evaluation of behavioural supersensitivity).
* $P < 0.05$ in comparison to respective saline pretreated group (Mann-Whitney U test).

and Naylor, 1975) and to elicit signs of extrapyramidal disorders by acting probably at mesolimbic level (Scatton et al. 1979). This inference is supported by our observations showing that (-)-sulpiride given even at low doses strongly inhibited LY 171555-induced hypermotility and apomorphine-elicited sedation (Vaccheri et al. 1986), which are mediated by dopaminergic system in the nucleus accumbens (Sharp et al. 1987, Radhakisum and Van Ree, 1987).

A possible explanation of the finding that repeated treatments with (-)-sulpiride plus SCH 23390, although displaying full neuroleptic activity, lacked to induced dopaminergic supersensitivity, could be due to the functional coupling of D1 and D2 receptors at striatal level (Stoof and Verherjden, 1986). Thus, low doses of SCH 23390, by blocking D1 receptors, could have greately potentiated the low potency of (-)-sulpiride in blocking D2 receptors in this area. The combined administration of (-)-sulpiride + SCH 23390 for 3 weeks fails to change the limbic characteristics of D1 or D2 receptor binding in membranes prepared from striata or nuclei accumbens (not shown), suggesting that these doses were ineffective for inducing behavioural supersensitivity.

REFERENCES

Arnt, J., 1985, Behavioral stimulation is induced by separate D1 and D2 receptor sites in reserpine-pretreated but not in normal rats, Eur. J. Pharmacol., 113:79.
Bartholini, G., 1976, Differential effect of neuroleptic drugs on dopamine turnover in the extrapiramidal and limbic system, J. Pharm. Pharmac., 28:429.
Breese, G. R., and Mueller, R. A., 1985, SCH 23390 antagonism of a D-2 dopamine agonist depend upon catecholaminergic neurons, Eur. J. Pharmacol., 113:109.
Burt, D. R., Creese, J., and Snyder, S. H., 1977, Antipsychotic drugs: chronic treatment elevates dopamine receptor binding in brain, Science, 196:326.
Christensen, A. V., Arnt, J., Hyttel, J., Larsen, J. J., and Svendsen, O., 1984, Pharmacological effects of a specific dopamine D-1 antagonist

SCH 23390 in comparison with neuroleptics, Life Sci., 34:1529.

Costall, B., and Naylor, R. J., 1975, Detection of the neuroleptic properties of clozapine, sulpiride and thioridazine, Psychopharmacologia
(Berlin), 43:69.

Creese, I., Burt, D. R., and Snyder, S. H., 1976, Dopamine receptor binding
predicts clinical and pharmacological potencies of antischizophrenic
drugs, Science, 192:481.

Fuxe, K., Ogren, S. O., Hall, H., Agnati, L. F., Andersson, K., Kohler, C.,
and Schwarcz, R., 1980, Effect of chronic treatment with l-sulpiride
and haloperidol on central monoaminergic mechanisms, Adv. Biochem.
Psychopharmacol., 24:193.

Iorio, L. C., Barnett, A., Leitz, F., Houser, V. P., and Korduba, C. A.,
1983, SCH 23390, a potential benzazepine antipsychotic with unique
interactions on dopaminergic system, J. Pharmacol. Exp. Ther., 226:462.

Jenner, P., Clow, A., Reavill, C., Theodorou, A., and Marsden, C. D., 1978,
A behavioral and biochemical comparison of dopamine receptor blockade
produced by haloperidol with that produced by substituted benzamide
drugs, Life Sci., 23:545.

Klawans, H. L., and Rubovitz, R., 1972, An experimental model of tardive
dyskynesia, J. Neural. Trans., 33:235.

Mielke, D. H., Gallant, D. M., and Kessler, C., 1977, An evaluation of a
unique new antipsychotic agent sulpiride: effects on serum prolactin
and growth hormone levels, Am. J. Psychiat., 134:1371.

Montanaro, N., Dall'Olio, R., Gandolfi, O., and Vaccheri, A., 1982, Differential enhancement of behavioral sensitivity to apomorphine following
chronic treatment of rats with (-)-sulpiride and haloperidol, Eur. J.
Pharmacol., 81:1.

Pugh, M. T., O'Boyle, K. M., Molloy, A. G., and Waddington, J. L., 1985,
Effects of the putative D-1 antagonist SCH 23390 on stereotyped behaviour induced by the D-2 agonist RU 24213, Psychopharmacology, 87:
308.

Radhakisun, F. S., and Van Ree, J. M., 1987, The hypomotility elicited by
smal doses of apomorphine seems exclusively mediated by dopaminergic
systems in the nucleus accumbens, Eur. J. Pharmacol., 136:41.

Rupniak, N. M. J., Jilpatrick, G., Hall, M. D., Jenner, P., and Marsden,
C. D., 1984, Differential alterations in striatal dopamine receptor
sensitivity induced by repeated administration of clinically equivalent
doses of haloperidol, sulpiride or clozapine in rats, Psychopharmacology, 84:512.

Scatton, B., Warms, P., Zivkovic, B., Depoortere, H., Dedek, K., and Bartholini, G., 1979, On the neuropharmacological spectra of "classical"
(haloperidol) and "atypical" (benzamide derivatives) neuroleptics, in:
"Sulpiride and Other Benzamides", P. F. Spano, M. Trabucchi, G. U. Corsini, and G. L. Gessa, eds., Italian Brain Research Foundation Press,
Milan, pp 53-66.

Seeman, P., 1980, Brain dopamine receptors, Pharmacol. Rev., 32:229.

Serra, G., Van Ree, J. M., and De Vied, D., 1983, Influence of classical
and atypical neuroleptics on apomorphine-induced behavioural changes
on extinction of a conditioned avoidance response, J. Pharm. Pharmacol.,
35:255.

Sharp, P., Zetterstrom, T., Ljungberg, T., and Ungerstedt, U., 1987, A direct comparison of amphetamine-induced behaviours and regional brain
dopamine release in the rat using intracerebral dialysis, Brain Res.,
401:322.

Stoof, J. C., and Verherjden, P. F. H. M., 1986, D-2 Receptor stimulation
inhibits cyclic AMP formation brought about by D-1 receptor stimulation in rat neostriatum but not in nucleus accumbens, Eur. J. Pharmacol., 129:205.

Tarsy, T., and Baldessarini, R. J., 1974, Behavioral supersensitivity to
apomorphine following chronic treatment with drugs which interfere with
synaptic function of cathecolamines, Neuropharmacology, 13:927.

Trabucchi, M., Memo, M., Battaini, F., Reggiani, A., and Spano, P. F., 1980, Effect of long-term treatment with haloperidol and sulpiride on different types of dopaminergic receptors, Adv. Biochem. Psychopharmacol., 24:275.

Vaccheri, A., Dall'Olio, R., Gandolfi, O., and Montanaro, N., 1986, Involvement of different dopamine receptors in rat diphasic motility response to apomorphine, Psychopharmacology, 89:265.

RESPONSES OF MOTOR- AND NONMOTOR-RELATED NEOSTRIATAL NEURONS

TO AMPHETAMINE AND NEUROLEPTIC DRUGS

George V. Rebec, John L. Haracz, JoAnn T. Tschanz, Zhongrui Wang
and Ilsun White

Program in Neural Science
Department of Psychology
Indiana University, Bloomington, IN   47405   U.S.A.

INTRODUCTION

The neostriatum has been implicated in the behavioral response to amphetamine and in the ability of neuroleptic drugs to attenuate this response (e.g., Rebec and Bashore, 1984). Recent efforts to identify the mechanisms underlying amphetamine-induced behavioral effects have focused on single-unit recordings of neostriatal activity in freely moving animals. When such recordings are obtained exclusively from neurons that are active in rats performing a locomotor task, amphetamine causes a further increase in firing rate (West et al., 1987). In contrast, neurons sampled when animals are resting quietly show a heterogeneous response to amphetamine that includes both excitations and inhibitions (Gardiner et al., 1988; Ryan et al., 1989). These divergent results may reflect a differential action of amphetamine on motor- and nonmotor-related neurons. In an initial test of this hypothesis, we found that cells activated during movement were significantly more likely to increase their firing rate to amphetamine than neurons showing activity unrelated to movement (Haracz et al., 1989). In this report, we extend this research and also compare the ability of haloperidol, a classical neuroleptic that blocks virtually all components of the amphetamine behavioral response, and clozapine, an atypical drug that reverses relatively few amphetamine-induced behaviors (Tschanz and Rebec, 1989), to block the neuronal response to amphetamine.

METHODS

Data were obtained from male, Sprague-Dawley rats (approximately 400 g) that were prepared for single-unit recording as previously described (Gardiner et al., 1988). Following a recovery period of at least 7 days, a varnish-insulated tungsten microelectrode was lowered into the anterior neostriatum, and single-unit activity (signal-to-noise ratio of 4:1 or more) was amplified and displayed by conventional means. Neuronal discharges were counted on-line and routed to a computer for subsequent analysis. Each animal was housed inside a sound-attenuating chamber equipped with a videotaping system that permitted direct

The Basal Ganglia III, Edited by G. Bernardi *et al.*
Plenum Press, New York

recording of open-field behavlor. Firing rates were compared between periods when the animal rested quietly and when the animal displayed spontaneous movements (e.g., locomotion, rearing, and head turning). Neuronal activity also was recorded during tactile stimulation of the rat's vibrissae, snout, and body and during periods when the animal was provoked to move its head, forepaws, or hindlimbs.

Following a 5-10 min period of quiet rest, each animal received a subcutaneous (sc) injection of 1.0 mg/kg d-amphetamine sulfate (free base) followed 30 min later by either haloperidol (0.1-1.0 mg/kg) or clozapine (5.0-30.0 mg/kg). In some cases, the animal received a second neuroleptic injection 30 min after the first. Neuronal activity and behavior were monitored continuously until 30 min after the last injection. The mean firing rate of each neuron was calculated for the baseline period of quiet rest and for successive 5-min periods thereafter. Baseline firing rate was defined as 100%, and drug-induced changes were calculated as a percentage of this value for each 5-min period. Upon completion of testing, each animal was anesthetized and current was passed through the electrode to mark the recording site. Following a transcardial perfusion, the brain was removed and prepared for histological analysis.

RESULTS

A total of 50 single units, having a mean baseline firing rate of 2.23 $\pm$ 0.68 (SEM) spikes/sec during quiet rest, was recorded from the anterior neostriatum. Included in this total were 38 motor-responsive cells that changed firing rate either during head movements (N=13) or general body movements (N=25). In 36 of these 38 cells, the response to movement was a frank excitation, typically manifest as bursts of activity coinciding with specific behaviors (e.g., bobbing or turning of the head, grooming, and rearing). Occasionally, however, we encountered motor-responsive cells that showed a tonic increase in firing rate beginning a few seconds before movement and ending shortly after movement stopped. Only 2 motor-responsive cells were inhibited during motor activity. The remaining 12 neurons in our sample showed spontaneous activity unrelated to movement. Both motor- and nonmotor-related neurons often changed their firing rate during the probing of specific body regions (e.g., face, limbs, or dorsum). All neurons displayed biphasic action potentials with peak-to-peak amplitudes of 800-1000 μV. Subsequent histological analysis revealed that both motor- and nonmotor-related cells were located in the neostriatum (N=44) or in the ventral neostriatum, commonly known as the nucleus accumbens (N=6).

Amphetamine (1.0 mg/kg) produced a characteristic increase in behavioral activation that included forward locomotion, head bobbing, rearing, and sniffing. Consistent with previous evidence (Haracz et al., 1989), these behaviors began within 5-15 min after injection and were accompanied by unidirectional increases (N=36) or decreases (N=13) in neuronal activity. Almost all motor-responsive cells (33 of 37) were excited by amphetamine, typically increasing their firing rate to above 300% of the baseline rate. In contrast, this response occurred in only 3 of 12 nonmotor-related neurons. This difference in frequency of excitations was significant: X^2 (1) = 26.92, p < 0.001. The remaining neurons in each category were inhibited by amphetamine, typically to below 20% of the baseline rate.

Motor-related neurons excited by amphetamine generally increased their activity during locomotion, head movements, and rearing. When neuronal activity during these discrete behaviors was compared before and after drug injection, firing rate after amphetamine was usually higher in each case, suggesting a primary effect of amphetamine on these cells rather than a neuronal response related solely to movement. In a neuron excited during locomotion, for example, statistical analysis of firing rates during comparable bouts of locomotion before and after amphetamine revealed a significantly greater excitation following amphetamine administration: t (10) = 6.71, p < 0.01.

Most animals were challenged at 30 and 60 min after amphetamine with haloperidol (0.1-1.0 mg/kg) and clozapine (5.0-30.0 mg/kg) in counterbalanced order. Haloperidol blocked all components of the amphetamine behavioral response and also inhibited 36 of 37 neurons, usually to a firing rate below the pre-amphetamine baseline. This reduction in firing rate occurred regardless of the direction of the response to amphetamine and whether clozapine had been administered previously. The only neuron activated by haloperidol was unusual in that it consistently showed phasic movement-related inhibitions of activity. The response to clozapine (5.0-30.0 mg/kg), on the other hand, was more variable. This neuroleptic selectively attenuated amphetamine-induced rearing and sniffing and produced both increases (N=13) and decreases (N=12) in neuronal activity. The direction of the

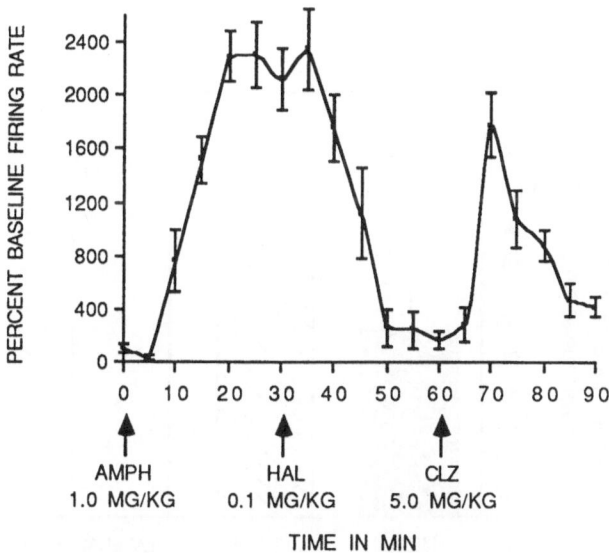

Fig. 1.   Response of a motor-related neuron to amphetamine, haloperidol, and clozapine. Baseline (100%) is defined as the firing rate during quiet rest for the 10-min period before amphetamine administration. Arrows indicate injection times. Note the opposing effects of haloperidol and clozapine. Data points represent means ± SEM of firing rates collected at 15-sec intervals.

clozapine response was unrelated to either the direction of the amphetamine response or the motor responsiveness of individual cells. When injected after haloperidol in 9 of 11 animals, clozapine reversed the haloperidol-induced depression. An example of a motor-related neuron activated by amphetamine, suppressed by haloperidol, and reactivated by clozapine is shown in Fig. 1.

The magnitude of the haloperidol-induced inhibition was examined in three anatomical subgroups : medial (N=9), which included placements in the medial neostriatum or nucleus accumbens, central (N=4), which was defined as a 0.5-mm-wide strip equidistant from the most medial and lateral extents of the neostriatum, and lateral neostriatum (N=7). As shown in Fig. 2, haloperidol inhibited cells in lateral neostriatum more strongly than medially located neurons: t (14) = 1.927, p < 0.05. The inhibition in lateral neostriatum typically reached 20% or less of the pre-amphetamine quiet baseline, whereas neurons in the medial group were inhibited only to 40% of their baseline rate.

In one animal in which we identified a motor-related neuron, we administered chloral hydrate (800 mg/kg, intraperitoneally), which produced immobility and some signs of general anesthesia (e.g., decreased flinching to ear pinch). After 90 min, the animal was challenged with amphetamine followed 50 and 80 min later by haloperidol and clozapine, respectively. The animal showed no overt motor activity in response to amphetamine. Surprisingly, the motor-related neuron in this case responded with a clear inhibition of firing rate that haloperidol rapidly converted to an excitation. The elevated firing rate was sustained following clozapine administration. This series of aberrant responses is shown in Fig. 3.

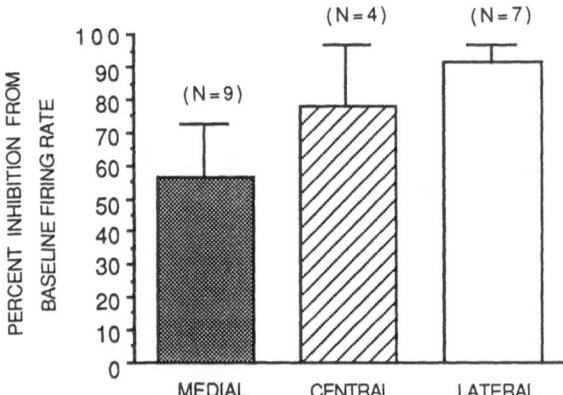

Fig. 2. The magnitude of haloperidol-induced inhibitions in three different anatomical regions (see text). The inhibitions are expressed as a percent decrease from the pre-amphetamine baseline firing rate during quiet rest. N refers to the number of neurons in each group and includes only those cells in rats treated with 1.0 mg/kg haloperidol as the first neuroleptic.

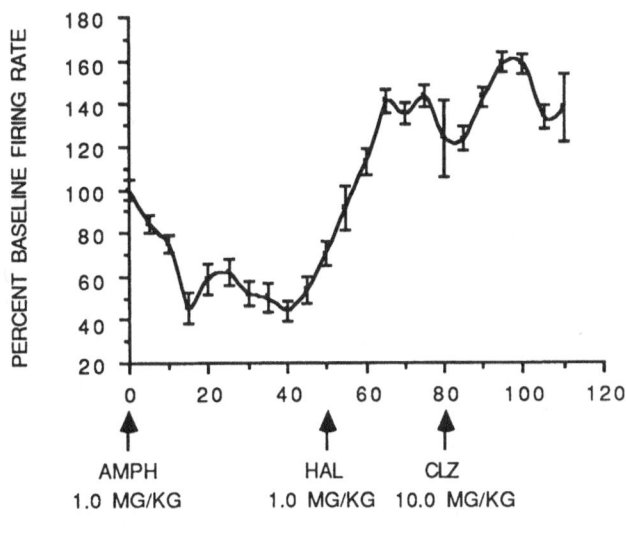

Fig. 3. Effect of chloral hydrate on the subsequent response of a motor-related neuron to amphetamine, haloperidol, and clozapine. Chloral hydrate was administered 90 min before amphetamine. Baseline (100%) is defined as the mean firing rate 10-min prior to amphetamine, and drug-induced changes in firing rate are plotted as in Fig. 1. Note that under chloral hydrate anesthesia amphetamine inhibited and haloperidol excited this motor-related neuron, which in the absence of anesthesia would be expected to show an amphetamine-induced excitation and a haloperidol-induced inhibtion.

DISCUSSION

Our results support previous evidence obtained from freely moving animals that amphetamine produces both increases and decreases in single-unit activity in both neostriatum and nucleus accumbens (Gardiner et al., 1988; Haracz et al., 1989; Ryan et al., 1989). Increases were typical of motor-responsive cells, whereas nonmotor-responsive neurons usually were suppressed. Although we cannot rule out a behavior-related contribution to these results, it seems unlikely that behavior alone is responsible for the changes in firing rate. Drug effects secondary to behavior can be controlled, at least in part, by comparing neuronal activity during similar behavioral responses before and after drug injection (e.g., Ranck et al., 1982). When controlled in this way, our results demonstrate a clear effect of amphetamine on neuronal activity.

Haloperidol inhibited the activity of virtually all the neurons tested in our sample (97%), whereas clozapine had mixed effects, either increasing or decreasing approximately equal numbers of cells. These results may reflect mechanisms underlying the differential behavioral effects of these drugs (e.g., Tschanz ahd Rebec, 1989). Our results with haloperidol and clozapine, however, contrast markedly with data obtained from immobilized or anesthetized rats, in which both neuroleptics, acutely administered with or without amphetamine pretreatment, elevated neuronal activity in neostriatum and nucleus accumbens (Rebec et al., 1979, 1980; Skirboll and Bunney, 1979; Napier et al., 1985; Hu and Wang, 1986). Thus, immobilization or anesthesia may alter the neuroleptic-induced changes in firing rate from taht recorded in behaving animals. In fact, in one case in which we pretreated an animal with chloral hydrate, the haloperidol response was reversed. Interestingly, the inhibitory response of the motor-related neuron in this animal to amphetamine also was opposite to that typically found in awake, behaving subjects. Immobilization or anesthesia also may alter amphetamine-induced changes in firing rate since single units show excitations most frequently in behaving rats (Haracz et al., 1989), whereas amphetamine-induced inhibitions predominate in neostriatum or nucleus accumbens of the immobilized preparation (Rebec et al., 1979).

We also found that haloperidol, which has a high affinity for D2-dopamine receptors (see Creese, 1987), inhibited lateral neostriatal neurons more strongly than neurons in medial neostriatum or nucleus accumbens. This result is consistent with evidence for a medial-lateral gradient in D2-dopamine receptor concentrations in neostriatum (Joyce and Marshall, 1987; Loopuijt, 1989). It also is noteworthy that lateral neostriatum receives considerable input from sensory-motor cortex (Dube et al., 1988; McGeorge and Faull, 1989). Thus, to the extent that lateral neostriatum plays a key role in sensory-motor processing (Schneider, 1987; West et al., 1987), a strong inhibition of neuronal activity in this area by haloperidol may contribute to the pronounced motor side effects of this drug. Clozapine, in contrast, is largely devoid of such effects, and also fails to produce a large-scale inhibition of neostriatal activity. In fact, we found that in most cases clozapine reversed the haloperidol-induced depression, which parallels preliminary clinical data that this drug often attenuates the motor side effects produced by haloperidol and other classical neuroleptics (Meltzer, 1986; Lieberman et al., 1989). Moreover, the ability of clozapine in some cases to enhance the neuronal effects of amphetamine suggests a potential mechanism for similar behavioral findings (Robertson and MacDonald, 1984; Tschanz and Rebec, 1989).

In summary, our results confirm the heterogeneous nature of the neuronal response to amphetamine in the neostriatum and suggest that the response of individual neurons to this drug is strongly related to the motor-response characteristics of these cells. We also found a strong inhibitory action of haloperidol, especially in lateral neostriatum, that contrasted with the more variable response to clozapine and with previous neuroleptic data obtained from immobilized or anesthetized animals. Collectively, our results underscore the importance of the freely moving preparation for elucidating the mechanisms underlying the behavioral response to amphetamine and neuroleptic drugs.

ACKNOWLEDGEMENTS

This research was supported by USPHS Grant DA 0241. Paul Langley provided expert technical assistance; Faye Caylor helped with the preparation of the manuscript.

REFERENCES

Creese, I., 1987, Biochemical properties of CNS dopamine receptors, in:
   "Psychopharmacology: The Third Generation of Progress," H. Y. Meltzer, ed., Raven,
   New York.

Dube, L., Smith A. D., and Bolam, J. P., 1988, Identification of synaptic terminals of
   thalamic or cortical origin in contact with distinct medium-size spiny neurons in the
   rat neostriatum, J. Comp. Neurol., 267:455.

Gardiner, T. W., Iverson, D. A., and Rebec, G. V., 1988, Heterogeneous responses of
   neostriatal neurons to amphetamine in freely moving rats, Brain Res., 463:268.

Haracz, J. L., Tschanz, J. T., Greenberg, J., and Rebec, G. V. ,1989, Amphetamine-induced
   excitations predominate in single neostriatal neurons showing motor-related activity,
   Brain Res., 489:365.

Hu, X. T., and Wang, R. Y., 1986, Differential effects of haloperidol and clozapine in the rat
   neostriatum and nucleus accumbens: microiontophoretic studies, Soc. Neurosci. Abstr.,
   12:1389.

Joyce, J. N., and Marshall, J. F., 1987, Quantitative autoradiography of dopamine D2 sites in
   rat caudate-putamen: localization to intrinsic neurons and not to neocortical afferents,
   Neuroscience, 20:773.

Lieberman, J. A., Saltz, B. L., Johns, C. A., Pollack, S., and Kane, J. M., 1989, Clozapine
   effects on tardive dyskinesia, Psychopharmacol. Bull., 25:57.

Loopuijt, L. D., 1989, Distribution of dopamine D-2 receptors in the rat striatal complex
   and its comparison with acetylcholinesterase, Brain Res. Bull., 22:805.

McGeorge, A. J., and Faull, R. L. M., 1989, The organization of the projection from the
   cerebral cortex to the striatum in the rat, Neuroscience, 29:503.

Meltzer, H. Y., 1986, Novel approaches to the pharmacotherapy of schizophrenia, Drug Dev.
   Res., 9:23.

Napier, T. C., Coyle, S., and Breese, G. R., 1985, Ontogeny of striatal unit activity and effects
   of single or repeated haloperidol administration in rats, Brain Res., 333:35.

Ranck, J. B., Jr., Kubie, J. L., Fox, S. E., Wolfson, S., and Muller, R. V., 1982, Single unit
   recording in behaving animals: bridging the gap between neural events and sensory
   behavioral variables, in: "Behavioral Contributions to Brain Research," T. Robinson,
   ed., Oxford University Press, Oxford.

Rebec, G. V., and Bashore, T. R., 1984, Critical issues in assessing the behavioral effects of amphetamine, Neurosci. Biobehav. Rev., 8:153.

Rebec, G. V., Bashore, T. R., Zimmerman, K. S., and Alloway, K. D., 1979, "Classical" and "atypical" antipsychotic drugs: differential antagonism of amphetamine- and apomorphine-induced alterations of spontaneous neuronal activity in the neostriatum and nucleus accumbens, Pharmacol. Biochem. Behav., 11:529.

Rebec, G. V., Bashore, T. R., Zimmerman, K. S., and Alloway, K. D., 1980, Neostriatal and mesolimbic neurons: dose-dependent effects of clozapine, Neuropharmacology, 19:281.

Robertson, A., and MacDonald, C., 1984, Atypical neuroleptics clozapine and thioridazine enhance amphetamine-induced stereotypy, Pharmacol. Biochem. Behav., 21:97.

Ryan, L. J., Young, S. J., Segal, D. S., and Groves, P. M., 1989, Antidromically identified striatonigral projection neurons in the chronically implanted behaving rat: relations of cell firing to amphetamine-induced behaviors, Behav. Neurosci., 103:3.

Schneider, J. S., 1987, Basal ganglia-motor influences: role of sensory gating, in: "Basal Ganglia and Behavior: Sensory Aspects of Motor Functioning," J. S. Schneider and T. I. Lidsky, eds., Hans Huber, Toronto.

Skirboll, L. R., and Bunney, B. S., 1979, The effects of acute and chronic haloperidol treatment on spontaneously firing neurons in the caudate nucleus of the rat, Life Sci., 25:1419

Tschanz, J. T., and Rebec, G. V., 1989, Atypical antipsychotic drugs block selective components of amphetamine-induced stereotypy, Pharmacol. Biochem. Behav., 31:529.

West, M. O., Michael, A. J., Knowles, S. E., Chapin, J. K. and Woodward, D. J., 1987, Striatal unit activity and the linkage between sensory and motor events, in: "Basal Ganglia and Behavior: Sensory Aspects of Motor Functioning," J. S. Schneider and T. I. Lidsky, eds., Hans Huber, Toronto.

THE ROLE OF THE DORSAL AND VENTRAL STRIATUM IN THE THERAPEUTIC AND EXTRAPYRAMIDAL SIDE EFFECTS OF NEUROLEPTIC DRUGS

Bart A. Ellenbroek and Alexander R. Cools

Psychoneuropharmacol. Res. Unit
Cath. University of Nijmegen
P.O. Box 9101
6500 HB Nijmegen, the Netherlands

INTRODUCTION

Ever since their first introduction into clinical practice in the mid 1950s the neuroleptic drugs have been the most effective drugs in various forms of psychoses, most notably the schizophrenic psychosis. Despite their widespread use, however, the neuronal mechanisms giving rise to the therapeutic and side effects is still not completely elucidated. Until the beginning of the sixties it was thought that neuroleptic drugs exert their action by directly blocking the noradrenergic neurotransmission. However, the studies performed by Carlsson & Lindqvist (1963) and van Rossum (1966) clearly showed that blockade of the dopaminergic transmission especially in the caudate nucleus (Cools & van Rossum, 1970) is the primary mode of action of neuroleptics. These findings were extended independently by two groups in 1976. Both Creese et al (1976) and Seeman et al (1976) showed that the potency of neuroleptic drugs for blocking the dopamine receptors was correlated with the clinical dose. Although these data had great impact, some critical comments should be made. First both studies were performed in vitro. This implies that the possible role of metabolites in the therapeutic effect was not taken into account. This is of importance since several neuroleptics, like thioridazine (Axelsson & Marterisson, 1978), fluphenazine (Hitzemann et al, 1984) and trifluoperazine (Javaid et al, 1980) have metabolites with antipsychotic efficacy. Moreover, it has recently been shown that schizophrenic patients resistant to the classical neuroleptic haloperidol show good improvement upon treatment with the atypical (see below) neuroleptic clozapine (Kane et al, 1988). This strongly suggests that clozapine does not work through a blockade of D_2 receptors, since haloperidol has a much higher affinity for these receptors than clozapine (Richelson & Nelson, 1984). Although a pharmacokinetic explanation cannot be fully ruled out Wolkin et al (1989) showed that the D_2 receptor occupancy was not different between patients responding and those not responding to haloperidol. These data indicate that at least haloperidol and clozapine seem to work through different receptor systems.

Another strong indication that there are differences between neuroleptic drugs has been the identification of the so called atypical neuroleptics. Up until the mid 1970s it was generally accepted that all neuroleptic drugs produce so called extrapyramidal side effects, like parkinsonian symptoms, akathisia, dystonia and (tardive) dyskinesia. Moreover it was even thought that extrapyramidal side effects were a prerequisite for a therapeutic action (Haasse & Janssen, 1985). Despite early indications that certain neuroleptics (most notably thioridazine) produced significantly less extrapyramidal side effects than others (NIMH, 1964), it was not until the introduction of clozapine that the concept of atypical neuroleptics was born (Pi & Simpson, 1983). Since then it has been repeatedly shown that clozapine combines an antipsychotic effect with almost no extrapyramidal side effects (Gerlach et al, 1974; Sayers & Amsler, 1977; Claghorn et al, 1983).

The concept of atypical neuroleptics not only has clinical relevance, but also important theoretical implications. It suggests that the neuronal mechanisms underlying the therapeutic effect and the extrapyramidal side effects are distinct. Indeed it has been suggested on the basis of electrophysiological experiments that the therapeutic effects of neuroleptic drugs are due to a specific action on the dopaminergic A10 cell group, whereas the extrapyramidal side effects occur as a result of the actions on the dopaminergic A9 cell group (Chiodo & Bunney, 1983; White & Wang, 1983). Although additional support was obtained for this hypothesis (Lane & Blaha, 1986; Skarsfeldt, 1988), several findings seem to be in conflict with this general idea. First of all, there are differences between the findings of Chiodo & Bunney and White & Wang. The latter authors found no effect of clozapine (either acute or chronic) on the number of active cells in the A9 cell region, whereas Chiodo & Bunney found an increased number of active cells in A9 after both acute and chronic clozapine treatment. In agreement with the latter finding Saller & Salama (1986) found an increase in 3MT accumulation in the dorsal striatum after acute administration of clozapine. Moreover O'Connor et al (1989) recently showed that clozapine increases the extracellular dopamine concentrations in the dorsal striatum. Likewise Haracz et al (1989) showed that clozapine influences the activity of cells in the dorsal striatum. Thus these data show that, at least after an acute injection, clozapine also influences the dopaminergic cells in the A9 area as well as the terminals of these cells in the dorsal striatum. This then implies that the original concept that schizophrenia is due to an overactivity of the dopamine transmission within the caudate nucleus (Randrup & Munkvad, 1972; Cools, 1975) may still be valid (see also the recent discussion by Carlsson, 1988). Especially since there appears to be an increase in dopamine D_2 receptors in the caudate and putamen of drug naive schizophrenic patients (Wong et al, 1986).

Furthermore, if the depolarisation inactivation of the A10 cell group is responsible for the therapeutic effects of neuroleptic drugs, it is difficult to understand that chronic morphine treatment also induces depolarisation inactivation of these cells (Gysling & Wang, 1983), since morphine is without marked antipsychotic effect. Moreover data from our own laboratory show that at least one of the so called extrapyramidal side effects, orofacial dyskinesias, is at least in part mediated by the mesolimbic dopamine system and its first order output system the ventral pallidum (Cools et al, 1976; 1989).

Taken all these data together, it seems that the neuronal mechanisms underlying the neuroleptic actions are still not fully understood and that at this moment the strict separation of therapeutic and extrapyramidal side effects as being mediated by respectively the A10 and the A9 dopamine cell groups is not unequivocal.

THE PAW TEST: AN ANIMAL MODEL WITH PREDICTIVE VALIDITY

The above discussion has made it clear that there is a need for an a-
nimal model in which the neuronal mechanisms underlying the therapeutic
and the extrapyramidal side effects of neuroleptic drugs can be studied
separately. Most of the classically used animal models (like blockade of
apomorphine induced stereotypy, blockade of self stimulation and blockade
of conditioned avoidance response) do not appear to fulfil the criteria
for predictive validity (see table 1, Ellenbroek 1988). For instance a
number of false positives occur in these models (Wauquier, 1976; Thomas &
Handley, 1978; Arnt, 1982).

Table 1. The criteria for assessing the predictive validity of animal
models for schizophrenia.

---

1) Neuroleptics of various chemical classes should be effective.

2) No false negatives should occur.

3) No false positives should occur.

4) Anticholinergic drugs should not reduce the effects of neuroleptic drugs.

5) Chronic treatment should not reduce the effects of neuroleptic
drugs.

6) There should be a relationship betweem the clinical potency
of neuroleptics and their potency in the model

Of more importance is the fact that most of these animal models viola-
te criteria 4 and 5, which are based on clinical practice (see Matthysse
1981; Ellenbroek, 1988). Therefore we have developed a new animal model:
the paw test. This model is based on EMG studies which indicated that the
dorsal and ventral striatum play a differential role in regulating forelimb and hindlimb muscle tone. More specifically we showed that the nucleus accumbens plays an important role in the regulation of forelimb
muscle tone, and the dorsal striatum plays an important role in the regulation of the hindlimb muscle tone (Ellenbroek et al, 1985; 1988). In the
paw test the animal's ability to retract its forelimbs and its hindlimbs
is analysed. The test box is a perspex platform (30 x 20 cm with a height
of 20 cm) and contains two holes for the forelimbs and two holes for the
hindlimbs as well as a slit for the tail (Ellenbroek et al, 1987). 30 Minutes after the injection of a drug the animals are placed in the test box
and the time to withdraw one forelimb (Forelimb Retraction Time, FRT) and
one hindlimb (Hindlimb Retraction Time, HRT) is scored.

Sofar a great variety of drugs have been tested in the paw test (see
also table 2) and the picture until now is rather clear. Thus, all neuroleptics tested influence the HRT, whereas the tested non neuroleptics do
not. Moreover, all tested classical neuroleptics influence FRT, whereas
the tested atypical neuroleptics do not. This suggests that an increase in
HRT is an animal model for the therapeutic efficacy of neuroleptic drugs,
whereas an increase in FRT is an animal model for so called extrapyramidal
side effects. Furthermore the paw test was submitted to the criteria as
defined in table 1 and all criteria were fulfilled. Thus the haloperidol
induced increase in FRT but not HRT was sensitive to treatment with sco-

polamine and to chronic haloperidol treatment (Ellenbroek & Cools, 1988).

For this reason, the paw test seems to provide a rapid screening model for new possible neuroleptic drugs. Moreover it seems to be an experimentally simple test for studying the neuronal mechanisms underlying the therapeutic and unwanted side effects of neuroleptics.

Table 2. The minimal effective dose (MED) for significantly increasing the forelimb retraction time (FRT) and the hindlimb retraction time (HRT) for a number of classical and atypical neuroleptics as well as of a number of non neuroleptic drugs.

DRUG	MED for FRT	MED for HRT
Classical Neuroleptics		
Chlorpromazine	7.5 mg/kg	7.5 mg/kg
cis-Flupenthixol	0.5 mg/kg	0.5 mg/kg
Fluphenazine	0.5 mg/kg	0.5 mg/kg
Haloperidol	0.15 mg/kg	0.15 mg/kg
Atypical Neuroleptics		
Clozapine	>50 mg/kg	10.0 mg/kg
Thioridazine	60 mg/kg	15.0 mg/kg
Non Neuroleptics		
Desipramine	>20 mg/kg	>20 mg/kg
Diazepam	> 5 mg/kg	> 5 mg/kg
Morphine	>20 mg/kg	>20 mg/kg
Promethazine	>20 mg/kg	>20 mg/kg
trans-Flupenthixol	> 5 mg/kg	> 5 mg/kg

THE ROLE OF THE DORSAL AND VENTRAL STRIATUM IN THE PAW TEST

The paw test was used to evaluate the role of the dorsal and ventral striatum in the effects of neuroleptic drugs. For this reason rats were stereotaxically implanted with two cannulae aimed at the dorsal or at the ventral striatum. At least one week after the operation the animals were injected intraperitoneally with either haloperidol (0.5 mg/kg) or clozapine (20 mg/kg) followed 15 minutes later by an injection into either the dorsal or the ventral striatum. The animal were then subjected to the paw test at 30, 40 and 50 minutes after the intraperitoneal injection and the FRT and HRT were determined as the mean of the three values at 30, 40 and 50 minutes. After the experiments the animals were killed by an overdose of sodium pentobarbital and their brains removed for histological verification. Only the data of animals with correctly placed cannulae are shown here.

Figure 1 shows the effects of the local administration of the selective D-2 agonist LY 171555 into the dorsal and ventral striatum on the effects of the classical neuroleptic drug haloperidol. First of all, this figure shows that after saline injections into either the dorsal and ventral striatum, haloperidol still led to an increase in both FRT and HRT.

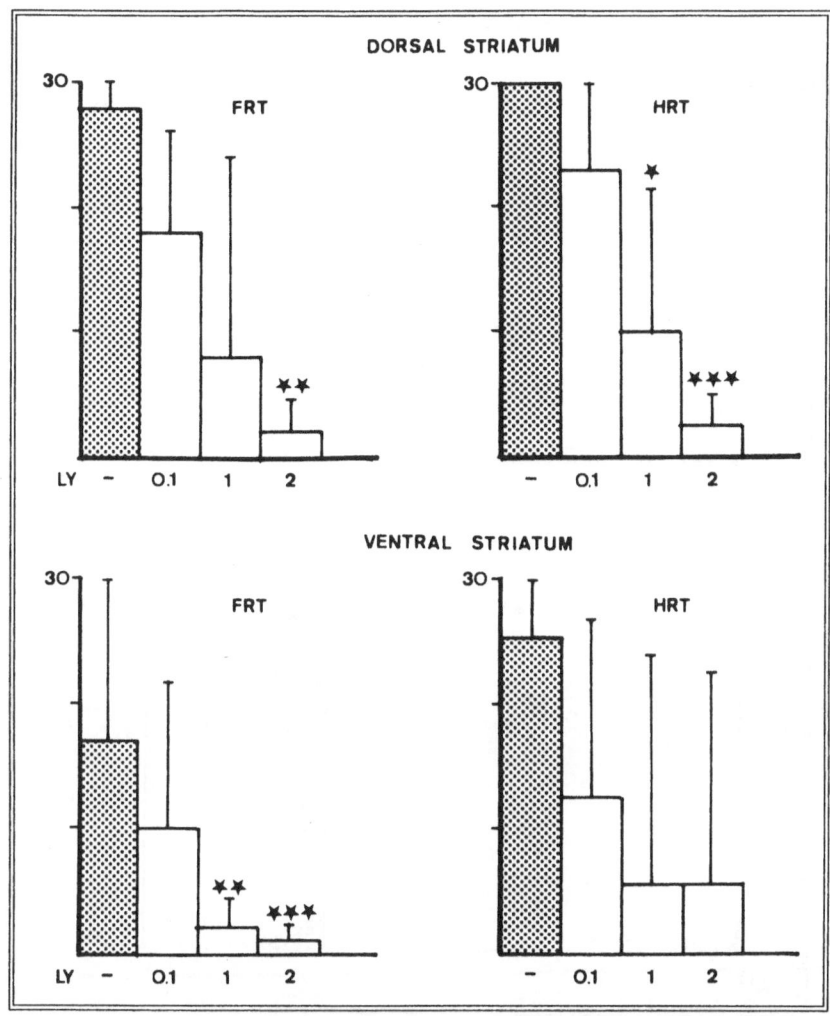

Figure 1. The effects of local application of LY 171555 into
the dorsal and ventral striatum on the effects of
haloperidol (0.5 mg/kg i.p) on the FRT and HRT in
the paw test. Represented are the median values
and the 75% point in seconds. * p<0.05; ** p<0.01;
*** p<0.005, Two tailed Mann Whitney U test vs haloperidol i.p plus control.

This indicates that, as was shown in previous experiments (Ellenbroek et
al, 1987) haloperidol indeed increases both FRT and HRT. Moreover the
data show that injections of LY 171555 into the ventral striatum antagonised the haloperidol induced increase in FRT but not in HRT. Furthermore
injections of LY 171555 into the dorsal striatum antagonised the haloperidol induced increase in both FRT and HRT.

Figure 2 shows the effects of the local administration of LY 171555 on
the effects of the atypical neuroleptic clozapine. From this figure it can
be seen that after control injections into the dorsal or ventral striatum
clozapine induced only an increase in HRT, which is in agreement with
earlier data (Ellenbroek et al, 1987). Moreover, the data show injections
of LY 171555 either into the dorsal or into the ventral striatum were unable to influence the effects of clozapine on the HRT.

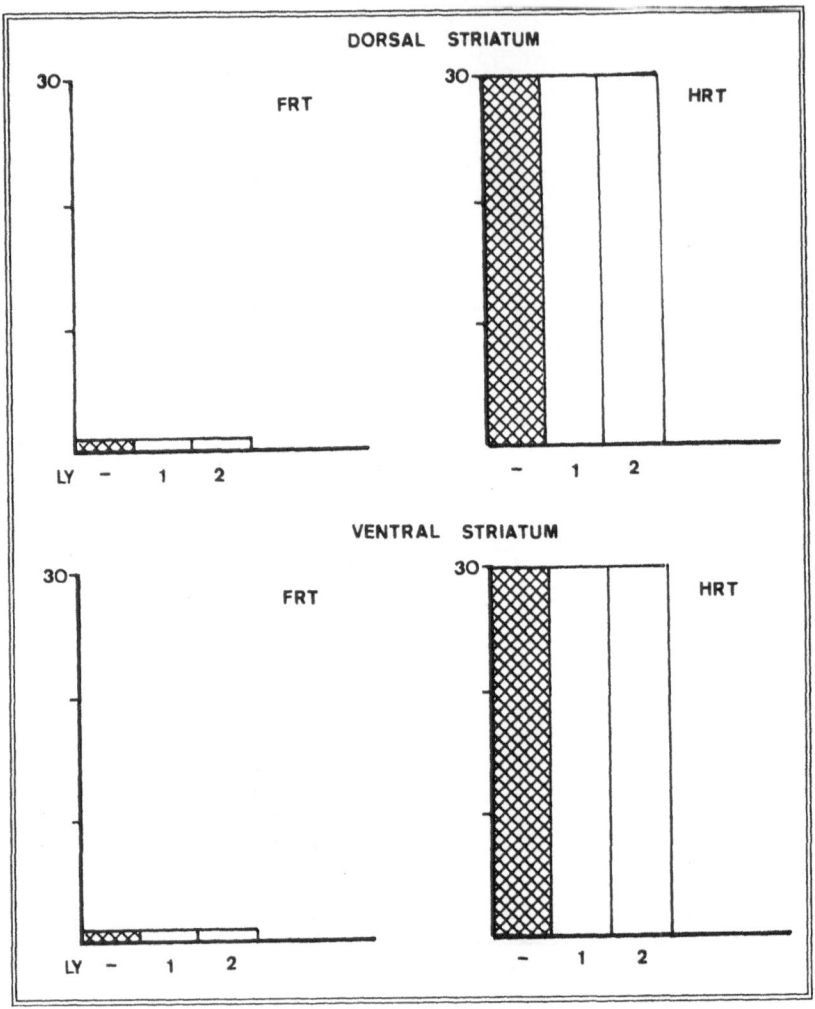

Figure 2. The effects of local application of LY 171555 into the dorsal and ventral striatum on the effects of clozapine (20 mg/kg i.p) on the FRT and HRT in the paw test. Represented are the median values and the 75% point in seconds.

DISCUSSION

The data presented in figures 1 and 2 suggest that both the D-2 receptors in the dorsal and the ventral striatum play a role in the effects of haloperidol on FRT, whereas the D_2 receptors in the dorsal striatum also plays a role in the effects of haloperidol on HRT. Moreover it appears that the role of the ventral striatum in regulating FRT is somewhat larger since LY 171555 already shows an effect at 1 µg/0.5 µl, whereas in the dorsal striatum the minimal effective dose of LY 171555 is twice as high. When these data are compared to the results on the tonic EMG activity there appears to be a difference with respect to the role of D_2 receptors in the ventral striatum. In the tonic EMG studies, the haloperidol induced rigidity of the triceps muscle could be blocked by the alpha-noradrenergic agonist phenylephrine but not by the dopaminergic agonist apomorphine injected into the ventral striatum (Ellenbroek et al, 1988). Although the effects of LY 171555 on the haloperidol induced tonic EMG activity were not studied, these data suggest that the D_2 receptors do not play a prominent role in this phenomenon. However, in the paw test the D_2 receptors do play a role in the retraction time of the forelimbs. This suggests that there are differences between the neuronal mechanisms underlying the tonic EMG rigidity in the triceps and the retraction time of the forelimbs. With respect to the tonic EMG rigidity of the gastrocnemius soleus muscle and the hindlimb retraction time this difference was not found. Thus both the rigidity (Ellenbroek et al, 1985) and the hindlimb retraction time (present study) were mediated via dopamine receptors in the dorsal striatum.

Figure 2 shows that, in contrast to haloperidol, D-2 receptors in the dorsal and ventral striatum appear not to play a role in the effects of clozapine. Although a general role for D-2 receptors in the effects of clozapine can not be ruled out, unpublished data from our laboratory seem to suggest that also intraperitoneal injections of LY 171555 is without effect on the clozapine induced increase in HRT. This then suggests that clozapine might work through another mechanism. Although it is as yet unclear which mechanism is involved, it is interesting to speculate on the possible role of the D-1 receptor. Chipkin & Latranyi (1987) already pointed to the resemblance between the effects of clozapine and SCH 23390, a selective D-1 antagonist. In the paw test, SCH 23390 also resembles clozapine (Ellenbroek et al, 1987) in that it more potently increases HRT. However, more experiments are needed to investigate the possible role of D-1 receptors in the effects of clozapine.

CONCLUSIONS

The present study shows that dopamine D_2 receptors play no role in the effects of the atypical neuroleptic clozapine on the parameters as measured in the paw test. On the other hand dopamine D_2 receptors seem to play an important role in the effects of the classical neuroleptic haloperidol ot the parameters as measured in the paw test. More specifically, the D_2 receptors in the dorsal striatum appear to play an important role in the regulation of the hindlimb retraction time, although a minor role in the regulation of the forelimb retraction time is als indicated by the present data. The D_2 receptors in the ventral striatum appear to play an important role in the regulation of the forelimb retraction time.

As discussed above, earlier studies indicated that the increase in hindlimb retraction time may be an animal model for the therapeutic effects whereas an increase in forelimb retraction time may represent an animal model for the extrapyramidal side effects of neuroleptic drugs. If this is true, the present data suggest that D_2 receptors in the dorsal striatum play a major role in the therapeutic effects of classical neuroleptics like haloperidol. Furthermore D_2 receptors in the ventral striatum seem to play a major role in the extrapyramidal side effects of classical neuroleptic drugs. Finally the data indicate that D_2 receptors do not seem to play a major role in the therapeutic effects of clozapine.

REFERENCES

Arnt,J., 1982, Pharmacological specificity of conditioned avoidance response inhibition in rats: Inhibition by neuroleptics and correlation to dopamine receptor blockade, Acta Pharmacol. Toxicol., 51: 321.

Axellson,R. and Martensson,E., 1978, Relationship between serum concentrations of thioridazine and its main nonconjugated metabolites and the clinical response in thioridazine-treated patients, Curr. Ther. Res., 24: 232.

Carlsson,A., 1988, The current status of the dopamine hypothesis of schizophrenia, Neuropsychopharmacol., 1:179.

Carlsson,A. and Lindqvist,M., 1963, Effects of chlorpromazine or haloperidol on formation of 3-methoxytyramine and normetanephrine in mouse brain, Acta Pharmacol. Toxicol., 20: 140.

Chiodo,L. and Bunney,B., 1983, Typical and atypical neuroleptics: Differential effects of chronic administration on the activity of A9 and A10 midbrain dopaminergic neyurons, J. Neurosci., 5: 2539.

Chipkin,R. & Latranyi,M., 1987, Similarity of clozapine and SCH 23390 in reserpinized rats suggests a common mechanism of action, Europ. J. Pharmacol., 136: 371-375.

Claghorn,J., Abuzzahab,F., Wang,R., Larsson,C., Gelenberg,A., Klerman,G., Tuason,V. and Steinbook,R., 1983, The current status of clozapine, Psychopharmacol. Bull., 19: 138.

Cools,A., 1975, An integrated theory of the aetiology of schizophrenia. Impairment of the balance between certain, in series connected dopaminergic, serotonergic and noradrenergic pathways within the brain, in: "On the origin of schizophrenic psychoses", H. van Praag, ed, De erven Bohn, Amsterdam.

Cools,A., Spooren,W., Cuypers,E., Bezemer,R, and Jaspers,R., 1989, Heterogeneous role of neostriatal and mesostriatal pathology in disorders of movement: A review and new facts, in: "Neural mechanisms in disorders of movement", A. Crossman and M. Sambrook, eds, Libbey, London.

Cools,A., Struyker-Boudier,H. and van Rossum,J., 1976, Dopamine receptors: Selective agonists and antagonists of functionally distinct types within the feline brain, Europ. J. Pharmacol., 37: 282.

Cools,A. and van Rossum,J., 1970, Caudate dopamine and stereotyped behaviour of cats, Arch Intern. Pharmacodyn. Ther., 187: 163.

Ellenbroek,B., 1988, "Animal models for schizophrenia and neuroleptic drug action, including a survey on the research and treatment of schizophrenia" unpublished PhD thesis, Krips, Meppel.

Ellenbroek,B. and Cools,A., 1988, The paw test: An animal models for neuroleptic drugs which fulfils the criteria for pharmacological isomorphism, Life Sciences, 42: 1205.

Ellenbroek,B., Peeters,B., Honig,W. and Cools,A., 1987, The paw test: a behavioural paradigm for differentiating between classical and atypical neuroleptic drugs, Psychopharmacol., 93: 343.

Ellenbroek,B., Schwarz,M., Sontag,K., Jaspers,R. and Cools,A., 1985, Muscular rigidity and delineation of a dopamine specific neostriatal subregion: tonic EMG activity in rats, Brain Res., 345: 132.

Ellenbroek,B., van den Hoven,J. & Cools,A., 1988, The nucleus accumbens and forelimb muscular rigidity in rats, Exp. Brain Res., 72: 299.

Gerlach,J., Koppelhus,P., Helweg,E. and Monrad,A., 1974, Clozapine and haloperidol in a single blind cross-over trial: therapeutic and biochemical aspects in the treatment of schizophrenia, Acta Psychiatr. Scand., 40: 410.

Gysling,K. and Wang,R., 1983, Morphine-induced activation if A10 dopamine neurons in the rat, Brain Res., 277: 119.

Haasse,H. and Janssen,P., 1985, "The action of neuroleptic drugs", Elsevier, Amsterdam.

Haracz,J., Tschanz,J., White,I., Wang,Z., Miller,D. and Rebec,G., 1989, Single-unit activity in the medial neostriatum and nucleus accumbens: Effects of amphetamine, haloperidol and clozapine in freely moving rats, Soc. Neurosci. Abs., 15: 1134.

Hitzemann,R., Garver,D., Mavroides,M., Hirschowitz,J. and Zemlan,F., 1986, Fluphenazine activity and antipsychotic actiity, Psychopharmacol., 90: 270.

Javaid,J., Pandey,G., Daslah,B., Hu,H. and Davis,J., 1980, Measurement of neuroleptic concentrations by GLC and radioreceptor assay, Comm. Psychopharmacol., 4: 467.

Kane,J., Honigfeldt,G., Singer,J. and Meltzer,H. 1988, Clozapine in treatment resistant schizophrenics, Psychopharmacol. Bull., 24: 63.

Lane,R. and Blaha,C., 1986, Electrochemistry in vivo: Application to CNS pharmacology, Ann. N. Y. Acad. Sci., 473: 50.

Matthysse,S., 1981, Nucleus accumbens and schizophrenia, 1980, in: "The neurobiology of the nucleus accumbens," R. Chronister and J. DeFrance, eds., Haer Institute for Electrophysiological Res., Maine.

National Institute of Mental Health, 1964, Phenothiazine treatment in acute schizophrenics, Arch. Gen. Psychiatr., 10: 246.

O'Connor,W., Drew,K. and Ungerstedt,U., 1989, Differences in dopamine release and metabolism in rat striatal subregionsa following acute clozapine using in vivo microdialysis, Neurosci. Lett., 98: 211.

Pi,E. and Simpson,G., 1983, Atypical neuroleptics: clozapine and the benzamides in the treatment of tardive dyskinesia, Mod. Probl. Pharmacopsychiatr., 21: 80.

Randrup,A. and Munkvad,I., 1972, Evidence indicating an association between schizophrenia and dopaminergic hyperactivity in the brain, Orthomol. Psychiatr., 1: 2.

Richelson,E. and Nelson,A., 1984, Antagonism by neuroleptic of neurotransmitter receptors of normal human brain in vitro, Europ. J. Pharmacol., 103: 197.

Saller,C. and Salama,A., 1986, 3-Metoxytyramine accumulation: Effects of typical neuroleptics and various atypical compounds, Naunyn Schmiedeb. Arch. Pharmacol., 334: 125.

Sayers,A. and Amsler,H., 1977, Clozapine, in: "Pharmacological and biochemical properties of drug substances, vol.1," M. Goldberg, ed., American Pharmaceutical Association.

Skarsfeldt,T., 1988, Differential effects after repeated treatment with haloperidol, clozapine, thioridazine and tefludazine on SNC and VTA dopamine neurons in rats, Life Sciences, 42: 1037.

Thomas,K. and Handley,S., 1978, Modulatiuon of dexamphetamine-induced compulsive gnawing – including the possible involvement of presynaptic aalpha-adrenoceptors, Psychopharmacol., 56: 61.

van Rossum,J., 1966, The significance of dopamine receptor blockade for the mechanism of action of neuroleptic drugs, Arch. Int. Pharmacodyn. Ther., 160: 492.

Wauquier,A., 1976, The influence of psychoactive drugs on brain self-stimulation in rats: A review, in: "Brain stimulation reward," A. Wauquier and E. Rolls, eds., Elsevier, Amsterdam.

White,F. and Wang,R., 1983, Differential effects of classical and atypical antipsychotic drugs on A9 and A10 dopamine neurons, Science, 211: 1054.

Wolkin,A., Barouche,F., Wolf,A., Rotrosen,J., Fowler,J., Shine,C., Cooper,T. and Brodie,J., 1989, Dopamine blockade and clinical rsponse: Evidence for two biological subgroups of schizophrenia, Am. J. Psychiatr., 146: 905.

Wong,D., Wagner,H., Tune,L., Dannals,R., Pearlson,G., Links,J., Tamminga, C., Brousolle.E,. Ravert,H., Wilson,A., Toung,J., Malat,J., Williams, J., O'Tuama,L., Snyder,S., Kuhar,M. and Gjedde,A., 1983, Positron emission tomography reveals elevated D-2 receptors in drug naive schizophrenics, Science, 234: 1558.

INTERACTION OF EAA- AND GABA-NEUROTRANSMISSION IN THE SNR OF CATS:

OPEN FIELD BEHAVIOUR

C. Heim, R. Jaspers, F. Block and K.-H. Sontag

Max-Planck-Institute for experimental Medicine
Hermann-Rein-Strasse 3
D-3400 Göttingen, FRG

INTRODUCTION

The pars reticulata of the substantia nigra (SNR) is considered to be an intercalated station between striatal GABAergic and subthalamic glutamatergic fibres (Dray, 1979; Kitai and Kita, 1987).
Injections of a GABA agonist or antagonist in the SNR of cats result in characteristic changes of behaviour (Wolfarth et al., 1981; Cools et al., 1983; Cools et al., 1984; Sontag et al., 1984; Heim et al., 1986). The application of 250-500 ng/0.5 μl of the GABA antagonist picrotoxin into the SNR induces an akinetic posture characterized by body freezing, contra-lateral static head turning and eye fixation. This effect can be explained by the fact that the GABAergic output neurons are disinhibited by a blocka-de of inhibitory striato-nigral GABAergic activity (Deniau et al., 1976).
Recently, it was reported that electrical stimulation of neurons in the subthalamic nucleus (STN) evokes monosynaptically EPSPs in SNR neurons (Kitai and Kita, 1987). Single STN neurons have branched excitatory acting axons to the globus pallidus (external segment), the nucleus entopeduncu-laris (internal segment) and the substantia nigra (Kitai and Kita, 1987). Lesioning of the STN induces a decrease in ^{3}H -D-aspartate uptake in the pallidum and the SNR ipsilateral to the lesion. In contrast, the uptake of ^{14}C -GABA is not changed in these structures (Brotchie and Crossman, 1989). These data indicate that the STN-SNR pathway is glutamatergic/ aspartatergic.
If electrical stimulation of STN neurons elicits monosynaptically EPSPs in the SNR and the pathway indeed is glutamatergic/aspartatergic, it can be expected that increased STN activity will excite SNR neurons. The result would be a condition comparable to a blockade of GABAergic striato-nigral neurotransmission, which disinhibits SNR GABAceptive output neurons. We therefore compared the effect of NMDA application with the effect of picrotoxine injections injected into the SNR.

METHODS

Cats were anaesthetized with sodium pentobarbital (45 mg/kg i.p. and bilaterally equipped with guide cannulae directed to the caudo-lateral SNR (A 2.0-3.0; L 5.0-6.5; H 2.5-3.5; Snider and Niemer, 1964, see Fig. 1). Micro-injections (0.5 μl) were carried out starting 7 d following sur-gery. Drugs were delivered with help of injection cannulae, at a rate of 0.2 μl/30 sec; maximally three injections on each side and within 6

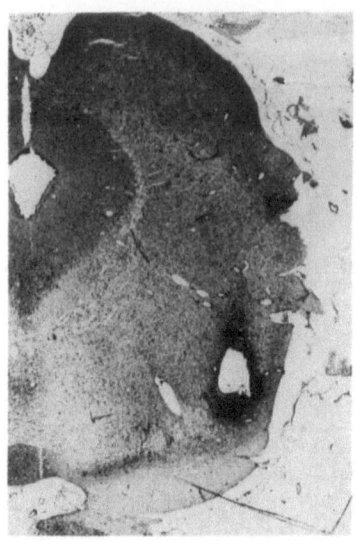

Fig. 1. Cross section of a cat brain illustrating the target point of the injections in the caudolateral SNR.

weeks were given. NMDA was used as a NMDA receptor agonist and injected in a dose of 200 ng. In order to investigate the receptor specifity of the NMDA induced behavioural effects, the NMDA receptor antagonist AP_7 (100 ng) was injected into the SNR 5 min prior to NMDA in a second series of experiments. In order to study the interaction of NMDA and GABA receptor activity, the GABA agonist muscimol (100 ng) was injected 15 min before NMDA in a third series of experiments. Open field behaviour of cats, pre-adapted to the observation cage (120 x 100 x 100 cm), was recorded with help of a video-system allowing subsequent analysis.

The following dependent variables were analyzed until 30 min after the NMDA injection:
1. Freezing: indicating a behavioural state in which the cat did not make any movement for a period of at least 7 sec.
2. Eye fixation: implying absence of any eye movement for a period of at least 7 sec.
3. Licking: licking the fur repeatedly, limited to particular parts of the body.
The results were statistically evaluated with help of Student's t-test.

RESULTS

Shortly after the application of NMDA (200 ng) the cats displayed freezing behaviour and eye fixation (Fig. 2). AP_7 (100 ng) significantly antagonized both freezing and eye fixation (Fig. 2). Muscimol significantly counteracted the NMDA induced freezing and eye fixation (Fig. 3). Single injections of AP_7 or muscimol resulted in stereotyped ipsi-as well as contralateral licking (Fig. 4).

DISCUSSION

Our results demonstrate that the microinjection of NMDA into the SNR induces a NMDA receptor specific effect comparable to picrotoxin when applicated at the same target point, i.e. the caudo-lateral SNR (see Heim et al., 1986). The effects of AP_7 and muscimol were also comparable: both elicit stereotyped licking. The present data suggest that blockade of the excitatory drive of the subthalamo-nigral glutamatergic/aspartatergic

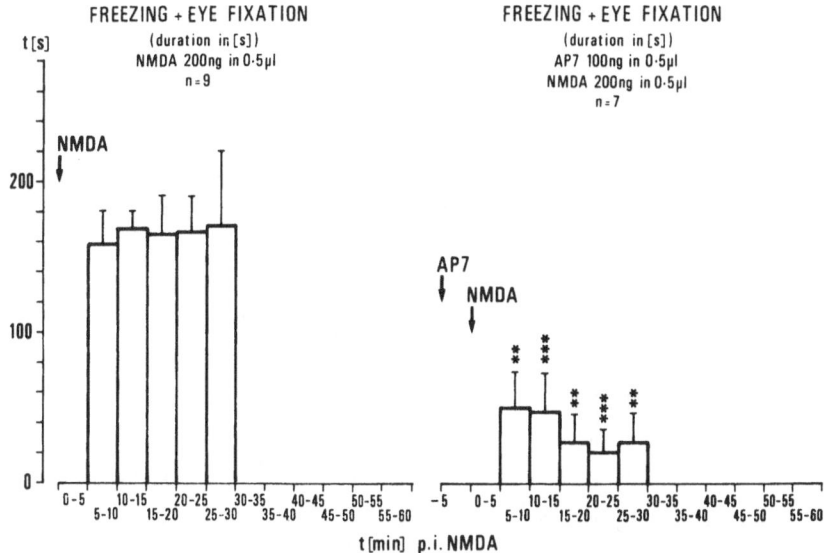

Fig. 2. Duration of Freezing behaviour plus Eye fixation (in s) following
unilateral SNR injections of NMDA (left) and after unilateral SNR
injections of NMDA preceded by ipsilateral injections of AP7
(right).

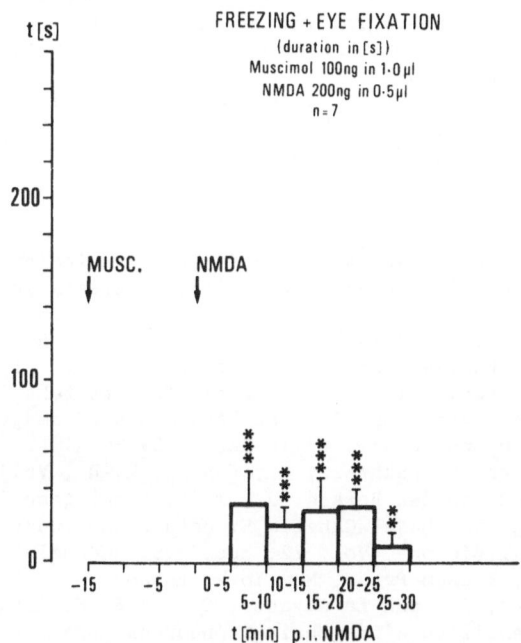

Fig. 3. Duration of Freezing behaviour plus Eye fixation (in s) following
unilateral SNR injections of NMDA preceded by ipsilateral injections of muscimol.

pathway fascilitates the striato-nigral inhibitory GABAergic pathway. The
experiments with intranigral injections of picrotoxin as reported pre-

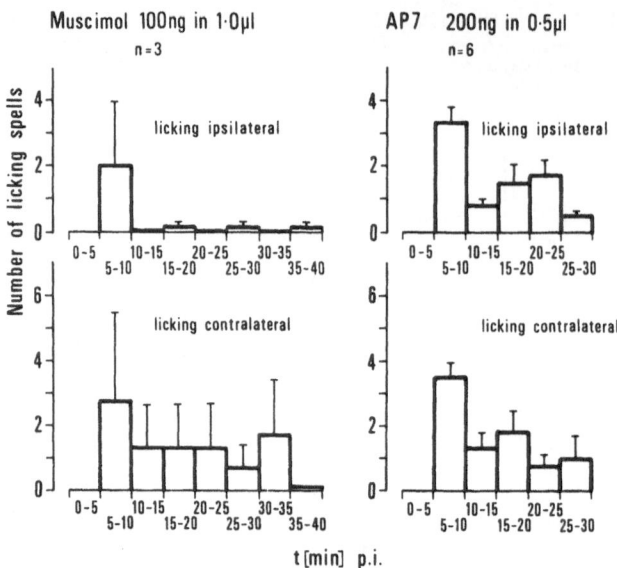

Fig. 4. Number of ipsi- and contralateral licking spells following unilateral SNR injections of muscimol (left) and AP7 (right).

viously (see Introduction) together with the present data concerning the
intranigral application of NMDA reveal a functional interaction between the
influence of the inhibitory GABAergic striato-nigral and the glutamatergic
/aspartatergic subthalamo-nigral system. In conclusion, the SNR appears to
modulate the outflow of the striato-nigral, striato-pallido-nigral and the
cortico-subthalamo/pallido-subthalamo-nigral system. The SNR as an output
nucleus of the basal ganglia system is influenced by a short disynaptic
descending influence via STN and a long polysynaptic loop via the striatum
(for review see Kitai and Kita, 1987).

REFERENCES

Brotchie, J.M., and Crossman, A.R., 1989, Evidence for excitatory amino
    acid (EAA) mediated subthalamic nucleus efferents in the rat. Basal
    Ganglia '89, Capo Boi, Cagliari, Italy, June 10-13
Cools, A.R., Jaspers, R., Kolasiewicz, W., Sontag, K.-H. and Wolfarth,
    S., 1983, Substantia nigra as a station that not only transmits,
    but also transforms, incomming signals for its behavioural
    expression: striatal dopamine and GABA mediated responses of pars
    reticulata neurons. Behav. Brain Res. 7:39
Cools, A.R., Jaspers, R., Schwarz, M., Sontag, K.-H., Vrijmoed de
    Vries, M., and van den Berken, J., 1984, Basal ganglia and switching
    behaviour, in: The Basal Ganglia Structure and Function, Advances
    in Behavioural Biology, Vol. 72, eds. J.S. McKenzie, R.E. Klemm,
    L.N. Wilcock, Plenum Press, New York, London
Deniau, J.M., Feger, J., and Le Guyader, C., 1976, Striatal evoked inhi-
    bition of identified nigro-thalamic neurons. Brain Res. 104:152
Dray, A., 1979, The striatum and substantia nigra: a commentary on their
    relationship. Neuroscience, 4:1407
Heim, C., Schwarz, M., Klockgether, T., Jaspers, R., Cools, A.R., and
    Sontag, K.-H., 1986, GABAergic neurotransmission within the reticular
    part of the substantia nigra (SNR): role for switching motor
    patterns and performance of movements. Exp. Brain Res. 63:375
Kitai, S.T., and Kita, H., 1987, Anatomy and physiology of subthalamic nuc-
    leus: a driving force, in: The Basal Ganglia II, Structure and

Function - Current Concepts Advances in Behavioural Biology, Vol.
32, eds. M.B. Carpenter, A. Jayaraman, Plenum Press, New York,
London

Snider, R.S. and Niemer, W.T., 1964, A Stereotaxic Atlas of the Cat Brain,
University of Chicago Press, Chicago

Sontag, K.-H., Heim, C., Schwarz, M., Jaspers, R., Cools A.R., and Wand,
P., 1984, Consequences of disturbed GABAergic transmission in substantia nigra pars reticulata in freely moving cats on their motor
behaviour and in anaesthetized cats on their spinal motor elements,
in: Basal Ganglia Structure and Function, Advances in Behavioural
Biology, Vol. 72, eds. J.S. McKenzie, R.E. Klemm, L.N. Wilcock,
Plenum Press, New York, London

Wolfarth, S., Kolasiewicz, W., and Sontag, K.-H., 1981, The effects of
muscimol and picrotoxine injections into the cat substantia nigra.
Naunyn-Schmiedeberg's Arch. Pharmacol. 317:54

RESPONSES OF MONKEY DOPAMINE NEURONS TO EXTERNAL STIMULI: CHANGES WITH LEARNING

Tomas Ljungberg, Paul Apicella and Wolfram Schultz

Institut de Physiologie
Université de Fribourg
CH-1700 Fribourg
Switzerland

SUMMARY

Previous studies have shown that dopamine (DA) neurons respond with a phasic activation to external stimuli triggering a behavioral reaction. In this study, we investigated responses of DA neurons to stimuli before, during and after they acquired a behavioral significance. Dopamine neurons with typical electrophysiological characteristics were recorded in midbrain areas A9, A8 and A10 in two *Macaca fascicularis* monkeys. Before behavioral reactions were demanded, the opening of a small door of a food box or the illumination of a small light did typically not activate DA neurons (N=173). In a second phase, a small morsel of food was presented behind the opening door, and the monkeys reached out for it. Most of 76 DA neurons thereafter recorded responded with a short burst of impulses to door opening, but lacked responses to the light. In a third phase, the light was used to trigger a similar reaching arm movement towards a lever for obtaining liquid reward. After acquisition of this task, most of 64 DA neurons responded to the light in a similar way as to door opening. During the transitory phase of task acquisition, an increased proportion of DA neurons were phasically activated by the liquid reward. This response largely disappeared when the behavioral reaction was fully established. Thus, the responses of DA neurons are related to the acquired significance of environmental stimuli in a behavioral context.

INTRODUCTION

The ascending dopamine (DA) systems in the brain are important for the performance of normal behavior. For example, an increased DA receptor stimulation induces hyperactivity, stereotyped movements of head and forelimbs, and compulsive licking and gnawing. These behavioral changes occur in parallel with a disruption of goal-directed adaptive behaviors, such as the performance of operantly conditioned tasks, food and water intake, exploratory and social behaviors (Randrup and Munkvad, 1970; Beninger, 1983; Robbins and Sahakian, 1983; Fibiger and Phillips, 1986). A decreased DA transmission causes in some respect a mirror image - the animals have difficulties in initiating and performing a wide variety of behaviors. Furthermore, intracranial self-stimulation is associated with an activation of DA systems (Wise and Rompré, 1989).

In man, the ascending DA systems have been implicated in three severe clinical syndromes; Parkinsonism, schizophrenia and drug abuse. In Parkinsonism, the typical

The Basal Ganglia III, Edited by G. Bernardi *et al.*
Plenum Press, New York

difficulties in initiating and performing movements are linked to a degeneration of ascending DA systems in the brain. In schizophrenia, the antipsychotic medication exerts its therapeutic effects via the blockade of forebrain DA receptors (Creese et al., 1976; Seeman et al., 1976) and, finally, drugs of abuse are thought to cause an activation of DA systems (Bozarth, 1986).

Despite the fundamental importance in behavior, details of the functions of DA systems are poorly understood. Electrophysiological analysis of single DA neurons in the behaving animal can help to provide a closer insight. Previous experiments have shown that the activity of single DA neurons is characterized by a remarkable stability during a variety of behavioral acts and during tonic sensory stimulation (Steinfels et al., 1981; DeLong et al., 1983), and by burst-like responses to certain phasic stimuli. Several types of stimuli evoke these responses, such as relatively strong and novel stimuli eliciting an orienting reaction (Steinfels et al., 1983), touch of primary reward (Romo and Schultz, 1990), or conditioned stimuli triggering immediate behavioral reactions (Schultz, 1986; Schultz and Romo, 1990).

In the present study, we investigated how a previously ineffective stimulus acquires the propensity to activate DA neurons after animals have learnt to use the stimulus to trigger a behavioral reaction. We presented animals with two different, behaviorally neutral, phasic stimuli which did not activate DA neurons. After recording data in this 'no task' situation, the behavioral meaning of one of the stimuli was changed by using the same stimulus to trigger a reaching movement for reward. The second stimulus remained neutral and was used to control for arousal and stimulus specificity. In a third phase, the second stimulus was also used to trigger a behavioral reaction leading to reward.

METHODS

Behavioral procedures

The completely enclosed experimental apparatus contained a touch-sensitive, immovable resting key, a food box, and a yellow light-emitting diode. Both the food box and the light were placed at eye level, at reaching distance of the forearm and at 27° lateral to the midsagittal plane in front of the animal. The door of the food box opened vertically upward within 20-22 ms. In later stages of experimentation, a small lever was placed immediately below the light. The touch-sensitive resting key was positioned so that the monkey could keep its hand relaxed on the key (elbow joint at approximately 90°).

Two *Macaca fascicularis* monkeys were conditioned to keep their hand on the resting key by rewarding them with a drop of fruit juice every 2.5-3.5 s. In the first stage of experimentation, the light-emitting diode was illuminated and the empty food box was opened once every 5-10 s in separate sessions, respectively ('no task' situation). In a second phase of experimentation, a small morsel of apple was placed in the food box. Animals immediately learned to reach out for it when the door of the box opened. The light-emitting diode continued to be phasically illuminated in separate sessions. In a third phase, the delivery of fruit juice for keeping the hand on the resting key was stopped. Instead, the monkey obtained a drop of fruit juice when it performed a reaching movement and touched the small lever below the light after it had been illuminated. Behavior was electronically monitored from standard electronic pulses generated from door opening, illumination of the light, release of the resting key, entering the food box and touching the lever.

Data acquisition and analysis

After animals were able to remain relaxed on the resting key for extended periods of time, they were implanted with a recording chamber and electrodes for monitoring eye movements and arm muscle activity. The substantia nigra was localized radiographically and electrophysiologically.

Using procedures described previously (Schultz, 1986), extracellular activity was recorded with movable microelectrodes from single DA neurons in the midbrain during contralateral performance of behavioral tasks. Neurons were defined as dopaminergic using the combined criteria of anatomical position, discharge frequency of <9/s and impulse duration of >2.0 ms. Signals from neuronal activity were conventionally amplified, filtered, displayed on oscilloscopes, passed through an adjustable Schmitt-trigger, and sampled together with eye movements, electromyograms (EMG) from extensor digitorum communis and biceps brachii and time markers from the behavioral performance by a laboratory computer. The behavioral relationships of neuronal discharges, muscle activity and eye movements were displayed in each trial on-line in the form of dot-displays and analog curves. Only results from neurons with at least 10 trials of a given test are reported.

Constant time windows were introduced in order to compare neuronal responses between different stages of experimentation. According to the latencies and durations of fully developed neuronal responses, time windows after stimulus presentation were for box opening: 80-216 ms, light: 92-228 ms, liquid reward: 176-312 ms. The average discharge frequency in the time window was compared with the average activity during 500 ms preceding stimulus presentation in each trial. A specifically implemented two-tailed Wilcoxon matched-pairs signed-rank test (Schultz, 1986) was used to statistically verify changes in activity following stimulus presentation, using a $p<0.01$.

Following termination of data collection, recording sites of neurons were reconstructed from small electrolytic marker lesions on cresylviolet-stained, 50 um thick, coronal sections of the brain.

RESULTS

In the no task situation, the empty food box was opened or the light illuminated without requiring the animal to react to the stimuli. Typically, DA neurons did not change their impulse activity in response to stimulus presentation during this stage of experimentation (Fig. 1). The figure also shows the absence of saccadic eye movements directed towards the stimuli. Neuronal responses and ocular saccades were observed in each animal during the first sessions after the stimuli were introduced (food box: 4 of 83 neurons, 5 %; light: 5 of 56 neurons, 9 %).

In the second phase of experimentation, a small morsel of apple was placed in the food box. The morsel became visible and available to the monkey when the door opened. The animal immediately learned to reach out for the apple, bring it to the mouth and consume it. The arm movement was regularly preceded by a saccadic eye movement towards the food box. A phasic response to door opening occurred in 41 of 76 (54 %) DA neurons thereafter recorded (Fig. 2, left). This response lacked temporal relationship to individual saccades and was similar to that previously described using a food box task (Schultz, 1986). In contrast, DA neurons remained largely unresponsive to the light which was still presented without a specific task (1 response in 34 DA neurons, 3 %; Fig. 2, right). Target-directed eye movements continued to be absent in this situation. This selectivity shows that the responses were not due to unspecific factors induced by the behavioral conditioning procedure, such as a general increase in neuronal responsiveness.

In the third phase of experimentation, liquid reward was made contingent upon touching a lever following presentation of the light. The lever was placed at the same position as the previously employed food box, thus making the reaching movements very similar. The conditioning of the animals lasted two days and was performed in steps which guided them from releasing the resting key (which no longer automatically gave a reward) to finally touching the lever. The duration of light illumination and the corresponding response period were successively shortened to 2 s. After animals acquired the correct behavioral reaction, the majority of DA neurons responded with a phasic activation to illumination of the light (37 of 64 neurons; 58 %). As with the food box, animals now regularly performed a saccadic eye movement. A typical response of a DA neuron is shown in Fig. 3. The fact that the light elicited the same type of response as door opening, demonstrates that the selective responses of the previous phase were not due to inherent differences in stimulus properties.

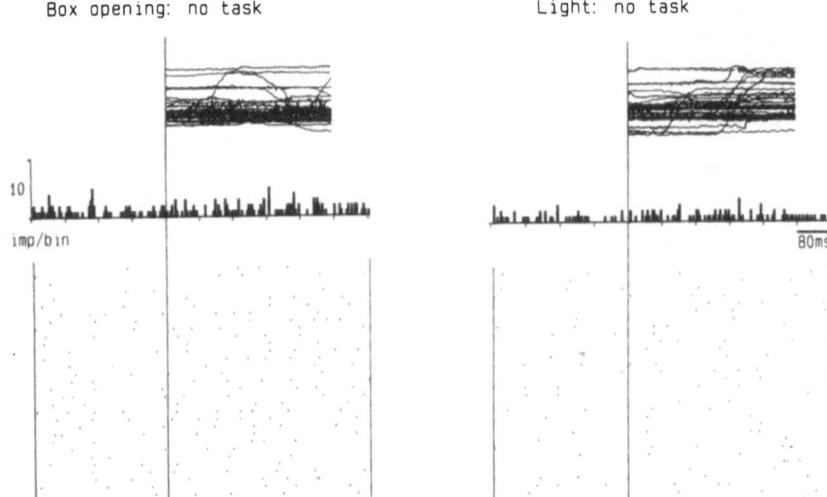

Fig. 1. Lack of response of one DA neuron to the presentation of two phasic stimuli, door opening and light, during the no task situation. In this and the following figures are shown, from above downward: superposed horizontal electrooculograms; perievent time histograms of neuronal impulses; dot display of neuronal impulses. Each dot represents the time of a neuronal impulse, and each line of dots represents one trial. The histograms are composed of neuronal impulses shown as dots below them. Eye movements and neuronal impulses were recorded simultaneously in the same trials. Bin width is 4 ms, and small markers below histograms indicate 20 bins.

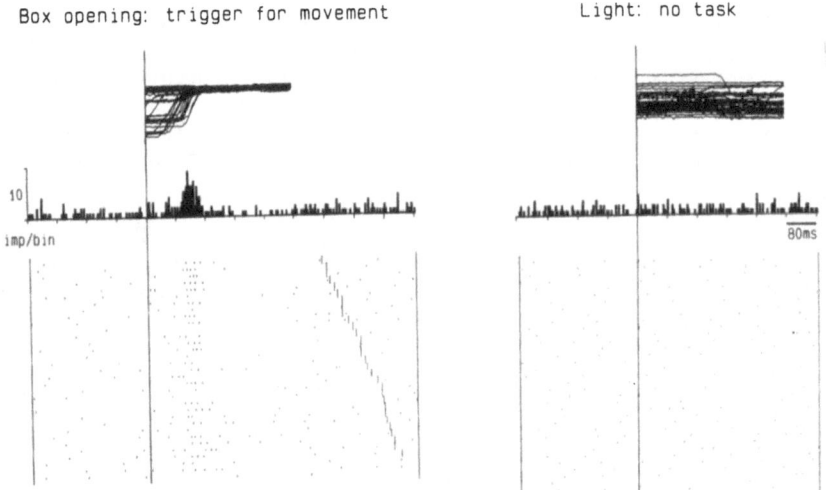

Fig. 2. Left: Response of a DA neuron to door opening used to trigger a reaching movement of the arm and an eye movement. Small bars in dot display represent the time of the hand arriving at the food box. The sequence of trials is rearranged according to the time intervals between door opening and arrival at the food box. **Right**: No response of the same neuron to light illumination when a particular behavioral reaction is not required.The horizontal components of saccades towards the right are shown by upward deflections.

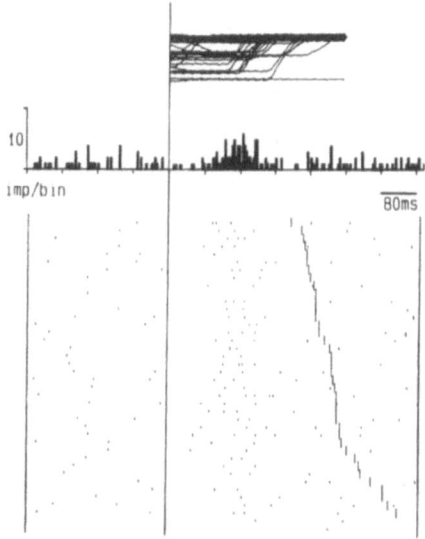

10
imp/bin
80ms

Fig. 3. Response of a DA neuron to the presentation of the phasic light stimulus used to trigger a reaching movement. Small bars in dot display indicate the time of the hand leaving the resting key. Trials are rearranged according to the time intervals between light illumination and key release.

Dopamine neurons were also recorded during the acquisition of the task. During this transitory phase, the proportion of DA neurons responding to light illumination was still low (8 of 25 neurons; 32 %). In contrast, more DA neurons responded with a phasic activation to the delivery of liquid reward (11 of 21 neurons; 52 %), as compared to previous phases during which liquid was administered automatically when the hand stayed on the resting key (18 of 138 neurons; 13 %). When the task was fully acquired, the proportion of DA neurons responding with an activation to the liquid reward was again low (11 of 63 neurons; 18%). These results are summarized in Fig. 4.

Histological reconstructions showed that in each of the three phases about 80% of neurons were recorded from the pars compacta of substantia nigra (area A9), while about 10% of neurons were located in each of areas A8 and A10. Dopamine neurons responding to the different stimuli were evenly distributed over areas A9, A8 and A 10.

DISCUSSION

The present results show that responses of DA neurons to phasic stimuli depend upon the behavioral context. The physical stimulus properties per se were not sufficient for a neuronal response. Chronologically, DA neurons responded when the door opening or light stimuli were applied during the first few sessions. These responses may be related to the potential interest for novel stimuli, as suggested by the presence of target-directed eye movements. With repeated exposure, animals lost overt behavioral reactions, the stimuli became neutral, and neuronal responses disappeared. Later, when the same stimuli elicited behavioral reactions, a strong phasic neuronal activation occurred. A similar context dependency has been observed in a different experiment in which DA neurons responded to trigger stimuli during performance of a behavioral task while being unresponsive to the same stimuli presented in a no task situation (Schultz and Romo, 1990). The present experiments further revealed responses of DA neurons to delivery of liquid reward during the

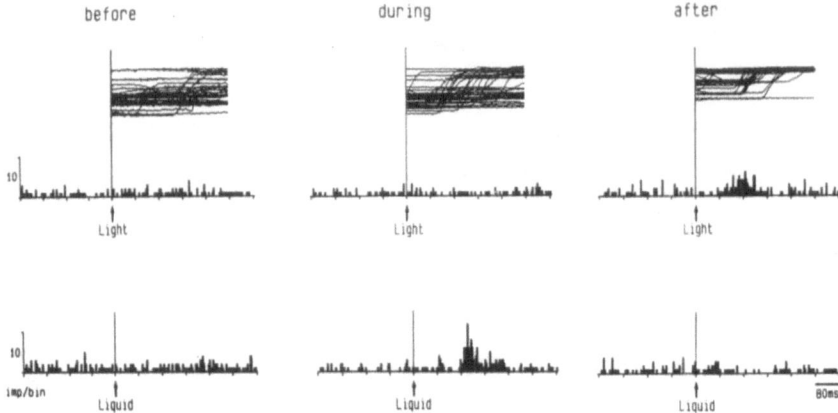

Fig. 4. Activity of three DA neurons following presentation of light and delivery of liquid reward, before, during and after establishment of the reaching movement triggered by light illumination. Shown from above downward are: superposed horizontal electrooculograms; perievent time histograms of neuronal impulses aligned to light illumination and delivery of liquid, respectively.

establishment of the lever-pressing task. This response was greatly diminished when the task was fully acquired.

Common properties of stimuli effective in activating DA neurons appear to be their behavioral significance. The animal needs to pay attention to stimuli that are either novel, indicate reward, or are used for triggering a behavioral reaction (see also Schultz, 1986 and Schultz and Romo, 1990). The earlier finding that responses are diminished when animals are being distracted (see Strecker and Jacobs, 1987) would be consistent with this attentional aspect. It may be inferred that responses of DA neurons are also related to an increased tendency to react to those environmental stimuli that are responsible for the activation and attract the animal's attention. Through the behavioral acts generated, the animal would adapt its behavior according to the information contained in the stimulus, for example, exploratory tendencies are initiated by novel stimuli. In an operant task, the conditioned stimulus would activate DA neurons and lead to initiation of the appropriate behavioral reaction for acquiring the reward. These considerations would correspond to previous suggestions that activation of DA systems causes motivational arousal (Bindra, 1974; Wise, 1982), has energizing properties (Fibiger and Phillips, 1986) or causes a psychomotor stimulation (Robbins and Sahakian, 1983; Wise and Bozarth, 1987). An excessive stimulation of DA receptors would overdrive this mechanism, causing the well known syndrome of hyperactivity and stereotypies (Randrup and Munkvad, 1970). A decrease in DA receptor stimulation results in the opposite, i.e. a difficulty in identifying important stimuli leading to initiation and performance of behavioral acts.

While animals learned the lever-pressing task, DA neurons were activated by the delivery of liquid reward. This was an important stimulus for the thirsty animal that no longer received liquid in an automatic manner as during the no task situation. A similar phasic activation to reward has previously been observed when animals touched an invisible morsel of food in self-initiated movements without preceding stimuli (Romo and Schultz, 1990). In the present experiment, the neuronal response to the reward stimulus could serve to activate the behavior of the animal and result in an appetizing effect leading to approach and exploratory behavior. Eventually, animals would learn that a lever has to be pressed after light presentation in order to receive the reward. When the task is acquired, the delivery of liquid has changed its significance as stimulus. Instead, the light sustains the behavioral reaction and activates the DA neurons. The selective response of DA neurons by reward during task acquisition could thus serve to direct the animal's behavior so that the appropriate behavioral reactions for reaching the goal can be established. The interpretation that responses of DA neurons subserve an appetizing function is compatible with previous views that activation of DA systems has reinforcing or hedonic properties responsible for electrical self-stimulation and self-administration of drugs (Wise, 1982; Bozarth, 1986; Wise and Rompré, 1989).

Recently, the incentive motivation theory has been advanced for explaining some of the functions of dopaminergic transmission in the brain (Beninger, 1983; Fibiger and Phillips, 1986). Its main tenet is that a state of expectancy or anticipation develops through the association of environmental stimuli with the availability of goal objects. Repeated pairing of these stimuli transforms a previously neutral stimulus into a reliable predictor of a primary reward, and by that the conditioned stimulus gains incentive value. Afterwards, presentation of the incentive stimulus alone is sufficient for initiating and sustaining behavior (Fibiger and Phillips, 1986). In the present experiments, the neutral stimuli of the no task situations did not elicit responses in DA neurons. After a conditioning procedure in which the stimuli were paired with reward, a neuronal response occurred together with the initiation of a specific behavioral reaction leading to reward. During acquisition of the task, the observed transient response to reward may have been transferred to the conditioned trigger stimulus. Although these results would be compatible with the incentive motivation theory, we found in subsequent experiments a decrease of responses of DA neurons to trigger stimuli in overtrained animals, despite an improvement in task performance. These data, as well as the responses to novel stimuli, would be at variance with predictions from the incentive motivation theory. This would not argue against a role of dopaminergic neurotransmission in incentive processes, but suggests a more general involvement of dopaminergic impulse activity in behavior, such as in activation and motivational arousal mentioned above.

It has previously been found that animals are particularly sensitive to the disrupting effects of decreased DA receptor stimulation during the acquisition of an operant task (see Beninger, 1983). The selective and pronounced responses of DA neurons found in the present study during the establishment of the task might explain this particular sensitivity.

During one stage of experimentation, the same DA neurons responded to door opening used for triggering a behavioral reaction while being unresponsive to the light presented in the no task situation. In a recent learning experiment, similar discriminative responses were observed in rabbit cingulate cortex (Gabriel et al., 1980). In contrast to our results in which these responses developed together with behavioral responding, Gabriel et al. found that the discriminative properties preceded task acquisition. The presently observed discriminative responses appear to be related to the arousing and directing properties of behaviorally significant stimuli and do not suggest a specific involvement of DA neurons in the acquisition of associative significance of conditioned stimuli.

ACKNOWLEDGEMENTS

We acknowledge the technical assistance of F. Tinguely. The study was supported by the Swiss NSF (grant 3.473-0.86) and the United Parkinson Foundation.

REFERENCES

Beninger, R. J., 1983, The role of dopamine in locomotor activity and learning, Brain Res. Rev., 6: 173.

Bindra, D., 1974, A motivational view of learning, performance, and behavior modification, Psychol. Rev., 81: 199.

Bozarth, M. E., 1986, Neural basis of psychomotor stimulant and opiate reward: evidence suggesting the involvement of a common dopaminergic system, Behav. Brain Res., 22: 107.

Creese, I., Burt, D. R., and Snyder, S. H., 1976, Dopamine receptor binding predicts clinical and pharmacological potencies of antischizophrenic drugs, Science, 192: 481.

DeLong, M. R., Crutcher, M. D., and Georgopoulos, A. P.,1983, Relations between movement and single cell discharge in the substantia nigra of the behaving monkey, J. Neurosci., 3: 1599.

Fibiger, H. C. and Phillips, A. G., 1986, Reward, motivation, cognition: psychobiology of mesotelencephalic dopamine systems, In: Handbook of Physiology-The Nervous System IV, Bethesda.

Gabriel, M., Foster, K., and Orona, E., 1980, Interaction of laminae of the cingulate cortex with the anteroventral thalamus during behavioral learning, Science 208: 1050.

Randrup, A., and Munkvad, I., 1970, Biochemical, anatomical and psychological investigations of stereotyped behavior induced by amphetamines, In: Amphetamines and related compounds, E. Costa, and S. Garattini, eds., Raven Press, New York.

Robbins, T. W., and Sahakian, B. J., 1983, Behavioral effects of psychomotor stimulant drugs: clinical and neuropsychological implications, In: Stimulants: neurochemical, behavioral, and clinical perspectives, I. Creese, ed., Raven Press, New York.

Romo, R., and Schultz, W., 1990, Dopamine neurons of the monkey midbrain: contingencies of responses to active touch during self-initiated arm movements, J. Neurophysiol., in press.

Schultz, W., 1986, Responses of midbrain dopamine neurons to behavioral trigger stimuli in the monkey, J. Neurophysiol., 56: 1439.

Schultz, W., and Romo, R., 1990, Dopamine neurons of the monkey midbrain: contingencies of responses to stimuli eliciting immediate behavioral reactions, J. Neurophysiol., in press.

Seeman, P., Lee, T., Chau-Wong, M., and Wong, K., 1976, Antipsychotic drug doses and neuroleptic/dopamine receptors, Nature, 261: 717.

Steinfels, G. F., Heym, J., and Jacobs, B. L.,1981, Single unit activity of dopaminergic neurons in freely moving animals, Life Sci., 29: 1435.

Steinfels, G. F., Heym, J., Strecker, R. E., and Jacobs, B. L.,1983, Behavioral correlates of dopaminergic unit activity in freely moving cats, Brain Res., 258: 217.

Strecker, R. E., and Jacobs, B. L., 1987, Dopaminergic unit activity during behavior, in: Neurophysiology of dopamine systems - current status and clinical perspectives, L. A. Chiodo and A. S. Freeman, eds., Lakeshore Publishing Company.

Wise, R. A., 1982, Neuroleptics and operant behavior: The anhedonia hypothesis, Behav. Brain Sci., 5: 39.

Wise, R. A., and Bozarth, M. A., A psychomotor stimulant theory of addiction, Psychol. Rev., 94 (4): 462.

Wise, R. A., and Rompré, P.-P., 1989, Brain dopamine and reward, Ann. Rev. Psychol., 40: 191.

THE BASAL GANGLIA: A ROLE IN MEDIATION OF MOTOR EQUIVALENCE

Ivan Divac

Institute of Neurophysiology
University of Copenhagen
School of Medicine

Numerous attempts have been made to describe the functions of the basal ganglia (BG; Divac and Oberg, 1979). The number and variability of proposed concepts reflect different approaches to the study of the BG, the complexity of the organization of the forebrain, as well as the imagination of the authors. I have contributed to the collection of vague notions by proposing that the BG "translate cognition to action" (Divac, 1977). At a later date I suggested that the BG may mediate the phenomena of motor equivalence (Divac, 1984). Now, I would like to discuss the evidence for the latter suggestion and relate it to the former one.

Motor equivalence refers to the ability of organisms to achieve the same goal by means of different movements. For example, a door in Europe can be opened by hand, elbow, knee, foot and other parts of the body. A decision to open a door, or to choose the left corridor of a maze, or to achieve any other goal, may be put into effect in a number of alternative movement patterns.

The reason to believe that the BG play a role in this transformation from a decision to movement is based on the following observations: First, anatomy suggests that the BG are situated as an interface between the cerebral cortex on the one hand, and the pyramidal and some extrapyramidal paths, on the other hand (Allen and Tsukahara, 1974; Divac, 1977; Kemp and Powell, 1971; Webster, 1975). Furthermore, different cortical areas are related to specific neostriatal regions so that the forebrain can be divided into vertical systems which include parts of the thalamus, cerebral cortex and the neostriatum (Divac, 1977). There are in theory as many systems as there are separable cortical areas with specific thalamic and neostriatal appendages (Divac, 1968, 1972, 1977, 1984). The efferents of the vertical systems partially overlap on pallidal and nigral neurons (Percheron et al., 1984) and ultimately feed to the thalamic regions connected to the motor cortical areas as well as to the mesencephalic regions from which some extrapyramidal paths originate (Carpenter, 1984).

Secondly, neurobehavioral evidence suggests that different "association" cortical areas mediate different complex functions (Kolb and Whishaw, 1985; Rosenkilde, 1979). In monkeys, for example, the prefrontal cortex and the related neostriatal region are parts of the prefrontal system (Rosvold and Szwarcbart, 1964) which was visualized by Divac and Diemer (1980) and shown to be indispensable for delayed response-type behavior (Divac, 1972; Rosvold and Szwarcbart, 1964). The mediation of delayed responding by the prefrontal system, notably, does not depend on the actual behavioral setup: lesions in this system interfere with delayed responding regardless of whether it is performed by moving only one arm while sitting in a primate chair, or moving both arm and body as in the Wisconsin General Testing Apparatus or the Nencki Testing Apparatus, or locomoting in a T-maze (e.g. Stamm, 1970; Divac and Warren, 1971). Even when performed by gaze shifting, delayed responding engages the prefrontal cortex (Funahashi et al., 1989). Other evidence indicates that the prefrontal system can make use of predelay information provided by any sensory modality (Gross and Weiskrantz, 1963). Thus, it appears that computations in the prefrontal system can utilize information provided by any sensory modality and have access to any set of the final common path neurons. The equivalent arrangement may exist in other cognitive vertical systems of the frontal, parietal and temporal lobes. These systems decide for example about the visual pattern to be chosen or about the moment to reverse a preference (Divac et al., 1967). Particular circumstances seem able to mobilize the appropriate vertical system (e.g. if one essentially depends on a predelay cue, one must rely on computations of the dorsolateral prefrontal system) which then makes decisions about what is to be done. The responsibility about how to achieve this goal motorically falls on the basal ganglia formations and the cerebro-bulbospinal paths. It seems, thus, that the basal ganglia complex serves to couple non-somatotopic, "what-to-do" vertical systems to somatotopic, "how-to-do" pathways while "translating cognition to action".

Obviously, this concept describes at best only one aspect of the organization of the forebrain. For example, relations between the limbic system and the BG are not mentioned. Many other problems (Divac, 1977) remain unsolved and new features of the BG structure and connections ought to be incorporated in a more comprehensive theory about BG functions. Especially burning problem is the relation between the cognitive vertical systems and the vertical system which includes the motor cortical areas. A hypothesis about this relation was proposed by Rolls and Williams (1984).

ACKNOWLEDGEMENTS

Among the foundations which supported my research in the past decade, the most generous have been Danish Medical Research Council, Fonden af 1870, Hasselblad Foundation, Ib Henriksens Fond and Wacherhausens Legat. Jesper Mogensen and R. Gunilla E. Oberg have contributed in many ways to the research which in part provided the basis of the present concept.

REFERENCES

Allen, G.I., and Tsukahara, N., 1974, Cerebrocerebellar communication systems, Physiol. Rev., 54:957-1006.
Carpenter, M.B., 1984, Interconnections between the corpus striatum and brain stem nuclei, in: "The basal ganglia: structure and function", J.S. McKenzie, R.E. Kemm, and L.N. Wilcox, eds., Plenum, New York, pp. 1-68.
Divac, I., 1968, The functions of the caudate nucleus, Acta Biologiae Experimentalis (Warsaw), 28:107-120.
Divac, I., 1972, Neostriatum and functions of prefrontal cortex, Acta Neurobiol. Exp., 32:461-477.

Divac, I., 1977, Does the neostriatum operate as a functional entity?, in: "Psychobiology of the striatum", A.R. Cools, A.H.M. Lohman, and J.H.L. van der Berken, eds., Elsevier, Amsterdam, pp. 21-30.

Divac, I., 1984, Functions of the basal ganglia, CIBA Foundation Symposium 107, Pitman, London, pp. 201-215.

Divac, I., and Diemer, N.H., 1980, The prefrontal system in the rat visualized by means of labelled deoxyglucose. Further evidence for functional heterogeneity of the neostriatum, J. Comp. Neurol., 190:1-13.

Divac, I., and Oberg, R.G.E., 1979, Current conceptions of neostriatal functions. History and a evaluation, in: "The Neostriatum", I. Divac, and R.G.E. Oberg, eds., Pergamon, Oxford, pp. 215-230.

Divac, I., Rosvold, H.E., and Szwarcbart, M., 1967, Behavioral effects of selective ablation of the caudate nucleus, J. Comp. Physiol. Psychol., 63:184-190.

Divac, I., and Warren, J.M., 1971, Delayed response by frontal monkeys in the Nencki Testing Situation, Neuropsychologia, 9:209-217.

Funahashi, S., Bruce, C.J., and Goldman-Rakic, P.S., 1989, Mnemonic coding of visual space in the monkeys dorsolateral prefrontal cortex, J. Neurophysiol., 61:331-349.

Gross, C.G., and Weiskrantz, L., 1964, Some changes in behavior produced by lateral frontal lesions in the macaque, in: "The frontal granular cortex and behavior", J.M. Warren, and K. Akert, eds., McGraw Hill, New York, pp. 74-101.

Kemp, J.M., and Powell, T.P.S., 1971, The connections of the striatum and globus pallidus: synthesis and speculation, Trans. Roy. Soc. B, 262:441-457.

Kolb, B., and Whishaw, I.Q., 1985, Fundamentals of human neuropsychology", Second ed., Freeman, San Francisco.

Percheron, G., Yelnik, J., and Francois, C., 1984, The primate striato-pallido-nigral system: an integrative system for cortical information, in: "The basal ganglia: structure and function", J.S. McKenzie, R.E. Kemm, and L.N. Wilcox, eds., Plenum, New York, pp. 87-105.

Rolls, E.T., and Williams, S.G.V., 1987, Sensory and movement-related neuronal activity in different regions of the primate striatum, in: "Basal ganglia and behavior: Sensory aspects of motor functioning", J.S. Schneider, and T.I. Lidsky, eds., Hans Huber, Bern, pp. 61-67.

Rosenkilde, C.E., 1979, Functional heterogeneity of the prefrontal cortex of the monkey: a review, Behavior. Neurol. Biol., 25:301-345.

Rosvold, H.E., and Szwarcbart, M.K., 1964, Neural structures involved in delayed-response performance, in: "The frontal granular cortex and behavior", J.M. Warren, and K. Akert, eds., McGraw Hill, New York, pp. 1-15.

Stamm, J.S., 1970, Dorsolateral frontal ablations and response processes in monkeys, J. Comp. Physiol. Psychol., 70:437-447.

Webster, K.E., 1975, The basal ganglia, Proc. Roy. Soc. Med., 68:203-210.

MODELS OF BASAL GANGLIA
PATHOLOGY

THE IMPORTANCE OF MPP+ LOCALIZATION FOR THE MANIFESTATION OF MPTP-INDUCED NEUROTOXICITY

Andrew Giovanni, Patricia K. Sonsalla and Richard E. Heikkila

Department of Neurology, University of Medicine and Dentistry of New Jersey, Robert Wood Johnson Medical School, Piscataway, N.J., 08854, U.S.A.

INTRODUCTION

1-Methyl-4-phenyl-1,2,3,6-tetrahydropyridine (MPTP) is a neurotoxin which is very effective in inducing a lesion of the dopaminergic nigrostriatal pathway in several species, including humans, monkeys and mice (Davis et al., 1979; Burns et al., 1983; Langston et al., 1984.; Heikkila et al., 1984a). MPTP provides a means with which to produce dopaminergic deficits in experimental animals which closely mimic those of human Parkinson's disease. However, the degree of dopaminergic neurotoxicity induced by a given dose of MPTP can vary greatly in these species. For example, rats and mice exhibit marked differences in sensitivity to MPTP, with the rat being relatively unaffected by a dose of MPTP which would produce an 80-90% loss of dopamine in the neostriata of mice. Additionally, there is a considerable difference in sensitivity to MPTP among several strains of mice (Sonsalla and Heikkila, 1986; Sundstrom et al, 1987), and even within a particular strain, older mice are affected to a greater extent than are younger mice by a given dose of MPTP (Ricaurte et al., 1987). In some cases, as with the strains of mice, the differential sensitivity to MPTP-induced neurotoxicity may be explained by the pharmacokinetics of MPTP. More specifically, the amount of the 1-methyl-4-phenylpyridinium species (MPP+), the neurotoxic oxidation product of MPTP, which is found in the neostriata of a particular strain of mice correlates positively with the degree of dopaminergic neurotoxicity induced by MPTP in that strain, relative to the other strains. These data suggest that the brain level, or perhaps more importantly the neostriatal level, of MPP+ is a determinant of MPTP-induced dopaminergic neurotoxicity. Moreover, MAO-B inhibitors have been shown to block the formation of MPP$^+$ in the brain of MPTP-treated animals and also to protect against MPTP-induced neurotoxicity (Heikkila et al., 1984b). In other studies, dopamine uptake inhibitors have been shown to protect against MPTP-induced neurotoxicity, presumably by preventing MPP$^+$ accumulation by dopaminergic nerve terminals (Javitch et al., 1985; Mayer et al., 1986).

However, the differential susceptibility to MPTP demonstrated by mice and rats cannot be explained solely by differences in the neostriatal content of MPP$^+$. Following a 60 mg/kg s.c. dose of MPTP, MPP$^+$ concentrations in the rat neostriata are even higher than those in the mouse, yet the rat exhibits little dopaminergic neurotoxicity while the mouse is greatly affected. In the present study, we have examined the relationship between the brain content of MPP$^+$ and the extent of neurotoxicity, following MPTP administration, in several strains of mice and in one strain of rat. Furthermore, we have investigated the role of the dopaminergic uptake system in the pharmacokinetics of MPTP and MPP$^+$ in the brain and neostriata of MPTP-treated animals. This was an attempt to determine what fraction of the total amount of MPP$^+$ found in the brain or neostriata of MPTP-treated animals could be attributed to being within the dopaminergic nerve terminals. Additionally, we have evaluated the influence of the dopamine storage vesicle, which has been shown to accumulate MPP$^+$, on the neurotoxicity and pharmacokinetics of MPTP and MPP$^+$. The data will show that although the brain content of MPP$^+$ is an important factor, the subcellular distribution of MPP$^+$ must also play a very important role in determining the degree of neurotoxicity induced by MPTP.

METHODS

Measurement of Monoamines and Metabolites

The method used has been previously described (Heikkila et al., 1984b). Briefly, mice or rats were stunned by a blow to the head, the brains removed, and the neostriata dissected free. The neostriata were homogenized in 0.2 M perchloric acid containing 3,4-dihydroxybenzylamine as an internal standard (10 mg wet weight tissue/ml). The homogenates were centrifuged at 27,000 x g for 15 min and the contents of dopamine (DA) and its metabolites were determined by high performance liquid chromatography with electrochemical detection.

The In Vivo Metabolism of MPTP

The amounts of MPTP and its two metabolites (MPDP$^+$ and MPP$^+$) present in the brains (or selected brain region) of mice or rats were determined at various time points after the systemic injection of MPTP. At the time of sacrifice, animals were decapitated, the brains rapidly removed and the brain areas of interest dissected free. Samples were immediately homogenized in 5 or 10 volumes of 0.2 M perchloric acid. The homogenate was centrifuged for 3 min. and the amounts of the parent compound and its metabolites in the supernatant were determined by HPLC with UV detection using a modification of the procedure described by Shinka et al. (1987). The results are given in nmoles/g tissue and are corrected for recovery.

RESULTS AND DISCUSSION

MPTP-Induced Dopaminergic Neurotoxicity in Four Strains of Mice

Four strains of mice (C57-bl, CF-W, CF-1 and CD-1) were given MPTP at 20 mg/kg intraperitoneally four times, once every 2 hours. MPTP differentially affected the strains, with the C57-bl strain being the most affected; the decrement in neostriatal dopamine (DA) content was greater than 80% as compared to untreated controls. The CF-W strain was somewhat less sensitive to MPTP and exhibited about a 60% decrement in neostriatal dopamine. The CF-1 and CD-1 strains were unaffected by this particular MPTP treatment and demonstrated no significant depletion of striatal DA. It should be noted, that at higher doses, MPTP caused dopaminergic deficits in all of these strains.

MPTP-induced neurotoxicity is dependent upon 1) MAO-B for the bioactivation of MPTP to MPP^+, and 2) the dopamine uptake system for the accumulation of MPP^+ into dopaminergic neurons. Thus, these two parameters were examined in the four strains of mice for a possible explanation of the varied susceptibility of the strains to MPTP. Additionally, the neostriatal content of MPP^+ was determined following the cumulative dosing paradigm. Of these three factors, the neostriatal level of MPP^+ most closely correlated with the sensitivity of the strains to MPTP. The more sensitive strains had higher levels of MPP^+ for longer periods of time than did the less sensitive strains. While the MAO-B content was highest in the neostriata of the C57-bl mice (the most sensitive strain), the CF-W mice had about 80% of the neostriatal MAO-B activity of the C57-bl strain, and the CF-1 mice, which were insensitive to MPTP, displayed only a slightly lower neostriatal activity of MAO-B than the C57-bl strain. Additionally, there was no correlation found between neostriatal uptake capacity and sensitivity to MPTP in the four strains of mice (all exhibited similar DA uptake capacity). Thus, it appears that although MAO-B catalyzed metabolism and DA uptake are essential for the expression of MPTP-induced neurotoxicity, these factors are not the sole determinant of the strains' relative sensitivity to MPTP.

As with the four strains of mice discussed above, rats and mice generally exhibit differential susceptibility to MPTP-induced dopaminergic neurotoxicity. For example, 60 mg/kg of

MPTP administered subcutaneously would produce an 80-90% depletion of DA in the neostriata of several strains of mice while Sprague-Dawley (S.D.) rats would display no signs of dopaminergic neurotoxicity. However, S.D. rats exhibited higher neostriatal levels of MPP^+ than did mice following this dosing regimen (fig. 1). Thus, unlike the situation with the four strains of mice where the neostriatal MPP^+ content correlates with strain sensitivity to MPTP, species sensitivity cannot be explained solely by neostriatal MPP^+ content following MPTP administration. MAO-B content and DA uptake capacity were examined in S.D. rats and Swiss-Webster (S.W.) mice with the following results. MAO-B levels were found to be higher in rat than in mouse neostriata, with V_{max} values of 5256 ± 236 and 3297 ± 618 nmole/g tissue/hour for rats and mice respectively. K_m values for MAO-B did not differ significantly between the S.D. rats and the S.W. mice. Additionally, neostriatal dopamine uptake capacity was essentially the same in these S.D. rats and S.W. mice.

MPTP Pharmacokinetics and MPP^+ Uptake

The remaining studies in this article focus on identifying the proportion of the neostriatal MPP^+ which is localized within dopamine nerve terminals, the proposed site of action of the neurotoxin. We thought that if we could determine the content of MPP^+ in the terminals, the physiologically relevant fraction of MPP^+, this information would help us interpret the results from studies which compare pyridinium content with MPTP-induced dopaminergic neurotoxicity in animals of differing ages, strains, species or under various pharmacological treatments. Several parallel experiments were carried out in which the pharmacokinetics of MPP^+, the neurotoxic metabolite of MPTP, were determined in the brains and neostriata of mice treated with MPTP (fig. 2). Figure 2a depicts the time course of MPP^+ content in the brain and neostriata of Swiss-Webster (S.W.) mice following a 30 mg/kg s.c. injection of MPTP. This dose of MPTP was chosen to produce a significant but intermediate depletion of neostriatal DA ($\approx$50% decrement, see fig. 3). The MPP^+ content found in the neostriata was consistently 2-3 times higher than that in the ROB (rest of the brain or whole brain without the neostriata). Initially, we attributed the higher MPP^+ content in the neostriata to the relatively high capacity, compared to other brain regions, of neostriatal dopaminergic nerve terminals to accumulate MPP^+ via the dopamine uptake system. However, the following studies indicate that this is not the case.

Pretreatment of mice with mazindol, a dopamine uptake inhibitor, effectively protects them from MPTP-induced dopaminergic neurotoxicity, presumably by inhibiting the accumulation of MPP^+ into dopaminergic neurons (Javitch et al., 1985; Mayer et al., 1986). However mazindol, at a dose which

504

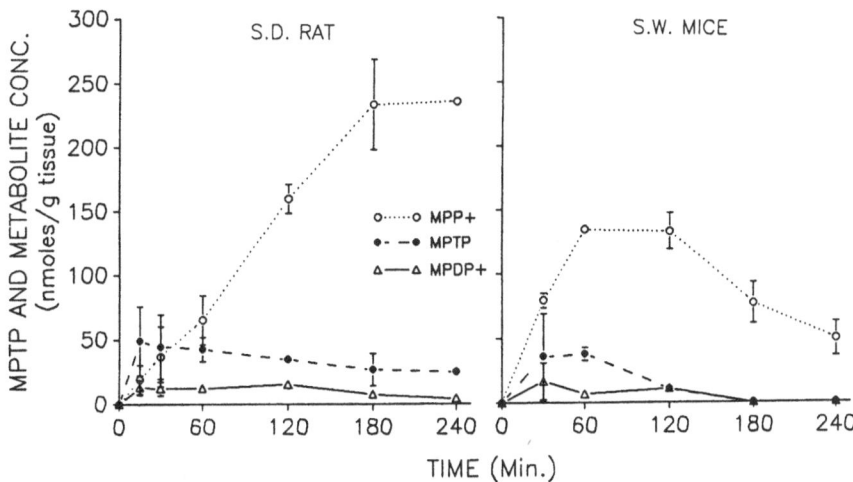

FIGURE 1. MPTP AND METABOLITE CONTENT IN NEOSTRIATUM FOLLOWING MPTP ADMINISTRATION One s.c. injection of MPTP (60 mg/kg) was administered to male Sprague-Dawley rats weighing 250-275 g and to mice weighing 25-30 g. The concentrations of MPTP and its major metabolites were measured in the neostriatum at various time points. Results in rats are the mean content in nmoles/g wet weight ± S.D. for 3 rats, except for the 3 and 4 hr. time points which are from 2 and 1 rats, respectively. Results in mice are the mean values for three samples (two neostriata per sample).

attenuated MPTP-induced dopaminergic neurotoxicity (see fig. 3), had little effect on the pharmacokinetics of MPP^+ in the brain or neostriata of MPTP-treated mice (fig 2b). Furthermore, in the presence of mazindol the neostriatal content of MPP^+ remained 2-3 times higher than that of the ROB for up to five hours after MPTP administration. The fact that mazindol pretreatment blocked the effects of MPTP, yet did not alter neostriatal MPP^+ content, indicates that the majority of the pyridinium found in the neostriata is not within the dopaminergic nerve terminals. Corroborating evidence for this observation is presented in figure 2c which depicts a parallel experiment in which the MPTP analog 2'n-propyl-MPTP is administered to mice. 2'n-propyl-MPTP is oxidized to its corresponding pyridinium, 2'n-propyl-MPP^+, primarily via MAO-A. Additionally, 2'n-propyl-MPP^+ is a potent inhibitor of mitochondrial respiration at Complex I, a requirement we believe to be essential for this class of compounds to possess neurotoxic potential (Youngster et al., 1989). However, 2'n-propyl-MPP^+ is a relatively poor substrate for the dopamine uptake system as compared to MPP^+ and other neurotoxic analogs of MPP^+ (Youngster and Heikkila., 1988) and is therefore not

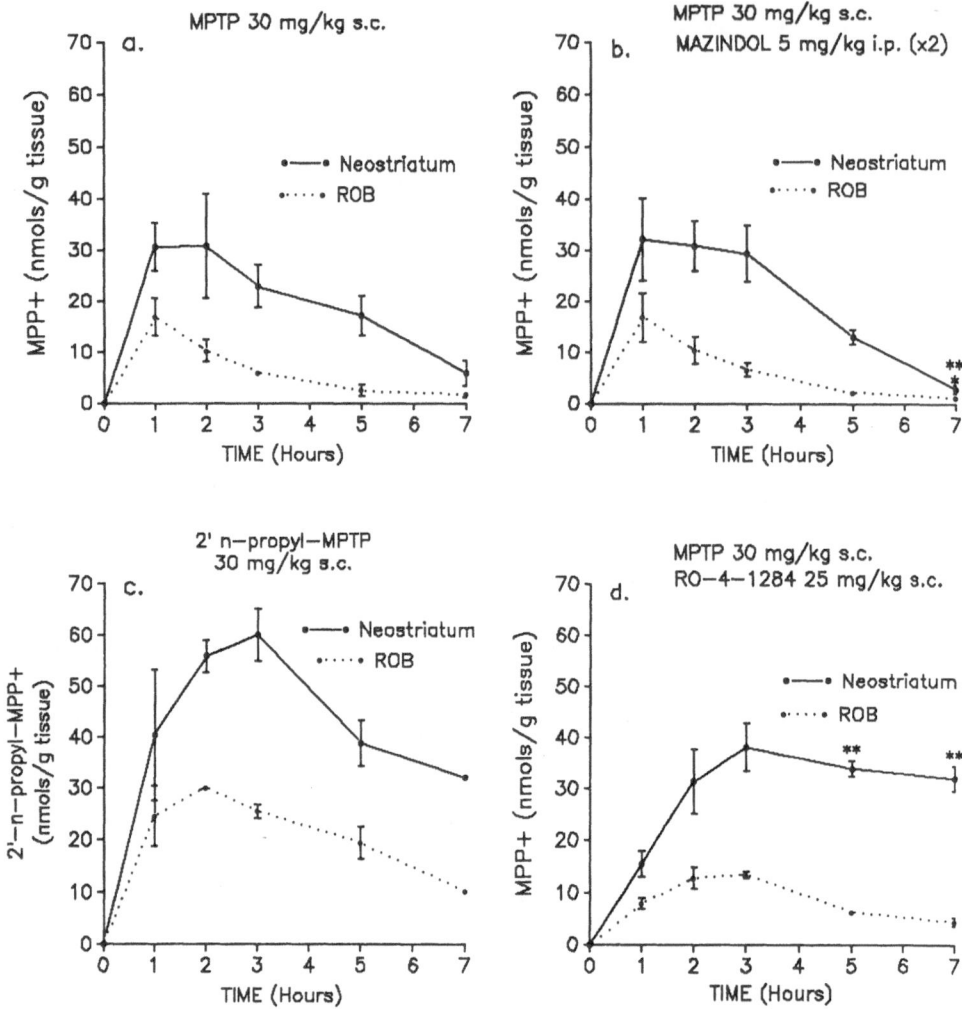

**FIGURE 2. MPP⁺ OR 2'n-PROPYL-MPP⁺ CONTENT IN NEOSTRIATUM AND
REST OF BRAIN AFTER MPTP OR 2'n-PROPYL-MPTP ADMINISTRATION**
2a: MPTP (30 mg/kg) was administered s.c. to Swiss Webster mice
(25 to 30g). Animals were sacrificed at various time points and
their neostriatal and rest of brain (ROB) MPP⁺ levels were
determined. **2b:** same as 2a but mice received an injection of
mazindol (5 mg/kg i.p.) 30 minutes prior to and 2 hr. after the
MPTP injections. **2c:** same as 2a but 30 mg/kg 2'n-propyl-MPTP
was administered instead of MPTP. **2d:** same as 2a but mice
received RO-4-1284 (25 mg/kg s.c.) 15 min. before and 2 hr.
after MPTP administration. Each data point represents the mean
± S.D. of 3 mice. *This data point was not significantly
different from the rest of the brain at the same time point
based on a Duncan comparison test with α=0.05. **These data
points were significantly different from the neostriata from
the MPTP alone group at the same time point based on a Duncan
comparison test with α=0.05.

506

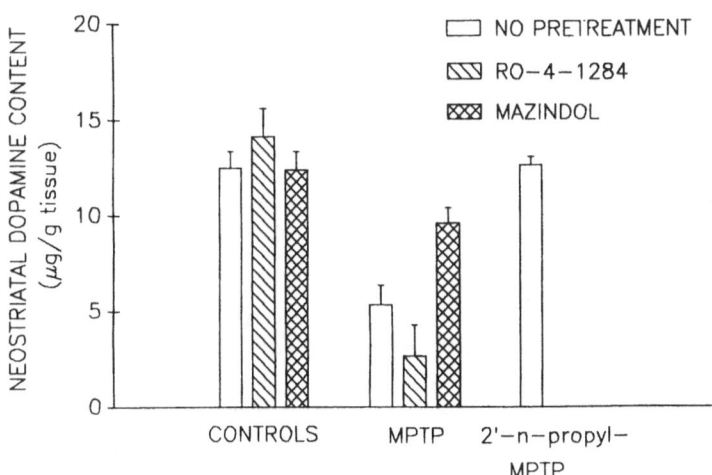

FIGURE 3. NEUROTOXICITY DATA Mice were pretreated with mazindol or RO-4-1284 before and after MPTP administration as described in figure 2. 2'n-propyl-MPTP was given as described in figure 2. Mice were sacrificed 5 days later and neostriatal dopamine levels were determined. Each group consists of 4 to 6 mice. Data represent the mean ± S.D.

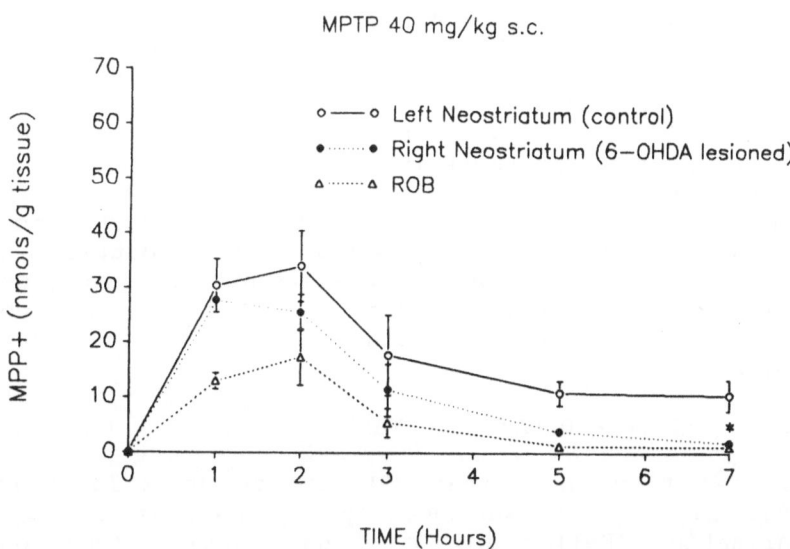

FIGURE 4. NEOSTRIATAL MPP⁺ LEVELS FOLLOWING MPTP ADMINISTRATION TO MICE WITH UNILATERAL LESIONS OF THE NIGROSTRIATAL PATHWAY Mice were lesioned unilaterally with an intrastriatal injection of 25 μg of 6-OHDA, one week prior to the s.c. administration of 40 mg/kg of MPTP. DA content in the lesioned neostriatum was less than 25% of that in the nonlesioned neostriatum. Lesioned and nonlesioned neostriata, respectively, were pooled from two animals. Data are the mean ± S.D. of 3 sets of pooled neostrata. *Significantly different from control neostriatum and not from the ROB based on Duncan multiple comparison test.

neurotoxic (see fig. 3). Figure 2c depicts the ROB and neostriatal content of 2'n-propyl-MPP$^+$ following the s.c. administration of 30 mg/kg 2'n-propyl-MPTP. Here the neostriatal levels of pyridinium are higher than those seen with the equivalent neurotoxic dose of MPTP (compare 2a and 2c). Also note that the neostriatal levels of pyridinium are 2-3 times greater than those of the ROB, yet this compound is not effectively accumulated into dopaminergic neurons.

The results from the studies discussed above prompted another experiment to examine the influence of the dopamine uptake system on the pharmacokinetics of MPP$^+$ in the brain and neostriata of MPTP-treated mice. In this case we chose to destroy the dopaminergic nerve terminals in the mouse neostriata via the stereotaxic administration of 6-hydroxy-dopamine (6-OHDA) several days prior to the administration of MPTP. 1μl containing 25 μg of 6-OHDA was injected stereo-taxically into the right neostriata of mice 7 days before the MPTP metabolic study was conducted. 6-OHDA administration resulted in the destruction of greater than 75% of striatal DA nerve terminals (greater than 75% loss of DA as compared to both the non-injected neostriatum and age-matched control mice). In this study MPTP was given at a dose of 40 mg/kg s.c. and the MPP$^+$ content was determined in the lesioned and non-lesioned neostriatum and in the ROB (see fig. 4). This was a higher dose of MPTP than used in figure 2, resulting in higher MPP$^+$ levels in the control neostriata and ROB. At the earliest time periods MPP$^+$ levels in the control and lesioned neostriata did not differ significantly, but by the 7 hour time point we found slightly but significantly less MPP$^+$ in the lesioned neostriatum as compared to the non-lesioned neostriatum. The difference in the pyridinium content between the 6-OHDA lesioned and non-lesioned neostriatum may reflect the amount of pyridinium that is inside the dopaminergic nerve terminals. However, since lesioning the neostriata is an invasive way of eliminating the dopamine uptake system, several controls should be carried out (such as sham stereotaxic injections) before this data can be considered any more than preliminary.

MPTP AND VESICULAR DOPAMINE UPTAKE

It has been shown that MPP$^+$ can be accumulated by rat striatal slices and subsequently released in response to depolarization (Keller and Da Prada, 1985). Additionally, Reinhard et al, (1988) have shown that inhibitors of vesicular catecholamine uptake potentiate MPTP-induced neurotoxicity in mice. In light of these findings, we have investigated the effect of the tetrabenazine-like vesicular catecholamine uptake inhibitor RO-4-1284 on the pharmacokinetics of MPP$^+$. We found a potentiation of MPTP-induced dopaminergic neurotoxicity by RO-4-1284 as indicated by the greater DA depletion in the neostriata of mice treated with RO-4-1284 and MPTP compared to that observed in mice treated with MPTP alone (see fig. 3).

Dopamine vesicles may provide protection from MPTP by supplying a storage site for MPP$^+$ within the dopaminergic nerve terminals. In animals in which these storage sites were eliminated with RO-4-1284 treatment, we expected to find reduced neostriatal pyridinium content following MPTP administration. What we found in these animals, compared to controls, was a dramatic increase in the neostriatal pyridinium content for up to 7 hours following MPTP administration (see fig. 2d). These results are surprising and difficult to rationalize but may be explained by the ability of the dopaminergic vesicles to release MPP$^+$. One possible explanation is that the accumulation and subsequent release of MPP$^+$ by the dopaminergic vesicle may actually aid in the elimination of MPP$^+$ from the mouse neostriata. These results are somewhat tentative, although similar results were obtained with reserpine pretreatment (data not shown).

SUMMARY AND CONCLUSIONS

From the studies outlined above it appears that most of the MPP$^+$ found in the neostriata of MPTP-treated mice is not within dopaminergic neurons. Furthermore, in both mice and rats, the brain or neostriatal MPP$^+$ content does not always correlate with MPTP-induced dopaminergic neurotoxicity. This implies that, although brain or neostriatal MPP$^+$ content is important for the expression of MPTP-induced dopaminergic neurotoxicity, the precise localization of the MPP$^+$ within the tissue must also play a role in determining the extent of neurotoxicity induced by MPTP. Therefore, we conclude that care should be taken when interpreting results from studies which relate the brain or neostriatal levels of MPP$^+$ with the extent of MPTP-induced dopaminergic neurotoxicity.

REFERENCES

Burns RS, LeWitt PA, Ebert MH, Pakkenberg H, Kopin IJ (1985) The clinical syndrome of striatal dopamine deficiency. Parkinsonism induced by 1-methyl-4-phenyl-1,2,3,6-tetrahydropyridine (MPTP). N.Engl.J.Med., 312, 1418-1421.

Davis GC, Williams AC, Markey SP, Ebert MH, Caine ED, Reichert CM, and Kopin IJ (1979) Chronic parkinsonism secondary to intravenous injection of meperidine analogs. Psychiatry Res., 1, 249-254.

Heikkila RE, Hess A, Duvoisin RC (1984a) Dopaminergic neurotoxicity of 1-methyl-4-phenyl-1,2,5,6-tetrahydropyridine in mice. Science, 224, 1451-1453.

Heikkila RE, Manzino L, Cabbat FS, Duvoisin RC (1984b) Protection against the dopaminergic neurotoxicity of 1-methyl-4-phenyl-1,2,5,6-tetrahydropyridine by monoamine oxidase inhibitors. Nature, 311, 467-469.

Heikkila RE, Manzino L, Cabbat FS, Duvoisin RC (1985) Studies on the oxidation of the dopaminergic neurotoxin 1-methyl-4-phenyl-1,2,5,6-tetrahydropyridine by monoamine oxidase B. J.Neurochem., 45, 1049-1054.

Javitch JA, D'Amato RJ, Strittmatter SM, Snyder SH (1985) Parkinsonism-inducing neurotoxin, N-methyl-4-phenyl-1,2,3,6-tetrahydropyridine: uptake of the metabolite N-methyl-4-phenylpyridine by dopamine neurons explains selective toxicity. Proc.Natl.Acad.Sci.U.S.A., **82**, 2173-2177.

Keller HH, and Da Prada, M (1985) Evidence for the release of 1-methyl-4-phenylpyridinium (MPP$^+$) from rat striatal neurons in vitro. Eur.J.Pharmacol., **119**, 247-250.

Langston JW, Ballard P, Tetrud JW, Irwin I (1983) Chronic Parkinsonism in humans due to a product of meperidine-analog synthesis. Science, **219**, 979-980.

Langston JW, Irwin I, Langston EB, Forno LS (1984) Pargyline prevents MPTP-induced parkinsonism in primates. Science, **225**, 1480-1482.

Mayer RA, Kindt MV, Heikkila RE (1986) Prevention of the nigrostriatal toxicity of 1-methyl-4-phenyl-1,2,3,6-tetrahydropyridine by inhibitors of 3,4-dihydroxy-phenylethylamine transport. J.Neurochem., **47**, 1073-1079.

Reinhard JF Jr, Daniels AJ, Viveros OH (1988) Potentiation by reserpine and tetrabenazine of brain catecholamine depletions by MPTP (1-methyl-4-phenyl-1,2,3,6-tetrahydropyridine) in the mouse; evidence for subcellular sequestration as basis for cellular resistance to the toxicant. Neurosci.Lett., **90**, 349-353.

Ricaurte GA, Irwin I, Forno LS, DeLanney LE, Langston E, Langston JW (1987) Aging and 1-methyl-4-phenyl-1,2,3,6-tetrahydropyridine-induced degeneration of dopaminergic neurons in the substantia nigra. Brain.Res., **403**, 43-51.

Shinka T, Castagnoli N Jr, Wu EY, Hoag MKP, Trevor AJ (1987) Cation-exchange high-performance liquid chromatography assay for the nigrostriatal toxicant MPTP and its monoamine oxidase B generated metabolites in brain tissues. J. Chromatog., **398**, 279-287.

Sonsalla PK, Heikkila RE (1986) The influence of dose and dosing interval on MPTP-induced dopaminergic neurotoxicity in mice. Eur.J.Pharmacol., **129**, 339-345.

Sundstrom E, Stromberg I, Tsutsumo T, Olson L, Jonsson G (1987) Studies on the effect of 1-methyl-4-phenyl-1,2,3,6-tetrahydropyridine (MPTP) on central catecholamine neurons in the C57Bl/b mouse. Comparison with three other strains. Brain Res. **405**, 26-38.

Youngster SK, Heikkila RE (1988) A biological evaluation of some 2'-substituted analogs of MPTP, in: "Progress in Parkinson Research," F. Hefti and W.J.Weiner, Eds., Plenum Press, New York.

Youngster SK, Nicklas WJ, Heikkila RE (1989) Structure-activity study of the mechanism of 1-methyl-4-phenyl-1,2,3,6-tetrahydropyridine (MPTP)-induced neurotoxicity. II. Evaluation of the biological activity of the pyridinium metabolites formed from the monoamine oxidase-catalyzed oxidation of MPTP analogs. J. Pharmacol. Exptl. Ther., **249**, 829-835.

METHAMPHETAMINE AND MPTP: SIMILARITIES AND DIFFERENCES IN MECHANISMS OF NEUROTOXIC ACTION

Patricia K. Sonsalla and Richard E. Heikkila

Department of Neurology, University of Medicine and Dentistry of New Jersey-Robert Wood Johnson Medical School, Piscataway, NJ 08854, USA

INTRODUCTION

The primary pathophysiological finding in the brains of those afflicted with Parkinson's Disease (PD) is an extensive loss of nigrostriatal dopaminergic neurons within the basal ganglia. Although considerable research has focused on the elucidation of the cause for this neurodegeneration, the substance or mechanism responsible for this neurodegeneration is unknown. A Parkinsonian syndrome can occur in individuals who are exposed acutely to low doses of 1-methyl-4-phenyl-1,2,3,6-tetrahydropyridine (MPTP; Ballard et al., 1985) or to those exposed chronically to manganese (e.g. manganese miners; see Barbeau, 1984). The symptoms seen in those patients who have ingested toxic quantities of MPTP are nearly identical to those seen in idiopathic PD whereas those seen in humans who have been exposed chronically to high levels of manganese are somewhat different than those in idiopathic PD and may reflect damage not only to nigrostriatal dopaminergic but to other neuronal systems as well. These findings have led to two different hypotheses regarding the etiology of PD. One is that a compound similar in structure or biochemical characteristics to MPTP, derived from exogenous or endogenous sources, is responsible for the neurodegeneration of nigrostriatal dopaminergic neurons in PD. Another is that the oxidation of DA results in the formation of oxygen-derived reactive species (quinones and/or superoxide or the hydroxyl radical) capable of cytotoxicity as is believed to occur in Mn^{2+} poisoning.

In experimental animals damage to neostriatal dopaminergic neurons can be produced by MPTP and manganese salts as well as by several other compounds, such as methamphetamine. There is considerable evidence which indicates that the mechanisms of neurotoxic action for these compounds are different. In this communication, we will summarize our findings and those of others regarding the neurotoxicity of these substances, particularly of MPTP and methamphetamine. These observations, summarized in table 1, will clearly demonstrate that although MPTP and methamphetamine have some features in common, they are more dissimilar than similar in their actions.

MPTP-INDUCED TOXICITY TO DOPAMINERGIC NEURONS

A rather selective degeneration of nigrostriatal dopaminergic neurons is observed after the systemic administration of MPTP to several animal species which include non-human primates (Burns et al., 1983) and mice (Heikkila et al., 1984a). It is not MPTP itself but rather its metabolite 1-methyl-4-phenylpyridinium (MPP$^+$) which is the actual neurotoxic substance. The major processes which are thought to be important for neurotoxicity are 1) the bioactivation of MPTP by monoamine oxidase-B (MAO-B) to a dihydropyridinium intermediate which is then spontaneously oxidized to the pyridinium metabolite (Chiba et al., 1984), 2) the accumulation of MPP$^+$ within dopaminergic neurons by the dopamine (DA) transporter (Javitch et al., 1985), 3) the concentration of MPP$^+$ within mitochondria (Ramsay et al., 1986) and the inhibition of mitochondrial respiration at Complex I (Nicklas et al., 1985). Thus compounds which either inhibit MAO-B activity (e.g. deprenyl) and thus prevent MPP$^+$ formation (Heikkila et al., 1984b) or compounds which inhibit the DA transporter (e.g. mazindol) and thus prevent the uptake of MPP$^+$ into DA neurons protect against MPTP-induced neurotoxicity (Javitch et al., 1985). Damage to mesolimbic and mesocortical dopaminergic pathways and noradrenergic neurons within the cortex and locus coeruleus can occur after the administration of MPTP to experimental animals although, in general, this damage is less than that to nigrostriatal dopaminergic neurons (Sundstrom et al., 1987). However, because these pathways are also damaged in idiopathic PD, the MPTP-treated animal may prove to be the best animal model of PD. For more information on the neurotoxicity of MPTP see reviews by Kopin, 1987; Schultz, 1988; and Langston, 1989.

SIMILARITIES IN NEUROTOXICITY INDUCED BY MPTP AND METHAMPHETAMINE

Like MPTP, the peripheral administration of methamphetamine to mice produces rather selective damage to striatal dopaminergic neurons and in this species methamphetamine is actually more potent than MPTP as a dopaminergic neurotoxin (Sonsalla and Heikkila, 1988). Interestingly, both methamphetamine and MPTP are more potent and more selective dopaminergic neurotoxins in mice than in rats. Although they are very resistant to the neurotoxic effects of MPTP, under those conditions in which there is an extensive lesion of the dopaminergic neurons in rats, there is also marked damage to the striatal serotonergic neurons as well (unpublished observations). There is also a pronounced loss of serotonergic nerve terminals in rats treated with methamphetamine (Hotchkiss and Gibb, 1980) whereas in methamphetamine-treated mice there is little or no damage to serotonergic nerve terminals (unpublished observations). The reasons for this species difference in susceptibility of the serotonergic neurons to damage by either MPTP or methamphetamine (or other amphetamine analogs) are unknown.

512

Table 1. Comparison of the Dopaminergic Neurotoxicity Induced by MPTP, Methamphetamine and Manganese

	MPTP	METHAMPHETAMINE	Mn^{2+}
Toxicity to DA neurons	+	+	+
Actual Toxic Species	MPP^+	?	?
Mechanism of Toxicity	Inhibition of Mitochondrial Respiration?	Oxidative Stress? Excito-toxicity?	Oxidative Stress?
Protection by:			
MAO B Inhibitors	+	−	−[a]
DA Uptake Inhibitors	+	+	?
DA Synthesis Inhibitors	−	+	+
DA Receptor Antagonists	−	+	?
NMDA Recptor Antagonists	−	+	?
Antioxidants	±	+	+

[a]It has been reported that pargyline pretreatment of rats actually potentiates Mn^{2+}-induced toxicity (Parenti et al., 1986).

It is not known how methamphetamine causes neuronal damage. Like MPP^+, the amphetamines utilize the DA transporter and cause the release of dopamine; presumably intracellular DA is transported out of the neuron by reversal of the DA transporter (see McMillen, 1983). Release of this pool of DA appears to play a very critical role in methamphetamine-induced neurotoxicity. If, for example, the DA transporter is blocked, the neurotoxic effects of methamphetamine on dopaminergic nerve terminals are prevented (Fuller and Hemrick-Luecke, 1982; Schmidt and Gibb, 1985). Thus DA uptake inhibitors such as mazindol will protect against neurotoxicity induced by either MPTP (by preventing the uptake of MPP^+) or by methamphetamine (by preventing the methamphetamine-facilitated release of DA). This protection by DA uptake inhibitors is about the extent of the similarities between the neurotoxic actions of MPTP and methamphetamine.

NEUROTOXICITY INDUCED BY METHAMPHETAMINE

It is apparent that the DA released by methamphetamine plays an important role in methamphetamine-induced neurotoxicity but the release of DA by MPTP is not critical for its neurotoxicity. If DA synthesis is inhibited by α-methyl-p-tyrosine, which blocks tyrosine hydroxylase activity and thus depletes the cytosolic pool of DA, methamphetamine-induced toxicity to dopaminergic as well as serotonergic neurons is prevented (Schmidt et al., 1985). In contrast, the pretreatment of mice with α-methyl-p-tyrosine (or with compounds which deplete the vesicular pool of DA) does not alter the toxicity of MPTP (Fuller and Hemrick-Luecke, 1985). It has been hypothesized that subsequent to this methamphetamine-facilitated release of DA there is oxidative stress which results in the formation of oxygen-derived species and that these reactive species are responsible for the damage incurred (Seiden and Vosmer, 1984). In support of this hypothesis, there is evidence which indicates that antioxidants such as Vitamin E, ascorbic acid or L-cysteine provide protection against methamphetamine-induced damage (DeVito and Wagner, 1989; Steranka and Rhind, 1987). In contrast, there is little or no protection by antioxidants against MPTP-induced neurotoxicity (Perry et al., 1985).

Methamphetamine-induced dopaminergic neurotoxicity depends on the release of the newly-synthesized cytosolic pool of DA. Although there is some evidence that oxidative stress may be a component of this toxicity, it does not appear as though this can be solely responsible for the neurotoxic effects observed. This is because the neurotoxic damage of methamphetamine but not of MPTP can be prevented by the pretreatment of animals with DA receptor antagonists or with antagonists of the N-methyl-D-aspartate (NMDA) receptor subtype of the glutamatergic receptors. In the case of protection by the DA receptor antagonists, there appears to be an even greater release and turnover of DA in animals treated with both the antagonist and METH than with METH alone. Thus, if the release and subsequent oxidation of DA are responsible for neurotoxicity, it is difficult to explain protection in a situation in which there is even more release of DA into the synaptic cleft. Protection against methamphetamine-induced dopaminergic damage by the DA receptor antagonists (e.g. haloperidol, sulpiride, SCH 23390) indicates that the interaction of DA with either D1 or D2 receptors is required for the neurotoxic actions of methamphetamine (Buening and Gibb, 1974; Sonsalla et al., 1986). Exactly how DA receptor activation contributes to neurotoxicity is a matter of considerable speculation.

We have found that noncompetitive NMDA receptor antagonists which act at the phencyclidine binding site within the receptor-linked ion channel (MK-801, phencyclidine, ketamine) protect against neurotoxicity induced by methamphetamine but not by MPTP (Sonsalla et al., 1989). In more recent experiments we have found that ifenprodil and SL 82.0715, noncompetitive NMDA antagonists which act at a site other than the phencyclidine site (possibly a polyamine site), also prevent methamphetamine-induced dopaminergic damage. Likewise, antagonists which are competitive at the NMDA site (CGS 19755, NPC 12626) also exert protective effects. Moreover, an intrastriatal infusion of NMDA potentiates the actions of peripherally administered methamphetamine (unpublished observations). Taken together, these observations indicate that the stimulation of NMDA receptors by excitatory amino acids is an important aspect of the neurotoxic process associated with methamphetamine. Although the amino acids glutamate and aspartate play a central role in excitatory synaptic transmission in the central nervous system, it is now thought that an overexposure of neurons to these transmitters results in cell death. Indeed, NMDA receptor antagonists have been found to be protective against the neuronal loss observed in various animal models of neurodegeneration. Our findings are consistent with the concept that excitatory amino acids may also be involved with the degeneration of dopaminergic neurons in idiopathic PD.

COMPARISON OF NEUROTOXICITY INDUCED BY METHAMPHETAMINE AND Mn^{2+}

A loss of neostriatal dopamine is observed in rats after the chronic administration of manganese salts in the drinking water or after an acute intrastriatal or intranigral infusion of manganese (see Barbeau, 1984; Parenti et al., 1986). It has been hypothesized that the neurotoxic actions of Mn^{2+} to the dopaminergic neurons are due to an enhancement by Mn^{2+} of the autoxidation of DA and the subsequent increased generation of free radicals (see Donaldson, 1987). In support of this hypothesis of a role for DA in this neurotoxicity is the observation that Mn^{2+}-induced toxicity is markedly attenuated in rats pretreated with α-methyl-p-tyrosine (Parenti et al., 1988). In this aspect, neurotoxicity produced by Mn^{2+} resembles that of methamphetamine; i.e. little or no dopaminergic neurotoxicity by Mn^{2+} or methamphetamine occurs when cytosolic DA stores are depleted. Moreover, Mn^{2+}-induced dopaminergic neurotoxicity is enhanced by pargyline or L-DOPA pretreatment, both of which increase the content of DA in the brain (Parenti et al., 1986). This enhancement of Mn^{2+}-induced neurotoxicity by pargyline contrasts markedly with the protective effect of pargyline against MPTP-induced neurotoxicity and the essential lack of effect of pargyline on methamphetamine-induced neurotoxicity. Further evidence of a role for oxidative stress comes from the observation that antioxidants such as Vitamin E protect against Mn^{2+}-induced neurotoxicity (Parenti et al., 1988).

SUMMARY

In summary, there clearly are important differences in the mechanisms of neurotoxic action of MPTP, methamphetamine and Mn^{2+}. Whereas the neurotoxicity of MPTP depends on the formation of MPP^{+}, a compound which is foreign to the brain, that of methamphetamine and Mn^{2+} appear to be dependent on the actions of endogenous substances which include the neurotransmitters DA and the excitatory amino acids. It remains to be determined if the degeneration of the nigrostriatal dopaminergic neurons seen in idiopathic PD occurs by any of the mechanisms associated with the neurotoxic actions of MPTP, methamphetamine or manganese. As has been discussed, some pharmacological agents can protect against damage induced by one of the neurotoxins but fail to protect or may even potentiate the damage induced by the others. Thus, it becomes crucial that the cause of neurodegeneration in PD be unraveled so that appropriate therapeutic measures may be developed to prevent or retard the degeneration of these dopaminergic neurons.

REFERENCES

Ballard P.A., Tetrud J.W., and Langston J.W.(1985)
Permanent human parkinsonism due to 1-methyl-4-phenyl-
1,2,3,6-tetrahydropyridine (MPTP): Seven cases. Neurology
35,949-956.

Barbeau A.(1984) Manganese and extrapyramidal disorders.
Neurotoxicology 5,13-36.

Buening M.K., and Gibb J.W.(1974) Influence of methamphetamine
and neuroleptic drugs on tyrosine hydroxylase activity.
Eur.J.Pharmacol. 26,30-34.

Burns R.S., Chiueh C.C., Markey S.P., Ebert M.H.,
Jacobowitz D.M., and Kopin I.J.(1983) A primate model of
parkinsonism: Selective destruction of dopaminergic neurons
in the pars compacta of the substantia nigra by
N-methyl-4-phenyl-1,2,3,6-tetrahydropyridine.
Proc.Natl.Acad.Sci.USA 80,4546-4550.

Chiba K., Trevor A., and Castagnoli N Jr.(1984) Metabolism of
the neurotoxic tertiary amine, MPTP, by brain monoamine
oxidase. Biochem.Biophys.Res.Commun. 120,574-578.

DeVito M.J., and Wagner G.C.(1989) Methamphetamine-induced
neuronal damage: A possible role for free radicals.
Neuropharmacology 28,1145-1150.

Donaldson J.(1987) The physiopathologic significance of
manganese in brain: Its relation to schizophrenia and
neurodegenerative disorders. Neurotoxicology 8,451-462.

Fuller R.W., and Hemrick-Luecke S.K.(1982) Further studies
on the long-term depletion of striatal dopamine in
iprindole-treated rats by amphetamine. Neuropharmacology
21,433-438.

Fuller R.W., and Hemrick-Luecke S.K.(1985) Mechanisms of
MPTP (1-methyl-4-phenyl-1,2,3,6-tetrahydropyridine)
neurotoxicity to striatal dopamine neurons in mice.
Prog.Neuropsychopharmacol.Biol.Psychiatry 9,687-690.

Heikkila R.E., Hess A., and Duvoisin R.C.(1984a)
Dopaminergic neurotoxicity of 1-methyl-4-phenyl-1,2,5,6-
tetrahydropyridine in mice. Science 224,1451-1453.

Heikkila R.E., Manzino L., Cabbat F.S., and Duvoisin
R.C.(1984b) Protection against the dopaminergic
neurotoxicity of 1-methyl-4-phenyl-1,2,5,6-
tetrahydropyridine by monoamine oxidase inhibitors. Nature
311,467-469.

Hotchkiss A.J. and Gibb J.W.(1980) Long-term effects of
multiple doses of methamphetamine on tryptophan hydroxylase
and tyrosine hydroxylase activity in rat brain.
J.Pharmacol.Exp.Ther. 214,257-262.

Javitch J.A., D'Amato R.J., Strittmatter S.M., and Snyder
S.H.(1985) Parkinsonism-inducing neurotoxin,
N-methyl-4-phenyl-1,2,3,6-tetrahydropyridine: Uptake of the
metabolite N-methyl-4-phenylpyridine by dopamine neurons
explains selective toxicity. Proc.Natl.Acad.Sci.USA
82,2173-2177.

Kopin I.J.(1987) MPTP: an industrial chemical and
contaminant of illicit narcotics stimulates a new era in
research on Parkinson's Disease. Environ.Health Perspect.
75,45-51.

Langston J.W.(1989) Mechanisms underlying neuronal
degeneration in Parkinson's Disease: An experimental and
theoretical treatise. Movement Disorders 4,S15-S25.

McMillen B.A.(1983) CNS stimulants: two distinct mechanisms of action for amphetamine-like drugs. Trends Pharmacol.Sci. 4,429-432.

Nicklas W.J., Vyas I., and Heikkila R.E.(1985) Inhibition of NADH-linked oxidation in brain mitochondria by 1-methyl-4-phenyl-pyridine, a metabolite of the neurotoxin, 1-methyl-4-phenyl-1,2,5,6-tetrahydro- pyridine. Life.Sci. 36,2503-2508.

Parenti M., Flauto C., Parati E., Vescovi A., and Groppetti A.(1986) Manganese neurotoxicity: Effects of L-Dopa and pargyline treatments. Brain Res. 367,8-13.

Parenti M., Rusconi L., Cappabianca V., Parati E.A., and Groppetti A.(1988) Role of dopamine in manganese neurotoxicity. Brain Res. 473,236-240.

Perry T.L., Yong V.W., Clavier R.M., Jones K., Wright J.M., Foulks J.G., and Wall R.A.(1985) Partial protection from the dopaminergic neurotoxin N-methyl-4-phenyl-1,2,3,6-tetrahydropyridine by four different antioxidants in the mouse. Neurosci.Lett. 60,109-114.

Ramsay R.R., Salach J.I., and Singer T.P.(1986) Uptake of the neurotoxin 1-methyl-4-phenylpyridine (MPP$^+$) by mitochondria and its relation to the inhibition of the mitochondrial oxidation of NAD$^+$-linked substrates by MPP$^+$. Biochem.Biophys.Res.Commun. 134,743-748.

Schmidt C.J., and Gibb J.W.(1985) Role of the dopamine uptake carrier in the neurochemical response to methamphetamine: Effects of amfonelic acid. Eur.J.Pharmacol. 109,73-80.

Schmidt C.J., Ritter J.K., Sonsalla P.K., Hanson G.R., and Gibb J.W.(1985) Role of dopamine in the neurotoxic effects of methamphetamine. J.Pharmacol.Exp.Ther. 233,539-544.

Schultz W.(1988) MPTP-induced parkinsonism in monkeys: mechanism of action, selectivity and pathophysiology. Gen.Pharmacol. 19,153-161.

Seiden L.S., and Vosmer G.(1984) Formation of 6-hydroxy-dopamine in caudate nucleus of the rat brain after a single dose of methylamphetamine. Pharmacol. Biochem.Behav. 21,29-31.

Sonsalla P.K., Gibb J.W., and Hanson G.R.(1986) Roles of D1 and D2 dopamine receptor subtypes in mediating the methamphetamine-induced changes in monoamine systems. J.Pharmacol.Exp.Ther. 238,932-937.

Sonsalla P.K., and Heikkila R.E.(1988) Neurotoxic effects of 1-methyl-4-phenyl-1,2,3,6-tetrahydropyridine (MPTP) and methamphetamine in several strains of mice. Prog.Neuropsychopharmacol.Biol.Psychiatry 12,345-354.

Sonsalla P.K., Nicklas W.J., and Heikkila R.E.(1989) Role for excitatory amino acids in methamphetamine-induced nigrostriatal dopaminergic toxicity. Science 243,398-400.

Steranka L.R. and Rhind A.W.(1987) Effect of cysteine on the persistent depletion of brain monoamines by amphetamine, p-chloroamphetamine and MPTP. Eur.J.Pharmacol. 133,191-197.

Sundstrom E., Stromberg I., Tsutsumi T., Olson L., and Jonsson G.(1987) Studies on the effect of 1-methyl-4-phenyl-1,2,3,6-tetrahydropyridine (MPTP) on central catecholamine neurons in C57BL/6 mice. Comparison with three other strains of mice. Brain.Res. 405,26-38.

RETROGRADE AXONAL TRANSPORT OF MPTP AFTER UNILATERAL CAUDATE NUCLEUS

INFUSION IN THE MONKEY

Hisamasa Imai, Toshiki Nakamura, Nobuo Miyashita, Katsunori
Nishi and Hirotaro Narabayashi

Department of Neurology, Juntendo University School of
Medicine, Tokyo, Japan

INTRODUCTION

1-Methyl-4-phenyl-1,2,3,6-tetrahydropyridine (MPTP) elicits selective
destruction of nigrostriatal dopamine neurons in human and non-human
primates along with clinical symptoms of Parkinson's disease (Langston et
al., 1983; Burns et al., 1983). Systemically administered MPTP is bio-
transformed into 1-methyl-4-phenylpyridium ion (MPP$^+$), which then enters
dopaminergic neurons via the dopamine uptake system to destroy nigral cells.
Although MPP$^+$ uptake is more prominent in nerve terminals than in nerve
cell bodies, terminal degeneration alone is not enough to lead to the death
of the neuron. Either MPP$^+$ is retrogradely transported to the cell bodies
after being taken up at the nerve terminals, or the dopamine uptake sites
on the cell bodies and their dendritic processes are responsible for the
toxin directly entering the neuron (Langston and Irwin, 1986; Snyder and
D'Amato, 1986).
Perhaps in favor of the latter, Snyder and D'Amato (1986) have
pointed out that the differential sensitivity of norepinephrine and dopamine cell bodies may stem from variations in the catecholamine innervation
of the locus ceruleus and substantia nigra, respectively. The locus
ceruleus is densely innervated by catecholamine terminals that should
accumulate MPP$^+$ and thus protect the locus ceruleus norepinephrine neurons
from neurotoxicity. In contrast, no such protection is available for the
dopaminergic cells of the substantia nigra that receive only a few catecholamine terminals.
In an attempt to clarify this question and to construct a model for
pure hemiparkinsonism, we administered MPTP directly into the unilateral
caudate nucleus of crab-eating monkeys via an Alzet osmotic minipump (Imai
et al., 1988).

MATERIALS AND METHODS

Adult male crab-eating monkeys (Macaca fascicularis) weighing 3-5 kg
were used for this experiment. Under anesthesia by intramuscular injection
of ketamine (30 mg/kg), each monkey was placed in a stereotaxic frame, and
stereotaxic ventriculography was done to improve the accuracy of cannula
placement. A stainless-steel, L-shaped cannula (0.6- or 0.8-mm outer
diameter) was implanted in the central part of the head of the unilateral
caudate nucleus. The cannula was connected by polyvinyl tubing to an Alzet

The Basal Ganglia III, Edited by G. Bernardi *et al.*
Plenum Press, New York

Table 1. Numbers of Monkeys Used for
Infusion Sites / Drug Concentractions

Infusion Site	MPTP HCl (mg)			
	4	1	0.4	0
Caudate Nucleus	5 $\begin{pmatrix} H & 5 \\ THH2 \end{pmatrix}$	1 (THC1)	5 $\begin{pmatrix} H & 3 \\ THC1 \end{pmatrix}$	1 (H 1)
Thalamus	2 (H 1)			

H: Histology
THH: TH Immunohistochemistry
THC: TH Chemistry

200-µl osmotic minipump that had been filled with 0.4-4 mg of MPTP HCl in physiological saline. After the pump had been wetted by placement in the subcutaneus tissue of the hind neck, it delivered its content for 14 days. The numbers of monkeys used are listed in Table 1. Three monkeys served as controls; one was infused with the vehicle alone in the unilateral caudate nucleus, and the other 2 were infused with 4 mg of MPTP HCl in the unilateral thalamus.

During up to 6 months following minipump implantation, neurological symptoms were checked frequently and recorded on videotape. Apomorphine (0.1-0.2 mg/kg) was administered intramuscularly, and its effects were also monitored by videotaping, especially for quantification of circling.

For histological studies, all monkeys were deeply anesthetized with pentobarbital sodium (50 mg/kg) intraperitoneally and perfused trans-cardially with heparinized physiological saline followed by perfusion with ice-cold 4% paraformaldelyde in phosphate buffer. The brain was rapidly removed, and placed and kept in the same cold fixative overnight. Frontally sliced blocks were dehydrated and embedded in paraffin, and cut at 4-6 µm. Two sets of tissue were obtained: one set for Nissl and Klüver-Barrera stains, and another set of adjacent sections for tyrosine hydroxylase (TH) immunohistochemistry. For demonstration of the immunoreactivity of TH, deparaffinized tissue sections were incubated with a rabbit antiserum against TH purified from bovine adrenal medulla. The sections were then processed by the avidin-biotin method employing diaminobenzidine, and were lightly counterstained with hematoxylin.

For neurochemical analysis, monkeys were deeply anesthetized with pentobarbital sodium given intraperitoneally and decapitated. The brain was removed and frozen rapidly in dry ice, and kept at -70°C in a freezer until used. The brain was sliced frontally into blocks of 5-6-mm thickness at 0°C. The required brain regions including caudate nucleus, nucleus accumbens and putamen were punched out with a special needle. Tissues were homogenized in ice-cold 0.32 M sucrose solution. An aliquot for determination of dopamine concentration was deproteinized and injected into an HPLC with a fluoredetector. TH activity was measured by the method of Nagatsu et al. (1979).

RESULTS AND DISCUSSION

Monkeys Administered 4 mg of MPTP into the Caudate Nucleus

Behavior. Within a week after the start of MPTP infusion, each monkey exhibited a flexed posture and hypokinesia of the contralateral limbs. Reaching for food was generally limited to the uninvolved upper

Fig.1.   Flexed posture of the right upper limb in a monkey after the
         left caudate nucleus infusion of 4 mg MPTP. The monkey was
         reluctant to use this limb, so he took food with the left hand.

limb (Fig.1). And spontaneous mild circling toward the MPTP-treated side
was frequently observed. After treatment with apomorphine a striking, dose-
dependent reversal of the circling motion occurred. These behavioral disturbances continued to increase for 3 months and then reached a plateau. No
parkinsonian tremor at test was observed in any of the monkeys. In spite
of the hemiparkinsonian motor impairments, all the monkeys were able to
feed themselves and maintained good general health without antiparkinsonian
medication throughout the observation period.

    Histology and Chemistry. The MPTP infusion site was initially
confirmed to be in the center of the head of the caudate nucleus.   On the
control side, homogeneous TH immunoreactivity, indicating the dopamine
nerve terminals, was seen throughout the striatum. TH staining of the
MPTP-treated side of the striatum, however, showed that there was almost
no immunoreactivity in the caudate nucleus and dorsal putamen, but moderate to nearly normal immunoreactivity in the region from the ventromedial
putamen to the nucleus accumbens. In the midbrain, the number of TH-
immunoreactive neurons in the substantia nigra (SN) of the treated side was
markedly reduced by approximately 90% along the entire rostrocaudal and
dorsoventral extent relative to that of the control side. On the other
hand, TH-immunoreactive neurons and fibers in the ventral tegmental area of
Tsai (VTA) on the treated side did not show any marked change. Both TH
activity and dopamine concentration in the caudate nucleus and putamen on
the MPTP-treated side were also reduced by approximately 70-95% relative to
those of the untreated side, but they were preserved in the nucleus accumbens, parallel to the histochemical findings.

    Discussion. The above-mentioned findings clearly showed that hemi-
parkinsonism was produced in the monkey after unilateral caudate nucleus
infusion of 4 mg of MPTP HCl. MPP$^+$ uptake at the dopamine nerve terminals
alone and then retrograde axonal transport to the cell bodies seems to be
sufficient to destroy nigral dopamine neurons in the primate. Why did a
diffuse nigral cell loss occur after a local intracaudate infusion?   Area
of intrastriatal diffusion of MPTP would depend on the amount of the toxin
infused.

Monkeys Administered 0.4 mg of MPTP into the Caudate Nucleus

Behavior. No clear hemiparkinsonism occurred in this series; no flexed posture of the contralateral upper limb, no hypokinesia of the limb and no spontaneous circling were observed. After treatment with apomorphine, however, a mild circling away from the MPTP-treated side, the same direction as that of 4 mg of MPTP, appeared.

Histology and chemistry. Both TH activity and dopamine concentration in the caudate nucleus on the drug-treated side were markedly reduced, but in the putamen they were reduced only near the infusion site. Histologically in the midbrain the SN on the treated side showed a partial cell loss, mainly rostrally, and clusters of spared cells. This histological finding supports the data of the SN afferents to the caudate nucleus and putamen in primate by Parent et al. (1983).

CONCLUSION

Persistent hemiparkinsonism was produced in the monkey after unilateral caudate nucleus infusion of MPTP. MPP^+ uptake at the dopamine nerve terminals alone and then retrograde axonal transport to the cell bodies seemed to be sufficient to destroy nigral dopamine cells in the monkey. Infusion of 4 mg of MPTP into the caudate nucleus produced an almost complete loss of dopamine nerve terminals in the caudate nucleus and dorsal putamen and an almost total dopaminergic cell loss in the ipsilateral SN. Dopamine nerve terminals in the nucleus accumbens and ventromedial putamen, however, remained fairly preserved, as did dopaminergic cells in the ipsilateral VTA. Infusion of 0.4 mg of MPTP into the caudate nucleus produced a partial nigral cell loss, mainly rostrally, and cluster of spared cells.

REFERENCES

Burns, R.S., Chiueh, C.C., Markey, S.P., Ebert, M.H., Jacobowitz, D.M. and Kopin, I.J., 1983, A primate model of parkinsonism: selective destruction of dopaminergic neurons in the pars compacta of the substantia nigra by N-methyl-4-phenyl-1,2,3,6-tetrahydropyridine, Proc. Natl. Acad. Sci. U.S.A., 80:4546.
Imai, H., Nakamura, T., Endo, K. and Narabayashi, H., 1988, Hemiparkinsonism in monkeys after unilateral caudate nucleus infusion of 1-methyl-4-phenyl-1,2,3,6-tetrahydropyridine (MPTP): Behavior and histology, Brain Res., 474:327.
Langston, J.W. and Irwin, I., 1986, MPTP: current concepts and controversies, Clin. Neuropharmacol., 9:485.
Langston, J.W., Ballard, P.A., Tetrud, J.W. and Irwin, I., 1983, Chronic parkinsonism in humans due to product of meperidine-analogue synthesis, Science, 219:979.
Nagatsu, T., Oka, K. and Kato, T., 1979, Highly sensitive assay for tyrosine hydroxylase activity by high performance liquid chromatography, J. Chromatogr., 163:247.
Parent, A., Mackey, A. and De Bellefeuille, L., 1983, The subcortical afferents to caudate nucleus and putamen in primate: a fluorescence retrograde double labeling study, Neuroscience, 10:1137.
Snyder, S.H. and D'Amato, R.J., 1986, MPTP: a neurotoxin relevant to the pathophysiology of Parkinson's disease, Neurology, 36:250.

CHRONIC ADMINISTRATION OF 1-METHYL-4-PHENYL-1,2,3,6-TETRAHYDROPYRIDINE TO MARMOSETS

Alberto Albanese, Maria Concetta Altavista, Carlo Colosimo
Anna Rita Bentivoglio, Paola Rossi, Giorgio Macchi

Istituto di Neurologia
Università Cattolica del Sacro Cuore
I-00168 Roma (Italy)

INTRODUCTION

It is well established that the administration of 1-methyl-4-phenyl-1,2,3,6-tetrahydropyridine (MPTP) to human beings or to non-human primates brings about a parkinsonian syndrome which closely resembles idiopathic Parkinson's disease (Snyder and D'Amato, 1986). With few noticeable exceptions (Langston and Ballard 1983), the human cases of accidental exposure to MPTP were caused by acute administration of this toxin (i.e., a single or few repeated injections; Langston, 1987). Similarly, in order to obtain experimental animals which are clinically affected, monkeys are usually injected with repeated doses of MPTP over 2 to 5 days. Regimen varies according to the species. In marmosets, 4-5 daily injections of 1-4 mg/kg/day i.p. are effective so as to obtain animals which are obviously parkinsonian, but who can maintain themselves (Jenner et al., 1984). The resemblance of MPTP-induced parkinsonism to Parkinson's disease is still an unsolved issue (Albanese 1989). Recently, Kish et al. (1988) found that, in patients affected by Parkinson's disease, dopamine loss is more severe in the putamen than in the caudate nucleus, while in monkeys poisoned acutely with MPTP dopamine loss is higher in the caudate nucleus than in the putamen.

Based on the consideration that Parkinson's disease is a chronic and slowly progressive clinical condition, which is associated with a degeneration of dopaminergic neurons, the present study was undertaken in order to evaluate the effect of a chronic administration of low doses of MPTP. The question as to whether a chronic regimen of MPTP administration is capable to induce the behavioural and pathological signs of parkinsonism is relevant for a better understanding of the putative pathogenetic role played by environmental toxins in the human pathology (Barbeau et al., 1986; Rajput et al., 1986). Indeed, if environmental pyridines do really play a role in the natural human disease, then they must be able to produce pathological and biochemical alterations by the additive effect of low doses over time. Aim of the present paper is to study the behavioural, morphological and biochemical effects brought about by the administration of MPTP to marmosets according to a chronic regimen.

The Basal Ganglia III, Edited by G. Bernardi *et al.*
Plenum Press, New York

TABLE 1.   Global Disability Scale for Marmosets

Clinical condition	Score
Normal	0
Slowing of movements, with normal balance and posture	1
Abnormal posture of the limbs, trunk or tail; abnormal balance when pushed or pulled	2
Very slow, unable to keep balance on a perch	3
Immobile, bound to the floor	4

MATERIALS AND METHODS

Thirty common marmosets (*Callithrix jaccus*) of either sex were used in the study. The animals were housed in couples; the cages were kept at 25° C with a natural daylight cycle.

Dose-response study

Twenty animals were divided in five experimental groups, each one consisting of 4 sex- and age-matched individuals. Sixteen monkeys were dosed i.p. twice a week with MPTP: **group A** received 0.25 mg/kg of the toxin in each injection; **group** B received 0.50 mg/kg; **group C** received 0.75 mg/kg; **group D** received 1.25 mg/kg. Four monkeys, belonging to **group E,** were injected twice a week with vehicle only (saline solution, 0.5 ml, i.p.). Assessment of parkinsonian disability was performed before starting the experiment, and at weekly intervals thereafter, by means of a disability scale for MPTP-treated marmosets (Jenner and Marsden, 1985; Table 1). In addition, the animal behaviour was observed for at least one hour every day and after each injection.

Three months after treatment, the animals were killed by decapitation, under deep general anaesthesia with a mixture of alphaxalone and alphadolone acetate (*Saffan*, 15 mg/kg). After extraction, the brain stem was fixed by immersion in 4% phosphate buffered paraformaldehyde for approximately 2 weeks, then transferred to 30% phosphate buffered sucrose for 3 days, before being processed for morphological analysis. Coronal sections were cut at 30 µm intervals, by means of a freezing microtome. Alternate adjacent sections were stained by means of cresyl violet or processed for tyrosine hydroxylase (TH) immunohistochemical staining, according to the following procedure. All antisera were diluted in phosphate buffered saline (PBS) containing 0.1% Triton X-100, 3% goat serum, and 1% (w/v) bovine serum albumin. Sections were treated for 30 min. in PBS containing 0.2% (v/v) hydrogen peroxide followed by repeated wash in PBS. Incubation in TH antisera was overnight at room temperature. Sections were then washed in PBS, incubated with goat anti-rabbit IgG serum (1:20), washed again and then incubated for 1 h at room temperature with rabbit peroxidase-antiperoxidase antiserum (Miles) diluted 1:100. After washing, the tissue sections were then exposed to the chromogen solution (1,3-diaminobenzidine, 0.05% w/v) in PBS containing 0.02% (v/v) hydrogen peroxide as substrate for approximately 10 min. Sections through the midbrain were studied by means of a microcomputer based image analyser, blindly, by a single observer. Cell counts were performed in the pars compacta: the relative density of TH-positive cell bodies was computed in medial, intermediate and lateral segments of this nucleus. In addition, the

Table 2. Analytical Rating Scale for Marmosets with Parkinsonian Symptoms

Item	Score
Alertness	0-1
Reaction to stimuli	0-3
Head motility	0-2
Eye motility	0-1
Posture alterations	0-4
Balance impairment	0-2
Motility (akinesia)	0-3
Vocalisation	0-2
Tremor	0-1
Fur condition	0-1
Total disability score	0-20

cross-sectional area of all neurons displaying a well-defined non stained nucleus was measured in the substantia nigra and in the ventral tegmental area. Cell counts of glial cells were performed on Nissl-stained sections. Data obtained from the five experimental groups were compared individually, by means of pooled t two-sample analysis.

Chronic administration and follow-up

Ten marmosets of either sex, aged between 2.5 and 7 years (average 6 years), were used in this experiment. Five animals were treated with 1 mg/kg MPTP, which was dissolved in 0.5 ml saline and was injected i.p., twice a week, consecutively for four months. Five monkeys, used as controls, were injected with vehicle only using the same schedule of administration. The monkeys were observed before each dose of MPTP was administered and for half an hour afterwards. Parkinsonian disability was assessed by means of a rating scale specifically devised for assessing parkinsonism in this animal species (Table 2). In each monkey, ratings were performed before starting the experiment and periodically for twelve months (i.e., for eight months after MPTP administration was discontinued). Each evaluation was performed by two observers, whose scores were averaged. The animals are still under observation.

RESULTS

Dose-response analysis

After chronic administration of different doses of MPTP it was observed that parkinsonian signs occurred in all the experimental groups. In groups C and D all monkeys were obviously affected, whilst some monkeys belonging to groups A and B were rated normal. As shown in figure 1, average disability scores increased with time in all groups; in addition, average scores were always higher in the experimental groups treated with higher doses of MPTP. Acute motor abnormalities, which commonly occur after acute administration of MPTP (Jenner et al., 1986) were not seen during chronic treatment. The earliest behavioural changes consisted in a progressive bradykinesia, which affected monkeys of groups C and D, starting less than one month after the beginning of treatment. Rigidity of the limbs, abnormalities of posture and balance and blepharospasm appeared later in the same experimental groups. Monkeys belonging to group A and B became bradykinetic only by the third month of treatment. Motor

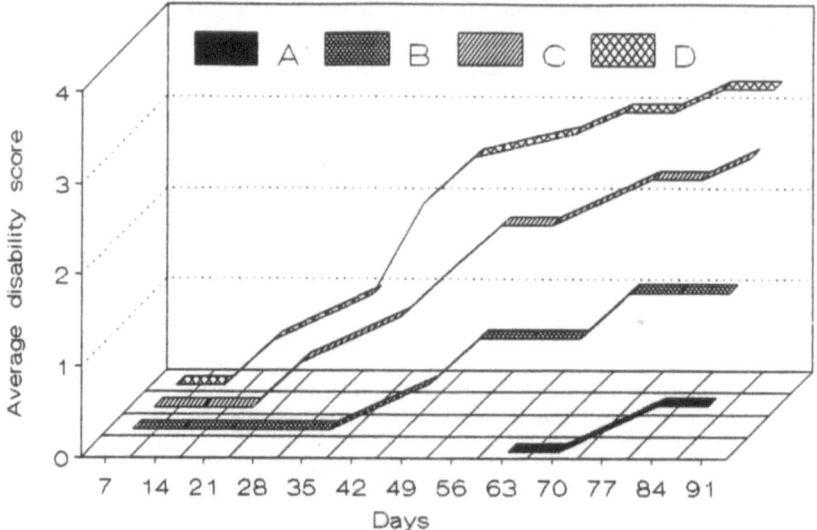

Fig 1. Dose-response curves show that marmosets treated higher doses of MPTP show more severe behavioural alterations. This experiment shows that a clear parkinsonian picture is observed in groups C and D after three months of MPTP administration.

symptoms were never as severe as those observed following an acute sublethal treatment. All the animals were able to feed themselves and to take care of themselves. After three months of poisoning, group C and D animals were grossly parkinsonian: they were all bradykinetic; most of them suffered from dystonia. Tremor only occurred in animals belonging to group D. Most monkeys of group B only showed a mild bradykinesia and some additional abnormalities, while most monkeys of group A were rated normal.

In all monkeys, histological analysis of the brain stem allowed to observe TH-containing neurons located in the substantia nigra and in the ventral tegmental area. In MPTP-treated monkeys, both regions were partially depleted of TH-stained perikarya. Morphometric analysis of the substantia nigra showed that, as compared to controls, the number of TH-stained neurons located in the substantia nigra was reduced in MPTP treated monkeys; however, no linear correlation could be found between morphometric data and the regimen of MPTP administration (Fig. 2). No differences were observed when the medial, intermediate and lateral segments of pars compacta were compared. Average cell density in control animals was 491.87 cells/μm^2; in group A it decreased by 40.16% to 294.35 cells/μm^2; in group B it was 263.18 cells/μm^2 (-46.49%); in group C it reached 256.29 cells/μm^2 (-47.90%); and in group D it was 283.88 cells/μm^2 (-42.29%). Morphometric analysis also showed that the size of TH-stained neurons located in the substantia nigra and in the ventral tegmental area was not affected by MPTP treatment (Fig. 3). In fact, between-group comparison showed that the cross-sectional area did not differ significantly among groups A, B, C, D and controls. Finally, when the number of glial nuclei located in the pars compacta was compared, no significant differences were found between MPTP-treated and control monkeys.

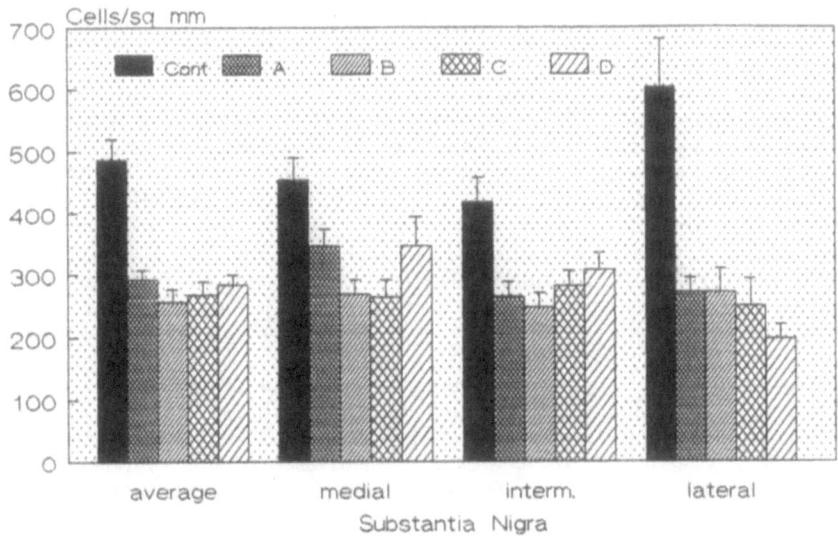

Fig 2. Chronic administration of MPTP brings about a reduction in the number of TH-stained neurons (mean ± SEM) which are observed in the
pars compacta of the substantia nigra. In monkeys belonging to
group D, cell depletion is particularly severe in the lateral part of
the nucleus.

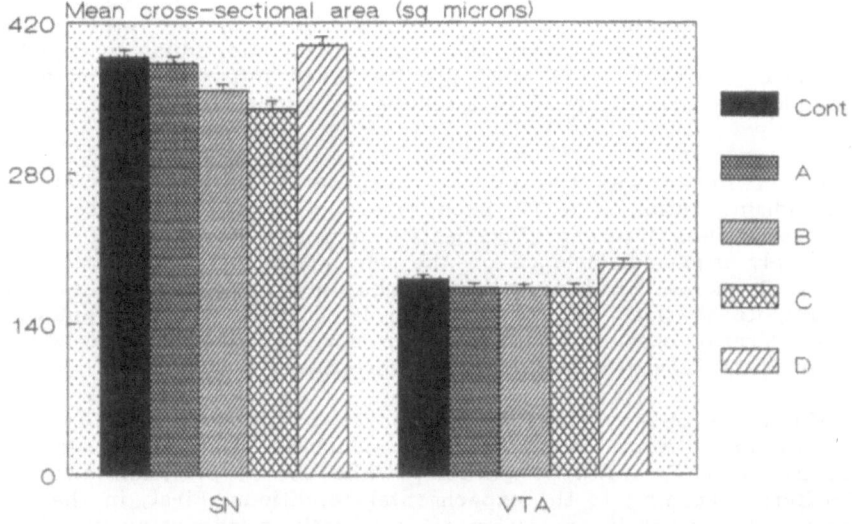

Fig 3. Chronic MPTP treatment did not affect the size (mean ± SEM) of
TH-stained neurons located in the pars compacta of substantia nigra
(SN) or in the ventral tegmental area (VTA).

Chronic administration and follow-up

The present study is in keeping with earlier observations reporting that chronic treatment with MPTP (1 mg/kg, twice a week) brings about a parkinsonism, that progressively worsens when subsequent doses of the toxin are administered (Albanese et al. 1990). The average of total disability scores showed that a marked and progressive worsening of symptoms occurred during the first month of treatment, and that clinical disability progressed at a much slower rate from the third month of treatment. Acute toxic effects related to the administration of MPTP were never observed. The highest disability average score was reached during the fifteenth week of treatment, shortly before the end of it. After discontinuance of chronic MPTP administration, parkinsonian features gradually recovered for few months and remained stable afterwards (Fig. 4). Behavioural recovery was particularly evident in two out of five monkeys.

The analysis of individual items of the disability scale showed that some scores were more commonly altered than others. Motility of the head and eyes, vocalisation, posture and the ability to react to external stimuli were constantly affected. On the other hand, tremor was constantly absent, and balance, alertness and self-caring (as shown by fur conditions) were always rated normal.

DISCUSSION

The present study shows that when marmosets are chronically exposed to MPTP a progressive parkinsonian syndrome occurs. The low doses, which are required for the chronic regimen, do not produce acute pharmacologic effects, which usually appear when repeated maximal doses of MPTP are administered (Jenner and Marsden, 1986). In addition to behavioural signs, chronic administration of MPTP also produces morphological alterations, which are non-linearly related with the total amount of MPTP given over time. Finally, the existence of individual susceptibilities to the toxic effects of MPTP must also be taken into account, as different clinical pictures are observed in monkeys belonging to a single experimental group.

Bradykinesia was the clinical sign most clearly associated with chronic administration of MPTP. Bradykinesia gradually progressed as toxin administration was pursued. Vocalisation and eye movements were always not severely impaired. Balance on the perch was mildly impaired in some monkeys. Dystonia (including blepharospasm) was also observed occasionally; its severity was clearly not related to the total amount of toxin injected. Tremor was only observed in some animals treated with 1.25 mg/kg of MPTP. Postural abnormalities were never observed.

When our observations are compared to those of monkeys treated with acute or subacute regimens, it is concluded that the basic parkinsonian syndrome is quite comparable, while the severity of each clinical sign differs according to the experimental conditions. First, in the chronic regimen motor disabilities, although gradually progressive, impair motor behaviour without threatening vital functions (e.g., feeding and self-caring). Second, chronic MPTP treatment neither brings about acute toxic reactions nor it is associated with such self-damaging behaviours, which are commonly seen after repeated treatments with high doses of MPTP. As a consequence, higher cumulative doses of MPTP can be given with the chronic regimen. Such high cumulative doses are required in order to

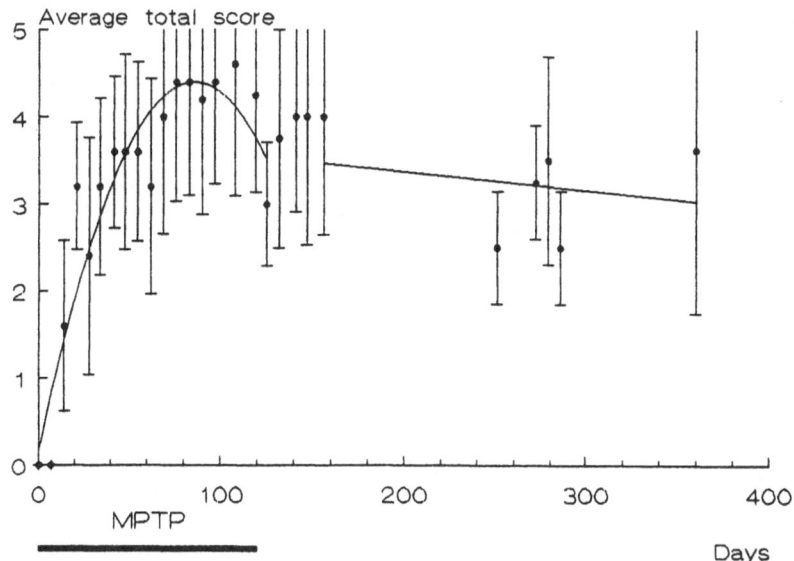

Fig 4. Chronic administration of 1 mg/kg MPTP induced a progressive parkinsonian syndrome. Average total scores (± SEM) are plotted
together with a second order polynomial curve fit. This indicates
that a gradual recovery of clinical signs occurs shortly after discontinuance of the toxin; afterwards, the neurological impairment becomes stable.

produce motor abnormalities comparable to those observed after acute
MPTP administration. Taking all these considerations into account, it can be
observed that, at variance with the syndrome produced by acute or subacute administration of MPTP, in chronically treated monkeys (as well as in
the human idiopathic condition) a slow progression of signs and symptoms,
which are mainly restricted to the motor system, is observed.

The existence of morphological abnormalities in the midbrain of marmosets treated chronically with MPTP is well in keeping with knowledge
that similar alterations are usually found in idiopathic and in MPTP-
induced parkinsonism. The present data show that, after chronic treatment,
loss of TH staining without cell shrinkage may occur in midbrain
dopamine-containing neurons. As shown by Lams et al. (1988), the lack of
staining for a metabolic marker does not necessarily imply that non stained
neurons are not viable; as a matter of fact, the derangement from a normal
metabolic balance is reversible to a large extent. Therefore, it may well be
the case that a loss of TH staining in the midbrain of marmosets treated
chronically with MPTP may be due to biochemical alterations brought about
by uptake (and, possibly, build-up) of toxic compounds into such neurons.
The existence of a metabolic reaction due to a direct effect of toxic compounds is also supported by the lack of a linear dose-response curve in
our immunocytochemical data. As shown in Fig. 2, the density of TH-stained
neurons located in the substantia nigra is significantly lower in treated
monkeys than in controls, but it does not change according to the dose
regimen of MPTP administration. This leads to the fundamental question as
to whether the parkinsonian signs of chronically treated marmosets are re-

lated to the occurrence of cell loss in the substantia nigra or they are a transient pharmacologic effect.

The occurrence of a behavioural recovery after discontinuance of MPTP treatment is in keeping with earlier observations based on acute or subacute administration of MPTP to monkeys (Eidelberg et al. 1986; Ueki et al. 1989). Since recovery of parkinsonian signs is a common phenomenon, the causes of which are still unknown (Ueki et al. 1989), the occurrence of a certain degree of recovery after discontinuance of chronic MPTP administration does not necessarily imply that in this case parkinsonian signs are a transient pharmacologic effect. Furthermore, it must be considered that, after few months of rapid recovery, parkinsonian signs remained stable for the following three months.

The morphological and biochemical study of these marmosets, which were treated chronically with MPTP for four months and which were followed-up for eight months, is expected to solve some still unanswered questions. First, the study of dopamine and of its metabolites contained in the neostriatum will allow to define whether in the chronic model dopamine depletion is more severe in the caudate nucleus or in the putamen. Second, data obtained from the morphological analysis of midbrain dopaminergic neurons will be compared to those obtained in the dose-response study, in which marmosets were killed soon after MPTP administration was discontinued. This will probably allow to clarify to what extent a damage produced by a chronic administration of MPTP for four months may be reversed by discontinuing the toxin.

ACKNOWLEDGEMENTS

This work was supported in part by CNR (Consiglio Nazionale delle Ricerche) grant 88.00416.04. The analytical scale for marmosets described in table 2 has been worked out by a team of researchers (including A.A.) visiting P. Jenner and C.D. Marsden at the Institute of Psychiatry.

REFERENCES

Albanese A., in press, Extrapyramidal system, motor ganglia and movement disorders, *Reviews in the Neurosciences.*

Albanese A., Altavista M.C., Gozzo S., Rossi P., Colosimo C., Bentivoglio A.R., Perretta G., Elia M., Monaco V., Macchi G., in press, Chronic administration of MPTP to marmosets, *Adv. Neurol.*

Barbeau A., Roy M., Cloutier T., Plasse L., Paris S, 1986, Environmental and genetic factors in the etiology of Parkinson's disease, *in:* "Parkinson's disease", M. Yahr, K.J. Bergmann, eds., Raven, New York, *Adv. Neurol.* 45: 299.

Eidelberg E., Brooks B.A., Morgan W.W., Walden J.G., Kokemoor R.H., 1986, Variability and functional recovery in the N-methyl-4-phenyl-1,2,3,6-tetrahydropyridine model of parkinsonism in monkeys, *Neuroscience* 18: 817.

Jenner P., Marsden C.D., 1986, The actions of 1-methyl-4-phenyl-1,2,3,6-tetrahydropyridine in animals as a model of Parkinson's disease, *J. Neural Transm. Suppl.* 20: 11.

Jenner P., Rupniak N.M.J., Rose S., Kelly E., Kilpatrik G., Lees A., Marsden C.D., 1984, 1-methyl-4-phenyl-1,2,3,6-tetrahydropyridine-induced parkinsonism in the common marmoset, *Neurosci Lett.* 50: 85.

Jenner P., Rose S., Nomoto M., Marsden C.D., 1986, MPTP-induced parkinsonism in the common marmoset: behavioral and biochemical effects, *in:* "Parkinson's disease", M.D. Yahr, K.J. Bergmann, eds., Raven, New York, *Adv. Neurol.* 45: 183.

Kish S.J., Shannak K., Hornykiewicz O., 1988, Uneven pattern of dopamine loss in the striatum of patients with idiopathic Parkinson's disease, *N. Engl. J. Med.* 318: 876.

Lams B.E., Isacson O., Sofroniew M.V., 1988, Loss of transmitter-associated enzyme staining following axotomy does not indicate death of brainstem cholinergic neurons, *Brain Res.* 475: 401.

Langston J.W., 1987, MPTP: the promise of a new neurotoxin, *in*: "Movement Disorders 2", C.D. Marsden, S. Fahn, eds., Butterworths, London.

Langston J.W., Ballard P.A., 1983, Parkinson's disease in a chemist working with 1-methyl-4-phenyl-1,2,3,6-tetrahydropyridine, *New Engl. J. Med.* 4: 310.

Rajput A.H., Uitti R.J., Stern W., Laverty W., 1986, Early onset Parkinson's disease and the childhood environment, *in*: "Parkinson's disease", M. Yahr, K.J. Bergmann, eds., Raven, New York, *Adv. Neurol.*, 45: 295.

Snyder S.H., D'Amato R.J., 1986, MPTP: a neurotoxin relevant to the pathophysiology of Parkinson's disease, *Neurology* 36:250.

L-DOPA-INDUCED CHOREA AND DYSTONIA IN MPTP-TREATED SQUIRREL MONKEYS

N.M.J. Rupniak, S. Boyce, M.J. Steventon and S.D. Iversen

Merck Sharp and Dohme Research Laboratories, Neuroscience
Research Centre, Terlings Park, Eastwick Road, Harlow
Essex, CM20 2QR, U.K.

Administration of L-DOPA to Parkinsonian patients frequently
induces a range of disabling involuntary movements of which chorea
and dystonia are the most common. Choreiform movements are often
associated with peak plasma levels of drug ('peak-dose dyskinesias';
Markham, 1974) suggesting an association with the antiparkinsonian
action of L-DOPA. In contrast, dystonias are less predictable,
being prevalent not only during the time of peak effect but also
while plasma L-DOPA levels are rising and falling ('diphasic
dyskinesias'; Barbeau, 1975; Muenter et al, 1977).

Dyskinesias represent a major complication of treatment with
antiparkinsonian drugs. The possibility that different underlying
mechanisms are responsible for chorea and dystonia raises the
possibility of development of drugs with reduced propensity to
induce dyskinesias and to develop novel antidyskinetic agents which
do not compromise beneficial effects of existing antiparkinsonian
medication.

Research into mechanisms responsible for drug-induced
dyskinesias has been made possible by the discovery of the
MPTP-induced parkinsonism model in experimental primates. In Old
World primates, such as macaques, MPTP induces a syndrome which
closely resembles idiopathic Parkinson's disease regarding
neuropathology, symptomatology and susceptibility to L-DOPA-induced
dyskinesias (Bedard et al, 1986; Clarke et al, 1987). Parkinsonism
has also been successfully induced in two New World species using
MPTP which are lower than macaques on the phylogenetic scale:
squirrel monkeys (Langston et al, 1984) and marmosets (Jenner et al,
1986). These species offer certain advantages over larger primates
because they can be housed in relatively large numbers and can be
handled more easily (for example, for the purpose of oral drug
administration). However, considerable differences in brain
specialisation and motor control exist between higher and lower
order primates (see Passingham, 1981). The most useful animal model
of Parkinson's disease for preclinical studies would be one which
responded to antiparkinsonian therapy in a similar manner to man.
Marmosets have become a popular species for research in recent
years, but there are no reports that marmosets exhibit dyskinesias
in response to L-DOPA or any other drug treatments. Whilst

The Basal Ganglia III, Edited by G. Bernardi *et al.*
Plenum Press, New York

dyskinesias in parkinsonian squirrel monkeys have not yet been
documented, this species has been widely employed in research into
neuroleptic-induced and idiopathic dyskinesias (Murphey and Dill,
1972; Neale et al, 1982; Neale et al, 1984). Thus, unlike rodents
or marmosets, there is good evidence that squirrel monkeys possess
the appropriate functional neuroanatomy required to exhibit similar
drug-induced motor syndromes to those seen in man. Recently we have
characterised in detail the dyskinesias induced by L-DOPA in
parkinsonian squirrel monkeys and have found a striking similarity
with those observed in Parkinson's disease.

   In 1987 we induced parkinsonism in seven adult male squirrel
monkeys (Saimiri sciureus) using once weekly injections of MPTP (2mg
i.p.). Treatment with MPTP was terminated once animals exhibited at
least 3 of the 4 following signs consistently throughout a 2-week
period: tremor, akinesia, bradykinesia and rigidity. The monkeys
showed considerable differences in individual susceptibility to
MPTP, requiring between 6 and 45mg to reach this end-point. Some
two years later the animals had undergone spontaneous recovery from
the acute effects of MPTP but remained akinetic, hunched and
bradykinetic by comparison with normal monkeys (stage I-II on the
Hoehn and Yahr scale, 1967). Whilst it has been common to employ
more aggressive treatment regimes to induce severe parkinsonism (eg
4mg/kg for 4 consecutive days), this procedure is associated with a
high mortality rate (Langston et al, 1984) and is neither suitable
nor necessary for long-term behavioural investigations in this
species. Prior to the experiments described here we had repeatedly
challenged monkeys with L-DOPA (20mg/kg p.o.) in order to establish
susceptibility to develop dyskinesias. Cumulative doses of up to
1.5g of L-DOPA were administered during this phase.

   For the purpose of characterisation of the behavioural effects
of L-DOPA, locomotor activity and dyskinesias were scored by direct
continuous visual observation for up to 3h. Dyskinesias were
recorded as chorea or dystonia and were scored separately in each
limb using a clinically-based rating scale of 0-4 depending on
frequency and severity (Guy, 1976). Chorea was defined as random
and chaotic flexion and extensions of the limbs. Chorea was often
superimposed upon more repetitive, stereotypic dyskinesias and could
not be scored independently from these otherwise identical
movements. Dystonia was characterised by brief or sustained
repetitive abnormal posturing associated with flexion or extension
of the limbs. L-DOPA was suspended in 0.5% methyl cellulose and
administered by oral gavage in the dose range 5-40mg/kg 60 min after
treatment with carbidopa in a fixed dose ratio of 1:2mg/kg p.o.

   Spontaneous locomotor activity in the parkinsonian squirrel
monkeys was reduced to around 30% of levels seen in normal animals.
Acute administration of L-DOPA at doses of 5mg/kg p.o. or higher
stimulated locomotor activity by comparison with the effects of
vehicle treatment (Figure 1A). Locomotor activity was
dose-dependently increased by higher doses up to 15mg/kg p.o. This
peak was not exceeded by administration of 20, 30 or 40mg/kg. The
reason for this plateau was the emergence of competing behaviours,
especially climbing (see below).

   Examination of the time course for motor stimulation by L-DOPA
revealed an onset after around 15min using doses of 5-15mg/kg, and a
peak between 60 and 110 min after treatment. At higher doses
(20-40mg/kg), the time course for locomotor activation was more
complex, first peaking between 35 and 100 min and again for a second

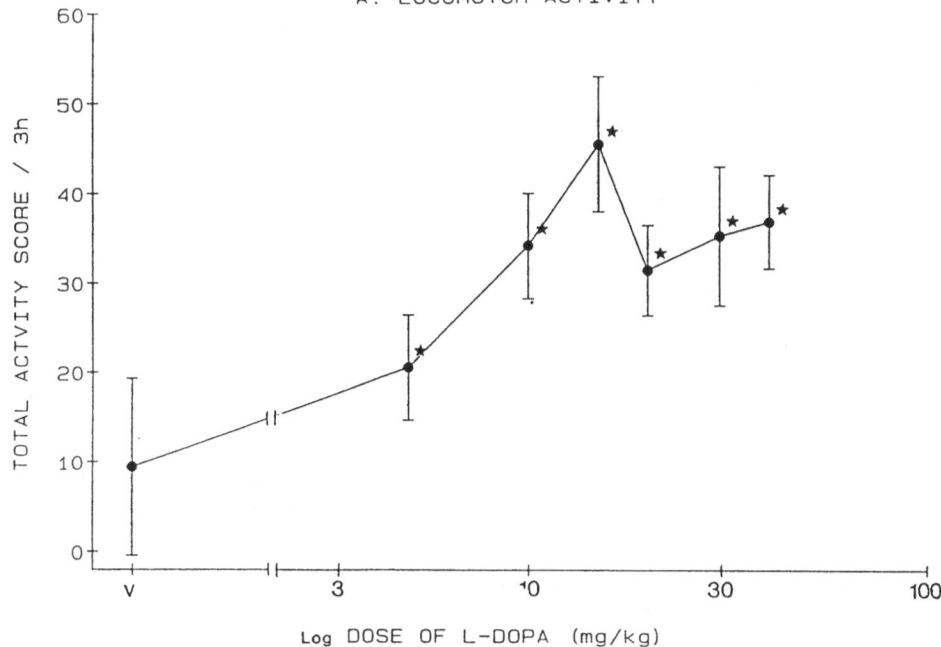

A. LOCOMOTOR ACTIVITY

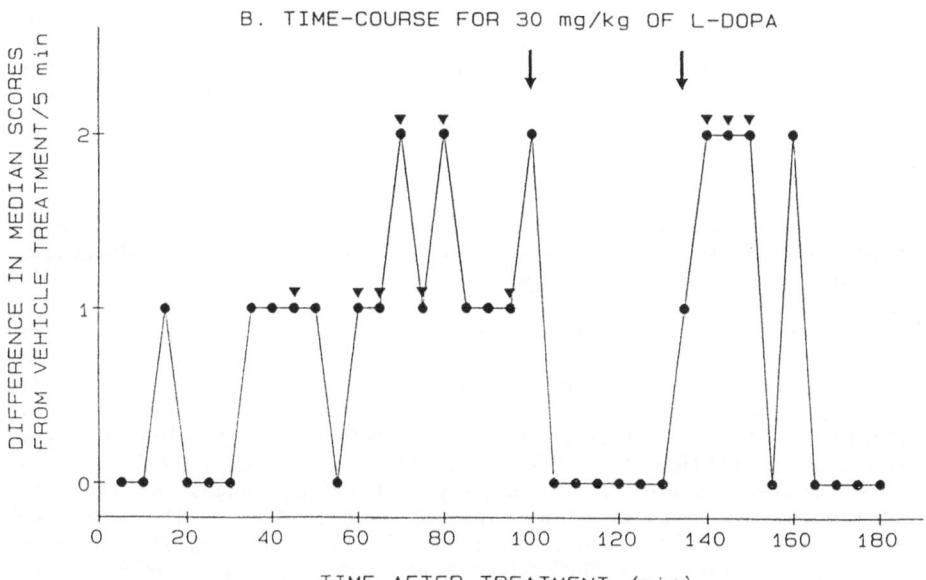

B. TIME-COURSE FOR 30 mg/kg OF L-DOPA

Figure 1. Locomotor activity induced by L-DOPA. (A) Dose-response curve for activity following oral administration of L-DOPA (5 to 40mg/kg) and (B) time-course data for 30mg/kg of L-DOPA. Dose-response curve data are expressed as the mean ($\pm$ 1 SEM) and time-course data as the median score obtained from 7 animals.
   * $p < 0.05$ compared to vehicle treatment, Dunnett's multiple comparison test.
   ▼ $p < 0.05$ compared to vehicle treatment at each time point, Wilcoxon matched-pair ranked-sign test.
Arrows denote the emergence of a compulsive climbing response.

time between 135 and 150 min (Figure 1B). The apparent reduction in locomotor activity between 100 and 135 min was associated with emergence of compulsive climbing. Animals would then jump onto the grid cage front and hang motionless at the top of the cage in a manner reminiscent of climbing induced by apomorphine in rodents (Protais et al, 1976).

In addition to stimulating locomotion, doses of 10mg/kg or higher of L-DOPA induced dyskinesias of mild to severe intensity and frequency, depending on dose. Dyskinesias were predominantly choreic in nature affecting both the arms and the legs, but not the oral or facial region. Chorea was generally more severe in the legs than in the arms. Choreiform movements were never observed spontaneously in parkinsonian squirrel monkeys or following treatment with carbidopa alone. The dose-response curves for chorea in the arms and legs were bell-shaped (Figure 2A). Administration of 10mg/kg of L-DOPA induced mild, intermittent choreiform movements. At higher doses (15, 20 and 30mg/kg p.o.) chorea became severe and virtually continuous, characterised by chaotic alternating flexion and extensions of the limbs, with adduction and abduction of the shoulder and hip and circling and twisting movements of the hands and feet. The time of peak effect for chorea was between 60 and 140min after treatment, coinciding with the time of peak locomotor activity.

L-DOPA-induced dyskinesias in squirrel monkeys also had a dystonic component. Dystonic-like postures were not as frequent or as severe as chorea and were almost exclusively confined to the legs; only one animal developed dystonia of the arm (Figure 2B). In the legs, dystonia consisted of either sustained flexion of the hip and knee with plantar-flexion of the ankle and eversion of the foot, accompanied by abduction of the hip, or extension of the hip and knee, with plantar-flexion of the ankle. Dystonic postures were repeated periodically typically affecting a particular limb, and were maintained briefly (around 5 sec). Unlike chorea, dystonias were observed at the time of peak effect (peak-dose dystonia) and as drug effects were diminishing (end-of-dose dystonia). Moreover, dystonia was occasionally observed spontaneously.

The selective distribution of dyskinesias in the limbs rather than the orofacial region probably reflects the age and severity of parkinsonism in our animals. Young onset parkinsonian patients also develop dyskinesias restricted to the extremities, unlike older patients whose dyskinesias are more widespread, including the head and oral parts (Gerlach, 1977). Dyskinesias typically affect those limbs or regions most affected by parkinsonism, suggesting that the spread and distribution of dyskinesias is dependent on the magnitude and location of the nigrostriatal lesion. In addition, the duration of prior exposure to L-DOPA may influence dyskinesias. MPTP-treated macaques, whose parkinsonism was assumed not to have progressed, gradually developed facial as well as limb dyskinesias during chronic L-DOPA treatment (Clarke et al, 1987; Boyce et al, 1990a).

As might be predicted from previous studies using neuroleptic drugs, we have found that parkinsonian squirrel monkeys develop similar choreic and dystonic movements to those described in man using L-DOPA. We have continued to explore this model in order to elucidate the role of D_1 and D_2 receptors in antiparkinsonian and dyskinetic activity (Boyce et al, 1990b) and to demonstrate a selective antidyskinetic effect of CCK-8S (Boyce et al, 1990c). The availability of such close animal models is unique for Parkinson's

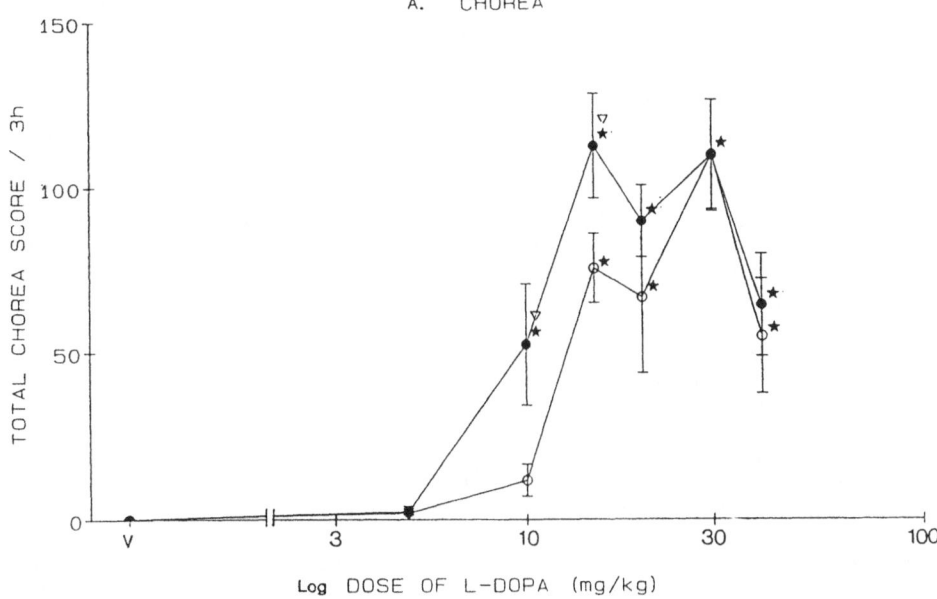

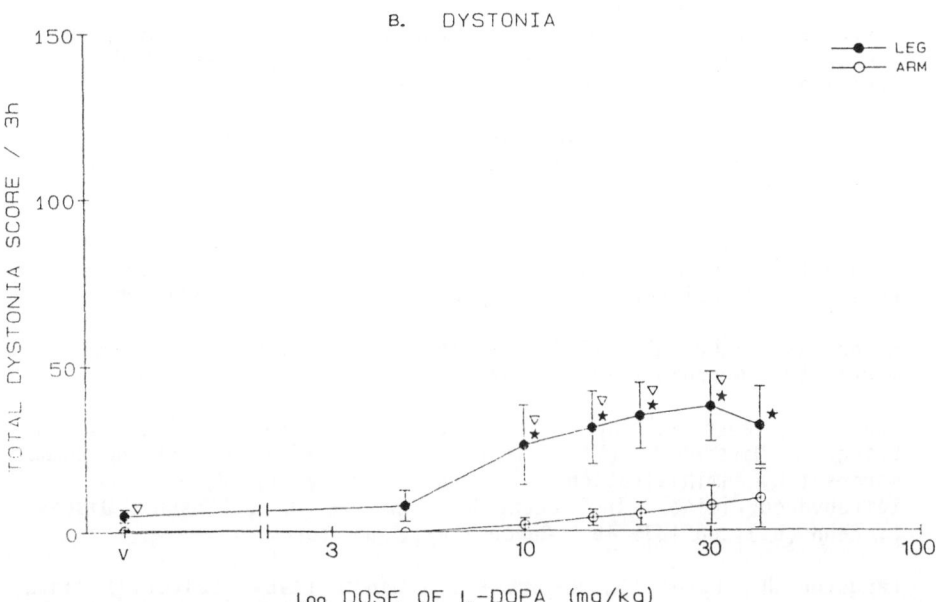

Figure 2. Dose-response curves for (A) chorea and (B) dystonia in MPTP-treated squirrel monkeys following oral administration of L-DOPA (5 to 40mg/kg). Values are expressed as the mean (± 1 SEM) obtained from 7 animals.

* p < 0.05 compared to vehicle treatment, Dunnett's multiple comparison test.

▽ p < 0.05 compared to equivalent dose in the arms, paired Student's t-test.

disease and will greatly increase the power of preclinical screening for novel therapeutic agents.

References

Barbeau A (1975) Diphasic dyskinesias during levodopa therapy. Lancet i: 756.

Bedard PJ, Di Paolo T, Falardeau P, Boucher R (1986) Chronic treatment with L-DOPA, but not bromocriptine induces dyskinesia in MPTP-parkinsonian monkeys. Correlation with ^{3}H-spiperone binding. Brain Res. 379: 294-299.

Boyce S, Clarke CE, Luquin R, Peggs D, Farmery SM, Robertson RG, Mitchell IJ, Sambrook MA, Crossman AR (1990a) Induction of chorea and dystonia in parkinsonian primates. Movement Disorders in press.

Boyce S, Rupniak NMJ, Steventon MJ, Iversen SD (1990b) Differential effects of D_1 and D_2 agonists in MPTP-treated primates: functional implications for Parkinson's disease. Neurology in press.

Boyce S, Rupniak NMJ, Steventon M, Iversen SD (1990c) CCK-8S inhibits L-DOPA-induced dyskinesias in parkinsonian squirrel monkeys. Neurology in press.

Clarke CE, Sambrook MA, Mitchell IJ, Crossman AR (1987) Levodopa-induced dyskinesia and response fluctuations in primates rendered parkinsonian with 1-methyl-4-phenyl-1,2,3,6-tetrahydro-pyridine (MPTP). J. Neurol. Sci. 78: 273-280.

Gerlach J (1977) Relationship between tardive dyskinesia, L-DOPA-induced hyperkinesia and parkinsonism. Psychopharmacology 51: 259-263.

Guy W (1976) Psychopharmacology Research Branch NIMH. Abnormal involuntary movement scale (AIMS) p 534-537 In: Guy W (ed) ECDEU Assessment Manual for psychopharmacology. DHEW, Rockville MD.

Hoehn NM, Yahr MD (1967) Parkinsonism: onset, progression and mortality. Neurology 17: 427-442.

Jenner P, Rose S, Boyce S, Kelly E, Kilpatrick G, Rupniak NMJ, Briggs R, Marsden CD (1986) Induction of parkinsonism in the common marmoset by administration of 1-methyl-4-phenyl-1,2,3,6-tetrahydropyridine. In: Recent Developments in Parkinson's Disease, S. Fahn (ed), pp 137-146. Raven Press, New York.

Langston JW, Forno LS, Rebert CS, Irwin I (1984) Selective nigral toxicity after systemic administration of 1-methyl-4-phenyl-1,2,3,6-tetrahydropyridine (MPTP) in the squirrel monkey. Brain Res. 292: 390-394.

Markham CH (1974) The choreoathetoid movement disorder induced by levodopa. Clin. Pharmacol. Therap. 12: 340-343.

Muenter MD, Sharpless TS, Tyce GM, Darley FL (1977) Patterns of dystonia ("I-D-I" and "D-I-D") in response to L-DOPA therapy for Parkinson's disease. Mayo Clin. Proc. 52: 163-174.

Murphey DL, Dill RE (1972) Chemical stimulation of discrete brain loci as a method of producing dyskinesia models in primates. Exp. Neurol. 34: 244-254.

Neale R, Gerhardt S, Fallon S, Liebman JM (1982) Progressive changes in the acute dyskinetic syndrome as a function of repeated elicitation in squirrel monkeys. Psychopharmacology 77: 223-228.

Neale R, Gerhardt S, Liebman JM (1984) Effects of dopamine agonists, catecholamine depletors, and cholinergic and GABAergic drugs on acute dyskinesias in squirrel monkeys. Psychopharmacology 82: 20-26.

Passingham RE (1981) Primate specialisation in brain and intelligence. Symp. Zool. Soc. Lond. 46: 361-388.

Protais P, Constentin J, Schwartz JC (1976) Climbing behaviour induced by apomorphine in mice: a simple test for the study of dopamine receptors in striatum. Psychopharmacology 60: 1-6.

ELECTROPHYSIOLOGICAL STUDY OF DYSKINESIA PRODUCED BY MICROINJECTION

OF PICROTOXIN INTO THE STRIATUM OF THE RAT

Shinichi Muramatsu, Mitsuo Yoshida
and Satoshi Nakamura

Department of Neurology, Jichi Medical School
Tochigi, Japan

INTRODUCTION

From the findings in human diseases, such as the main pathological change in Huntington's disease is an atrophy of caudate neurons, striatum has been related to dyskinesia. It was, however, difficult to produce an animal model of dyskinesia by electrostimulation or lesioning of the striatum. In the 1970's, by pharmacological manipulation of the striatum, a model of dyskinesia was reported; that is, injection of picrotoxin (PTX), a selective antagonist of γ-aminobutyric acid (GABA), into the striatum of the rat and cat produced dyskinesia similar to choreiform or myoclonic movements (McKenzie and Viik, 1975; Tarsy et al., 1978). The underlying neuronal mechanism of the dyskinesia, however, is not yet fully understood. In this paper, we performed the following experiments to define more precisely the nature of the dyskinesia and to elucidate the underlying neuronal mechanism. 1) We studied behaviorally the influence of the dopaminergic system on the dyskinesia, since dopamine is closely related to choreiform dyskinesia as L-DOPA induced dyskinesia. 2) We investigated electrophysiologically changes in neuronal activities in the striatum and its output structures during dyskinesia.

MATERIAL AND METHODS

Behavioral study

The methods used in this study was almost same as those in the previous report (Slater and Dickinson, 1982) . In brief, twenty Wistar albino rats weighing 280-300g were implanted with the guide cannula under stereotaxic control over the left caudoputaminal unit(CPU). In 6 rats, 8 μg of 6-OH dopamine with 0.8 μg of ascorbic acid in 0.4 μg of saline were injected into the pars compacta of substantia nigra (coordination in the atlas (Pellegrino et al., 1979); P -3.0 mm, L 2.0 mm from bregma, D 8.5 mm from dura) to destroy the dopaminergic neurons. After a 7-day post-operative recovery interval, 1 μg of PTX dissolved in 1 μl of saline was injected into the midpoint of the CPU (A 1.4 mm, L 3.5 mm from bregma, D 5.0 mm from dura) through the guide cannula. Apomorphine (0.5 μg in 0.5 μl of saline) was given intrastriatally to 3 rats two minutes before PTX injection and bromocriptine (Sandoz co. 10

The Basal Ganglia III, Edited by G. Bernardi *et al.*
Plenum Press, New York

mg/kg in 1 ml of saline) was administrated intraperitoneally to another 4 rats 1 hour before PTX injection. The frequency of the repetitive jerks that occurred in the right forelimb was recorded and intensity was measured on a 0-4 scale: 0 = no jerking; 1 = weak jerking; 2 = sustained jerking of forelimb; 3 = pronounced jerking also involving upper part of body; 4 = 3 plus short periods of myoclonic seizures.

Electrophysiological study

1) Preparation of animals

Sixty Wistar albino rats were used. Slight or moderate anesthesia was maintained by subcutaneous injection of pentobarbital sodium (15-20 mg/kg) or ketamine hydrochloride (30-50 mg/kg). The rat's head was fixed in a stereotaxic apparatus (Narishige Co., Tokyo). The left cranium was removed and the left cerebral hemisphere was widely exposed. In some animals (n=10), the cerebral cortex(CX), except for the ventral cortex, was ablated as widely as possible until the corpus callosum (CC) could be seen. Afferent fibers of the CPU from the thalamus and the pars reticulata of substantia nigra (SNr) were also cut by the mediolateral movement of a spatula that had been deeply inserted between the thalamus and CPU.

2) Microinjection of PTX

One mg of PTX was dissolved in 1 ml of saline, and 1 µl of the solution was drawn into a Hamilton syringe (outer diameter: 350 µm). The needle was inserted vertically into the midpoint of the CPU. PTX (1 µg) was injected into the CPU at a rate of 0.5 µl/min, and the syringe was kept in situ for 2 minutes after the injection was completed. In some animals (n=10), angled injection of PTX was performed by declining the syringe by 30 degree rostrocaudally in order to avoid damaging the cortical region just above the CPU.

3) Reversible and irreversible lesion of CX

To test involvement of the CX in the production of dyskinesia, we poured ice water for 10 to 30 seconds onto the surface of the cortex by using an injection syringe. And for production of an irreversible lesion in the cerebral cortex, a small amount of silver nitrate was placed on the desired area.

4) Recording

Elgiloy electrodes were used for recording of field potentials as well as extracellular single unit spikes of the CPU and also for recording of single unit spikes of SNr (P -3.0 mm, L 2.5 mm, D 8.0 mm) and the ventromedial nucleus of thalamus (VM)(P -0.8 mm, L 1.5 mm, D 7.0 mm). Potentials of the CX were recorded by placement of silver ball-electrodes with diameter of 0.5 mm on the surface of the sensorimotor CX overlying the CPU. An Ag-AgCl electrode was placed on the right ear bar and was used as an indifferent electrode. Electromyograms (EMG) were recorded differentially between 2 needles inserted into the right biceps brachii muscle or masseter muscles. The signals were displayed on a 4-channel oscilloscope (VC-10, Nihon Kohden, Tokyo) through a 4-channel conventional preamplifier.

5) Identification of recording sites

To identify recording sites, we made electrolytic lesions by passing negative DC current of 3V for 20 sec. into the sites. The brain was fixed in 10% formal saline, and frontal sections of 100 μm thickness were cut and stained by the Klüver-Barrera method.

RESULTS

The results of behavioral study were summarized in Tables 1 and 2. Apomorphine increased the frequency of dyskinesia and bromocriptine intensified the dyskinesia. The dyskinesia reduced, on the contrary, in the 6-OH dopamine pretreated rats.

Changes in extracellularly recorded single unit spikes in the CPU in association with production of dyskinesia

Extracellular single unit spikes of the CPU were recorded during development of PTX-induced cortical spike potentials, as shown in Fig.1. Before and immediately after PTX injection the unit spikes fired tonically (See Rec 2 in A). Although it is possible that there existed CPU neurons which did not show any spontaneous firing (see Rec 1 in A-E, Fig.1), spontaneous firing of the single unit spikes of the CPU was usually low, ranging from 0.1 to 7 Hz (m±SD, 2.4±1.8, n=100). The firing pattern became gradually irregular (B) and further showed a tendency of grouping (C and D). The grouped firing also extended spatially and could be recorded from the other electrode (Rec 1 in D). Finally, synchronous with the grouped firing of the unit spikes of the CPU, fully developed spike potentials were observed in the CX, accompanied by manifestation of EMG activity in the contralateral forelimb (E).

Table 1. Effects of dopaminergic modification on the frequency of the dyskinesia

Drugs	Picrotoxin 1 μg (control)	Apomorphine 0.5 μg in the CPU	Bromocriptine 10 mg/kg i.p.	6-OH dopamine 8 μg in the SNc
n	7	3	4	6
m ± SE[*]	54.6±4.9	71.0±5.9[**]	50.3±2.4	42.0±2.7[**]

*: mean ± standard error   **: P<0.05

Table 2. Effects of dopaminergic modification on the intensity of the dyskinesia

Drugs	Picrotoxin 1 μg (control)	Apomorphine 0.5 μg in the CPU	Bromocriptine 10 mg/kg i.p.	6-OH dopamine 8 μg in the SNc
n	7	3	4	6
m ± SE[*]	3.4±0.2	3.6±0.3	4.0[**]	2.8±0.4

*: mean ± standard error   **: P<0.05

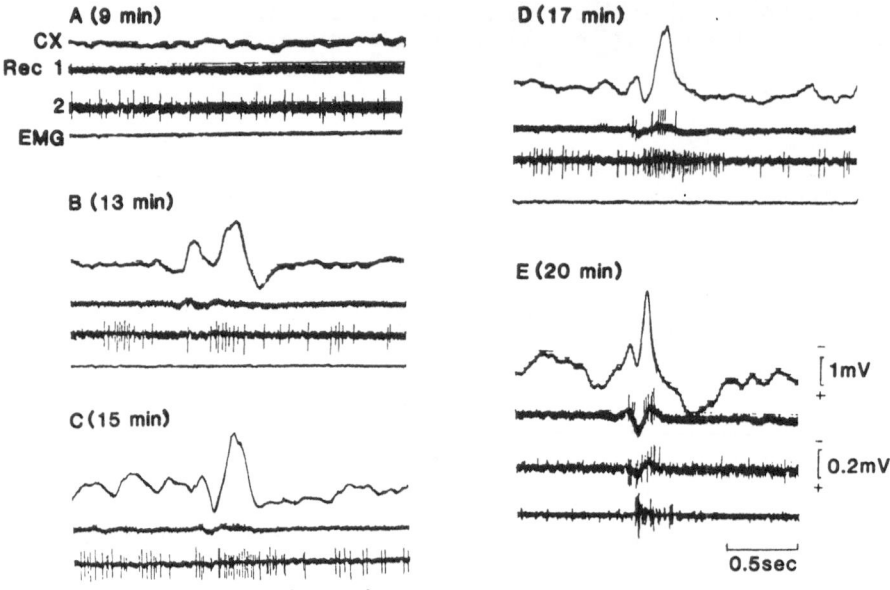

Fig.1. Sequential changes in single unit activity in CPU during development of spike potentials. A-E: 9, 13, 15, 17, and 20 minutes after PTX injection into left CPU. CX: left sensorimotor cerebral cortex. Rec 1: record of single unit activity by an Elgiloy electrode. Rec 2: record by another electrode 2 mm from electrode of Rec 1. The injection site lay approximately midway between the two electrodes. EMG: electromyogram from right biceps brachii muscle. Low-pass filter: 0.1 Hz for the CPU and the CX, and 0.1 KHz for EMG. (From Muramatsu et al., 1989.)

Sequential recordings of field potentials also revealed that the initial potential change occurred within the CPU and that dyskinesia did not appear until cortical spike potentials had fully developed.

Effect of cooling or irreversible lesioning of CX on dyskinesia

Dyskinesia, as well as corresponding EMGs, disappeared 30 seconds after initiation of pouring of ice water onto the surface of the CX, as shown in Fig.2A. Concomitantly cortical spike potentials markedly diminished or disappeared, while CPU spike potentials remained without any visible change in amplitude although a slight change in shape and decrease in frequency were observed (A-2). Three minutes after cessation of cooling, small spike potentials reappeared in the CX (A-3); and the potential was fully recovered and EMG activities also appeared again (A-4). On the other hand, an irreversible lesion made in the CX by silver nitrate resulted in disappearance of cortical spike potentials and dyskinesia as well (Fig.2B). A change in shape and decreased frequency of CPU spike potentials were also observed.

Effect of deafferentation of the CPU

In order to isolate the CPU from the influence of the major afferents, we ablated the CX as widely as possible by suction; and thalamo-striatal as well as nigro-striatal fibers were cut with a spatula. Furthermore, efferent fibers of the CPU were also cut in this

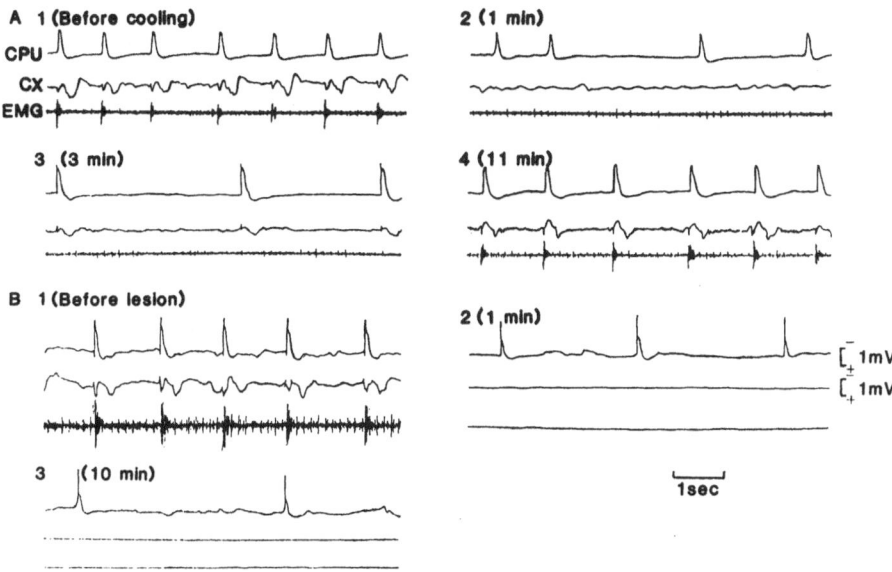

Fig.2. Effects of cooling and irreversible lesioning of CX on CPU spike potentials. A: effect of cooling of the CX. Ice water was put onto the left CX for 30 seconds. 1: before cooling. Fully developed spike potentials of CPU and CX, and of EMG, are seen. 2: 1 minute after cessation of cooling. Spike potentials of CX diminished remarkably and EMG disappeared, while CPU spike potentials remained. 3: 3 minutes after cooling. Small spike potentials of CX reappeared. 4: 11 minutes after cooling. Fully recovered spike potentials of CX with EMG are seen. B: effect of lesioning of CX with silver nitrate. 1: before lesioning. 2 and 3: 1 and 10 minutes respectively after lesioning. Neither spike potential of CX nor EMG is seen. CPU: left caudoputaminal unit. CX: left sensorimotor cerebral cortex. EMG: electromyogram of right biceps brachii muscle. Low pass filter: same as in Fig.1. (From Muramatsu et al., 1989.)

procedure. Periodical generation of the spike potential was clearly preserved within the isolated CPU.

Changes in the firing patterns of SNr and VM neurons in association with CPU field spike potentials

Spontaneous firing of extracellular unitary spikes of the SNr ranged from 20 to 70 Hz (m±SD, 42±10, n=149). The relationship between the firing pattern of the SNr neurons and the CPU field spike potential was analyzed.

Extracellular unitary spikes of the SNr were recorded in association with CPU field spike potentials (Fig.3A-C). All 147 neurons, which were histologically identified as SNr cells, showed marked inhibition in correspondence with the CPU spike potentials. No recognizable excitation of the SNr neurons could be observed in association with the CPU spike potential; however, spindle-shaped, grouped discharges, consisting of many small spikes and with a peak latency of 40-50 msec from the onset of the CPU spike potential, were often recorded synchronously with the CPU spike potentials from the same recording site, as seen in A-C of Fig.3.

In contrast to the inhibitory response of SNr neurons, all 110 VM

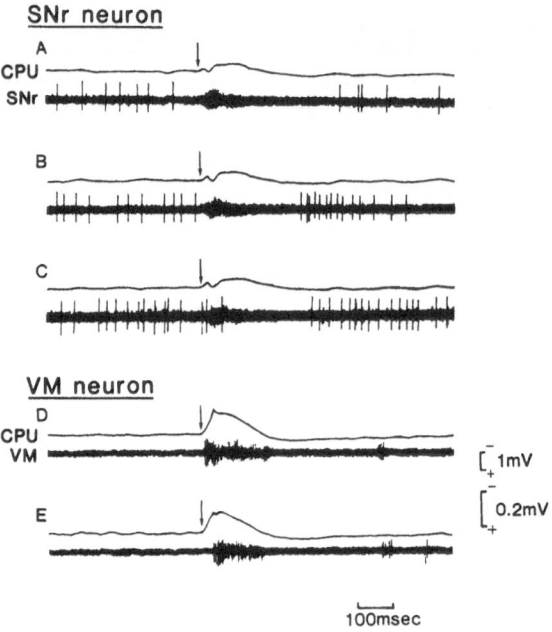

SNr neuron

A

CPU

SNr

B

C

VM neuron

D

CPU

VM

E

100msec

Fig.3. Changes in single unit spike activities of SNr and VM in association with CPU spike potentials. Upper records of A-E: CPU spike
potentials. Lower records single unit spikes. A-C: single unit spike
activities of 3 different SNr neurons. D and E: single unit spike activities of 2 different VM neurons. Arrows indicate onset of CPU spike
potentials. (From Muramatsu et al., 1989.)

neurons recorded showed excitation in relation to the CPU spike potential, as shown in D and E of Fig.3.

DISCUSSION

Modulatory effects of dopamine on dyskinesia

Dyskinesia such as choreiform or myoclonic movements was produced
in the contralateral forelimb of rats by injection of PTX, a specific
GABA antagonist, into the CPU. Our results of behavioral study were as
clinically expected: that dopamine potentiate dyskinesia. It is well
known that while L-DOPA often causes dyskinesia in Parkinsonian
patients, dopamine antagonists attenuate dyskinesia in Huntington's
chorea. Caudate neurons show either inhibitory or excitatory response
to dopamine mediated by D_1 or D_2 receptors respectively(Akaike et al.,
1987; Calabresi et al., 1987). It is possible that some group of CPU
efferent neurons is excited even if direct action of dopamine is inhibitory by inhibiting inhibitory interneurons in the CPU or by reducing
effect of lateral inhibition from another group of efferent neurons.

Mechanism of spike potential generation within the CPU

Since periodical field spike potentials were generated in the CPU even when the CPU was isolated from the major inputs, the CPU spike potential is probably generated by some intrinsic mechanism of the CPU per se.

As the site of action of PTX, which blocks GABAergic inhibitory synapses and gives rise to the excitatory effect, the following two possible sites are so far presumed: Inhibitory GABAergic recurrent axon collaterals inhibit CPU efferent fibers, and blocking of the inhibition by PTX would result in increased firing activity of the efferent fibers. Also, GABAergic interneurons are reported, and removal of the inhibitory effect of the interneurons on CPU efferent fibers by PTX could also be involved in excitation of CPU efferent neurons.

On the other hand, some inhibitory process should also be involved for the CPU spike potential to be generated not continuously but periodically. Firstly, inhibitory influences from CPU areas not reached by PTX may be exerted on the site activated by PTX. Secondly, excessive firing of neurons produces summation of after-hyperpolarization, resulting in decreased firing of the neurons themselves.

The role of the CX

In our experiments, dyskinesia did not appear until full development of the cortical spike potential and reversible or irreversible dysfunction of the CX resulted in disappearance of dyskinesia without visible changes in amplitude of CPU spike potentials. It can, therefore, be concluded that CX is necessary for the manifestation of the dyskinesia, although the primary originator is the CPU.

There was a report that involvement of small lesions of the sensorimotor cortex made by the injection needle was necessary for production of dyskinesia (Tarsy et al, 1978). Our results, however, disclosed that lesioning of a wide area of the cerebral cortex always resulted in disappearance of dyskinesia. The geometry of the injection such as angled or vertical, therefore, seems to be not crucial for production of dyskinesia. In fact, dyskinesia never occurred in some experiments (n=5) regardless of whether the PTX injection was angled or vertical. Analysis of these abortive results by recording of spike potentials of CPU and CX, and also EMG simultaneously, revealed that in some cases full development of spikes of either CPU or CX did not occur, resulting in failure to produce dyskinesia.

Relationship of the firing patterns between neurons of the CPU, SNr, and VM

In contrast to electrical stimulation, which would uniformly activate brain structures, pharmacological activation of the neuronal components adopted in the present study would probably influence neuronal elements unevenly, resulting in activation of somewhat different CPU efferent neurons. This lack of uniformity might have been the principal reason for the production of variable latencies in the inhibition of the SNr neurons and excitation of the VM neurons occurring in association with each CPU field spike potential.

Since caudato-nigral as well as putamino-nigral fibers are inhibitory in nature (Yoshida and Precht, 1971), it is reasonable that in our experiments the SNr neurons were strongly inhibited in correspondence with the CPU field spike potentials accompanied by high-frequency grouped firing of the CPU neurons.

In the SNr, spindle-shaped, grouped discharges consisting of small spikes were frequently observed to be synchronous with the CPU spike potential (see Fig.3). It is possible that this grouped discharge represents the excitation of SNc neurons through the inhibition of inhibitory interneurons of the SNr (Grace and Bunny, 1979), or electrotonic reflection of the excited neurons or axons surrounding the SNr, such as those of the midbrain reticular formation (MRF) or of the cerebral peduncle.

The following possible mechanisms may be offered for the excitation of VM neurons. First, the SNr neuron exerts monosynaptic inhibitory effects on VM neurons (Ueki, 1983), and the latter neurons generally receive tonic inhibition from the inhibitory SNr. Thus, transient removal of the SNr inhibition could lead to the excessive excitation of VM neurons. Although it is hard to prove these disinhibitory mechanisms in our experiments because of the difficulty in determining strict time relations, they are reported to be operative in other structures of the basal ganglia (Chevalier et al., 1985; Deniau et al., 1985; Hikosaka and Sakamoto,1986). Second, since inhibitory efferent fibers of the SNr also project to the MRF, excitation of the MRF might occur through a disinhibitory mechanism. Then the MRF would excite VM neurons. Third, VM neurons may be excited secondarily by the cortical spike potentials via the cortico-thalamic pathway. This is not likely, however, since ablation of the CX did not eliminate the VM excitation in our experiment (unpublished observation).

ACKNOWLEDGMENTS

We are grateful to Dr. Akira Ueki for his collaboration in the initial part of the experiment. This work was supported by a Grant-in-Aid (No. 01623003) for Scientific Research on Priority Areas from the Japanese Ministry of Education, Science, and Culture.

REFERENCES

Akaike, A., Ohno Y., Sasa M. and Takaori S., 1987, Excitatory and inhibitory effects of dopamine on neuronal activity of the caudate nucleus neurons in vitro, Brain Res., 418:262-272.

Calabresi P., Mercuri N., Stanzione P., Stefani A., and Bernardi G., 1987, Intracellular studies on the dopamine-induced firing inhibition of neostriatal neurons in vitro: Evidence for D1 receptor involvement, Neuroscience., 20:757-771.

Chevalier, G., Vacher, S., Deniau, J. M., and Desban, M., 1985, Disinhibition as a basic process in the expression of striatal functions.I. The striato-nigral influence on tecto-spinal/tecto-diencephalic neurons, Brain Res., 334:215-226.

Deniau, J. M., and Chevalier, G., 1985, Disinhibition as a basic process in the expression of striatal functions.II. The striato-nigral influence on thalamocortical cells of the ventromedial thalamic nucleus, Brain Res., 334:227-233.

Grace, A. and Bunny, B., 1979, Paradoxical GABA excitation of nigral dopaminergic cells: indirect mediation through reticurata inhibitory neurons, Eur. J. Pharmacol., 59:211-218.

Hikosaka, O. and Sakamoto, M., 1986, Cell activity in monkey caudate nucleus preceding saccadic eye movements, Exp. Brain Res., 63: 659-662.

McKenzie, G. M, and Viik, K., 1975, Chemically induced choreiform activity: Antagonism by GABA and EEG patterns, Exp. Neurol., 46: 229-243.

Muramatsu, S., Yoshida, M. and Nakamura, S., 1989, Electrophysiological study of dyskinesia produced by microinjection of picrotoxin in the striatum of the rat, Neurosci. Res. in press.

Pellegrino, L. J., Pellegrino, A. S., and Cushman, A. J., 1979, "A stereotaxic atlas of the rat brain", 2nd ed., Plenum, New York.

Slater, P., and Dickinson, S. L., 1982, Role of acetylcholine and dopamine in myoclonus induced by intrastriatal picrotoxin, Neurosci. Lett., 28:253-257.

Tarsy, D., Pycock, C. J., Meldrum, B. S., and Marsden, C. D., 1978, Focal contralateral myoclonus produced by inhibition of GABA action in the caudate nucleus of rats, Brain, 101:143-162.

Ueki, A., 1983, The mode of nigro-thalamic transmission investigated with intracellular recording from cat thalamus, Exp. Brain Res., 49:116-124.

Yoshida, M. and Precht W., 1971, Monosynaptic inhibition of neurons of the substantia nigra by caudate-nigral fibers, Brain Res., 32: 225-228.

NEUROPHYSIOLOGICAL DEVELOPMENT OF FETAL NEOSTRIATAL NEURONS TRANSPLANTED INTO

ADULT NEOSTRIATUM

John P. Walsh, Chester D. Hull, Carlos Cepeda, Michael S. Levine
and Nathaniel A. Buchwald

Mental Retardation Research Center, University of California, Los
Angeles, California   90024-1759

INTRODUCTION

Behavioral studies of rats with excitotoxic neostriatal lesions have demonstrated functional recovery of locomotor activity, motor symmetry, skilled motor tasks and performance in maze learning tasks after grafting of fetal neostriatal tissue (Dunnett et al, 1988; Deckel et al, 1983; Isacson et al, 1986). Histological analysis of the grafted neostriatal tissue suggests that these effects are mediated by the survival and growth of neurons which for the most part have been classified as the medium-sized spiny cell of the neostriatum (McAllister et al, 1985; Clarke et al, 1988, DiFiglia et al, 1988). It has been further proposed that the grafted neurons function via a reinstatement of the neurocircuitry normally found in the neostriatum (Dunnett et al, 1988; for review see Norman et al, 1988). A number of techniques have been used to show that nigral, cortical, thalamic and raphe afferents grow into the grafted neostriatal tissue (Clarke et al, 1988; Wictorin et al, 1989; Pritzel et al, 1986; Wilson et al, 1987). Conversely, grafted neostriatal tissue has been shown to reinstate gamma amino butyric acid (GABA) release in the globus pallidus and substantia nigra in lesioned animals (Sirinathsinghji et al, 1988).

However, there is evidence that the grafted neostriatal tissue does not fully develop all of the morphological and synaptic features normally present in the neostriatum. Differences have been observed in the morphology of the cell types that survive and in the distribution of afferents innervating the graft (Zhou et al, 1989; DiFiglia et al, 1988; Walker et al, 1987; Pritzel et al, 1986; Wictorin et al, 1988; 1989a,b).

In order to determine if grafted neostriatal cells develop normal physiological and morphological characteristics we examined the physiology and single cell morphology of transplanted neostriatal neurons (TSNs) 2-4 weeks after transplantation (Walsh et al, 1988). The physiological and morphological differences observed between TSNs and host neostriatal neurons were tentatively assigned to the developmentally immature state of the TSNs. The present study extends this work to include analyses of properties of neurons from older grafts (up to 5 months post-transplantation) and a parallel analysis of neostriatal neurons from rat pups age-matched to the grafted neurons. Our findings indicate that the physiological differences previously noted in young TSNs continue to be expressed in older TSNs.

The Basal Ganglia III, Edited by G. Bernardi *et al.*
Plenum Press, New York

MATERIALS AND METHODS

Fetal neostriatal tissue was grafted into the intact (nonlesioned) neostriatum of adult host Sprague-Dawley rats. Donor neostriatal tissue (embryonic day (E)13 - E14) was prepared for grafting into the host neostriatum by the cell suspension technique of Schmidt et al. (1981). The resulting tissue suspension was stereotaxically injected into anesthetized adult rats using the following coordinates: 1 mm anterior to bregma, 3.5 mm lateral to midline, and 5.0 mm below the dural surface.

Host rats were sacrificed from 2 weeks to 5 months post-transplantation in order to examine the physiological and morphological development of the grafted neostriatal neurons. The brains of host animals were rapidly removed and prepared for in vitro brain slice analysis (see Walsh et al, 1988 for details). In order to compare the development of TSNs with neostriatal neuronal development in intact animals, neurons obtained from rat pups age-matched to the post-transplantation times studied were examined in parallel studies using similar brain slice techniques (postnatal day 7 to adult). The equivalent postnatal age of TSNs was estimated by subtracting 7 days from the post-transplant age of the grafted tissue at the time of the experiment (donor tissue was E14 at the time of grafting).

Intracellular recordings were made with glass pipettes filled with 3 M potassium acetate with DC resistance ranging from 50–100 MΩ. Intracellular signals were amplified with an Axoclamp 2A amplifier and stored on video cassette tape for later analysis. Synaptic potentials were elicited by extracellular stimulation with a bipolar electrode (0.2 mm silver wires placed 1–2 mm from the recording microelectrode, tip separation 200–500 μM). Stimulation intensities ranged from 50–500 μA.

Lucifer yellow (LY) was also injected into TSNs and pup neurons to examine the morphological development of neostriatal cells. Microelectrodes used for intracellular labelling were filled with 5–10% LY dissolved in distilled water and 1.0 M LiCl. LY was injected by passing 2–5 nA of negative DC current for 2–5 min. Following injection, slices were transferred to fixative and kept overnight at 4°C. They were washed with buffer, treated with alcohols, cleared in xylene, and mounted on glass slides in Depex mounting media.

RESULTS

For electrophysiological and morphological analyses TSNs and rat pup neurons were separated into two age groups: a young group, 2–4 weeks after grafting or postnatal day(P)7–P21 in pups and an older group, 5–8 weeks after grafting or P30–P60 in age-matched intact rats.

Input resistance. The input resistance (R_{in}) of neostriatal neurons was determined from the change in membrane potential produced in response to injection of different intensities of hyperpolarizing intracellular current pulses. Measures were made at the peak of the response using a range of membrane potentials that was positive to the activation potential of the anomalous rectifying current (Kita et al, 1984). The average R_{in} of TSNs examined in the young group (2–4 weeks after grafting) was 36±5 MΩ (±SEM)(n=7). The average R_{in} of the age-matched pups (P7–P21) was 39±4 MΩ (n=31). The difference between these average values was not statistically significant (t<1, df=36). Examples of the recordings from neurons in each group are illustrated in Fig. 1. The younger cells (both TSNs and pup neurons) also displayed the low amplitude, long duration action potentials characteristic of immature neurons (Misgeld et al, 1986; Schwartzkroin and Altschuler, 1977; McCormick and Prince, 1987). The average R_{in} of TSNs in the older group (5–8 weeks post-transplantation) remained high (47±4 MΩ, n=13).

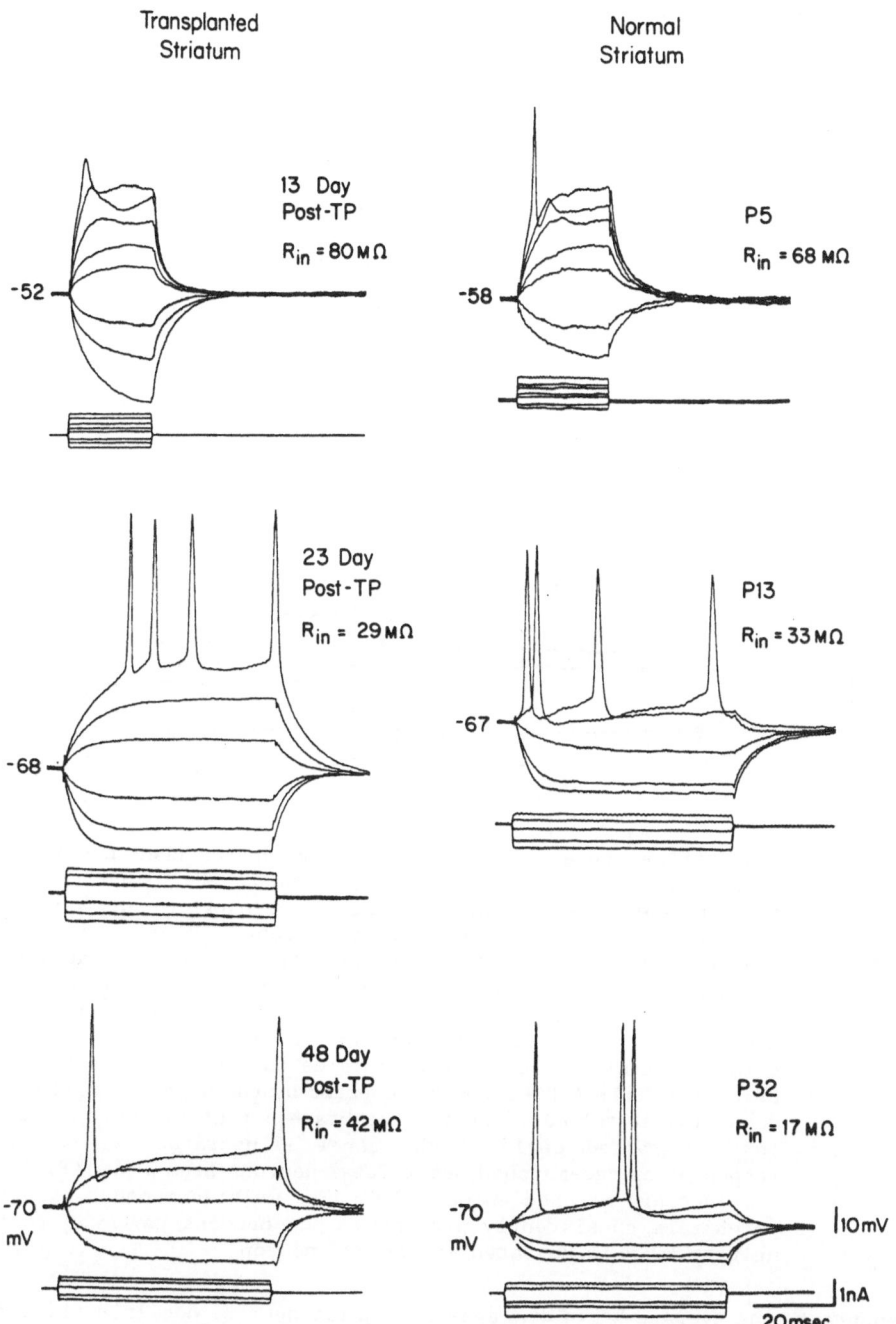

Fig. 1.   Neuronal responses to injections of current. The left panel shows
responses of 3 TSNs. The youngest neuron illustrated is 13 days
post–transplant. R_{in} is indicated to the right of each set of
traces. The right panel shows responses of neurons from pups.

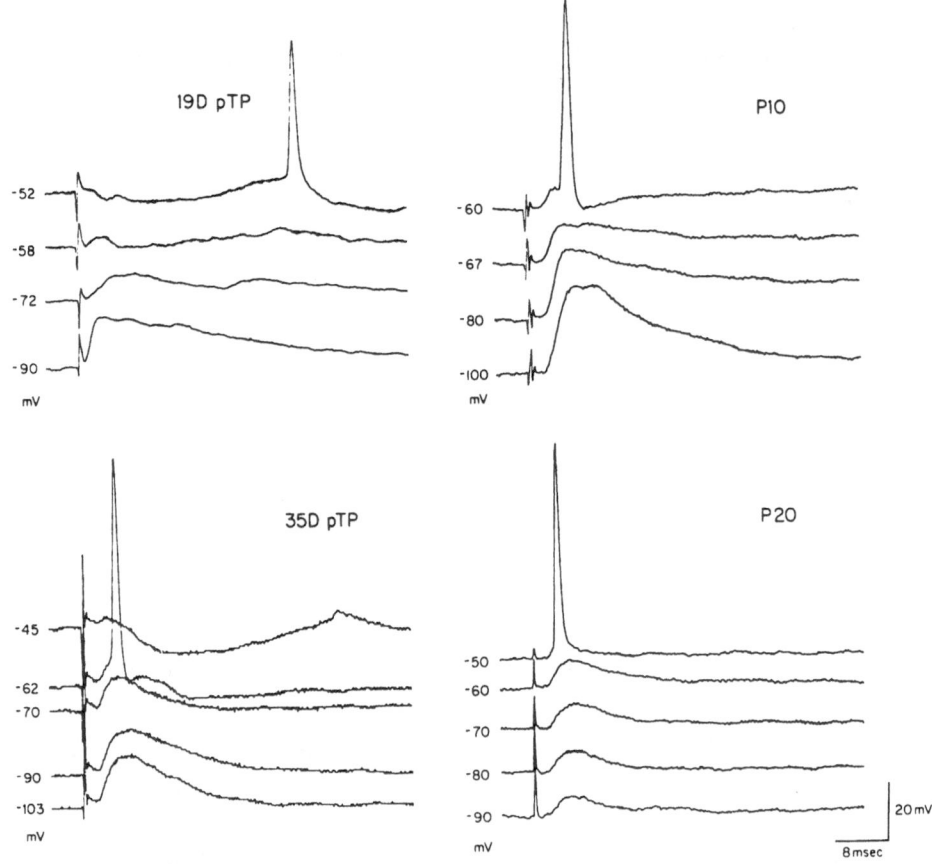

Fig. 2.   Synaptic responses evoked in TSNs and age—matched neostriatal
neurons from young and mature age groups. For each cell the
membrane potential (numbers to left of each trace) was varied by
intracellular injection of direct current to test the voltage
sensitivity of the synaptic responses. An identical extracellular
stimulation pulse was delivered at each level of membrane
potential. Left panel illustrates the responses of two grafted
neurons (19 days post transplant (D pTP) and 35D pTP) to
stimulation of the host tissue outside the graft. In both cases,
EPSPs and IPSPs (IPSPs are represented by the negative deflection
occurring at the most positive membrane potential) were produced
in the grafted cells. Right panel illustrates the synaptic
responses of age—matched neostriatal neurons from a pup (P10) and
an older age—matched animal (P20). The multi-component synaptic
potentials consistently present in pup neurons gave way to the
unitary EPSP illustrated in the P20 neuron.

In contrast, the average R_{in} of the age—matched rat neurons declined (15±1 MΩ,
n=10) (Fig. 1). The difference between these average values was statistically
significant (t=6.72, df=21, p<.005). The oldest TSNs examined (5 months after
grafting) continued to express high R_{in} values (51±4 MΩ, n=3). The R_{in} of
neurons obtained from intact adult rats was 14±1 MΩ (n=9). The difference
between these values was also statistically significant (t=12.84, df=10,
p<.005).

Synaptic potentials. Synaptic potentials were typically evoked in TSNs
by extracellular stimulation of the host tissue 1—2 mm from the border of the

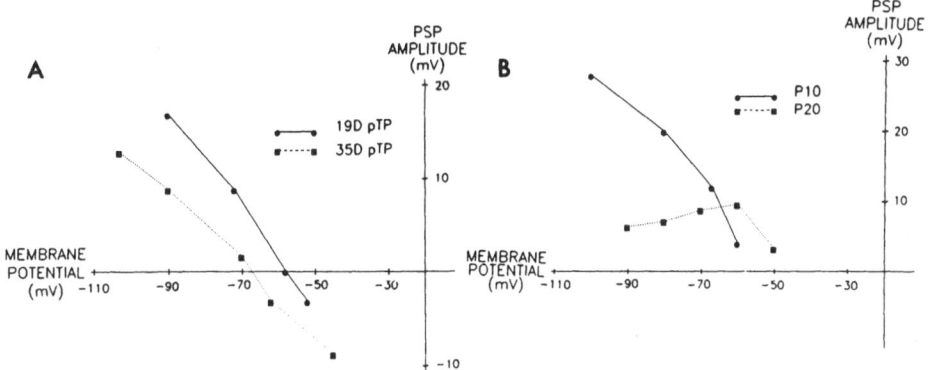

Fig. 3.   A. Plots of peak amplitude of the later synaptic response
           illustrated in Fig. 2 versus the membrane potential of the
           transplanted cells. The synaptic responses had reversal
           potentials which equaled the equilibrium potential for Cl⁻ in
           mammalian neurons. B. Plots of the peak amplitude of the synaptic
           responses versus the membrane potential of the age-matched
           neurons shown in Fig. 2.

transplant (left panel, Fig.2). Identical stimulation parameters and
electrode placement distances were used for the age-matched pup neurons.
Multi-component long duration synaptic potentials evoked in young age-matched
neurons gradually changed to short-duration, rapidly occurring excitatory
post-synaptic potentials (EPSPs) in animals P20 or older (right panel, Fig.
2). The synaptic potentials elicited in TSNs were consistently of long
duration and were composed of multiple components at all post-transplantation
times examined. In some cases, large complex synaptic potentials up to 400
msec in duration could be evoked. The synaptic activity generally consisted
of multi-component EPSPs followed by an inhibitory postsynaptic potential
(IPSP) (Fig. 2). The reversal potential of the IPSP was consistent with the
Cl⁻ reversal potential (Fig. 3) (Mueller et al, 1984; Ben-Ari et al, 1989).
TSNs also displayed considerable spontaneous synaptic potentials in most of
the animals examined. In contrast, spontaneous synaptic potentials were
rarely observed in the youngest of the age-matched rat pups and never in
animals older than P10.

    Morphology. Injections of LY revealed that TSNs possessed morphological
features that resembled the medium-sized spiny neurons filled in age-matched
pups (Fig. 4). The cell bodies of TSNs were, in general less than 15 μM in
diameter and spiny specializations were present over most of their dendritic
surface. A number of differences were observed, however, between TSNs and
age-matched neostriatal neurons. Measurements of cross-sectional diameter and
dendritic field radius indicated that TSNs had significantly larger somas and
longer dendrites than age-matched neurons in the older group. TSNs examined
2-4 weeks after transplantation had average soma diameters of 10.8±0.7 μM
(n=7) and average dendritic field radii of 113±11 μM (n=7). Neostriatal
neurons from age-matched rat pups (P7-P21) had average soma diameters of
9.9±0.3 μM (n=56) and average dendritic field radii of 93±8 μM (n=6). During
this age period differences between these average values were not
statistically significant (t=1.02, df=61 for soma diameter, t=1.36, df=11 for
dendritic field radius). TSNs examined 1-5 months after grafting had average
soma diameters of 14.2±1.2 μM (n=16) and average dendritic field radii of
189±15 μM (n=12). Neostriatal neurons from age matched pups (P21-adult) had
average soma diameters of 11.5±0.6 μM (n=18) and average dendritic field
radii of 133±9 μM (n=18). During this age period the differences between the
average values were statistically significant (t=2.09, df=32, p<.025 for soma
diameter, t=3.38, df=28, p<.005 for dendritic field radius). Dendrites of

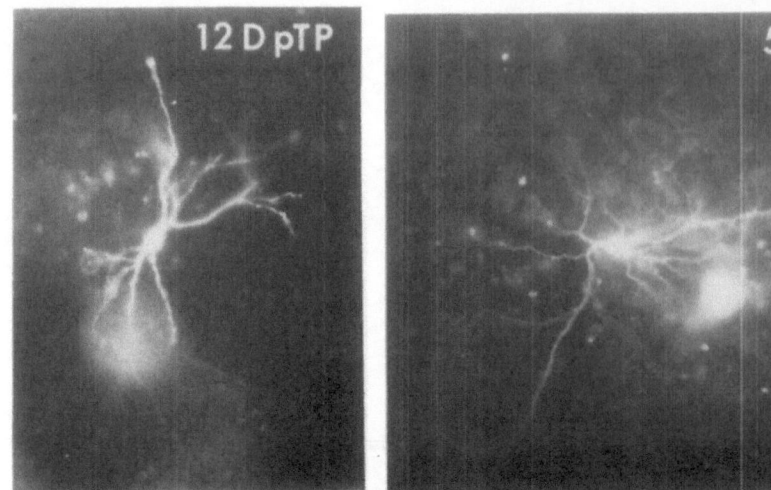

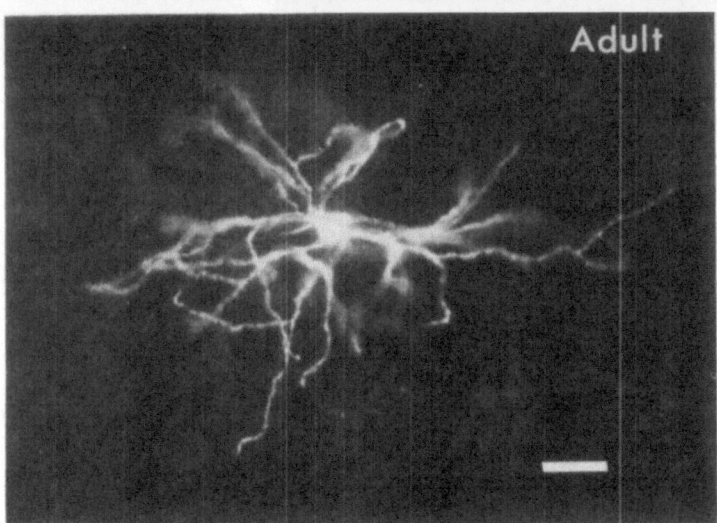

Fig. 4.   TSNs and a neostriatal neuron from an adult rat filled with LY.
Top left: 12 D pTP neuron filled with LY. Top right: 5 month (mo)
pTP neuron filled with LY. The distal processes of this neuron
projected beyond the field of view of this photomicrograph.
Bottom: adult neostriatal neuron from an intact rat. TSNs tended
to have longer but less complex dendrites than neostriatal
neurons from intact age-matched animals. The spine density on
dendrites of TSNs tended to be lower than that of age-matched
neostriatal cells. Calibration bar is 50 μm and refers to all
panels.

TSNs tended to have lower spine densities than those of age-matched pups. In
several case (n=5) TSN dendrites and/or axons crossed the transplant border
and grew into the surrounding host neostriatum.

DISCUSSION

The results of this study demonstrates that TSNs develop the ability
to generate passive and active neuronal responses. They receive synaptic
input from host tissue. Morphologically, recorded TSNs have many of the

properties of medium-sized spiny neostriatal neurons. However, a number of significant developmental differences were observed between the electrophysiological and morphological properties of TSNs and neostriatal neurons obtained from intact age-matched rats.

The lack of developmental decline in R_{in} for TSNs contrasts with the significant decrease in R_{in} during early maturation observed in neostriatal cells and forebrain neurons in other areas in intact animals (Misgeld et al, 1986; Schwartzkroin and Altschuler, 1977; McCormick and Prince, 1987). The difference between TSNs and age-matched cells in intact animals could be due to the action of a number of factors which potentially influence R_{in}. We observed, as have others, that TSNs tend to have a lower spine density than their age-matched counterparts (Walsh et al, 1988; McAllister et al, 1985). The occurrence of spines on dendrites of neostriatal cells contributes significantly to reducing the measurable R_{in} by providing additional pathways of current spread (Wilson, 1984; Wilson, 1988). DiFiglia et al (1988) have shown that the density of synapses on TSNs is lower than those on intact neostriatal cells. The potential lack or diminution of tonic synaptic influence could decrease the resting steady state conductance of the TSNs. Lastly, we observed that the primary branch points of TSN dendrites tended to occur further from the soma and fewer branches occurred more distally. Both of these phenomena would be expected to increase R_{in}.

In agreement with previous studies in the neostriatum, hippocampus and neocortex, we found that evoked synaptic responses obtained from intact rats underwent developmental changes (Misgeld et al, 1986; McCormick and Prince, 1987; Ben-Ari et al, 1989). Long duration mixed EPSP-IPSPs observed in rat pups gradually gave way to short duration unitary EPSPs in animals older than P20. By contrast, complex multi-component EPSP-IPSPs were present in TSNs at all of the post-transplantation ages examined. Even in grafts examined 5 months after transplantation, long duration complex synaptic activity was observed. TSNs also frequently displayed a steady barrage of multicomponent spontaneous synaptic activity. Spontaneous synaptic input was only rarely observed in the youngest of age-matched rat pups. The unusual synaptic activity observed in the TSNs may reflect aberrant connections made by inputs intrinsic to the graft and/or connections resulting from sprouting of host fibers into the graft. Sprouting of fibers in response to hippocampal lesions also produces abnormal synaptic activity (Tauck and Nadler, 1985). In agreement with the present findings, Buzsaki et al (1989) and Freund and Buzsaki (1988) have observed similar phenomena in hippocampal transplants, where the synaptic potentials become exaggerated to the point of developing seizure-like activity.

In general, TSNs had highly branched dendrites and spiny specializations characteristic of medium spiny neurons of the neostriatum (Chang et al, 1982; Chronister et al, 1976). There were many morphological details in TSNs, however, which differed from age-matched neostriatal neurons. In agreement with DiFiglia et al (1988), we found TSNs to have larger cross-sectional diameters than their age-matched counterparts. The dendrites of TSNs also tended to extend further from the soma than the age-matched neostriatal neurons from intact rats. These dendrites also lacked the symmetry around the soma typically seen in neostriatal neurons. As indicated above, dendrites of TSNs also generally had fewer spines than age-matched neurons. These observations were apparent in the oldest grafts indicating that either the TSNs expressed incomplete morphological development or that the abnormal exposure to host brain and the grafting process caused the TSNs to follow their own unique path of development.

The inadvertent grafting of pallidal and neocortical neurons could explain some of the differences observed between the properties of TSNs and age-matched neostriatal neurons (Wictorin et al, 1989a,b). Even if this potential confound occurred, the predominant neuron grafted still appeared to be of neostriatal origin. Our sample of recorded and labelled cells was consistent. The results cannot be explained by a selective sampling or survival of non-neostriatal neurons. Neurons having morphological properties of neocortical pyramidal neurons or globus pallidus cells were never encountered.

We have provided physiological and morphological evidence for communication between host and transplant. This communication expressed synaptically suggests, however, that abnormal connections may have been formed. Such findings could be a reflection of the exposure of TSNs to potentially inappropriate developmental cues during the ontogeny of the grafted tissue. Together with the many behavioral studies demonstrating recovery of function, our findings indicate that normal physiological activity may not be a prerequisite for recovery of behavioral function in response to grafting (Dunnett et al, 1988; Deckel et al, 1983). These results and those of Buzsaki et al (1988) suggest, however, that caution should be used in the selection of grafting as a procedure for the alleviation of neurological disorders.

ACKNOWLEDGEMENTS

This research was supported by USPHS Grant HD05958.

REFERENCES

Ben-Ari, Y. Cherubini, E., Corradetti, R. and Gaiarsa, J.L., 1989, Giant synaptic potentials in immature rat CA3 hippocampal neurones. J. Physiol., 416:303-325.

Buzsaki, G., Bayardo, F., Miles, R., Wong, R.K.S. and Gage, F.H., 1989, The grafted hippocampus: An epileptic focus. Exp. Neurol., 105:10-22.

Chang, H.T., Wilson, C.J. and Kitai, S.T., 1982, A golgi study of rat neostriatal neurons: Light microscopic analysis. J. Comp. Neurol., 208:107-126.

Chronister, R.B., Farnell, K.E., Marco, L.A. and White, L.E., Jr., 1976, The rodent neostriatum: A golgi analysis. Brain Res., 108:37-46.

Clarke, D.J., Dunnett, S.B., Isacson, O., Sirinathsinghji, D.J.S. and Bjorklund, A., 1988, Striatal grafts in rats with unilateral neostriatal lesions - Ultrastructural evidence of afferent synaptic inputs from the host nigrostriatal pathway. Neurosci., 24: 791-801.

Deckel, A.W., Robinson, R.G., Coyle, J.T. and Sanberg, P.R., 1983, Reversal of long-term locomotor abnormalities in the kainic acid model of Huntington's disease by day fetal 18 striatal implants. Eur. J. Pharmac., 93:287-288.

DiFiglia, M., Schiff, L. and Deckel, A.W., 1988, Neuronal organization of fetal striatal grafts in kainate- and sham-lesioned rat caudate nucleus: Light and electron-microscopic observations. J. Neurosci., 8:1112-1130.

Dunnett, S.B., Isacson, O., Sirinathsinghji, D.J.S., Clarke, D.J. and Bjorklund, A., 1988, Striatal grafts in rats with unilateral neostriatal lesions - III. Recovery from dopamine dependent motor asymmetry and deficits in skilled paw reaching. Neurosci., 24:813-820.

Freund, T.F., and Buzsaki, G., 1988, Alterations in excitatory and GABAergic inhibitory connections in hippocampal transplants. Neurosci., 27:373-385.

Isacson, O., Dunnett, S.B., and Bjorklund. A., 1986, Behavioral recovery in an animal model of Huntington's disease. Proc. Natl. Acad. Sci. (USA), 83:2728-2732.

Kita, T., Kita, H., and Kitai, S.T., 1984, Passive electrical membrane properties of rat neostriatal neurons in an in vitro slice preparation. Brain Res., 300:129-139.

McAllister, J.P., Walker, P.D., Zemanick, M.C., Weber, A.B., Kaplan, L. and Reynolds, M.A., 1985, Morphology of embryonic neostriatal cell suspensions transplanted into adult neostriatum. Dev. Brain Res., 23:282-286.

Misgeld, U., Dost, H.U. and Frotscher, M., 1986, Late development of intrinsic excitation in rat neostriatum: an in vitro study. Dev. Brain Res. 27:59-67.

McCormick, D.A., and Prince, D.A., 1987, Postnatal development of electrophysiological properties of rat cerebral cortical pyramidal neurones. J. Physiol., 393:743-762.

Mueller, A.L., Traub, J.S., and Schwartzkroin, P.A., 1984, Development of hyperpolarizing inhibitory postsynaptic potentials and hyperpolarizing response to r-aminobutyric acid in rabbit hippocampus studied in vitro. J. Neurosci., 4:860-867.

Norman, A.B., Lehman, M.N. and Sanberg, P.R., 1988, Functional effects of fetal striatal transplants. Brain Res. Bull., 22:163-172.

Pritzel, M., Isacson, O., Brundin, P., Wiklund L. and Bjorklund, A., 1986, Afferent and efferent connections of striatal grafts implanted into the ibotenic acid lesioned neostriatum. Expl. Brain Res., 65:112-126.

Schmidt, R.H., Bjorklund, A., and Stenevi, U., 1981, Intracerebral grafting of dissociated tissue suspensions: A new approach for neural transplantation to deep brain sites. Brain Res., 218:347-356.

Sirinathsinghji, D.J.S., Dunnett, S.B., Isacson, O., Clarke, D.J., Kendrick, K. and Bjorklund, A., 1988, Striatal grafts in rats with unilateral neostriatal lesions - II. In vivo monitoring of GABA release in globus pallidus and substantia nigra. Neurosci., 24:803-811.

Schwartzkroin, P.A. and Altschuler, R.S., 1977, Development of kitten hippocampal neurons. Brain Res., 134:429-444.

Tauck, D.L. and Nadler, J.V., 1985, Evidence for functional mossy fiber sprouting in hippocampal formation of kainic acid-treated rats. J. Neurosci., 5:1016-1022.

Walker, P.D., Chovanes, G. and McAllister, J.P., 1987, Identification of acetylcholine-reactive neurons and neuropil in neostriatal transplant. J. Comp. Neurol., 259:1-12.

Walsh, J.P., Zhou, F.C., Hull, C.D., Fisher, R.S., Levine, M.S. and Buchwald, N.A., 1988, Physiological and morphological characterization of striatal neurons transplanted into the striatum of adult rats, Synapse, 2:37-44.

Wictorin, K., Clark, D.J., Bolam, J.P. and Bjorklund, A., 1989a, Host corticostriatal fibers establish synaptic connections with grafted striatal neurons in the ibotenic acid lesioned striatum. Eur. J. Neurosci., 1:189-195.

Wictorin, K., Ouimet, C.C. and Bjorklund, A., 1989b, Intrinsic organization and connectivity of intrastriatal striatal transplants in rats as revealed by DARPP-32 immunohistochemistry: Specificity of connections with the lesioned host brain. Eur. J. Neurosci., 1:690–701.

Wictorin, K., Isacson, O., Fisher, W., Nothias, F., Peschanske, M., and Bjorklund, A., 1988, Connectivity of striatal grafts implanted into the ibotenic acid-lesioned striatum – I. Subcortical afferents. Neurosci., 27:547–562.

Wilson, C.J., Emson, P. and Feler, C., 1987, Electrophysiological evidence for the formation of a corticostriatal pathway in neostriatal tissue grafts. Soc. Neurosci. Abst., 13:11.

Wilson, C.J., 1984, Passive cable properties of dendritic spines and spiny neurons. J. Neurosci., 4:281–297.

Wilson, C.J., 1988, Cellular mechanisms controlling the strength of synapses. J. Electron Microsc., 10:293–313.

Zhou, F.C., Buchwald, N.A., Hull, C.D., and Towle, A., 1989, Neuronal and glial elements of fetal neostriatal grafts in the adult neostriatum. Neurosci., 30:19–31.

A PRIMATE MODEL OF HUNTINGTON'S DISEASE : UNILATERAL STRIATAL LESIONS
AND NEURAL GRAFTING IN THE BABOON (Papio papio)

D. Riche, P. Hantraye*, O. Isacson**, and M. Maziere*

Laboratoire de Physiologie Nerveuse, Equipe de Neuroanatomie
Fonctionnelle, CNRS, 91198 Gif-sur-Yvette, France
* URA 1285, CNRS, Service Hospitalier Frédéric Joliot, CEA
4 place du Général Leclerc, 91406 Orsay, France
** The Regeneration Laboratory, Harvard Medical School
Mailman Res. Ctr., McLean Hospital, Belmont, MA 02178, USA

INTRODUCTION

A striking neuropathological similarity has been shown between
excitatory striatal lesions in the rat and Huntington's disease (HD).
Thus, excitatory lesions in rats have served extensively as
experimental models of HD (Coyle and Schwartz, 1983; Isacson et al.,
1985; Beal et al., 1986; Sanberg et al., 1989). A large body of data
now supports the notion that this lesion is similar to the striatal
lesion observed in HD and that this rat model can mimic the
neurological dysfunction associated with the striatal degeneration in
Huntingtonian patients. However, motor manifestations in lesioned
rodents are restricted to changes in locomotor activity and do not
include dyskinesias or chorea-like movements. The therapeutic
predictions from the rodent model are therefore limited compared to
data that could be obtained from other species, especially primates.
In order to study the neurological basis for dyskinesias associated
with striatal dysfunction in the primate and to test new therapeutic
strategies for the treatment of HD such as neural transplantation, we
have developed a striatal excitotoxic lesion model in primates using
ibotenic acid (IA) as excitotoxin. In this study data are presented
showing : 1) the occurrence in the primate of "chorea-like" movements
relevant to HD following excitotoxic lesion of the caudate nucleus and
putamen. 2) the survival and the anatomical and functional
characteristics of fetal rat striatal cells following their
implantation into the neuron-depleted striatum of the baboon.

METHODS

Excitotoxic lesion. Thirteen baboons were used. Nine baboons received
IA injections into the right caudate-putamen (CP) corresponding to a
total IA dose of either 350 µg or 700 µg. Four non-operated animals
served as controls (for details, see Hantraye et al., 1989, 1990).

Behaviour. Before lesion and 2 to 39 weeks after lesion, the baboons were repeatedly tested over a 60 min period for spontaneous and drug-induced (apomorphine 1 mg/kg i.m.) behaviours. All test sessions were video-recorded and the baboons were rated for incidence of abnormal movements. Different types of abnormal motor behaviours were observed under apomorphine testing and categorized into discrete classes to allow quantification of the observations. The incidence of these categories of motor behaviours was recorded for each individuals and noted as present (1) or absent (0) during each time-period of 10 min of the 60 min test-session. Parametric statistical comparison of the occurrence of these symptoms in the controls versus the IA-lesioned baboons was done by one-way analysis of variance with post-hoc Neuman Keuls test and Scheffe F-test.

Transplantation procedure and immunosuppressive treatment. Four to 6 weeks after lesion, 5 lesioned animals were prepared for xenogeneic transplantations of embryonic rat striatal tissue. Baboons to be grafted were treated daily with i.m. injections (10-15 mg/kg) of cyclosporin-A (kindly provided by Sandoz-France). Cyclosporin-A treatment was initiated 2-3 days prior grafting. The day of the transplantation, cell suspensions were prepared from primordial striatum taken from rat fetuses of 14-15 days gestational age following the method of Schmidt et al. (1981).

Histology. After completion of the behavioural testing, animals were deeply anaesthetized and perfused transcardially with heparinized saline (1 l) and 3-5 l of 4% paraformaldehyde containing 0.1% glutaraldehyde in phosphate buffer (pH 7.4). Frozen sections (40 μm thick) were serially cut, and collected in series of 10 adjacent sections in tris buffer containing azide. In each series, one section was stained with cresyl violet, one section processed for localization of acetylcholinesterase (AChE), the remaining being kept for immunocytochemical localization of glial fibrillary acidic protein (GFAP), tyrosine hydroxylase (TH), serotonin (5-HT) and Leu-enkephalin (Enk). Technical details have been given elsewhere (Isacson et al., 1989).

Fig. 1. Frontal sections from a same animal receiving 700 μg IA into the right CP. a) At the level of the striatum, there is no acetylcholinesterase (AChE) reaction into the dorsal part of the caudate nucleus (NC). It is reduced in the medial part of the putamen (P) compared to the normal left side. There is also a decrease in the ipsilateral globus pallidus (GP). The enlargement of the lateral ventricle (Vl) is obvious. (CC: corpus callosum; CA: anterior commissure; CI: internal capsule; Hy: hypothalamus). b) Section at the level of the substantia nigra (SN), stained for AChE (cl: nucleus centralis lateralis; GL: lateral geniculate; md: n. dorsalis medialis; rt: n. reticularis; VPI: n. postero- inferior; St: subthalamus; nr: n. ruber; PC: cerebral peduncle). Compared to the left side, some flattening of the SN can be observed. c) Section at the same level processed for TH-immunoreactivity (IR) and slightly counterstained with cresyl violet. TH-IR is observed in the SN and in the ventral tegmental area (vt). The flattening of the right SN seems to be due to a reduction of volume of the pars reticulata rather than to a reduction in the number of TH-positive cells in the pars compacta.

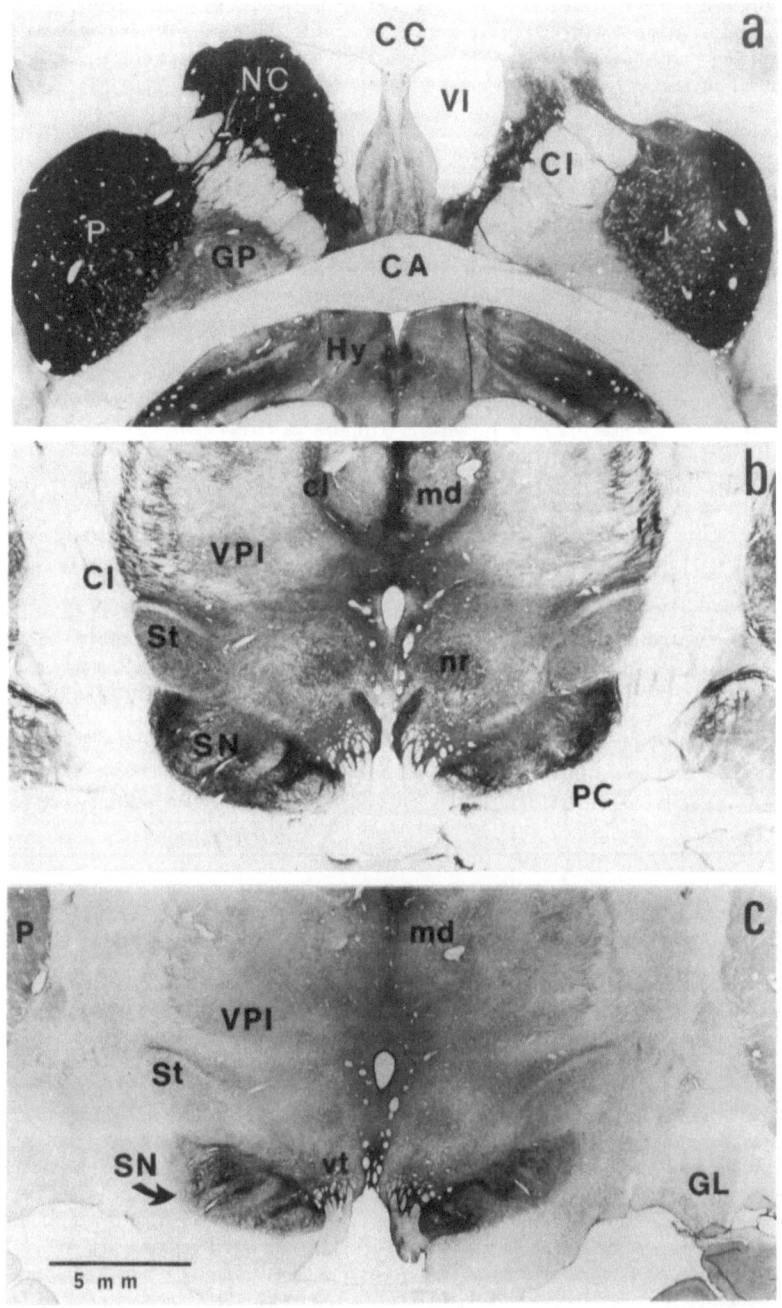

Fig. 1.

OBSERVATIONS IN THE LESIONED ANIMALS

1. Behavioural studies

Spontaneous behaviours : Two to five days after surgery the baboons recovered their usual behaviour. Few spontaneous abnormal movements were observed. Rare dyskinetic head movements occurred in some of the lesioned animals, but were never seen in controls.

Dopamine agonist-induced behaviours : Given the previous findings that dopamine-stimulating drugs can induce choreatic movements in presymptomatic Huntingtonian patients or aggravate existing symptoms in HD patients (Paulson, 1976), we used this pharmacological method to elicit symptoms in baboons with unilateral lesions of the CP. The dopamine-agonist drug apomorphine was administered at doses ranging between 0.5-2 mg/kg to observe possible facilitation of abnormal behaviours in locomotion and movements. The most dramatic features of the response were various stereotypic and dyskinetic movements. These motor behaviours included postural asymmetries and repetitive dystonic movements of hands and head, that may be analogous to choreic and involuntary movements in humans. Intact animals never showed any of the dyskinesias, asymmetric postures or increased locomotion described, while they occurred to a degree in all the lesioned animals. Abnormal movements were most marked contralateral to the lesion, but most often involved complex bilateral integration of movements of the whole body. When the trunk of the body was involved, the general locomotion, frequency and intensity of movements increased. Stereotypic jaw movements also occurred in association with biting behaviours. About 5-10 min after the apomorphine injection, episodes of explosive movements were seen. These abnormal movements were brief (<10 seconds) and included bursts of jumps and random ballistic movements. The highest locomotor activity occurred in animals showing strong postural asymmetry and movements were associated with continuous ipsilateral circling behaviour. These dyskinesias occurred for several months without any sign of recovery.

2. Anatomical studies

In the animals receiving IA injections, cresyl violet or AChE stained sections revealed the presence of a lesion, with a size related to the amount of excitotoxin injected. The injection site appeared to consist of three different areas, concentric to the needle track. The center of the lesion in the vicinity of the needle and where the amount of toxin was maximal, showed some necrosis and a complete cell loss, even with the lowest dose of IA used (350 µg). Around this central zone was a transition zone characterized by marked neuronal loss. Finally, between the outer limits of the transition zone and the remaining striatum which appeared normal was a region with intense reactive astrocytosis. In AChE sections, these regions showed low reaction (Fig. 1a). Usually, the shrinkage of the CP on the lesioned side was accompanied by an enlargement of the lateral ventricle, as previously observed in the rat (Isacson et al., 1984). At the level of the substantia nigra (Fig. 1b,c), some shrinkage of this structure was observed, ipsilateral to the lesion. In sections processed for immunocytochemical localization of TH, this phenomenon seemed to be due not to a loss of neurons in the pars compacta but rather to a flattening of the pars reticulata (Fig. 1c). Using the antiserum against GFAP, morphological variations in the astrocytic population have been observed from 4 days to 6 months after lesion.

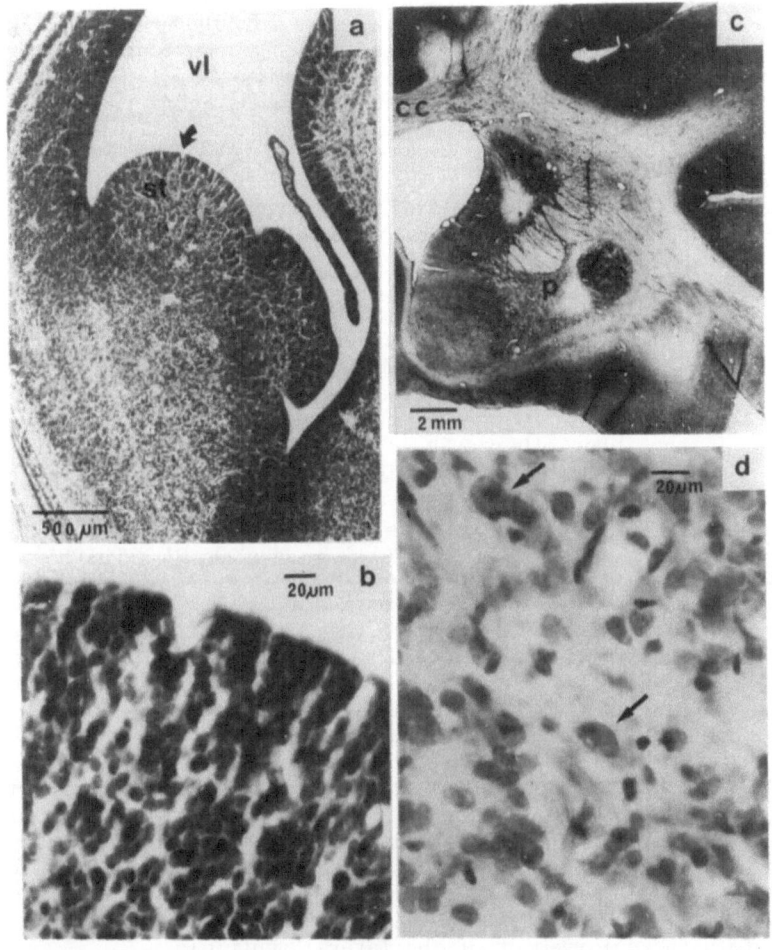

Fig. 2. Embryonic tissue and development of the graft.

a) Paraffin section (10 μm thick) cresyl violet stained from the brain of a 14 days embryo rat showing the striatum primordium (st), (vl: lateral ventricle). b) Higher magnification of the area denoted in (a) by the arrow. Cells appear densely packed and slightly shrunken. c) Frozen section from baboon T3 (survival: 2 1/2 months). The implant is visible in the putamen (P). Some neuronal loss appears colorless in the CP. d) Higher magnification in the graft. The size of rat cells is increased compared to (b). They are different neuronal types, including large neurons (arrows).

1. Behavioural results

Four IA-lesioned and transplanted baboons have been challenged with apomorphine testing. All these baboons showed a gradual decline both of ipsilateral half-turns and dyskinesias at test-sessions up to several months after implantation. In particular, the oro-facial dyskinesias and dystonic/chorea-like movements were greatly compensated, while body asymmetry and turning behaviour were partially corrected. The general locomotor hyperactivity was also reduced by the striatal implants. These preliminary results of a beneficial effect of the grafts require extensive substantiation and elucidation by additional experiments. However, it must be said that no recovery or alteration of the dyskinetic response to apomorphine has been observed in any of the lesioned baboons without transplants for up to 2 years after lesion (Hantraye et al., in preparation).

2. Anatomical results

In the first transplanted animal (T1, survival time : 6 months) which received 350 µg IA, no graft could be found. In the second animal (T2) and in the absence of a specific marker for rat cells, we had only doubts about the presence of some rat cells along the needle tracks. In the three other animals (T3,T4,T5), rat cells survived and were found morphologically identical to those observed in rats by previous investigators using the same protocol (Isacson et al., 1984, 1985, 1987 ; Clarke et al., 1988 ; McAllister et al., 1985) (Fig. 2a,b). In cresyl violet stained sections, neurons of apparently normal morphology were observed aggregated in lobules or patches (Fig. 2c), showing no signs of increased infiltration by macrophages but a density of cells higher (2-3 times) than in normal host striatum. As described in a preliminary report (Isacson et al., 1989), these rat cells seem to have differenciated (Fig. 2d) and to have grown to their normal phenotypic size. The total volume of each implant was 14, 24 and 29 mm3 in T3, T4 and T5, respectively. As the total volume of cells injected was 70µl, in each animal, the volume of the grafts found in the host brains varies from 20 to 40% of the amount of cells injected. This volume is less than expected from similar studies in the rat. Fig. 3 shows the development of the graft in T4, 2 1/2 months after transplantation. In this case, the implant was more developed in the putamen than in the caudate nucleus. In one case (T5), the graft has developed in both the caudate and the putamen, compensating almost completely the enlargement of the lateral ventricle.

On sections stained for AChE, the grafted tissue showed a typical patchy appearance with a dense network of fibers and some cells presumably of the interneuron type. Sections stained for GFAP showed a strong reactive astrocytosis surrounding the graft, with some long fibers infiltrating through the lobules. Some small sized glial cells bearing long and thin processes were seen in the same regions, representing probably donor cells. TH-labeled fibers were observed in the implant, less dense than in the lesioned host surrounding tissue, with a small number of TH-positive neurons. Their presence in the graft can be considered as normal because TH-IR neurons have been recently described in the normal rat striatum (Tashiro et al., 1989). Among other antibodies tested, Leu-Enk immunoreactivity (IR) has shown, in the normal dorsal part of the striatum thin labeled terminals and a more intense labeling in ventral parts of the caudate and putamen with some IR-cells. In the globus pallidus, the labeling was conspicuous with thick fibers as in the substantia innominata and in the perforated substance. In the transplant, Enk-IR fibers and terminals were also seen among the lobules. Some cell bodies can be

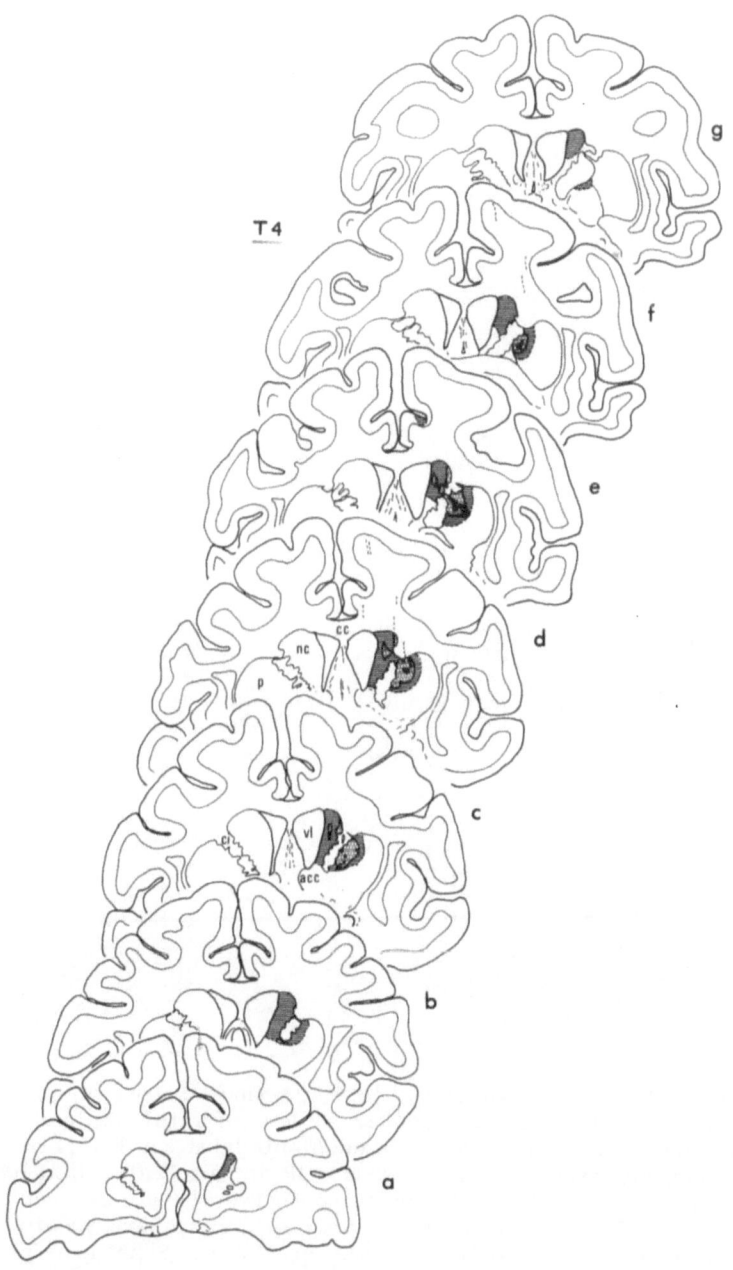

Fig. 3. Schematic drawings of frontal sections through the striatum of baboon T4, in the rostro-caudal direction (a-g). Horizontal hatched areas: lesioned tissue. Vertical broken hatched areas: implanted tissue (acc: n. accumbens; cc: corpus callosum; nc: n. caudatus; p: putamen; vl: lateral ventricle).

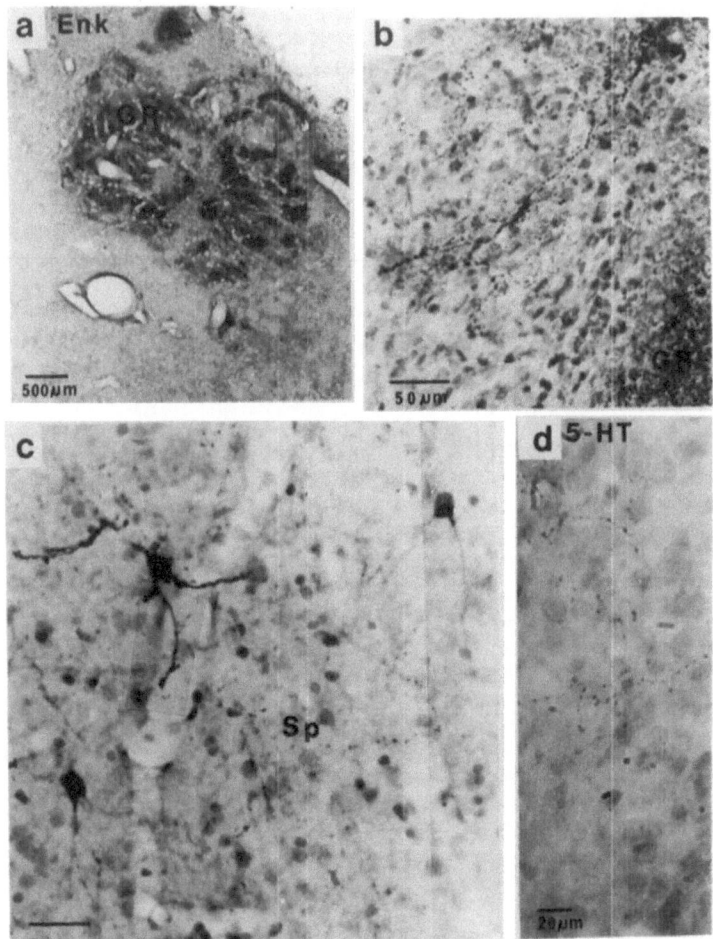

Fig. 4. Leu-enkephalin (a,b,c) and serotonin-immuno-
reactivity (d) in the grafts. a) Rat striatal implant (GR)
in the caudate nucleus of transplanted baboon T5. Note the
patchy appearance of the Leu-Enk IR in the graft. b) Higher
magnification: Leu-Enk reactive nerve terminals can be
observed in the transplant as well as long fibers running in
the immediate vicinity of the graft. c) Some Leu-Enk-IR
cells with long processes and thin fibers (scale bar: 50 μm)
at the level of the septum (Sp). d) 5HT-IR fibers in the
graft. These very thin fibers with small fusiform
varicosities, can be considered as small varicose axons.

distinguished (Fig. 4a,b,c). The lesioned tissue around the graft was devoid of labeling. With the antibody against 5-HT, an accumulation of labeled fibers was observed in the lesioned striatum and around the graft. Some thin fibers with varicosities were observed in the graft.

DISCUSSION

Ibotenic acid lesions : behavioural deficits and histopathological modifications

As previously described, the prominent features of the response to apomorphine administration in the IA-lesioned baboons was the various behavioural manifestations resembling those found in HD patients. Thus, clearly recognizable "choreic-like" movements were observed, always associated with the striatal lesion. These abnormal movements, unpredictables in nature, generally involved both upper and lower limbs as well as the neck, face and trunk of the animals. Typically, series of dyskinesias of the neck, limbs and trunk were followed by rapid jerky movements involving the lower limbs and the trunk of the animal. Twisting movements of the forearms or hind limbs resembling dystonias were also frequently noted. Accompanying these symptoms, oro-facial dyskinesias were observed as well as, in some animals, a typical turning behaviour, the animal rotating toward the lesioned side. This latter symptom, when present, was rather stereotyped and persisted for almost the entire testing period. In the past, various attempts were made to reproduce in the primate the involuntary movements observed in HD. A number of experiments were undertaken by means of electrolytic lesions in the striatum. Unfortunately, these lesions that not only destroy striatal neurons but also affect striatal afferents and fibers en passant failed to produce "choreic-like" movements in the primate. As an alternative, striatal excitotoxic lesions were developed in the rat using stereotaxic injections of various endogenous or exogenous glutamate receptor agonists (Coyle and Schwarcz, 1983; Beal et al., 1986). The present study shows that such excitotoxic lesions can produce neuropathological changes similar to those encountered in HD brains, including the relative sparing of striatal afferents and of certain striatal cell types also preserved in the neurodegenerative disease. However, even if these excitotoxins can partly simulate the biochemical and neuropathological alterations of HD, their direct application into the CP is not enough to produce dyskinesias or dystonias in excitotoxically lesioned primates (Kanazawa et al., 1986; Hantraye et al., 1989). The combined use of excitotoxic lesion and systemic administration of dopamine-agonist drugs is necessary to produce such symptoms in the primate. Since neither the excitotoxic striatal lesion or the administration of apomorphine alone can induce the choreic-like movements, the neurophysiological and neuropharmacological basis for the observed symptoms may therefore be attributed to the combined effects of striatal excitotoxic lesion and excessive stimulation of CNS dopaminoceptive neurons. As suggested in the present paper and in previous studies (Ellison et al., 1987), the excitotoxic lesion of the striatum produces a denervation of two main striatal output structures (globus pallidus and substantia nigra) and therefore remove the GABAergic inhibitory control of the striatum on these structures. The decreased GABAergic inhibition led to a decrease in the pallidothalamic inhibition exerted via the subthalamic nucleus and the medial part of the globus pallidus (Penney and Young, 1983). Then, an excessive dopamine receptor activation by apomorphine will activate motor associated circuitry in the spared ventral striatum as well as in the cerebral cortex and some areas of the basal ganglia. This activation may elicit abnormal movements due to the decrease in the pallidothalamic inhibition in lesioned but not in control animals. Consistent with the idea that the ventral striatum is an important component in the production of choreic movements is the reduction in the incidence of the dyskinesias and locomotor hyperactivity found after IA-lesion of the nucleus accumbens in one baboon previously lesioned in the CP (unpublished observation).

Striato-striatal grafts : morphological and immunohistochemical characteristics

Our results show that xenografts of rat striatum primordium implanted into the lesioned CP of the baboon can survive and express a number of biochemical and morphological properties characteristic of normal striatal neurons. Survival of the xenograft was achieved by using the immunosuppressive agent, cyclosporin-A. Two 1/2 months after transplantation into the lesioned CP, striatal implants were identified as a well-delineated tissue mass in the lesioned area. By improving the transplantation protocole and the immunosuppressive treatment, the volume of the graft was brought from 14 mm3 to 30 mm3. With AChE histochemistry, each graft has a patchy appearance showing intense reaction in fibers and neurons. Similarly, by detection of Leu-Enk-immunoreactivity, a patchy labeling of fibers and terminals was observed within the transplant while no labeling was seen in the surrounding lesioned striatum. Peptide immunoreactivity has been found depleted within choreic brains (Waters et al., 1988), as well as a severe shrinkage of the substantia nigra but with a total number of pigmented neurons (showing TH-IR) within the normal range. This is in agreement with our observations in the baboon and consistent with biochemical observations that TH activity or dopamine concentration are normal in the basal ganglia of choreic patients (Bird and Iversen, 1974). With 5HT-immunohistochemistry, small varicose axons could be observed within the transplant. As no 5HT-positive neurons were observed in the graft, these fibers may be derived from the host, arising from midbrain raphe nuclei. Accumulation of 5HT-positive fibers was observed in lesioned regions of the striatum in lesioned animals and in areas surrounding the graft in transplanted cases. This, further suggests that IA excitotoxicity is mainly directed toward the intrinsic neuronal somata and spare afferent fibers.

In agreement with previous observations in the rat, after 2 1/2 months survival, fetal catecholamine- and peptides-containing neurons survive transplantation and show immunocytochemical and morphological characteristics suggesting that some reconstruction of the neuronal circuitry occurs between the rat neural transplant and the baboon host brain. The observation of viable striatal implants growing into severely lesioned CP provide an anatomical basis for the behavioural compensatory effects observed in our transplanted animals.

ACKNOWLEDGMENTS

The authors gratefully acknowledge S. Mirman, J.P. Bouillot, H. Hryn for skillful technical assistance, and J. Cayla and his staff for excellent primate care. This work was partially supported by NATO Grant CRG 890583, BRSG Grant RRO5484 awarded by National Institutes of Health and DRET Grant 88/053.

REFERENCES

Beal, M.F., Kowall, N.W., Ellison, D.W., Mazurek, M.F., Swartz, K.J., Martin, J.B., 1986, Replication of the neurochemical characteristics of Huntington's disease by quinolinic acid, Nature, 321:168-171.

Bird, E.D. and Iversen, L.L., 1974, Post mortem measurements of glutamic acid decarboxylase, choline acetyltransferase and dopamine in basal ganglia. Brain, 97:457-472.

Clarke, D.J., Dunnett, S.B., Isacson, O., Sirinathsinghji, D.J.S. and Björklund, A., 1988, Striatal grafts in rats with unilateral neostriatal lesions. I. Ultrastructural evidence of afferent synaptic inputs from the host nigrostriatal pathway. Neurosci., 24:791-801.

Coyle, J.T. and Schwarcz, R., 1983, The use of excitatory amino acids as selective neurotoxins, in: "Handbook of Chemical Neuroanatomy, Vol. 1, A. Björklund & T. Hökfelt, eds, Amsterdam:Elsevier, pp. 508-527.

Ellison, D.W., Beal, M.F., Mazurek, M.F., Malloy, J.R., Bird, E.D. and Martin, J.B., 1987, Aminoacid neurotransmitter abnormalities in Huntington's disease and the quinolinic acid animal model of Huntington's disease. Brain, 110:1657-1673.

Hantraye, P., Riche, D., Maziere, M., Maziere, B., Loc'h, C. and Isacson, O., 1989, Anatomical, behavioural and positron emission tomography studies of unilateral excitotoxic lesions of the baboon caudate-putamen as a primate model of Huntington's disease, in: "Neural Mechanisms in Disorders of Movement", A.R. Crossman & M.A. Sambrook, eds, John Libbey, London, pp. 183-193.

Hantraye, P., Riche, D., Maziere, M. and Isacson, O., 1990, A primate model of Huntington's disease: behavioral and anatomical studies of unilateral excitotoxic lesions of the caudate-putamen in the baboon. Exp.Neurol., in press.

Isacson, O., Brundin, P., Kelly, P.A.T., Gage, F.H. and Björklund, A., 1984, Functional neuronal replacement by grafted neurons in the ibotenic acid-lesioned striatum, Nature, 311:458-460.

Isacson, O., Brundin, P., Gage, F.H. and Björklund, A., 1985, Neural grafting in a rat model of Huntington's disease: progressive neurochemical changes after neostriatal ibotenate lesions and striatal tissue grafting, Neurosci., 16:799-817.

Isacson, O., Dawbarn, D., Brundin, P., Gage, F.H., Emson, P.C. and Björklund, A., 1987, Neural grafting in a rat model of Huntington's disease: striosomal-like organization of striatal grafts as revealed by acetylcholinesterase histochemistry, immunocytochemistry and receptor autoradiography, Neurosci., 22:481-497.

Isacson, O., Riche, D., Hantraye, P., Sofroniew, M.V. and Maziere, M., 1989, A primate model of Huntington's disease: cross-species implantation of striatal precursor cells to the excitotoxically lesioned baboon caudate-putamen, Exp.Brain Res., 75:213-220.

Kanazawa, I., Tanaka, Y. and Cho, F., 1986, "Choreic" movement induced by unilateral kainate lesion of the striatum and L-dopa administration in monkey, Neurosci. Lett., 71:241-246.

McAllister, J.P., Walker, P.D., Zemanick, M.C., Weber, A.B., Kaplan, L.J. and Reynolds, M.A., 1985, Morphology of embryonic neostriatal cell suspension transplanted into adult neostriata, Dev. Brain Res., 23:282-286.

Paulson, G.W., 1976, Predictive tests in Huntington's disease, in: "The Basal Ganglia", M.D. Yahr, ed., Raven Press, pp. 317-329.

Penney, J.B. and Young, A.B., 1983, Speculations on the functional anatomy of basal ganglia disorders, Ann. Rev. Neurosci., 6:73-94.

Sanberg, P.R., Calderon, S.F., Giordano, M., Tew, J.M. and Norman, A.B., 1989, The quinolinic acid model of Huntington's disease: locomotor abnormalities, Exp. Neurol., 105:45-53.

Schmidt, R.H., A. Björklund, A. and Stenevi, U., 1981, Intracerebral grafting of dissociated CNS tissue suspensions: a new approach for neuronal transplantation to deep brain sites, Brain Res., 218:347-356.

Tashiro, Y., Sugimoto, T., Hattori, T., Uemura, Y., Nagatsu, I., Kikuchi, H. and Mizuno, N., 1989, Tyrosine hydroxylase-like immunoreactive neurons in the striatum of the rat, Neurosci. Lett., 97:6-10.

Waters, C.M., Peck, R., Rossor, M., Reynolds, G.P. and Hunt, S.P., 1988, Immunocytochemicals studies on the basal ganglia and substantia nigra in Parkinson's disease and Huntington's chorea. Neurosci., 25:419-438.

[14C]-2-DEOXYGLUCOSE AUTORADIOGRAPHIC STUDIES OF DOPAMINERGIC DRUGS IN

AN ANIMAL MODEL OF PARKINSON'S DISEASE

Joel M. Trugman and G. Frederick Wooten

Department of Neurology, Box 394
University of Virginia Health Sciences Center
Charlottesville, Virginia 22908

INTRODUCTION

Parkinson's disease is unique among neurological disorders in that all of its cardinal signs and symptoms can be attributed to brain dopamine deficiency and corrected by dopamine replacement. Although great progress has been made, drug treatment of patients with advanced Parkinson's disease is severely limited by declining efficacy of dopaminergic drugs, the emergence of drug-induced dyskinesias, and the occurrence of complex clinical fluctuations. Two important questions remain regarding dopaminergic therapy: 1) Which dopamine receptors (i.e. anatomic location, receptor subtype) need to be stimulated to achieve an antiparkinson effect? 2) Subsequent to dopamine receptor stimulation, which neural circuits are activated to mediate these effects? Substantive answers to these questions may suggest new treatment strategies for Parkinson's disease, or alternatively, more rational use of currently available drugs.

We have addressed these questions by studying the effects of systemically administered dopaminergic drugs upon regional cerebral glucose utilization (RCGU) in rats with unilateral 6-hydroxydopamine (6-OHDA) lesions of the substantia nigra, a classical animal model of Parkinson's disease (Ungerstedt et al., 1973). Since glucose is the primary energy-producing substrate in the brain and the major expenditure of brain energy is for pumping of ions across membranes, the autoradiographic measure of RCGU provides a functional map of brain physiological activity (Sokoloff et al., 1977; Mata et al., 1980). The primary aim of these studies has been to use the 2-deoxyglucose (2-DG) method to identify which brain regions are activated when dopaminergic drugs are delivered to a dopamine-depleted basal ganglia. The results demonstrate that selective D1 and D2 agonists produce different effects on glucose utilization in the entopeduncular nucleus (EP) and the substantia nigra pars reticulata (SNr), two critical basal ganglia output nuclei. D1 receptor stimulation markedly increases RCGU in these regions whereas D2 stimulation does not (Trugman and Wooten, 1987). Experiments using drugs of varying receptor selectivity in combination with D1 and D2 antagonists suggest that D1 dopamine agonist effects can be quantitatively assessed in vivo using 2-DG autoradiography (Trugman et al., 1989). In this paper we review these studies and discuss the implications for the treatment of Parkinson's disease.

The Basal Ganglia III, Edited by G. Bernardi et al.
Plenum Press, New York

METHODS

Animals and Lesions

Male Sprague-Dawley rats weighing 275-325 grams were pretreated with desipramine hydrochloride 25 mg/kg, i.p., and lesions were performed as previously described (Wooten and Collins, 1981). Six-hydroxydopamine hydrobromide (9.6 μg/2.4 μl of 0.1% ascorbic acid in normal saline) was injected over 3 minutes into the left perinigral region. Rats were tested 10 days later with apomorphine hydrochloride, 0.5 mg/kg, i.p., to determine lesion efficacy. Rats that turned contralaterally a minimum of 5 rotations/min at the peak action of apomorphine were used in subsequent 2-DG studies 4-8 weeks after lesion.

2-Deoxyglucose Autoradiography

[14C]-2-DG autoradiography was performed according to the method of Sokoloff et al. (1977). On the experimental day, after rats had been fasted overnight, a central venous catheter was inserted into the right external jugular vein under 1% halothane anesthesia. After 3 hours of recovery, rats were injected intravenously with the experimental drug followed by [14C]-2-DG (8 μCi/100 g b.wt. in 0.3-0.4 ml saline; 55 mCi/mM; American Radiolabeled Chemicals, Inc., St. Louis, MO) 2 min later. Forty-five minutes after the [14C]-2-DG injection, the rats were killed with a 50 mg/ml intravenous bolus of sodium pentobarbital, and the brains were rapidly removed and prepared for autoradiography. Film autoradiographs were analyzed with a Leitz variable aperture micro-densitometer. Data for glucose utilization were expressed as ratio of gray matter optical density (OD) to white matter OD.

RESULTS AND DISCUSSION

Effects of Selective D1 and D2 Agonists

To determine the functional metabolic consequences of selective D1 and D2 receptor stimulation, we examined the effects of the D1 agonist SKF-38393 and the D2 agonist quinpirole (LY-171555) on RCGU in rats with unilateral nigral lesions (Trugman and Wooten, 1987). SKF-38393 (0.5-25 mg/kg) and quinpirole (0.01-5.0 mg/kg) produced indistinguishable behavioral responses, including vigorous contralateral rotation. Treatment with each drug similarly increased glucose utilization, dose-dependently, in the parafascicular thalamus, subthalamic nucleus, and deep layers of the superior colliculus ipsilateral to the nigral lesion; RCGU was decreased in the ipsilateral lateral habenula. In contrast, D1 and D2 agonists differentially altered RCGU in the EP and SNr (Fig. 1). SKF-38393, 5.0 and 25.0 mg/kg, increased 2-DG uptake 127 and 275%, respectively, in the SNr ipsilateral to the lesion. Quinpirole, 1.0 and 5.0 mg/kg, caused maximal contralateral turning yet did not significantly alter 2-DG uptake in the ipsilateral SNr. The glucose utilization response of the ipsilateral EP paralleled that of the SNr, demonstrating large increases following the administration of SKF-38393 and minimal change following quinpirole. These data are consistent with the hypothesis that, upon systemic administration, SKF-38393 selectively stimulates striatoentopeducular and striatonigral neurons, a population of nerve cells known to express a high density of D1 receptors (Barone et al., 1987).

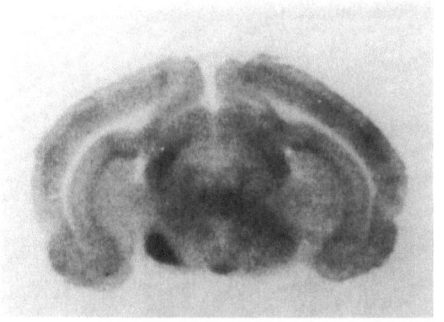
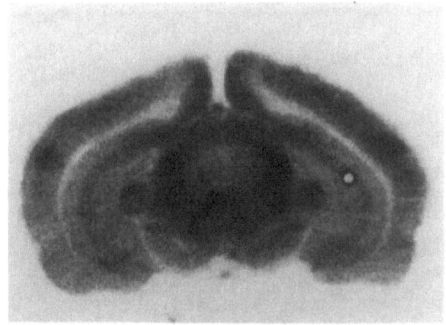

SKF 38393
25 mg/kg

Quinpinole
1.0 mg/kg

Fig 1.   [14C]-2-DG autoradiographs depicting RCGU patterns in the
unilateral nigral-lesioned rat treated with SKF-38393 (D1
agonist) and quinpirole (D2 agonist). The 6-OHDA lesion is on
the left side. SKF-38393, but not quinpirole, markedly
increased RCGU in the SNr ipsilateral to the lesion (from
Trugman and Wooten, 1987).

Glucose Utilization Increases Measure D1 Agonist Activity In Vivo

The results with selective D1 and D2 agonists suggested that in
this model, the ability of a dopamine agonist to increase RCGU in the EP
and SNr may represent a new in vivo assay of D1 agonist activity. To
test further this hypothesis, we compared the effects of bromocriptine,
pergolide, and (+)-4-propyl-9-hydroxynaphoxazine (PHNO), three dopamine
agonists used for the treatment of Parkinson's disease (Trugman et al.,
1989). Pergolide, in contrast to bromocriptine or PHNO, increased RCGU
in the SNr ipsilateral to the lesion in a dose-dependent manner (0.04
mg/kg, up 52%; 0.4 mg/kg, up 111%), resulting in asymmetric glucose
utilization on the dopamine-denervated and intact sides of the brain.
Pretreatment with a selective D1 antagonist (SCH 23390, 0.5 mg/kg)
blocked completely the RCGU increase elicited by pergolide (0.4 mg/kg)
whereas pretreatment with a selective D2 antagonist (eticlopride, 1.0
mg/kg) only mildly attenuated this increase (Figs. 2 and 3). The effect
of drug treatments on RCGU in the EP was similar to that in the SNr.
These results demonstrate that the large RCGU increase in the EP and SNr
after pergolide administration is dependent primarily on D1 receptor
stimulation. Administration of bromocriptine and PHNO minimally altered
RCGU in the ipsilateral EP and SNr and did not result in significant
left/right asymmetry. Of the three agonists studied, only pergolide has
D1 stimulatory properties in vitro (Goldstein et al., 1980). These
results suggest that, in this model, the magnitude of the RCGU increase
in the EP and SNr elicited by a dopamine agonist, above the modest
effects produced by selective D2 stimulation, represents a quantitative
measure of D1 agonist effect in vivo. The results support a non-
selective D1/D2 stimulatory effect of pergolide (0.04-0.4 mg/kg) and a
selective D2 action of both bromocriptine and PHNO.

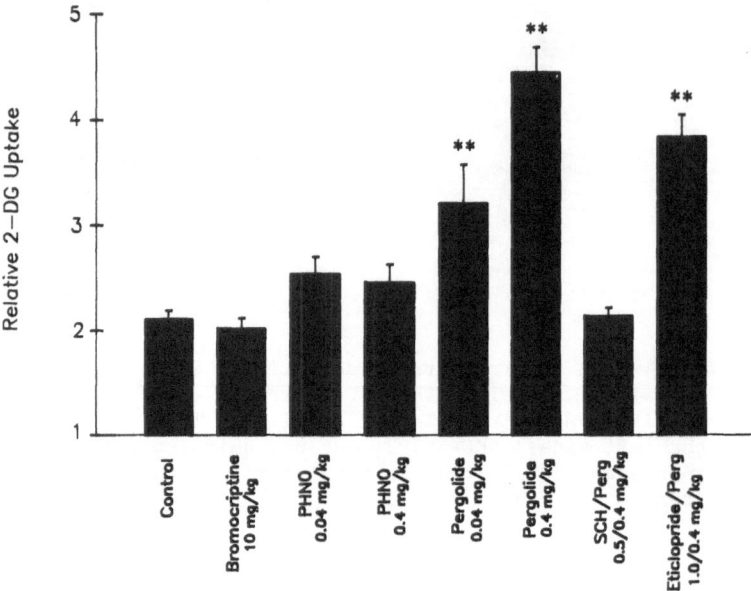

Substantia Nigra Pars Reticulata

Fig. 2. Glucose utilization in the SNr ipsilateral to the 6-OHDA
substantia nigra lesion. Values for pergolide (0.04 mg/kg),
pergolide (0.4 mg/kg) and eticlopride/pergolide (1.0/0.4 mg/kg)
differed from each other as well as all other groups (P<.05,
Duncan multiple range test) (from Trugman et al., 1989).

Metabolic Effects of Levodopa

Levodopa, the mainstay of the drug treatment of Parkinson's
disease, acts via conversion in the brain to dopamine. Following
pretreatment with carbidopa (25 mg/kg, i.p.), levodopa (10-50 mg/kg,
s.c.) markedly increased RCGU in the ipsilateral EP and SNr (Trugman and
Wooten, 1986). In contrast, levodopa did not alter RCGU in the globus
pallidus, consistent with the hypothesis that dopamine has different
effects on striatopallidal neurons as compared with striatoentopeduncular and striatonigral neurons. Moderate RCGU increases were
observed in the ipsilateral subthalamic nucleus, lateral midbrain
reticular formation, and deep layers of the superior colliculus, all
regions that receive direct projections from the globus pallidus, EP, or
SNr. The data suggest that the midbrain reticular formation and the
deep layers of the superior colliculus are functionally activated during
levodopa-induced turning and that nigroreticular and nigrocollicular
projections may mediate, in part, the effects of levodopa in humans.

To test the hypothesis that the RCGU increases in the EP and SNr
following levodopa administration are mediated primarily by D1 receptor
activation, we employed a paradigm of selective D1 and D2 antagonist
pretreatment. SCH 23390 (1.0 mg/kg) completely blocked the levodopa-

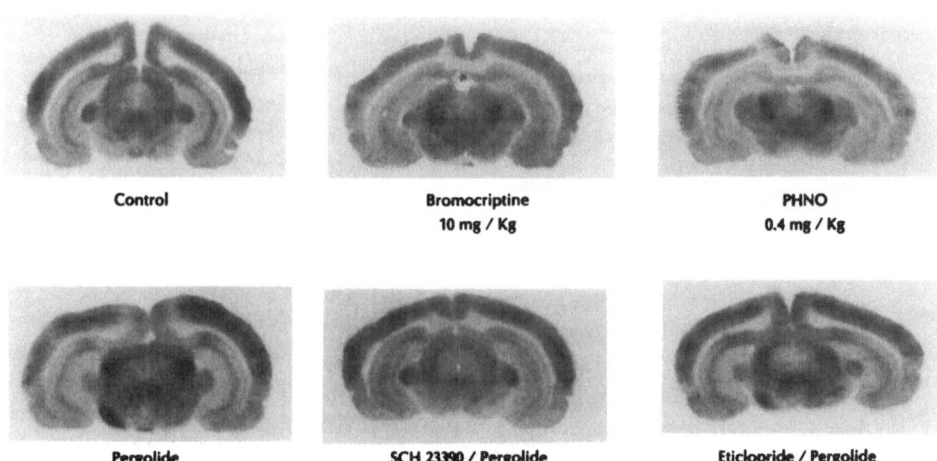

Fig. 3. 2-DG autoradiographs depicting RCGU patterns after dopamine
agonist treatment of rats with unilateral, left sided, 6-OHDA
substantia nigra lesions. Pergolide (0.4 mg/kg) increased RCGU
markedly in the SNr ipsilateral to the lesion, an effect which
was blocked by pretreatment with a selective D1 antagonist (SCH
23390), but only mildly attenuated by pretreatment with a
selective D2 antagonist (eticlopride). Bromocriptine and PHNO
failed to significantly increase glucose utilization in the SNr
ipsilateral to the lesion, consistent with a selective D2
agonist effect (from Trugman et al., 1989).

induced RCGU increases in the EP and SNr, returning RCGU values to
control levels (Fig. 4). Pretreatment with the selective D2 antagonist
eticlopride (2.0 mg/kg) was far less effective at attenuating levodopa-
induced changes, with RCGU values remaining significantly above control
levels. The data suggest that dopamine formed following decarboxylation
of levodopa stimulates both D1 and D2 receptors in vivo and that the
major metabolic effects of levodopa are critically dependent on D1
receptor stimulation.

CONCLUSIONS

     The administration of dopamine agonists to rats with unilateral
6-OHDA substantia nigra lesions produces two distinct but overlapping
patterns of RCGU corresponding to D1 and D2 receptor stimulation. Both
D1 and D2 agonists increase RCGU in the ipsilateral subthalamic nucleus,
deep layers of the superior colliculus, and parafascicular thalamus, and
decrease RCGU in the lateral habenula. The two patterns differ in that
D1 agonists, but not D2 agonists, markedly increase RCGU in the
ipsilateral EP and SNr. The results obtained with drugs of varying
D1/D2 selectivity have proven this distinction consistent and have

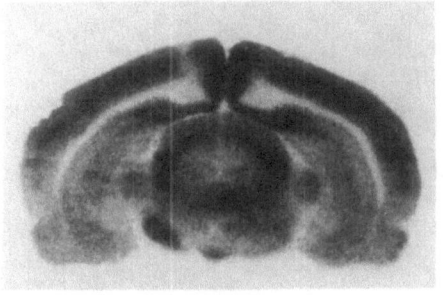
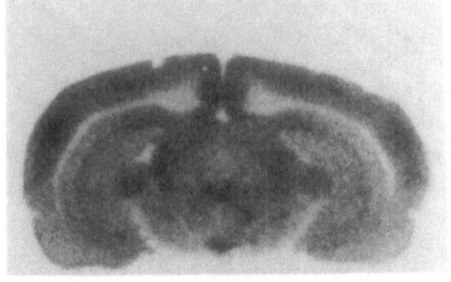

Levodopa
25 mg/kg

SCH 23390 / Levodopa
1.0 / 25 mg/kg

Fig. 4. Levodopa markedly increased RCGU in the SNr ipsilateral to the
6-OHDA lesion. Pretreatment with the D1 antagonist, SCH 23390,
completely blocked this RCGU increase.

suggested that, in this model, RCGU increases in the EP and SNr
ipsilateral to the 6-OHDA lesion can be used as a quantitative in vivo
index of D1 agonist activity. As such, 2-DG autoradiography may prove
to be a useful procedure for screening new compounds for D1 effects in
vivo.

Interpretation of RCGU changes requires anatomical, biochemical,
and electrophysiological correlation. Experimental evidence has
suggested that RCGU reflects predominantly activity in nerve terminals
as opposed to cell bodies (Mata et al., 1980; Kadekaro et al., 1985).
In 6-OHDA-lesioned rats, systemically administered apomorphine and SKF-
38393 consistently inhibit the firing of SNr output neurons whereas the
selective D2 agonist quinpirole is much less effective in producing
inhibition (Waszczak et al., 1984; Weick and Walters, 1987). We have
therefore proposed that D1-mediated RCGU increases in the EP and SNr
represent increased physiological activity in axon terminals of
GABAergic inhibitory striatoentopeduncular and striatonigral neurons.
The selective localization of D1 receptors on striatonigral neurons
(Harrison et al., 1989) suggests that dopamine may directly stimulate
this population of striatal cells.

Regarding the therapy of Parkinson's disease, the data suggest that
dopamine formed from levodopa stimulates both D1 and D2 receptors in
vivo. Therefore, to mimic the effects of levodopa with direct acting
drugs should require a combination of D1 and D2 agonists. The RCGU
effects of levodopa in the EP and SNr, however, are blocked completely
by D1 antagonist pretreatment but only mildly attenuated by D2
antagonist pretreatment, suggesting that the metabolic effects of D1 and
D2 stimulation by endogenous dopamine are separable. While the effects
of selective D1 and D2 stimulation in humans have not been clarified,
these results support the possibility of separating the adverse effects
from the therapeutic effects of antiparkinson drugs by a titrated
manipulation of the D1/D2 dopamine system.

ACKNOWLEDGEMENTS

This work was supported by a Cotzias Fellowship from the American
Parkinson Disease Association to J.M. Trugman, NIH-K08-NS01174, and The
Harrison Endowment of the University of Virginia.

REFERENCES

Barone, P., Tucci, I., Parashos S.A. and Chase, T.N.: D-1 dopamine receptor changes after striatal quinolinic acid lesion. Eur. J. Pharmacol. **138**: 141-145, 1987.

Goldstein, M., Lieberman, A., Lew, J.Y., Asano, T., Rosenfeld, M.R. and Makman, M.H.: Interaction of pergolide with central dopaminergic receptors. Proc. Natl. Acad. Sci. **77**: 3725-3728, 1980.

Harrison, M.B., Wiley, R.G. and Wooten, G.F.: Selective localization of D1 receptors to striatonigral neurons in the rat. Soc. Neurosci. Abs. **15**: 585, 1989.

Kadekaro, M., Crane, A.M. and Sokoloff, L.: Differential effects of electrical stimulation of sciatic nerve on metabolic activity in spinal cord and dorsal root ganglion in the rat. Proc. Natl. Acad. Sci. U.S.A. **82**: 6010-6013, 1985.

Mata, M., Fink, D.J., Gainer, H., Smith, C.B., Davidsen, L., Savaki, H., Schwartz, W.J. and Sokoloff, L.: Activity-dependent energy metabolism in rat posterior pituitary primarily reflects sodium pump activity. J. Neurochem. **34(1)**: 213-215, 1980.

Sokoloff, L., Reivich, M., Kennedy, C., Des Rosiers, M.H., Patlak, C.S., Pettigrew, K.D., Sakurada, O. and Shinohara, M.: The [14C] deoxyglucose method for the measurement of local cerebral glucose utilization: theory, procedure and normal values in the conscious and anesthetized albino rat. J. Neurochem. **28**: 897-916, 1977.

Trugman, J.M. and Wooten, G.F.: The effects of L-DOPA on regional cerebral glucose utilization in rats with unilateral lesions of the substantia nigra. Brain Res. **379**: 264-274, 1986.

Trugman, J.M. and Wooten, G.F.: Selective D1 and D2 dopamine agonists differentially alter basal ganglia glucose utilization in rats with unilateral 6-hydroxydopamine substantia nigra lesions. J. Neurosci. **7(9)**: 2927-2935, 1987.

Trugman, J.M., Arnold, W.S., Touchet, N. and Wooten, G.F.: D1 dopamine agonist effects assessed in vivo with [14C]-2-deoxyglucose autoradiography. J. Pharmacol. Exp. Ther. **250**: 1156-1160, 1989.

Ungerstedt, U., Avemo, A., Avemo, E., Ljungberg, T. and Ranje, C.: Animal models of parkinsonism. Adv. Neurol. **3**: 257-271, 1973.

Weick, B.G. and Walters, J.R.: Effects of D1 and D2 dopamine receptor stimulation on the activity of substantia nigra pars reticulata neurons in 6-hydroxydopamine lesioned rats: D1/D2 coactivation induces potentiated responses. Brain Res. **405**: 234-246, 1987.

Wooten, G.F. and Collins, R.C.: Metabolic effects of unilateral lesion of the substantia nigra. J. Neurosci. **1**: 285-291, 1981.

EFFECTS OF TRANSIENT GLOBAL ISCHEMIA ON THE BASAL GANGLIA OF RAT

R. Schmidt-Kastner, W. Paschen, K.-A. Hossmann

Max-Planck-Institut für neurologische Forschung
Dep. Experimental Neurology, Ostmerheimer Str. 200
D-5000 Köln 91, FR Germany

SUMMARY

The present studies focussed upon the effect of transient global brain ischemia on basal ganglia of rat. A modified four-vessel occlusion (4VO) model was used to induce forebrain ischemia. Regional cerebral blood flow was severely depressed in the striatum during 4VO. Biochemical measurements revealed a massive depletion of energy-rich metabolites in the striatum during ischemia which was followed by a transient recovery of energy metabolism. Neuropathological studies demonstrated early and massive lesions in the striatum. Damage increased from ventral to dorso-lateral areas of the striatum which may be related both to variations in the microcirculatory disturbances and to differences in intrinsic neuronal organization. Immunohistochemical studies showed a depression of staining for astrocytes and formation of vasogenic edema in the dorso-lateral striatum.

INTRODUCTION

Pathophysiological mechanisms of ischemic brain damage are under intense experimental investigation (reviewed in 1-4). Histological analysis showed that brain damage following global ischemia manifests preferentially in selectively vulnerable brain regions (5). Experimental global ischemia lasting 10 min or more produces neuronal lesions in the striatum (caudate putamen) of rats (5-7). Ischemic lesions in the striatum manifest in the early recirculation period (5, 8) wheras neuronal damage in other brain regions, as in the CA1 sector of the hippocampus, is delayed by 2 days (5, 9, 10). The present study focussed upon the specific features of ischemic damage in the basal ganglia using blood flow measurements, biochemical investigation of energy metabolism (11) and morphological techniques (10).

MATERIAL AND METHODS

Ischemia model: Forebrain ischemia was produced in adult male Wistar rats using a modification of the four-vessel occlusion model (12) as described in detail elsewhere (11). In short, animals were anesthetized with 1.5-2.5 % halothane in 70% nitrous oxide / 30% oxygen. On day 1, both vertebral arteries were coagulated. On day 2, both common carotid arteries were carefully exposed and occluded after halothane blow-off when the animals were still unresponsive. Ischemia was maintained for 30 min. Flattening of the electroencephaologram (EEG) was used as a selection criterion.

Biochemistry: Control rats with vertebral artery occlusion, animals subjected to ischemia of 30 min, and animals recirculated after ischemia for 8 and 24 hours (n=6 each) were re-anesthetized and brains were frozen in situ under mechanical ventilation (11). Tissue probes was dissected from the striatum for analysis of ATP, phosphocreatine, glucose and lactate content using standard enzymatic techniques.

Regional cerebral blood flow: Blood flow was measured by the ^{14}C-iodoantipyrine technique and quantitative autoradiography in controls (n=2) and animals during 4VO with a flat EEG (n=3) (11). Autoradiograms were taken at several frontal section levels and matched to a standard atlas of the rat brain (13). Residual blood flow which occurs during 4VO was calculated as percent of control.

Histology: Control rats (n=2-4 for each method) and post-ischemic animals (n = 3-10 at 3, 8, 24, and 72 hours survival) were perfusion-fixed through the left ventricle with 4% neutral-buffered formaldehyde under deep anesthesia. For conventional neuropathological analysis blocks containing the basal ganglia were embedded in paraplast, sectioned at 4-6 μm and stained by combined cresyl-violet/luxol-fast blue. Neuronal damage was estimated using a histological score with: 0 = normal, 1 = up to 10% of cells damaged, 2 = up to 50% of cells damaged and 3 = more than 50% of cells damaged. For spatial analysis of these changes the striatum was subdivided into four quadrants: dorso-lateral (dl), dorso-medial (dm), ventro-medial (vm), ventro-lateral (vl). Immunohistochemical staining was performed on closely spaced sections using the standard PAP-technique. Antibodies were directed against the astroglia-specific proteins, glial fibrillary acidic protein (GFAP) and S-100 protein. Serum-proteins which enter the brain tissue only under pathological conditions were visualized using antibodies against rat whole serum-proteins or rat serum-albumin. Other brains (n=8) were perfusion-fixed, sections were cut at 40 μm on a cryostat and reacted free-floating for acetylcholinesterase (AChE)-histochemisty using the method of Hedreen et al. (14).

RESULTS

Biochemistry: The changes of energy metabolites in the striatum are summarized in Tab. 1. During ischemia the energy metabolism was severely depressed with loss of high-energy phosphates or glucose and an increase of lactate. At 8 h recirculation, ATP and phosphocreatine were in the normal range whereas lactate was increased. By 24 h, metabolism deteriorated with a loss of ATP and further rise in lactate. (For further details of biochemical pathology of the striatum in 4 VO, see ref. 15, 16).

Regional cerebral blood flow: Four-vessel occlusion induced a massive reduction of blood flow to the striatum. Residual values ranged between 4-5% of control in vm and vl striatum and 1-2% in dl and dm striatum. Differences between dorsal and ventral areas were significant (t-test, p<0.001). Residual flow was higher in globus pallidus (15%) and substantia nigra (25%).

Table 1. Energy metabolism in striatum before, during, and 8 and
24 hours after 30 min of ischemia induced by 4VO

	Control*	Ischemia	8 h	24 h
PCr	3.43 ± 1.13	0.27 ± 0.67*	2.81 ± 1.00	2.48 ± 0.41
ATP	2.86 ± 0.27	0.19 ± 0.12*	2.34 ± 0.29	1.61 ± 0.75*
Glucose	1.52 ± 0.51	0.10 ± 0.20*	3.57 ± 0.67*	5.16 ± 0.82*
Lactate	2.29 ± 0.75	15.16 ± 1.16*	7.02 ± 2.74*	10.06 ± 2.20*

PCr = phosphocreatine. #) = Animals with vertebral artery occlusion alone.
*) = Significance at $p < 0.01$ (modified t-tests).

Neuropathology of the striatum: Control sections stained with cresyl
violet/luxol-fast blue revealed the typical cytology of the striatum with
about 98% of neurons of the medium-sized and 2% of the large variety
(Fig.3a). Immunohistochemical control staining for GFAP was low and patchy
in the normal striatum, but high and homogenous for S100-protein. All
pathological changes after ischemia appeared with a gradient of damage
along the dorso-ventral axis of the striatum which is reflected in the
histological scores for striatal subfields shown in Fig.1. Also, lesions
regularly extended along the striatum in the rostro-caudal axis.

At 3 h recirculation (Fig.3b) the neuropil in dl and vl striatum was
vacuolated leading to compression of capillaries. Several medium-sized
neurons appeared compressed and stained darkly. Large neurons appeared
normal or were also compressed. Astrocytic staining for GFAP or S100-
protein was as in control material. In some sections, staining for serum-
proteins surrounded larger vessels forming a halo of diffuse staining.

At 8 h recirculation (Fig.3c) dl, vl and dm striatum was massively
affected. The neuropil was highly vacuolated, and the vast majority of the
medium-sized neurons were severely distorted and shrunken into triangular,
dark-staining forms. Large neurons were found with a normal nucleus and
compressed cytoplasm. The underlying neuropil stained blue with luxol fast
blue and gave a peculiar refraction. Staining of the myelin-bundles was
suppressed. Large vessels were open whereas smaller vessels appeared
compressed. In all sections tissue damaged in the described way was mixed
in a patchy manner with islands of normal neurons. Glial staining was as in
controls. Staining for serum-proteins appeared in the lateral areas and
covered the neuropil in a diffuse manner leaving the damaged neurons and
vacuoles unstained.

At 24 h recirculation (Fig.3d) severe damage of the dl,vl and dm
striatum was already apparent on macroscopic inspection presenting as a
marked paleness. On microscopic examination the tissue showed a
progression of neuronal damage. A prominent feature was the marked
patchiness of the damage. Strongly damaged tissue was intermingled with
normal areas. Vacuolation had diminished in intensity whereas ischemic
neuronal death had progressed. Most of the medium-sized neurons were

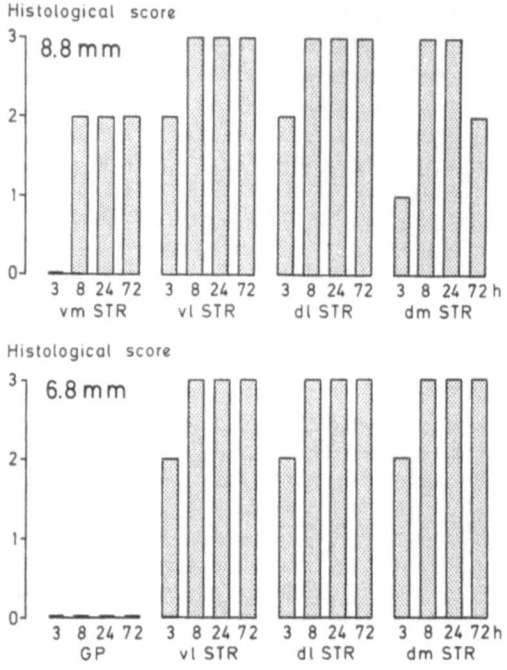

Figure 1. Neuronal damage in different areas of the striatum (STR) and in globus pallidus (GP). A histological score was used to estimate neuronal lesions in different areas with 0 = no damage, 1 = up to 10% , 2 = up to 50%, 3 = more than 50% of neurons damaged. Two frontal section levels were evaluated at 8.8 and 6.8 mm with reference to intraaural line (ref.13). Groups of n=3-6 animals survived for 3, 8, 24 and 72 hours after 30 min of ischemia. The striatum was divided into four quadrants: vm = ventro-medial, vl = ventro-lateral, dl = dorso-lateral, dm = dorsomedial.

necrotic and showed up as small irregular remnants of nuclei. Several large neurons had a well-defined nucleus and some vacuolated cytoplasm. The vessels were largely preserved but some enlargement of the endothelial cells was noted. GFAP-and S100-protein staining were normal in medial areas whereas the lateral areas revealed suppression of glial staining. Residues of the neurons reacted positive for the S-100 protein presumably indicating diffusion and adhesion of this protein. Staining for serum-proteins further increased in the neuropil and necrotic remnants of neurons were strongly stained. Staining for AChE was not different from control material.

At 72 h recirculation (Fig.3e) the picture was even more heterogenous with remnants of necrotic neurons mixed with persisting large neurons, reactive astrocytes, polymorphic phagocytes, microglial cells and perivascular proliferative elements. The most severe damage was in the lateral areas as shown in Fig.2. At this stage, the topographic differentiation in the GFAP-staining was even more pronounced (Fig.3f,g; Fig.2b). Ventral and medial areas, and also the most dorsal aspects showed very strongly stained, reactive astrocytes with thick processes (Fig.3f). Lateral areas showed depressed staining in most animals (Fig.3g). In the same areas, S-100 protein staining also showed depression whereas the ventro-medial belt showed enlarged reactive astrocytes. Serum-protein staining filled up the neuropil of the lateral striatum (Fig.3h; Fig.2c). Necrotic neurons were very strongly stained (Fig.3h). Persisting large neurons or reactive astrocytes might also show slight positivity. Staining for AChE was maintained with a lower intensity as controls.

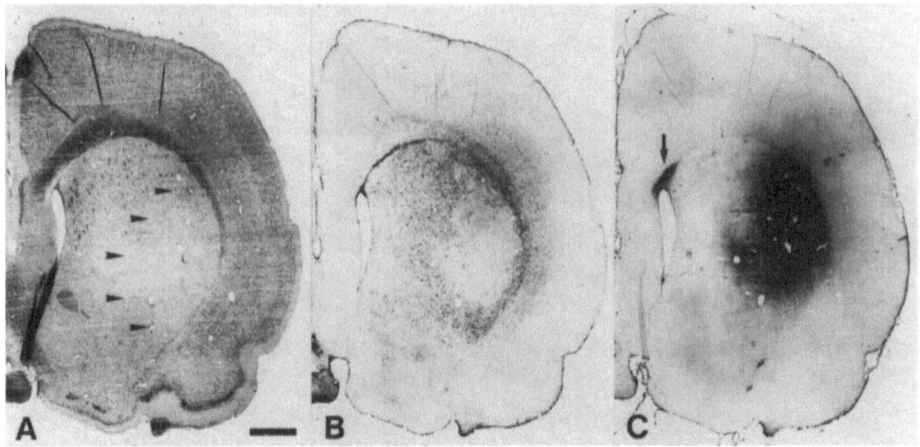

Figure 2. Striatal damage at 72 h after ischemia. A) Cresyl-violet stained section with area of massive neuronal lesions outlined by arrowheads. B) Section stained for glial fibrillary acidic protein (GFAP) with immunohistochemical technique. Note loss of staining in the lateral area contrasting with the marked increase in GFAP-stained astrocytes in surrounding areas. C) Immunohistochemical staining for serum-proteins in the lateral area. Note spreading of edema fluid along the corpus callosum towards the lateral ventricle as indicated by arrow. Bar = 1 mm.

Neuropathology of other basal ganglia nuclei: The dorsal margin of nucleus accumbens was variably involved in continuity with stiatal damage. Occasionally, lesions spread into the fundus striati. About half of animals revealead irregular foci of neuronal necrosis in the olfactory tubercle. The globus pallidus, nucleus basalis of Meynert, entopeduncular nucleus, subthalamic nucleus were nearly always intact. Two animals surviving 24 h revealed a peculiar oval lesion situated in pars reticulata of substantia nigra which was characterized by shrunken neurons, depressed glial staining and serum-protein extravasation.

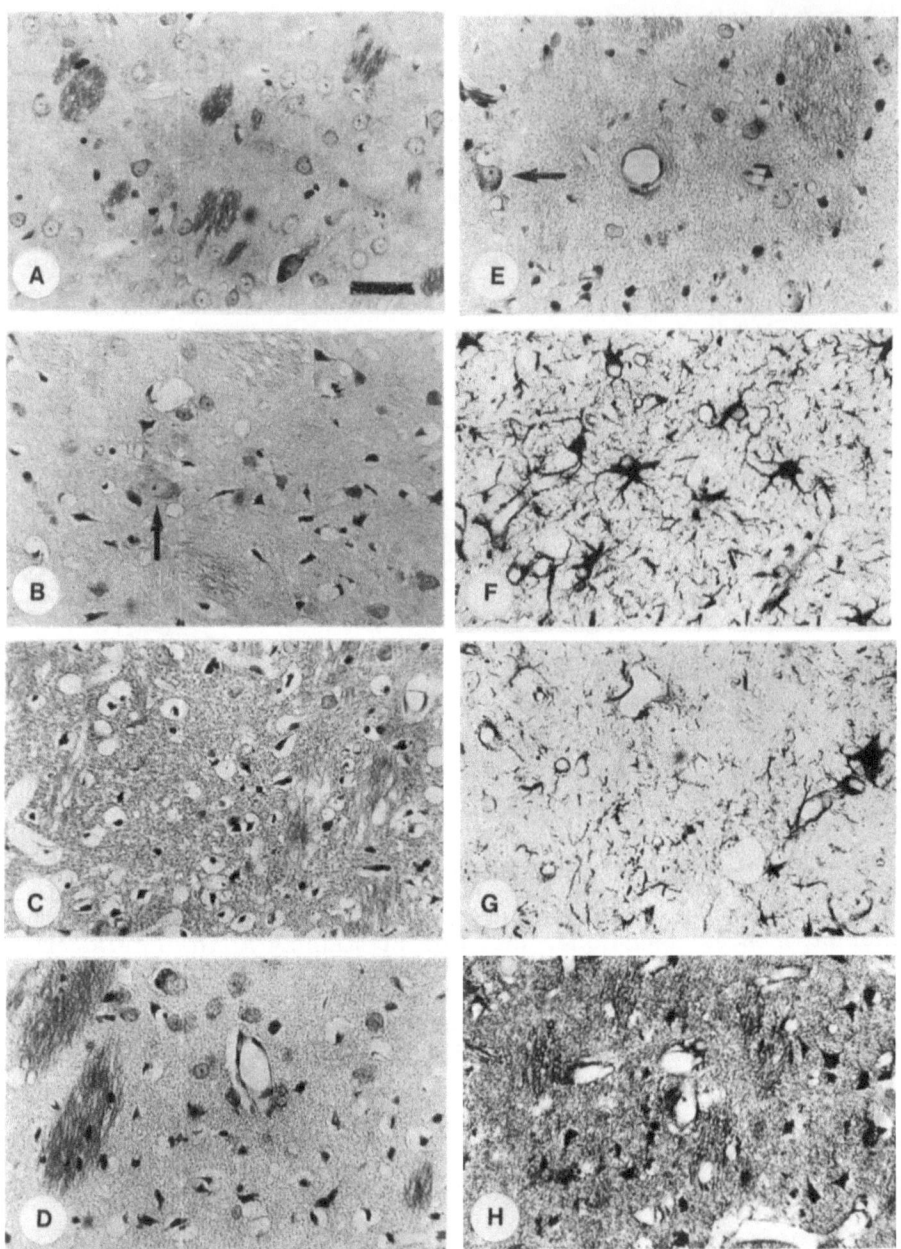

Figure 3. Neuropathology of the striatum at different survival times following 30 min of ischemia. Combined cresyl-violet/luxol fast blue staining in A) to E), immunohistochemistical staining in F) to H). A) Control striatum. B) 3 hours after ischemia. C) 8 hours survival, very strong vacuolation and refraction of neuropil giving high background staining. D) 24 hours survival, note mixture of damaged, shrunken neurons and intact cells. E) 72 hours survival, near-complete disappearance of medium-sized neurons, persistance of large neurons (arrow) and astrocytosis. F) GFAP-staining in the ventro-medial aspect of the striatum at 72 hours survival, strongly enlarged astrocytes with massive increase of GFAP-content. G) GFAP-staining in lateral area with absence of reactive astrocytosis. H) Serum-protein immunohistochemistry in lateral area, massive uptake of edem-fluid into necrotic neurons and in background. Bar = 50 μm.

DISCUSSION

Ischemic damage of the brain is caused by a variety of pathophysiological factors that interact in a complex manner (1-4). In the recovery period after ischemia persisting disturbances in blood flow and energy balance influence neuronal recovery. In the later recirculation period, biochemical deterioration as seen in striatum reflects neuronal death (11). In addition, the intrinsic organization of different brain regions determines the region-specific type of injury. The following discussion will be focussed upon specific features of ischemic damage found in the basal ganglia of rat.

In the present study the most severely affected structure was the dorso-lateral striatum confirming previous reports (5, 6, 8). In fact, in this region massive tissue injury was seen which included most neurons and possibly astroglial cells. Thus, the pathology of the lateral striatum could excede the stage of selective neuronal necrosis and then approached (sub-)infarction as noted in earlier studies (8). Interestingly, in the hippocampus the same dose of ischemia produces selective necrosis of neurons whereas astrocytes remain intact and react (10). One reason may be that the intrinsic resistance of astroglial cells and the neuron-glial interactions differ between hippocampus and striatum. Alternatively, the early and massive death of the majority of striatal neurons secondarily leads to involvement of astrocytes. Opening of the blood-brain barrier as observed with antibodies against serum-proteins was in proportion to the profound tissue damage.

Gradients in blood flow in different vascular territories might cause a stronger damage of dorso-lateral than ventro-medial areas. The striatum of the rat is supplied by arteries arising in a ventro-dorsal fashion, and vessels feeding the dorsal striatum are functional endarteries (17). Small differences in residual flow measured during ischemia are unlikely to account for such gradients of damage. After ischemia the striatum undergoes a period of post-ischemic hypoperfusion (18, 19) and uncoupling between blood flow and metabolism (6). However, in a recent study no regional differences within striatum were detected during the post-ischemic hypoperfusion phase in the 4VO model (19). Most likely, qualitative changes in the microcirculation existed and led to inhomogenous perfusion changes which escaped detection by autoradiographic measurements. Evidence for perfusion changes may be provided by the inhomogenous, patchy type of neuronal damage in striatum. On the other hand, the striatal neuropil is anatomically organized into "patch" and "matrix" compartments of neurotransmitters and fiber connections (20). In the present study the patchiness of ischemic damage does not resemble these basic anatomical compartments in size or distribution which is in line with recent immunohistochemical studies after 4VO (21). Several gradients exist along the dorso-ventral axis of the striatum in the anatomical and neurochemical organization (22-24). The distribution of the dopamine receptors (24) may have pathophysiological significance since a role for dopamine as a mediator of striatal ischemic injury has been proposed (19, 25, 26). Since gradients appear both in the vascular organization and in the intrinsic neuronal anatomy and biochemistry, the striatum is not suited to decide which are the leading pathogenetic factors in ischemic injury.

Another striking feature of ischemic injury is the selective vulnerability of different striatal cell classes. After ischemia a class of large neurons persisted even in the most severely lesioned areas (27). Evidence form several sources indicates that these neurons are equivalent to the cholinergic interneurons. Quantitative biochemical measurements of the enzyme, cholineacetyltransferase (ChAT), in post-ischemic striatum showed maintained levels (27). The persistence of AChE-staining (28) in the present study would support this observation. Similarly, neurons stained by antibodies against ChAT were found spared in ischemic lesions of the striatum induced by combined hypoxia/ischemia in newborn rats (29). It remains to be studied in more detail if additional neuronal classes resisted ischemic injury. Given the near-complete destruction of the medium-sized neurons as the typical projecting cells of the striatum the functional significance of resistant interneurons may be neglegible. However, the neurobiological principles underlying this differential vulnerability are of general interest. Large interneurons have a low number of spines on their dendrites in difference to projection neurons (30). It is proposed that dendritic surface specialization and/or specific receptors may be critical factors rendering projection neurons specifically vulnerable in a situation of relative energy-deficiency. Further, the resistance of certain striatal neurons is shared by ischemia with several other pathologies including Huntington`s disease (31) and intra-striatal lesions induced by excitotoxic substances (32).

A further example of selective vulnerability within the basal ganglia is provided by the resistance of the pallidum contrasting with the marked lesions of the striatum. However, in addition to intrinsic differences of neuronal organization the higher residual flow to the pallidum must be considered as protective factor. On the other hand, a peculiar type of common neuronal and glial damage developed in pars reticulata of substantia nigra despite high residual flows (7). The nucleus basalis of Meynert and the pars compacta of substantia nigra were resistant to acute ischemic injury. On the other hand these nuclei are vulnerable in degenerative diseases of the human brain, namely Alzheimer`s disease and Parkinson`s disease. This observation suggests that these cell groups do not reveal an outstanding vulnerability to acute ischemia. Consequently, ischemic events would not appear as major pathogenetic factors in degenerative processes of these areas.

REFERENCES

1. Siesjö, B.K., 1981, Cell damage in the brain: a speculative synthesis, J. Cereb. Blood Flow Metabol. 1 : 155-185.

2. Hossmann, K.-A., 1982, Treatment of experimental cerebral ischemia, J. Cereb. Blood Flow Metabol. 2: 275-297.

3. Plum, F.,1983, What causes infarction in ischemic brain?: The Robert Wartenberg Lecture, Neurology 33: 222-233.

4. Kogure, K., Hossmann, K.-A., Siesjö, B.K., Welsh, F.A., 1985, Molecular mechanisms of ischemic brain damage, Prog. Brain Res. 63.

5. Pulsinelli, W.A., Brierley, J.B., Plum, F. 1982, Temporal profile of neuronal damage in a model of transient forebrain ischemia, Ann. Neurol. 11: 491-498.

6. Ginsberg, M.D., Graham, D.I., Busto, P., 1985, Regional glucose utilization and blood flow following graded forebrain ischemia in the rat: correlation with neuropathology, Ann.Neurol. 18: 470-481.

7. Smith, M.-L., Kalimo, H., Warner, D.S., Siesjö, B.K., 1988, Morphological lesions in the brain preceding the development of postischemic seizures, Acta Neuropathol. 76: 253-264.

8. Petito, C.K., Pulsinelli, W.A., Jacobson, G., Plum, F., 1982, Edema and vascular permeability in cerebral ischemia: comparison between ischemic neuronal damage and infarction, J. Neuropathol. Exp. Neurol. 41: 423-436.

9. Kirino, T, 1982, Delayed neuronal death in the gerbil hippocampus following ischemia, Brain Res. 239: 57-69.

10. Schmidt-Kastner, R., Hossmann, K.-A., 1988, The distribution of ischemic neuronal damage in the dorsal hippocampus of rat, Acta Neuropathol. 76: 411-421.

11. Schmidt-Kastner, R., Paschen, W., Grosse Ophoff, B., Hossmann, K.-A., 1989, A modified four-vessel occlusion model for inducing incomplete forebrain ischemia in rats, Stroke 20: 938-946.

12. Pulsinelli, W.A., Brierley, J.B., 1979, A new model of bilateral hemispheric ischemia in the unanesthetized rat, Stroke 10: 267-272.

13. Pellegrino, L.J., Pellegrino, A.S., Cushman, A.J., 1979, A stereotaxic atlas of the rat brain, 2nd ed., Plenum Press, New York London.

14. Hedreen, J.C., Bacon, S.J., Price, D.L., 1985, A modified histochemical technique to visualize acetylcholinesterase-containing axons, J. Histochem. Cytochem. 33: 134-140.

15. Paschen, W., Djuricic, B., Mies, G., Schmidt-Kastner, R., Linn, F., 1987, Lactate and pH in the brain: Association and dissociation in different pathophysiological states, J. Neurochem. 48: 154-159.

16. Paschen, W., Schmidt-Kastner, R., Djuricic, B., Meese, C., Linn, F., Hossmann, K.-A., 1987, Polyamine changes in reversible cerebral ischemia, J. Neurochem. 49: 35-37.

17. Rieke, G.K., Bowers, D.E., Penn, P., 1981, Vascular supply pattern to rat caudatoputamen and globus pallidus: scanning electron-microscopic study of vascular endocasts of stroke-prone vessels, Stroke 12: 840-847.

18. Pulsinelli, W.A., Levy, D.E., Duffy, T.E., 1982, Regional cerebral blood flow and glucose metabolism following transient forebrain ischemia, Ann. Neurol. 11: 499- 509.

19. Globus, M.Y.-T., Ginsberg, M.D., Harik, S.I., Busto, R., Dietrich, W.D., 1987, Role of dopamine in ischemic striatal injury. Metabolic evidence, Neurology 37: 1712-1719.

20. Gerfen, C.R., Herkenham, M., Thibault, J., 1987, The neostriatal mosaic: II. Patch- and matrix directed mesostriatal dopaminergic and non-dopaminergic systems, J.Neurosci. 7: 3915-3934.

21. Freund, T.F., Buszaki, G., Leon, A., Baimbridge, K.G., Somogyi, P., 1990, Relationship between neuronal vulnerability and calcium binding protein immunoreactivity in ischemia, submitted.

22. Rea, M.A., Simon, J.R., 1981, Regional distribution of cholinergic parameters within the rat striatum, Brain Res. 219: 317-326.

23. Beal, M.F, Martin, J.B, 1985, Topographical dopamine and serotonin distribution and turnover in rat striatum, Brain Res. 358: 10-15.

24. Joyce, J.N., Loeschen, S.K., Marshall, J.F., 1985, Dopamine D-2 receptors in rat caudate-putamen: the lateral to medial gradient does not correspond to dopaminergic innervation, Brain Res. 338: 209-218.

25. Globus, M.Y.-T., Ginsberg, M.D., Dietrich, W.D., Busto, R., Scheinberg, P., 1987, Substantia nigra lesion protects against ischemic damage in the striatum. Neurosci. Lett. 80: 251-256.

26. Globus, M.Y.-T., Busto, R., Dietrich, W.D., Martinez, E., Valdes, I., Ginsberg, M.D., 1988, Effect of ischemia on the in vivo release of striatal dopamine, glutamate, and gamma-aminobutyric acid studied by intracerebral microdialysis, J. Neurochem. 51: 1455-1464.

27. Francis, A., Pulsinelli, W., 1982, The response of GABAergic and cholinergic neurons to transient cerebral ischemia, Brain Res. 243: 271-278.

28. Lynch, G.S., Lucas, P.A., Deadwyler, S.A., 1972, The demonstration of acetylcholinesterase containing neurons within the caudate nucleus of the rat, Brain Res. 45: 617-621.

29. Johnston, M.V., Hudson, C., 1987, Effects of postnatal hypoxia-ischemia on cholinergic neurons in the developing rat forebrain: choline acetyltransferase immunocytochemistry, Dev. Brain Res. 34: 41-50.

30. Chang, H.T., Wilson, C.J., Kitai, S.T., 1982, A Golgi study of rat neostriatal neurons: light microscopic analysis, J.Comp.Neurol. 208: 107-126.

31. Ferrante, R.J., Kowall, N.W., Beal, M.F., Richardson, E.P., Bird, E.D., Martin, J.B., 1985, Selective sparing of a class of striatal neurons in Huntington's disease, Science 230: 561-563.

32. Beal, M.F., Kowall, N.W., Swartz, K.J., Ferrante, R.J., Martin, J.B., 1989, Differential sparing of somatostatin-neuropeptide Y and cholinergic neurons following striatal excitotoxin lesions, Synapse 3: 38-47.

ROLE OF D1 AND D2 DOPAMINE RECEPTORS IN PILOCARPINE-INDUCED SEIZURES

Paolo Barone*, Vincenzo Palma*, Sotirios A. Parashos$
Thomas N. Chase$ and Giuseppe Campanella*

(*) Institute of Neurology, II Faculty of Medicine, University of
Naples, Italy and ($) Experimental Therapeutics Branch, NINCDS
Bethesda, Md, USA

INTRODUCTION

Both substantia nigra (SN) and caudate-putamen (CP) are brain regions
documented to be involved in the regulation of seizure susceptibility (Gale,
1984). Nigral microinjection of gamma-amino-butyric-acid (GABA) agonists or
antagonists respectively attenuates or potentiates seizures in several
experimental models of epilepsy (Iadarola and Gale, 1982; Turski et al., 1986,
1987a). Similarly, pilocarpine induced-seizures convulsions are prevented by
chemical stimulation of striatal cells but facilitated by lesioning of the caudate-
putamen (Turski et al., 1987b).

Most studies focus on the role of striatal GABA-ergic projections, while
little is known about the contribution of dopaminergic system to the
propagation of epileptic activity. Certain dopaminergic agonists are reported to
exert anticonvulsant activity in epilepsy induced by electroshock,
pentylenetetrazol and photically induced myoclonus (Kilian and Frey, 1973;
Meldrum et al., 1975; McNamara et al, 1984). On the other hand, experimental
evidence indicates that pentylenetetrazol-induced convulsions are inversely
proportional to striatal dopamine (DA) content indicating that DA depletion
might be a good indicator of seizure resistance (Fariello et al., 1987). These
discrepancies may reflect differences in DA involvement in various seizure
models.

The understanding of DAergic mechanisms in motor seizures propagation
is further complicated by the existence of two dopamine receptors, D1 and D2
(Kebabian and Calne, 1979). These receptors are generally recognized by their
contrasting effects on adenylate cyclase: D1 receptor stimulation increases,
while D2 receptor stimulation inhibits the enzyme activity (Stoof and Kebabian,
1981). However, behavioral and neurophysiological studies do not support such
opposing actions: for example, classical motor behaviors in intact rat requires
the concurrent D1 and D2 receptor stimulation (Braun et al., 1986). Conceivably,
functional interaction between D1 and D2 receptors may be crucial in the
DAergic mechanisms of seizures. Interestingly, both D1 and D2 dopamine
receptors are abundant in CP and SN (Dawson et al., 1988).

The Basal Ganglia III, Edited by G. Bernardi *et al.*
Plenum Press, New York

Both CP and SN are involved in the mechanisms of propagation rather than generation of epileptic discharges (Iadarola and Gale, 1982; Turski et al., 1987a). Similarly, a possible DA involvement in epilepsy might be revealed by epilepsy models studying the spread of seizure. For this purpose, we used the pilocarpine-induced epilepsy which is a reliable model to evidence the process of secondary generalization and to evaluate drug effects on the propagation of convulsive activity (Turski et al., 1983).

The present article discusses a series of investigations in which agonists and antagonists with different selectivity for D1 and D2 receptors were tested for their effects on thresholds for pilocarpine-induced seizures in rats. The following drugs were used in this study: the D1 agonist SKF 38393 (Smith, Kline and French, Philadelphia, PA), the D2 agonist LY 171555 (Eli Lilly and Co., Indianapolis, IN), the D1 antagonist SCH 23390 (Schering Plough, NJ) and the D2 antagonist raclopride (Astra, Sweden).

MATERIALS AND METHODS

Male Sprague-Dawley rats (180-200 g) were placed in individual plastic observation cages and allowed to habituate 30-40 min before drug administration. In order to reduce peripheral cholinergic effects, methylscopolamine (1 mg/kg; s.c.) was injected in all animals 30 min prior to pilocarpine.

The first group of experiments investigated the effects of dopaminergic drugs on epilepsy induced by a subconvulsant dose of pilocarpine. Animals received one of the following treatments: SKF 38393 (0.1-10 mg/kg; i.p.); LY 171555 (0.1-10 mg/kg; i.p.); SCH 23390 (0.01-1 mg/kg; i.p.); raclopride (3-5 mg/kg; i.p.); saline (i.p.). After 5 min, the rats received 200 mg/kg pilocarpine (i.p.), a dose which produces stereotypies but no convulsions (Turski et al., 1983).

The second group of experiments evaluated the effect of D1 receptor blockade on SKF 38393-induced convulsions. Prior to pilocarpine (200 mg/kg) injection, groups of 8 rats for each dose were challenged with the selective D1 antagonist SCH 23390 (0.1-mg/kg; i.p.) plus a fixed dose of SKF 38393 (1 mg/kg) which had previously been found to be the ED_{50} for convulsions.

The third group of experiments evaluated the ability of SCH 23390 (0.01-0.1 mg/kg; i.p.) to block convulsions induced by a convulsant dose of pilocarpine (400 mg/kg; i.p.).

Behavior recording

The presence of the following pilocarpine-induced behavioral alterations was eventually recorded: akinesia, tremor, stereotypies, motor limbic seizures and status epilepticus (Turski et al., 1983). Behavioral differences between experimental treatments and different doses for each treatment were evaluated by two-way analysis of variance.

EEG recording

Under chloral hydrate anesthesia (400 mg/kg, i.p.), 3 to 6 animals for each treatment group received electrodes in the dorsal hippocampus and over the occipital cortex for EEG recording. The electrodes were connected to an amplification apparatus (SAN-EI, Japan; time constant 0.3 sec, high cut-off filter 35 Hz, paper speed 15 mm/sec); electrical activity was continuously recorded from 30 min before to 2 hours after drug administration.

RESULTS

Behavioral and electrographic results are reported in Table 1 and Fig 1, respectively. In the saline pretreated rats pilocarpine (200 mg/kg) induced akinesia, tremor and automatisms including sniffing, chewing and teeth chattering, but no convulsive activity. These animals completely recovered 120-150 min after pilocarpine injection.

TABLE 1. Effect of pretreatment with DA-agonists and DA-antagonists on epilepsy induced by a subconvulsant dose of pilocarpine (200mg/kg).

treatment	Dose m g / k g	N. Rats	Motor Limbic Seizures	Status Epilepticus occurrence %	Lethality
Saline	-	18	0	-	0
SKF 38393	0.1	8	37.5 [a]	25.0	0
	1	8	62.5 [a]	50.0 [a]	25.0
	5	11	90.9 [a]	90.9 [a]	72.7 [a]
	10	8	100 [a]	100 [a]	100 [a]
SCH 23390	0.05	6	0	0	0
	0.1	6	0	0	0
SKF 38393 + SCH 23390	1 0.1	8	12.5 a,b	12.5 a,b	12.5 a,b
LY 171555	1	6	0	0	0
	5	6	0	0	0
	10	6	0	0	0
Raclopride	3	10	60 [a]	40 [a]	40 [a]
	5	5	100 [a]	100 [a]	100 [a]

Both agonists and antagonists were injected 5 min prior to pilocarpine. Status epilepticus was defined as continuous convulsions for a period longer than 30 min. Mortality was calculated 24 hs after pilocarpine injection. [a] = $p<0.01$ vs saline-treated animals;
[b] = $p<0.05$, vs corresponding dose of SKF 38393 (ANOVA).

SKF 38393 pretreatment, on the other hand, induced generalized convulsive activity including both motor limbic seizures and eventually status epilepticus in a dose dependent manner. Motor limbic seizures occurred 30 min after pilocarpine injection. No rats survived at higher doses of SKF 38393. Electrical activity closely paralleled the behavioral alterations: a well synchronized, high voltage, spiking activity in the hippocampus and cortex occurred at the same time of the motor limbic seizures. In most rats these electrographic seizures resulted in status epilepticus within approximately 40 min after pilocarpine injection.

SCH 23390 pretreatment did not affect the behavioral response to pilocarpine: animals developed automatisms but neither convulsive behavior nor electrographic seizure was observed . On the other hand, in experiments of combined treatments, SCH 23390 inhibited the convulsive activity induced by both pilocarpine (200 mg/kg) plus SKF 38393 (1 mg/kg) (Fig. 1A). Furthermore, pretreatment with D1 antagonist prevented seizures induced by the convulsant dose of pilocarpine (400 mg/kg) (Fig. 1B).

Pretreatment with LY 171555 did not alter the pilocarpine-induced automatisms. Like the D1 antagonist, the D2 agonist pretreatment exerted neither convulsive activity nor electrographic seizure.

On the other hand, pretreatment with selective D2 antagonist raclopride significantly reduced the threshold for pilocarpine-induced convulsions.

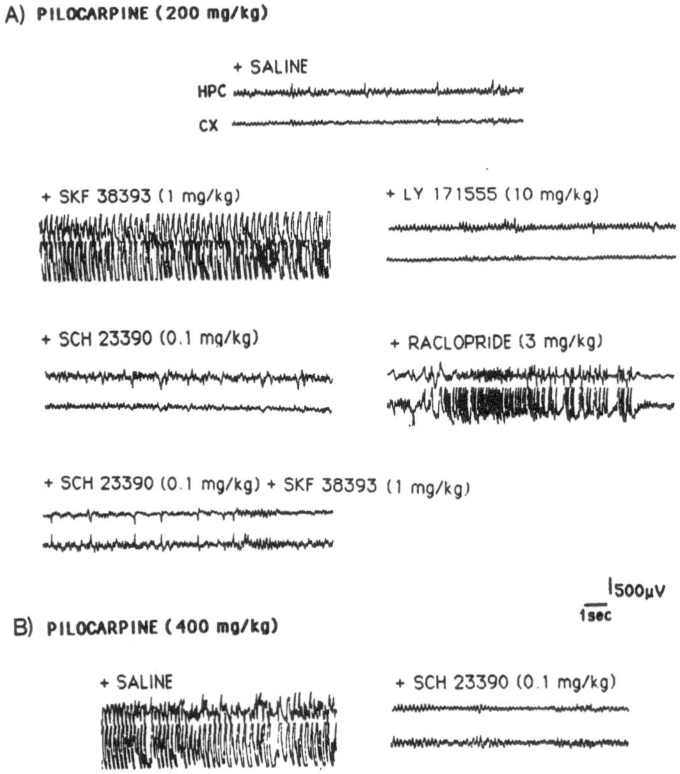

Fig. 1. Electrographic recordings illustrating the effects of dopamine agonists and/or antagonists on epilepsy induced by either a subconvulsant (A) or a convulsant (B) dose of pilocarpine. Recordings were taken at 30-40 min after pilocarpine injection. Pretreatment with either saline or D2 agonist or D1 antagonist did not alter basal activity; no electrographic seizure was observed. On the other hand, seizures occurred in animals pretreated with either the D1 agonist or the D2 antagonist. Conversely, D1 receptor blockade prevented seizures induced by either 200 mg/kg pilocarpine plus SKF 38393 (A) or 400 mg/kg pilocarpine, given alone (B). HPC, hippocampus; CX, cortex.

DISCUSSION

The present results demonstrate that dopamine system exerts a control on seizure propagation and convulsive activity. Stimulation of D1, but not D2, receptors potently reduced the threshold for pilocarpine-induced convulsions in the rat. Furthermore, D1 receptor blockade prevented convulsive activity, while D2 receptor blockade facilitated pilocarpine-induced convulsions. Neither DAergic agent altered the pilocarpine induced stereotypies, indicating that DA receptors are primarily involved in the spread rather than in the generation of epileptic discharges.

The existence of opposing roles for the two DA receptor subtypes in modulating convulsive activity is confirmed by a series of experimental observations. Anticonvulsant effects have been observed with D2 receptor stimulation in seizures induced by electroshock, pentylenetetrazol (Kilian and Frey, 1973;McNamara et al, 1984), pilocarpine (Turski et al., 1988), photically induced myoclonus (Meldrum et al., 1975) and amygdala-kindling in rats as well as air blast stimulation in gerbils (Loscher, 1985; Loscher and Czuczwar, 1986). Conversely, D2 antagonist administration is known to display convulsant effects in both animal epilepsy models and human epilepsy. Although SKF 38393 produced no effect in certain epilepsy models (Loscher and Czuczwar, 1986), the present study clearly demonstrate that D1 receptor stimulation exerts a very potent effect in reducing the threshold for pilocarpine-induced seizures. It is possible that DA system is differently involved in various seizure models.

Conceivably, D1 receptors are primarily involved in the diffusion of the epileptic discharge. High density of D1 receptors has been found in SNR (Dawson et al., 1988), a brain region indicated to be crucial for epilepsy propagation (Iadarola and Gale, 1982). In particular, nigral D1 receptors are localized on the terminals of GABA neurons, suggesting a modulatory role for these receptors in the regulation of GABA release (Barone et al., 1987). It is possible that D1 agonists facilitate the pilocarpine-induced convulsions by reducing GABA release in the SNR, mimicking the effect produced by the nigral injection of GABA antagonists (Iadarola and Gale, 1982; Turski et al., 1986, 1987a). At the present time, however, there is no direct evidence that D1 receptor stimulation is capable of reducing GABA release (Arbilla et al., 1981; Girault et al., 1986).

Interestingly, in the present study D1 receptor blockade prevented convulsions induced by a convulsant dose of pilocarpine (400 mg/kg) but did not affect the pilocarpine-induced stereotypies. These results suggest that D1 receptor blockade might be effective in preventing the secondary generalization of seizures in human epilepsy.

REFERENCES

Arbilla, S., Kamal, L.A., and Langer, S.Z., 1981, Inhibition by apomorphine of the potassium evoked release of [3H]-gamma-aminobutyric acid from the rat substantia nigra, Br. J. Pharmacol. ,74:389.

Barone, P., Tucci, I., Parashos, S.A. ,and Chase T.N., 1987, D1 dopamine receptor changes after striatal quinolinic acid lesion, Eur. J. Pharmacol. 138:141.

Braun, A.R., Barone, P., and Chase, T.N., 1986, Interaction of D1 and D2 dopamine receptors in the expression of dopamine agonist induced behaviors, in :

"Neurobiology of central D1 dopamine receptors", G.R. Breese and I. Creese , eds Plenum Publishing Corp, New York.

Dawson, T.M., Barone, P., Sidhu, A., Wamsley, J.K., and Chase, T.N., 1988, The D1 dopamine receptor in the rat brain: quantitative autoradiographic localization using an iodinated ligand, Neuroscience , 26:83.

Fariello, R.G., DeMattei, M., Castorina, M., Ferraro, T.N., and Golden, G.T., 1987, MPTP and convulsive responses in rodents, Brain Res. , 426:373.

Gale K., 1984, Role of the substantia nigra in the anticonvulsant actions of GABAergic drugs, in "Neurotransmitters, Seizures and Epilepsy", R.G.Fariello,ed., Raven Press, New York.

Girault, J.A., Spampinato, U., Savaki, H.E., Glowinski, J., and Besson, J.D. ,1986, In vivo release of [3H]-gamma-aminobutyric acid in the rat neostriatum. I.: Characterization and topographical heterogeneity of the effects of dopaminergic and cholinergic agents, Neuroscience, 19:1101.

Iadarola, M.J., and Gale, K., 1982, Substantia nigra: site of anticonvulsant activity madiated by GABA, Science, 218:1237.
J. Neurosci. , 4:2410.

Kebabian, J.W., and Calne, D.B., 1979, Multiple receptors for dopamine, Nature, 260:257.

Kilian, M. ,and Frey, H. H., 1973, Central monoamines and convulsive thresholds in mice and rats, Neuropharmacol. ,12:681.

Løscher, W., 1985, Influence of pharmacological manipulation of inhibitory and excitatory neurotrasmitter systems on seizure behavior in the mongolian gerbil, J. Pharmacol. Exp. Ther. ,233:204.

Løscher, W., and Czuczwar, S.J., 1986, Studies on the involvment of dopamine D1 and D2 receptors in the anticonvulsant effect of dopamine agonists in various rodent models of epilepsy, Eur. J. Pharmacol. ,128:55.

McNamara, J.O., Galloway, M.T., Rigsbee, L.C., and Shin, C. ,1984, Evidence implicating substantia nigra in regulation of kindled seizure thresholds,

Meldrum, B., Anlezark, G., and Trimble, M., 1975, Drugs modifying dopaminergic activity and behaviour, the EEG and epilepsy in Papio-Papio, Eur. J. Pharmacol. , 32: 203.

Sovner, R. ,and DiMascio, A., 1978, Extrapyramidal syndromes and other neurological side effects of psychotropic drugs, in "Psychopharmacology: a generation progress", M. A. Lipton, A. DiMascio and K. F. Killiam, eds, Raven New York.

Stoof, J.C., and Kebabian, J.W., 1981, Opposing roles for D1 and D2 dopamine receptors in efflux of cAMP from rat neostriatum, Nature, 294:366.

Turski, L., Cavalheiro, E.A., Bortolotto, Z.A., Ikonomidou-Turski, C., Kleinrok, Z., and Turski, W.A., 1988, Dopamine-sensitive anticonvulsant site in the rat striatum. J. Neurosci. , 8:4027.

Turski, L., Cavalheiro, E.A., Schwarz, M., Turski ,W.A., De Moares Mello, L., Bortolotto, Z.A., Klockgether, T., and Sontag K.H., 1986, Susceptibility to seizures produced by pilocarpine in rats after microinjection of isoniazid or vinyl- GABA into the substantia nigra, Brain Res. , 370:294.

Turski, L., Cavalheiro, E.A., Turski, W.A. ,and Meldrum, B.S., 1987a, Excitatory neurotrasmission within substantia nigra pars reticulata regulates thresholds for seizures produced by pilocarpine in rats: effects of intranigral 2-amino-7- phosphonoheptanoate and N-methyl-D-aspartate, Neuroscience , 18:61.

Turski, L., Meldrum, B. S., Cavalheiro, E.A., Calderazzo-Filho, L.S., Bortolotto, Z.A., Ikonomidou-Turski, C., and Turski, W.A., 1987b, Paradoxical anticonvulsant activity of the excitatory amino acid N-methyl-D-aspartate in the rat caudate-putamen, Proc. Natl. Acad. Sci. , 84:1689.

Turski, W.A., Cavalheiro, E.A., Schwar, M., Czuczwar, S.J., Kleinrok, Z., and Turski, L., 1983, Limbic seizures produced by pilocarpine in rats: behavioural, electroencephalographic and neuropathological study, Behav. Brain Res. 9:315.

DOPAMINERGIC DYSFUNCTIONS IN NEONATAL HYPOTHYROIDISM

Andrea Vaccari and Zvani L. Rossetti

"B.B. Brodie" Department of Neuroscience
Chair of Toxicology, Medical School
Via Porcell 4, 09124 Cagliari, Italy

BACKGROUND

It is well known that the nervous and the endocrine systems themselves communicate. Consistently, there exists a consensus that thyroid dysfunction may play a role in mental disorders.[1] On the ground that impairment of central dopaminergic (DA) pathways may be relevant to the origin of some psychiatric disorders, the regulation of brain DA by thyroid hormones is generating much interest. In this connection, the present article dealing with the latter argument, will devote special attention to the early postnatal period, when the thyroid gland is necessary for brain maturation.

It must preliminarly be remarked that DA-alterations are not the only province of perinatally-provoked disorders of thyroid hormones. Every aspect of neurotransmission may, indeed, be impaired.[2]

Generally speaking, location and severity of transmitter impairments due to thyroid malfunction do strictly obey the rule of reflecting the time of exposure, and the extent of maturation in different brain regions, through fetal and/or early postnatal age.[3,4] Expectedly, fetal and/or neonatal hypothyroidisms will affect only those brain regions or neuronal subpopulations whose adult state has not yet been achieved when the fetal thyroid becomes functionally competent. Thus, early-developing brain structures are little sensitive to postnatal-onset hypothyroidism. From the foregoing, it is generally accepted that neonatal hypothyroidism impairs the morphogenesis of developing neural networks, and adult-onset hypothyroidism may elicit more subtle biochemical alterations in neuronal transmission, while leaving unchanged the underlying structures. Thus, perinatally-provoked and adult-onset thyroid deprivations may provoke qualitatively different effects on morphological, biochemical, and behavioral indexes of the central DA-system.

The Basal Ganglia III, Edited by G. Bernardi *et al.*
Plenum Press, New York

Fetal thyroid secretion in the rat begins at the 18th day of pregnancy, and in humans at the beginning of the second trimester of pregnancy. Therefore, in the rat, brain regions such as the cerebellum, undergoing an almost entirely postnatal development, and the striatum, whose maturation is only partially prenatal, are highly dependent on thyroid homeostasis. The fact that the adult shape and connections of the human striatum are attained at the middle of fetal life, much later than the onset of fetal thyroid secretion, supports the possibility that teratogenic lesions of the striatum may occur in congenitally hypothyroid children too.

The newborn striatum in rats is characterized by the presence of immature synapses and by the appearance of the first varicosities in DA-fibres through the first days of age (cf.[5] for references). Furthermore, DA-neurons in the striatal matrix are innervated by the substantia nigra consistently later than corresponding neurons segregated in the patches.[6] Fetal and/or neonatal hypothyroidism may, indeed, disrupt the timing of striatal maturation. In fact, caudate neuronal proliferation is delayed, and the density of synaptic contacts decreases.[7,8] Generalizing from the above considerations, it is to be expected that early hypothyroidism may reflect on the ontogenetic increase (cf.[5] for references) of DA-functions, synthesis and catabolism of DA, DA uptake and release, DA receptors and second messengers included.

EXPERIMENTAL MODELS

Neonatal hypothyroidism is routinely obtained either with the chronic administration of the antithyroid drugs methimazole (MMI) or propylthiouracyl (PTU) to pregnant rats and then to their suckling pups, or with the only postnatal treatment of newborns. Thyroid deprivation must occur during the sensitive period of brain maturation, which in the rat is thought to extend from the middle of gestation to at least 24 days after birth,[9] or for a shorter time.[10-12]

Additional procedures involve surgical parathyroidectomy, which cannot be performed before the first week of age, or the injection of high concentrations of ^{131}I to 1 day-old rats.

All methods suffer limitations, due to extrathyroidal effects of surgical- and radio-thyroidectomy, and to some possibility that the thiol goitrogens used may cause both maternal and fetal hypothyroidism when given in utero,[13] as well as may directly affect synaptic membranes.[14]

The congenitally hypothyroid hyt/hyt mice seem now to represent a well exploitable new model of neonatal hypothyroidism.[15-17]

THE DOPAMINERGIC SYSTEM IN HYPOTHYROIDISM

Presynaptic events

Early, fluorometry-based reports indicated that neonatal radiothyroidectomy or MMI-administration to rats decreased at postnatal day 30 the levels of brain DA by almost 50% of the control values, in terms of g tissue.[18,19] In contrast, overall

Table 1. Effects of neonatal hypothyroidism on presynaptic dopaminergic markers in the rat striatum

Marker	Site labeled		Euthyroid	Hypothyroid
[^{3}H]DA[a] uptake	neuronal + vesicular transporters	V_{max} K_m	13.1 ± 1.7 46.9 ± 2.1	4.9 ± 1.0** 25.1 ± 3.3§
[^{3}H]MAZ[b] binding	neuronal transporter	B_{max} K_D	2.6 ± 0.4 6.9 ± 1.8	2.1 ± 0.1 4.1 ± 0.8
[^{3}H]PTA[a] binding	vesicular transporter	B_{max} K_D	3.7 ± 0.2 6.8 ± 0.6	2.2 ± 0.3** 4.5 ± 0.5*

Pregnant rats received 50 mg/kg methimazole (MMI) in the drinking water from day 15 to delivery. Their pups were daily, s.c. injected with 20 mg/kg MMI from birth to day 10, and 30 mg/kg MMI from then to postnatal day 30. Control and hypothyroid rats were sacrificed at day 32. Results are means ± SEM from 5-6 rats, and are expressed as pmoles/mg protein (B_{max} for Mazindol and p-Tyramine binding), pmoles/mg protein/min (V_{max} for DA uptake), or nM (K_D or K_m). * P<0.02; ** P<0.005; § P<0.0001 vs euthyroids, one-way ANOVA.
[a] From.[5]   [b] From.[25]

mesencephalic contents of DA and DOPAC in 20 day-old mice having received PTU pre- and postnatally, were similar to that in controls.[20] At our hands, pre-and postnatal administration of MMI did not affect striatal concentrations of DA in 32 day-old rats, though it decreased by approximately 50% DOPAC and HVA contents, as expressed in terms of mg proteins.[5] An obvious explanation was, of course, inhibition of DA turnover as adaptive reaction to lower availability of DA, or the reflection of minor needs of DA by the hypoactive neuron. Reasons for this finding can be hardly related to a diminished DA catabolism. In fact, MAO activities in the developing hypothyroid brain do not show clearly-cut tendency to be either increased or decreased.[21] The same is true for brain or striatal tyrosine hydroxylase activity.[20,22] Thus, other causes must be found, such an hypothyroidism-stimulated wash-out of DA metabolites from the newborn brain, which in normal conditions differs from the adult in its efficiency at eliminating organic acids from cerebrospinal fluid.[23]

The administration of PTU starting at postnatal days 2-3 reportedly provoked a small deficit in the total amount of[^{3}H]DA taken-up by the hypothyroid striatum, perhaps accounting for some loss of DA terminals.[22] We have obtained a by far more impressive decrease in synaptosomal [^{3}H]DA uptake when the antithyroid treatment with MMI was started at day 15 of gestation.[5] In light of the introductory considerations, the in utero-plus neonatal exposure to the goitrogen influenced both prenatal and postnatal steps of striatal maturation, whereas in the previous study, the final developmental features had only been impaired. The V_{max} for [^{3}H]DA uptake was decreased by 63%, and there was a 46% decrease of K_m values, i.e. the affinity for the substrate was greater than in controls (Table 1).

Taken together, latter findings clearly indicate that neonatal, pharmacologically-produced hypothyroidism preferentially affects the storage step for DA, rather than its transport across the neuronal membrane.[5] Inasmuch as [^{3}H]DA uptake is considered a rough index for the density of functional dopaminergic endings in the striatum,[27] some retardation in the attainment of adult configuration of striatal afferents may be inferred. Less DA terminals may imply less vesicles available for DA, as well as a normal number of organelles bearing decreased amounts of carrier units. On the other hand, the finding of normal contents of [^{3}H]mazindol-labeled, neuronal transporter units, and the changes in affinity values for DA-uptake and PTA-binding processes, may mainly reflect alterations in chemico-physical characteristics of the vesicular transporter, rather than a pure loss of functional DA-endings. Of course, it is difficult to correlate almost normal contents of striatal DA with a deficient storage system, unless the contents of residual vesicles and/or carrier units are still enough to warrant the storage of endogenous DA produced.

In conclusion, these data allow us to suggest that neonatal and, to a greater extent, combined pre- and postnatal antithyroid treatment of rats may retard the DA-turnover, when the maturation of striatal innervation is impaired and, more specifically, the vesicular transport of DA is deficient.

Since the DA-uptake process reflects both translocation of DA through the neuronal membrane, and its approach to storage vesicles, we wished to ascertain whether these two compartments could be differently affected by neonatal hypothyroidism. Relevant to this purpose was the finding that the number of binding sites for [³H]tyramine ([³H]PTA), a trace amine which primarily (by 85%) labels the vesicular carrier for DA,[24,25] was decreased by 40%, and its corresponding binding affinity was greater than in controls (Table 1). In contrast, fetal/neonatal hypothyroidism did not affect striatal high-affinity binding of [³H]mazindol, an almost pure marker for the neuronal DA-transporter[25,26] (Table 1).

Table 2. Effects of neonatal hypothyroidism on postsynaptic dopaminergic markers in the rat striatum

Marker	Site labeled		Euthyroid	Hypothyroid
[³H]SPI[a] binding	D_2	B_{max}	2.81 ± 0.02	2.17 ± 0.01*
		K_D	0.61 ± 0.06	0.79 ± 0.10
[³H]SCH 23390[b] binding	D_1	B_{max}	0.30 ± 0.05	0.15 ± 0.02*
		K_D	0.53 ± 0.08	0.34 ± 0.04
DA-stimulated[b] adenylate cyclase	D_1	V_{max}	79.4 ± 4.2	50.6 ± 4.3**
		K_m	3.7 ± 0.9	1.9 ± 0.3

Neonatal hypothyroidism was provoked as described in Table 1. B_{max} for Spiperone (SPI) and SCH 23390 binding, and V_{max} for DA-stimulated cAMP accumulation values are expressed as pmoles/mg protein; K_D = nM, K_m = μM. Means $\pm$ SEM from 6-11. 32 day-old rats.
* $P<0.05$; ** $P<0.001$ vs euthyroids, one-way ANOVA.
[a] From.[28]   [b] From.[5]

Postsynaptic events

A marked impairment in the number of striatal, [³H]spiperone-labeled D_2-type DA-receptors has previously been shown, decreases ranging from 23-27% in 31 day-old hypothyroid rats, depending on the sex,[28] to 36-40% in 3 or 6 week-old newborns[22] (Table 2). There were no changes in corresponding affinity values. In the light of a putative location of D_2-receptors on cholinergic axon terminals rather than on dendrites of striatal interneurons-,[29] the development of fewer cholinergic nerve terminals during thyroid deficiency has been also consistently demonstrated.[22]

Alterations extend also to D_1-type receptors. In fact, the B_{max} for [³H]SCH-23390 binding to striatal membranes was impaired by 49% in samples from hypothyroid rats (Table 2). Deficits in numbers of D_1-sites were well reflected by a consistent decrease in D_1-mediated, DA-stimulated adenylate cyclase activity. V_{max} values for cAMP production were 36% smaller than in euthyroid counterparts (Table 2).

The putative location of D_1-sites on intrinsic striatal neurons, rather than on nigrostriatal terminals,[30,31] and the presence of D_2-receptors on cholinergic interneurons, may well suggest that intrinsic neurons in the neonatal hypothyroid striatum contain significantly less dopamine receptors. Thus, alterations in DA-mediated behaviours were to be expected.

Table 3. Effects of neonatal hypothyroidism on D_1-mediated or D_1-facilitated behaviours of adult rats[a]

Thyroid state	Acute treatment	Yawns	Penile erections
		(number/30 min)	
Euthyroid	saline	0.37 ± 0.18	0.12 ± 0.12
	apomorphine	8.50 ± 1.52	1.83 ± 0.11
Hypothyroid	saline	0.36 ± 0.20	0.36 ± 0.28
	apomorphine	4.93 ± 0.94*	1.13 ± 0.19**

Neonatal hypothyroidism was provoked as described in Table 1. MMI administration was continued up to 49-59 days of age, 24 hours before behavioral testings. The number of D_1-mediated penile erection episodes, or of D_2-mediated, D_1-facilitated yawning responses as elicited by a s.c. injection of 50 μg/kg apomorphine, or saline, was scored starting 5 min after injections. Values are means ± SEM from 8-15 animals.* $P<0.05$; ** $P<0.001$, compared with apomorphine-treated, euthyroid controls; one-way ANOVA. [a] From.[35]

Dopaminergic behaviours in neonatal hypothyroidism

The behavioural effects associated with changes in dopaminergic function (stereotypy, hyperlocomotion, catalepsy etc.) have historically been associated with alterations in D_2-receptor activity.[32] Normally, changes in receptor numbers may be related with modifications in behavioural changes evoked by stimulation or antagonism of these receptors. The only information available on D_2-mediated behaviours in neonatal hypothyroidism is based on a study of Overstreet et al.[33] Lactating dams and then, their weaned pups, were placed on a low-iodine diet until postnatal day 130, when they exhibited increased incidence of sniffing and locomotor activities after apomorphine injection, compared to iodine-repleted counterparts. An enhanced sensitivity of hypothyroids to D_2- (and D_1) stimulation, fit-ting with increased density of striatal D_2-receptors was, thus, suggested. The latter finding goes counterwise to the repeatedly demonstrated trend of striatal D_2-sites to decrease, when binding assays were performed at much younger age in neonatal hypothyroids.[22,28,34] Besides the different procedure used to provoke early-onset hypothyroidism, and variability of results inherent to different methods,[2] further experiments will be necessary in order to exclude adaptive responses of DA-parameters towards long-lasting antithyroid treatments, once full adulthood is reached.

We were here particularly interested to ascertain whether the hypothyroidism-related deficit in density and biochemical coupling of D_1-receptors could have behavioral consequences. As a matter of fact, the number of apomorphine-evoked penile erection (P.E.) and yawning (Y) episodes was consistently smaller (38% and 42%, respectively) in adult, neonatally-rendered hypothyroid rats, as compared with apomorphine-challenged euthyroid counterparts[35] (Table 3). The induction of P.E. responses is thought to reflect the D_1-component of apomorphine activity,[36] and Y seems to be mediated by D_2-receptors functionally correlated to D_1-sites.[37] Thus, a smaller availability of striatal D_1- and D_2-receptors (see above) may likely underlie the reduced ability of apomorphine at provoke P.E. and Y responses in hypothyroidism.

CONCLUDING REMARKS

Present results strongly suggest that thyroid deprivation during fetal and early postnatal ages, when the corpus striatum in rats undergoes maturation, provokes a widespread depression of the dopaminergic nigrostriatal system. As stated before, hypothyroidism-related impairments may extend to additional neurotransmitter pathways in the striatum, as well as to other brain regions.[2]

The most consistent features here shown are some loss of striatal innervation and, more important, a selective impairment of the vesicular transport of DA. The DA-turnover may, thus, be adaptively reduced, as reflected by the smaller contents of DOPAC and HVA metabolites. The loss in the numbers of striatal dopamine receptors, and deficit in their biochemical transduction have significant consequences on those behaviours which are D_1/D_2-mediated.

A further involvement of DA-receptor deficiency in the cause of additional disturbances such as hyperlocomotion and gait abnormality commonly found in both hypothyroid newborn rats[38-40] and children[41] cannot be discounted. Therefore, present results issue the challenge of ascertaining, whenever possible, whether some part of neurobehavioral pathology in children suffering severe, replacement-resistant, or late-treated congenital hypothyroidism, may have dopaminergic origin.

Acknowledgements

This work was supported by grants from the Italian Ministry of Education (40% and 60%, years 1987-1988) to A.V.

REFERENCES

1. A. J. Prange, J. C. Garbutt, and P. T. Loosen, The hypothalamic-pituitary-thyroid axis in affective disorders, in: "Psychopharmacology: the Third Generation of Progress," H. Y. Meltzer, ed., Raven Press, New York (1987).
2. A. Vaccari, Teratogenic mechanisms of dysthyroidism in the central nervous system, in: "Biochemical Basis of Functional Neuroteratology," G. J. Boer et al., eds, Elsevier, Amsterdam (1988).
3. J. M. Manson, Teratogens, in: "Casarett and Doull's Toxicology," C. D. Klaassen et al., eds, MacMillan Publ. Co., New York (1986).
4. M. J. Ellenhorn and D. G. Barceloux, "Medical Toxicology,"Elsevier, New York (1988).
5. A. Vaccari, Z. L. Rossetti, G. De Montis, E. Stefanini, E. Martino, and G. L. Gessa, Neonatal hypothyroidism induces striatal dopaminergic dysfunction, Neuroscience (in press).
6. C. R. Gerfen, K. G. Baimbridge, and J. Thibault, The neostriatal mosaic: III. Biochemical and developmental dissociation of patch-matrix mesostriatal systems, J.Neurosci. 7:3935 (1987).
7. E. J. Lu and W. J. Brown, The developing caudate nucleus in the euthyroid and hypothyroid rat, J.Comp.Neurol. 171:261 (1977).
8. E. J. Lu and W. J. Brown, An electron microscopic study of the developing caudate nucleus in euthyroid and hypothyroid states, Anat.Embryol. 150:335 (1977).
9. J. T. Eayrs and S. Levine, Influence of thyroidectomy and subsequent replacement therapy upon conditioned avoidance learning in rats, J.Endocrinol. 25:505 (1963).
10. A. Ruiz-Marcos, F. Sanchez-Toscano, M. J. Obregon, F. E. Del Rey, and G. M. De Escobar, Thyroxine-treatment and recovery of hypothyroidism-induced pyramidal cell damage, Brain Res. 239:559 (1982).
11. A. Ruiz-Marcos, J. Salas, F. Sanchez-Toscano, F. E. Del Rey, and G. M. De Escobar, Effect of neonatal and adult onset hypothyroidism on pyramidal cells of the rat auditory cortex, Devl Brain Res. 9:205 (1982).
12. R. Hebert, J. M. Langlois, and J. H. Dussault, Permanent defects in rat peripheral auditory function following perinatal hypothyroidism: determination of a critical period, Devl Brain Res. 23:161 (1985).
13. G. M. De Escobar and F. Escobar Del Rey, Thyroid hormone and the developing brain, in: "Congenital Hypothyroidism," J.H. Dussault and P. Walker, eds, Academic Press, New York (1983).
14. R. Biassoni and A. Vaccari, Selective effects of thiol reagents on the binding sites for imipramine and neurotransmitter amines in the rat brain, Br.J.Pharmacol. 85:447 (1985).

15. W. G. Beamer, E. M. Eicher, L. J. Maltais, and J. L. Southard, Inherited primary hypothyroidism in mice, Science 212:61 (1981).

16. W. G. Beamer and L. A. Creswell, Defective thyroid ontogenesis in fetal hypothyroid (hyt/hyt) mice, Anat.Rec. 202:387 (1982).

17. P. M. Adams, S. A. Stein, M. Palnitkar, A. Anthony, L. Gerrity, and D. R. Shanklin, Evaluation and characterization of the hypothyroid hyt/hyt mouse I: somatic and behavioral studies, Neuroendocrinol. 49:138 (1989).

18. R. L. Singhal, R. B. Rastogi, and P. D. Hrdina, Brain biogenic amines and altered thyroid function, Life Sci. 17:1617 (1975).

19. R. B. Rastogi, Y. La Pierre, and R. L. Singhal, Evidence for the role of brain biogenic amines in depressed motor activity seen in chemically thyroidectomized rats, J.Neurochem. 26:446 (1976).

20. J. Puymirat, Effects of dysthyroidism on central catecholaminergic neurons, Neurochem.Int. 7:969 (1985).

21. A. Vaccari, R. Biassoni, and P. S. Timiras, Selective effects of neonatal hypothyroidism on monoamine oxidase activities in the rat brain, J.Neurochem. 40:1019 (1983).

22. R. N. Kalaria and A. K. Prince, Effects of thyroid deficiency on the development of cholinergic, GABA, dopaminergic and glutamate neuron markers and DNA concentrations in the rat corpus striatum, Int.J. Devl.Neurosci. 3:655 (1985).

23. C. Atack, N. H. Bass, and P. Lundberg, Mechanisms for the elimination of 5-hydroxyindoleacetic acid for brain cerebrospinal fluid of the rat during postnatal development, Brain Res. 77:111 (1974).

24. A. Vaccari, High affinity binding of [^{3}H]-tyramine in the central nervous system, Br.J.Pharmac. 89:15 (1986).

25. A. Vaccari and G. L. Gessa, [^{3}H]Tyramine binding: a comparison with neuronal [^{3}H]dopamine uptake and [^{3}H]mazindol binding processes, Neurochem.Res. 14:949 (1989).

26. J. A. Javitch, R. O. Blaustein, and S. H. Snyder, [^{3}H]-Mazindol binding associated with neuronal dopamine uptake sites in corpus striatum membranes, Eur. J.Pharmacol. 90:461 (1983).

27. R. E. Heikkila, B. S. Shapiro, and R. C. Duvoisin, The relationship between loss of dopamine nerve terminals, striatal [^{3}H]spiroperidol binding and rotational behaviour in unilaterally 6-hydroxydopamine-lesioned rats, Brain Res. 211:285 (1981).

28. A. Vaccari and P. S. Timiras, Alterations in brain dopaminergic receptors in developing hypo- and hyperthyroid rats, Neurochem.Int. 3:149 (1981).

29. J. Lehmann and S. Z. Langer, The striatal cholinergic interneuron: synaptic target of dopaminergic terminals? Neuroscience 10:1105 (1983).

30. S. E. Leff, L. Adams, J. Hyttel, and I. Creese, Kainate lesion dissociates striatal dopamine receptor ligand binding sites, Eur.J.Pharmac. 70:71 (1981).

31. F. M. Filloux, J. K. Wamsley, and T. M. Dawson, Presynaptic and postsynaptic D$_1$ dopamine receptors in thenigrostriatal system of the rat brain: a quan-

titative autoradiographic study using the selective D_1 antagonist [^{3}H]SCH 23390, <u>Brain Res.</u> 408:205 (1987).

32. J. Arnt, Behavioral studies of dopamine receptors: evidence for regional selectivity and receptor multiplicity, <u>in</u>: "Dopamine Receptors," I. Creese and C.M. Fraser, eds, Alan R. List, Inc., New York (1987).

33. D. H. Overstreet, A. D. Crocker, C. A. Lawson, G. M. McIntosh, and J. M. Crocker, Alterations in the dopaminergic system and behaviour in rats reared on iodine-deficient diets, <u>Pharmacol.Biochem.Behav.</u> 21:561 (1984).

34. M. R. Del Cerro, G. Somoza, S. Segovia, and A. Guillamon, Effects of neonatal thyroidectomy on neurotransmitter receptors in several regions of the rat brain, <u>IRCS Med.Sci.</u> 14:92 (1986).

35. A. Vaccari, M. Collu, and G. Serra, Dopamine-mediated yawning and penile erections in neonatally-rendered hypothyroid rats: effects of GM_1 ganglioside, (submitted).

36. G. Serra, M. Collu, P. D'Aquila, and G. L. Gessa, SKF 38393, a selective D_1 DA agonist, induces penile erections in rats, <u>Pharmacol.Res.Comm.</u> 20:247 (1988).

37. G. Serra, M. Collu, and G. L. Gessa, Yawning is elicited by D_2 dopamine agonists but is blocked by the D_1 antagonist SCH 23390, <u>Psychopharmac.</u> 91:330 (1987).

38. W. S. Schwark, Cretinism animal model: neonatal hypothyroidism in the rat, <u>Am.J.Pathol.</u> 87:437 (1977).

39. C. E. Hendrich, W. J. Jackson, and S. P. Porterfield, Behavioral testing of progenies of Tx (hypothyroid) and growth hormone-treated Tx rats: an animal model for mental retardation, <u>Neuroendocrinol.</u> 38:429 (1984).

40. C. P. Comer and S. Norton, Behavioral consequences of perinatal hypothyroidism in postnatal and adult rats, <u>Pharmacol.Biochem.Behav.</u> 22:605 (1985).

41. B. S. Hetzel and I. D. Hay, Thyroid function, iodine nutrition and fetal brain development, <u>Clin.Endocrinol.(Oxf.)</u> 11:445 (1979).

ADVANCES IN THE UNDERSTANDING OF NEURAL MECHANISMS IN MOVEMENT DISORDERS

I.J. Mitchell, J.M. Brotchie, W.C. Graham, R.D. Page, R.G. Robertson
M.A. Sambrook and A.R. Crossman

Experimental Neurology Group
Department of Cell and Structural Biology
University of Manchester
Manchester M13 9PT, U.K.

INTRODUCTION

Relatively little is known about how various primary pathologies within the basal ganglia manifest themselves in the appearance of movement disorders. For example, both Parkinson's disease and Huntington's disease are characterised by dysfunction of the striatum. In the case of Parkinson's disease, loss of the ascending dopamine systems results in disordered striatal activity, whereas in Huntington's disease, parts of the striatum degenerate. It is implicitly assumed that the abnormal movements result from the disordered striatal activity acting on lower motor centres, presumably via the intermediary of the thalamus and its connections with the cortex. However, the pathophysiological processes by which the abnormal striatal output acts upon the intervening basal ganglia nuclei are yet to be defined. In an attempt to address these issues we have developed primate models of a spectrum of movement disorders and then attempted to both elucidate the neural circuitry which is responsible for mediating them and to define the underlying characteristic changes in neural activity in each of the affected basal ganglia nuclei.

1. BALLISM(US)

The first primate model of a movement disorder that we developed was of hemiballismus. This was based on the classic work of Carpenter who had demonstrated that lesions of the subthalamic nucleus in the monkey produced wild flinging movements of the contralateral limbs, known as hemiballismus or ballism (1). We were able to reproduce the movements in a reversible manner in the macaque by the injection of bicuculline, a GABA antagonist, into the subthalamic nucleus (2). We subsequently investigated the neural mechanisms that underlie this condition using the 2-deoxyglucose (2-DG) uptake procedure (3).

The Basal Ganglia III, Edited by G. Bernardi *et al.*
Plenum Press, New York

The 2-DG uptake technique is based on the premise that changes in the firing rate of a neuron will result in corresponding changes in glucose consumption. Thus if the rate of glucose utilization can be established it can be used as an index of neuronal activity. This can be achieved by administering a radiolabelled analogue of glucose, 2-DG, which can only be partially metabolised and so remains trapped inside the neuron. The localisation and quantification of label can then be determined autoradiographically.

Application of the 2-DG uptake technique to animals with active ballism resulted in autoradiographs which showed decreased 2-DG uptake in both segments of the globus pallidus and the ventral anterior/ventral lateral thalamic complex ipsilateral to the injection of bicuculline (4). There is now considerable evidence to show that the regional changes in 2-DG uptake, which occur as a result of experimental manipulation are probably due to changes in the activity of nerve terminals within the affected structure, rather than primarily reflecting the activity of intrinsic neuronal cell bodies (5,6,7). Assuming the changes in 2-DG uptake seen in ballism are due to changes in terminal activity, the decreased uptake in the globus pallidus probably reflects decreased neuronal activity in the projection to it from the subthalamic nucleus (1). This suggests that the injection of bicuculline inhibits the subthalamic nucleus neurons, probably by a depolarising blockade. The decrease in 2-DG uptake seen in the thalamus is probably due to decreased terminal activity in the input from the medial pallidal segment (1). Such underactivity in the medial pallidal segment implies that the structure must normally be excited by the subthalamic nucleus. This result thus provided the first behaviourial evidence that the transmitter at the subthalamopallidal pathway must be excitatory and not inhibitory as had previously been thought.

2. CHOREA

Chorea can be induced in a reversible manner in the primate by the injection of bicuculline into the region where the putamen borders onto the lateral pallidal segment (8). The dyskinesia consisted of writhing movements of the limbs contralateral to the injection. Examination of 2-DG autoradiographs from such animals revealed a marked regional increase in 2-DG uptake in the ipsilateral subthalamic nucleus, the area of increased uptake being restricted to the dorsolateral tip of the nucleus (9). Decreased uptake of 2-DG was seen in both the medial segment of the globus pallidus and the ventral anterior/ventral lateral thalamic complex, this distribution of label being the same as that seen in ballism.

If the increase in 2-DG uptake in the subthalamic nucleus was due to a change in terminal activity, it is most likely due to increased activity in the afferent input to it from the lateral pallidal segment (1). This hypothesis was tested using anatomical tracing techniques. The tracer, HRP was injected at a site which had previously given dyskinesia and the anterograde transport of label studied. Terminal label was seen in the subthalamic nucleus, but was restricted to the dorsolateral portion of the structure, ie., that part of the nucleus which was labelled in the 2-DG experiment (9). The remarkable correspondence between the two patterns of label strongly suggests that the increase in 2-DG uptake seen in the subthalamic nucleus of the choreic brain is due to increased activity in the pallidosubthalamic pathway. The pallidosubthalamic pathway is known to be GABAergic. It would therefore appear that the chorea arises from increased neuronal activity in the pallidosubthalamic pathway which results in increased inhibition of the subthalamic nucleus. In view of the similarity of 2-DG uptake in both the medial pallidal segment and the thalamus of the ballistic and choreic brains it is tempting to speculate that this may also be the case in other hyperkinetic conditions.

3. TARDIVE DYSKINESIA

In collaboration with Prof. L. Gunne (Uppsala, Sweden), we have extended these studies to a third primate model of dyskinesia, namely tardive dyskinesia. Cebus monkeys were treated with depot neuroleptics for several years. This procedure resulted in the appearance of tardive dyskinesia in some of the animals but not in others (11). The neuroleptic-treated animals thus serve as a good control for the neuroleptic treated dyskinetic animals. Preliminary data has demonstrated that the pattern of 2-DG uptake in the tardive dyskinetic animal is similar to that seen in experimental chorea and ballism. Thus, the tardive dyskinetic animal showed decreased 2-DG uptake in the medial pallidal segment compared with that seen in the non-dyskinetic animal but increased uptake in the subthalamic nucleus (12). Thus, it would appear that overactivity of the pallidosubthalamic pathway and underactivity of subthalamopallidal pathway are fundamental characteristics chorea, ballism and tardive dyskinesia.

4. LEVODOPA-INDUCED DYSKINESIA

Chronic treatment of parkinsonian patients with dopamine replacement therapy is known to result in the appearance of hyperkinetic side effects at peak dose. We have been able to reproduce some of these motor abnormalities in the MPTP-treated parkinsonian primate. Thus, administration of therapeutic doses of either levodopa or direct dopamine agonists such as apomorphine results in the appearance of apparent restlessness of the lower limbs after as little as 5 doses. Following 4-10 weeks of regular levodopa therapy, animals developed "peak-dose" choreiform movements in the lower limbs which spread, with time, to involve the upper limbs and oral/facial musculature. With further treatment (5-21 months), the animals developed peak dose dystonia in the lower limbs (13). The induction of peak-dose chorea and dystonia in the parkinsonian primate is dependent on the length of exposure to dopaminergic agents, such that chorea develops after 1-3 months while dystonia requires much longer periods of treatment. We consider these conditions to represent novel models of levodopa-induced chorea and dystonia in man since they depend upon the same underlying neuropathology and treatment regimen as their human counterparts.

Preliminary data from 2-DG uptake experiments suggests that the neural mechanisms that mediate levodopa-induced chorea are the same as those which underlie other forms of experimentally-induced hyperkinesias. Thus, levodopa-induced chorea appears to be mediated by underactivity in the subthalamopallidal and pallidothalamic pathways. The pallidothalamic pathway is thought to utilise GABA as its transmitter (10). Underactivity in this pathway would therefore be expected to result in loss of inhibition of the thalamus. If this hypothesis is correct it would be predicted that lesions of the overactive thalamus should alleviate hyperkinesias. This hypothesis was tested in the levodopa-induced choreic primate.

Anatomical tracing experiments were initially performed in order to define the area of thalamus which receives the output from the medial segment of the globus pallidus. HRP-WGA was injected stereotactically into the medial segment of the globus pallidus and its anterograde transport studied. Terminal label was observed in a restricted portion of the ventral anterior/ventral lateral thalamic complex in the area defined as VApc and dc by Ilinsky (14). This region lies just posterior to that receiving afferents form the substantia nigra and anterior to the thalamic region which has been implicated in the mediation of tremor.

The effects of localised thalamic lesions on hyperkinesia were then studied in primates with levodopa-induced chorea, (see figure 1). The results of this work clearly demonstrate that lesions which destroy the majority of the terminal fields of the pallidothalamic pathway completely alleviate chorea of the contralateral limbs. Smaller lesions which spared some of the terminal area had only a transitory effect, the chorea reappearing after several weeks. In contrast, lesions which were located medial to the area of pallidal termination had no effect on dyskinesia. These data strongly suggest that levodopa-induced chorea is mediated via underactivity of the pallidothalamic pathway which results in disinhibition of the thalamus. Furthermore, since we have concluded that a spectrum of hyperkinetic disorders are mediated by a common mechanism, the data suggests that thalamotomy may represent a useful therapeutic approach for the alleviation of dyskinesias which arise as a result of a variety of different pathologies.

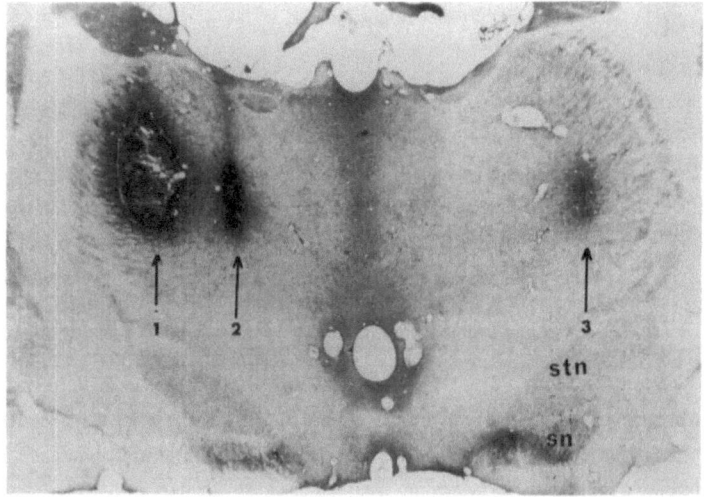

FIGURE 1

Histological demonstration of radiofrequency lesions in the thalamus of a monkey with dopamine agonist induced chorea. Lesion 1 (4 weeks premortem) was correctly placed in the terminal field of the pallidal input to the thalamus and completely abolished chorea of the contralateral limbs. Lesion 2 (9 weeks premortem) was placed medial to the pallidothalamic terminals and had almost no effect on dyskinesia. Lesion 3 (7 weeks premortem) partially destroyed the pallidothalamic terminals and had a transient effect.

5. PARKINSONISM

We have also applied the 2-DG uptake technique to monkeys rendered parkinsonian by the systemic administration of the neurotoxin 1-methyl-4-phenyl-1,2,3,6-tetrahydropyridine, (MPTP). The autoradiographs from these animals revealed increased 2-DG uptake in the globus pallidus and thalamus (15). These data are interpreted as suggesting that the major afferent inputs to these structures, namely the putamen, subthalamic nucleus and medial pallidal segment are overactive in experimental parkinsonism. In contrast, decreased 2-DG uptake was seen in the subthalamic nucleus. This finding suggests that the major input to the nucleus, namely the lateral pallidal segment is underactive in the MPTP-treated monkey.

In collaboration with Dr. P. Emson's research group, (Babraham, U.K.), we have tried to confirm some of these findings using in situ hybridization techniques. To this end we have studied enkephalin mRNA expression in the striatum of the parkinsonian monkey. The results of this experiment demonstrated that there is a dramatic increase in enkephalin mRNA expression in the parkinsonian putamen compared with control values (16). These enkephalinergic striatal cells are known to project to the lateral pallidal segment. These data are therefore in agreement with the conclusion drawn from the 2-DG experiment that the projection from the putamen to the lateral pallidal segment is over active in parkinsonism.

6. DYSTONIA

As mentioned in the previous section, chronic dopamine replacement therapy of both parkinsonian patients and primates can result in the appearance of dystonia at peak dose. Dystonia is defined as: "a syndrome of sustained muscle contractions, frequently causing twisting and repetitive movements, or abnormal postures" (17). Very little is known about the primary pathology which underlies the disorder nor the pathophysiological mechanisms that mediate it. However, secondary dystonia is often associated with basal ganglia dysfunction. It can occur as a symptom of Huntington's disease, as a result of basal ganglia infarcts, and in parkinsonism, where it can occur both at peak dose and prior to levodopa treatment.

We have investigated the neural mechanisms that mediate levodopa induced dystonia in MPTP-treated parkinsonian primates using the 2-DG uptake technique. The figure below shows an autoradiograph from an animal that was rendered hemiparkinsonian by a unilateral injection of MPTP into the right common carotid artery. During the period of active 2-DG uptake the animal showed torticollis and prolonged periods of adduction of the left leg which was flexed at the hip and knee. The pattern of 2-DG uptake seen is a combination of that seen in the choreic brain and that seen in the parkinsonian brain, (see figure 2). Thus, there is increased uptake in the subthalamic nucleus and decreased uptake in the thalamus as seen in chorea, but increased uptake in the globus pallidus as seen in parkinsonism (18). A parsimonious explanation for the pattern of uptake is that;

i. The increased uptake in the subthalamic nucleus is due to increased activity in the input from the lateral pallidal segment.

611

ii. This would inhibit the subthalamic nucleus which would result in loss of excitation of the medial pallidal segment.

iii. Decreased neuronal activity in the medial pallidal segment would then account for the decreased 2-DG uptake in the thalamus.

However, this would predict that the medial pallidal segment would show decreased uptake when it fact it shows increased uptake. This increased uptake could most easily be accounted for by postulating an increase in the activity in the putaminopallidal pathway. This data suggests that dopamine agonist-induced dystonia is characterised by; increased activity in the putaminopallidal and pallidosubthalamic pathways but decreased activity in the subthalamopallidal and pallidothalamic pathways.

This conclusion is supported by an anecdotal observation that manipulating GABAergic transmission in the medial pallidal segment of the monkey can elicit acute dystonia. These observations, however, are not supported by data from thalamotomy studies. In contrast to levodopa-induced chorea, lesions of the ventral anterior/ventral lateral thalamic complex appear to have little effect on alleviating levodopa-induced dystonia.

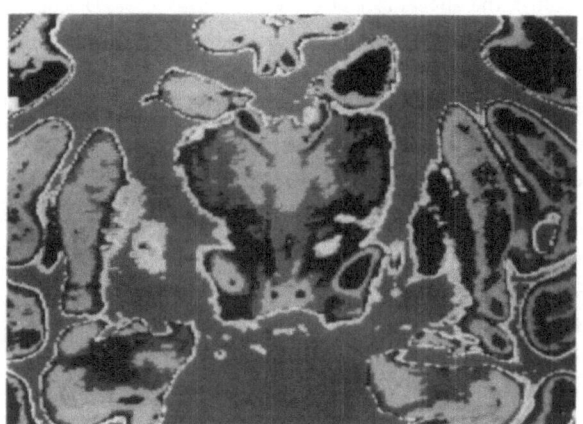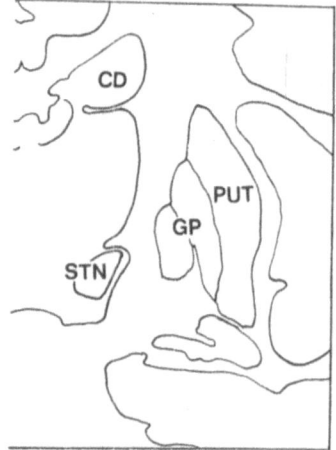

FIGURE 2

Autoradiograph from the brain of a monkey with unilateral dystonia showing asymmetries in 2-DG uptake between the side of the brain giving rise to dystonic symptoms (right side) and the control side. Note the increased 2-DG uptake in the right caudate nucleus (CD), putamen, (P) globus pallidus (GP) and subthalamic nucleus (STN) compared with the left (control) side.

CONCLUSIONS DRAWN FROM 2-DG STUDIES OF PRIMATE MODELS OF MOVEMENT DISORDERS

Taken together, the results of these studies have allowed us to draw the following conclusions;

1. The subthalamopallidal and pallidothalamic pathways are critically involved in a variety of hyperkinetic disorders including, ballism, chorea and tardive dyskinesia. Both pathways are underactive in these conditions.

2. Stereotactically placed lesions of the area of termination of the pallidothalamic pathway can alleviate levodopa induced chorea.

3. The subthalamopallidal and pallidothalamic pathways also play an important role in the mediation of parkinsonism, both pathways being overactive in this condition. These conclusions have been supported by electrophysiological data from the laboratories of Delong (19) and Filion (20).

4. The projection from the putamen to the lateral pallidal segment is overactive in experimental parkinsonism.

5. Levodopa induced dystonia (and possibly other forms of dystonia) arise from an interaction of abnormal inputs to the medial pallidal segment from the striatum, such that; the direct projection from the putamen to the medial pallidal segment is overactive whilst the input which arrives indirectly via the lateral pallidal segment and the subthalamic nucleus is ultimately underactive.

6. The transmitter in the subthalamopallidal pathway is excitatory. Evidence from immunohistochemical studies from the laboratories of Parent (21) and Kitai (21) have suggested that it is an excitatory amino acid (EAA).

EXCITATORY AMINO ACID (EAA) TRANSMISSION AND MOVEMENT DISORDERS

If the conclusions presented above are correct, it follows that modulation of EAA transmission at the subthalamopallidal synapse should have a dramatic effect on the control of movement. We have begun to investigate this possibility using a number of different experimental approaches.

1. NEUROCHEMICAL DEMONSTRATION OF EAA UPTAKE SITES ON SUBTHALAMIC NUCLEUS TERMINALS IN THE GLOBUS PALLIDUS

Firstly, we have attempted to use neurochemical techniques to confirm that the transmitter in question really is an EAA. To this end transmitter uptake systems were studied in rats with unilateral quinolinic acid lesions of the subthalamic nucleus. A double label procedure was used which permitted assessment of EAA and GABA uptake in the same tissue. The assay measured sodium dependent [^{3}H]-D-aspartate and [^{14}C]-GABA uptake in neuronal membrane preparations. The pharmacokinetic properties of the assay were consistent with radioligand accumulation specifically by the EAA and GABA uptake sites respectively used to remove the transmitters from synaptic junctions. We therefore believe that the assay provides a means of quantifying pre-synaptic terminals utilising EAA and GABA transmitters.

The assay revealed a dramatic decrease (-86%) in [3H]-D-aspartate uptake in the globus pallidus ipsilateral to the lesion compared with that seen in the contralateral structure. In contrast no significant changes were seen in the uptake of GABA (21). This loss of EAA uptake sites suggests that there must be an EAA input to the globus pallidus from the subthalamic nucleus.

2. CHOREA ELICITED BY EAA BLOCKADE IN THE MEDIAL PALLIDAL SEGMENT

On the basis of these results we have tried to induce dyskinesia in the monkey by blocking EAA transmission in the medial pallidal segment. This experiment involved infusing the broad spectrum EAA antagonist, Kynurenic acid, unilaterally into the medial pallidal segment of the conscious monkey. This resulted in the appearance of choreic movements of the limbs contralateral to the injection site when the injection was centred in the medial pallidal segment but not in the surrounding structures (24). In order to confirm that the manipulation had resulted in decreased pallidal output, 2-DG uptake was studied in some of these animals. The resultant autoradiographs revealed decreased uptake of 2-DG in the ventral anterior/ventral lateral thalamic complex ipsilateral to the kynurenic acid injection. This result strongly suggests that the dyskinesia induced by kynurenic acid results from loss of excitation in the pallidothalamic pathway.

3. ALLEVIATION OF AKINESIA BY EAA BLOCKADE IN THE MEDIAL PALLIDAL SEGMENT HOMOLOGUE IN RODENT PARKINSONIAN MODELS

From the premises that blocking glutamate transmission at this synapse elicits chorea, and that the pathway plays a very important role in mediating parkinsonism, it follows that manipulating glutamate transmission may alleviate some of the symptoms of Parkinson's disease.

In order to investigate this possibility a series of experiments have been conducted using 6OHDA lesioned rats. The rats received bilateral injections of 20 μg 6OHDA directly into the substantia nigra. This treatment rendered the rats extremely akinetic, presumably because of the loss of striatal dopamine. EAA antagonists were then injected unilaterally into the entopeduncular nucleus, the rat homologue of the medial pallidal segment, in an attempt to alleviate the akinesia. Saline was found to have no effect on movement, whereas kynurenic acid (20 μg), the broad spectrum EAA antagonist, gave a dramatic but short lasting reversal of the akinesia (25).

Delineation of the EAA receptor subtype involved in this effect has been undertaken in rats rendered akinetic by systemic injections of reserpine (5mg/kg, i.p.). Dose-response studies with a wide range of EAA antagonists revealed that compounds blocking NMDA receptor associated transmission (MK801, CPP, AP7 and HA-966) were more effective than the broad spectrum antagonist kynurenic acid and the non-NMDA preferring CNQX, (see figure 3). These compounds were only efficacious in a narrow band of doses, being anaesthetics at high doses. These results suggests that manipulating NMDA glutamate receptors in the medial pallidal segment of the globus pallidus could be used to reverse some of the symptoms of parkinsonism.

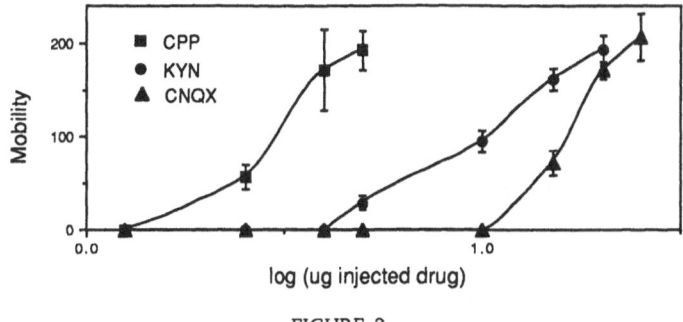

FIGURE 3

Dose response curves for the effects on mobility of injecting EAA antagonists into the entopeduncular nucleus of reserpine-induced akinetic rats. The mobility score is a measure of total forelimb movement during the 15 minute period immediately post-injection. Note that CPP, the NMDA receptor specific antagonist, was the most effective drug in reversing the akinesia. Kynurenic acid, the broad spectrum EAA antagonist, and CNQX, which has specificity for non-NMDA receptor subtypes, were less potent.

CONCLUSIONS

The results of the experiments presented in this paper point to the importance of the medial pallidal segment in mediating movement disorders. Underactivity in this structure is characteristic of a multitude of experimentally-induced hyperkinetic conditions including ballism, chorea and tardive dyskinesia. In contrast, overactivity of this structure characterises the parkinsonian brain. We have demonstrated that the activity of the medial pallidal segment is regulated by an EAA input from the subthalamic nucleus. Preliminary data strongly suggest that manipulations of EAA transmission at these synapses can have a profound effect on the control of movement and may represent an avenue for novel therapeutic approaches to the alleviation of disorders of movement.

ACKOWLEDGEMENTS

This work was financially supported by grants from the Medical Research Council, The Wellcome Trust, the Parkinson's Disease Society, Action Research for the Crippled Child and the Dystonia Medical Research Foundation.

REFERENCES

1. CARPENTER, M.B. & SUTIN J. (1983) Human Neuroanatomy, Williams & Wilkins, Baltimore/London.
2. CROSSMAN, A.R., SAMBROOK, M.A. A& JACKSON, A. (1984) Brain, 107, 579-596.
3. SOKOLOFF L., REIVICH M., KENNEDY C., DES ROSIERS M. H., PATLACK C. S., PETTIGREW K. D., SAKURADA O. AND SHINOHARA M. J. Neurochem., 1977; 28, 897-916.

4. MITCHELL I.J., SAMBROOK M.A. AND CROSSMAN A.R.
   Brain, 1985; 108, 405-422.

5. AUKER, C.R, MESZLER, R.M. AND CARPENTER, D.O.
   J. Neurophysiol., 1983; 49:1504

6. MATA M., FINK D.J., GAINER H., SMITH C.B., DAVIDSEN L., SAVAKI H., SCHWARTZ W. J. &
   SOKOLOFF L. J. Neurochem., 1980; 34, 213-215.

7. SCHWARTZ W.J., SMITH C.B., DAVIDSEN L., SAVAKI H. AND SOKOLOFF L. Science, 1979; 205,
   723-725.

8. CROSSMAN, A.R., MITCHELL, I.J., JACKSON, A. & SAMBROOK, M.A. (1988) Brain, 111, 1211-
   1233.

9. MITCHELL, I.J., JACKSON, A., SAMBROOK, M.A. AND CROSSMAN, A.R.
   Brain, (In press).

10. PARENT, A., (1986) Comparative Neurobiology of the Basal Ganglia, John Wiley.

11. GUNNE, L.M., HAGGSTROM, J.E. & SJOQUIST, B. (1984) Nature, 309: 347-349.

12. MITCHELL, I.J., PAGE, R.D., ROBERTSON, R.G., CROSSMAN, A.R. & GUNNE, L.M. Br. J.
    Pharmacol., (In press)

13. BOYCE, S., CLARKE, C.E., LUQUIN, R., PEGGS, D., FARMERY, S.M., ROBERTSON, R.G.,
    MITCHELL, I.J., SAMBROOK, M.A. AND CROSSMAN, A.R. Movement Disorders (In press)

14. ILINSKY, I.A & KULTAS-ILINSKY 1987 J. Comp. Neurol. 262: 331-364.

15. MITCHELL, I.J., CLARKE, C.E., BOYCE, S., ROBERTSON, R.G., PEGGS, D., SAMBROOK, M.A. AND
    CROSSMAN, A.R. Neuroscience (In press).

16. AUGOOD, S.J., EMSON, P.C., MITCHELL, I.J., BOYCE, S., CLARKE, C.E. AND CROSSMAN, A.R.
    1989 Mol. Brain Res., 6, 85-92.

17. Classification of Extrapyramidal Disorders J. Neurol. Sci., 1981; 51, 311-327.

18. MITCHELL, I.J., LUQUIN, R., BOYCE, S., CLARKE, C.E., ROBERTSON, R.G., SAMBROOK, M.A. &
    CROSSMAN, A.R. Movement Disorders, (In press)

19. MILLER, W.C. & DELONG, M.R. (1987) In: Advances in Behavioural Biology, 32, The Basal
    Ganglia II. Structure and Function - Current Concepts (Carpenter M.B. & Jayaraman A., eds.)
    Plenum Press, New York, 395-403.

20. FILION, M., TREMBLAY, L. & BEDARD, P.J. (1988) Brain Res., 444: 165-176.

21. SMITH, Y. & PARENT, A. (1988) Brain Res., 453: 353-356.

22. KITA, H. & KITAI, S.T. (1987) J. Comp. Neurol. 260: 435-452.

23. BROTCHIE, J.B., CROSSMAN, A.R. Br. J. Pharmacol. (In press)

24. ROBERTSON, R.G., FARMERY, S.M., SAMBROOK, M.A. & CROSSMAN, A.R. (1989) Brain Res.
    476: 317-322

25. BROTCHIE, J.B., MITCHELL, I.J. & CROSSMAN, A.R. Eur. J. Pharmacol (Submitted)

STUDIES OF DIRECT CLINICAL
INTEREST

NEUROPEPTIDE mRNA EXPRESSION IN HUMAN BASAL GANGLIA

G.Mengod[1], E.Ruberte[1,*], A.Probst[2], and J.M.Palacios[1]

1 Preclinical Research, Sandoz Pharma AG, Basel
2 Institut für Pathologie, Abteilung Neuropathologie
Universität Basel, Basel. Switzerland. * Present
address: Lab. Génétique Moléculaire CNRS, Faculté
Médecine. Strasbourg, France

SUMMARY

We have examined the distribution in human basal ganglia of cells expressing mRNAs for several neuropeptides in control and diseased postmortem material by in situ hybridization histochemistry using ^{32}P-labelled oligonucleotides as hybridization probes. In control cases, enkephalin mRNA showed a patchy-like distribution. Cells containing somatostatin mRNA and neuropeptide Y mRNA were scattered throughout the caudate and putamen in a very similar pattern. Cholecystokinin mRNA was not found in the striatum. In Parkinson's disease, our preliminary results showed a decrease in the hybridization signal with the enkephalin probe. In contrast, somatostatin mRNA levels and neuropeptide Y levels did not exhibit significant changes. In Huntington's chorea, we have observed a marked decrease in the levels of enkephalin mRNA in both caudate and putamen but not in the nucleus accumbens. In contrast, no alteration of the hybridization signal was observed for both somatostatin and neuropeptide Y mRNAs. These results show that in situ hybridization histochemistry is very useful for the study of the molecular anatomy of the human basal ganglia and their pathology.

INTRODUCTION

Advances in our understanding of the functions of basal ganglia in normal and pathologic brains are related to the development of new techniques allowing for a deeper analysis at the anatomical and molecular levels. Because of the high cellular complexity of this brain region and the limited resolution of classical biochemical techniques, histochemical methods have been developed to provide cellular and subcellular analysis with a high degree of molecular specificity. Histochemical, immunohistochemical and autoradiography procedures take advantage of the existence of antibodies and radioactive markers for brain molecules (Björklund and Hökfelt , 1983).

Many of these molecules survive the postmortem conditions of the human brain and are amenable to analysis using biochemical or histochemical procedures (Bird and Iversen, 1982, Palacios et al., 1986). The systematic study of samples from diseased brains and their comparison with "control" populations has lead to the discovery of selective biochemical deficits in some diseases of the human brain. Classical examples are the dopamine deficit in Parkinson's disease (Hornykiewicz, 1972) and the acetylcholine deficit in Alzheimer's disease (Rossor, 1982). An extensive catalogue of biochemical alterations in these and other diseases is now available (Palacios et al., 1986).

Several types of neurons have been recognized in the mammalian basal ganglia, including the human, on the basis of their morphological characteristics (Graveland et al, 1985). Immunohistochemical studies have allowed the identification of the main transmitters used by these different neurons. Recent studies have also described a subnuclear organization of the striatum in two main compartments which have been named islands, striosomes or patches and matrix. Several neuropeptides appear to be characteristic ofsome of these neuronal populations. Striatal spiny neurons contain enkephalin and substance P, while somatostatin and neuropeptide Y have been localized to aspiny neurons. Neuropeptides are also contained in afferents to the striatum, an example being cholecystokinin.

Until now, most of the available information on the brain neuropeptide content, cellular localization and alteration with disease and drug treatment has been obtained thanks to the application of immunological techniques such as radioimmunoassay and immunohistochemistry. These techniques, however, present some limitations related to the sensitivity and specificity of the antibodies used. Successful visualization by immnunohistochemistry may depend either on the procedure used in the fixation of the tissue or on the necessity of pretreatment of the animals with colchicine to allow the visualization of cell bodies. Some of these problems are difficult to overcome when human postmortem tissues are investigated. The cloning and sequencing of genes coding for neuropeptides offer an alternative approach for their study. For example, visualization of cells containing neuropeptide mRNAs in the rat brain was accomplished by the application of the in situ hybridization histochemistry technique (Valentino et al., 1987). Numerous studies using this technique have been published on the distribution of neuropeptide mRNA containing cells and the regulation of neuropeptide gene expression by deafferentation or drug treatment.

In this paper we will review the use of in situ hybridization histochemistry for the study of neuropeptide gene expression in human basal ganglia. Particular emphasis will be put on the assessment of the specificity of the hybridization signal obtained with this material. The influence of parameters such as postmortem delay, age, gender and disease will also be discussed. The basal ganglia are particularly attractive due to the wealth of information already available on its chemical, cellular and regional organization (Graybiel and Ragsdale, 1984, Beal and Martin, 1986; Gerfen et al., 1987a,b; Parent, 1987) and because of their well documented involvement in diseases such as Parkinson, Huntington's Chorea and Schizophrenia (Emson et al 1986). The goal of our studies is to examine the distribution of the cells expressing neuropeptides in the healthy human brain and possible changes as a consequence of the different types of neuronal degeneration characteristic of diseases affecting the basal ganglia.

METHODS

in situ HYBRIDIZATION HISTOCHEMISTRY

The oligonucleotides used as probes in the in situ hybridization experiments were the following: NPY/1, complementary to the nucleotide sequence corresponding to amino acids 29-45 and NPY/2 to amino acids 84-98 of human neuropeptide Y precursor mRNA (Minth et al., 1984). SOM/1, complementary to amino acids 88-98 and SOM/3 to amino acids 22-37 of human somatostatin mRNA sequence (Shen et al., 1982). CCK/1, complementary to amino acids 80-95 and CCK/2 to amino acids 1-16 of human cholecystokinin mRNA (Takahashi et al., 1985). ENK/1, complementary to amino acids 133-146 and ENK/2 to amino acids 1-15 of human preproenkephalin A mRNA (Noda et al., 1982). They were labelled with ^{32}P-αdATP at their 3' end and hybridized to tissue sections as described by Mengod et al., 1989.

Several control experiments were carried out with all the probes used in the present study in order to determine the specificity of the hybridization signal obtained. To check the specificity of the oligonucleotides used, Northern analyses were performed for each probe. We found that they hybridized with single mRNA species of sizes corresponding to those originally described and only in brain regions where the peptides are known to be expressed. Figure 1 shows such an analysis for the enkephalin (A) and cholecystokinin (B) probes. While the enkephalin probe hybridized with an mRNA of 1500 nucleotides from striatum (A, lane 2), no signal could be detected in frontal cortex RNA (A, lane 1). The cholecystokinin probe hybridized to cortical mRNA (B, lane 1) with an expected size of 800 nucleotides, but no hybridization was obtained with RNA from the striatum (B, lane 2).

For each mRNA analyzed, at least two oligonucleotides were synthesized and used separately as hybridization probes. The hybridization pattern obtained in tissue sections was the same for both (not shown). Hybridization could be blocked by competing with 20-50-fold excess of unlabelled oligonucleotide (not shown). The thermal stability of the hybrids was found to be close to the theoretical Tm (not shown). In general, good agreement was found between the hybridization pattern obtained with all the oligomers and the available immunohistochemical results. Background levels were defined as those of tissue areas where no cellular signal was observed, for example in the cerebellum, and they were comparable to those obtained after competition of the labelled probe with excess of unlabelled probe.

Microdensitometric quantification of the hybridization signal for some of these peptides did not reveal any significant correlation between the level of hybridization and parameters such as postmortem delay, age, or gender in the control population examined.

RESULTS

Anatomical distribution in basal ganglia

We examined the distribution of the cells containing mRNA coding for enkephalin, cholecystokinin, somatostatin and neuropeptide Y, in several control cases without reported neurological diseases. A heterogeneous distribution of these cells was observed (Figure 2). Preproenkephalin A mRNA (figure 2A) presented a "patchy" distribution in the caudate and putamen nuclei. In contrast, cells containing somatostatin mRNA (figure 2B) were scattered throughout the caudate and putamen nuclei. In several striatal areas, they were organized around the patches of high preproenkephalin A mRNA content (data not shown). Neuropeptide Y mRNA containing cells (figure 2D) were distributed in the basal ganglia with a pattern similar to the one observed for somatostatin mRNA. This co-distribution was particularly evident in the basal ganglia but not observed in other areas such as the anterior thalamus, where high levels of somatostatin mRNA contrasted with the apparent absence of hybridization signal for neuropeptide Y mRNA. No significant hybridization was observed for cholecystokinin mRNA whereas a high density of cells containing this mRNA were observed in adjacent cortical regions and in the claustrum (figure 2C), indicating that striatal cholecystokinin immunoreactivity is localized to processes from cells extrinsic to this nucleus.

Figure 3 illustrates the relative abundance of cells containing transcripts for preproenkephalin A, somatostatin and neuropeptide Y in the human putamen, as observed using liquid emulsion autoradiography. While neuropeptide Y and somatostatin mRNA were expressed by a small number of cells (B,C), a larger number of cells containing preproenkephalin A mRNA were observed throughout this region (A).

Neuropeptide mRNA changes in Parkinson's and Huntington's diseases

In order to obtain more information on the cellular localization of these neuropeptides in the human striatum we examined tissues from diseased human brains where lesions of the striatal afferents and efferents are well described.

Parkinson's disease represents a model of a central dopaminergic denervation in man to study the relationships between neuropeptides and dopamine-containing neurons. Such interactions have been evidenced from investigations on the corresponding animal model, the 6-hydroxydopamine (6-OHDA)-induced degeneration of the rat dopaminergic ascending pathways. The levels of the peptide enkephalin are decreased in the striatum in Parkinson's disease (Agid and Javoy-Agid,

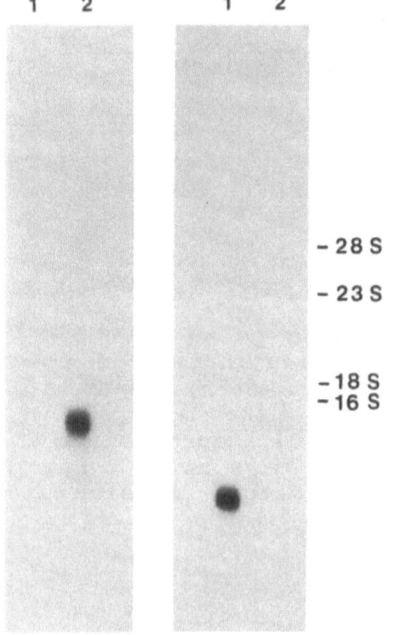

Figure 1 Autoradiogram of Northern blots of human poly(A)$^+$ RNA hybridized with the preproenkephalin A (A) and cholecystokinin (B) labelled oligonucleotide probes. Lane 1: 10 μg of poly(A)$^+$ RNA from cortex. Lane 2: 10 μg of poly(A)$^+$ RNA from putamen. The size of the hybridizing mRNA was calculated in comparison with molecular weight standards loaded on adjacent lanes. Reproduced from Mengod et al. 1990 with permission.

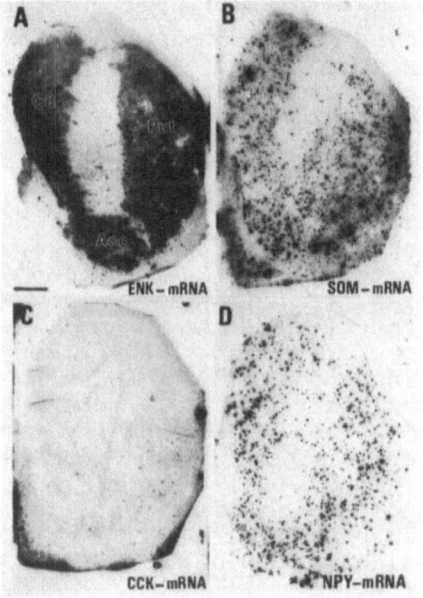

Figure 2 Neuropeptide mRNA visualization in human basal ganglia. The pictures are photographs from film-autoradiograms obtained by hybridizing sections from the same control case with ^{32}P-labelled oligonucleotides complementary to the mRNAs coding for: (A): preproenkephalin A (ENK); (B): somatostatin (SOM); (C): cholecystokinin (CCK); and (D): neuropeptide Y (NPY). Dark areas are rich in the corresponding mRNA. Films were exposed for 14 days at -70°C with intensifying screens. Cd, caudate nucleus. Put, putamen. Acc, nucleus accumbens. Bar: 5mm. Reproduced from Mengod and Palacios, 1990 with permission.

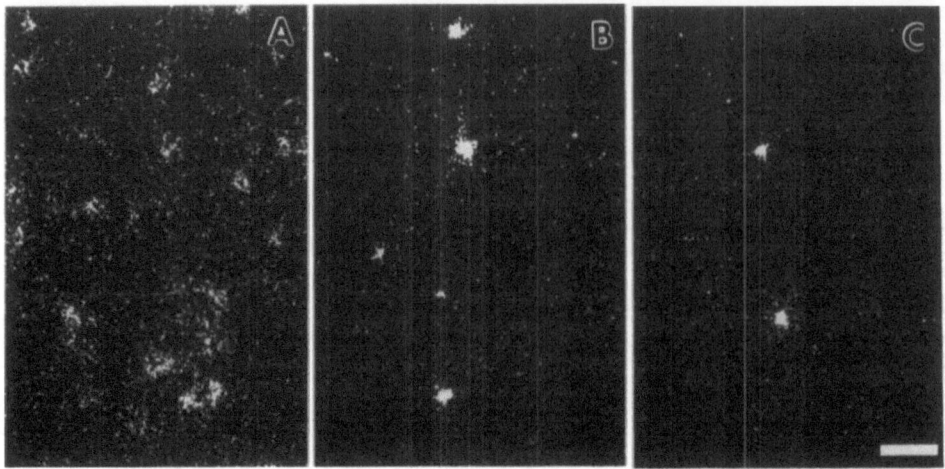

Figure 3 Relative abundance of cells expressing neuropeptides in the human putamen. Dark-field illumination photomicrographs obtained from liquid emulsion autoradiograms, where hybridization signal is seen as white dots. (A): preproenkephalin A, (B): somatostatin and (C): neuropeptide Y mRNA containing cells. Exposure time was 14 days at 4°C. Bar: 100μm. Reproduced from Mengod et al., 1990 with permission.

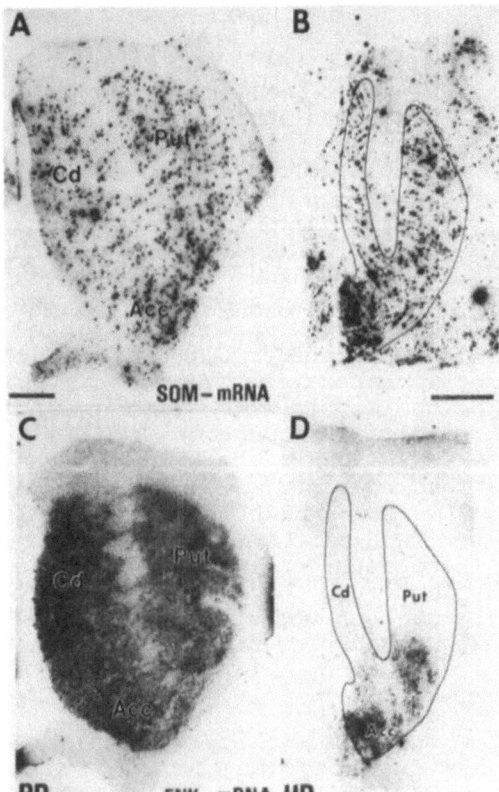

Figure 4 Alterations in the levels of neuropeptide mRNAs in Parkinson's and Huntington's disease basal ganglia. Levels of somatostatin mRNA were preserved throughout the striatum in Parkinson's (A) and Huntington's (B) striatum, while those of enkephalin mRNA were decreased in a Parkinson's case (C), compared to a control case in Figure 2A. The decrease was more pronounced in the Huntington's striatum (D), some ENK mRNA can be seen in the ventral parts. Cd, caudate nucleus. Put, putamen. Acc, nucleus accumbens. Bar: 5mm. Reproduced from Mengod and Palacios, 1990 with permission.

624

1985). In our preliminary results a decrease of the hybridization signal with the enkephalin probe was observed in the basal ganglia from all the analyzed cases dying with Parkinson's disease when compared to control cases (Figure 4C). These results are in contradiction with those obtained with 6-OHDA lesioned rats, where a marked increase in the level of enkephalin mRNA can be observed in the lesioned side (Young et al, 1988, Savasta et al, 1989). In contrast, neither somatostatin (Figure 4A) nor neuropeptide Y (not shown) mRNA levels exhibited significant changes in the Parkinsonian striatum. We were unable to observe any alterations in the levels of cholecystokinin mRNA in the regions analyzed. A differential regulation of somatostatin and neuropeptide Y mRNA levels by the ascending dopaminergic input has been recently documented immunohistochemically in the experimental animal (Kerkerian et al., 1986). The ability to analyze similar changes in Parkinson's disease at the mRNA level is now possible using in situ hybridization.

In Huntington's chorea some neurons of the caudate nucleus and putamen are selectively affected (Kowall et al, 1987), namely the spiny neurons. While the concentrations of a number of neurotransmitters which are known to be expressed by intrinsic cells projecting outside the striatum are decreased (as for example acetylcholine, GABA, substance P and enkephalins), other neurotransmitters such as dopamine, serotonin and neurotensin, which are localized to extrinsic afferents, are preserved. Interestingly, some intrinsic transmitters like somatostatin and neuropeptide Y and some cholinergic interneurons are also preserved. Using in situ hybridization we have observed a marked decrease in the levels of enkephalin mRNA in both caudate and putamen of the Huntington's chorea cases analyzed, but not much change in the nucleus accumbens (figure 4D). No alterations were seen in the cholecystokinin mRNA content (not shown). In contrast, an increase in the hybridization signal was observed for both somatostatin (figure 4B) and neuropeptide Y (not shown) mRNAs in Huntington's chorea striatum . The distribution of the neurons expressing somatostatin and neuropeptide Y was similar to that seen in control. However, because of the marked shrinkage of the basal ganglia in this disease, an apparent increase in the mRNA levels for these two neuropeptides was observed.

CONCLUSIONS

The most important finding of these investigations is the demonstration that the mRNA for several neuropeptides can be visualized at the microscopic level in human postmortem brain tissues using in situ hybridization histochemistry and synthetic oligonucleotides as probes. The specificity of the hybridization signal detected in each case is supported by several criteria such as Northern blot analysis, use of two oligomers complementary to different regions of the same target mRNA, competition between labelled and excess unlabelled oligoprobes, and melting curve analysis. Furthermore, factors such as age, postmortem delay, or gender did not show influence on the hybridization signal.

The results obtained for the different neuropeptides examined are, in general, in good agreement with the available information on their distribution and cellular localization as determined by radioimmunoassay or immunohistochemistry.

In the basal ganglia, we have found that numerous cells contained preproenkephalin A mRNA. These cells showed heterogeneous distributio throughout the caudate and putamen nuclei. This distribution is in good agreement with the proposed localization of enkephalin to spiny projection neurons (Pickel et al., 1980; Graybiel and Ragsdale, 1984; Beal and Martin, 1986). Somatostatin and neuropeptide Y mRNA containing cells were less abundant and presented a very similar distribution throughout the striatum, in agreement with the histochemical colocalization of these two neuropeptides to the same neuronal population, the aspiny interneurons (Graybiel and Ragsdale, 1984). The distribution agrees well with that of both neuropeptides in the cat striatum as revealed by immunohistochemical techniques (Chesselet and Graybiel, 1986). No detectable levels of cholecystokinin mRNA could be seen in the striatum, indicating that striatal cholecystokinin immunoreactivity is localized to processes from cells extrinsic to this nucleus.

In Parkinson's disease we have not detected significant alterations of the mRNA content for somatostatin and neuropeptide Y, whereas the number of preproenkephalin A mRNA containing cells was reduced to some extent in the caudate-putamen nucleus in some of the cases examined until now.

We have been able to show that there is a marked decrease in the levels of preproenkephalin mRNA in both caudate and putamen of the Huntington's chorea cases analyzed, while in the nucleus accumbens this mRNA content was less affected. No alterations were seen in the hybridization signal for somatostatin and neuropeptide Y mRNA, in accordance with the immunohistochemical studies (Kowall et al., 1987).

In conclusion, these results show that in situ hybridization histochemistry provides a new way to examine neuropeptide gene expression in the human brain. This method allows the visualization of cell bodies expressing the genes for the different neuropeptides. In combination with image analysis, it could be used to study disease or drug induced changes in the levels of expression of these genes.

REFERENCES

Agid, Y., Javoy-Agid, F., 1985, Peptides and Parkinson's disease. Trends Neurosci., 1:30.
Beal, M.F., Martin, J.B., 1986, Neuropeptides and neurological disease. Ann Neurol., 20:547.
Bird, E.D., Iversen, L.L., 1982, Human brain postmortem studies of neurotransmitter and related markers. In: Handbook of Neurochemistry, Lajtha A., ed., vol 2 . Plenum Press, London, New York ., pp. 225.
Björklund, A., Hökfelt, T., (eds), 1983, In: Handbook of Chemical Neuroanatomy, vol 1. Elsevier Press. Amsterdam.
Chesselet, M.F., Graybiel, A.M., 1986, Striatal neurons expressing somatostatin-like immunoreactivity: evidence for a

peptidergic interneuronal system in the cat. Neurosci. 17:547.

Emson, P.C., Rossor, M.N., Tohyama, M., (eds), 1986, Neuropeptides and neurodegenerative diseases. Progress in Brain Research, vol 66. Elsevier Science Publishers B.V. Amsterdam

Gerfen, C.R., Herkerham, M., Thibault, J., 1987a, The neostriatal mosaic: II.Patch and matrix directed mesostriatal dopaminergic and non-dopaminergic systems. J. Neurosci., 7:3915.

Gerfen, C.R., Baimbridge, K.G., Thibault, J., 1987b, The neostriatal mosaic: III. Biochemical and developmental dissociation of patch-matrix mesostriatal systems. J. Neurosci., 7:3935.

Graveland, G.A., Williams, R.S., DiFiglia, M., 1985, A Golgi study of the human neostriatum: Neurons and afferent fibers. J. Comp. Neurol., 234:317.

Graybiel, A.M., Ragsdale, Jr. C.W., 1984, Biochemical anatomy of the striatum. In: Chemical Neuroanatomy. P.C. Emson, ed., Raven Press, New York, pp. 427-504

Hornykiewicz, O., 1972, Dopamine and its physiological significance in brain function. In: The Structure and Function of Neurons Systems. vol 6. G.M. Bournes ed., Academic Press, New York, pp. 367-415

Kerkerian, L., Bosler, O., Pelletier, G., Nieoullon, A., 1986, Striatal neuropeptide Y neurons are under the influence of the nigrostriatal dopaminergic pathway: immunohisto-chemical evidence. Neurosci. Lett., 66:106.

Kowall, N.W., Perrante, R.S., Martin, J.B., 1987, Patterns of cell loss in Huntington's disease. Trends Neurosci., 10:24.

Mengod, G., and Palacios, J.M.,(1990) Molecular neuro-pathology: The study of transmitter and receptor expression in human postmortem materials by in situ hybridization and receptor autoradiography. In: Neuropsychopharmacology, edited by W.E. Bunney jr., H. Hippius, G. Laakmann and M. Schmauss. Springer Verlag, Berlin. In press.

Mengod, G., Charli, J.-L., and Palacios, J.M., (1990) The use of in situ hybridization histochemistry for the study of neuropeptide gene expression in the human brain. Cel. Mol. Neurobiol.. In press.

Mengod, G., Martinez-Mir, M.I., Vilaró, M.T., and Palacios, J.M. 1989 Localization of the mRNA for dopamine D receptor in the rat brain by in situ hybridization histochemistry. Proc. Natl. Acad. Sci., USA., 86:8560.

Minth, C.D., Bloom, S.R., Polak, J.M., and Dixon, J.E., 1984, Cloning, characterization, and DNA sequence of a human cDNA encoding neuropeptide tyrosine. Proc. Natl. Acad. Sci., USA, 81:4577.

Noda, M., Teranishi, Y., Takahashi, H., Toyosato, M., Notake, M., Nakanishi, S., and Numa, S., 1982, Isolation and structural organization of the human preproenkephalin gene. Nature 297:431.

Palacios, J.M., Probst, A., and Cortés, R., 1986, Mapping receptors in the human brain. Trends Neurosci., 9:284.

Parent, A., 1987, In: Comparative Neurobiology of the basal ganglia. J. Wiley and Sons. New York.

Pickel, V.M., Sumal, K.K., Beckley, S.C., Miller, R.J., and Reis, D.J. 1980, Immunocytochemical localization of enkephalin in the neostriatum of rat brain: a light and electron microscopic study. J. Comp. Neurol. 189:721.

Rossor, M.N., 1982, Dementia (Neurotransmitters and CNS Disease), Lancet, ii:1200.

Shen, L-P., Pictet, R.L., and Rutter, W.J., 1982, Human somatostatin I: sequence of the cDNA. Proc. Natl. Acad. Sci., USA ,79:4575.

Savasta, M., Ruberte, E., Palacios, J.M., and Mengod, G., 1989, The colocalization of cholecystokinin and tyrosine hydroxylase mRNAs in mesencephalic dopaminergic neurons in the rat brain examined by in situ hybridization. Neurosci., 29:363.

Takahashi, Y., Kato, K., Hayashizaki, Y., Wakabayashi, T., Ohtsuka, E., Matsuki, S., Ikehara, M., and Matsubara, K., 1985, Molecular cloning of the human cholecystokinin gene by use of a synthetic probe containing deoxyinosine. Proc. Natl. Acad. Sci., USA, 82:1931.

Valentino, K.L., Eberwine, J.H., and Barchas, J.D. (eds) 1987, In: In situ hybridization. Applications to neurobiology. Oxford University Press. New York, Oxford.

Young, III W.S., Bonner, T.I., Brann, M.R., 1986, Mesencephalic dopamine neurons regulate the expresion of neuropeptide mRNAs in the rat forebrain. Proc. Natl. Acad. Sci. USA, 83:9827.

NEUROPEPTIDES RECEPTORS CHANGES IN HUNTINGTON'S CHOREA BASAL

GANGLIA ARE UNRELATED TO CHANGES IN NEUROPEPTIDES LEVELS

Chinaglia G.[1], Alvarez F.J.[2], Probst A.[1] and Palacios J.M.[2]

[1]Department of Pathology, Division of Neuropathology, University of Basel, Basel Switzerland and [2]Preclinical Research, Sandoz Pharma Ltd., Basel, Switzerland

SUMMARY

We have studied the distribution and densities of receptors for somatostatin, neurotensin and substance P in basal ganglia of patients affected by Huntington's chorea using autoradiographic techniques. Somatostatin binding sites density appear to be decreased in neostriatum of these patients, probably because they are located on neurons sensitive to the disease process of Huntington's chorea. Neurotensin and substance P receptors densities were found to be comparable to those of control cases suggesting that they are located on surviving neurons. Previous reports stated that in the same region the concentration of the peptide somatostatin is greatly increased, substance P is decreased and neurotensin is in the control range. Consequently, changes in receptors appear largely unrelated to alterations in transmitter levels. Our results suggest that receptors are expressed by cells which are different from those expressing the transmitter.

INTRODUCTION

Huntington's disease (HD) is an autosomal dominant disorder characterized by marked neuronal loss and astrogliosis in neostriatum[1]. The neurodegenerative process does not affect all kinds of neostriatal neurons. In fact while the spiny projection neurons (classified according to the Golgi's method[2]) degenerate[3], two populations of aspiny interneurons, one containing the peptides somatostatin (SS) and neuropeptide Y (NPY) and the enzyme nicotinamide adenine dinucleotide phosphate diaphorase (NADPH-d)[4] and the other characterized by the enzyme acetylcholinesterase (AChE)[5], are relatively well preserved. The levels of neurotransmitters such as gamma-aminobutyric acid (GABA)[6], enkephalines or substance P (SP)[7,8], expressed by the spiny neurons, have been found markedly decreased. In contrast, levels of

The Basal Ganglia III, Edited by G. Bernardi et al.
Plenum Press, New York

629

dopamine (DA)[9] and serotonin (5-HT)[10], expressed by extrinsic afferents, are unaltered or slightly elevated. Finally, SS and NPY, localized in the surviving interneurons, are greatly increased[11,12]. Several studies have been performed on neostriatal receptors in HD. Reductions of muscarinic cholinergic, GABA, DA and 5-HT receptors[13,14,15] probably reflects their localizations on degenerating neurons. The neostriatum contains considerable amounts of SS (SS1 type), NT and SP binding sites in humans[16,17,18]. Kainic acid treated rats exhibit a complete loss of SP binding sites in neostriatum, suggesting their localization on projection neurons, which are known to be destroyed by this neurotoxin[19]. A study performed with various neurotoxins and lesions suggest a localization of NT receptors on different cell populations in rat neostriatum[20]. The aim of the present study was to investigate SS1, SP and NT receptors in the basal ganglia of HD patients using radiohistochemical methods. The influence of neuroleptics, commonly administered to HD patients, on receptor densities, was taken into account by including in our study a group of schizophrenic (SCH) patients, all chronically treated with the same category of drugs.

MATERIALS AND METHODS

Frozen postmortem brain tissues from 7 patients with Huntington's Chorea grade 3[1] (5 females and 2 males, age range: 26-74 years, postmortem interval: 1- 42 hours) were provided by Prof. E. Bird, Brain Tissue Resource Center of the McLean Hospital, Belmont, Massachussetts and from Dpt. of Pathology of the University of Basel. Samples from 15 control human brains (5 males and 10 females, age range: 22-86 years, postmortem delay: 2-15 hours) and 7 SCH patients (3 males and 4 females, aged 61-89 years, postmortem delay 10-30 hours) were provided from Dpt. of Pathology of the University of Basel.

Autoradiographic studies were performed according to previously published procedures[21,22,23]. Briefly, 10 μm sections were obtained using cryostat-microtome (Leitz 1720, from Leitz, Wetzlar, F.R.G.) from frozen tissue blocks and mounted onto gelatin-coated glass slides. The stable somatostatin octapeptide analogue [^{125}I]204-090 (specific activity 2000 Ci/mmol, kindly provided by Dr. C. Bruns, Sandoz Ltd.) was used to visualize the somatostatin SS1 type receptors, the [^{125}I]Bolton-Hunter SP (BHSP) (2000 Ci/mmol; Amersham, Buckingamshire, U.K.) for SP receptors and monoiodo ([^{125}I]Tyr$_3$)NT (2200 Ci/mmol, NEN Research Products, Boston, MA) for NT receptors.

Autoradiograms were generated by apposing the labeled sections to [^{3}H]-sensitive Ultrofilm (LKB, Sweden) for a convenient exposure time. Plastic [^{125}I]-standards (Amersham, UK) were exposed along with the labeled tissues. Autoradiograms were analyzed using a computer-assisted image analysis system (MCID Imaging Research Inc, St. Catharines, Ontario, Canada). Statistical analysis were carried out using an ANOVA test. For anatomical mapping, adjacent sections to those used in the autoradiographic experiments were stained with cresyl violet and for acetylcholinesterase activity.

630

RESULTS

Moderate to high densities of SS1, NT and SP receptors were measured in caudate nucleus, putamen and nucleus accumbens in control cases. The distribution of each of these receptors was heterogeneous with patches of higher density alternating with areas of lower binding density. Slight but statistically not significant decreases of NT receptors were observed in the neostriatum and substantia nigra (SN) of SCH patients. The nucleus accumbens too was characterized by a diminution of NT binding sites (p<0.01) in the same group of patients. In contrast, SS1 and SP receptors did not display any modification in SCH brains.

A marked reduction of SS1 receptors density was observed in the caudate and putamen in all HD cases studied (p<0.0005). The determination of SS1 receptor density in areas like the nucleus accumbens, insular cortex and claustrum showed densities comparable to those found in controls (Figure 1).

SP and NT receptors were slightly (not statistically significant) decreased in the caudate of HD cases. SP and NT receptor densities similar to control values were found in the putamen, accumbens, insular cortex, claustrum. NT receptor density in SN of HD patients was in the control range (Figure 2).

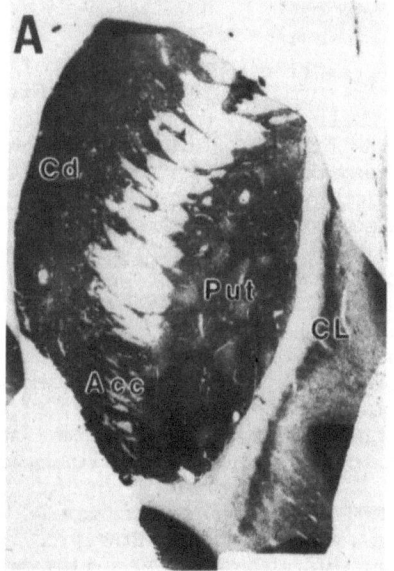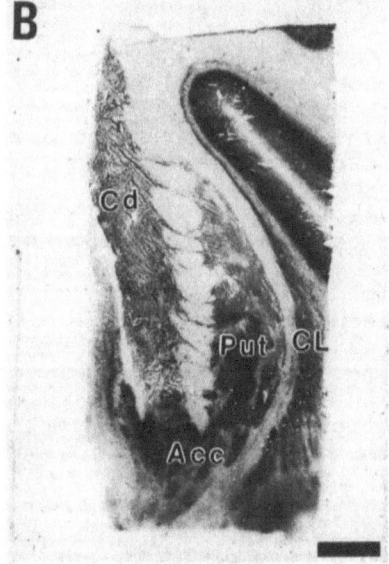

Figure 1. Photomicrographs from autoradiograms showing the distribution of the somatostatin analogue [125I]-204-090 binding sites in the basal ganglia of a control (A) and Huntington's chorea (B) cases. In the Huntington's chorea case, somatostatin (SS1) receptor binding was preserved in tha nucleus accumbens (Acc), ventral aspect of the anterior third of putamen (Put), claustrum (CL) and insular cortex, but markedly reduced in the caudate nucleus and in the dorsal part of putamen. Bar: 2.5mm.

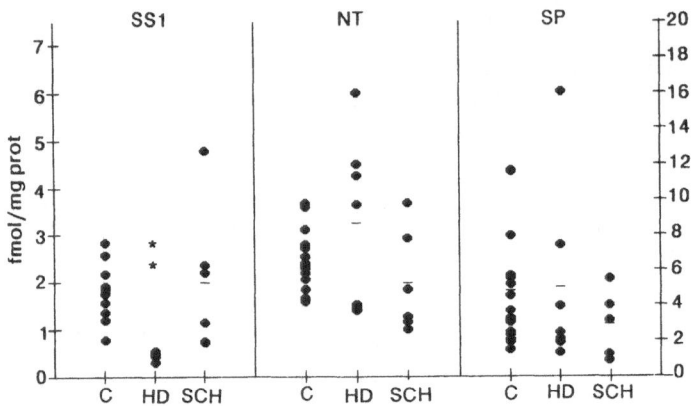

Figure 2. Scattergrams displaying the densities of [^{125}I] somatostatin SS1 subtype (SS1), neurotensin (NT) and substance P (SP) binding sites in the middle third of putamen in control (C), Huntington's chorea (HD) and schizophrenic (SCH) brains. Each point represents a value obtained for one subject and the small horizontal bars indicate mean values. Units are fmol/mg protein. In HD brains a significant decrease (** p<0.0005) is observed in SS1 but not in NT and SP receptor densities. Note: different scale for SP scattergram.

DISCUSSION

Our results demonstrate a marked reduction of SS1 receptor density in the neostriatum of HD patients. Different hypothesis can be proposed to explain it. A decreased density of SS1 binding sites could result from a reduced binding of the radioligand to the somatostatin receptors, due to receptor occupation by the endogenous peptide. This is, however, unlikely since tissue sections were preincubated in order to remove the endogenous ligand. In addition incubation in the presence of a 5-fold concentration of ligand did not result in a significant increase in specific binding. Furthermore, gray areas surrounding basal ganglia, like insular cortex or claustrum, showed normal levels of receptor binding. A down-regulation of SS1 receptors may conceivably result from the increased endogenous peptide. The reverse situation has been described in the experimental animal where an up-regulation of somatostatin receptors follows a treatment with cysteamine, a depleting agent for somatostatin[24]. A reduction of SS1 binding sites could also result of chronic neuroleptic treatment. However, the possible role of neuroleptics can be ruled out by the normal SS1 receptor densities in SCH patients. Finally, SS1 receptor loss in HD could be explained by their localization on neuronal populations which degenerate in this disease. Receptors losses unrelated to alterations in presynaptic markers are not uncommon in HD. Examples are those for GABA, acetylcholine, DA, substance P or cholecystokinin[13,25].

SP and NT receptor densities were essentially unaltered in spite of the evident shrinkage of neostriatum in HD[1].

However, NT receptor density was decreased in neostriatum and accumbens of SCH patients. This could be explained by a down-regulation mechanism of NT receptors. Previous studies in rat brain have demonstrated that haloperidol increases the NT-like immunoreactivity in the same regions [28,29]. On the other hand increased NT content was reported in the caudate of HD patients, but not in SCH patients[30]. Regarding the result in SCH individuals, the authors do not specify if the patients were treated with neuroleptics. In HD the elevated NT concentration could result from increased density of NT containing neurons due to shrinkage. The neostriatal shrinkage in HD but not in SCH patients could explain the discrepancy in our results between NT receptor density in SCH and HD patients. Thus, a slight NT receptor loss probably exists in HD subjects, although masked by the striatal atrophy.

Regarding SP receptors, there is a lack of correlation between our results and those obtained in experimental animals treated with kainic acid[19]. It is important to underline that this neurotoxic lesion is no longer considered a good model for HD. In fact it results in a decrease of somatostatin containing cells, which are spared in HD[26]. It is also worth mentioning that SP receptors have been localized to astrocytes in culture[27]. As a reactive astrogliosis occurs in HD, the question arises about a possible expression ofSP receptors by striatal astocytes in HD. However, SP receptor density is in control range although the amount of glial cells is considered to increase manyfold in the neostriatum of HD patients[1].

Hence the nature of the cell populations bearing the receptors remains problematical in the basal ganglia. If the receptor is localized to a cell which degenerates in HD, the receptor will be lost in spite of the preservation of the transmitter. What physiological role the transmitter could have in the absence of the receptor is difficult to understand. A further implication of these findings is that both pre- and postsynaptic parameters have to be examined before a therapeutical strategy is designed.

ACKOWLEDGEMENTS

The authors gratefully aknowledge Prof. E. Bird from the Brain Tissue Resource Center, Mc Lean Hospital, Belmont, MA, USA, for kindly supplying the Huntington's disease specimens. This center is supported in part by PHS grant number MH/NS 31862.

References

1. J.P. Vonsattel, R.H. Myers, T.J. Stevens, R.J. Ferrante, E.D. Bird, and E.P. Richardson, Jr, Neuropathological classification of Huntington's disease, <u>J. Neuropathol. Exp. Neurol</u>. 44: 559-577 (1985).
2. G.A. Graveland, R.S. Williams, and M. Di Figlia, A Golgi study of the human neostriatum: Neurons and afferent fibers. <u>J. Comp. Neurol</u>. 234: 317-333 (1985).

3. A. Reiner, R.L. Albin, K.D. Anderson, C.J. D'Amato, J.B. Penney, and A.B. Young, Differential loss of striatal projection neurons in Huntington disease, *Proc. Natl. Acad. Sci.* 85: 5733-5737 (1988).

4. R.J. Ferrante, N.W. Kowall, M.F. Beal, J.B. Martin, E.D. Bird, and E.P. Richardson, Jr., Morphologic and histochemical characteristics of a spared subset of striatal neurons in Huntington's disease, *J. Neuropathol. Exp. Neurol.* 46: 12-27 (1987).

5. R.J. Ferrante, M.F. Beal, N.W.Kowall, E.P. Richardson, Jr, and J.B. Martin, Sparing of acetylcolinesterase-containing striatal neurons in Huntington's disease, *Brain Res.* 411: 162-166 (1987).

6. T.L. Perry, S. Hansen, and M. Kloster, Huntington's chorea: Deficiency of τ-aminobutyric acid in brain, *N. Engl. J. Med.* 288: 337-342 (1973).

7. P.C. Emson, A. Arregui, V. Clement-Jones, B.E.B. Sandberg, M. Rossor, Regional distribution of methionine-enkephalin and substance P-like immunoreactivity in normal human brain and in Huntington's disease, *Brain Res.* 199: 147-160 (1980).

8. M.F. Beal, D.W. Ellison, M.F. Mazurek, K.J. Swartz, J. R. Malloy, E.D. Bird, and J.B. Martin, A detailed examination of substance P in pathologically graded cases of Huntington's disease, *J. Neurol. Sci.* 84: 51-61 (1988).

9. E.D. Bird, and L.L. Iversen, Huntington's Chorea. Post-mortem measurement of glutamic acid decarboxylase, choline acetyltransferase and dopamine in basal ganglia, *Brain*, 97: 457-472 (1974).

10. H. Bernheimer, and O. Hornykiewicz, Brain amines in Huntington's chorea, *Adv. Neurol.* 1: 525-531 (1973).

11. N. Aronin, P.E. Cooper, L.J. Lorenz, E.D. Bird, S.M. Sagar, S.E. Leeman, and J.B. Martin, Somatostatin is increased in the basal ganglia in Huntington's disease, *Ann. Neurol.*, 13 (1983) 519-526.

12. D. Dawbarn, M.E. De Quidt, and P.C. Emson, Survival of basal ganglia neuropeptide Y-somatostatin neurones in Huntington's Disease, *Brain Research*, 340: 251-260 (1985).

13. J.B. Penney, Jr., and A.B.Young, Quantitative autoradiography of neurotransmitter receptors in Huntington's disease, *Neurology*, 32: 1391-1395 (1982).

14. P.J. Whitehouse, R. Trifiletti, B.E. Jones, S. Folstein, D.L. Price, S.H. Snyder, and M.J. Kuhar, Neurotransmitter receptor alterations in Huntington's disease: autoradiographic and homogenate studies with special reference to benzodiazepine receptor complexes, *Ann. Neurol.*, 18: 202-210 (1985).

15. C. Waeber and J.M. Palacios, Serotonin-1 receptor binding sites in the human basal ganglia are decreased in Huntington's chorea but not in Parkinson's disease: a quantitative in vitro autoradiography study, *Neuroscience*, 32: 337-347 (1989).

16. J.C. Reubi, R. Cortés, R. Maurer, A. Probst and J.M. Palacios, Distribution of somatostatin receptors in the human brain: an autoradiographic study, *Neuroscience*, 18: 329-346 (1986).

17. A. Sarrieau, F. Javoy-Agid, P. Kitabgi, M. Dussaillant, M. Vial, J.P. Vincent, Y. Agid and W.H. Rostène, Characterization and autoradiographic distribution of neurotensin binding sites in the human brain, Brain Res., 348: 375-380 (1985).

18. J.C. Beaujouan, Y. Torrens, M. Saffroy and J. Glowinski, Quantitative autoradiographic analysis of the distribution of binding sites for [^{125}I]Bolton Hunter derivatives of eledoisin and substance P in the rat brain, Neuroscience, 18: 857-875 (1986).

19. J.K. Ritter, D.R. Gehlert, J.W. Gibb, J.K. Wamsley, and G.R. Hanson, Neuronal localization of substance P receptors in rat neostriatum, Eur. J. Pharmacol., 109: 431-432 (1985).

20. M. Goedert, K. Pittaway, and P.C. Emson, Neurotensin receptors in the rat striatum: lesion studies, Brain. Res., 299: 164-168 (1984).

21. J.C. Reubi, A. Probst, R. Cortés, and J.M. Palacios, Distinct topographical localization of two somatostatin receptor subpopulations in the human cortex, Brain Res., 406: 391-396 (1987).

22. J.M. Palacios, A. Pazos, M.M. Dietl, M. Schlumpf and W. Lichtensteiger, The ontogeny of brain neurotensin receptors studied by autoradiography, Neuroscience, 25: 307-317.

23. M.M. Dietl, M. Sanchez, A. Probst and J.M. Palacios, Substance P receptors in the human spinal cord: decrease in amyotrophic lateral sclerosis, Brain Res. 483, 39-49.

24. C.B.Srikant and Y.C. Patel, Cysteamine-induced depletion of brain somatostatin is associated with up-regulation of cerebrocortical somatostatin receptors, Endocrinol., 115: 990-995 (1984).

25. G. Mengod and J.M. Palacios, Molecular neuropathology: the study of transmitter and receptor expression in human postmortem materials by in situ hybridization and receptor autoradiography, in: "Neuropsychopharmacology", W.E. Bunney Jr, H. Hippius, G. Laakman and M. Schmauss, eds., Springer Verlag, Berlin (1990) in press.

26. M.F. Beal, N. W. Kowall, K.J. Swartz, R.J. Ferrante and J.B. Martin, Differential sparing of somatostatin-neuropeptide Y and cholinergic neurons following striatal excitotoxin lesions, Synapse, 3: 38-47 (1989).

27. Y. Torrens, J.C. Beaujouan, M. Saffroy, M.C. Daguet de Montety, L. Bergstrom and J. Glowinski, Substance P receptors in primary cultures of cortical astrocytes from the mouse, Proc. Natl. Acad. Sci. U.S.A., 83:9216-9220 (1986).

28. K.W. Eggerman and D.S. Zahm, Numbers of neurotensin-immunoreactive neurons selectively increased in rat ventral striatum following acute haloperidol administration, Neuropeptides, 11: 125-132 (1988).

29. S. Govoni, J.S. Hong, H.Y.T. and E. Costa, Increase of neurotensin content elicited by neuroleptics in nucleus accumbens, J. Pharmacol. Exp. Therap., 215: 413-417 (1980).

30. C.B. Nemeroff, W. W. Youngblood, P.J. Manberg, A.J. Prange, Jr. and J.S. Kizer, Regional brain concentrations of neuropeptides in Huntington's chorea and schizophrenia, Science 221: 972-975 (1983).

NEURAL ACTIVITY OF THE BASAL GANGLIA IN PARKINSON'S DISEASE STUDIED BY

DEPTH RECORDING AND PET SCAN

Chihiro Ohye, Tohru Shibazaki, Masafumi Hirato
Yasuhiro Kawashima, Masaru Matsumura and Takashi Shibasaki

Department of Neurosurgery
Gunma University School of Medicine
Maebashi, Gunma, JAPAN

INTRODUCTION

The basal ganglia is certainly one of the major centers of the central nervous system controlling voluntary movement and posture[1], but its precise role is still far from understanding. Only experimental studies in monkeys with various conditioned voluntary movements or with experimentally produced involuntary movements were carried out to observe neural activities of the basal ganglia associated with these particular conditions.

To elucidate the neural mechanism or on-going changes underlying movement disorders of Parkinson's disease and choreic movement, the neural activity of the basal ganglia (caudate nucleus and internal segment of globus pallidus) was recorded during the course of stereotactic thalamotomy, and correlated with local glucose metabolism studied by positron emission tomography (PET). Preliminary parts of this work were already reported[2-4].

SUBJECTS AND METHODS

Seven cases of parkinson's disease (tremor type: 5, rigid type:2) and a case of hemichorea after asphyxia were studied. Further, for comparison, additional 2 cases of essential tremor and a normal volunteer were examined.

For PET study, Hitachi PCT-H1 (128 detectors, space resolution 8 mm, slice thickness 16 mm, 7 slices at a time) was used[5]. In our institution, PET scanner and XCT (Hitachi) scanner are settled in the same room and the patient is transferred from one scanner to the other without changing his head position. Therefore, images obtained by both scanners are almost comparable in position, facilitating the determination of the region of interest. A babycyclrotron (Japan Steel Work) supplies 18F-fluorodeoxyglucose (18FDG). Thus, 5 mci of 18FDG was injected intravenously, and the scanning started from 60 min after injection. PET study was performed always several days before stereotactic surgery.

Using Leksell's stereotactic apparatus, stereotactic thalamotomy was performed on all subjects except for a normal control. We use microrecording method to find out exactly the target area,

The Basal Ganglia III, Edited by G. Bernardi et al.
Plenum Press, New York

637

thalamic ventralis intermedius (Vim) nucleus in such cases[6,7]. As our trajectory toward the Vim nucleus from prefrontal area almost always passes the caudate (Cd) nucleus, and not infrequently also traverses a medial most part of the internal segment of globus pallidus, we take such opportunities to observe neural activity of basal ganglia in human case. In fact, stereotactic CT scan or stereotactic MRI method that we have developed recently can verify the passage of electrode through basal ganglia structures[8]. Bipolar concentric needle type of electrode (outer diameter 0.3 mm, tip about 10 um, interpolar distance 10-20 um, electrical resistance about 100 kohms) was used. A pair of microelectrodes is used in a position of stylet of coagulation needle (outer diameter, 2 mm), and introduced into the brain with the aid of a micromanipulator (step wise motor drive, remote controlled system).

Electrical activity was demonstrated on a oscilloscope by a conventional method and recorded on a running paper by thermalcorder.

RESULTS

Electrical Activity Of The Basal Ganglia

In the stereotactic thalamotomy guided by microrecording, as already described, the recording electrode usually passes the

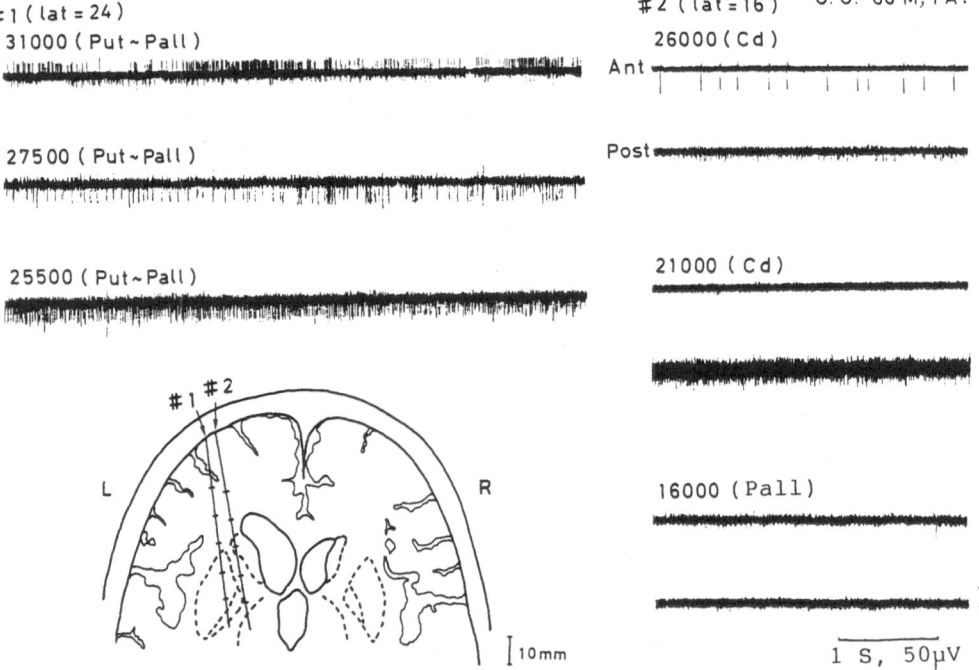

Fig. 1. Electrical activity of the basal ganglia in a case of tremor type parkinson's disease. Left side (#1), recording from the putaminal region. Right side (#2), recording from the Cd-GP by a pair of electrodes (ant,post). Each tracking is shown on the left lower figure made from stereotactic MRI. Numbers in each trace denote distance in microns from tentative zero, at lower border of the thalamic ventrointermedius nucleus.

cortical gray matter, caudate nucleus (Cd), again white matter and
then comes into the thalamus. In some cases, especially when the
electrode is introduced with more inclined angle than usual, the
electrode may pass the corner of the internal segment of th globus
pallidus (GP) after passing through the Cd nucleus and before coming
into the thalamus. As each subcortical structure exhibits
characteristic spontaneous electrical activity, it is not difficult
to identify the Cd nucleus, which shows rather high background
activity with about 30 Hz fast oscillation. When a single spike
discharge is isolated, sporadic low frequency spontaneous discharge
is seen[9].

In a case with tremor type PA, as shown in Fig. 1, in the Cd
nucleus, a typical Cd spike discharge at a certain point (in this
particular case, 26000 um) and also a high spontaneous background
activity at slightly deeper point (21000 um) were observed, for
example. Further advancing the electrode, background activity
increased again, and this point (16000 um) was thought to be in the
GP, if compare to the XCT image as shown on the left. The background
activity in this supposed GP area was less than that in the Cd
region. Thalamic activity was encountered in this case, from about
8000 um (not shown in the figure). In other tremor type PA, it was
always the case that Cd spontaneous activity was higher than that in
GP. In this particular case shown in Fig. 1, the first tracking
passed through putamen-pallidum area by incidental miss calculation

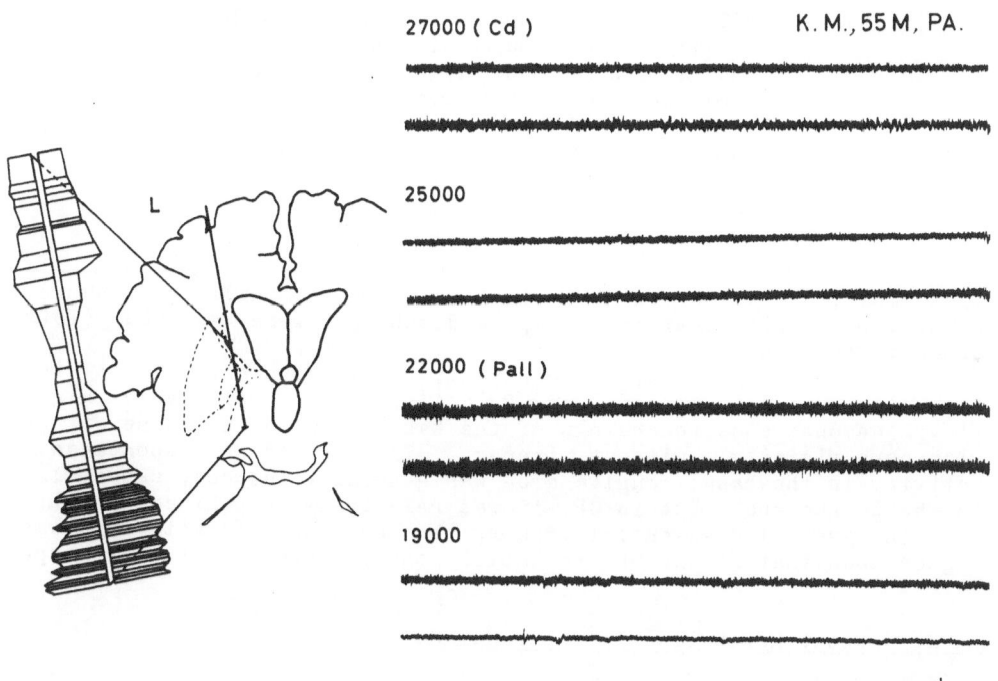

Fig. 2. Electrical activity of the basal ganglia in a case of rigid
type parkinson's disease. Recording track is shown on the
left figure with histograms of neural activity. A pair of
histograms from a pair of electrodes. Several samples of
electrical activities from related areas are shown on the
right. A pair of traces from a pair of electrodes.

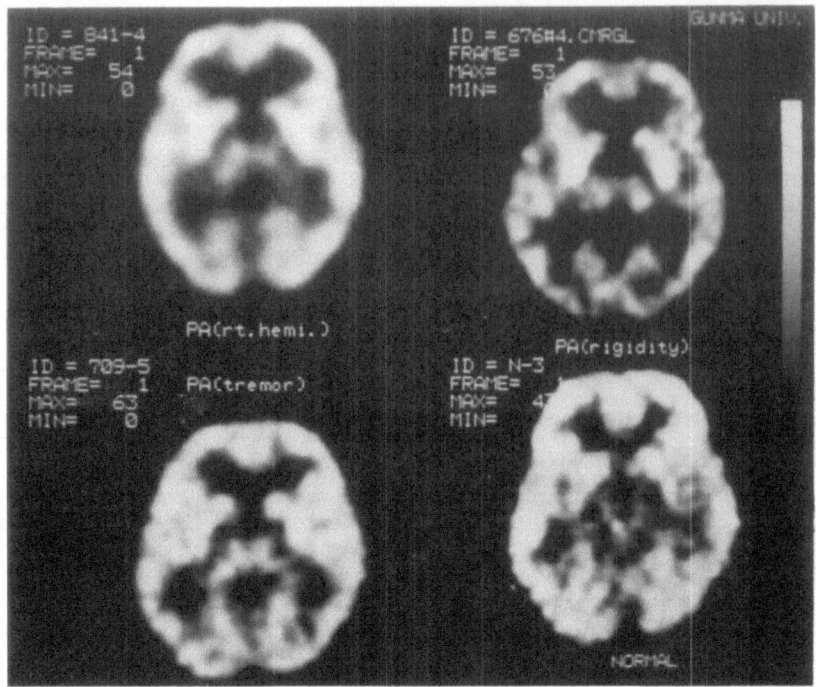

Fig. 3. PET images of glucose metabolism at the level of
cerebral cortex and basal ganglia. Upper left:
tremor type PA. Upper right: rigid type PA.
Lower left: essential tremor. Lower right:
normal person. Intensity of white color is parallel to intensity of glucose accumulation.
(Copy from original color illustration)

(certainly, it was verified after checking X-ray film), and there,
several particular patterns of spike discharges were recorded (left
upper three traces).

In rigid type PA, in contrast, spontaneous activity in the
Cd nucleus was not intense, and GP activity was relatively high.
Fig. 2 demonstrates recordings of the basal ganglia in a case of PA
with pure rigidity, without tremor. In this case, spontaneous
activity in the basal ganglia area was generally reduced, especially
in the Cd nucleus. But in GP, it was relatively maintained.

In cases with essential tremor, the Cd nucleus activity was
higher than that of the GP, resembling the pattern of tremor type
PA.

Local Glucose Metabolism In The Basal Ganglia

Local glucose metabolism was studied by means of PET scan
using 18FDG. Results were summarized in Fig. 3. In a normal control
case(Fig. 3, lower right), glucose accumulation was high in frontal
and parietal cortex, the Cd area and a small spot in the thalamus.
In cases with essential tremor (Fig. 3, lower left) glucose distribution pattern was almost similar to that in normal case. However,
in cases with PA, glucose accumulation was generally decreased,
especially in cortical areas. This tendency was most conspicuous in
cases with rigid type PA (Fig. 3, upper right), where only a slight

glucose activity was noticed in frontal area. But in contrast, glucose accumulation was relatively preserved in the basal ganglia region, probably being more in lenticular nucleus than Cd nucleus. And it was often the case that an active spot in the Cd nucleus was lacking. In cases with tremor type PA (Fig. 3, upper left), glucose accumulation pattern looked like partly to that of essential tremor and partly to that of rigid type PA: the glucose activity in basal ganglia was relatively high but cortical glucose activity was reduced slightly.

In A Case Of Hemichorea

In a case with right hemichorea after asphyxic accident, XCT showed a slit like low density area in bilateral putamen. During the course of stereotactic left thalamotomy, depth recording revealed reduced spontaneous electrical activity in the Cd, but exaggerated grouped discharges were often encountered in the GP. Such grouped discharge was never found in GP of parkinson's disease. In this case, abnormal grouped discharges were also encountered inside the thalamus, especially in anterodorsal part, probably corresponding to the ventrooral nucleus. The grouped discharges were not necessarily synchronous with the peripheral choreic movement, but sometimes it looked like to be synchronous. PET study demonstrated that the glucose accumulation in the GP was higher (about 16%) in the affected side, supporting the results of depth recording.

DISCUSSION AND SUMMARY

The present observation revealed, for the first time , on-going changes in the human basal ganglia associated with parkinson's tremor and rigidity. And for further information, cases with the essential tremor and choreic movement were also studied. The most interesting point is that the electrical activity of the basal ganglia in rigid type PA and that in tremor type PA are different. In rigid type PA, the electrical activity in the Cd nucleus is very much reduced probably associated with reduced cortical metabolic activity, while in tremor type PA, the electrical activity in the Cd nucleus is high, together with high cortical metabolism. As the electrical activity in the Cd nucleus in essential tremor is also high, being comparable to tremor type PA, high spontaneous activity in the Cd seems to be the usual feature. Apparently, the results are somewhat incompatible with the simple application of the concept that the nigrostriatal connection is inhibitory, because in PA nigrostriatal dopaminergic system is dysfunctioning. But probably, the striatum has stronger excitatory source from cortical motor and premotor areas[10], and this excitatory input is lacking in PA, as shown by extremely low glucose metabolism in the cortex of PA, especially in rigid type PA, revealed by PET scan.

In the experimental study on animal model of parkinson's disease in monkeys, it was postulated that activity in the GP was elevated. Using autoradiographic method, it was claimed that glucose uptake was high in GP of monkeys with parkinsonian like rigidity and akinesia due to MPTP administration[11-13]. On the other hand, the same kind of parkinsonian model in monkeys has been used to elucidate the neural activity of the basal ganglia, and it was found that the electrical activity of pallidal neurons were exaggerated, with abnormal discharge pattern of irregular bursting and unusual responses to peripheral natural stimuli (bilateral response, for example)[14,15]. Also our PET study revealed that glucose metabolism was relatively increased in putamen-GP region, supporting the results of animal experiment.

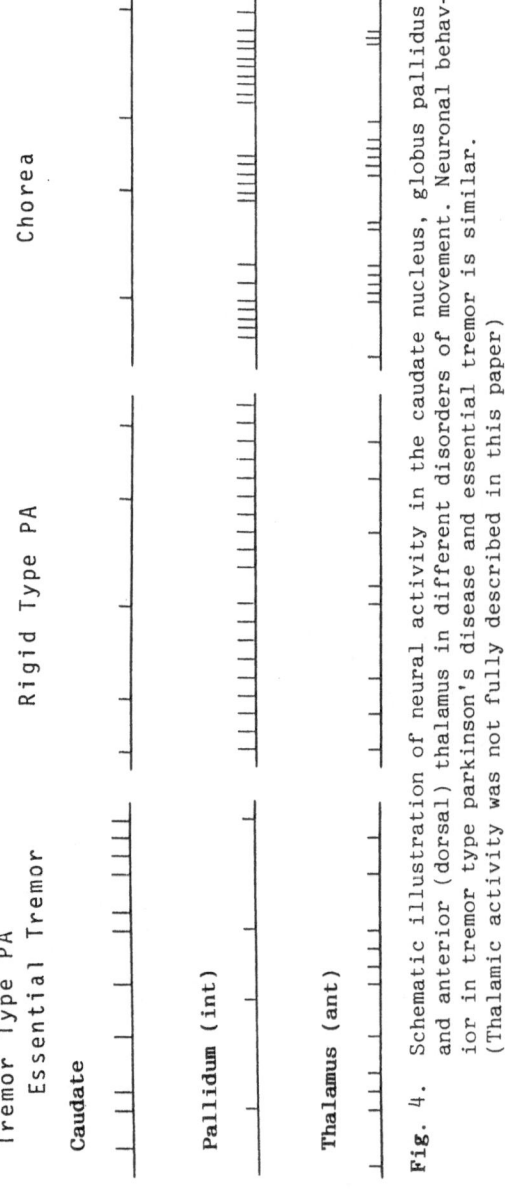

Fig. 4. Schematic illustration of neural activity in the caudate nucleus, globus pallidus and anterior (dorsal) thalamus in different disorders of movement. Neuronal behavior in tremor type parkinson's disease and essential tremor is similar. (Thalamic activity was not fully described in this paper)

In animal model of choreic movement as well as in humans, activity of GP is considered to be the opposite of parkinson's disease, namely pallidal activity being reduced, contrary to our particular case with choreic movement[16]. Results of our depth recording are schematically summarized in Fig. 4. It is understood that the electrical activity of the Cd and GP is reciprocally connected.

Anyway, in our observation, on the cases with parkinsonian tremor, rigidity, essential tremor ana choreic movement, electrical activity of basal ganglia and glucose metabolism are very well correlated. Where electrical activity is high, metabolic rate is also high and vice versa. In this sense, it seems likely that PET study on glucose metabolism reflects rather faithfully general energy level in that particular region. Therefore such correlative study of depth recording and PET study may be promising for further elucidation of neural mechanisms of movement disorders in humans.

REFERENCES

1. M.D. DeLong and A.P. Georgopoulos, Motor function of of the basal ganglia. in; "Handbook of Physiology, The Nervous System II", V.B. Brooks, ed., Amer. Physiol. Soc., Bethesda (1981) pp1017-1061.
2. C. Ohye, A new aspect of parkinsonian rigidity and tremor (in japanese). Internal Med., 63:831-836 (1989).
3. M. Hirato, Y. Kawashima, T. Shibasaki, T. Shibazaki, and C. Ohye, Cerebral metabolism in parkinson's disease-Comparison between rigidity and tremor (in japanese). Stereotactic Surgery, (in press).
4. M. Hirato, Y. Kawashima, T. Shibazaki, T. Shibasaki and C. Ohye, Pathophysiology of parkinsonian rigidity (in japanese). Stereotactic Surgery, (in press).
5. T. Shibasaki, Analysis of human cerebral function using positron emission tomography (in japanese). Progr. Neurosci., 28:211-228 (1986).
6. C. Ohye, Selective Thalamotomy for Movement Disorders: Microrecording Stimulation, Techniques and Results, in; Modern Stereotactic Surgery" , L.D. Lunsford, ed., Martius Nijhoff Pub., Boston, (1988) pp315-331.
7. C. Ohye, T. Shibazaki, T. Hirai, H. Wada, M. Hirato and Y. Kawashima, Further physiological observations on the ventralis intermedius neurons in the human thalamus. J. Neurophysiol., 61:488-500 (1989).
8. Y. Kawashima, M. Matsumura, M. Hirato, T. Shibazaki, C. Ohye and H. Nakajima, Application of CT and MRI for stereotactic functional neurosurgery (in japanese). Stereotctic Surgery, (in press).
9. C. Ohye, T. Shibazaki, M. Matsumura, Y. Kawashima and M. Hirato. Activity of the caudate neurons in humans. in: "Basal Ganglia II", M.B. Carpenter and A. Jayaraman, eds., Plenum, New York, (1987) pp483-488.
10. A. Dray, Physiology and pharmacology of mammalian basal ganglia. Progr. Neurobiol., 14:221-335 (1980).
11. A.R. Crossman, Primate model of dyskinesia-the experimental approach to the study of basal ganglia-related involuntary movement disorders. Neuroscience, 21:1-40 (1987).
12. M.A. Sambrook, A.R. Crossman, I. Mitchell, R.G. Robertson, C.E. Clarke and S. Boyce, The basal ganglia mechanisms mediating primate models of movement disorders. in: "Neural mechanisms

in disorders of movement", A.R. Crossman and M.A. Sambrook, eds., John Libbey, London-Paris, (1989) pp123-144.

13. I. Mitchell, C.E. Clarke, S. Boyce, R.G. Robertson, D. Peggs, M.A. Sambrook and A.R. Crossman, Neural mechanisms underlying parkinsonian symptoms based upon regional uptake of 2-deoxy-glucose in monkeys exposed to 1-methyl-4-phenyl-1,2,3,6-tetrahydropyridine. Neuroscience, 32:213-226 (1989).

14. M. Filion, L. Tremblay and P.J. Bedard, Abnormal influence of passive limb movement on the activity of globus pallidus neurons in parkinsonian monkeys. Brain Res., 444:165-176 (1988).

15. M. Filion, L. Tremblay and P.J. Bedard, Excessive and unselective response of medial pallidal neurons to both passive movement and striatal stimulation in monkeys with MPTP-induced parkinsonism. in: "Neural mechanisms in disorders of movement. A.R. Crossman and M.A. Sambrook, eds., John Libbey, London-Paris, (1989) pp157-164.

16. D.J. Brooks and R.S.J. Frackowiak, PET and movement disorders. J. Neurol. Neurosurg. Psychiat. special suppl.:68-77 (1989).

DIFFERENTIAL ASPECTS OF PARKINSONIAN AKINESIA AS REVEALED BY LIMB MOVEMENT STUDIES

F. Viallet*, E. Trouche**, E. Legallet**, P. Apicella**, and R. Khalil*

*Service de Neurologie - CHU La Timone - 13385 MARSEILLE CEDEX 5
France
**LNF 3 - CNRS - 31, chemin Joseph Aiguier - 13402 MARSEILLE CEDEX 9
France

SUMMARY

Parkinsonian akinesia is known to be a complex symptom. In the present study, some of its phenomenological aspects, such as difficulty in initiating movement and slowness of movement execution, are reassessed by comparing, on the one hand, data obtained in two stimulus-triggered reaction time (RT) tasks involving either aiming (pointing to a visual target) or no aiming (no spatial goal) and, on the other hand, two goal directed task conditions which could be performed either in a closed loop (normal vision of the movement) or in an open loop mode (without visual feedback). The data show that in parkinsonians, the RTs were more prolonged in the no aiming task : this deficit was partly compensated by dopatherapy. Furthermore, these patients exhibited a hypometric tendency in pointing movements when the visual reafferents were lacking. These results show that some of the factors contributing to the expression of akinesia in Parkinson's disease play a leading role, namely : motivation for the task, expectation that sensory feedback will be used during movement execution, muscle energizing deficits (mainly revealed by the hypometric tendency).

INTRODUCTION

Among the classical triad of symptoms associated with Parkinson's disease, the last to be determined was akinesia (WILSON, 1925); Wilson described parkinsonian patients as showing a reduction in the need for or impulse to perform muscular action, so that their "will to act" is impaired : he also measured the reaction time (R.T.) preceding the muscular response to a single visual stimulus and obtained the figures of 0.24 sec in the case of normal individuals and 0.36 sec in that of parkinsonians. Since then, the significance of this akinesia has gradually become recognized. One of its most prominent aspects consists of slowness in preparing for movement (JOUBERT and BARBEAU, 1969) and difficulty in initiating movement (NARABAYASHI et al., 1976) as in the case of the freezing phenomenon, for example. Nowadays, the general term akinesia is used to cover various aspects of the motor disability observed in Parkinson's disease including a loss of spontaneous associated movements, an increase in the RT, a slowness of movement and a tendency to make smaller movements than required (ROTHWELL, 1987). The mechanisms of akinesia have been taken to consist of two main components which are the muscles energizing deficit and the prolongation of RT (HALLETT, 1989) : as mentioned by MARSDEN (1989), parkinsonian akinesia consists of an inability in patients to hold or deliver a prepared motor program in response to their internal plan of action : consequently, they are less successful than normal subjects at performing predictive motor action (FLOWERS, 1978; STERN et al., 1983; DAY et al., 1984) and more dependent on visual feedback to accomplish their movements (FLOWERS, 1976; COOKE et al., 1978). The aim of the present study was to investigate these aspects of akinesia in parkinsonian patients by means of a limb movement analysis. The characteristics of the forthcoming movement are studied as a factor of akinesia in the first part by comparing two RT tasks involving either aiming or no aiming, i.e. the presence or the absence of a definite spatial goal. In the second part, the role of the visual feedback about the movement itself in the expression of akinesia is reassessed by comparing the subjects'performances in a goal-directed task under two conditions namely the closed loop or open loop mode, depending on whether visual feedback was present or not during the ongoing movement.

METHODS

The experimental arrangement has been described previously (VIALLET et al., 1987). The subjects were placed on a seat designed so as to maintain a standardized posture facing a vertical board (Fig. 1). On the lower medial part of this board, a platform was set on which the subjects had to keep their hand during a variable unpredictable foreperiod (P.P.) randomly distributed between four values (0.5; 1.0; 1.5, 2.0 sec) until the onset of a luminous signal. This signal (L.E.D. : 5 mm in diameter) was presented in the center of a square screen (33 x 33 cm in area) which was placed on the upper part of the board.

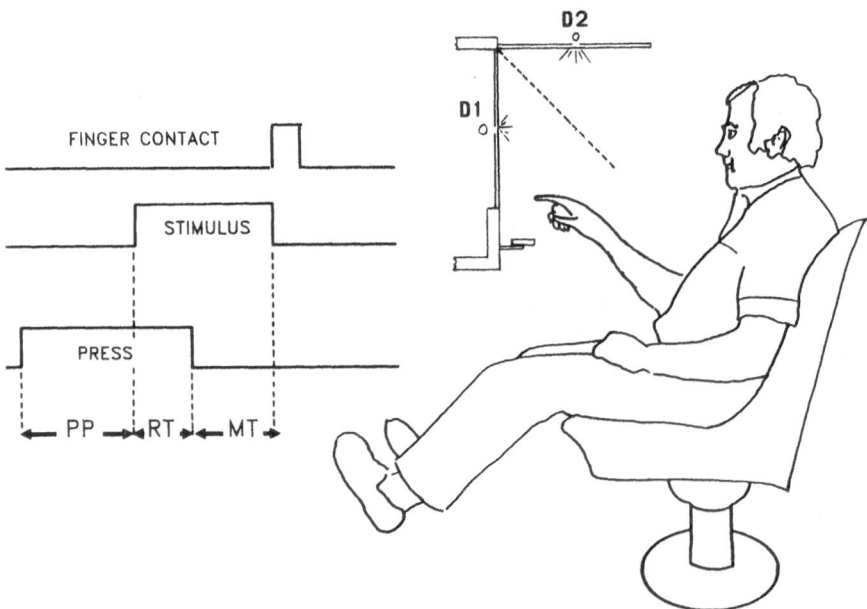

Fig. 1 . Experimental arrangement (see text for description).

Experiment 1 : Aimed vs non aimed RT tasks

Each subject performed four blocks of trials (each block consisting of a series of 48 trials). Both RT tasks (aimed and non aimed) were carried out with each hand alternately and successively in a random order from one subject to another. In the aimed RT task, which has been called the pointing task, the subjects were required to point at the luminous signal as soon as it appeared with their index finger as rapidly and as accurately as possible. In the non aimed one, which has been called the no aiming task, the subjects had only to release the platform as rapidly as possible when the signal appeared : the no aiming movement has been found by kinematic analysis to reproduce exactly the initial part of the pointing trajectory.

Thirteen patients with idiopathic Parkinson's disease participated in this study. All were predominantly right-handed (Edinburgh handedness inventory). Their mean age was 63.5 (range 50 to 73 years) and 7 were males and 6 females. On the Hoehn and Yahr scale (1967), 6 patients were stage 2, 2 stage 3 and 5 stage 4. All were tested both under their usual antiparkinson drugs (mainly L-DOPA) and either before starting the treatment or after withdrawal of the treatment ("drug holidays"). Ten age-matched controls were chosen from hospital inpatients with no brain damage; their mean age was 68.2 (range 58 to 79 years) and 6 were males and 4 females.

The RTs were taken to be the time interval between the onset of the luminous signal and the release of the platform. Statistical comparisons were made between aimed and non aimed mean RTs in the 3 groups of subjects (controls, parkinsonians without treatment and with treatment) using ANOVA Var 3 (ROUANET and LEPINE, 1977).

Experiment 2 : Closed vs open loop pointing movement

Each subject performed four blocks of trials as in experiment 1, both visual conditions and both hands being tested successively and alternately in a random order. Under the visual closed loop conditions, the visual reafferents were available over the entire hand trajectory. Under the visual open loop conditions, the luminous signal D2 was presented in the target location D1 in the form of a mirror image without any change in the experimental environment (see Fig. 1) : the pointing movement was thus performed without any visual control of the moving hand.

Fourteen patients with idiopathic Parkinson's disease, all predominantly right-handed, were tested. Their mean age was 62.8 (range 36 to 74 years) and 9 were males and 5 females. On the Hoehn and Yahr scale, 5 patients were stage 2, 7 stage 3 and 2 stage 4. All are tested only under their usual antiparkinson treatment. Nine age-matched controls were chosen from inpatients with no brain damage; their mean age was 64.8 (range 56 to 75 years) and 5 were males ans 4 females.

The parameters recorded were as follows : R.T. defined as in Experiment 1, movement time (M.T.) which was taken to be the time interval between releasing the platform and touching the screen, pointing error which focused on the vertical error (ERY) corresponding to the vertical distance between the target and the first finger contact on the screen. Statistical comparisons were made between closed loop and open loop mean RTs, MTs and ERYs in the 2 groups of subjects (controls, parkinsonians), again using ANOVA Var 3.

RESULTS

Experiment 1 : Aimed vs non aimed RT tasks

The mean RT values are presented on histograms (Fig. 2). The combined results show that there was no significant difference between hands either in the controls (F = 0.49, df 1,9, p > 0.01) or the parkinsonians (F = 2.45, df 1,24, p > 0.01). In parkinsonians not under treatment, the RTs were significantly longer in both tasks than in the controls (pointing : F = 11.93, df 1,21, p < 0.005; No aiming : F = 15.65, df 1,21, p < 0.001). In the patients with Parkinson's disease, dopatherapy improved the RTs significantly in the no aiming task (F = 9.39, df 1,24, p < 0.01) and non-significantly in the pointing task (F = 6.19, df 1,24, p > 0.01). Comparison between both tasks showed the existence of a significantly greater RT impairment in the no aiming than in the pointing movement in the patients with Parkinson's disease (F = 36.13, df 1,24, p < 0.0005), whereas no significant difference was observed in the controls (F = 5.52, df 1,9, p > 0.01). This differential RT impairment was more marked in parkinsonians not undergoing treatment (F = 23.18, df 1,12, P < 0.0005) than when undergoing treatment (F = 14.10, df 1,12, p < 0.005).

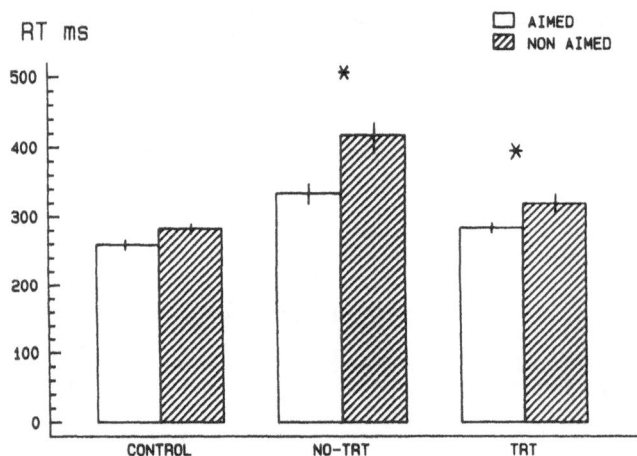

Fig. 2 . RTs in msec (mean ± SEM) in the 3 groups.
(control parkinsonian without treatment (no TRT),
parkinsonian with treatment (TRT) in both tasks (*:p < 0.01).

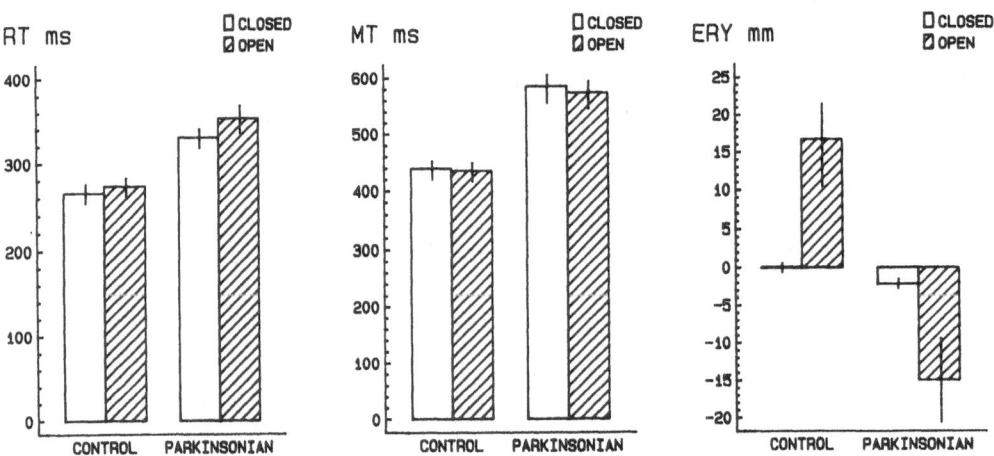

Fig. 3 . RTs, MTs and ERYs (mean ± SEM) in both groups
(controls, parkinsonians) in both conditions.

Experiment 2 : Closed vs open loop visual conditions

The mean RT, MT, and ERY values are given in the histograms (Fig. 3). The combined results show that in the parkinsonians the RTs were not significantly longer than in the controls under either condition (F = 7.06, df 1,21, p > 0.01), whereas the MTs were significantly longer in the parkinsonians than in the controls (F = 8.67, df 1,21, p<0.01). Comparisons between the ERYs also showed that a significant difference existed between controls and parkinsonians (F = 8.15, df 1,21, p<0.01). The latter result is mainly due to the effect of the open loop condition since the ERYs, which lay in the same range under closed loop conditions in both groups (F = 4.91, df 1,44, p>0.01), differed significantly between the control and parkinsonian groups under the open loop conditions (F = 13.84, df 1,44, p<0.001) : actually, under the open loop conditions, the ERYs showed a trend towards a positive value in the control group corresponding to hypermetria (mean ERY value = +16.7 mm), whereas the reverse trend towards a negative value was observed in the parkinsonian group, corresponding to hypometria (mean ERY value : -15 mm).

DISCUSSION

Akinesia is difficult to assess objectively, even using standardized movement analysis, because it can have several different facets in Parkinson's disease as well as varying from one patient to another (TERAVAINEN and CALNE, 1980). The main interindividual factors involved in akinesia include not only the stage of the disease and the effect of the treatment but also the clinico-pathological pattern of the disease and the motor strategy used to perform movements. However, beyond these interindividual variations, it seemed to be interesting to look at the general trends shown by a well-defined group of parkinsonians in comparison with controls, in order to determine some of the consequences of akinesia in the initiation and execution of motor tasks. Actually, the most characteristic features of akinesia which have been put forward in the literature, are the prolonged RT prior to initation of a movement and the slowness of movement execution (DRAPER and JOHNS, 1964; BARBEAU and DE GROOT, 1966; HALLETT and KHOSHBIN, 1980). The results obtained here in both experiments 1 and 2 provide some complementary answers to the questions which have been raised as to the real nature of akinesia. The first point analyzed in experiment 1 was the differential expression of akinesia in aimed vs non aimed RT tasks : it emerged clearly that the RTs were longer in patients with Parkinson's disease in both tasks, which is a classical result (HEILMAN et al., 1976; MARSDEN, 1984; VIALLET et al., 1984; SHERIDAN et al., 1987). But here the RTs were found to be more severely affected in the noaiming than in the pointing task in parkinsonian patients, whereas no significant difference was observed between tasks in the controls. The latter result suggests that the expression of akinesia, as reflected in the RT measurements, may depend on the characteristics of the forthcoming movement. It is generally held that the greater the amount of information to be programmed, the longer the latency of a movement will be (BERNSTEIN, 1967) : this means that the RTs should have been longer in the case of the pointing movement in which the informational load seems to be, a priori, greater than in that of the no aiming task. Now, the fact that this was not so in the controls, while the reverse was even observed in the patients suggests that the informational load cannot have been the decisive factor. In fact, these two RT tasks differed simply in that the pointing movement was aimed at reaching a visual target whereas the non aimed movement involved no spatial goal. From this point of view, the greater RT impairment shown by the parkinsonians in the no aiming task can be interpreted partly in terms of the motivational aspects; actually, it has already been mentioned that motor performance in Parkinson's disease can be influenced by motivational factors (SCHWAB and ZIEPER, 1965) : moreover, motivated behaviour aimed at satisfying basic internal or conditioned needs is known to be regulated by the dopaminergic system (WISE, 1982); in this context, it is noteworthy that dopatherapy improved the non aimed more than the pointing movements in parkinsonian patients. Besides these motivational aspects, it can also be assumed that, in Parkinson's disease, the expectation of that some sensory feedback will be available for use during the movement execution might influence the RT performance : in this framework, non aimed movements, during which less sensory feedback is expected, will then be more severely impaired than pointing movements which can involve the use of visual feedback information about the goal

(HUMPHREY, 1979). The compensatory role of the visual feedback in the control of movement execution in Parkinson's disease was studied more specificaly in experiment 2. The main finding here was the impairment in the spatial accuracy which was observed along with a general hypometric tendency in the parkinsonian group. This contrasts with the overshooting of the target shown by the control group. Hypometria has already been observed in Parkinsonians performing other motor tasks (FLOWERS, 1976; COOKE et al., 1978). This aspect of akinesia may have resulted from a reduction in the activation of forebrain dopaminergic motor systems (MEADOR et al., 1986), which connect the basal ganglia and the supplementary motor area (SMA), as mentioned by GOLDBERG (1985). In this context, it has also been shown that the amplitude of the so called Bereitschftspotential (BP), a readiness potential preceding a self-paced voluntary movement, is reduced in Parkinsonians (DEECKE et al., 1977; SHIBAZAKI et al., 1978) : moreover, further studies have demonstrated that the earlier component of the BP was abnormal in Parkinson's disease when the dopatherapy was withheld (DICK et al., 1987; 1989); similar data have also been obtained on patients with Parkinson's disease as regards the slow electronegative brain potential (CNV) occurring prior to a stimulus-triggered movement : the amplitude of this potential was lower in these patients when off that when on treatment (AMABILE et al., 1986). In the absence of visual feedback, the poor activation of the SMA by the basal ganglia output may therefore have resulted in the insufficient energizing of the initial part of the movement, which could not be compensated for by visual guidance during the execution phase (VIALLET et al., 1987). However, this hypothesis suggesting the existence of a direct correlation between the level of SMA activation and the amplitude of the initial EMG burst triggering the actual movement still remains to be confirmed in further studies.

REFERENCES

AMABILE, G., FATTAPPOSTA, F., POZZESSERE, G., ALBANI, G., SANARELLI, L., RIZZO, P.A. and POROCUTTI C., 1986), Parkinson's disease : electrophysiological (CNV) analysis related to pharmacological treatment. Electroencephal. Clin. Neurophysiol., 64: 521-524.

BARBEAU, A. and DE GROOT, J.A., 1966, The problem of measurement of akinesia. J. Neurosurg., 24: 331-334.

BERNSTEIN, N.A., 1967, The coordination and regulation of movements. Pergamon Press, Oxford.

COOKE, J.D., BROWN, J.D. and BROOKS, V.B., 1978, Increased dependence on visual information for movement control in patients with Parkinson's disease. Can. J. Neurol. Sci., 5: 413-415.

DAY, D.L., DICK, J.P.R. and MARSDEN, C.D., 1984, Patients with Parkinson's disease can employ a predictive motor strategy. J. Neurol. Neurosurg. Psych., 47: 1299-1306.

DEECKE, L., ENGLITZ, H.G., KORNHÜBER, H.H. and SCHMITT, G., 1977, Cerebral potentials preceding voluntary movements in patients with bilateral or unilateral Parkinson akinesia. Progr. Clin. Neurophysiol., 1: 151-163.

DICK, J.P.R., CANTELLO, R., BURUMA, O., GIOUX, M. BENECKE, R., DAY, B.L., ROTHWELL, J.C., THOMPSON, P.D. and MARSDEN, C.D., 1987, The Bereitschaftspotential, L-DOPA and Parkinson's disease. Electroenceph. Clin. Neurophysiol., 66: 263-274.

DICK, J.P.R., ROTHWELL, J.C., DAY, B.L., CANTELLO, R., BURUMA, O., GIOUX, M., BENECKE, R., BERARDELLI, A., THOMPSON, P.D. and MARSDEN, C.D., 1989, The Bereitschaftpotential is abnormal in Parkinson's disease. Brain 112: 233-244.

DRAPER, I.T. and JOHNS, R.J., 1964, The disordered movement in parkinsonism and the effect of drug treatment. Bull. Johns Hopkins Hosp. 115: 465-480.

FLOWERS, K.A., 1976, Visual "closed-loop" and "open-loop" characteristics of voluntary movement in patients with parkinsonism and intention tremor. Brain 99: 269-310.

FLOWERS, K.A., 1978, Lack of prediction in the motor behaviour of parkinsonism. Brain 101: 35-52.

GOLDBERG, G., 1985, Supplementary motor area. Structure and function : Review and hypotheses. Behav. Brain Sci. 8: 567-616.

HALLETT, M., 1989, Clinical neurophysiology of akinesia, in: Proceedings of the international meeting of the french neurological society. From posture to initiation of movement. 7.

HALLETT, M. and KHOSHBIN, S., 1980, A physiological mechanism of bradykinesia. Brain 103: 301-314.

HEILMAN, K.M., BOWERS, D., WATSON, R.T. and GREER, M., 1976, Reaction times in Parkinson's disease. Arch. Neurol. 33: 139-140.

HOEHN, M.M. and YAHR, M.D., 1967, Parkinsonism : onset, progression and mortality. Neurology 17: 427-432.

HUMPHREY, D.R., 1979, On the cortical control of visually directed reaching : contributions by non precentral motor areas, in: Posture and movement, R.E. Talbott and D.R. Humphrey, eds., Raven Press, New York, pp. 51-112.

JOUBERT, M. and BARBEAU, A., 1969, Akinesia in Parkinson's disease, in: Progress in Neurogenetics, A. Barbeau and J.R. Brunette, eds., International Congress Series, 175: 366-376, Excerpta Medica, Amsterdam.

MARSDEN, C.D., 1984, Which motor disorder in Parkinson's disease indicates the true motor function of the basal ganglia?, in: Functions of the basal ganglia. Ciba Foundation symposium, 107: 225-241, Pitman, London.

MARSDEN, C.D., 1989, Slowness of movement in Parkinson's disease. Movement Disorders 4: 526-537.

MEADOR, F.J., WATSON, R.T., BOWERS, D. and HEILMAN, K.M., 1986, Hypometria with hemispatial and limb motor neglect. Brain 109: 293-305.

NARABAYASHI, H., IMAI, H., YOKOCHI, M., HIRAYAMA, K. and NAKAMURA, R., 1976, Cases of pure akinesia without rigidity and tremor and with no effect by L-DOPA therapy, in: Advances in Parkinsonism, W. Birkmayer and O. Hornyckiewicz, eds., Roche, Basle, pp. 335-342.

ROTHWELL, J.C., 1987, Control of human voluntary movement, Croom Helm, London & Sydney.

ROUANET, H. and LEPINE, D., 1977, L'analyse des comparaisons pour le traitement des données expérimentales. Informatique et Sciences Humaines 33: 9-123.

SCHWAB, R.S. and ZIEPER, I., 1965, Effects of mood, motivation, stress and alertness on the performance in Parkinson's disease. Psychiatr. Neurol. 150: 345-357.

SHERIDAN, M.R., FLOWERS, K.A. and HURRELL, J., 1987, Programming and execution of movement in Parkinson's disease. Brain 110: 1247-1271.

SHIBASAKI, H., SHIMA, F. and KUROIWA, Y., 1978, Clinical studies of the movement-related cortical potential (M.P.) and the relationship between the dentatorubrothalamic pathway and readiness potential (R.P.). J. Neurol. 219: 15-25.

STERN, Y., MAYEUX, R., ROSEN, J. and ILSON, J., 1983, Perceptual motor function in Parkinson's disease : a deficit in sequential and predictive voluntary movement. J. Neurol. Neurosurg. Psych. 46: 145-151.

TERAVAINEN, H. and CALNE, D., 1980, Quantitative assessment of Parkinsonian deficits, in: Parkinson'disease. Current progress, problems and management, U.K. Rinne, M. Klingler and G. Stamm, eds., Elsevier, North-Holland, pp. 145-164.

VIALLET, F., TROUCHE, E., BEAUBATON, D., LEGALLET, E. and KHALIL, R., 1984, Impairment of a goal-directed movement performed by Parkinsonians : an objective index of bradykinesia. Neurosci. Lett. Suppl. 18: S.85.

VIALLET, F., TROUCHE, E., BEAUBATON, D., LEGALLET, E. and KHALIL, R., 1987, Visual feedback and motor performance in human and animal basal ganglia dysfunction, in: Basal ganglia and behavior : sensory aspects of motor functioning, J.S. Schneider and T.I. Lidsky, eds., Hans Huber, Toronto-Lewiston-New York-Bern-Stuttgart, pp. 71-82.

WILSON S.A.K., 1925, The Croonian Lectures on some disorders of motility and of muscle tone, with special reference to the corpus striatum : Lecture I. The Lancet, 1-10.

WISE, R.A., 1982, Neuroleptics and operant behavior : the anhedonian hypothesis. Behav. Brain Sci. 5: 39-87.

RETINAL VISUAL DYSFUNCTION, IN PARKINSON'S DISEASE, IS ENHANCED BY LOW CONTRAST STIMULI

P. Stanzione, M. Tagliati, A. Peppe*, G. Sancesario, and F. Pierelli*

Clinica Neurologica II Università, Roma
*Istituto di Clinica delle Malattie Nervose de Mentali Università- La Sapienza, Roma

INTRODUCTION

Electrophysiological studies, performed in the past several years, (Bodis-Wollner and Yahr 1978) have demonstrated visual alterations in Parkinson's disease (PD) patients. Further studies confirmed the delay of the pattern reversal Visual Evoked Potential's (VEP) (Sollazzo 1984; Tartaglione et al. 1984; Bodis-Wollner et al. 1986) major positive wave (P100). Research has proved that the cause of this delay also dependens on several parameters of the visual stimulus: the type of visual pattern (Tartaglione et al. 1984), the spatial (Onofrj et al. 1986) and temporal frequency (Marx et al. 1987), and the contrast level (Gottlob et al. 1987)(for a review see Bodis-Wollner et al. 1988) . Namely, an enhanced P100 delay has been observed in PD patients, when sinusoidal grating stimuli at high spatial frequency (2-4cpd) and at low contrast levels (less then 50%) are utilized (Bodis-Wollner 1988). The physiopathologic relationship with the dopaminergic deficit of PD has been proved by several studies emphasizing the recovery of this delay during L-DOPA therapy (Bodis-Wllner and Yahr 1978; Sollazzo 1984; Bodis-Wollner et al. 1988).

A retinal origin of VEP delay has been proposed on the basis of the reported presence of dopaminergic innervation of the retina (Frederick et al. 1982). The retinal origin of the reported visual dysfunctions has been studied by utilizing Pattern Electroretinogram (PERG) recordings. This potential is a low voltage (1-2 uV) one, recordable, in humans, by an electrode positioned in the lower cantus. It is composed of a sequence of three waves, the most prominent of which (b-wave) is considered to be generated at the ganglion cell layer (Maffei and Fiorentini 1981). Therefore the potential can be regarded as an index of the

---

Author for correspondence : P. Stanzione, Clinica Neurologica, Dpt. Sanita' Pubblica e Biologia Cellulare, II Universita' di Roma - Via O. Raimondo 8- 00173 Roma ITALY.

Key Words:PERG,Parkinson's disease,contrast sensitivity

electrical activity performed by the retina at its last stage.

In a previous study (Stanzione et al. 1989), performed with the aim of exploring the possible retinal origin of VEP delay, VEP and PERG were simultaneously recorded in PD patients. This study demonstrated an increase of both PERG b-wave and VEP P100 latencies suggesting a retinal contribution to the VEP delay.

Furthermore both PERG and VEP latency differences, between the PD patients and the Standard group. proved larger and true at a higher significance level, utilizing higher spatial frequency stimuli (2cpd) .

In the present study we attempted to analyze the influence of low contrast stimuli on PERG parameters in PD patients as also shown for VEP alterations, by previous electrophysiological (Bodis-Wollner et al. 1988) studies. Therefore PERGs were recorded in 10 Parkinson's disease patients, utilizing different contrast levels (100%, 50%, 20%, 10%) , before and during dopaminergic monotherapy. The results were compared to a standard group of 8 normal subjects, homogeneous in age.

The PERG b-wave latency, progressively increased as the contrast level was decreased from 100% to 10%. This was observed both in healthy subjects and in parkinsonian patients without therapy. The latter showed a greater extent of b-wave latency delay at every contrast level with a striking prolongation in recordings obtained by a 10% contrast stimulus.

The dopaminergic monotherapy produced a recovery of the reported PERG abnormalities.

We conclude that low contrast stimuli enhance PERG sensitivity to the visual dysfunction of Parkinson's disease patients. Moreover the observed effects after therapy confirm the dopaminergic nature of these electrophysiological alterations.

SUBJECTS AND METHODS

10 parkinsonian patients (PD)(mean age 63 +/-8.1 years) and 8 normal subjects, matched for age (mean age 60.4+/-5.3 years), were examined. The patients were classified according to the Hoen and Yahr into II and III stage (Hoen and Yahr 1967). All the patients reacted positively to dopaminergic therapy.

The PERGs were elicited by a reversing grating pattern on a television screen . The entire pattern subtended a 17 degrees visual angle. The mean luminance was 60 cd /m2 . Four different contrast levels (100%, 50%, 20%, 10%) calculated according to Hess and Baker (Hess and Baker 1984) (C= [Lmax-Lmin]/[Lmax+Lmin])were tested in each subject. The spatial frequency of the grating was 2cpd; the rate counter phase modulation was 1 Hz.

PERG responses were recorded with a silver-wire electrode placed in the lower cantus of the eye, a silver-cup reference electrode placed over the homolateral temple and a ground electrode on the mastoid. Details of the electrophysiological methods used are reported in Ref.11.

Monocular right and left recordings were obtained in all the normal subjects and in all the PD patients not earlier than 10 days after therapy withdrawal. The same recordings were performed in PD patients 2 weeks after L-DOPA monotherapy (range 375-750 mg/die). This second recording was performed two hours after the midday oral ingestion of 125 or 250 mg of L-DOPA plus a peripheral inhibitor, as this is considered the peak-time of the blood concentrations after oral ingestion (Fahn 1985).

In the PERG analysis we took into account the b-wave latency. The data obtained from the standard group and from the PD patients, before and during therapy, were compared utilizing the Bonferroni t statistic Test for multiple samples. Within each group the statistical analysis was performed utilizing the analysis of variance (ANOVA one way) corrected by Tukey test for the Critical Difference. All the subjects participating in this study gave their informed consent.

RESULTS

PERG b-wave latencies were studied in 16 eyes of normal standard subjects and in 20 eyes of PD patients, before and during dopaminergic therapy (Fig. 1), utilizing four different contrast levels: 100%, 50%, 20%, 10%. In 3 eyes of PD patients , a reliable and reproducible b-wave was not recorded in control conditions at 10% of contrast level, but the b-wave was recorded in the same eyes after therapy. Therefore their values are shown separately (Fig. 3, Fig. 4), while the statistical analysis was performed on 17 eyes of PD patients before and during therapy (Fig. 1, Fig. 2).

Evaluation of b-wave latency at different contrast levels within the Standard Group and within the PD patients' group before and during therapy

In the Standard Group, the b-wave latency progressively increased as the contrast level was decreased. The percentage difference, in this group, between the mean latency at 100% and 10% was 8.5%. At a first approximation, a linear relationship existed between b-wave latencies and the contrast levels. The increase in latency through the decrease of contrast from 100% to 10%, was true at the significant level of p< 0.001 (ANOVA one-way). However the single difference between the mean latency at 100% and 10% was not statistically significant if analysed by ANOVA corrected by Tukey test (Fig. 2A).

In the group of PD patients before therapy, a b-wave progressive latency increase (32% difference in the mean latencies between 100% and 10%), when lowering the contrast level, was seen. This progressive increase was true at p< 0.001,(ANOVA one-way). Contrary to the Standard group, the difference between the mean latency at 100% and 10% was statistically significant in the PD patients' group before therapy (p= 0.01 ANOVA corrected by Tukey test, Fig. 2A). Moreover, the higher latency increase, at a contrast level of 10%, produced a non-linear behaviour of the b-wave latency (Fig. 2A-B).

In the same PD patients during treatment the mentioned difference between the mean latencies at 100% and 10% was no longer statistically significant when analysed by ANOVA corrected by Tukey test, as preaviosly apparent in the Standard group (Fig. 2B). The tendency to a latency

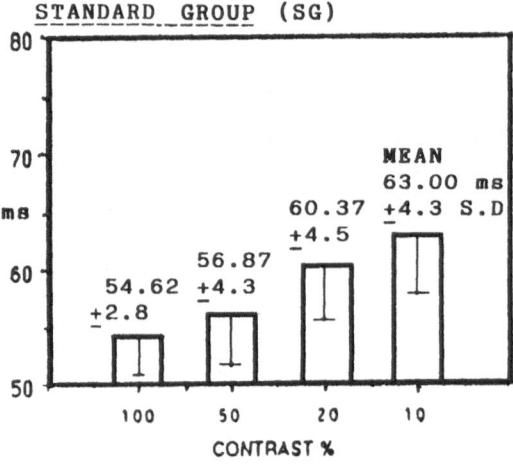

STANDARD GROUP (SG)

** p 0.01 /* p=0.05
Bonferroni Test

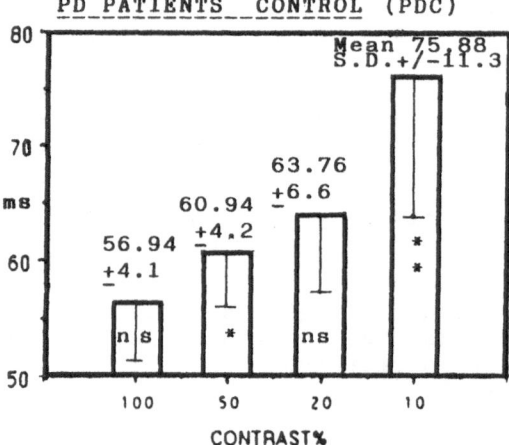

PD PATIENTS CONTROL (PDC)

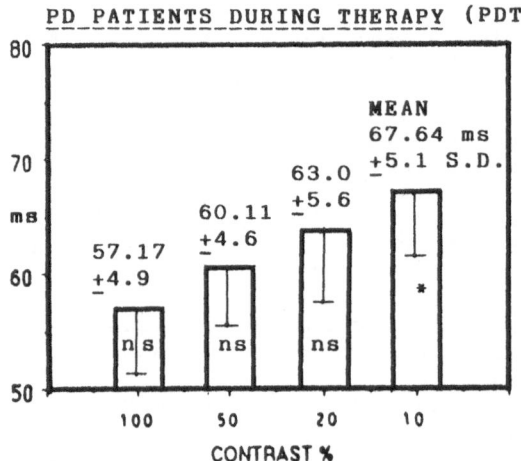

PD PATIENTS DURING THERAPY (PDT)

increase from 100% to 10%, was true at the significance level of p< 0.001 (ANOVA one-way). Moreover the b-wave latency delay, again appeared to be in linear relationship to the contrast level, as observed in the standard group.

F I G. 1

Mean b-wave latencies in the Standard Group SG, upper frame) and in the PD patients group, before (PDC, middle frame) and during dopaminergic therapy (PDT, lower frame).

In each frame the bars indicate the mean b-wave latency at each contrast level in that population. Contrast levels are shown on the x axis and the latency values on the y axis. The lines inside the bars, indicate the Standard Deviations (S.D.). On the top of each bar, the mean and the S.D. are given. The asterisk in the middle frame indicates the significance of the difference between the corresponding mean latencies of the PDC and SG groups. In the lower frame, the asterisk indicates the significance in the differences between the corresponding means in the PDC and PDT groups. ns: not significant. Note that 10% of contrast level produces between PDC and SG differences true at higher significant level, which recovers during therapy.

b-wave latency in the Standard group (SG)
and in PD patients before therapy(PDC)

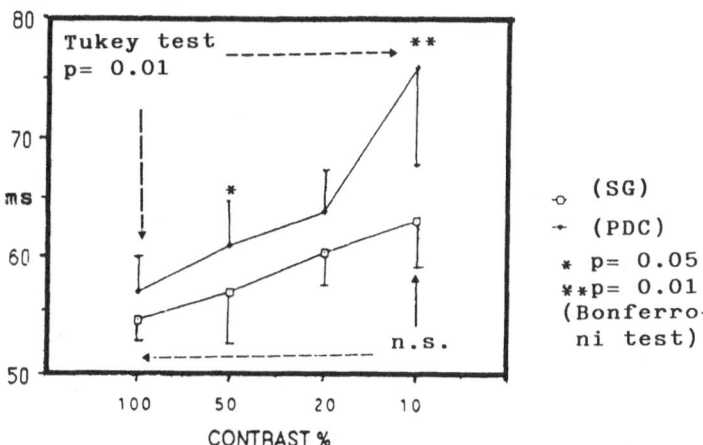

(SG) ⊶
(PDC) ⊷
* p= 0.05
**p= 0.01
(Bonferro-
ni test)

F I G. 2

Comparison of the
b-wave mean
latencies between
and within
groups.
The upper frame
shows the
comparison
between the SG
and the PDC
groups. Note that
in the PDC group
only the
difference
between the mean
latency at 100%
and 10% is
statistically
significant
because of the
non-linearity of
the PERG response
at 10%. In the
middle frame the
PDC is compared
with the PDT
group. Note the
recovery of the
linearity of the
response at 10%
in the PDT group.
In the lower
frame the data
obtained from the
SG and the PDT
groups are
compared. Note
the absence of
any statistically
significant
difference
between the two
groups.

b-wave latency in PD patients before
(PDC)and during therapy (PDT).

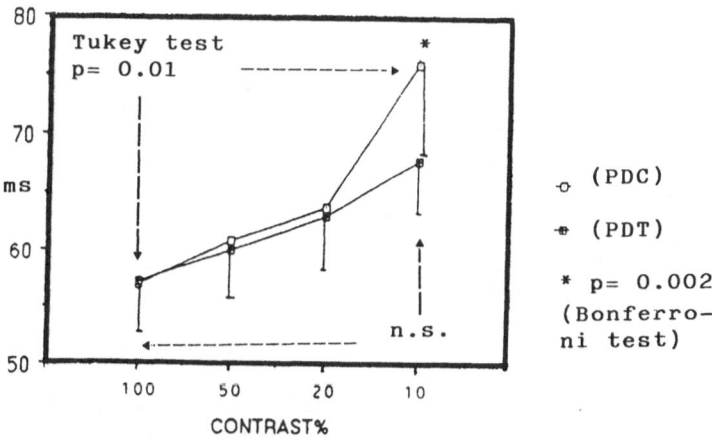

(PDC) ⊶
(PDT) ⊷

* p= 0.002
(Bonferro-
ni test)

b-wave latency in the Standard Group(SG)
and in PD patients during therapy(PDT)

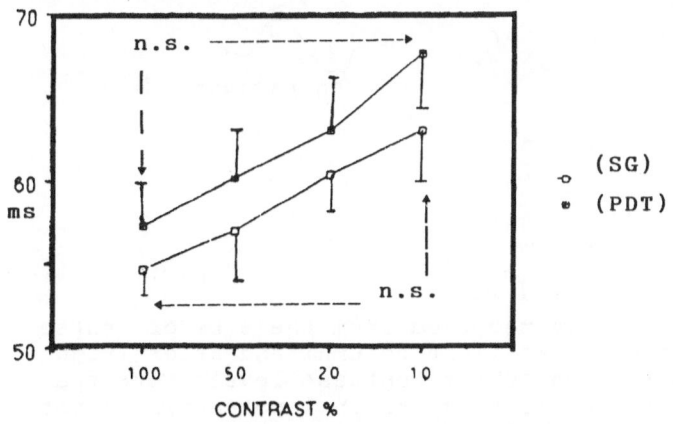

(SG) ⊶
(PDT) ⊷

657

Comparison of the mean b-wave latencies between the three Groups

The comparison of the mean latencies of the b-wave between the Standard Group and the PD patients before therapy showed a statistically significant difference at 50% (p = 0.05) and a higher statistically significant difference at 10% (p < 0.01, Bonferroni t-Test, Fig. 2A). The b-wave mean latencies during therapy showed a general decrease which was statistically significant at 10% (p= 0.05, Bonferroni t-Test) if compared to the same group before therapy. No statistically significant difference was observed (Fig. 2B) comparing the mean latencies of the b-wave in PD patients during therapy and in the standard group (Fig. 2C).

The three eyes which did not show a retinal potential at 10% of contrast level in control conditions.

Three eyes of three different PD patients did not show any clear cut retinal potential when they were stimulated at the lowest (10%) contrast level in control conditions (fig. 3 left column).

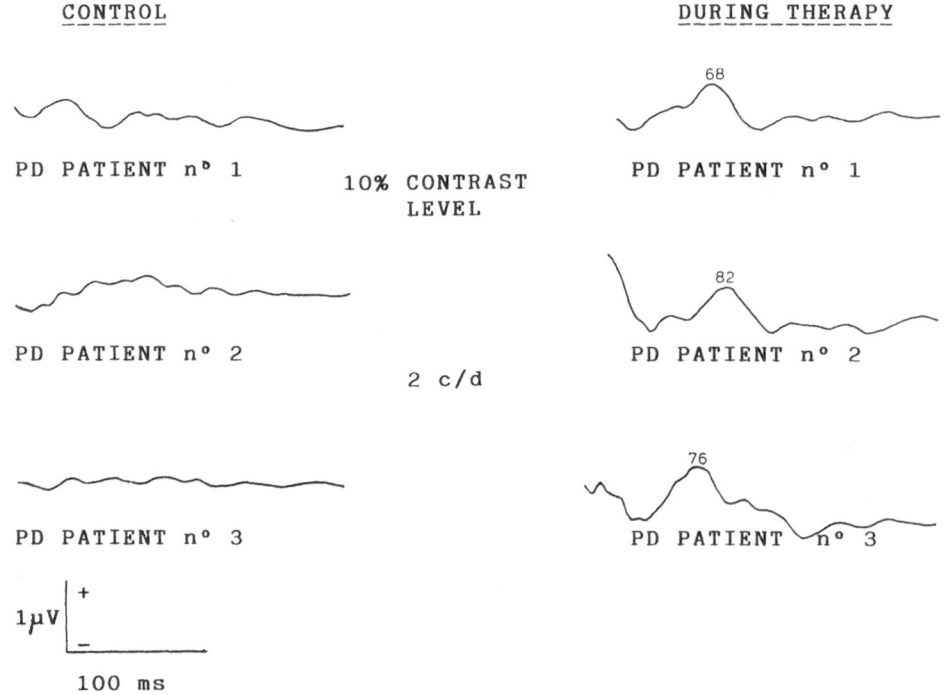

CONTROL   DURING THERAPY

PD PATIENT n° 1   10% CONTRAST   PD PATIENT n° 1
                            LEVEL

PD PATIENT n° 2   2 c/d   PD PATIENT n° 2

PD PATIENT n° 3   PD PATIENT n° 3

1μV

100 ms

F I G. 3

The fig. shows the traces recorded from the eyes of three different PD patients before (left column) and after (right column) therapy, utilizing 10% of contrast level. Note that in control condition, a clear cut retinal potential is not recorded. In the left column, during therapy, a low amplitude but clear b-wave appears in all the eyes.

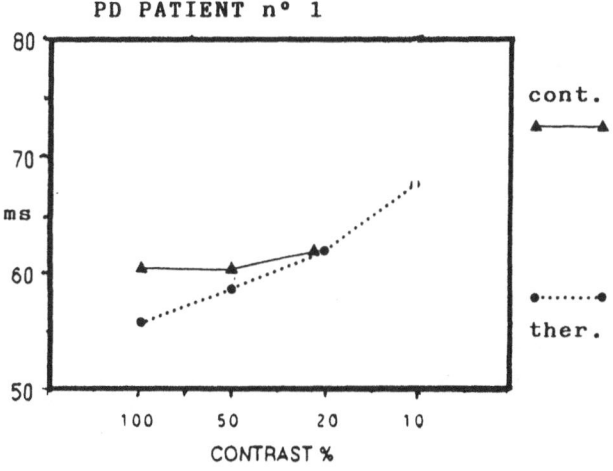

PD PATIENT n° 1

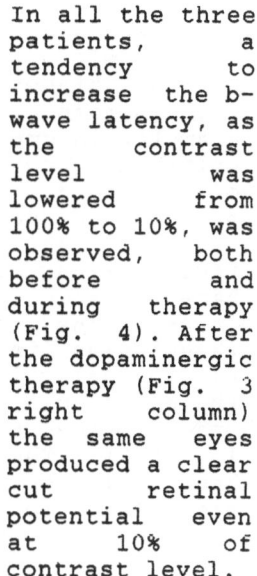

cont.

ther.

In all the three patients, a tendency to increase the b-wave latency, as the contrast level was lowered from 100% to 10%, was observed, both before and during therapy (Fig. 4). After the dopaminergic therapy (Fig. 3 right column) the same eyes produced a clear cut retinal potential even at 10% of contrast level.

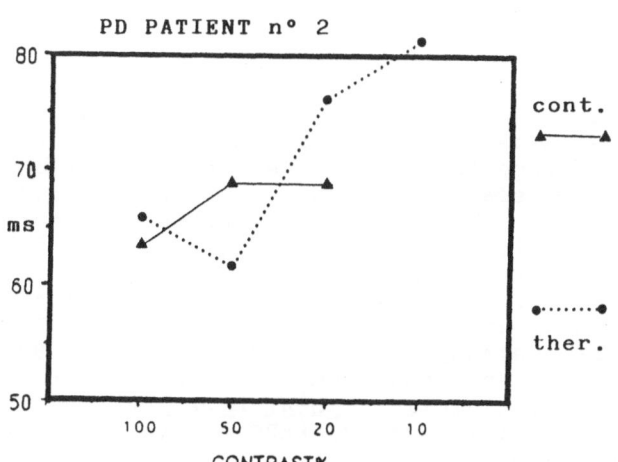

PD PATIENT n° 2

cont.

ther.

F I G. 4

The single values of b-wave latency recorded at each contrast level in these three patients are plotted before and after therapy, in each frame. Note the absence of the response at 10% in the control and the appearance of a responce at the same contrast level, after therapy.

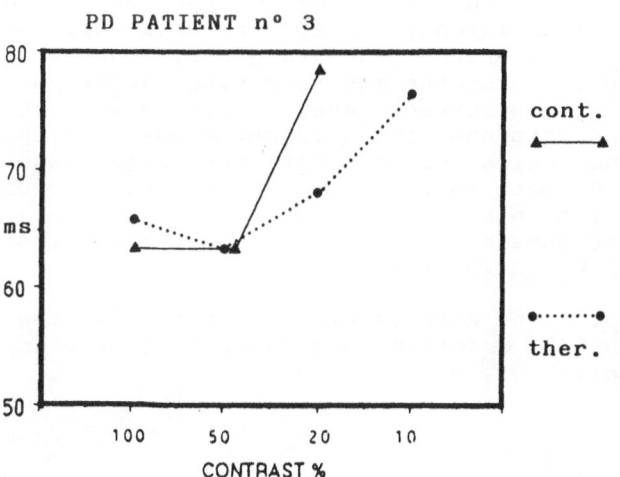

PD PATIENT n° 3

cont.

ther.

DISCUSSION

The main finding of this study was that b-wave mean latencies are increased at each contrast level in PD patients without therapy compared to the Standard group, but the highest significance was at a contrast level of 10%. Therefore a low contrast level seems to be a critical parameter in emphasizing the visual deficit of PD patients, confirming previous psychophysiological studies (Bodis-Wollner et al. 1987) on the same disease. Alterations in the contrast sensitivity functions have been related to the impairment of dopaminergic trasmission in PD patients (Bodis-Wollner et al. 1987). The recovery observed during therapy strenghtens the importance of the relationship between the contrast response and the dopaminergic transmission in the visual system. The retinal origin of the P100 delay in PD patients is supported by histochemical studies showing the presence of dopaminergic receptors on human retina horizontal and amacrine cells (Frederick et al. 1982). The dopaminergic receptors' activation on both these cells could modulate the signal trasmission to the " neighbouring neurons" (Vaney 1985) as it is demonstrated to do in the horizontal cells of the turtle retina (Piccolino et al. 1984). The impairment of this system could lead to a spreading of the signal in the "neighbouring neurons" at the inner and outer plexiform layer levels, activating, from one, several "vertical lines" (photoreceptor-bipolar-ganglion cells). Therefore the voltage generated by light activation could be spatially and temporally less concentrated, producing a decrease of the noise-to-signal ratio. This could partially explain the visual dysfunction observed in PD patients by the VEP recordings.

Previous studies in normal humans (Vaney 1985) reported a linear decrease of the steady state PERG amplitude (induced by a steady-state stimulation with 8Hz of temporal frequency) as a function of the contrast levels. Our study reports a similar linear behaviour of the transient (1 Hz of temporal frequency) PERG b-wave latency in the standard population.

A further relevant finding in PD patients was the non-linear behaviour of the responses in the untreated group. The b-wave mean latency in PD patients without therapy compared to that of the Standard group, increased at each contrast level, but the most striking difference was observed at the 10% contrast level. This causes the non- linearity of the response, which is not present in the standard group, and which accounts for the significant difference in latency between the 100% and 10% levels of contrast in the PD patients' population.

All these data thus suggest the use of low contrast stimuli in studing PERG in PD patients.

Acknowledgement. The authors express their gratitude to Roberta Losacco for the excellent technical assistance and for revising the English style.

REFERENCES

Bodis-Wollner I. 1988, Altered Spatio-Temporal Contrast Vision in Parkinson's disease and MPTP treated monkey: The role of Dopamine. In: Bodis-Wollner I. and Piccolino M. eds. Dopaminergic mechanisms in Vision. New York: Liss,43: 205-20.

Bodis-Wollner I.,Mark M.S., Mitra S., Bobak P., Mylin L. and Yahr M. 1987, Visual dysfunction in Parkinson's disease. Brain 110: 1675-98.

Bodis-Wollner I., Onofrj M.C., Marx M.S. and Mylin L.H. Visual Evoked Potential in Parkinson's disease: spatial frequency, temporal rate, contrast and the effect of dopaminergic drugs. 1988 In: Bodis-Wollner I. and Cracco J. eds. Evoked Potentials; Alan R. Liss, New York, 307-19.

Bodis-Wollner I. and Yahr M.D. 1978 Measurement of Visual Evoked Potential in Parkinson's disease. Brain, 101: 661-71

Fahn S. 1982, Fluctuations in disability in Parkinson's disease: pathophysiology. In: Mardsen C.D. and Fahn S. eds. Movement Disorders, London: Butterworths, 123-145.

Frederick J.M., Rayborn M.E., Laties A.M., Lam D.M.K. and Hollyfield J.G. 1982, Dopaminergic Neurons in the human retina. J. Comp. Neurol. 210: 65-79.

Gottlob I., Schneider E., Heider W. and Skrandies W. 1987, Alteration of Visual Evoked Potentials and Electroretinograms in Parkinson's disease. Electroenceph. Clin. Neurophysiol. 66: 349-57.

Hess F. and Baker L. jr. 1984, Human Pattern-Evoked Electroretinogram. J.Neurophysiol. 51 (5): 939-51

Hoen M.M. and Yahr M.D. 1967, Parkinsonism: onset, progression and mortality. Neurology 17: 427-442.

Maffei L. and Fiorentini A. 1981, Electroretinographic responses to alternating gratings before and after section of the optic nerve. Science 211,953-54.

Mark M., Bodis-Wollner, Bobak P., Harnois C., Mylin L. and Yahr M.1986, Temporal Frequency-dependent VEP changes in Parkinson's disease. Vision Res. 26: 185-93.

Onofrj M., Ghilardi M.F., Basciani M. and Gambi D.1986, Visual Evoked Potentials in Parkinson's disease and dopamine blockade reveal stimulus dependent dopamine function in humans.J.Neurol. Neurosurg. Psychiat. 49: 1150-59.

Piccolino M., Neyton J. and Gerschenfeld H.M. 1984,Decrease of GAP junction permeability induced by dopamine and cyclic adenosine 3': 5'- monophosphate in horizontal cells of turtle retina. J. Neurosci. 4(10): 2477-88.

Sollazzo D. 1984, Influence of L-Dopa/carbidopa on pattern reversal VEP. Behavioural difference in primary and secondary parkinsonism. Elactroenceph. Clin. Neurophysiol. 47: 305-7.

Stanzione P., Pierelli F., Peppe A., Stefano E., Rizzo P.A., Morocutti C. and Bernardi G. 1989, Pattern Visual Evoked Potentials Abnormalities in Parkinson's Disease: Effects of L-DOPA Therapy Clin. Vision Sci. 4(2): 115-27.

Tartaglione A., Pizio N., Bino G., Spadavecchia L. and Favale E. 1984, VEP changes in Parkinson's disease are stimulus dependent. J.Neurol. Neurosurg. Psychiat., 47: 305-307.

Vaney D.I. 1985, The morphology and topographic distribution of A II amacrine cells in the cat retina. Proc. R. Soc. Lond. B. 224: 475-88.

RAPID EYE MOVEMENTS DURING STAGE REM ARE MODULATED BY NIGROSTRIATAL

DOPAMINE NEURONS?

Masaya Segawa and Yoshiko Nomura

Segawa Neurological Clinic for Children
Tokyo, Japan

Rapid eye movements (REMs), the symbolic parameter of stage REM (sREM), are saccadic, conjugate eye movements during sleep and are known to be modulated by cholinergic neurons in the brainstem (Sakai, 1984; Hobson et al., 1974) but their neural network has remained unclarified. On the other hand, voluntary saccadic eye movements have been shown to be controlled by the pathway from the frontal eye field via the basal ganglia to the superior colliculus through the caudate nucleus and pars reticulata of the substantia nigra with double GABAergic inhibitory systems (Hikosaka and Wurtz, 1983; Hikosaka et al.,1989). From polysomnographical examinations on various kinds of basal ganglia diseases including the levodopa responsive fluctuating dystonia, that is, hereditary progressive dystonia with marked diurnal fluctuation (HPD), we have revealed that the nigrostriatal dopamine (NS-DA) neurons modulate the number and direction of REMs, probably through the caudate-nigra-superior colliculus pathway, the same as the voluntary saccade.

SUBJECTS AND METHODS

Four cases of HPD diagnosed by clinical features, one with unilateral Parkinson's disease, four with unilateral basal ganglia lesion revealed by brain imaging studies and six normal persons were subjected to this study.

The polysomnography (PSG) was performed with a 21 channel electroencephalograph according to the method previously reported (Segawa et al., 1987). Two channels were used to record the electrooculogram (EOG) of the horizontal eye movement; one for the original wave to evaluate the REMs with a time constant of 2 seconds, and the other with DC for velocity wave for the evaluation of the number and direction of REMs. Eye movements exceeding the rising angle of 30° under the paper speed of 15 mm per second were defined as REMs. Sleep stages were evaluated according to the criteria of APSS (Rechtschaffen and Kales, 1968). Two kinds of body movements, gross movement (GM) and twitch movement (TM), were also evaluated by assessing the EMG activities of muscles of the trunk and extremities, recorded by surface electrodes placed on 6 muscles on each side. GMs are diffuse sequential EMG activities including those of the rectus abdominalis and last more than two seconds. TMs are short EMG activities localized to one muscle and last less than 0.5 second.

The Basal Ganglia III, Edited by G. Bernardi *et al.*
Plenum Press, New York

The number of REMs that appeared in stage REM was calculated. Besides the total number, those towards the right (R) or left (L) were calculated separately and the ratio of the former against the latter (R/L ratio) was assessed. The numbers of GMs and TMs against each sleep stage were also counted and their sleep stage dependent modulations were evaluated. The ratio of the total number of TMs of the mentalis muscle in stage REM against the total number of REMs in stage REM (ment TM sREM/REMs) was assessed.

These phasic parameters of sleep were shown to be specifically controlled by the NS-DA neurons, parts of the basal ganglia and the brainstem neurons, respectively (Segawa et al., 1987). Furthermore, they were shown to be uninfluenced by the first night effects (Segawa et al., 1987). In this study, these parameters were examined to evaluate whether the abnormalities detected were dependent on the dysfunction of the NS-DA neurons, one or more parts of the basal ganglia or the brainstem neurons.

PSGs were performed twice on patients with HPD and case T.M. with lesions in the left putamen, before and after levodopa. For the rest of the patients and normal subjects the results of the one-night PSG were analyzed.

RESULTS

Clinical Features

Four cases with HPD had clinical characteristics described previously (Segawa et al., 1986). As to the grade of dystonia, three of them had side predominance to the left and one to the right. All of them responded markedly to levodopa and were normalized within 8 weeks with a maximum dosis of about 20mg/kg/day and their effects have been continuing for more than five years. The case with unilateral Parkinsionism (A.H.) was a 55 years old female with clinical onset at 54 years of age and had symptoms with left side predominance. The four cases with unilateral basal ganglia lesions were thought to be nonprogressive and to be caused by focal vascular lesions or asphyxia (Table 1); case J.S., a 5-year-old boy with mild hemiplegia of the right extremities which

Table 1. Cases with Unilateral Lesions

Subject	Age (yrs)	Sex	Disorder
A.H.	55	F	Lt Hemiparkinsonism
J.S.	5	M	Rt Hemiparesis – caudal portion of the PUTAMEN and lateral segment of the lt PALLIDUM
T.M.	7	M	Rt Hemidystonia – infarction of the lt PUTAMEN
T.H.	39	M	Sequelae of the Lt PUTAMINAL hemorrhage
K.S.	11	M	Lt Hemidystonia – low density area in the rt CAUDATE body

occurred at the age of 7 months, whose CT scan showed a focal low density area in the caudalo-medial portion of the left putamen expanding to the lateral portion of the lateral segment of the pallidum; case T.M., a 7-year-old boy with right sided hemidystonia, with CT revealing a low density area in the left putamen which developed around 10 months of age; case T.H., a 37-year-old male with MRI showing a focal lesion in the left putamen, due to hemorrhage; and case K.S., an 11-year-old boy with left hemidystonia, with a low density area in the right caudate body revealed by the CT scan. Case T.M. did not reveal any definite improvement after levodopa, except for the PSG changes described below.

Polysomnographical findings

In the 4 cases of HPD, GMs showed abnormal sleep stage dependent modulation with increase in the rate of its occurrence in stage 2 and the decrease in stage 1 and REM. TMs decreased in number in all sleep stages but without alteration of sleep stage dependent modulation and this feature was prominent in muscles of the more affected side. These abnormalities improved after levodopa. In A.H. a marked decrease in number of both movements was observed; only a few GMs were observed in stage 2 without any GMs in the other sleep stages. TMs also reduced markedly and no TMs were detected at all throughout the recording in the muscles on the right which was the more affected side. The four cases with unilateral basal ganglia lesion showed a different sleep stage dependent modulation of GMs specific for each; cases T.M. and T.H. revealed features fundamentally similar to those of HPD, however, case K.S. had different features with increase in rate during stage 4 and decrease of it in stage REM, while J.S. had a normal pattern. TMs showed marked decrease in number in the muscles of the affected side in three patients. However, in case K.S. only a minor variation was observed.

In HPD and unilateral Parkinsonism, the ratio of ment TM sREM/REMs was reduced significantly from the normal values of 0.4 ± 0.2 (Segawa, 1985). Among cases with unilateral basal ganglia lesions, the ratio showed significant decrease in cases T.M. and T.H., while in the others it was below normal range.

In case T.M., levodopa alleviated the abnormal modulation of GMs and the ratio of ment TM sREM/REMs (Segawa et al., 1987). However, TMs of the normal side increased in number markedly, though no alteration was observed in TMs of the affected side (Segawa et al., 1987).

Table 2   Direction of REMs in Normal Children

Subject Age (yrs)	Sex	REMs (number/hour in sREM) Towards the right	Towards the left	R/L ratio
4	M	85.6	64.4	1.32
6	F	377.0	451.1	0.83
8	F	66.3	47.3	1.40
9	M	153.7	110.8	1.37
10	F	212.0	183.3	1.15
12	M	234.8	156.5	1.50
		188.2 + 114,0	168.9 + 146.7	1,26 + 0.24

Direction of REMs

The number of REMs showed a marked variation among subjects, both normal and diseased. However, in normal subjects, the direction showed a side preference to the right and the R/L ratio revealed only a small variation with an average of 1.26 ± 0.24 (Table 2).

In pathologic cases, REMs revealed a particular side preference in relation to the side of the affected or predominantly affected extremities (Fig. 1). That is, 3 of 4 cases with HPD showed significant lower values in the R/L ratio, revealing an increase in the rate of REMs towards the left. In these cases the left extremities were more affected. After levodopa, the R/L ratio of these cases showed an increase, while the case with a slight decrease in the ratio had a right side preference clinically, and showed further decrease in the ratio after levopoda (Fig. 2).

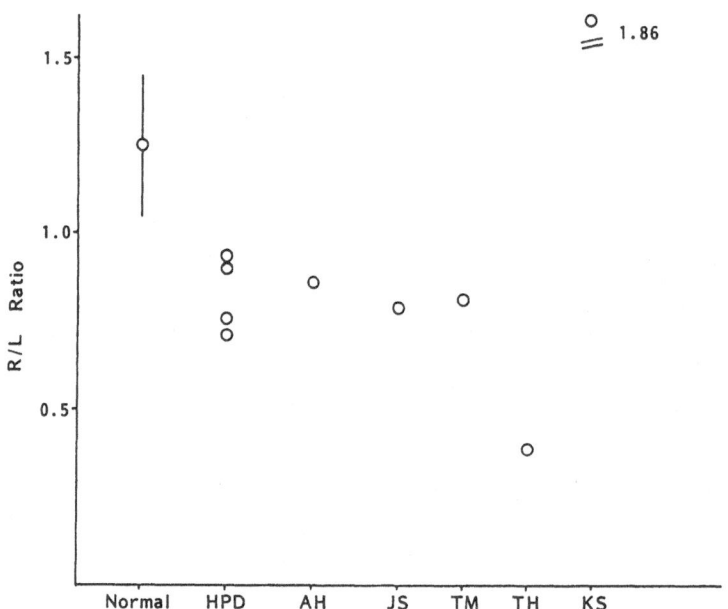

Fig. 1. R/L ratio in cases with HPD and cases with unilateral basal ganglia lesions. AH, JS, TM, TH and KS are listed in Table 1.

The case with left hemiparkinsonism showed increase in the ratio of REMs towards the left with marked decrease in the R/L ratio. Cases with unilateral basal ganglia lesion on the left (cases J.S., T.M. and T.H.) showed preference of REMs towards the left. But the case with the lesion in the right basal ganglia (case K.S.) showed increase in the rate of REMs towards the right, resulting in significant deviation, that is, increase in the R/L ratio. In T.M. the ratio increased to normal range after levodopa (Fig. 2).

DISCUSSION

The pathologic cases subjected to this study had asymmetry in the clinical symptoms, which was considered to be due to the asymmetrical involvement of the basal ganglia. In PSG this side difference correlated with the reduction of number of TMs of the affected muscles in all cases though the grade of severity differed with each case. However, the side preference of REMs in relation to the side of the affected limbs differed among cases. That is, in HPD and hemiparkinsonism the REMs preferred to direct towards the more affected side while in cases with unilateral basal ganglia lesions, the REMs tended to direct more towards the normal side. As the pathophysiology of HPD is considered to be the deficiency of DA at the terminal of the NS-DA neurons (Segawa et al., 1986), the difference in side preference of REMs observed between these groups depended on whether the lesion was presynaptically in the NS-DA neurons (HPD and hemiparkinsonism) or postsynaptically in the basal ganglia (cases with unilateral basal ganglia lesions).

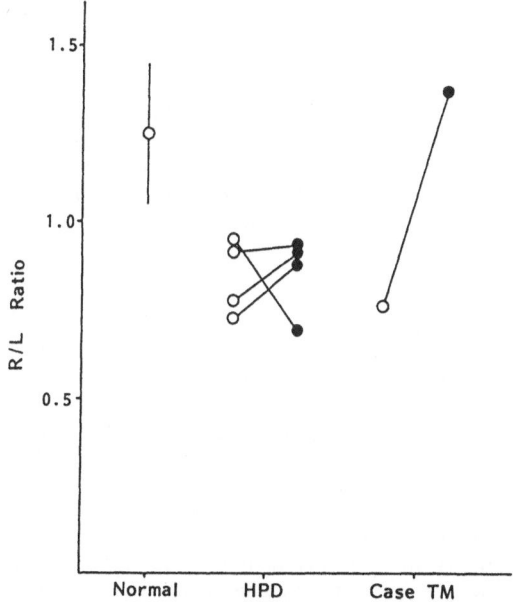

Fig. 2. Alteration of R/L ratio after levodopa observed in cases with HPD and case TM with unilateral basal ganglia disease. Open circles indicate values before levodopa and closed circles indicate values after levodopa. Average and standard deviation of normal subjects are shown on the left.

Our previous studies (Segawa et al., 1987; Segawa 1985), have shown that the NS-DA neurons and the basal ganglia have specific control of muscle activities during sleep; for GMs, they regulate the number and sleep stage dependent modulation via the efferent pathway to the thalamus from the medial portion of the pallidum, and for TMs they control the number via the putative efferent descending pathway directly to the brainstem from the basal ganglia. However, they do not control the sleep stage dependent modulation of TMs which is regulated by brainstem neurons.

On the other hand our PSG examinations on HPD have already suggested that NS-DA neurons regulate the number of REMs via the caudate-nigra (pars reticulata) superior colliculus pathway (Segawa et al. 1987) the same as the voluntary saccades (Hikosaka and Wurtz, 1983; Hikosaka et al. 1989). That is, the reduction in activities of the NS-DA neurons finally causes exaggeration of the saccade neurons in the superior colliculus, by disinhibition of GABAergic neurons in the caudate nucleus, which inhibit other GABAergic neruons in the pars reticulata of the substantia nigra (Segawa et al., 1987). So hyper- and hypo-activities of NS-DA neurons cause particular abnormalities in these phasic parameters during sleep, and their hypo-activities decrease the number of TMs and increase that of REMs, which are reflected in the reduction in rate of ment TM sREM/REMs.

The modulation of muscle activities observed in the PSG of HPD and hemiparkinsonism was identical to that observed in the hypofunction of NS-DA neurons (Segawa et al., 1987; Segawa 1985), and the features of TMs revealed that the hypoactivity was more prominent in the basal ganglia contralateral to the more affected limbs. Among cases with unilateral basal ganglia diseases, two with large lesion in the putamen (T.M. and T.H.) revealed abnormalities implicating DA hypoactivity on the side of the lesion. This was also confirmed by a marked decrease in the ratio of ment TM sREM/REMs of these cases.

TMs of the case with lesion in the posterior part of the putamen (J.S.) revealed abnormalities suggesting DA hypofunction, but the pattern of GMs was normal. However, abnormalities of these body movements in cases with lesion in the body of the caudate nucleus (K.S.) was not those observed in DA deficiency. In these cases the ratio of ment TM sREM/REMs reduced only slightly.

In animal experiments, REMs, the saccadic eye movements during sleep, are known to be generated by neurons of the pontine regions (Sakai, 1984; Hobson et al., 1974), particularly those caudal to the locus coeruleus and rostral to the abducens nucleus (Siegal, 1985). However, in this study the results of PSGs revealed no abnormalities in these neurons. Thus the side preference of REMs must be explained by the disturbance of the NS-DA neurons or the modulation of the neural structures of the basal ganglia.

As most cases with HPD have left side predominance (Segawa et al., 1986) the NS-DA neurons on the right are less active, causing ipsilateral saccade neurons to be more active than the contralateral ones and increase the REMs towards the left with a decrease in the R/L ratio. After levodopa this situation normalized with clinical improvement (Fig.3). Similar features were present in the case with hemiparkinsonism.

In cases with unilateral basal ganglia lesions, the lesions in the striatum may cause hypofunction of the GABAergic neurons in the caudate which disinhibit GABAergic neurons of the substantia nigra of pars reticulata and cause suppression of the saccade neurons in the superior colliculus on the same side. Thus in these cases the direction of REMs showed side preference towards the side of the lesioned basal ganglia or the side of the less affected limbs. This feature is shown in the schematical illustration of case T.M. (Fig.4). In this case the direction of REMs normalized after levodopa. This might have been due to the hyperactivity of the right DA neurons, while no alteration was observed on the left, probably because of the irreversible lesion in the left striatum which was shown by the side difference in modulation of TMs after levodopa (Segawa et al, 1987) (Fig.4).

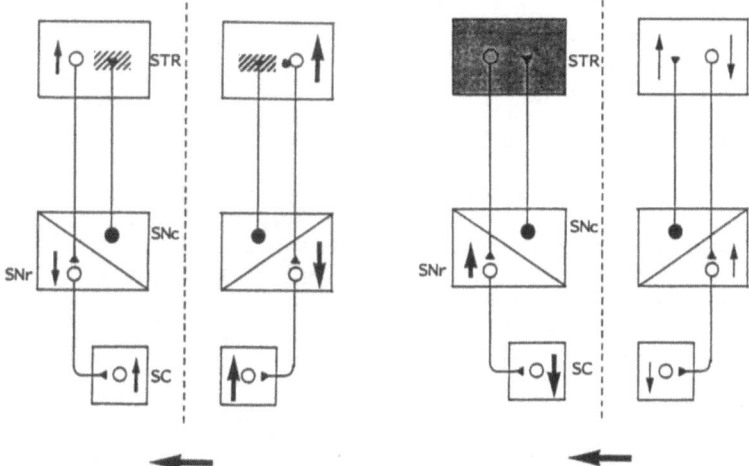

Fig.3. Schematic diagram of the pathophysiology of HPD and the direction of REMs before (left half) and after (right half) levodopa. Hatched bars at the terminal of NS–DA neurons indicate the locus of lesion and the size of the arrows indicate the severity of the hypoactivity.
STR: striatum; SNc: substantia nigra pars comparta; SNr: substantia nigra pars reticulata; SC: superior colliculus. Closed circles indicate nigrostriatal dopamine neurons and open circles indicate saccade neurons.

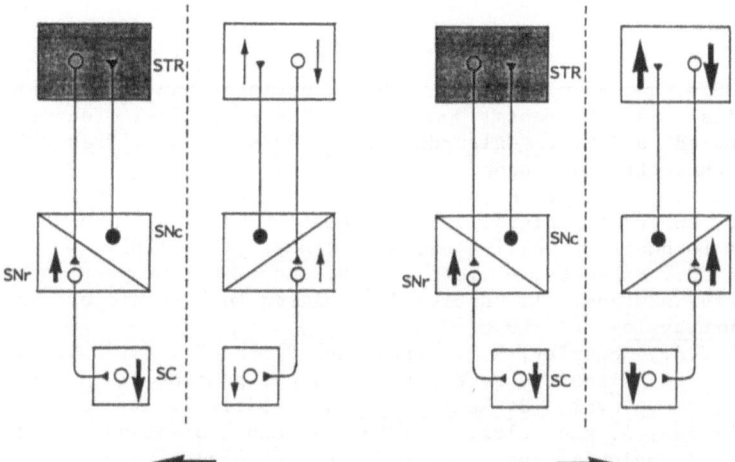

Fig. 4. Schematic diagram of the pathophysiology and preference direction of REMs of case TM before (left half) and after (right half) levodopa. Closed bars indicate the locus of lesion.
STR: striatum; SNc: substantia nigra pars comparta; SNr: substannigra pars reticulata; SC: superior colliculus. Closed circles indicate nigrostriatal dopamine neurons and open circles indicate saccade neurons.

The side preference towards the right seen in normal subjects reflected the asymmetry in activities of NS-DA neurons observed in neurohistochemical examinations (Glick et al., 1984). One case with HPD with right side predominance in symptoms, had R/L ratio slightly below the normal range and only in this case the ratio further reduced after levodopa with improvement in the symptoms. In this case the normal values of R/L ratio might have been set in the lower levels.

These evidences further confirm that the NS-DA neurons and the basal ganglia modulate REMs and further suggest that it is through the caudate-nigra-superior colliculus pathway, the same as for voluntary saccade. For REMs the pallidofugal thalamic pathway might not be involved because the PSG of unilateral basal ganglia lesions showed no definite relation between the direction of REMs and the pattern of GMs, which are modulated by this pathway (Segawa et al., 1987).

The lesions in the striatum observed in cases subjected to this study varied among cases (Table 1). The direction of REMs was not influenced by the variation, while the features of TMs differed with case or locus of lesion (Segawa et al., 1987). So the neurons involved in the modulation of REMs might pass through the striatum rostrocaudally from the caudate to the substantia nigra pars reticulata via the putamen without any synaptic changes and connection with the interneurons. On the contrary, the striatofugal fibers modulating TMs might have synaptic changes or connection with the interneurons in the striatum. This might result in the decrease of TMs in the affected limbs with decrease of REMs by the suppression of the ipsilateral saccade neurons in cases with unilateral striatal lesion, showing quite a contrast to the features observed in the lesion in the NS-DA neurons, presynaptic to the striatum in which REMs and TMs showed reciprocal response to DA activity. Thus estimations of REMs and TMs by PSG are necessary for the evaluation of the pathophysiology of basal ganglia diseases.

REFERENCES

Glick, S.W. and Shapiro, R.M., 1984, Functional and neurochemical asymmetries, in "Cerebral Dominance - The Biological Foundation," N. Geschwind and A.M. Galburda, eds, Harvard Univ. Press, Cambridge, Massachusetts and London.
Hikosaka, O., Wurtz, R.,H., 1983, Visual and oculomotor functions of monkey substantia nigra pars reticulata. IV. Relation of substantia nigra to superior colliculus. J Neurophysiol 49:1285.
Hikosaka, O., Sakamoto, M., Usui, 1989, Functional properties of monkey caudate neurons. I. Activities related to saccadic eye movements, J Neurophysiol 61:780.
Hobson, J.A., McCarley, R.W., Freedman, R. and Pivick, R.T., 1974, Time course of discharge rate changes by cat pontine brain stem neurons during sleep cycle, J. Neurophysiology, 37:1297.
Rechtschaffen, A. and Kales, A.,1968, "A Manual of Standardized Terminology, Techniques and Scoring System for Sleep Stages of Human Subjects," U.S. Government Printing Office, Washington, DC,
Sakai, K., 1984, Central mechanisms of paradoxical sleep, in "Sleep Mechanisms," Experimental Brain Research, Suppl. 8, A. Borbely and J.L. Valatx, eds, Springer-Verlag, Berlin-Heidelberg, New York, Tokyo.
Segawa, M., Nomura, Y., Hikosaka, O., Soda, M., Usui, S. and Kase, M., 1987, Roles of the basal ganglia and related structures in symptoms of dystonia, in: "Basal Ganglia II- Structure and Function," M.B. Carpenter and A. Jayaraman eds, Plenum Press, New York.
Segawa, M., Nomura, Y., Kase, M., 1986, Diurnally fluctuating hereditary progressive dystonia, in: "Handbook of Clinical Neurology, Extrapyra-

midal Disorders," Vol.49 (Revised Series 5), P.J. Vinken, G.W. Bruyn, H.L. Klawans, eds, Elsevier Science Publishers, Amsterdam.

Segawa, M., 1985, Body movements during sleep: its significance in neurology (in Japanese), <u>Neurological Medicine (Tokyo)</u>, 22:317.

Siegal, J.M., 1985, Ponto-medullary interaction in the generation of REM sleep, in: "Brain Mechanism of Sleep," D.J. McGinty, R. Drucker-Colin, A. Morrison and P.L. Parmeggian, eds, Raven Press, New York.

IN VIVO NEUROCHEMICAL ANALYSES OF EXOGENOUSLY ADMINISTERED

L-DOPA: IMPLICATIONS FOR TREATMENT OF PARKINSON'S DISEASE

Elizabeth D. Abercrombie and Michael J. Zigmond

Depts. of Behavioral Neuroscience and Psychiatry
University of Pittsburgh
Pittsburgh, PA 15260 (USA)

INTRODUCTION

The fundamental therapeutic actions of L-DOPA in Parkinson's disease are attributed to its ability to be taken up into the brain where it is converted to dopamine (DA) by the enzyme aromatic amino acid decarboxylase (AADC) (Birkmayer and Hornykiewicz, 1962; Lloyd et al., 1975; Melamed et al., 1984). Thus, it is believed that by replenishing the depleted stores of DA in the basal ganglia caused by the disease-related loss of DA neurons, the neurological deficits characteristic of Parkinson's disease are ameliorated. However, although many insights regarding the therapeutic mechanism of action of L-DOPA in Parkinson's disease have been gained, several important questions remain largely unresolved. Two of the issues which we have addressed in our laboratory are: 1) Is the release of DA formed from exogenous L-DOPA regulated in any way or does it merely reflect a non-specific spillover of DA from the site of formation? 2) Do the characteristics of the release of DA formed from exogenous L-DOPA in the intact striatum differ from those in the striatum lacking a normal dopaminergic innervation as in Parkinson's disease?

We have attempted to answer these questions by using in vivo microdialysis to examine extracellular DA levels in striatum following L-DOPA administration. This is a most useful technique, developed relatively recently (Imperato and Di Chiara, 1984; Ungerstedt, 1984), which permits the study of alterations in the extracellular concentration of substances in the brain of intact rats. Briefly, a probe that incorporates a small piece of dialysis membrane is implanted into the brain under anesthesia and the animal is then allowed to recover from surgery. The dialysis probe is perfused at a very slow rate with an artificial cerebrospinal fluid solution. Neurotransmitter molecules gaining access to the extracellular space diffuse through the pores in the dialysis membrane down their concentration gradients and the solution is collected and analyzed. In this way we have successfully studied changes in extracellular DA level in striatum in response to a variety of behavioral and pharmacological manipulations (Figure 1; Abercrombie et al., 1989a; 1989b; 1990).

The Basal Ganglia III, Edited by G. Bernardi *et al.*
Plenum Press, New York, 1990

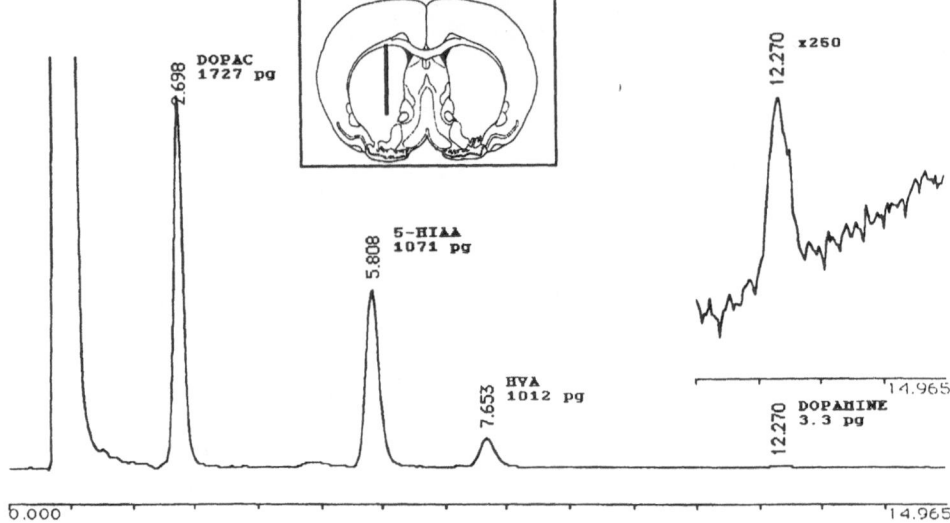

Fig. 1.   Dialysis probes were implanted into the medial portion of the
          rostral striatum as indicated by the dark bar in the inset.
          The chromatogram was obtained by analyzing 20 ul of striatal
          dialysate using HPLC-EC (Waters 460 detector, E_{work}=+0.60 V).
          Note that the level of the metabolites DOPAC and HVA are much
          greater than that of dopamine itself (shown enlarged 250X).

EXOGENOUS L-DOPA AND EXTRACELLULAR DOPAMINE IN STRIATUM

All animals were pretreated with the peripheral AADC inhibitor
RO4-4602 (50 mg/kg, i.p.) 30 min prior to administration of L-DOPA
methyl ester (25-200 mg/kg, i.p.). L-DOPA decreased extracellular
DA at 25 mg/kg (n=4) and 50 mg/kg (n=4) and increased it at 100
mg/kg (n=7) and 200 mg/kg (n=4). No behavioral effects of L-DOPA
were observed with the 25 mg/kg or 50 mg/kg doses. At 100 mg/kg
L-DOPA occasionally produced some locomotor activity and sniffing
behavior whereas 200 mg/kg consistently produced marked increases
in locomotion that included a great deal of sniffing and rearing.

Thus, L-DOPA affected extracellular DA in the striatum in a
complex, dose-dependent manner (Figure 2). The decrease in
extracellular DA following the lower doses of L-DOPA may be
attributable to activation of presynaptic inhibitory autoreceptors
on DA neurons. In support of this hypothesis, it has been
demonstrated that L-DOPA administration can abolish the
electrophysiological activity of dopaminergic neurons following
conversion to DA via a direct action on DA receptors (Bunney et
al., 1973). We presume that either the temporal resolution of the
microdialysis technique is insufficient to detect the brief
increase in DA efflux which must necessarily precede a decrease
mediated by autoreceptor stimulation or that the magnitude of this
increase is such that there is no detectable overflow from the
synapse and therefore a net decrease in extracellular DA is
measured. The facilitatory effect of L-DOPA on DA efflux at the
higher doses may indicate that higher doses of L-DOPA can elicit
the release of DA which is independent of action potentials. Such

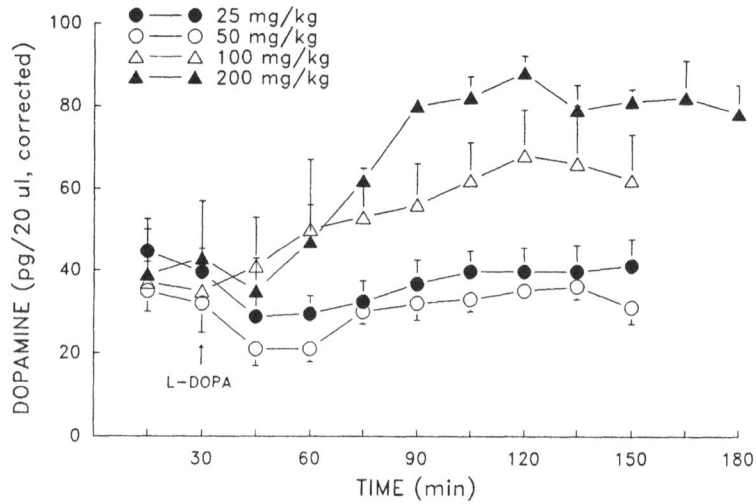

Fig. 2. Dose-response effects of i.p. administration of L-DOPA on extracellular DA level in striatum of intact rats. Results are mean ± SEM. From Abercrombie et al. (1990).

release could result from spillover of cytoplasmic DA either from dopaminergic terminals after saturation of the vesicular storage mechanism and/or monoamine oxidase or from non-dopaminergic sites (Melamed and Dafni, 1982; Melamed et al., 1985). The increase in 3-methoxytyramine formation following a high dose of L-DOPA is not affected by administration of direct DA agonists such as apomorphine, supporting the view that basal DA release is not under the control of impulse activity under these conditions (Ponzio et al., 1983).

IMPULSE DEPENDENCE OF DOPAMINE RELEASE AFTER EXOGENOUS L-DOPA

The above data highlight the possibility that low doses of L-DOPA enhance the storage pool of DA in dopaminergic neurons so that stimulation-evoked release of DA is selectively enhanced whereas higher doses of L-DOPA lead to impulse-independent spillover of DA (see also Globus and Melamed, 1983; Hefti et al., 1981a). We have tested the latter proposal by examining the extent to which DA release observed in response to 100 mg/kg L-DOPA is affected by concomitant application of tetrodotoxin (TTX), an agent that inhibits impulse activity by blocking voltage-sensitive sodium channels.

After the basal level of DA in the striatal dialysates was established, the perfusate solution was switched to one containing 10^{-5} M TTX. Addition of TTX to the perfusate typically reduced extracellular DA level to below the detection limit of the assay (0.5 pg DA) within 60 - 90 min. At this time, L-DOPA was administered either systemically (100 mg/kg, i.p.) or via the perfusion medium (5 x 10^{-5} M).

Under control conditions, systemic L-DOPA produced a net increase in DA of 20 ± 3 pg/sample relative to the basal level of 52 ± 8 pg/sample (n=5; all values are corrected for relative probe recovery). In the presence of TTX, L-DOPA increased DA from non-detectable levels to 23 ± 3 pg/sample (n=6). Thus, the absolute increase in extracellular DA produced by systemic L-DOPA administration did not differ significantly in the two groups, providing evidence that DA formed from exogenous L-DOPA in striatum is released in an impulse-independent manner (Figure 3). Qualitatively similar results were obtained when L-DOPA was administered directly via the dialysis probe, controlling for the possibility that the DA measured in the presence of TTX after systemic L-DOPA had been released outside the sphere of influence of the TTX.

METABOLISM OF L-DOPA IN DOPAMINE-DEPLETED STRIATUM

There is reason to hypothesize that the fate of DA formed from exogenous L-DOPA may differ in the intact and DA-depleted striatum. For example, the increase in tissue DA level measured in striatum after L-DOPA administration is lower in animals with large 6-HDA-induced depletions of striatal DA than in control animals (Hefti et al., 1981a; Langelier et al., 1973; Lloyd et al., 1975; Rinne et al., 1971). However, administration of L-DOPA to 6-HDA treated animals produces significant behavioral activation whereas relatively few, if any, behavioral effects of L-DOPA are observed in intact animals (Ervin et al., 1977; Hollister et al., 1979; Schoenfeld and Uretsky, 1973). Thus, we have examined the response to L-DOPA in animals with depletions of DA in striatal tissue produced by treatment with 6-HDA. Depletion of striatal DA with 6-HDA in the rat has proven to be a valuable animal model of Parkinson's disease (Zigmond and Stricker, 1989).

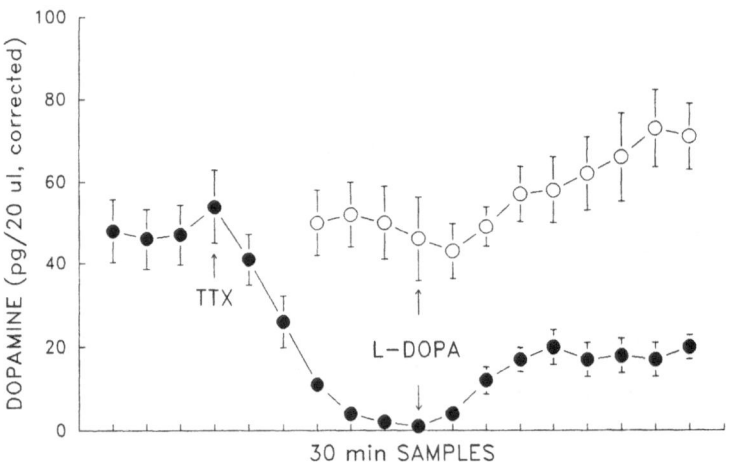

Fig. 3. Effect of 100 mg/kg i.p. L-DOPA administered alone (open circles) and in the presence of 10^{-5} M TTX (filled circles) on extracellular DA in striatum. The absolute increase in extracellular DA level did not differ significantly in the two conditions. Results are mean ± SEM.

Extracellular DA was evaluated after administration of L-DOPA (100 mg/kg) in rats with severe (>80%) striatal tissue DA depletion. Animals that had received i.c.v. 6-HDA bilaterally (n=6; mean tissue DA depletion=87%) were compared to a group of intact control animals (n=7) whereas the DA-depleted striatum of animals treated unilaterally with 6-HDA was compared to the intact striatum of the same animal (n=6; mean tissue DA depletion=96%).

L-DOPA produced a significantly greater increase in extracellular DA in animals with bilateral DA depletions than in controls. DA was increased from 37±5 pg/sample to 68 ±11 pg/sample in control animals compared to an increase from 8 ± 3 pg/sample to 266 ± 60 pg/sample in the 6-HDA-treated animals. The behavioral response to L-DOPA also was greatly augmented in the 6-HDA-treated animals. Responses included extreme hyperactivity with sniffing and rearing as well as frequent periods of intense stereotyped motor patterns. A similar phenomenon was observed in animals with large unilateral depletions of striatal DA. L-DOPA produced significant increases in extracellular DA on both the intact side (from 28 ± 11 pg/sample to 61 ± 8 pg/sample) and on the DA-depleted side (from 7 ± 4 pg/sample to 245 ± 67 pg/sample). However, the absolute magnitude of the increase was significantly greater on the lesion side (Figure 4). The animals treated unilaterally with 6-HDA displayed pronounced contralateral circling bahavior in response to 100 mg/kg L-DOPA, with an average peak of 24 ± 6 full turns per 5 min.

Thus, in rats sustaining severe depletions of striatal tissue DA (>80%), administration of 100 mg/kg L-DOPA produced an increase in extracellular DA that was greater, in absolute terms, than that observed in intact rats. This phenomenon was observed regardless of whether the tissue DA depletion was bilateral or unilateral (Abercrombie et al., 1990). Tissue DA levels following L-DOPA administration were not measured in the present study. Previous

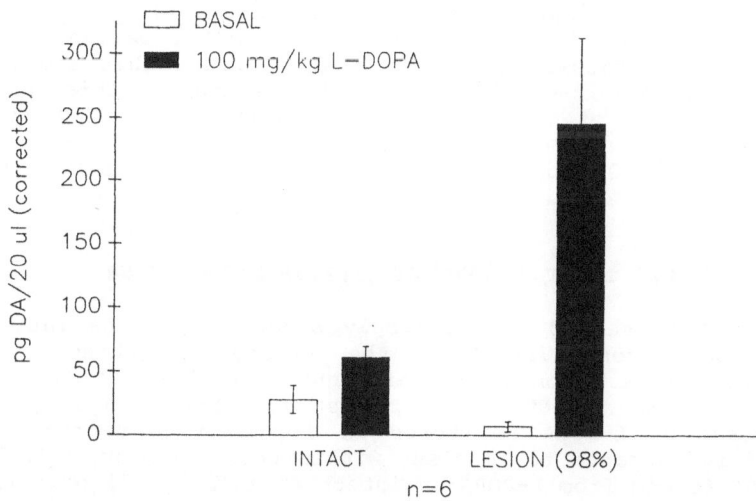

Fig. 4. Effect of L-DOPA on extracellular DA in the intact and DA-depleted striata of animals with unilateral depletions of striatal DA produced with 6-HDA. L-DOPA elicited a significant increase in DA on both sides. The increase on the DA-depleted side, however, was significantly greater than that observed on the intact side. From Abercrombie et al. (1990).

reports have shown, however, that L-DOPA increases absolute levels of tissue DA by less in Parkinsonian or 6-HDA-treated striata than in intact striata (see above; also Yahr et al., 1972). In contrast, the present data show that extracellular DA was increased by more, in absolute terms, in DA-depleted striata than in intact striata. Taken together, these data support the concept that a greater proportion of the DA that is formed from L-DOPA in the DA-depleted striatum gains access to the extracellular fluid. In agreement with the results obtained with TTX (see above), these observations would be predicted if the extracellular DA observed after L-DOPA in DA-depleted striata was not stored in vesicles but rather was released in an impulse-independent fashion. We hypothesize that this spillover of DA into extracellular fluid, along with the loss of high-affinity DA uptake sites associated with the decreased number of dopaminergic nerve terminals, combine to produce supranormal levels of extracellular DA following L-DOPA in DA-depleted striata.

ROLE OF HIGH AFFINITY DOPAMINE UPTAKE IN THE EFFECTS OF L-DOPA

We have further examined the hypothesis that a major factor in the large elevation of extracellular DA obtained following L-DOPA administration in DA-depleted striata is the loss of high affinity DA uptake sites. To do so, we measured extracellular DA in intact rats after the administration of L-DOPA (100 mg/kg, i.p.) and/or nomifensine (5 mg/kg, i.p.), an inhibitor of high affinity DA uptake. An increase in extracellular DA was observed after administration of nomifensine. This increase was significant both when the drug was administered alone and when administered in combination with L-DOPA (Figure 5). Indeed, the increase in extracellular DA produced by nomifensine plus L-DOPA in intact animals and the peak effect of L-DOPA alone in the severely DA depleted animals did not differ significantly from one another although both were significantly greater than the effects observed in intact animals given either L-DOPA alone or nomifensine alone. The present data show, therefore, that partial blockade of high affinity DA uptake in combination with L-DOPA treatment produces increases in extracellular DA in intact striata that are similar to those observed in 6-HDA-treated striata. This provides evidence in favor of the hypothesis that the loss of high affinity DA uptake sites is a major factor in the differential effect of a high dose of L-DOPA on extracellular DA level in intact vs. DA-depleted striata.

IMPLICATIONS FOR THE TREATMENT OF PARKINSON'S DISEASE

The data presented here strongly suggest that the fate of DA formed from exogenous L-DOPA in the striatum may change over the course of degeneration of DA afferents to that structure. We propose that in intact animals, most DA formed in response to lower doses of L-DOPA is stored in dopaminergic neurons and is released in response to impulses within these neurons. At higher doses, DA formed from L-DOPA in intact animals spills over from DA neurons or from non-dopaminergic striatal elements and rapidly is inactivated by the large population of sites for high affinity DA uptake. As the number of DA nerve terminals declines, a larger proportion of DA formed from L-DOPA is derived from cells that do not have the capacity to store DA and although the absolute amount of DA formed is less, the appearance of DA in the extracellular

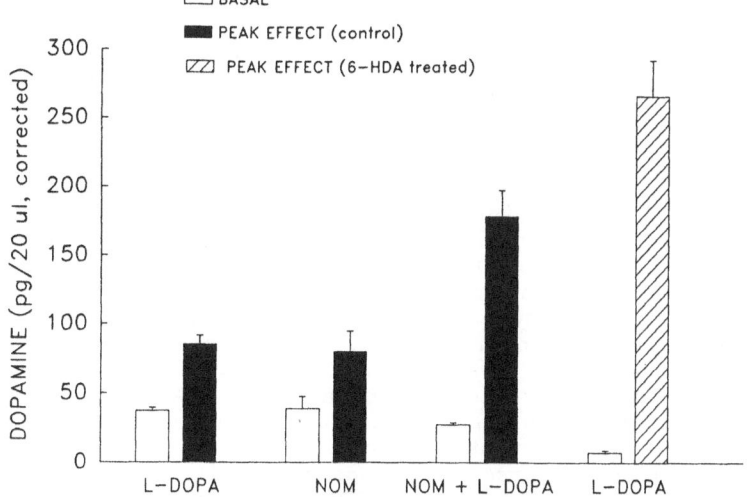

Fig. 5.   Effect of blockade of high affinity DA uptake with nomifensine
(NOM; 5 mg/kg, i.p.) on the extracellular DA response to 100
mg/kg L-DOPA in intact rats. Nomifensine plus L-DOPA produced
an increase in extracellular DA in intact rats (n=5) that was
greater than that seen after L-DOPA (n=7) or nomifensine (n=4)
alone and that was not significantly different from the
increase observed after L-DOPA alone in in bilaterally 6-HDA-
treated animals (n=6). From Abercrombie et al. (1990).

fluid is enhanced due to decreased inactivation by high affinity
DA uptake.   Under these conditions, DA release is uncoupled from
dopaminergic neuronal activity.   This model is consistent with the
proposals of several other investigators (Melamed, 1988; Ng et
al., 1972; Trugman and Wooten, 1986).

If the above model is correct, then several observations in
the literature are explained.   In this and previous studies,
behavioral activation is observed following administration of L-
DOPA to DA-depleted animals whereas few behavioral effects of
L-DOPA are observed in intact animals.   In general, these results
previously have been attributed to postsynaptic supersensitivity
of striatal DA receptors in the lesioned animals.   However, we
have observed that behavioral supersensitivity to L-DOPA occurs
both sooner and at smaller depletions of striatal DA than do
increases in postsynaptic receptor sensitivity (Zigmond and
Stricker, 1980).   Thus we believe that a shift in the distribution
of DA formed from L-DOPA in favor of the extracellular compartment
in DA-depleted animals together with the loss of high affinity DA
uptake sites also contributes to this phenomenon.   This difference
in the fate of DA derived from exogenous L-DOPA in intact striatum
as compared with DA-depleted striatum also may partially explain
the results of Trugman and Wooten (1986) and of Porrino and coworkers (1987).   These investigators examined alterations in
regional cerebral glucose utilization in rats with unilateral
lesions of the substantia nigra and in MPTP treated monkeys,
respectively, and observed that in many structures the rate of
glucose utilization following L-DOPA administration in DA depleted
animals was increased to levels that <u>exceeded</u> rates measured in

normal animals. Importantly, in the latter study, this effect was observed following chronic administration of L-DOPA and thus probably cannot be attributed to postsynaptic receptor supersensitivity since such treatments have been reported to cause receptor subsensitivity (Guttman and Seeman, 1985; Reches et al., 1984).

The supranormal extracellular levels of DA that were observed in DA-depleted striata in response to L-DOPA administration in the present study also may have important implications for understanding the loss of therapeutic efficacy that occurs with long-term L-DOPA treatment in Parkinson's disease. It has been suggested that this decline in therapeutic efficacy is associated not with the severity of the disease, but rather with the duration of L-DOPA treatment (Barbeau, 1969; Lesser et al., 1979). Abnormally elevated extracellular levels of DA in response to L-DOPA administration, coupled with inadequate means of inactivating this DA, could have potentially toxic effects. The catabolism of DA by monoamine oxidase is associated with the formation of hydrogen peroxide, a compound that can contribute to cellular destruction. Moreover, it is known that catabolic pathways exist for catecholamines in addition to oxidative deamination by monoamine oxidase and O-methylation by catechol-O-methyltransferase and that these oxidative processes can lead to the formation of toxic species (Graham, 1978). Indeed, there is increasing evidence that the formation of free radicals may contribute to the primary pathology of Parkinson's disease (Dexter et al., 1989; Riederer et al., 1989; Spina and Cohen, 1989). Although one study has demonstrated that long-term administration of L-DOPA does not damage dopaminergic neurons in the mouse (Hefti et al., 1981b), this study was conducted in intact animals in which the ability to inactivate DA via high affinity uptake was not impaired. More recently, it was reported that chronic L-DOPA treatment decreased the viability of fetal DA neuron grafts placed in the DA-depleted striatum of rats (Steece-Collier et al., 1989).

We propose that a pronounced loss of DA inactivation may enhance the oxidative stress produced by elevated extracellular DA levels after L-DOPA administration in the Parkinsonian striatum and that L-DOPA therapy may therefore have two distinct consequences: (1) a rapid, short-term improvement of the symptoms of Parkinson's disease due to elevated levels of striatal DA, and (2) a gradual, long-term exacerbation of the pathology due to enhancement of oxidative stress in the striatum.

In summary, we have observed that L-DOPA administration in an animal model of Parkinson's disease leads to increases in striatal extracellular DA level, as measured with microdialysis, that are greater than the elevations observed in intact animals. This phenomenon is attributed to the formation of DA from L-DOPA in residual DA terminals and in non-dopaminergic striatal elements that is released in an impulse-independent fashion coupled with a decrease in the capacity for efficient DA inactivation due to the loss of high affinity DA uptake sites. It is possible that such a phenomenon may produce oxidative stress in the Parkinsonian striatum that contributes to the loss of therapeutic efficacy of L-DOPA with long-term use.

Acknowledgements: We thank Alfred Bonatz, Sandra Castro, Kristen Keefe, and Holly Morris for technical assistance. This work was supported in part by USPHS grants NS-19608, MH-00058, MH09658, and the National Alliance for Research on Schizophrenia and Depression.

REFERENCES

Abercrombie, E.D., Keefe, K.A., DiFrischia, D.S., and Zigmond, M.J., 1989a, Differential effect of stress on in vivo dopamine release in striatum, nucleus accumbens, and medial frontal cortex, J. Neurochem., 52:1655.

Abercrombie, E.D., Keefe, K.A., and Zigmond, M.J., 1989b, Evidence that nerve terminal density is an important contributor to apparent differences in the activation of central dopamine systems, Behav. Pharmacol., 1(Suppl.):26.

Abercrombie, E.D., Bonatz, A.E., and Zigmond, M.J., 1990, Effects of L-DOPA on extracellular dopamine in striatum of normal and 6-hydroxydopamine-treated rats, Brain Res., in press.

Barbeau, A., 1969, L-DOPA therapy in Parkinson's disease: A critical review of nine years' experience, Canad. Med. Assoc. J., 101:59.

Birkmayer, W., and Hornykiewicz, O., 1962, Der 1-dioxyphenylalanin (L-DOPA-Effekt beim Parkinson-syndrom des menschen: Zur pathogenese und behadlung der Parkinson-akinese, Arch. Psychiat. Nervenkr., 203:560.

Bunney, B.S., Aghajanian, G.K., and Roth, R.H., 1973, Comparison of effects of L-DOPA, amphetamine and apomorphine on firing rate of rat dopaminergic neurones, Nature, 245:123.

Dexter, D.T., Carter, C.J., Wells, F.R., Javoy-Agid, F., Agid, Y., Lees, A., Jenner, P., and Marsden, C.D., 1989, Basal lipid peroxidation in substantia nigra is increased in Parkinson's disease, J. Neurochem., 52:381.

Ervin, G.N., Fink, J.S., Young, R.C., and Smith, G.P., 1977, Different behavioral responses to L-DOPA after anterolateral or posterolateral hypothalamic injections of 6-hydroxydopamine, Brain Res., 132:507.

Globus, M., and Melamed, E., 1983, Combined administration of direct dopamine agonists and L-DOPA does not interfere with utilization of exogenous L-DOPA in rat corpus striatum, Prog. Neuropsychopharm. Biol. Psychiatry, 7:211.

Graham, D.G., 1978, Oxidative pathways for catecholamines in the genesis of neuromelanin and cytotoxic quinones, Mol. Pharmacol., 14:633.

Guttman, M., and Seeman, P., 1985, L-DOPA reverses the elevated density of D_2 dopamine receptors in Parkinson's diseased striatum, J. Neural. Trans., 64:93.

Hefti, F., Melamed, E., and Wurtman, R.J., 1981a, The site of dopamine formation in rat striatum after L-DOPA administration, J. Pharmacol. Exp. Ther., 217:189.

Hefti, F., Melamed, E., Bhawan, J., and Wurtman, R.J., 1981b, Long-term administration of L-DOPA does not damage dopaminergic neurons in the mouse, Neurology, 31:1194.

Hollister, A.S., Breese, G.R., and Mueller, R.A., 1979, Role of monoamine neural systaems in L-dihydroxyphenylalanine-stimulated activity, J. Pharmacol. Exp. Ther., 208:37.

Imperato, A., and Di Chiara, G., 1984, Trans-striatal dialysis coupled to reverse-phase high performance liquid chromatography with electrochemical detection: A new method for the study of the in vivo release of endogenous dopamine and metabolites, J. Neurosci., 4:966.

Langelier, P., Roberge, A.G., Boucher, R., and Poirier, L.J., 1973, Effects of chronically administered L-DOPA in normal and lesioned cats, J. Pharmacol. Exp. Ther., 187:15.

Lesser, R.P., Fahn, S., Snider, S.R., Cote, L.J., Isgreen, W.P., and Barrett, R.E., 1979, Analysis of the clinical problems in parkinsonism and the complications of long-term levodopa therapy, Neurology, 29:1253.

Lloyd, K.G., Davidson, L., and Hornykiewicz, O., 1975, The neurochemistry of Parkinson's disease: Effect of L-DOPA therapy, J. Pharmacol. Exp. Ther., 195:453.

Melamed, E., 1988, Mechanism of action of exogenous L-DOPA: Is it a physiological therapy for Parkinson's disease?, in: "Parkinson's Disease and Movement Disorders," J. Jankovic and E. Tolosa, eds., Urban & Schwarzenberg, Baltimore-Munich.

Melamed, E., and Dafni, N., 1982, Effect of electrical stimulation of nigrostriatal dopaminergic neurons on utilization of exogenous L-DOPA in rat corpus striatum. J. Pharm. Pharmacol., 34:820.

Melamed, E., Hefti, G., Bitton, V., and Globus, M., 1984, Suppression of L-dopa-induced circling in rats with nigral lesions by blockade of central dopa-decarboxylase: Implications for mechanism of action of L-dopa in Parkinsonism, Neurology, 34:1566.

Melamed, E., Globus, M., Uzzan, A., and Rosenthal, J., 1985, Is dopamine formed from exogenous L-DOPA stored within vesicles in striatal dopaminergic nerve terminals: Implications for L-DOPA's mechanism of action in Parkinson's disease, Neurol., 35 (Suppl.):118.

Ng, K.Y., Chase, T.N., Colburn, R.W., and Kopin, I.J., 1972, L-DOPA in parkinsonism: A possible mechanism of action, Neurology, 22:688.

Ponzio, F., Achilli, G., Perego, C., Rinaldi, G., and Algeri, S., 1983, Does acute L-DOPA increase active release of dopamine from dopaminergic neurons?, Brain Res., 273:45.

Porrino, L.J., Burns, R.S., Crane, A.M., Palombo, E., Kopin, I.J., and Sokoloff, L., 1987, Local cerebral metabolic effects of L-dopa therapy in 1-methyl-4-phenyl-1,2,3,6-tetrahydropyridine-induced parkinsonism in monkeys, Proc. Nat. Acad. Sci. USA, 84:5995.

Reches, A., Wagner, H.R., Jackson-Lewis, V., Yablonskaya-Alter, E., and Fahn, S., 1984, Chronic levodopa or pergolide administration induces down-regulation of dopamine receptors in denervated striatum, Neurology, 34:1208.

Riederer, P., Sofic, E., Rausch, W.-D., Schmidt, B., Reynolds, G.P., Jellinger, K., and Youdim, M.B.H., 1989, Transition metals, ferritin, glutathione, and ascorbic acid in Parkinsonian brains, J. Neurochem., 52:515.

Rinne, U.K., Sonninen, V., and Hyyppa, M., 1971, Effect of L-DOPA on brain monoamines and their metabolites in Parkinson's disease, Life Sci., 10:549.

Schoenfeld, R.I., and Uretsky, N.J., 1973, Enhancement by 6-hydroxydopamine of the effects of DOPA upon the motor activity of rats, J. Pharmacol. Exp. Ther., 186:616.

Spina, M.B., and Cohen, G., 1989, Dopamine turnover and glutathione oxidation: Implications for Parkinson's disease, Proc. Nat. Acad. Sci. USA, 86:1398.

Steece-Collier, K., Collier, T.J., Sladek, C.D., and Sladek, J.R., 1989, Chronic L-DOPA treatment decreases the viability of grafted and cultured embryonic rat mesencephalic dopamine neurons, Soc. Neurosci. Abstr., 15:1354.

Trugman, J.M., and Wooten, G.F., 1986, The effects of L-DOPA on regional cerebral glucose utilization in rats with unilateral lesions of the substantia nigra, Brain Res., 379:264.

Ungerstedt, U., 1984, Measurement of neurotransmitter release by intracranial dialysis, in: "Measurement of Neurotransmitter Release In Vivo," C.A. Marsden, ed., Wiley, New York.

Yahr, M.D., Wolf, A., Antunes, J.-L., Miyoshi, K., and Duffy, P., 1972, Autopsy findings in parkinsonism following treatment with levodopa, Neurology, 20 (Suppl.):56.

Zigmond, M.J., and Stricker, E.M., 1980, Supersensitivity after intraventricular 6-hydroxydopamine: Relation to dopamine depletion, Experientia, 36:436.

Zigmond, M.J., and Stricker, E.M., 1989, Animal models of Parkinsonism Using Selective Neurotoxins: Clinical and Basic Implications, Int. Rev. Neurobiol., 31:1.

THE PHARMACOKINETICS OF L-DOPA IN PLASMA AND CSF OF THE MONKEY

John P. Hammerstad, William R. Woodward
Perry Gliessman, Brian Boucher, and John G. Nutt

Oregon Health Sciences University, Portland, Oregon
Oregon Regional Primate Research Center, Beaverton
Oregon

INTRODUCTION

L-DOPA is the single most effective therapeutic agent in the treatment of Parkinson's disease (PD). After years of treatment, however, its efficacy is mitigated by the development of a fluctuating response ("on-off") that may be as disabling as the disease. There are no changes in the peripheral pharmacokinetics of L-DOPA to explain the development of fluctuations[1]. Therefore, a change in the central pharmacokinetics or pharmacodynamics of L-DOPA and dopamine are postulated to underlie the appearance of fluctuations[2-6]. When a constant plasma concentration of L-DOPA is maintained by intravenous infusion in Parkinsonian patients, the motor fluctuations may be abolished for hours to days, indicating that the clinical response is dependant on the plasma level and that pharmacokinetic mechanisms contribute to the clinical fluctuations[7]. The short plasma half-life of L-DOPA and the regulation of its entry into brain by carrier-mediated transport may become critical factors when the decreased striatal storage of dopamine due to the loss of dopaminergic nerve terminals makes motor function dependant on the constant delivery of L-DOPA to striatum to support the continuous synthesis of dopamine[8].

L-DOPA reaches its site of action in the striatum via a saturable carrier-mediated transport system that resides in the capillary endothelium and is shared with other large neutral amino acids[9]. A similar transport system is present in choroid plexus[10], which is thought to regulate amino acid levels in cerebrospinal fluid (blood-CSF barrier)[11]. There are no barriers between brain extracellular fluid (ECF) and CSF, and the two are thought to be essentially identical in composition[12]. Because of these similarities in the transport of amino acids at the blood-brain and blood-CSF barriers and the importance of the CSF in both regulating and reflecting the neuronal microenvironment, the pharmacokinetics of L-DOPA in CSF in Rhesus monkeys (Macaca mulata) was studied as a more direct indication of its disposition in brain. The kinetics in

The Basal Ganglia III, Edited by G. Bernardi *et al.*
Plenum Press, New York

cisternal and lumbar CSF were compared to determine the
feasibility of using lumbar CSF to study the central kinetics
of L-DOPA in patients.

METHODS

Two spayed, adult, female Rhesus monkeys were used. The
animals were anesthetized with Halothane and nitrous oxide.
Percutaneous canulae were placed in the cisterna magna for
withdrawal of CSF and in both saphenous veins, one for infusion
of L-DOPA and the other for sampling blood. When lumbar CSF
was sampled an additional cannula was placed in the lumbar
theca. A constant intravenous infusion of L-DOPA was employed
in order to duplicate the clinical studies previously reported
by our group[1,4,8]. Different rates of infusion produced a range
of L-DOPA levels that approximated the low therapeutic range in
patients.

RESULTS AND DISCUSSION

A typical profile of L-DOPA levels in plasma and in
cisternal and lumbar CSF is illustrated in Figure 1. Plasma
levels reached a plateau within 30-60 minutes, whereas
cisternal CSF levels rose more slowly, achieving stable levels
within 60-90 minutes. In contrast to plasma and cisternal CSF,
L-DOPA levels in lumbar CSF were still rising at 180 minutes.

The precise regulation of CSF secretion and solute
exchange by the brain capillary endothelium and by the choroid
plexus epithelium serve to maintain a stable composition of
brain ECF and CSF. The transport of amino acids including
L-DOPA from CSF to blood by choroid plexus is considered an

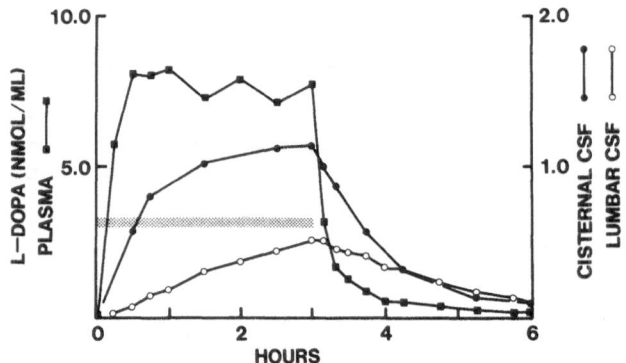

Figure 1. Concentrations of L-DOPA
(nmol/ml)in plasma (filled squares, left
ordinate), in cisternal CSF (filled
circles, right ordinate),and lumbar CSF
(open circles, right ordinate) during a 3
hour infusion at a rate of 8.5 mol/kg/hr.
The stippled bar represents the duration of
the infusion.

important mechanism for maintaining the low concentration of amino acids in CSF and, by inference, brain extracellular fluid[11,13]. An example of this regulation of solute exchange is seen in the low concentration of L-DOPA in CSF relative to plasma (17±6%, n=14, for cisternal and 7±4%, n=3, for lumbar CSF, Fig. 1). The L-DOPA infusions produced plateau levels in plasma and cisternal CSF that were proportional to the infusion rate. Moreover, there was a good correlation between plateau levels of L-DOPA in cisternal CSF and in plasma (Fig. 2). This relationship between plasma and CSF levels implies that at steady state the plasma concentrations are a reasonable index of cisternal CSF L-DOPA.

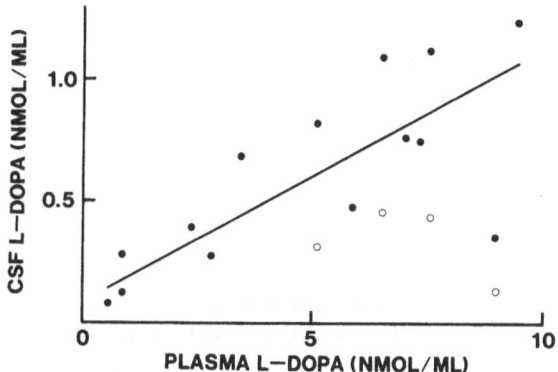

Figure 2. Correlation of the plateau concentrations of L-DOPA in plasma with those in cisternal CSF (filled circles, r^2=0.92) and in lumbar CSF (open circles). Each point represents a separate infusion, with 7 experiments each in the two animals. A correlation coefficient could not be estimated for lumbar CSF.

The disappearance of L-DOPA from cisternal CSF could be resolved into two kinetic components, a rapid distribution phase and a slower elimination phase. The distribution and elimination of L-DOPA in cisternal CSF approach that of plasma (Table 1). These kinetics are very similar to the kinetics of L-DOPA in ventricular CSF after an oral dose of L-DOPA in MPTP-treated monkeys[14]. Moreover, the ventricular CSF L-DOPA levels paralleled the motor effects of the drug in these animals.

Several related implications may be drawn from these observations. The rapid turnover of L-DOPA in ventricular and cisternal CSF may be representative of L-DOPA disposition in brain. Although there is almost a two-fold difference in the kinetics of distribution between plasma and CSF, the absolute disparity of a few minutes is probably not enough to dampen the oscillations in the plasma level and the delivery of L-DOPA to the brain. Therefore if the amount of synaptic dopamine is dependant on the moment to moment availability of L-DOPA, the

TABLE 1

Pharmacokinetics of L-DOPA in Plasma and CSF[1]

Compartment

	Plasma	Cisternal CSF	Lumbar CSF
Half-Lives			
Distribution (min)	4.9±1.6 (13)	8.9±3.5 (9)[2]	- - -
Elimination (min)	33.2±13.2 (13)	49.2±13.4 (9)	100.4±26.6 (3)[3]
Clearance (L/kg/hr)	1.7±0.7 (14)		
Apparent Volume of Distribution (L/kg)	0.49±0.19 (14)		

1 Values are expressed as the mean ± $sd_{(n-1)}$. Number of experiments are given in parentheses.

2 Distribution half-life in cisternal CSF was significantly greater than that in plasma (p=0.001; df=20).

3 Elimination half-life in lumbar CSF was significantly greater than that in cisternal CSF (p=0.001; df=10).

short plasma half-life of L-DOPA could account for the short-term fluctuations in the motor responses to the orally administered drug in patients with severe PD. Finally, because the pharmacokinetics of L-DOPA in plasma are similar to those of L-DOPA in cisternal CSF, there may be no advantage in monitoring cisternal CSF over measuring plasma drug levels in clinical studies.

The marked difference in the kinetics of L-DOPA in cisternal and lumbar CSF (Table 1) indicates that L-DOPA enters the cerebral ventricular CSF and is transported by bulk flow to the lumbar space. This is in accord with evidence that CSF is produced only within the cranial compartment. While there is substantial agreement that the majority of CSF is secreted by choroid plexus, there is less certainty about the proportion of CSF and ECF contributed by extrachoroidal sources, which by some estimates is as high as 40%, most coming from the parenchymal capillary bed (for further discussion of the evidence see Davson et. al[12]). The contribution to CSF production from capillaries in pia-arachnoid is uncertain but is probably negligible since there appears to be no CSF production in the spinal subarachnoid space[15]. Therefore, spinal pia-arachnoid and cord parenchyma is not a source for CSF and a difference in bulk of the tissue compartment through which L-DOPA might flux before reaching the CSF would not explain the marked contrast in kinetics. Although there is evidence of saturable transport of another LNAA, phenylalanine, by pia-arachnoid[16], the slow appearance of L-DOPA in lumbar CSF probably reflects simple diffusion from the cranial compartment, the site of CSF formation and solute exchange. A similar delayed appearance of radioactive L-DOPA in lumbar CSF in comparison to cisternal CSF after an intravenous bolus injection has been noted in an earlier study[17]. The removal of leucine[11] out of the spinal subarachnoid space by a process consistent with transport has been demonstrated, but the slow rate of disappearance is against a more rapid rate of removal by transport or metabolism as a reason for the lower concentration and slower equilibration. The clinical implication is that lumbar CSF concentrations of L-DOPA will not reflect dynamic changes in ventricular CSF nor, presumably, brain extracellular fluid, and therefore lumbar CSF will not be suitable for studying the central pharmacokinetics of L-DOPA in patients.

REFERENCES

1. Gancher ST, Nutt JG, Woodward WR. Peripheral pharmacokinetics of levodopa in untreated,stable, and fluctuating parkinsonian patients. Neurology 1987;37:940-944.
2. Shoulson I, Glaubiger GA, Chase TN. On-off response: clinical and biochemical correlations during oral and intravenous levodopa administration in parkinsonism patients. Neurology 1975;24:1144-1148.
3. Hardie RJ, Lees AJ, Stern GM. On-off fluctuations in Parkinson's disease: a clinical and neuropharmacological study. brain 1984;107:487-506.
4. Nutt JG, Woodward WR, Hammerstad JP, Anderson JL, Carter JH. The "on-off" phenomenon in Parkinson's disease: relation to levodopa absorption and transport. N Engl J Med 1984;310:483-488.
5. Quinn NP, Parkes D, Marsden CD. Control of on/off phenomenon by continuous intravenous infusion of levodopa. Neurology 1984;34:1131-1136.
6. Juncos J, Serrati C, Fabbrini G, Chase TN. Fluctuating levodopa concentrations and Parkinson's disease. lancet 1985;2:440-440.
7. Marsden CD. "On-off" phenomenon in Parkinson's disease. In: Rinne UK, Klinger M, Stamm G, eds. Parkinson's disease: Current progress, problems and management. Amsterdam: Elsevier, 1980:241-254.
8. Nutt JG, Woodward WR. Levodopa pharmacokinetics and pharmacodynamics in fluctuating parkinsonian patients. Neurology 1986;36:739-744.
9. Pardridge WM. Kinetics of competitive inhibition of neutral amino acid transport across the blood-brain barrier. J Neurochem 1977;28:103-108.
10. Lorenzo AV, Cutler RWP. Amino acid transport by choroid plexus in vitro. J Neurochem 1969;16:577-585.
11. Snodgrass SR, Cutler RWP, Kang BS, Lorenzo AV. Transport of neutral amino acids from feline cerebrospinal fluid. Am J Physiol 1969;217:974-980.

12. Davson H, Welch K, Segal MB. Physiology and Pathophysiology of the Cerebrospinal Fluid. Edinburgh:Churchill Livingstone, 1987:

13. Lindvall M, Hardebo JE, Owman CH. Barrier mechanisms for neurotransmitter monoamines in the choroid plexus. Acta Physiologica Scandinavica 1980;108:215-221.

14. Burns RS, Chiueh CC, Kopin IJ. Brain kinetics of L-dopa in the MPTP-treated monkey model of Parkinson's disease. Neurology 1985;35 (Suppl 1):224-224.

15. Hammerstad JP, Lorenzo AV, Cutler RWP. Iodide transport from the spinal subarachnoid space in the cat. Am J Physiol 1969;216:353-358.

16. Levin E, Sepulveda FV, Yudilevich DL. Pial vessels transport of substances from cerebrospinal fluid to blood. Nature 1974;249:266-267.

17. Pletscher A, Bartholini G, Tissot R. Metabolic fate of L-[^{14}C]dopa in cerebrospinal fluid and blood plasma of humans. Brain Res 1967;4:106-109.

TREATMENT OF MOTOR FLUCTUATIONS IN PARKINSON'S DISEASE:

CONTROLLED RELEASE PREPARATIONS

T. Caraceni, G. Geminiani, S. Genitrini. P. Giovannini, F. Girotti, D. Oliva and F.Tamma

Centre for the Study of Parkinson's Disease and Extrapyramidal Diseases, Istituto Neurologico Nazionale "C. Besta", Milan, Italy

INTRODUCTION

The fundamental problem in the treatment of Parkinson's disease is the appearance of motor fluctuations after long-term drug therapy (long-term L-dopa treatment syndrome, LTS, Marsden and Parkes, 1976). The clinical manifestations of LTS are complex, consisting of a varying association of akinetic symptoms and dyskinesias whose pathogenesis has still not been clarified.

From the therapeutic point of view it is useful to classify LTS phenomena according to their presumed correlation with dopaminergic stimulation, since most of the akinetic and hyperkinetic symptoms can be correlated with fluctuations in L-dopa plasma levels (Hardie et al, 1984); although other LTS manifestations (the random off phenomenon, lack of response to a single dose and paradoxical akinesia) appear independent of drug concentration in blood (Table 1).

TABLE 1. Classification of long-term l-dopa
         treatment symptoms
----
(1) Phenomena correlated with increased plasma L-dopa:
     (a) peak dose hyperkinesia
     (b) beginning of dose hyperkinesia
(2) Phenomena correlated to reduction in plasma L-dopa:
     (a) end-of-dose deterioration or wearing-off
     (b) off phase of predictable on-off
     (c) nocturnal akinesia
     (d) end of dose dyskinesias
     (e) nocturnal dystonia and early morning dystonia
(3) Phenomena unrelated to changes in L-dopa plasma levels
     (a) random off
     (c) lack of response to a single dose
     (d) akinesia paradoxica

Stabilisation of dopaminergic stimulation is expected to ease the symptoms related to oscillations in drug plasma levels, so that this objective is often pursued for selected patients in a clinical setting. Currently there are three ways by which a relative stabilisation of dopaminergic stimulation is achieved:

(i) intravenous, subcutaneous or transcutaneous administration of dopaminagonist drugs (Obeso et al, 1983);

(ii) use of long-acting dopaminagonists;

(iii) using controlled (or slow) release levodopa preparations.

We report here our experience with the controlled release preparation Madopar HBS. Each capsule of Madopar HBS contains 100 mg of levodopa and 25 mg of benserazide in a formulation which, after dissolution of the capsule wall in the stomach, forms a "hydrodynamically balanced system" (HBS) - a mass of hydrated gelatin. The mass is buoyant in gastric fluid and remains in the stomach a long time. At low pH the drugs are released slowly after diffusion through the hydrated layer (Erni and Held, 1987). The bioavailability of levodopa in Madopar HBS on oral administration is lower than that from standard Madopar, most probably because of incomplete absorption.

Several clinical studies have already been conducted with Madopar HBS: either using a single administration (Marion et al, 1987), or completely substituting standard Madopar (Chouza et al, 1987; Jensen et al, 1988; Ludin 1987; Nordera et al, 1987; Rondot et al, 1987), or combining both standard and HBS preparations (Quinn, 1987; Rinne, 1987; Lees, 1987). Other workers have investigated the controlled release preparation Sinemet CR which contains levodopa plus carbidopa (Friedman and Lannon, 1989; Cedarbaum et al, 1987; Hutton et al, 1988; Junkos et al, 1987). These studies have demonstrated that slow release formulations are able to reduce off periods without greatly increasing hyperkinesia in more than 50% of the subjects treated, though they do entail an increase in the mean daily levodopa dosage - by about 60% when HBS completely substitutes standard Madopar. Dose-related fluctuations are improved by the treatment, a delay in the onset of the clinical effect is noticed, and the response lasts longer but is less predictable.

In the light of these considerations we investigated the efficacy of controlled release Madopar HBS in an open short-term study in patients with Parkinson's disease complicated by akinetic-type motor fluctuations correlatable with reduced levodopa plasma levels.

PATIENTS AND METHODS

Subjects. Thirty-nine patients (20 men and 19 women) with idiopathic Parkinson's disease complicated by LTS took part; twenty-one were taking levodopa plus PDI, ten were taking dopaminagonists as well, six were also on anticholinergics and six were on regimes incorporating L-deprenyl. Average age was 62.8 (±8.60) years, mean duration of illness was 10.7 (±5.4) years, Hoehn and Yahr clinical stage was 3.3 (±0.7) and average length of treatment was 8.6 (±4.3) years.

On the basis of past experience three LTS symptoms were identified as potentially responding to controlled release preparations and patients were therefore divided into three groups according to their target symptom to be treated:

- Group I.   Nine patients with wearing off
- Group II. Fourteen patients with on-off
- Group III. Eleven patients with nocturnal akinesia

Clinical data on these groups is compared in Table 2. Additionally 2 subjects presented early morning dystonia as target symptom and 3 others had insomnia not correlated with nocturnal akinesia: they were treated as in group III to probe a possible effect of dopaminergic stimulation on sleep.

Methods.   Patients were maintained on a stabilised therapeutic regime for the month prior to beginning the study. In the week prior to commencement with Madopar HBS patients were asked to self-assess their target symptoms.

Those with wearing off and on-off (groups I and II) completed a diary every day for the week, noting, on an hourly basis, the off periods. After identifying relatively constant appearance times and durations, Madopar HBS was administered three hours before onset of the expected off. Initial dosage was 125 mg (1 capsule) with possible subsequent increase to a maximum of 375 mg (3 capsules). After one month of treatment there followed a week-long self assessment period.

The patients with nocturnal akinesia (group III), early morning dystonia and insomnia self-assessed the severity of their target symptoms on a 0 - 4 scale over a week. Madopar HBS treatment then began, taking 125 or 250 mg before going to bed. After one month self-assessment was repeated. During the study period the underlying therapeutic regime remained unchanged.

Table 2.   Clinical data for three groups
of Parkinson patients with LTS.

Group	I (n=9) Wearing off	II (n=14) On-off	III (n=11) Nocturnal akinesia
Age (years)	64.67 ± 9.94[b]	56.50±4.01[c]	66.27±7.31
Illness duration (years)	8.89±4.14	12.29±4.68	9.55±6.80
Stage (Hoehn & Yahr)	3.00±0.71[a]	3.57±0.65	3.24±0.47
Treatment period (years)	7.94±4.54[c]	10.57±4.16	7.45±4.23
Duration of LTS (years)	2.00±0.71	5.64±2.69	3.55±2.88
Levodopa dose + PDI (mg)	703.13±109.53	901.79±320.30[b]	647.73±284.05

[a] $p < 0.059$;   [b] $p \leq 0.05$;   [c] $p < 0.001$

RESULTS

The assessment one month after initiation of treatment was carried out for all subjects, although 11 patients dropped out after a month: 6 because of side effects and 5 at their own request because of the inefficacy of the treatment.

The mean Madopar HBS dose above basal therapy was 244.2 (±113.3) mg.

In patients with wearing off and on-off, treatment efficacy was evaluated from the number of off hours per week; those experiencing a reduction in off hours greater than 20% were considered responders.

Group I. Of the 9 patients with wearing off, 7 experienced an hourly reduction of between 27 - 54%. There was one dropout because of increased hyperkinesia. Mean dose was 222.2 (± 83.33) mg. The off hours before and after treatment are given for each patient in Table 3.

Group II. Of the 14 patients presenting on-off, 10 reported a reduction in off hours per week of between 24% and 50% There were 5 dropouts, 3 for increased hyperkinesia, 1 for perceptual disturbances and 1 for inefficacy of treatment. Mean Madopar HBS dose was 305.26 (± 149.09) mg. Off hours data are given in Table 4.

Group III. Of the 11 patients with nocturnal akinesia, early morning dystonia or insomnia, 9 received appreciable benefit from the treatment. There were 2 dropouts for lack of benefit. Mean Madopar HBS dose was 215.91 (± 58.39) mg. Of the 2 early morning dystonia patients, one experience a reduction in the phenomenon, while in the other it disappeared (Madopar HBS dosages were 250 and 125 mg respectively). None of the three patients with insomnia non-correlatable with nocturnal akinesia received any benefit from the treatment, which was therefore discontinued; one of these complained of increased night-time disquiet.

Table 3 Percentage reductions in OFF hours post-Madopar HBS treatment in group I (wearing off) patients.

Case	Off Hours Pre-Treatment	Post-Treatment	% Improvement
1	25.5	9	47.2
2	45.5	42.5	6.5
3	52	22	57.7
4	42	23	45.2
5	39.5	29	27.5
6 *	47	46	2.1
7	52	35	32.7
8	37	16.7	54.1
9	34	16	53
Average	41.6±8.68	26.58±12.55	36.27±20.61

* drop-out; p = 0.001

Table 4 Percentage reductions in OFF hours post-Madopar HBS treatment in group II (on-off) patients.

Case	Off Hours Pre-Treatment	Post-Treatment	% Improvement
1 *	54	37	31.5
2	41	25	39
3	34	28	17.7
4	44	27	38.7
5 *	35	19	45.8
6 *	52	42	19.2
7	37	28	24.3
8	44	44	0
9 *	42	20.5	50
10 *	69	60	13
11	38	24	36.8
12	34	20	41.2
13	58.5	34	42.4
14	50	37	26
Average	45.18±10.36	31.82±11.42	30.40±14.24

* drop-outs; p < 0.00001

CONCLUDING REMARKS

Before discussing the results we wish to emphasise a general point. Therapeutic trials with parkinsonians require very careful selection of patients based on the presumed pathogenesis of the symptoms to be treated. Our study evinced good responses to attempts to stabilise L-dopa plasma levels precisely because subjects were very carefully chosen.

Patients in the on-off subgroup were younger than those in the other two subgroups. This is consistent with others' findings on young onset Parkinson's disease (Quinn et al, 1987; Narabayashi et al, 1986; Giovannini et al, 1990) One possible explanation is that differing L-dopa kinetics in young people may result in higher plasma L-dopa levels being established quicker than in older patients (Yokochi, 1979); it is also possible that a young brain is more sensitive to L-dopa and hence more prone to developing fluctuations.

As the excellent results obtained for patients with wearing off demonstrate, controlled release L-dopa preparations seem above all to prolong the effect of a single administration; also - as shown by the good results for patients with on-off - they able to overcome off periods without excessive accentuation of hyperkinetic movements. Nocturnal akinesia and early morning dystonia improve when the slow release formulation is taken the night before; the effect is constant and fairly prolonged. The absence of any effect on insomnia confirms that slight but prolonged dopaminergic stimulation specifically affects nocturnal akinesia.

In summary our study confirms that, by a relatively modest increase in the daily dose of levodopa, slow release levodopa formulations are effective in relieving dose-related akinetic fluctuations without significantly increasing dyskinesias. This encouraging result is, we feel, largely due to the prior identification of certain aspects of LTS

(Nocturnal akinesia, well-defined off periods) as specific
targets for therapy. The therapy is easy administer, but
requires a little more attention on the part of the physician,
and greater compliance by patients, in order to be successful.

REFERENCES

Cedarbaum, J. M., Breck, L., Kutt, H., and McDowell, F. H.,
    1987. Controlled release levodopa/carbidopa. Sinemet
    CR4 treatment of response fluctuations in Parkinson's
    disease, Neurology, 37:1607.

Chouza, C., Romero, S., Medina, O. de, Aljanati, R.,
    Scarmelli, A., Caamano, J. L., Panizza, V. G., 1987,
    Substitution of standard Madopar by Madopar HBS in
    Parkinsonians with fluctuations, Eur.Neurol.,
    27: suppl. 1, 59.

Erni,W., and Held, K., 1987, The hydrodynamically Balanced
    System: a novel principle of controlled drug release,
    Eur. Neurol., 27: suppl. 1, 21.

Friedman, J. H., and Lannon, M. C., 1989, An open trial of
    controlled release carbidopa/L-dopa (Sinemet CR) for the
    treatment of mild-to-moderate Parkinson's disease,
    Clinical. Neuropharmacol., 12:220.

Giovannini, P., Piccolo, T., Genitrini, S., Soliveri, P.,
    Girotti, F., Geminiani, G., and Caraceni, T., 1990, Early
    onset Parkinson's disease: a personal experience,
    Movement Dis., in press.

Hardie, R. J., Lees, A. J., and Stern, G. M., 1984, On-off
    fluctuations in Parkinson's disease. A clinical and
    pharmacological study, Brain, 107:487.

Hutton, J. T., Morris, J. L., Roman, G. C., Imke, S. C., and
    Elias, J. M., 1988, Treatment of chronic Parkinson's
    disease with controlled-release carbidopa/levodopa,
    Arch. Neurol., 45: 861.

Jensen, N. O., Dupont, E., Hansen, E., Mikkelsen, B., and
    Mikkelsen, B. O., 1988, A controlled release form of
    Madopar in Parkinsonian patients with advanced disease and
    marked fluctuations in motor performance, Acta Neurol.
    Scand., 77: 422.

Junkos, J. L., Fabbrini, G., and Mouradian, M. M., 1987,
    Controlled release levodopa treatment of motor
    fluctuations in Parkinson's disease, J. Neurol. Neurosurg.
    Psychiatry, 50: 194.

Lees, A. J., 1987, A sustained-release formulation of L-Dopa
    (Madopar HBS) in the treatment of nocturnal and early-
    morning disabilities in Parkinson's disease, Eur. Neurol.,
    27: suppl.1, 126.

Ludin, H. P., 1987, Open clinical study of Madopar HBS, Eur.
    Neurol., 27: suppl.1,73. Malcom, S. L., Allen, J. G.,
    Bird, H., Quinn, N. P., Marion, M. H., Marsden, C. D.,
    O'Leary, C. G., 1987, Single-dose Pharmacokinetics of
    Madopar HBS in Patients and effect on food and antacid on
    the absorption of Madopar HBS in volunteers, Eur. Neurol.,
    27: suppl. 1, 28.

Marion, M. H., Stocchi, F., Malcom, S. L., Quinn, N. P.,
    Jenner, P., and Marsden, C. D., 1987, Single dose
    studies of a slow release preparation of Levodopa and
    benserazide (Madopar HBS) in Parkinson's disease, Eur.
    Neurol., 27: suppl.1, 54.

Marsden, C. D., and Parkes, J. D., 1976, On-off effects in patients with Parkinson's disease on chronic levodopa therapy, Lancet, i: 292.

Narabayashi, H., Yocochi, M., Iizuka, R., and Nagatsu, T., 1986, Juvenile parkinsonism, in "Handbook of Clinical Neurology, Extrapyramidal Disorders", P. J. Vinken and G. W. Bruyn, ed., Elsevier Science Publisher, Amsterdam.

Nordera, G. P., Lorizio, A., Lion, P., Durisotti, C., D'Andrea, G., and Ferro-Milone, F., 1987, Treatment of parkinsonian conditions with a controlled release form of Levodopa – Preliminary study, Eur. Neurol., 27: suppl. 1, 76.

Obeso, J. A., Luquin, M. R., Martinez-Lage, J. M., 1983, Lisuride infusion in Parkinson's disease, Ann. Neurol., 14: 134.

Quinn, N. P., Critchley, P., and Marsden, C. D., 1987, Young onset Parkinson's disease, Movement Dis., 2, 2: 73.

Quinn, N. P., Marion, N. H., and Marsden, C. D., 1987, Open study of Madopar HBS, a new formulation of Levodopa with benserazide, in 13 patients with Parkinson's disease and "on-off" fluctuations, Eur. Neurol., 27: suppl. 1, 105.

Rinne, U. K., 1987, Madopar HBS in the long-term treatment of Parkinsonian patients with fluctuations in disability, Eur. Neurol., 27: suppl. 1, 120.

Rondot, P., Ziegler, M., Aymard, N., and Holzer, J., 1987, Clinical trial of Madopar HBS in Parkinsonian patients with fluctuating drug response after long-term levodopa therapy, Eur. Neurol., 27: suppl.1, 114.

Yokochi, M., 1979, Juvenile Parkinson's disease: Pt. 2. Pharmacokinetic study, Adv. Neurol. Sci., 23: 1060.

COURSE OF MOTOR AND COGNITIVE IMPAIRMENT IN

HUNTINGTON'S CHOREA

F. Girotti, D. Oliva, V. Fetoni, D.Testa
P. Soliveri, P. Giovannini, G. Geminiani, and
T. Caraceni

Istituto Neurologico C. Besta, Milan, Italy

INTRODUCTION

Motor impairment, mental decay and psychopathological abnormalities all contribute to functional disability in Huntington's Chorea (HC) (1-3) though Mayeux et al (4) concluded that a major role is played by intellectual derangement.

As the illness progresses Parkinson-like signs and dystonia become evident too (5-6). Recent electrophysiological studies have shown that akinesia is a typical sign of Huntington disease which is present even in mildly affected or in very hyperkinetic patients (7-8).

Well known measures to assess akinesia are reaction time paradigms (9). In a previous study (10), we found significant Reaction Time (RT) and Movement Time (MT) elongation in non-demented Huntingtonians compared to normal controls and even to Parkinsonian patients.

The aim of the present study was to examine the progression of motor and cognitive disorders in patients followed up 2.5 years after initial examination.

PATIENTS AND METHODS

We were able to re-examine HC 17 patients after 2.5 years of disease. All except one, who had the akinetic-rigid variant, were affected by the hyperkinetic form. Clinical features are summarised in Table 1.

No patients were taking neuroleptic drugs on either examination. A few had been on mild neuroleptics but withdrew from the treatment ten days before re-examination. Functional ability was evaluated by the Shoulson and Fahn Scale (11); the number of hyperkinesias was counted over 2 minutes in each of the following parts of the body: head and face, upper and lower limbs and trunk (12); akinesia and bradykinesia were

Table 1. General Synopsis of HD Patients

No. of Cases	17
Sex	12M, 5F
Age (y)	48 (12.33)
Disease Duration (y)	7.8 (3.27)
Education (y)	8.1 (4.09)
Demented (n)	12

y = years, n = number,
figures in brackets = SD

assessed by the computerised analysis of RT and MT used in our previous study (10). Cognitive functions were assessed by the WAIS Scale (13) and based on DSM III criteria (14) twelve patients were found demented on second evaluation (Table 2.).

In one patient not all the motor and cognitive test batteries could be performed due to severe motor and mental derangement; in this patient we were only able to count the number of choreic movements. Statistical analysis used the paired Student t-test.

RESULTS

The data showed that number of hyperkinesias and cognitive performance did not change significantly, while RT and MT and especially functional ability worsened considerably over the period considered.

Twelve patients were found demented on second evaluation with mean VIQ unchanged and slight worsening of PIQ. Comparison between first and second examination results is displayed in Table 2.

A great variability in RT and MT performance was found, though the former had more markedly deteriorated in nearly all patients. The patient with the akinetic-rigid variant behaved differently in that he had the longest RT and MT though these did not change with respect to the first evaluation; PIQ considerably worsened in this patient.

DISCUSSION

Our data show, briefly, an obvious functional deterioration and a worsening of RT and MT with disease progression. Over the 2.5 years we considered, mental decay seems to decline at a different rate to motor impairment, progressing slowly overall.

Verbal functions were relatively unchanged being possibly more resistant to HC pathological alterations (15-17). Mayeux et al, (4) who examined 33 patients over a few years also found a partial integrity of cognitive function. On the other hand, since mental decay in HC seems to evolve more slowly than in other dementias (18) a longer longitudinal study than ours, employing tests better suited to evaluate memory and mental

Table 2. Motor and Cognitive Performance Scores
at First (A) and Second (B) Examination

	A		B	
HYPERKINESIAS	249	± 122	264.9 ± 155	
RT (msec)	524	± 171*	814	± 275
MT (msec)	590	± 131**	744	± 128
PIQ	80	± 15**	73	± 24
VIQ	87	± 14	87	± 15
SHOULSON SCORE	1.12 ±	0.34*	1.93±	0.68

* $p < 0.00001$
** $p < 0.003$
*** $p < 0.04$

flexibility (19-21), might be necessary to reveal significant cognitive alterations.

The marked worsening of RT and MT with disease evolution indicates that akinesia, also typical of HC (4,20,21) may be a good marker of disease progression. Stimulus identification problems and difficulties in the response selection and preparation phases may be responsible for the greater changes in RT compared to MT (22); RT alterations in HC are probably dependent on attention and emotional orientation disorders (5) as well as on other basal ganglia lesions.

We could not determine whether there were distinct subgroups of HC patients due to the small sample size, though the single patient with akinetic-rigid chorea had severe intellectual impairment which distinguished him from all the other patients. Further studies on larger groups of patients over a longer period are required.

REFERENCES

1. Caine E.D Ebert MJ, Weingartner H. An outline for the analysis of dementia: the memory disorder of huntington's disease. Neurology 27: 1087-1092, 1977.

2. Brandt J, Strauss ME, Larus J et al: Clinical correlates of dementia and disability in Huntington's disease. J Clin Neuropsychol 6(4): 401-412 (1984).

3. Caine ED, Hunt RD, Weingartner H, Ebert MJ: Huntington's dementia. Arch Gen Psychiatry 35:377-384 (1978).

4. Mayeux R, Stern Y, Herman A, Greenbaum L, Fahn S. Correlates of early disability in Huntington's disease. Ann Neurol 20: 727-731 (1986).

5. Bruyn GW. Huntington's chorea: historical, clinical and laboratory synopsis, in: "Handbook of Clinical Neurology", Vol 6, Vinken PJ, Bruyn JW eds. Amsterdam: North-Holland, pp 298-378 (1968).

6. Shoulson I. Care of patients and families with Huntington's disease. in: "Movement Disorders", Marsden CD and Fahn S eds. London: Butterworth, pp 277-290 (1982).

7. Hefter H, Homberg V, Lange HW, Freund HJ. Impairment of rapid movement in Huntington's disease. Brain 110: 585-612 (1987).

8. Thompson PD, Berardelli A, Rothwell JC, Day BL, Dick JPR, Benecke R, Marsden CD. The coexistence of bradykinesia and chorea in Huntington's disease and its implications for theories of basal ganglia control movement. Brain 11: 223-244 (1988).

9. Marsden CD. Slowness of movement in Parkinson's disease. Movement disorders. 4 (Suppl 1): 26-37 (1989).

10. Girotti F, Marano R, Soliveri P, Geminiani G, Scigliano G. Relationship between motor and cognitive disorders in Huntington's disease. J Neurol 235: 454-457 (1988).

11. Shoulson I, Fahn S. Huntington disease: clinical care and evaluation. Neurology 29: 1-3 (1979).

12. Girotti F, Carella F, Scigliano G, Grassi MP, Soliveri P, Giovannini P, Parati E, Caraceni T. Effect of neuroleptic treatment on involuntary movements and motor performances in Huntington's disease. J Neurol Neurosurg Psychiatry 47: 848-852 (1984).

13. Wechsler D. "The Wechsler Adult Intelligence Scale". New York Psychological Corporation (1955).

14. American Psychiatric Association "Diagnostic and Statistical Manual of Mental Disorders". 3rd ed. American Psychiatric Association. Washington D.C. (1980).

15. Aminoff MJ, Marshall J, Smith F. Pattern of intellectual impairment in Huntington chorea. Psychol Med 5: 169-172 (1975).

16. Norton JC. Patterns of neuropsychological test performance in Huntington's disease. J Nerv Ment Dis 161: 276-279 (1975).

17. Fedio P, Cox CS, Neophytides A, Canal-Frederick G, Chase PN. Neuropsychological profile of Huntington's disease: patients and those at risk, in: "Adv Neurol", Vol 23 Chase TN, Wexler NS, Barbeau A, eds. . New York: Raven Press. pp 449-556 (1979).

18. Brandt J, Folstein SE, Folstein MF. Differential cognitive impairment in Alzheimer's disease and Huntington's disease. Ann Neurol 23: 555-561 (1988).

19. Butters N, Sax D, Montgomery K, Tarlow S. Comparison of the neuropsychological deficits associated with early and advanced Huntington's disease. Arch Neurol 35: 585-589 (1978).

20. Josiassen RC, Curry LM, Mancall EL. Development of neuropsychological deficits in Huntington's disease. <u>Arch Neurol</u> 40: 791-796 (1983).

21. Huber SJ, Paulson GW. Memory impairment associated with progression of Huntington's disease. <u>Cortex</u> 23; 27-283 (1987).

22. Kerr B. Task factors that influence selection and preparation for voluntary movements, <u>in</u>: "Informartion Processing in Motor Control and Learning. Stelmach GE eds. New York, London: Academic Press, pp 55-69 (1978).

IMPAIRMENT OF THE COGNITIVE FUNCTIONS IN PARKINSON'S

DISEASE: A BRIEF SURVEY

A. G. Carlesimo, U. Nocentini and C. Caltagirone

Institute of Neurology
II University of Rome

INTRODUCTION

Research on the cognitive functions of subjects suffering from Parkinson's disease (PD) is both of clinical and theoretical interest.

At the clinical level, one must be aware that the occurrence of intellectual imairment in PD patients normally determines a global worsening of the functional disability. In some cases, disturbances in the cognitive domain may be related to the administration of anticholinergic drugs[1]. In other cases onset of cognitive deterioration appears as part of the normal course of the disease[2].

From a theoretical point of view, PD has been considered as a particularly revealing model in the study of the role played by subcortical structures in cognition[3]. However, this possible relationship between impaired cognition and modifications in dopaminergic pathways in PD has been dampened by the growing evidence of pathologic changes affecting structures different from the traditionally described ones. In fact, neuronal loss in the locus coeruleus[4] and in the nucleus basalis of Meynert[5] and the occurrence of degenerative modifications of the Alzheimer type at the cortical level[6] have been observed. A functional reduction of cholinergic transmission has also been more recently reported[7].

As for the occurrence of dementia in PD, authors generally point to a prevalence of dementia among Parkinsonians higher than for the same age group in the rest of the population[2]. Moreover, PD subjects not suffering from dementia nevertheless frequently display deficiencies of single cognitive functions[8]. It has been suggested that a sort of continuity exists between the two conditions, i.e. selective deficits of single cognitive functions could represent the first stage of the more general impairment displayed by subjects suffering from dementia[9]. However, while difficulties affecting certain skills (e.g. conceptual reasoning) are shown by the majority of patients even at very early stages of the disease[10], only in a relatively small proportion of the cases does this develop into actual dementia[11].

The Basal Ganglia III, Edited by G. Bernardi *et al.*
Plenum Press, New York

Bearing these uncertainties in mind, we shall approach the problem with an initial survey of studies ascertaining the prevalence and the qualitative features of dementia in PD. Moreover we shall proceed to consider research aiming at the identification of specific cognitive deficiencies in non demented Parkinsonian patients. In conclusion we shall offer an interpretation of the relationships between intellectual disturbances and neuropathological and neurochemical changes observable in PD.

PREVALENCE OF THE DEMENTIAL SYNDROME IN PARKINSON'S DISEASE

Today most authors agree that demential syndromes occur among Parkinsonian subjects at a significantly higher rate than that shown by the same age group in the rest of the population. The most recent studies offer an extimate of the prevalence of dementia among Parkinsonian patients ranging from 10%[12,13] to 40%[14]. Brown and Marsden[11] have raised various methodological objections concerning the selection of samples, suggesting 20% as a more realistic value of prevalence in Parkinsonians. These authors argue that the lack of uniformity in criteria adopted by various researchers in the diagnosis of dementia and the inadequate control of concomitant factors account for the variability of results reported in these studies.

As for the criteria used in the diagnosis of dementia we must distinguish between studies based on clinical observations and researches involving neuropsychological assessments. As shown in the table, clinical studies yield lower values for the occurrence of dementia than do studies of the neuropsychological type. In general, clinical studies have the advantage of being carried out on a very large number of cases (802 subjects in the study by Hoehn and Yahr[15]), and the patients often are followed over long periods of time. However, these studies present the disadvantage of using nonstandardized methods to assess cognitive deterioration. This point is further evidenced by the fact that the lowest values for the occurrence of general cognitive deterioration (10% and 11% respectively) were recorded in two studies[12,13] based on a retrospective review of case records. A further point to bear in mind in this respect is that diagnostic criteria of the clinical type, as those proposed in the DSM III[16], tend to emphasize the functional autonomy of patients. Such criteria are, however, difficult to apply to Parkinsonians, who invariably display motor disabilities.

Only a few studies in the area of neuropsychologically oriented research provide data on the prevalence of cognitive deterioration in PD. A general feature of such studies is the application of extensive batteries of tests able to assess a large range of cognitive functions and provide figures on the percentage of subjects showing diffusely deficient performances. Values of prevalence of diffuse cognitive deterioration derived from neuropsychological studies range from 20%[17] to 40%[14].

It is also important to bear in mind the clinical features displayed by the sample of Parkinsonian patients under consideration. In fact, some authors argue that cognitive impairment in Parkinsonian patients is correlated with the duration of the illness and/or the gravity of motor symptoms[14]. Moreover, the onset of dementia would appear to be more frequent in arteriosclerotic or postencephalitic Parkinsonians than in patients affected by idiopathic forms of the disease[14]. Finally, it has also been noted that the administration of anticholinergic drugs is associated with memory deterioration in Parkinsonian patients[1].

Neuropsychological assessment of Parkinsonian patients should therefore be accompanied by careful clinical assessment of the course of the illness, gravity of symptoms and the type of therapy applied, thus providing the opportunity to identify and possibly isolate predisposing and concomitant and/or causal factors.

In one of the studies recently carried out by our group[19] the cognitive functions of a large group of Parkinsonians were assessed. Neuropsychological assessment was performed by means of a battery of tests exploring memory, language, constructional praxia and general intelligence that, in a previous work, had revealed a high level of diagnostic reliability in discriminating between healthy subjects and patients affected by general intellectual deterioration[20]. In this study, about 40% of the Parkinsonian patients showed neuropsychological signs of dementia. Cognitive deterioration proved to be correlated with the duration of the illness and the administration of anticholinergic drugs. On the other hand, efficiency in cognitive performance did not appear to be significantly affected by the mean age of the patients, age at the onset of the symptoms, duration of dopaminergic treatment or main symptoms (tremor or bradykinesia).

NEUROPSYCHOLOGICAL FEATURES OF THE DEMENTIAL SYNDROME IN PARKINSON'S DISEASE

The findings in subjects affected by PD of neuropathological changes resembling those of patients suffering from Alzheimer's disease (AD) has led various authors[5,6,21] to hypothesize that cases of demential syndromes occurring in Parkinsonian patients could be attributed to a concomitant Alzheimer disease.

This hypothesis has, however, been challenged by other authors[22], who have pointed out that the occurrence of Alzheimer type histopathological changes in a number of Parkinsonian patients is not in itself sufficient to justify the assertion that association between the two illnesses occurs more frequently than chance alone could determine. In fact, the simple evidence of Alzheimer-like pathological changes in the cerebrum of subjects over 65 is not a specific finding and the diagnosis of AD is based on the finding of high number of senile plaques and neurofibrillary tangles[23,24].

One of the most interesting fields of research in clinical neuropsychology is the attempt to identify methods of cognitive assessment that can reveal differential patterns of impairment in subjects suffering from demential syndromes of various aetiologies. Taking dementia as a syndrome, it could be postulated that patterns of impairment could be detected which differ qualitatively according to the underlying process. If similar neuropsychological impairment patterns were observed in AD patients and Parkinsonians suffering from dementia, there would be some justification for hypothesizing one pathological process underlying the two conditions, while differential patterns of cognitive impairment would point to two essentially independent pathological processes. As matter of fact, one general feature that appears to distinguish the demential syndromes shown by the two conditions is the lesser degree of cognitive impairment shown by Parkinsonian subjects[25]. Confirmation of this has come from research carried out in our laboratory[26] on individuals affected by PD and AD respectively. The occurrence of general cognitive deterioration was assessed on the basis of a wide neuropsychological battery, while the gravity of such deterioration was assessed on the basis of the number of pathological performances shown by individual patients in the various tests forming the battery. In over 40% of the cases, Parkinsonian subjects showed neuropsychological features pointing to dementia. On examining the extent of cognitive impairment, 70% of the Parkinsonians with dementia were affected by mild forms of cognitive decline, while only 30% of the AD subjects showed this level of dementia.

Albert[27] suggested that, at the clinical level, PD display the specific characteristics of subcortical dementias, the features of which differ from forms, such as AD, which are characterized by a prevalent cortical location of pathological changes. According to the description by Albert et al.[28], cognitive impairment in subcortical dementias (Progressive Supranuclear Palsy, Huntington's chorea, Wilson's Disease, PD, etc.) appears to be characterized by slowness of mental processing, forgetfulness, impaired ability to manipolate acquired knowledge, altered personality with apathy or depression. The main feature distinguishing from cortical dementia is the preservation in subcortical forms of the linguistic, praxic and gnostic functions.

Attempts to identify different patterns of neuropsychological impairment in cortical and subcortical forms of dementia have not led to conclusive results. Cummings and Benson[29] have attempted to explain this in terms of the particular difficulty involved in detecting modifications in variables concerned with arousal, attention and affective behaviour, which are primarily regulated by the subcortical structures. A further methodological objection frequently raised concerns inadequate control over variables such as the gravity of the cognitive deterioration. In fact, in studies comparing Parkinsonian and AD patients without paying specific attention to the gravity of the demential syndrome[30,31,32] PD patients obtained better performances in all the fields examined. Indeed, few studies make comparison between PD and AD subjects after having ascertained that the groups are really comparable in terms of the gravity of cognitive deterioration. In a first study by Mayeux et al.[33] a modified version of the Mini Mental State (MMS) was administered to three groups of subjects with AD, Huntington's chorea (HD) and PD. The subjects in the three groups were in turn classified affected by mild, medium or severe forms of dementia according to the level of cognitive impairment. The AD, HD and PD subjects within each level of mental deterioration showed no significant differences for any of the MMS items. These results suggest, therefore, that subjects affected with the same degree of cognitive deterioration show identical patterns of neuropsychological impairment regardless of the type of underlying pathological process. It must, however, be borne in mind that the MMS generally used for rapid screening of intellectual deterioration does not seem suitable to supply detailed data on single cognitive activities and, therefore, to detect possible qualitative differences in the patterns of neuropsychological impairment. Subsequent studies[26,34,35] applying extensive batteries of neuropsychological tests have in fact generally revealed differential cognitive impairment patterns for PD and AD patients. More specifically, the AD patients show, in a context of global cognitive deficiency, a prevalent impairment in memory tests, while the Parkinsonians give their worst performances in tests exploring verbal fluency, formation of concepts and mental flexibility known to assess the efficiency of the frontal lobes. Caltagirone et al.[26] compared the results obtained by Parkinsonians affected by dementia and AD subjects in a large battery of neuropsychological tests. The subjects were in turn subdivided into two categories on the basis of the presence of a mild or a severe form of cognitive impairment. As it can be seen in figure 1, both comparisons between mildly and severely deteriorated PD and AD patients point to a twofold dissociation: in fact, while the AD patients obtain significantly worse results than the PD subjects in memory tests (and particularly in longterm verbal memory tests), PD patients give their worst performances, even worse than the AD patients' performances, in the verbal fluency test normally used in clinical assessment of frontal lobes functioning.

COGNITIVE DEFICITS IN NON DEMENTED PARKINSONIANS

It has frequently been observed that Parkinsonian subjects unaffected by dementia may show specific impairment in certain cognitive areas[10]. However, it must be pointed out that in many studies criteria
used for the selection of Parkinsonian subjects and identification of PD patients showing no signs of dementia are either not clearly stated or appear open to methodological objections. The uncertainties regarding the composition of PD groups in these studies makes difficult to correctly interpret the differences between PD and control groups. In fact, it can not be excluded that deficient performances shown by PD groups are due to the presence of an high number of demented subjects within the sample considered.
The cognitive areas most frequently studied in PD relate to visuo-spatial functions, memory and functions subsumed by the frontal lobes.
The hypothesis of selective deficiency in the visuo-spatial functions is based on the observation, borne out by many studies[18,36], that the results obtained by Parkinsonians in performances subtests of the Wechsler's intelligence scales are significantly lower than those obtained for verbal subtests. This is further confirmed by studies revealing the deficient performance of Parkinsonian subjects in tests assessing perception of the orientation of lines[36,37], the orientation of their own bodies in space[36], the ability to follow a route drawn on a map[38] and awareness of their own body schema[39].However, Brown and Marsden[8] point out that various methodological problems arise when the object of study is the visuo-spatial abilities of Parkinsonian patients. First of all, it is by no means easy to interpret performances of PD subjects on Wechsler's intelligence scales. Apart from possibly disturbed visuo-perceptual analysis, the low scores obtained in performance subtests may also be due to the reduced manual dexterity and/or speed characteristic of Parkinsonian patients.

Moreover, some of the tests used in these researches[38,39] require continual modifications of the mental representations. These tests may thus prove deficient for patients with limited conceptual flexibility, apart from the effect of possible disturbances of a visuo-spatial nature.

In fact, in studies in which a careful selection of patients non affected by general cognitive deterioration has been made and based on tests exclusively demanding the processing of visuo-spatial data, Parkinsonian subjects generally come up with performances comparable to those of the control group[40,41].Many studies have been dedicated to assessment of memory in PD. On the whole, these studies have shown generally good performance in tests, such as digit span, usually considered reliable for the assessment of immediate memory[10,42].

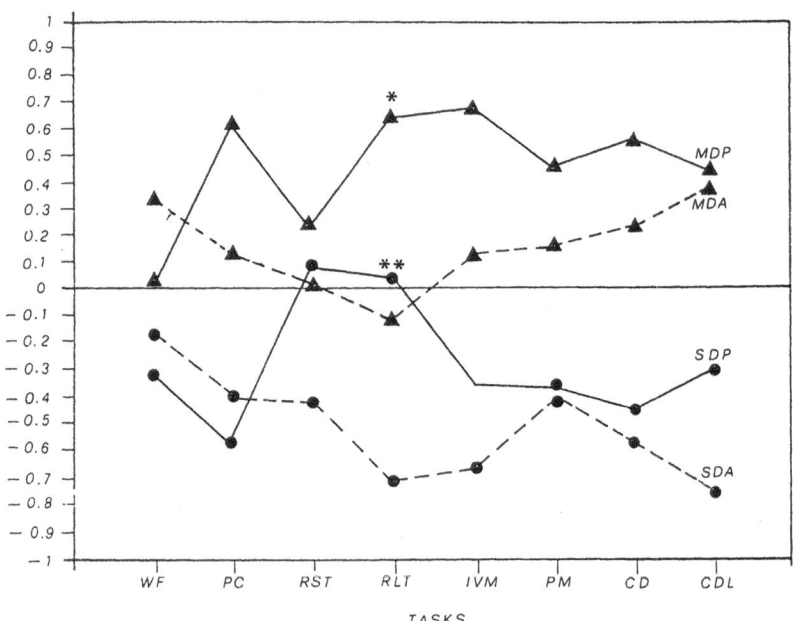

Fig. 1. Performance patterns obtained in the various tasks of the Mental Deterioration Battery tests by Parkinsonian and by Alzheimer's patients affected respectively by mild and severe forms of mental deterioration. The different scores are expressed as z scores. MDP = mildly deteriorated PD patients; MDA = mildly deteriorated AD patients; SDP = severely deteriorated PD patients; SDA = severely deteriorated AD patients. WF = Verbal Fluency; PC = Phrase Construction; RST = Short-term verbal memory; RLT = Long-term verbal memory; IVM = Immediate visual memory; PM = Raven '47 progressive matrices; CD = Simple copying; CDL = Copying with landmarks; * p = .06 in the comparison between mildly deteriorated PD and AD patients; ** p = .03 in the comparison between severely deteriorated PD and AD patients.

Generally deficient results are, however, obtained in tests assessing short-term recall of verbal data and visuo-spatial data[9,44], or longterm recall of verbal materia[19,45]. In contrast, recognition tests on verbal data[10,45] or visuo-spatial data[46] yield normal results. In short, these results would appear to demonstrate that Parkinsonians have normal information storage capacities but experience difficulty in the retrieval of this information from the long and short-term memory[8].

A recent study[47] attempts to define the mechanisms governing mnemonic deficiencies in Parkinsonian patients, comparing their performance with the performance of amnesic subjects in tests assessing differential aspects of memory, distinguishable both functionally and anatomically. A contrast was drawn between tests assessing the so-called declarative and procedural aspects of memory. In fact, Squire and his colleagues[48] suggested that declarative memory should be taken to cover information relating to specific facts, dates and incidents, while procedural memory should cover information relating to skills and procedures. In other words, they postulated a distinction between "know what" and "know how". The results obtained by Saint-Cyr et al.[47] would seem to point to a twofold dissociation, since Parkinsonians prove deficient in procedural but not in declarative memory, while amnesic patients show the reverse pattern. The authors suggest that this may seen as the result of a deficiency in PD subjects of the "heuristic strategy" necessary for procedural learning, and go on to suggest that this situation might be considered as the result of an impairment of the striatalfrontal pathways.

In fact, according to a considerable number of studies Parkinsonian subjects show deficient performance in tests assessing the functioning of the frontal lobes[10,39,49,50,51].

Caltagirone et al.[52] argue that "frontal deficiency" shown by Parkinsonian subjects is in fact selective in nature, i.e. also displayed by subjects whose intellectual efficiency is demonstrably reasonably well preserved. In fact, after excluding from a nonselected population of PD patients those patients who displayed the neuropsychological features of general cognitive deterioration, it was still possible to demonstrate differences with respect to the control group in tests revealing specific cognitive impairment. Thus the Parkinsonians unaffected by dementia showed results similar to those obtained by a healthy control group in tests assessing verbal and visual memory, constructional praxia, linguistic skills and general intelligence, showing deficienct performances only in the Wisconsin Card Sorting Test, where they experienced difficulty in the formation of concepts and persevered with the ready formulated concepts despite repeated situational feedback. The results of this research would appear to demonstrate quite clearly that deficiencies shown in "frontal tests" do not derive from a more general intellectual impairment but in fact reveal the features of a selective impairment in PD patients.

CONCLUSIONS

James Parkinson wrote in what was in fact the first description of the disease[53] that a major feature of the "shaking palsy" is motor disturbance, while "the senses and intellect" were spared. In the last few years, however, it has become increasingly accepted that impairment of the cognitive functions is a somewhat common feature in cases of PD. It has, in fact, been demonstrated that the occurrence of dementia (i.e. general deterioration of the cognitive functions) reaches levels of at least 15-20%, according to the most cautious extimates, compared with the 7-10% shown by the same age group in the population as a whole8. Another point now well documented is that, even in the early stages of the illness, Parkinsonians clearly not suffering from dementia nevertehless show specific deficiencies in individual cognitive abilities[10,47,52].

Researches in the direction of defining the morphofunctional changes underlying the demential syndrome in PD have not as yet designed a coherent framework. The observations made by some authors[5,6,21] that Alzheimer-like alterations at level of cortex and nucleus basalis of Meynert are a frequent finding in PD and pratically the rule for patients who have shown demential syndromes in their lives, have not been confirmed. In fact, other authors have documented no hystopathological changes of the Alzheimer type in the brain of PD patients suffering from dementia[7,22,54,55]. One possible interpretation of these data is that the occurrence of demential syndromes in parkinsonians subjects is subsumed by various types of morpho-functional alterations in the cortical and subcortical structures. In some cases the cognitive deterioration may derive from the characteristic hystopathological modifications occurring in AD[56]. Considering that both PD and AD are senile or presenile diseases, however, it is hardly surprising to find both illnesses affecting the same patient in a certain number of cases. It is yet to be established, however, whether the onset of AD occurs with equal or greater frequency among Parkinsonians than in the rest of the population and whether, therefore, PD can be taken as a factor predisposing subjects for the onset of an Alzheimer syndrome.

The hystopathological alterations described in some cases of PD with dementia recall the so-called "Lewy's body disease", characterized at the cortical level by the spread of cytoplasm inclusions generally observed in midbrain pigmented neurons[57,58]. Moreover, observations have also been recorded of reduced concentrations of cholinergic markers at the cortical level correlated with neuronal loss in the nucleus basalis of Meynert[7,59] and reduced dopaminergic and noradrenergic activity at the cortical level accompanied by hystological and neurochemical alterations in the ventral-tegmental area of the midbrain and the locus coeruleus[60].

The extent of deficiency in the superior cortical functions shown by parkinsonian subjects even when unaffected by dementia has attracted a great deal of attention, since it suggests that the basal ganglia play a role in the cognitive functions. There no longer seems to be any question of the fact that even in the early stages of the disease, parkinsonian patients often show selective deficiency in tests exploring functions governed by the frontal lobes. It has in fact emerged from various research studies that Parkinsonians not suffering from dementia do worse in tests of verbal fluency10, the Wisconsin Card Sorting Test[10,39,51,52] and other classification and conceptual flexibility tests[49,50].

It is, moreover, undoubtedly significant that the neuropsychological pattern displayed by Parkinsonian subjects with dementia appears characterized by prevalent impairment in functions governed by the frontal lobes, even in a context of general cognitive deterioration[26,35].

This neuropsychological evidence appears to receive support from data on the structural links and functional relationships between the diencephalic grey nuclei and the frontal cortical areas. De Long et al.[61] recently summarized the data available in the literature on the existence of two distinct circuits which connect the striatum and frontal cortex and retain structural and functional independence although following largely parallel courses. The first circuit ("motor loop") originates in the motor, premotor and somatosensorial areas and uses the putamen as primary relay station. Efferents leave from the putamen and, through the globus pallidus, the pars reticulata of the substantia nigra and the thalamic ventrolateral nuclei arrive at the supplementary motor area in the frontal lobes. In contrast, the second circuit ("complex loop") originates at the level of the associative areas of the frontal, temporal and parietal lobes. In this case, the relay structure in the striatum is represented by the caudate nucleus. The intermediary stations are represented by the globus pallidus, the substantia nigra and the anterior and middorsal ventral nuclei of the thalamus, while the cortical area closing the loop is the prefrontal cortex. Clearly, these structural links must create a close functional connection between the striatum and the frontal lobes. More specifically, the hypothesis has been advanced that the "motor loop" is involved in the control of voluntary motility, while the "complex loop", which originates from the complex of cortical associative areas but terminates at the level of the prefrontal areas, regulates the more complex behaviour functions governed by the frontal lobes.

Some authors[8,51] have suggested that damage to the "motor loop" may be the cause of motor disorders, while damage to the "complex loop" may trigger off the "frontal" features of the cognitive disturbance displayed by Parkinsonian subjects. Further evidence in support of selective impairment of the frontal structures is supplied by other aut who have observed reduced dopaminergic activity in PD at the level of the frontal areas, probably due to an impairment of the meso-cortical-limbic pathway leading from the ventraltegmental midbrain area to the frontal lobes.

REFERENCES

1. K. Syndulko, E. R. Gilden, E. C. Hansch, A. R. Potvin, W. W. Tourtellotte, J. H. Potvin, Decreased verbal memory associated with anticholinergic treatment in Parkinson's disease patients, Intern. J. Neurosc. 14:61-66 (1981).
2. J. A. Mortimer, K. J. Christensen, D. Webster, Parkinsonian dementia, in: "Handbook of Clinical Neurology, Vol. 2 (46): Neurobehavioral Disorders", J. A. M. Frederiks, ed., Elsevier Science Publishers, Amsterdam, New York, 371-384 (1985).
3. C. D. Marsden, Function of basal ganglia as revealed by cognitive and motor disorders in Parkinson disease, Can. J. Neurol. Sci. 11:129-135 (1984).

4. L. S. Forno, Pathology of Parkinson's disease, in:"Movement Disorders, 2", C. D. Marsden, S. Fahn, eds., Butterworths, London, 25-40 (1982).

5. P. J. Whitehouse, J. C. Hedreen, C. L. White, D. L. Price, Basal forebrain neurons in the dementia of Parkinson disease, Ann. Neurol. 13:243-248 (1983).

6. F. Boller, T. Mizutani, U. Roessmann, P. Gambetti, Parkinson's disease,dementia and Alzheimer disease: clinico-pathological correlations, Ann. Neurol. 1:329-55 (1980).

7. E. K. Perry, M. Curtis, D. J. Dick, J. M. Candy, J. R. Atack, C. A. Bloxham, G. Blessed, A. W. Fairbairn, B. E. Tomlinson, R. H. Perry, Cholinergic correlates of cognitive impairment in Parkinson's disease: comparisons with Alzheimer's disease, J. Neurol. Neurosurg. Psychiatry. 48:413-421 (1985).

8. R. G. Brown, C. D. Marsden, Neuropsychology and cognitive functions in Parkinson's disease: an overview, in: "Movement Disorders 2", C. D. Marsden, S. Fahn, eds., Butterworth, London, 99-123 (1987).

9. F. J. Pirozzolo, E. C. Hansch, J. A. Mortimer, D. D. Webster, M. A. Kuskowski, Dementia in Parkinson's disease: a neuropsychological analysis, Brain Cogn. 1:71-83 (1982).

10. A. R. Lees, E. Smith, Cognitive deficits in early stages of Parkinson's disease, Brain 106:257-70 (1983).

11. R. G. Brown, C. D. Marsden, How common is dementia in Parkinson's disease?, Lancet ii:1262-65 (1984).

12. A. H. Rajput, K. Offord, C. M. Beard, L. T. Kurland, Epidemiological survey of dementia in parkinsonism and control population, in: "Advances in Neurology, Vol. 40", R. G. Hassler, J. F. Christ, eds., Raven Press, New York, 229-234 (1984).

13. R. Mayeux, R. Rosenstein, Y. Stern, L. Cote, S. Fahn, The prevalence and risk of dementia in idiopathic Parkinson's disease, Ann. Neurol. 20:128 (1986).

14. G. C. Celesia, W. M. Wanamaker, Psychiatric disturbances in Parkinson's disease, Dis. Nerv. Syst. 3:577-583 (1972).

15. N. M. Hohen, M. D. Yahr, Parkinsonism: onset, progression and mortality, Neurology 17:427-442 (1967). 16. American psychiatric association committee on nomenclature and statistics, "Diagnostic and statistical manual of mental disorders, ed. 3", American psychiatric association, Washington, DC,1980.

17. R. H. S. Mindham, S. W. A. Ahmed, C. G. Clough, A controlled study of dementia in Parkinson's disease, J. Neurol. Neurosurg. Psychiatry 45:969-974 (1982).

18. A. W. Loranger, H. Goodel, F. H. McDowell, J. E. Lee, R. D. Sweet, Intellectual impairment in Parkinson's syndrome, Brain 95:405-412 (1972)..

19. C. Caltagirone, C. Masullo, N. Benedetti, G. Gainotti, Dementia in Parkinson's disease: possible specific involvement of the frontal lobes, Intern. J. Neurosc. 26:15-26 (1985).

20. C. Caltagirone, G. Gainotti, C. Masullo, G. Miceli, Validity of some neuropsychological tests in the assessment of mental deterioration, Acta. Psych. Scand. 60:50-56 (1979).

21. A. M. Hakim, G. Mathieson, Dementia in Parkinson's disease: a neuropathologic study, Neurology 29:1209-1214 (1979).

22. L. L. Heston, Dementia associated with Parkinson's disease: a genetic study. J. Neurol. Neurosurg. Psychiatry 43:846-848 (1980).

23. B. E. Tomlinson, G. Blessed, M. Roth, Observations on the brains of non-demented old people, J. Neurol. Sci. 7:331-356 (1967). 24. B. E. Tomlinson, G. Blessed, M. Roth, Observations of the brains of demented old people, J. Neurol. Sci. 11:205-242 (1970).

25. R. Portin, M. Laine, P. Molsa, U. K. Rinne, Cognitive impairment and cerebral atrophy in patients with Encephalopathy, Parkinson disease and Alzheimer disease, "Deauville, 5th INS European Conference", 1982.

26. C. Caltagirone, A. Carlesimo, U. Nocentini, S. Vicari, Differential aspects of cognitive impairment in patients suffering from Parkinson's and Alzheimer's disease: a neuropsychological evaluation, Intern. J. Neurosc. 44:1-7 (1989a).

27. M. L. Albert, Subcortical dementia in: "Alzheimer's disease: senile dementia and related disorders. Aging, vol. 7", R. Katzman, R. D. Terry, K. L. Bick, eds., Raven Press, New York, 173-180 (1978).

28. M. L. Albert, R. G. Feldman, A. L. Willis, The subcortical dementia of progressive sopranuclear palsy, J. Neurol. Neurosurg. Psychiatry 37:121-130 (1974).

29. J. L. Cummings, F. Benson, Subcortical dementia, Review of an emerging concept, Arch. Neurol. 41:874-879 (1984).

30. G. Gainotti, C. Caltagirone, C. Masullo, G. Miceli, Patterns of neuropsychological impairment in various diagnostic groups of dementia, in: "Aging of the brain and dementia", L. Amaducci, A. N. Davison, P. Antuono eds., Raven Press, New York, 1980.

31. S. Bentin, R. Silverberg, H. W. Gordon, Asymmetrical cognitive deterioration in demented and Parkinson patients, Cortex 17:533-544, 1981.

32. S. J. Huber SJ, E. C. Shuttleworth, G. W. Paulson, M. J. Bellchambers, L. E. Clapp, Cortical vs subcortical dementia. Neuropsychological differences, Arch. Neurol. 43:392-394 (1986). 33. R. Mayeux, Y. Stern, J. Rosen, D. F. Benson, Is "subcortical dementia" a recognizable clinical entity?, Ann. Neurol. 14:278-283 (1983).

34. M. Freedman, M. Oscar-Berman, Comparative neuropsychology of cortical and subcortical dementia, Can. J. Neurol. Sci. 13:410-414 (1986).

35. B. Pillon, B. Dubois, F. Lhermitte, Y. Agid, Heterogeneity of cognitive impairment in progressive supranuclear palsy, Parkinson's disease and Alzheimer's disease, Neurology 36:1179-1185 (1986).

36. F. Proctor, M. Riklan, I. S. Cooper, H. Teuber, Judgment of visual and postural vertical by parkinsonian patients, Neurology 14:287-293 (1964).

37. G. Danta, R. C. Hilton, Judgment of the visual vertical and horizontal in patients with parkinsonism, Neurology 25:43-47 (1975).

38. F. P. Bowen, M. M. Hoehn, M. D. Yahr MD, Parkinsonism: alterations in spatial orientation as determined by a route-walking test, Neuropsychologia 10:355-361 (1972). 39. F. P. Bowen, R. S. Kamienny, M. M. Burns, M. D. Yahr, Parkinsonism: effects of levodopa on concept formation, Neurology 25:701-704 (1975).

40. R. G. Brown, C. D. Marsden, Visuospatial function in Parkinson's disease, Brain 109:987-1002 (1986).

41. S. Della Sala, G. Di Lorenzo, A. Giordana, H. Spinnler, "Directional forecast": a specific visuo-spatial impairment of parkinsonians?, "Paper presented at the joint meeting of the Polish and Italian Societes of Neurology", Rome, Italy, 1985.

42. D. Asso, W.A.I.S. scores in a group of Parkinson patients, Brit. J. Psychiat. 115:555-556 (1969).

43. R. Portin, U. K. Rinne, Neuropsychological responses of Parkinsonian patients to long-term levodopa therapy, in: "Parkinson's disease: Current Progress, Problems and Management", M. Klinger, G. Stamm, eds., Elsevier/North Holland, Amsterdam, 271-304 (1980).

44. F. P. Bowen, M. M. Burns, E. M. Brady, M. D. Yahr MD, A note on alterations of personal orientation in parkinsonism, Neuropsychologia 14:425-429, 1976.

45. H. Weingartner, S. Burns, R. Diebel, P. A. Le Witt, Cognitive impairment in Parkinson's disease: distinguishing between effort-demanding and automatic cognitive processes, Psychiat. Res. 11:223-235 (1984). 46. K. A. Flowers, I. Pearce, J. M. S. Pearce, Recognition memory in Parkinson's disease. J. Neurol. Neurosurg. Psychiatry 47:1174-1181 (1984).

47. J. A. Saint-Cyr, A. E. Taylor, A. E. Lang, Procedural learning and neostriatal dysfunction in man, Brain 111:941-959 (1988).

48. N. J. Cohen, L. R. Squire, Preserved learning and retention of pattern-analyzing skill in amnesia: dissociation of knowing how and knowing that, Science 210:207-210 (1980).

49. A. R. Cools, J. H. L. Van Den Bercken, M. W. I. Horstink, K. P. M. Van Spaendonck, H. J. C. Berger, Cognitive and motor shifting aptitude disorder in Parkinson's disease, J. Neurol. Neurosurg. Psychiatry 47:443-453 (1984).

50. K. A. Flowers, C. Robertson, The effect of Parkinson's disease on the ability to maintain a mental set, J. Neurol. Neurosurg. Psychiatry 48:517-529 (1985).

51. A. E. Taylor, J. A. Saint-Cyr, A. E. Lang, Frontal lobe dysfunction in Parkinson's disease. The cortical focus of neostriatal outflow, Brain 109:845-883 (1986).

52. C. Caltagirone, A. Carlesimo, U. Nocentini, S. Vicari, Defective concept formation in parkinsonians is independent from mental deterioration. J. Neurol. Neurosurg. Psychiatry. 52:334-337 (1989b).

53. J. Parkinson, "An essay on the Shaking Palsy", 1817.

54. D. M. A. Mann, P. O. Yates, Pathological basis for neurotransmitter changes in Parkinson's disease, Neuropathol. Appl. Neurobiol. 9:3-19 (1983).

55. M. J. Ball, The morfological basis of dementia in Parkinson's disease, Can. J. Neurol. Sci. 11:180-184 (1984).

56. P. Gaspar, F. Gray, Dementia in idiopathic Parkinson's disease. Acta Neuropathol. 64:43-52 (1984).

57. A. W. Clark, J. Lehmann, Dementia with widespread Lewy bodies: studies of the neocortical cholinergic system. "(Abstract), Canadian Association of Neuropathologists, 23rd Annual Meeting", Banff, Alberta, September 1983.

58. M. Yoshimura, Cortical changes in the parkinsonian brain: a contribution to the delineation of "diffuse Lewy body disease", J. Neurol. 229:17-32 (1983).

59. B. Dubois, J. J. Hauw, M. Ruberg, M. Serdaru, F. Javoy-Agid, Y. Agid, Demence et maladie de Parkinson:

correlations biochimiques et anatomo-cliniques, Rev. Neurol. 141, 3:184-193 (1985).

60. O. Hornykiewicz, S. J. Kish, Neurochemical basis of dementia in Parkinson's disease. Can. J. Neurol. Sci. 11:185-190 (1984).

61. M. R. De Long, A. P. Georgopoulos, M. D. Crutcher, Cortico-basal ganglia relations and coding of motor performance, Exp. Brain Res. Suppl.7:30-40 (1983).

62. F. Javoy-Agid, Y. Agid, Is the mesocortical dopaminergic system involved in parkinson disease?, Neurology 30:1326-30 (1980).

63. B. Scatton, L. Rouquier, F. Javoy-Agid, Y. Agid, Dopamine deficiency in the cerebral cortex in Parkinson disease, Neurology 32:1039-1040 (1982).

ENVIRONMENTAL RISK FACTORS IN PARKINSON'S DISEASE

William C. Koller

University of Kansas Medical School
Kansas City, Kansas, 66103

INTRODUCTION

The clinical syndrome of parkinsonism can be caused by a variety of insults to the basal ganglia [1]. Idiopathic parkinsonism or Parkinson's disease is a distinct clinical entity with a defined neuropathology and neurochemistry. Neurons of the substantia nigra are degenerated and contain inclusion bodies called Lewy bodies [2]. There is disruption of neurotransmission in the dopaminergic nigrostriatal pathway because of nigral cell death. This results in a deficiency of dopamine in the striatum which is the neurochemical basis of Parkinson's disease. Parkinson's disease is a neurodegenerative condition for which the cause is unknown. Why do cells of the substantia nigra selectively die? New therapies or even preventative measures could be introduced if the etiology of Parkinson's disease could be found.

ETIOLOGIC CONSIDERATION

The possibility that Parkinson's disease is genetic in origin has long been considered [3]. While there is no clear genetic pattern to Parkinson's disease, it has often been noted by clinicians that it is not unusual for several members of a pedigree to have parkinsonism. The most definitive studies in this area have been investigations of monozygotic and dizygotic twin pairs in which one member had Parkinson's disease [4-6]. These studies have found a low concordance rate in twins which suggests a limited role for heredity in Parkinson's disease. However, it is still possible that there is a genetic predisposition to the disease.

If Parkinson's disease is not primarily genetic then it has to be acquired. An infectious etiology has been considered since parkinsonism was a part of the encephalitic and post-encephalitic syndrome of the viral epidemic of 1919 to 1923. However, currently there is no evidence that infection plays a role in Parkinson's disease. Pathologic examination

The Basal Ganglia III, Edited by G. Bernardi et al.
Plenum Press, New York

of parkinsonian brains has failed to reveal viral particles or other signs of infection [2].

The relationship of aging to the cause of Parkinson's disease has received some attention [7]. It is known that there is age-related decline in the number of nigral cells and loss of striatal dopamine [8]. However, these declines do not reach the necessary level (70-80 % loss of dopamine) to cause clinical symptoms. It has been suggested that an insult to the nigra occurs on the backdrop of an aging substantia nigra [7]. Thus aging could play a contributing role.

ENVIRONMENTAL CONSIDERATIONS

If Parkinson's disease is acquired, it is possible that an agent in the environment could be responsible. This hypothesis had little theoretical support until the chemical MPTP was discovered. This toxin produces a clinical syndrome almost identical to Parkinson's disease and causes selective damage to the substantia nigra with a resultant loss of striatal dopamine [8,9]. Could a similar toxin or putative environmental agent also cause nigral cell death and induce parkinsonism? Epidemiological studies have attempted to investigate this issue. Barbeau and coworkers [10] estimated the prevalence of Parkinson's disease in Quebec from public health records of l-dopa sales and found a higher prevalence in rural areas that had exposure to herbicides and pesticides. Several other studies have also reported that living in a rural environment increases the risk of developing Parkinson's disease. Rajput et al. [11] reported that all of the fifteen life-long Saskatchewan residents in his clinic with onset of Parkinson's disease at less than age 40 had lived exclusively in communities with populations less than 140 people. In 100 consecutive parkinsonian outpatients, Tanner [12] found those with onset at age less than 48 years to be more likely to have lived in a rural environment (town with a population less than 1000). In a mail survey of 1100 patients with Parkinson's disease, those with onset before age 48 were more likely to have lived in a rural environment than those with onset age 54 [13]. Ho and coworkers [14], in a study in Hong Kong, found that subjects with a long duration (greater than 40 years) in rural areas had a five-fold risk of getting Parkinson's disease. In a study in the state of Kansas, we also found that living in a rural residence (versus urban residence) doubled the odds of getting Parkinson's disease [15]. The number of years spent in the rural environment was signifi-cantly increased regardless of the age at onset (figure 1). Furthermore it has been reported from post-mortem material that non-parkinsonian rural residents had lower substantia nigra cell counts than urban residents of a similar age [16]. Despite the methodological differences in these studies, the persistence of rural living and Parkinson's disease suggest that some aspect of rural living increases the risk for Parkinson's disease.

Other studies have also pointed to a possible environ-mental cause of Parkinson's disease. Aquilonius and Hartvig [17] employed pharmaceutical records of l-dopa use to identify a regional distribution of parkinsonism in Sweden. Saw and paper mills and steel alloy industries were prevalent in the county where the most cases of Parkinson's disease occurred. They postulated that heavy metals were possible toxins causing Parkinson's disease. Data from death rate prevalences indicated that the disease is greater in the northern states than the southern [18]. It also appears that there is a lower prevalence of Parkinson's disease in nonindustrialized countries [19]. These observations suggest that environmental factors may predispose to the development of parkinsonism.

Some investigations have implicated herbicides and pesticides in the etiology of Parkinson's disease. Barbeau [10] found a higher incidence of parkinsonism in areas of Quebec that specialized in market gardening and pulp milling which utilized herbicides and pesticides. A correlation between Parkinson's disease and the use of these chemicals was found in studies in Madrid, Spain [20] and Hong Kong [14]. However, a case controlled investigation of 150 patients and controls in the state of Kansas reported no correlation [15]. In China, a case-control study found an increased risk for Parkinson's disease associated with exposure to industrial chemicals including, but not limited to, pesticides and herbicides [21].

The drinking of well water has also been implicated in the etiology of Parkinson's disease. Five epidemiological studies have implicated well water as a possible risk factor [11,12,14,15,20]. Several of these studies were performed in young onset Parkinson's disease since exposure to a putative toxin may have been greater in this group. These patients were more likely both to have drunk well water and lived in a rural environment. We also found the parkinsonians were more likely to have drunk well water (figure 2). Rajput et al. [11] analyzed well water for 23 metals and found no difference in metal composition of water from case and controls.

If environmental toxins are involved in the etiology of Parkinson's disease, it is likely that only certain chemicals are responsible for the disease, therefore, contradictory results may reflect variations in farming practices in different areas with consequent exposure to different chemicals or chemical combinations. Toxicity is determined by a large number of factors, such as dose and duration of exposure, body temperature, level of exertion, and metabolic and nutritional state. Exposure can occur by a variety of mechanisms, such as through drinking well water, ingesting exposed foods, or aerial spraying. Variations in any of these factors may explain differences in results in different investigations.

CONCLUSION

The search for an environmental toxin which may cause Parkinson's disease has just begun. If a chemical as simple as MPTP can cause the pathological and clinical features of Parkinson's disease, it is likely that a chemical with a similar neurotoxin mode of action may also exist. The identification of these putative neurotoxins will not be an easy task. Several unanswered questions make this search even more formidable. Is there a critical time period in which a neurotoxin can damage the substantia nigra? Is this early in life or is total life exposure more important? Does genetic

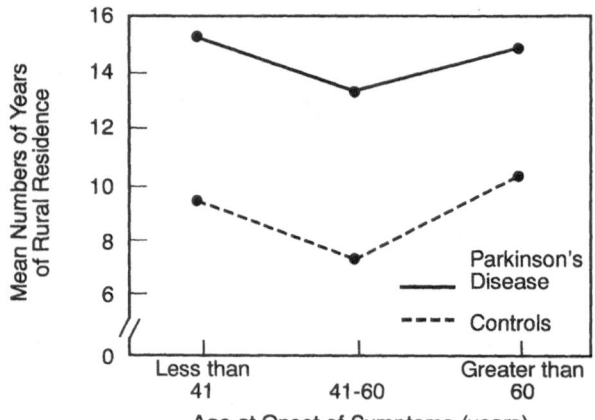

Figure 1. Mean number of years of rural residence for Parkinson's disease and controls prior to and including the 40th year of life for age of onset of parkinsonism less than 40 yr (N = 11), between 41 - 60 yrs. (n = 63), and greater than age 60 yrs. (N = 76).

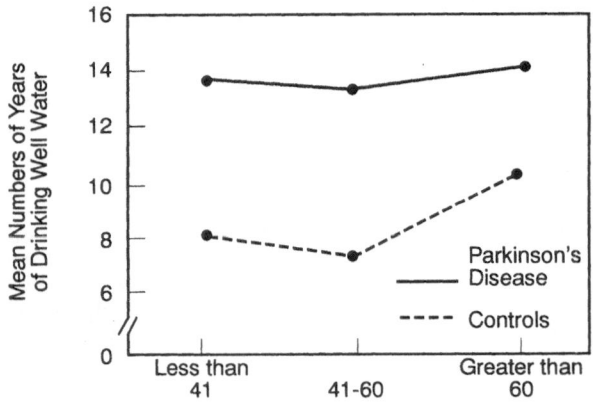

Figure 2. Mean number of years of drinking well water for Parkinson's disease and controls prior to and including the 40th year of life for age of onset of parkinsonism less than 40 yr (N = 11), between 41 - 60 yrs. (n = 63), and greater than age 60 years (N = 76).

predisposition play a major role in determining the neurotoxicity of putative toxins? Is there a long preclinical phase
of the illness? Hopefully future research will answer these
questions.

REFERENCES

1. Koller W.C. Classification of Parkinson's disease, in:
   Handbook of Parkinson's Disease, Koller W.C., ed. Marcel-
   Dekker, New York (1989).

2. Alvord E.C., Farno L.S., Kusske J.A., Kauffman R.J.,
   Rhodes J.S., Goetonski C.R., The pathology of
   parkinsonism. A comparison of degeneration in cerebral
   cortex and brainstem. *Adv. Neurol.* 5:175 (1974).

3. Lang A.E. Genetics, in: *Handbook of Parkinson's disease*,
   Koller W.C., ed., Marcel-Dekker, New York, 99, (1987)

4. Ward C.D., Duvoisin R.C., Ince S.E., Nutt J.D., Eldridge
   R., Calne D.B. Parkinson's disease in 65 pairs of twins
   and in a set of quadruplets. Neurology 33:815 (1983).

5. Marsden C.D., Parkinson's disease in twins. J. Neurol.
   Neurosurg. Psychiat. 30:105 (1987).

6. Marttila R.J., Kaprio J., Koskenvuo M., Rinne U.K.,
   Parkinson's disease in a nationwide twin cohort.
   Neurology 38:1217 (1988).

7. Calne D.B., Langston J.W., On the etiology of Parkinson's
   disease. Lancet 2:1457 (1983).

8. Ballard P.A., Tetrud J.W., Langston J.W., Permanent human
   parkinsonism due to MPTP. Neurology 35:969 (1985).

9. Burns R.S., LeWitt P., Ebert M.H., Pakkenberg H., Kopin
   I.J. The clinical syndrome of striatal dopamine
   deficiency; parkinsonism induced by MPTP. N. Engl. J.
   Med. 312:1418 (1985).

10. Barbeau A., Roy M., Cloutier T., Plasse L., Paris S.,
    Environmental and genetic factors in the etiology of
    Parkinson's disease. Adv. Neurol. 45:299 (1987).

11. Rajput A.H., Uitti R.J., Stern W., Laverty W., O'Donnell
    K., O'Donnell D., Yuen W.K. Dua A., Geography, drinking
    water chemistry, pesticides and herbicides and the
    etiology of Parkinson's disease. Can. J. Neurol. Sci.
    14:414 (1987).

12. Tanner C.M., Chen B., Wang W.Z., Peng M.L., Liu Z.L.,
    Liang X.L., Kao L.C., Gilley D.W., Schoenberg B.S.,
    Environmental factors in the etiology of Parkinson's
    disease. Can. J. Neurol. Sci. 14:419 (1987).

13. Tanner C.M., The role of environmental toxins in the
    etiology of Parkinson's disease. Trends in Neuroscience
    121:49 (1989).

14. Ho S.C., Woo J., Lee C.M., Epidemiologic study of Parkinson's disease in Hong Kong. Neurology 39:1314 (1989).

15. Koller W.C., Weldon G., Dolezal J., Chin T., Hassanein R., Environmental risk factors in Parkinson's disease. Neurology 39:363 (1989).

16. Thiesen B., Munoz D., Rajput A.H., Desel H., Substantia nigra neuronal counts in normal, rural, and urban populations. Neurology 38(Suppl):348 (1988).

17. Aquilonius S.M., Hartvig P., Utilization of antiparkinsonian drugs in Sweden. Upsala J. Med. Sci. 43:93 (1986).

18. Lux W.E., Kurtzke J.F., Is Parkinson's disease acquired? Evidence from a geographic comparison with multiple sclerosis. Neurology 37:467 (1987).

19. Langston J.W., Etiology of Parkinson's disease, in: *Handbook of Parkinson's Disease*, Koller W.C. ed., Marcel-Dekker: New York, (1986).

20. Jimenez-Jimenez F.J., Gonzales D.N., Jimenez-Rpiden S., Proceedings of the Ninth International Symposium on Parkinson's disease, World Congress of Neurology, p 118, (1988).

21. Tanner C.M., Chen B., Wang W-Z., Pang M., Liu Z., Liang X., Kao L.C., Gilley D.W., Goetz C.G., Schoenberg B.S., Environmental factors and Parkinson's disease (PD): A case-control study in China. Neurology 39:660 (1989).

NEW THERAPEUTIC STRATEGIES IN PARKINSON'S DISEASE: INHIBITION OF MAO-B BY Ro 19-6327 AND OF COMT BY Ro 40-7592

M. Da Prada, G. Zürcher, R. Kettler and A. Colzi

Pharmaceutical Research Department
F. Hoffmann-La Roche Ltd
CH-4002 Basel, Switzerland

INTRODUCTION

About three decades ago, it was observed for the first time that DOPA (levodopa, 3,4-dihydroxyphenyl-L-alanine) dramatically improved akinetic Parkinson's patients[1]. Several direct acting dopamine (DA) D_2 agonists were subsequently investigated but none proved to be sufficiently effective to be used routinely as the sole agent in the control of motor fluctuations in Parkinson's disease (PD). Motor fluctuations also frequently complicate DOPA therapy and limit its therapeutic benefit. When used as adjuvants to DOPA, DA agonists may modestly diminish the fluctuations, but a relatively high incidence of side-effects, chiefly psychotoxic, has limited their use[2]. It is noteworthy that chronic treatment of PD with DOPA is devoid of neurotoxic effects and does not produce damage to the nigro-striatal dopaminergic neurones[3]. Therefore, treatment with DOPA in combination with peripheral L-amino acid decarboxylase (AADC) inhibitors will remain the standard pharmacological therapy for PD at least for the next decade. However, new strategies are being actively investigated and some have been proposed to ameliorate dyskinesias and motor fluctuations occurring in patients under therapy with DOPA combined with a peripheral AADC inhibitor, e.g. benserazide (Madopar®) or carbidopa (Sinemet®)[2].

Many of the problems encountered during the long-term therapy of PD with DOPA are closely related to the short plasma half-life (about 1 h) of this amino acid. Thus, the rapid absorption of DOPA can lead to "peak-dose dyskinesia" followed by a bradykinetic state (wearing-off effect), due to its rapid elimination from the plasma.

In the attempt to ameliorate the pharmacokinetic behaviour of DOPA and to reduce peak plasma concentrations and fluctuations of the plasma DOPA levels, Madopar <HBS>[4] and Sinemet CR4[5], two controlled-release formulations have been recently developed. Although being advantageous in the treatment of PD, DOPA slow-release formulations have only partially solved the problem[4,6].

Inhibitors of monoamine oxidase type B (MAO-B) and catechol-O-methyltransferase (COMT) allow for novel strategies in the reduction of motor

The Basal Ganglia III, Edited by G. Bernardi *et al.*
Plenum Press, New York

fluctuations, which so frequently complicate DOPA therapy and limit its therapeutic benefit[2]. MAO-B inhibitors prolong the effect of DA at the receptor sites by reducing its extraneuronal deamination and possibly by suppressing hydrogen peroxide formation in oxidative deamination and the ensuing neuronal damage induced by free radicals[7,8]. COMT inhibitors improve DOPA bioavailability and, possibly, its plasma half-life by reducing the conversion of DOPA into 3-O-methyldopa (3-OMD)[9,10]. In the following, the preclinical neurochemical characteristics of a recently discovered inhibitor of the MAO-B (Ro 19-6327) and those of the COMT inhibitor, Ro 40-7592, will be highlighted. Both compounds are currently under clinical trials.

The molecular structure of the reversible MAO-B inhibitor Ro 19-6327 (N-[2-aminoethyl]-5-chloro-2-pyridine carboxamide HCl) and that of the reversible COMT inhibitor Ro 40-7592 (3,4-dihydroxy-4'-methyl-5-nitro-benzophenone) are shown in Fig. 1

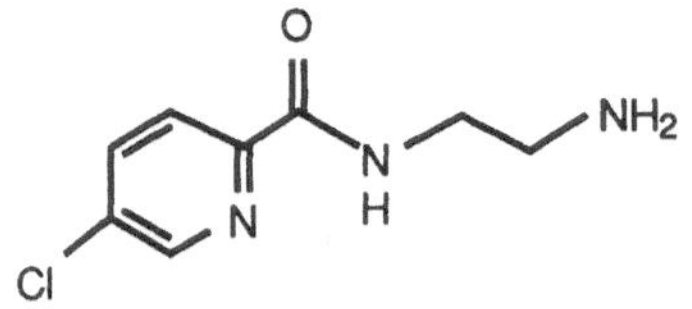

Ro 19-6327 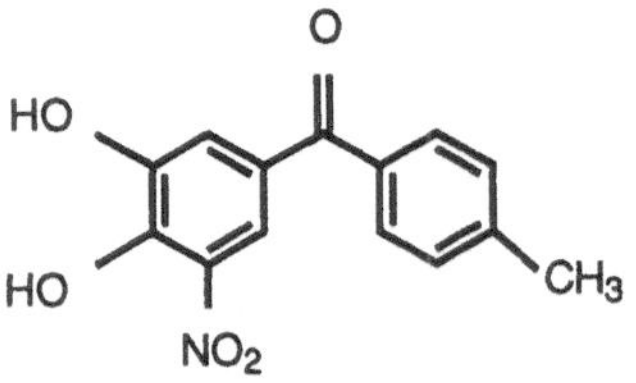   Ro 40-7592

Fig. 1. Molecular structures of MAO-B (left) and COMT (right) inhibitors

Ro 19-6327: A NOVEL, REVERSIBLE, HIGHLY SELECTIVE MAO-B INHIBITOR

To date, (-)deprenyl (selegiline) is the sole MAO-B inhibitor used clinically, as adjunct to DOPA and peripheral AADC inhibitors (e.g. Madopar® or Sinemet®) to improve symptom fluctuations in PD[11]. Moreover, as previously suggested[12], it was recently confirmed that an early selegiline therapy could delay the requirement for antiparkinsonian medication possibly by slowing the degenerative process underlying PD[13,14]. However, after repeated administration of relatively low doses (20 mg daily), selegiline did not only inhibit MAO-B but also MAO-A[15], with the consequent risk of provoking hypertensive crises (cheese-effect) by ingestion of food with high tyramine content. Furthermore, selegiline is metabolized to (-)-methamphetamine and (-)-amphetamine[15], and thus it remains an open question whether its beneficial effects in the therapy of PD are due to (-)-amphetamine production or to MAO-B inhibition. Ro 19-6327 belongs to a new chemical class of nontoxic, mechanism-based, fully reversible and highly selective MAO-B inhibitors both *in vitro* and *ex vivo*. In the rat brain *ex vivo*, the ED_{50} values (μmol/kg p.o. 2 h) obtained using PEA and 5-HT as substrates were: Ro 19-6327 0.3 and > 1200 vs selegiline 15 and > 500.

As shown in Fig. 2, Ro19-6327 almost completely inhibited MAO-B of rat liver at very low doses (1 μmol [0.23 mg]/kg p.o., 2 h). However, the compound, even at very high doses (1000 μmol/kg p.o.) did not affect MAO-A activity. In comparison selegiline was about 100 times less potent than Ro 19-6327 in inhibiting MAO-B and displayed much less selectivity than Ro 19-6327 (see Fig. 2). In fact, liver MAO-A activity was markedly inhibited 2 h after 100 μmol/kg selegiline p.o. (Fig. 2). The selectivity of Ro 19-6327 was further confirmed in experiments where the compound was administered orally to rats once daily for 2 weeks (Fig. 3).

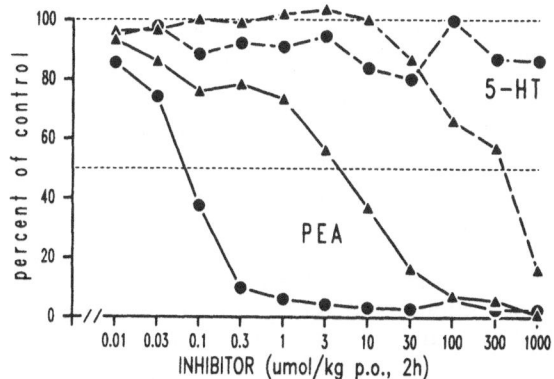

Fig. 2. Dose-dependent effects of Ro 19-6327 (circles) and selegiline (triangles) on the deamination of 5-hydroxytryptamine (5-HT) and phenylethylamine (PEA) in rat liver. MAO activities (means of triplicate determinations, N= 3-6, SEM < 10%) measured radiochemically[17] and expressed as percentage of corresponding controls (= 100%, N= 3-6), which, in nmol/mg tissue/h were: liver MAO-A, 19.1 ± 0.6; MAO-B, 11.9 ± 0.6.

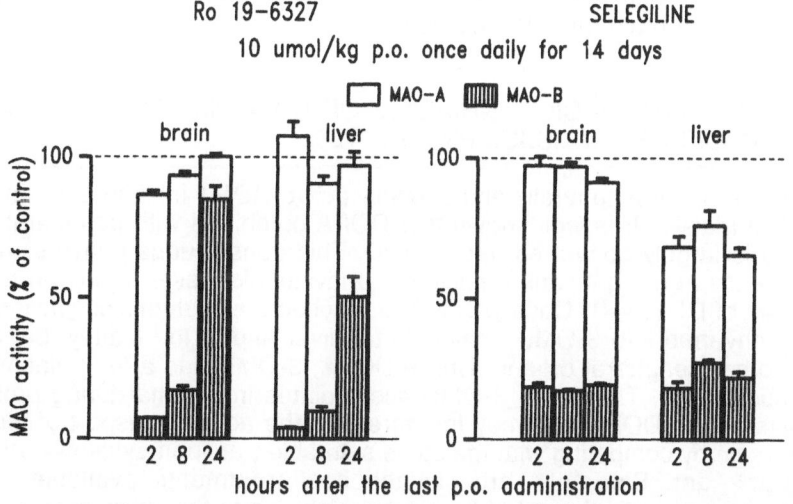

Fig. 3. MAO activity at various times after two weeks treatment with Ro 19-6327 or selegiline. MAO-A (□) and -B (▥) activities were measured in brain and liver of rats 2, 8 and 24 hours after the last dose of Ro 19-6327 or selegiline 10 μmol [about 2.3 mg]/kg p.o. each, once a day for two weeks.

These results show that even after repeated oral administration of a relatively high dose of Ro 19-6327, MAO-A activity was not affected (Fig. 3). In contrast, after repeated administration of equimolar amounts of selegiline, MAO-A activity of rat liver was markedly inhibited. As expected, after Ro 19-6327 but not after selegiline, MAO-B activity had largely recovered 24 h after the last dose both after single or repeated treatment (Fig. 3).

Previous experiments in black mice have shown that Ro 19-6327 prevented the 1-methyl-4-phenyl-1,2,5,6-tetrahydropyridine (MPTP)-induced depletion of brain DA and the formation of its quaternary derivative 1-methyl-4-phenyl-pyridinium (MPP^+) in the striatal tissue[16,17].

As shown previously and in contrast to selegiline, Ro 19-6327, which per se did not affect mean arterial blood pressure (MAP) even at high doses, did not potentiate the tyramine-induced increase in MAP in freely moving rats[16,17] and cannot be converted into amphetamine-like metabolites.

In healthy subjects, Ro 19-6327 is well tolerated even at the high dose of 200 mg and completely inhibits platelet MAO-B for 12 h after the dose of 40 mg[18]. In a recent study in healthy subjects brain MAO-B activity was measured by positron emission tomography (PET) using ^{11}C-selegiline i.v. as marker[19]. This PET study demonstrated that already 658 µg/kg Ro 19-6327 p.o. markedly inhibited (at least 90%) access of selegiline to brain MAO-B for up to 12 hours after dosing and provided indirect evidence that Ro 19-6327 potently inhibits human brain MAO-B activity *in vivo*. In these PET scan experiments an excellent correlation was found between brain K_3 (3-compartmental model analysis) and platelet MAO-B activity[19]. Therefore it can be concluded that in subjects treated with Ro 19-6327, platelet MAO-B inhibition can be used as a valid measure of brain MAO-B inhibition.

REVERSIBLE AND SPECIFIC INHIBITION OF THE COMT IN THE CNS AND IN THE EXTRACEREBRAL TISSUE BY Ro 40-7592

To date, no peripherally and centrally active COMT inhibitors are available for clinical trials[20]. It is well known that DOPA combined with peripheral AADC inhibitors is largely converted into 3-OMD. This conspicuous conversion takes place already in the gut wall, resulting in unavoidable waste and even reduced absorption of DOPA[9,21]. Once DOPA has reached the systemic circulation it can still be converted into 3-OMD, mainly in the liver and in the kidney, but also in various other peripheral organs. Unlike DOPA, 3-OMD has a long plasma half-life of about 15 h. Therefore, 3-OMD accumulates in plasma during prolonged administration of DOPA and may interfere with the active transport of the latter into the brain by competing with the same saturable transport system[21]. Although Madopar® and Sinemet® are still the best treatments available for PD, fluctuations in response, need of frequent dosing due to the short plasma half-life of DOPA, erratic pharmacokinetics, and considerable inter-individual variability in response to a given dose of DOPA, all cause problems in the management of these patients. Since we were convinced that part of these problems might be minimized by inhibition of COMT, we initiated a random biochemical screening in 1982 which, through the identification of the leads Ro 1-2812[22] and Ro 41-0960[23], culminated in the discovery of Ro 40-7592 (3,4-dihydroxy-4'-methyl-5-nitro-benzophenone)[24]. In contrast to the chemically closely related derivative OR-462, which inhibits COMT virtually only in the gastrointestinal tract[10,25], Ro 40-7592 inhibits COMT both in the CNS and in the periphery[25]. Ro 40-7592 is a competitive (K_i= 30 nM) inhibitor of COMT of intermediate duration of action, very potent in vivo (rat liver IC_{50}= 36 nM). As shown in Fig. 4,

Ro 40-7592 adminstered orally, inhibited COMT activity ex vivo in rat kidney, heart and liver (ED$_{50}$ [mg/kg p.o., 1 h]= 1.2, 3.3 and 6.3, respectively) in a dose-dependent manner and more potently than in rat brain (ED$_{50}$= 28 mg/kg p.o., 1 h).

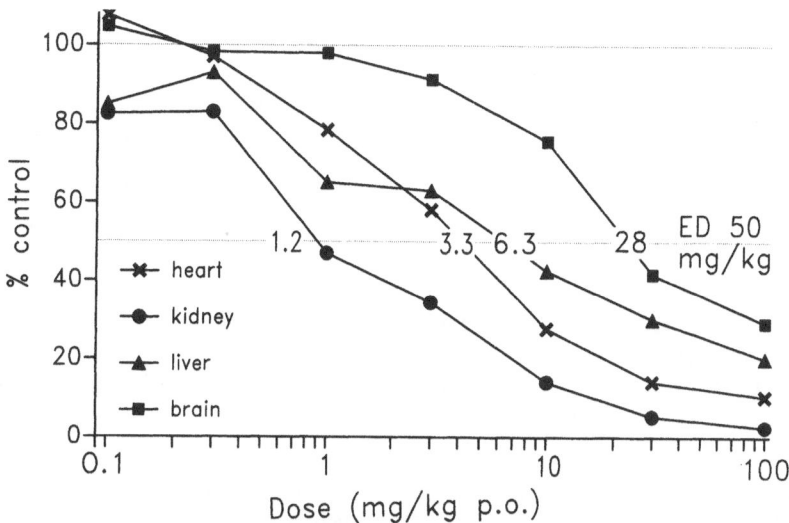

Fig. 4. Dose-dependent inhibition of COMT activity by Ro 40-7592 in various rat tissues. ED$_{50}$= (mg/kg p.o. 1 h). Absolute values for COMT activity in the rat (nmol/h/mg wet tissue, mean ± SEM, N= 3): kidney, 19 ± 1.2; heart, 0.4 ± 0.002; liver, 24 ± 2.2; ; brain, 0.9 ± 0.03. COMT activity was measured radiochemically[26].

It should be noted that the inhibitory potency expressed by ED$_{50}$ values determined *ex vivo* is underestimated because of the reversible nature of the COMT inhibition. Hence, a more reliable assessment of its COMT inhibitory effect is obtained by measuring, 2 h after Ro 40-7592 (30 mg/kg p.o.), the decrease of O-methylated catechol derivatives, e.g. 3-OMD, homovanillic acid (HVA) and 3-hydroxy-4-methoxy-phenylethyleneglycol (MOPEG); control values were attained about 16 h after dosing. OR-462 (100 mg/kg p.o.) failed to modify the level of HVA or MOPEG in rat brain[23].

The results in Fig. 5 show that Ro 40-7592 decreased in a dose-dependent manner the levels of the 3-O-methylated metabolites of DA, 3-methoxytyramine (3-MT) and HVA. During COMT inhibition by Ro 40-7592, the level of endogenous monoamines remained unaltered (not shown) whereas that of 3,4-dihydroxyphenylacetic acid (DOPAC) increased markedly (Fig. 5). As observed in rats[25] and monkeys[27], the bioavailability of DOPA, which was increased when administered in combination with benserazide, was further markedly increased when DOPA and benserazide were coadministered with Ro 40-7592, which blocked the conversion of DOPA into 3-OMD[25]. This increase was entirely due to COMT inhibition and to the consequent low converison of DOPAC into HVA. Free behaviour observations in rats have shown that Ro 40-7592 even at high doses (100 mg/kg p.o.) is devoid of stereotypic or cataleptic effects.

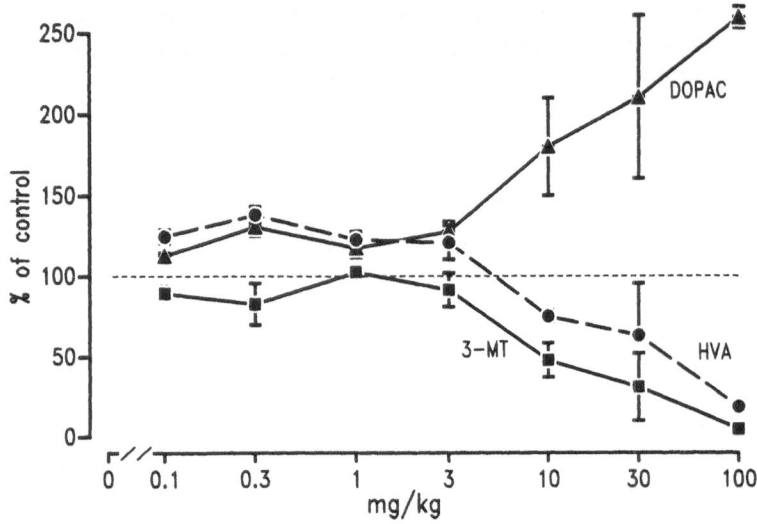

Fig. 5. Dose-dependent effect of Ro 40-7592 on the levels of the dopamine metabolites 3,4-dihydroxyphenylacetic acid (DOPAC), homovanillic acid (HVA) and 3-methoxytyramine (3-MT) in rat whole brain 1 h after p.o. administration (mean ± SEM, N= 6).

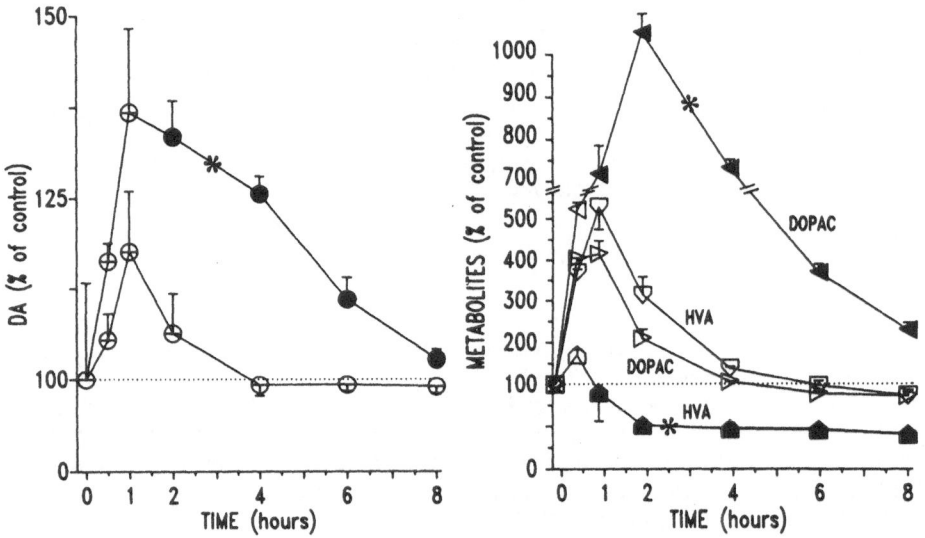

Fig. 6. Time course of changes in the concentration of dopamine (DA, left) and of its main deaminated metabolites homovanillic acid (HVA) and 3,4-dihydroxyphenylacetic acid (DOPAC) in rat brain. DOPA plus benserazide (10 mg/kg p.o., each) were administered without or with Ro 40-7592 (30 mg/kg p.o., curves marked with an asterisk) at time zero. Values are means ± SEM, N= 4. Statistical significance DOPA combined with benserazide and Ro 40-7592 versus DOPA plus benserazide: filled symbols, at least p < 0.05 (Mann-Whitney test). Absolute values (ng/g wet weight, mean, N= 4) DA= 957, DOPAC= 90, HVA= 91. Brain DA and its main metabolites were measured by HPLC with electrochemical detection.

The experiments in Fig. 6 show that the marked increase in DOPA bioavailability and the prolongation of the half-life of DOPA in plasma[25] produced in turn a marked and long-lasting increment of DA and DOPAC in the rat brain. DOPA coadministered with benserazide (10 mg/kg p.o., each) produced in the rat brain only a moderate and short-lasting increase of DA (Fig. 6, left) and of their main deaminated metabolites DOPAC and HVA (Fig. 6, right). In contrast, the administration of the COMT inhibitor Ro 40-7592 (30 mg/kg p.o.) in combination with DOPA and benserazide markedly increased for more than 6 h the level of DA and DOPAC in the rat brain (Fig. 6). In these conditions, as a result of the COMT inhibition, the brain level of HVA was significantly less than that measured in control rats in spite of the fact that the animals received DOPA (Fig. 6, right). Altogether, the experiments in Fig. 6 clearly show that in rats receiving relatively low doses of DOPA plus benserazide the central and peripheral blockade of the COMT induced by Ro 40-7592 resulted in a dramatic and long-lasting increment of the brain DA.

CONCLUDING REMARKS

The novel <u>MAO-B inhibitor Ro 19-6327</u> is a compound of high inhibitory potency and selectivity. Moreover, this compound is not converted to methamphetamine, does not interact with receptors or storage and reuptake mechanisms of the monoaminergic system, is devoid of hepatotoxicity and has no effect on the levels of rat brain monoamines and their metabolites or on gross behaviour in several animal species (rat, dog, Rhesus monkey). In healthy subjects, Ro 19-6327 was well tolerated even at doses (200 mg/kg p.o.) which produced virtually complete platelet MAO-B inhibition for more than 24 h.

From the present neurochemical comparative investigations with selegiline, Ro 19-6327 emerges as a novel, reversible, highly potent and selective MAO-B inhibitor. Whether the complete and selective inhibition of the MAO-B induced by Ro 19-6327 will be beneficial in the treatment of PD, especially retarding the progression of the disease and/or to alleviate "wearing-off" in advanced PD, is currently under clinical investigation. Additionally, the highly selective inhibitory effect on MAO-B makes Ro 19-6327 an excellent tool to clinically assess whether complete and selective inhibition of this enzyme protects dopaminergic neurones from age-associated degeneration, possibly by reducing the formation of free radical or the conversion of MPTP-like compounds into neurotoxins in the CNS.

In conclusion, Ro 19-6327 promises to be a safe adjuvant in the long-term therapy of PD and possibly of other degenerative brain disorders, e.g. Alzheimer's disease.

In basic biological research, [3H]Ro 19-6327, being a highly specific ligand of MAO-B, is also an ideal probe in binding studies for indirectly measuring MAO-B activity in tissue homogenates and for a precise localization of this enzyme in tissue slices in the CNS as well as in extracerebral tissues[28].

The preclinical profile of <u>the reversible COMT inhibitor Ro 40-7592</u> provides evidence that this compound is the most potent and selective inhibitor reported to date. Ro 40-7592, in contrast to old[20] or novel[10] COMT inhibitors, is able after oral administration to block markedly extracerebral as well as brain COMT activity. Ro 40-7592 alone or in combination with Madopar® has been shown to be well tolerated in clinical trials already under way. It might be anticipated that Ro 40-7592, by reducing the conversion of DOPA into 3-OMD, will have a marked DOPA sparing effect and, concomitantly, will reduce the utilization of the universal methyldonor S-adenosyl-L-methionine.

PET experiments in Rhesus monkeys (D.B. Calne, personal communication) demonstrated that Ro 40-7592, by inhibiting COMT activity, largely improved the visualization of the striatal dopaminergic nerve terminals labelled with ^{18}F-6-fluorodopa.

As indicated by our rat experiments, by reducing the formation of 3-OMD, Ro 40-7592 is expected to improve the therapeutic efficacy of DOPA by augmenting its brain availability. Therefore, by increasing DOPA bioavailability and its plasma half-life, Ro 40-7592, combined with Madopar® or Sinemet® standards or with their slow release formulations (Madopar <HBS>, Sinemet CR$_4$) offers a unique approach to a more refined drug treatment of PD.

REFERENCES

1. W. Birkmayer and O. Hornykiewicz, Der L-3,4-Dioxyphenylalanin (+ DOPA)- Effekt bei der Parkinson-Akinesie, Wien. Klin. Wschr. 73:787 (1961).
2. D.B. Calne, Drugs for the treatment of Parkinson's disease, in: "Handb. Exp. Pharm. 88", Springer Verlag (1989).
3. E. Melamed, Role of the nigrostriatal dopaminergic neurons in mediating the effect of exogenous L-dopa in Parkinson's disease, Mount Sinai J. Med. 55:35 (1988).
4. C.D. Marsden, U.K. Rinne, W.P. Koella and R. Dubuis, (eds), <Madopar.HBS, International Workshop on the "on-off"-phenomenon in Parkinson's disease. New possibilities for its management. Agno, Europ. Neurol. 27 [Suppl.]:1 (1985).
5. J.M. Cederbaum, L. Breck, H. Kutt and F.H. Dowell, Controlled-release levodopa/carbidopa. II. Sinemet CR4 treatment of response fluctuations in Parkinson's disease, Neurology 37:1607 (1987).
6. J.M. Cederbaum, The promise and limitations of controlled-release oral levodopa administration, Clin. Neuropharm., 12:147 (1989).
7. M. Sandler and V. Glover, Monoamine oxidase inhibitors in Parkinson's disease, in: "Drugs for the Treatment of Parkinson's Disease", D.B. Calne, ed., Handb. Exp. Pharm. 88, Springer Verlag, Berlin (1989).
8. P. Riederer, C. Konradi, G. Hebenstreit, and M.B.H. Youdim, Neurochemical perspectives to the function of monoamine oxidase, Acta Neurol. Scand., 126:41 (1989).
9. M. Da Prada, H.H. Keller, L. Pieri, R. Kettler and W.E. Haefely, The pharmacology of Parkinson's disease: basic aspects and recent advances, Experientia, 40:1165 (1984).
10. P.T. Männistö and S. Kaakkola, New selective COMT inhibitors: useful adjuncts for Parkinson's disease? TIPS, 10:54 (1989).
11. W. Birkmayer, P. Riederer, M.B.H. Youdim and W. Linauer, Potentiation of antiakinetic effect after L-DOPA treatment by an inhibitor of MAO-B - deprenyl, J. Neural Transm., 36:303 (1975).
12. W. Birkmayer, J. Knoll, P. Riederer, M.B.H. Youdim, V. Hars, and J. Marton, Increased life expectancy resulting from addition of L-deprenyl to Madopar® treatment in Parkinson's disease: a long-term study, J. Neural Transm., 64:113 (1985).
13. J.W. Tetrud, and J.W. Langston, The effect of deprenyl (selegiline) on the natural history of Parkinson's disease, Science, 245:519 (1989).

14. The Parkinson Study Group, Effect of deprenyl on the progression of disability in early Parkinson's disease, N. Engl. J. Med., 321:1364 (1989).
15. R. Schultz, K.-H. Antonin, E. Hoffmann, M. Jedrychowski, E. Nilsson, C. Schick, and P. Bieck, Tyramine kinetics and pressor sensitivity during monoamine oxidase inhibition by selegiline, Clin. Pharmacol. Ther., 46:528 (1989).
16. W. Haefely, R. Kettler, H.H. Keller, and M. Da Prada, Ro 19-6327, a reversible and highly selective monoamine oxidase B inhibitor: a novel tool to explore the MAO-B function in man, Adv. Neurol., 1990 (in press).
17. M. Da Prada, R. Kettler, H.H. Keller, and W.P. Burkard, Ro 19-6327, a reversible, highly selective inhibitor of type B monoamine oxidase, completely devoid of tyramine-potentiating effects: comparison with selegiline, in: "Progress in Catecholamines Research, Part B: Central Aspects, Alan R. Liss, Inc. (1988).
18. R. Kettler, and M. Da Prada, Platelet MAO-B activity in humans and stumptail monkeys: in vivo effects of the reversible MAO-B inhibitor Ro 19-6327, in: "Early Diagnosis and Preventive Therapy in Parkinson's Disease", H. Przuntek and P. Riederer, eds., Springer Verlag, Wien (1989).
19. G.W. Price, C.J. Bench, J.C. Cremer, S.K. Luthra, D.R. Turton, A.A. Lammertsma, R. Kettler, M. Da Prada, N. Wood, V. Jamieson, G. McClelland, and F.S.J. Frackowiak, Inhibition of human brain monoamine oxidase B by Ro 19-6327 - in vivo measurement using positron emission tomography, XIth Int. Congr. Pharmacol., Amsterdam, 1-6 July, 1990.
20. A. Reches, and S. Fahn, Catechol-O-methyltransferase and Parkinson's disease, Adv. Neurol., 40:171 (1984).
21. J.G. Nutt, Pharmacokinetics of levodopa, in: "Handbook of Parkinson's disease, W.C. Koller, ed., Marcel Dekker Inc., New York (1987).
22. G. Zürcher, H.H. Keller, H. Bruderer, J. Borgulya, and M. Da Prada, Caratteristiche neurochimiche di una nuova classe di inibitori della COMT attivi per via orale: livelli plasmatici di DOPA e 3-OMD nel ratto trattato con DOPA e benserazide, in: "Morbo di Parkinson e Demenze: Metodologie Diagnostiche", A. Agnoli and L. Battisti, eds., Pubbl. <<D. Guanella>> srl, Roma (1987).
23. A. Colzi, G. Zürcher, and M. Da Prada, Plasma concentrations of endogenous DOPA and 3-O-methylDOPA in rats administered benserazide and carbidopa alone or in combination with the reversible COMT inhibitor Ro 41-0960, in: "Early Diagnosis and Preventive Therapy in Parkinson's Disease", H. Przuntek and P. Riederer, eds., Springer Verlag, Wien (1989).
24. J. Borgulya, H. Bruderer, K. Bernauer, G. Zürcher, and M. Da Prada, Catechol-O-methyltransferase-inhibiting pyrocatechol derivatives: synthesis and structure-activity studies, Helv. Chim. Acta, 72:952 (1989).
25. G. Zürcher, H.H. Keller, R. Kettler, J. Borgulya, E.P. Bonetti, R. Eigenmann, and M. Da Prada, Ro 40-7592, a novel, very potent and orally active inhibitor of catechol-O-methyltransferase: a pharmacological study in rats, Adv. Neurol., 1990 (in press).
26. G. Zürcher, and M. Da Prada, Rapid and sensitive single-step radiochemical assay for catechol-O-methyltransferase, J. Neurochem., 38:191 (1982).

27. N. Penafiel, M. Da Prada, G. Zürcher, M.A. Mena, and J.G. de Yebenes, The effect of Ro 40-7592 on the peripheral metabolism of levodopa in the monkey, 1st Int. Congress of Movement Disorders, Washington, April 25-27 (1990).

28. M. Da Prada, R. Kettler, A.M. Cesura, and J.G. Richards, Reversible, enzyme-activated monoamine oxidase inhibitors: new advances, Pharmac. Res. Comm., 20:21 (1988).

CLINICAL AND NEUROENDOCRINOLOGICAL EFFECTS OF ESTROGENS IN

POSTMENOPAUSAL WOMEN WITH PARKINSON'S DISEASE

U. Bonuccelli, P. Piccini, A. Napolitano, A. Cagnacci*
G.B. Melis*, G.U. Corsini**, and A. Muratorio

Institute of Clinical Neurology, Institute of Obstetrics
and Gynecology*, Institute of Pharmacology**
University of Pisa, Italy

INTRODUCTION

Clinical and experimental evidence indicates a modulatory role of estrogens on the activity of nigrostriatal dopaminergic neurons (Hruska, 1985; Van Hartesveld & Joyce, 1986). However, conflicting data make the precise effects of estrogens on nigrostriatal dopaminergic system unclear, suggesting either a facilitatory (Nausieda et al., 1979; Hruska & Silbergeld., 1980; Chiodo et al., 1981) or an inhibitory (Bedard et al., 1978; Chiodo et al., 1979; Euvrard et al., 1979; Gordon, 1980) role.

In addition to the well known damage of the nigrostriatal dopaminergic system, a decreased dopamine content (Rinne, 1979; Javoy-Agid et al., 1984; Hornykiewicz & Kish, 1986) and the presence of Lewis's bodies (Den Hartog & Bethlem, 1960; Langston & Forno, 1978) have been demonstrated in the hypothalamus of Parkinson's disease (PD) subjects.

Moreover, some experimental data suggest a modulatory role of estrogens on the activity of hypothalamic neurons (Eikemburg et al., 1977; Raymond et al., 1978.)

Hypothalamic dopaminergic tonus influences the gonadotropin hypophyseal secretion, namely the luteinizing hormone (LH), by stimulating gonadotropin-releasing hormone (GnRH) secretion from the median eminence (Rasmussen et al., 1986) and modulating the activity of the endogenous opioid system (EOS) which in turn exerts an inhibitory control on LH secretion (Quigley & Yen, 1980,). This EOS action is strictly related to the circulating estrogen: in fact, the administration of an opioid receptor antagonist, such as naloxone, to women in fertile age, can increase LH plasma levels only in the late follicular phase and in the midluteal phase of the ovarian cycle, where a high estrogen secretion is present (Quigley & Yen, 1980). On the other hand, this LH response to naloxone is absent in hypogonadal states and

The Basal Ganglia III, Edited by G. Bernardi *et al.*
Plenum Press, New York

during postmenopausal age, but it can be restored by the administration of estrogen in both of these conditions (Reid et al, 1983; Melis et al., 1984; Melis et al., 1985; Shoupe et al., 1985).

LH secretion has not been investigated accurately in PD, thus we studied basal and naloxone stimulated LH secretion in postmenopausal women with PD before and after estrogen administration.

MATERIALS AND METHODS

We studied 6 women, mean age 64.4±6.4 years with idiopatic PD. The mean disease duration was 3.8±3.1 years, the Hoehn & Yahr stage was 1.8± ±0.7, and the Unified Parkinson's Disease Rating Scale (UPDRS)(Fahn et al., 1987) mean score was 33.6±10.2. Dementia and depression were ruled out by means of the Mini Mental State (Folstein et al., 1975) and the Hamilton Depression Rating Scale (1960) respectively. Four out of six patients had never received antiparkinsonian drugs; the remaining two patients had been treated with anticholinergic drugs until one month before the study. Six age-matched healthy women served as controls. All subjects had been in menopause for at least 10 years before the study and did not show any symptoms or signs of endocrinological or metabolic disorders.

In two consecutive days, at 9.00 am after an overnight fast, a 4-hour infusion of saline or saline plus naloxone (1.6 mg/h) has been performed. The subjects were supine and were not allowed to eat, drink or sleep during the test. Motor performances, during the infusions, were monitored by means of subitem III of the UPDRS and the tapping test (Nutt ed al., 1984) Blood samples, collected into heparinized plastic tubes every 10 minutes, were immediately centrifuged and plasma frozen until assayed. Plasma concentrations of prolactin (PRL), estrone (E1), estradiol (E2), progesterone (P), and testosterone (T) were measured in the first three samples by RIAs. Plasma LH and follicle-stimulating hormone (FSH) levels were measured in all samples by RIAs. Reagents were supplied by Sorin Biomedica (Vercelli, Italy).

All subjects were then treated with conjugated estrogens (CE) (1.25 mg/day bid) for 20 days, and the same clinical and endocrinological evaluation was then replicated.

All results are expressed as mean±SD, and statistical analyses of the results were performed by analysis of variance (ANOVA). Linear regression analysis was performed using the least squares method.

RESULTS

In all subjects basal FSH, PRL, E1, E2, P, and T levels were in the normal range for postmenopausal women and and similar in controls and in PD women (tab. 1).

Mean LH plasma levels were significantly (p<0.05) lower in PD women compared to controls (Fig.1); consequently the FSH/LH ratio was significantly higher in PD women; in this group mean LH plasma levels correlated significantly (r=-0,72, p<0.01) with the UPDRS score.

During naloxone infusion we did not observe any modifications of LH levels in both control and PD groups (Fig.2); motor score values were not changed during naloxone infusion in PD women when compared to the pre-infusion values.

TABLE 1. Mean (±SD) plasma concentrations of follicle -stimulating-hormone (FSH), prolactin (PRL), estrone (E1), estradiol (E2), progesterone (P) and testosterone (T) in 6 normal postmenopausal women and in 6 postmenopausal women with Parkinson's disease, before and after 20 days of conjugated estrogen (CE) administration (1.25 mg/day).

	CONTROLS		PARKINSONIANS	
	before CE	after CE	before CE	after CE
FSH (IU/L)	89.1 ±14.5	51.6** ± 8.9	79.7 ±13.3	36.0** ± 8.3
PRL (mcg/L)	8.1 ± 1.2	6.5 ± 1.8	10.6 ± 1.8	9.3 ± 1.1
E1 (pmol/L)	90.2 ±15.8	352.2** ±30.1	88.7 ±19.2	369.9** ±53.9
E2 (pmol/L)	42.0 ± 5.3	200.5** ±25.0	41.1 ± 4.2	220.2** ±18.3
P (nmol/L)	1.5 ± 0.1	2.22 ± 0.6	1.9 ± 0.2	1.6 ± 0.3
T (nmol/L)	1.8 ± 0.3	2.5 ± 0.5	2.1 ± 0.3	2.8 ± 0.5

** p<0.001 vs before CE

After CE administration in PD women a slight, although not significant, reduction of UPDRS score (29.0±8.8 vs 33.6±10.2) was observed, while tapping values were similar before and after CE.

CE administration increased E1 and E2 levels and reduced FSH levels in both groups (tab.1); by contrast CE treatment reduced significantly LH levels in controls but not in PD women (data not shown).

Naloxone infusion induced a significant (p<0.001) increase of LH levels in controls but not in PD women after estrogenization (Fig. 2); motor score values in estrogenized PD women were not modified by naloxone infusion when compared to the preinfusion score values.

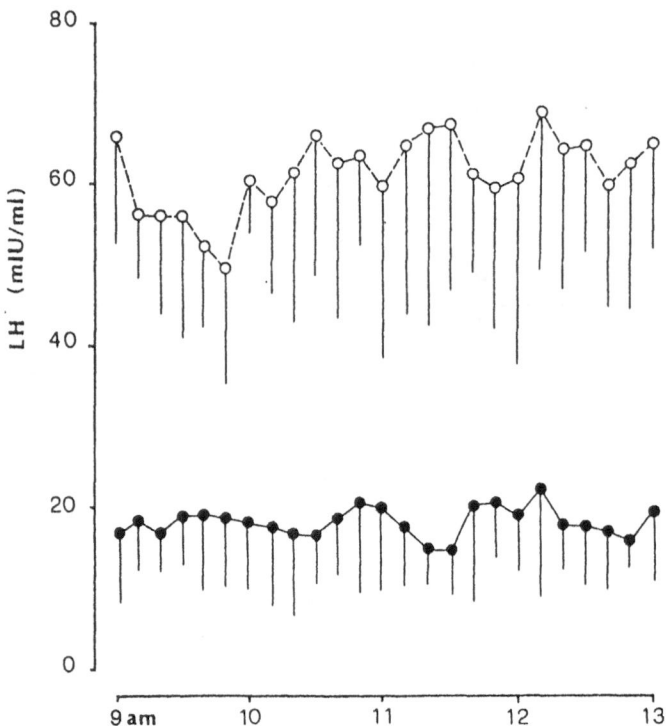

Fig. 1. Mean (± SD) luteinizing hormone (LH) plasma levels in 6 normal (O-- O) and 6 Parkinson's disease (PD) (●——●) postmenopausal women. LH levels were significantly lower in PD women compared to controls (p<0.05, ANOVA).

DISCUSSION

A selective reduction of LH secretion with a normal FSH secretion in PD women was found. Estrogen treatment was not able to significantly modify both basal and naloxone stimulated LH secretion in PD women. Motor performances were not modified in PD group by estrogen treatment.

The lack of a significant modification of motor score, after estrogen therapy, is in accordance with Giovannini et al.'s, (1981) study, in which a similar dosage of CE was given to 5 PD postmenopausal women. Bedard et al. (1977;1979), Koller et al. (1982) reported a worsening of motor performances but the severity of the disease was greater than in our study. More recently, Sandyk (1989) has reported

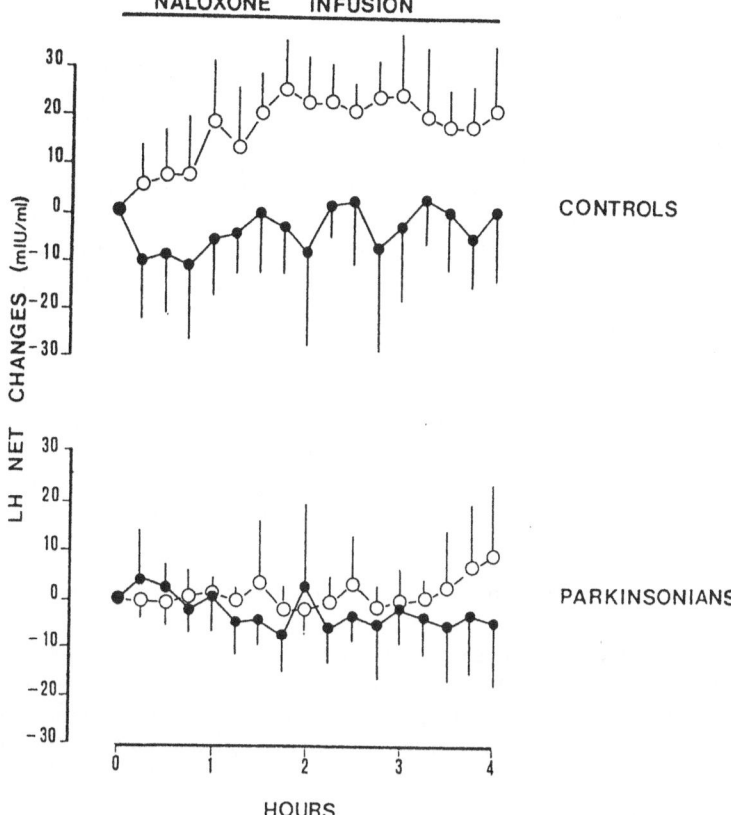

Fig. 2. Mean (± SD) net change in plasma luteinizing hormone (LH) levels during naloxone infusion (1.6 mg/h) in 6 controls (upper panel) and 6 Parkinson's disease (lower panel) postmenopausal women before (●—●) and after (o—o) conjugated estrogens (CE) administration (1.25 mg/day for 20 days). After CE administration, naloxone induced a significant increase of LH levels (p<0.001, ANOVA) in normal women.

that the suspension of estrogens in 2 PD postmenopausal women determined a worsening of PD symptoms. Taken together, these data suggest that estrogens can act differently in relation to the degree of disease severity. More PD patients need to be evaluated after estrogen treatment to fully clarify the effects of this hormone on motor performances in PD.

LH and FSH secretion is stimulated by hypothalamic GnRH and is modulated by gonadal peptides and steroids at peripheral level (Rivier et al., 1986; Vale et al., 1986; Ling et al., 1986). In our study the influence of peripheral factors can be excluded because of the levels of gonadal steroids were similar in PD patients and in controls and the secretion of ovarian peptides is presumably absent in postmenopausal women. Thus the decrease in LH we observed in PD women may be related to a disfunction of the central mechanisms which regulate its secretion. A reduced GnRH stimulation of LH secretion can be invoked to explain our data since LH is more dependent than FSH on the GnRH stimulus (Gross et al., 1987; Hall et al., 1988). An impaired hypothalamic dopaminergic function in PD may be responsible for a decreased GnRH stimulation and the abnormally low LH plasma levels we found. In fact, in PD patients, levo-dopa (Brown et al., 1978) or bromocriptine treatments (Hyyppa et al., 1978) have been reported to increase LH plasma levels; on the other hand, veralipride, a D2 receptor blocking agent is able to blunt LH plasma levels in postmenopausal women (Fioretti et al., 1989).

Because the EOS exerts an inhibitory control on LH secretion (Quigley & Yen, 1980) the reduced LH levels we found in this study, could be also explained by an hyperactivity of this system. However, during naloxone infusion we did not observe modifications of LH levels and a reduction of opioid peptides has been found in some cerebral areas (Agid & Javoy-Agid, 1985) and in cerebrospinal fluid (Nappi et al., 1985) of PD subjects, thus making a hyperactivity of EOS unlikely in this disease.

In normal postmenopausal women, chronic CE administration blunts LH levels probably by activating the hypothalamic EOS since naloxone infusion increases LH levels only after estrogenization (Melis et al., 1984). In PD women we observed no effect of CE on LH secretion and a lack of LH response to naloxone not only before, but also after estrogenization, thus indicating a complete unresponsiveness of the EOS hypothalamic system in PD.

It may be pointed out that at the hypothalamic level the activities of both the EOS and dopaminergic systems are strictly related: in fact, dopamine stimulates the release of ß-endorphin from the human hypothalamus in vitro (Rasmussen et al., 1987) and in normal postmenopausal women a chronic administration of bromocriptine can restore the LH response to naloxone (Melis et al., 1988). So far, no data concerning the possible restorative effect of dopamine agonists on LH response to naloxone in PD postmenopausal women are available. However, it is conceivable that an impaired activity of both opioid and dopaminergic systems at hypothalamic level can account for the abnormal LH secretion we observed in PD postmenopausal women. A few data indicating that levodopa (Brown et al., 1978) and bromocriptine (Hyyppa et al., 1978) are able to increase LH secretion in PD subjects, stand with this hypothesis which needs further support. Nonetheless, the usefulness of LH secretion as a biological marker of hypothalamic damage in PD seems very attractive because the diagnosis of PD depends at present entirely on recognition of clinical features.

Moreover there at present no useful early diagnostic markers of PD. Frank motor symptoms of the disease usually emerge only when at least 70-80% of striatal dopamine content has been reduced, indicating a long term pre-clinical phase of the disease, because several neurochemical compensatory mechanisms occur in the nigrostriatal system. In experimental animals a reduction of about 50% in hypothalamic dopamine content does not determine neuroendocrine modifications (Conte Devolx et al., 1981), thus suggesting that compensatory mechanisms also occur for hypothalamic dopaminergic functions. If the defect of LH secretion in PD postmenopausal women we found determines an abnormal clinical course of perimenopausal phenomena, it may represent a possible clinical window for early detection of PD, similar to depressive symptoms or reduced addictability to cigarette smoking both of which antedate the overt clinical onset of PD.

REFERENCES

Agid, Y., Javoy-Agid, F., 1985, Peptides and Parkinson's disease, TINS, 8:30-36.

Bedard, P., Langelier, P., Villeneuve, A., 1977, Oestrogens and extrapiramidal system. Lancet, II: 1367-1368.

Bedard, P., Dankova, J., Boucher, R., Langelier, P., 1978, Effect of estrogens on apomorphine-induced circling behaviour in the rat. Can. J. Physiol. Pharmacol., 65: 538-541.

Bedard, P., Langelier, P., Dankova, J., Villeneuve, A., Di Paolo, T., Barden, N., Labrie, F., Boissier, J.R., Euvrard, C., 1979, Estrogens, progesterone, and the extrapyramidal system, Adv. Neurol., 24: 411-422.

Brown, E., Brown, G.H., Kofman, O., Quarrington, B., 1978, Sexual function and affect in parkinsonian men treated with 1-Dopa. Am. J. Psychiatry, 135: 1552-1555.

Chiodo, L.A., Caggiula, A.R., Saller, C.F., 1979, Estrogen increases both spiperone-induced catalepsy and brain levels of ^{3}H spiperone in the rat. Brain Res., 172: 360-366.

Chiodo, L.A., Caggiula, A.R., Saller, C.F., 1981, Estrogen potentiates the stereotypy induced by dopamine agonists in the rat, Life Sci, 28: 827-835.

Conte-Devolx, B., Giraud, B., Castanas, E., Boudouresque, F., Orlando, M., Gillioz, P., Oliver, C., 1981, Effect of neonatal treatment with monosodium glutamate on the secretion of alpha-MSH, beta-endorphin and ACTH in the rat, Neuroendocrinology, 33: 207-211.

Den Hartog, W.A., Bethlem, J., 1960, The distribution of Lewy bodies in the central and autonomic nervous system in idiopathic paralisis agitans, J. Neurol. Neurosurg. Psychiatry, 23: 283-90.

Eikenburg, D.C., Ravitz, A.J., Gudelsky, G.A., Moore, K.E., 1977, Effects of estrogen on prolactin and tuberoinfundibular dopaminergic neurons, J. Neural. Transm., 40: 235-244.

Euvrard, C., Labrie, F., Boissier, J.R., 1979, Effect of estrogen on changes in the activity of striatal cholinergic neurons induced by DA drugs, Brain Res., 169: 215-220.

Fahn, S., Elton, R.L., Agid, Y., Barbeau, A., Calne, D.B., Duvoisin, R.C., Hoehn, M.M., Jankovic, J., Klawans, H.L., Lang, A.E., Lataste, X., Liebermann, A.N., Marsden, C.D., Markham, C.H., Mayeux, R., Rinne, U.K., Stern, G.M., Teychenne, P., Yahr, M.D., 1987, Unified Parkinson's disease rating scale, in: "Recent development in Parkinson's disease (Vol. II)", S. Fahn, C.D. Marsden, D.B. Calne, M. Goldstein, eds., Mcmillan Healthcare information, New York.

Fioretti, P., Cagnacci, A., Paoletti, A.M., Gambacciani, M., Soldani, R., Mauro, G.A., Spinetti, A., Melis, G.B., 1989, Effects of the antidopaminergic drug veralipride on LH and PRL secretion in postmenopausal women, J. Endocrinol. Invest., 12: 295-301.

Folstein, M.F., Folstein, S.E., McHugh, P.R., 1975, Mini Mental State, a practical method for grading the cognitive state of patients for the clinician, J. Psychiatry Res., 12: 189-193.

Giovannini, P., Martinez-Campos, A., Scigliano, G., Caraceni, T., 1981, Morbo di Parkinson ed estrogeni, in: "Atti della 7^ Riunione della Lega Italiana per la Lotta contro il Morbo di Parkinson e le Malattie Extrapiramidali", A. Agnoli, G. Bertolani, eds., D. Guanella, Roma.

Gordon, J.H., 1980, Modulation of apomorphine-induced stereotipy by estrogen: time course and dose response, Brain Res. Bull., 5: 679-682.

Gross, K.M., Matsumoto, A.M., Bremner, W.J., 1987, Differential control of luteinizing hormone and follicle-stimulating hormone secretion by luteinizing hormone-releasing hormone pulse frequency in man, J. Clin. Endocrinol. Metab., 64: 675-680.

Hall,J.E., Brodie, T.D., Badger, T.M., Rivier, J., Vale, W., Conn, P.M., Shoenfeld, D., Crowley, W.F., 1988, Evidence of differential control of FSH and LH secretion by gonadotropin-releasing hormone (GnRH) from the use of a GnRH antagonist, J. Clin. Endocrinol. Metab., 63: 524-531.

Hamilton, M., 1960, A rating scale for depression, J. Neurol. Neurosurg. Psychiatry, 23: 56-62.

Hornykiewicz, O., Kish, S.J., 1986, Biochemical pathophysiology of Parkinson's disease, Adv. Neurol., 45: 19-34.

Hruska, R.E., Silbergeld, E.K., 1980, Increased dopamine receptor sensitivity after estrogen treatment using the rat rotation model, Science, 208: 1466-1467.

Hruska, R.E., 1985, Sex hormone exposure and nonreproductive behaviour, Int. J. Ment. Health., 14: 112-134.

Hyyppa, M.T. Langvik, V.A., Rinne U.K., 1978 Plasma pituitary hormones in patients with Parkinson's disease treated with bromocriptine, J. Neural. Transm., 42: 151-157.

Javoy-Agid, F., Ruberg, M., Pique, L., Bertagna, X., Taquet, H., Studler, J.M., Cesselin, F., Epelbaum, J., Agid, Y., 1984, Biochemistry of hypothalamus in Parkinson's disease, Neurology, 34: 672-675.

Koller, W.C., Barr, A., Biary, N., 1982, Estrogen treatment of dyskinetic disorders, Neurology, 32: 547-549.

Langston, J.W., Forno, L.S., 1978, The hypothalamus in Parkinson's disease, Ann. Neurol., 3: 129-133.

Ling, N., Ying, S.Y., Ueno, N., Shimasaki, S., Esh, F., Hotta M., Guillemin, R., 1986, Pituitary FSH is released by a heterodimer of the ß-subunits from the two forms of inhibin, Nature, 321: 779-782.

Melis, G.B., Paoletti, A.M., Gambacciani, M., Mais, V., Fioretti, P., 1984, Evidence that estrogens inhibit LH secretion trough opioids in postmenopausal women, Neuroendocrinology, 39: 60-63.

Melis, G.B., Gargiulo T., Pallotti, A., Gambacciani, M., Cagnacci, A., Paoletti, A.M., Petacchi, F.D., Fioretti, P., 1985, Disappearance of opioid control of LH secretion in short term ovariectomized women, J. Endocrinol. Invest., 8 (suppl. 2), 41-46.

Melis, G.B., Cagnacci, A., Gambacciani, M., Paoletti, A.M., Caffi, T., Fioretti, P., 1988, Chronic bromocriptine administration restores luteinizing hormone response to naloxone in postmenopausal women, Neuroendocrinology, 47: 159-163.

Nappi, G., Petraglia, F., Martignoni, E., Facchinetti, F., Bono, G., Genazzani, A.R., 1985, ß-endorphin cerebrospinal fluid decrease in untreated parkinsonian patients, Neurology, 35: 1371-1374.

Nausieda, P.A., Koller, W.C., Weiner, W.J., Klawans, H.L. 1979, Modification of postsynaptic dopaminergic sensitivity by female sex hormones, Life Sci., 25: 521-526.

Nutt, J.G., Woodward, W.R., Hammerstad, J.P., Carter, J.H., Anderson, J.L., 1984, The on-off phenomenon in Parkinson's disease. Relation to levodopa absorption and transport, N. Eng. J. Med., 310: 483-488.

Quigley, M.E., Yen, S.S.C., 1980, The role of endogenous opiates on LH secretion during the menstrual cycle, J. Clin. Endocrinol, Metab., 51: 179-181.

Raymond, V., Beaulieu, M., Labrie, F., Boissier, J.R., 1978, Potent antidopaminergic activity of estradiol at the pituitary level on prolactin release, Science, 200: 1173-1175.

Rasmussen, D.D., Liu, J.H., Wolf, P.K., Jen, S.S.C., 1986, Gonadotropin-releasing hormone neurosecretion in the human hypothalamus. In vitro regulation by dopamine, J. Clin. Endocrinol., Metab., 62: 479-483.

Rasmussen, D.D., Liu, J.H., Wolf, P.L., Yen, S.S.C., 1987, Neurosecretion of human hypothalamic immunoreactive ß-endorphin: in vitro regulation by dopamine, Neuroendocrinology, 45: 197-200.

Reid, R.L., Quigley, M.E., Yen, S.S.C., 1983, The disappearance of opioidergic regulation of gonadotropin secretion in postmenomausal women, J. Clin. Endocrinol. Metab., 57: 1107-1110.

Rinne, U.K., 1979, Neuroendocrine and extrapyramidal dopamine neurons in Parkinson's disease, in: "Neuroendocrine Correlates in Neurology and Psychiatry", E.E. Muller, A. Agnoli, eds., Elsevier, Amsterdam, pp. 119-125.

Rivier, C., Rivier, J., Vale, W., 1986, Inhibin-mediated feedback control of follicle-stimulating hormone secretion in the female rat, Science, 234: 205-208.

Sandyk, R., 1989, Estrogens and the pathophysiology of Parkinson's disease, Intern. J. Neurosci., 45: 119-122.

Shoupe, D., Montz, F.J., Lobo, R.A., 1985, The effects of estrogens and progestins on endogenous opioid activity in oophorectomized women, J. Clin. Endocrinol. Metab., 60: 178-183.

Vale, W., Rivier, J., Vaughan, J., McClintock, R., Corrigan, A., Woo, W., Karr, D., Spiess, J., 1986, Purification and characterization of an FSH releasing protein from porcine ovarian follicular fluid, Nature, 321, 776-779.

Van Hartesveldt, C., Joyce, J.N., 1986, Effects of estrogen on the basal ganglia, Neurosci. Biobehav. Rev., 10: 1-14.

VISION AND VISUAL DISCRIMINATION SUFFER IN PARKINSON'S DISEASE: EVOKED POTENTIAL EVIDENCE

I. Bodis-Wollner,[1,2] S. Pang,[1] A. Falk,[1] M.F. Ghilardi,[1] and L. Mylin[1]

The VEP Laboratory, Box 1052, Department of Neurology[1] and Department of Ophthalmology,[2] Themount Sinai School of Medicine, CUNY, New York, NY 10029

Introduction

Patients affected by Parkinson's Disease (PD) not infrequently complain of some visual difficulty yet their visual acuity is intact. Although on careful questioning "double vision" may clarify itself as perceiving uncertain borders of images, more complex complaints cannot be easily classified as "visual" or as "cognitive" defects. In fact, both have been demonstrated in PD. While spatial frequency dependent visual defects probably result from retinal dopaminergic (DA) deficiency, the anatomy and neuropharmacology of cognitive changes in PD is uncertain.

In 1978 we (2) reported that half of advanced PD patients have abnormal visual evoked potentials (VEP). While those engaged in the study of movement disorders expressed scepticism, many subsequent studies demonstrated that the VEP and in addition the electroretinogram (ERG) is also abnormal in PD (3). These changes are most pronounced to grating or checkerboard stimuli consisting of medium coarse pattern element sizes (4,5,6). Furthermore both measures return to normality following Dopamine precursor therapy (5,6,7,8). Psychophysical tests of contrast sensitivity also revealed that the major visual deficit in PD affects the detection of medium sized patterns without substantially degrading acuity (9). Most of these visual deficits in PD could be explained by DA deficiency affecting DA amacrine cells, well described in the primate retina (10).

Despite the evidence implicating retinal DA deficiency as a cause of visual dysfunction, one can't conclude that all visual defects in PD are caused by retinal dopaminergic deficiency alone. Some remarkable observations do not fit this simple scheme. For one, it has been noted (11,12) that Parkinson's Disease patients suffer from visual deficits which are more evident for patterns in one and not in the other orientation. Neurons in the retina and LGN do not distinguish between orientation of stimuli, thus a "neural astigmatism" implies cortical pathology. One additional experimental reason to evaluate the relationship between retinal and retrochiasmatic changes in the visual system in Parkinson's Disease, emerges from neuropsychological data. These show that the behavioral response to visual orientation and the well known neuropsychological test called the Benton Line Orientation Test (13) are commonly affected in PD. Patients show defects in judging angular relations (14). It is therefore of interest to evaluate primary and later "cognitive" components of the visual evoked potential obtained simultaneously using identical stimuli. We used as stimuli vertically oriented and tilted grating patterns randomly interspersed. This stimulus presentation paradigm, often called "odd-ball," elicits late "cognitive" EPs. The best known is the P300 (15). This late positive component of the EP only occurs to the rarely presented target. Massive experimental evidence suggests that P300 indexes stimulus evaluation and decision by the observer. Although the origin of the cognitive components of the VEP is unknown, we postulated that if all visual changes in Parkinson's Disease occur as a result of a downstream retinal dopaminergic deficiency, then we should find perfect correspondence between abnormal primary and late evoked potentials. Otherwise we have to assume additional pathology. Here we report our exploratory EP study of Parkinsonian patients using concurrent measures of primary and cognitive visual EPs, using oriented sinusoidal grating patterns presented in an "odd-ball" paradigm as test targets. We obtained both primary and late VEP components (VP3) during the same task.

Methods

1. Stimuli

Sinusoidal grating patterns in two orientations were presented in an "odd-ball" paradigm. Both had the same spatial frequency: 2.3 cpd. The "frequent" stimulus was a vertical pattern presented 90% of time while the target was an oblique pattern presented rarely, 10% of the time. The patient's task was to count the number of target occurrences, typically 50, randomly interspersed between nontarget stimuli. The oblique targets were tilted from the vertical by 30 degrees or were horizontal. The screen subtended nine degrees at the eye and the patterns had a contrast of 55%. The mean luminance of the screen was 300 cd/m^2 and the observer's distance was 1.44 cm from the screen. Viewing was monocular, with the eye which had the better acuity. The stimulus screen was surrounded by an equiluminous surround.

2. Recordings

Scalp electrodes were fixed on the scalp in an eight channel (20 patients) or 16 channel (8 patient) montage. Common recording sites were the occipital mid-line electrode, (Z5); two lateral occipital electrodes labelled RO and LO; high occipital electrode at 23% of the inion nasion distance (Z23), right and left parietal electrodes, a vertex electrode

(Z50) and the midfrontal electrode (Z63). The reference electrode was Z70 for the VEP. The ground electrode was placed on the forehead. For the primary visual evoked potentials, traces were routinely evaluated at Z5 whereas for evaluating late cognitive components, the Z50 and the Z63 to linked mastoid montage was used. The signals were amplified and averaged either with Dantec Evomatic (8 channels) or with a Dantec Neuroscope (16 channels). Filters were set between 0.3 and 50 Hz. Primary visual evoked potentials were measured for the latency of N70 (17), P100 and 140 while VP3 and the preceding negative component N2 were measured both for amplitude and for latency. Amplitudes were measured both as N2 to P3 amplitude as well as individually from baseline. Baseline was established as the average voltage during 50 msec interval at the beginning of the onset of the stimulation.

3. Experimental paradigm

Stimulus presentation rate was 1.4 Hz, that is the duration of either rare or frequent targets was 700 msec. The onset and offsets were abrupt and between stimuli the screen remained equiluminant. The observers were asked to keep a mental count of the number of targets designated as rare.

4. Observers

The mean age of the 29 patients was 62.7 years with a range of 36 to 75. The average duration of PD was 120.4 months. Their staging (Hoehn and Yahr (16)) was from two to four, most classified in stage three. The average duration of levodopa treatment was 92 months with a range of 10 to 168 months. All patients were screened ophthalmologically. Patients with visual acuity better than 20/30 were selected for this study. There were eight age matched normal controls without Parkinson's Disease, and two non-Parkinsonian neurological patients without any neuropsychological impairment.

5. Control Recordings

In some patients we varied target stimulus orientation and used horizontally tilted patterns. Secondly, in an occasional patient and age-matched control we counterbalanced the stimulus presentation such that the rarely presented target was designated as the vertical pattern and the frequently presented stimulus was the oblique pattern. Lastly, in all controls and in some patients without a tendency to blepharospasm we recorded vertical and horizontal eye movement by placing electrodes above, below and on each side of one orbit. When we evaluated how eye movements may influence a presumed normal or abnormal P300 latency or amplitude, we found little overlap in time.

Results

The transient onset visual evoked potential to a stimulus with the spatial frequency of 2.3 cpd shows a small N70 component which is recorded best at the midline. There is an evident P100 and an evident N140. The normal values for N70

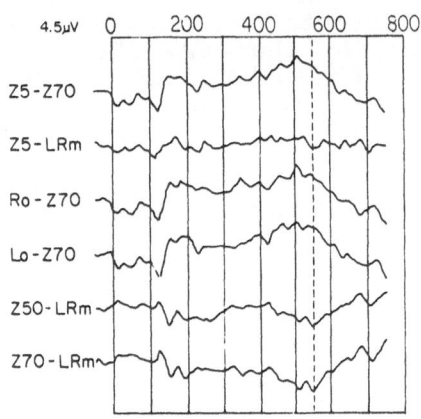

Fig. 1 shows primary VEPs (traces Z5, R0 and L0) and delayed (550 msec) VP3 to a target, in a 75 year old PD patient.

in our laboratory for the stimulus of 2.3 cpd in age matched controls is 85 msec ± 4.3 msec. For the P100 it is 126 msec ± 7.2 msec. Figure 1 shows responses to the (rear target) oblique stimulus. A VP3 does occur only to the rare stimulus, while a primary VEP occurs to both. Primary EPs are best seen in the montage from the occiput to the front (Z5 to Z70 electrode) and the VP3 is best seen in the montages from Z50 and from Z63 to linked mastoids. The latency of the normal VP3 based on our age matched normals for this paradigm, is 440 msec ± 37 msec. The "VP3 montage" reveals in addition to the VP3 also the inverted polarity primary evoked potential (Fig. 1).

Eye movements are <u>suppressed</u> during target presentation. The reason why we found minimal intrusion of eye movements in this task may be because the observer has to look at the center of the screen in order to judge pattern orientation. (In the auditory P300 paradigm there is evident automatic shift of the eyes as a result of target tones.)

The correlation between N70, the first postsynaptic cortical VEP component (17), and VP3 could not be evaluated in each patient. Unfortunately a precise measurement of N70 latency was not always possible due to its small amplitude. We then found the following results (see Tables).

Table 1

N70	P300 NORMAL	P300 ABNORMAL	
NORMAL	16	7	23
ABNORMAL	7	18	25
	23	25	

P100	P300 NORMAL	P300 ABNORMAL	
NORMAL	15	5	20
ABNORMAL	8	20	28
	23	25	

Abnormal potentials where defined as SD½1.5 for primary EPs and an SD½25 for P300. As the Tables show, of the examined patients, 16 eyes had normal N70 and VP3 and 18 had both abnormal. However, there were 7 discordant pairs: that is normal N70 and abnormal VP3 or vice versa. The correlation of P100 with VP3, there were 8 discordant pairs showing abnormal P100 with normal VP3 and only 5 patients with normal P100 had abnormal VP3s. Both P100 and VP3 were abnormal in 20 patients. Thus some patients may have an abnormal P100 and still normal VP3 and conversely, some may have a normal P100 and abnormal VP3. In summary, our results for this group of patients suggest that on the whole, there is a correspondence between an abnormal primary visual evoked potential and an abnormal VP3 obtained using simultaneously the same paradigm. The correspondence however is not perfect. Furthermore, the concordance between primary and cognitive VEPs is higher for P100 than for N70.

Discussion

The retinal effect of DA depletion was first described in rats by Malmfors in 1963 (18) and it was shown that the VEP can also be delayed by retinal dopaminergic blockade (19,20). Dopaminergic deficiency induced by either ocular 6-OH-DA treatment (21) or by systemic 1-methyl-4-phenyl-1-2-3-6-tetrahydropyridine (MPTP) causes abnormal ERGs and reduced retinal dopamine content in monkeys (22). Furthermore, the visual effect of DA deficiency in monkeys is similar to the visual deficits seen in PD. However, dopaminergic blockade does not lead to <u>loss</u> of vision: it leads to a change in normal visual functions regarding the <u>contrast sensitivity</u> and <u>time locked</u> responses (23) to particular sizes of retinal images. Are visual complaints in PD the consequence of combined retinal and basal ganglia DA deficiency?

While visuo-spatial functions and short term memory are often affected in PD (24), the role of the basal ganglia in visuo-spatial aspects of cognitive changes in PD is not well established (25). One attractive theory links basal ganglia structures to cognitive functions through thalamocortical loops. At least one of these is probably relevant to the visuo-spatial dysfunction in DA deficiency. This loop connects the caudate and the dorsolateral prefrontal cortex. Studies in the monkey indicate that dopaminergically innervated neurons of the dorsolateral prefrontal cortex (26,27) are involved in mnemonic coding of visual location in space. DA neurotransmission in fact is involved in linking frontal and parahippocampal regions. This is noteworthy because generator sources of a "cognitive" evoked potential component, the P300 have been inferred in or near the hippocampal formation in humans (28). In general P300 is generated as a response to the rear and random presentation of a target in a discriminatory task involving two slightly differing auditory stimuli. It is known that P300 latency grows with aging. P300 also becomes delayed in a number of neurodegenerative diseases. Several studies have suggested that auditory P300 changes occur in Parkinson's Disease (29,30,31) and there is a correlation between an abnormal P300 and certain subsets of a neuropsychological battery, foremost visual-spatial impairment in short-term memory (32). We have presented evidence that a late positive component of the visual evoked potential we labelled VP3, is elicited using simple sinusoidal grating stimuli. For eliciting the late components of the visually evoked potential, we used the presentation of the stimuli, designated target and non-target in the so-called "odd-ball" paradigm. Our previous studies (33) show that VP3 amplitude varies inversely and its latency directly with the probability of the stimulus. Hence the VP3 does fit into the category of late "cognitive" EP components. We have shown that a significant number of PD patients have abnormal VP3.

Is there a relationship between abnormal primary processing and visual cognitive changes in PD? The present exploratory study in PD suggests that the primary VEP and the VP3 obtained simultaneously may not be tight coupled since an abnormal P300 was not absolutely predictable from an abnormal primary visual evoked potential. However, since we did not have premorbid EPs, it is conceivable that patients classified as normal in reality could have been abnormal had we known their premorbid VEP. Our current longitudinal study in a larger population addresses coupling of P100 and VP3 as disease advances.

In summary, our study describes a new method to obtain primary and cognitive visual P300 potentials in humans and the results show that both were affected in a significant number of patients.

References

1. I. Bodis-Wollner, 1990, Visual deficits related to dopamine deficiency in experimental animals and Parkinson's Disease patients, Trends in Neuroscience, in press.
2. I. Bodis-Wollner and M.D. Yahr, 1978, Measurement of visual evoked potentials in Parkinson's disease, Brain, 101:661-671.
3. I. Bodis-Wollner, 1990, The visual system in Parkinson's Disease, in: "Vision and the Brain," B. Cohen and I. Bodis-Wollner, eds., Raven Press, New York.
4. A. Tartaglione, N. Pizio, I. Bo, L. Spadavecchia, and E. Favale, 1985, Spatial properties of pattern as determinants of visual evoked potential changes in Parkinson's syndrome, in: "Evoked potentials: neurophysiological and clinical aspects," C. Morocutti and P.A. Rizzo, eds., Elsevier, Amsterdam.
5. M. Onofrj, M.F. Ghilardi, M. Basciani, and Gambi D., 1986, Visual evoked potentials in Parkinson's disease and dopamine blockade reveal a stimulus-dependent dopamine function in humans, J Neurol Neurosurg Psychiatry, 49:1150-1159.
6. P. Stanzione, F. Pierelli, A. Peppe, P.A. Rizzo, and C. Morocutti, 1989, Pattern visual evoked potentials and electroretinogram abnormalities in Parkinson's disease: effects of L-dopa therapy, Clin Vis Sci, 4:115-128.
7. P.A. Bhaskar, S. Vanchilingam, E.A. Bhaskar, A. Devaprabhu, and R.A. Ganesan, 1988, Effect of L-dopa on visual evoked potentials in patients with Parkinson's disease, Neurology, 36:1119-1121.

8. I. Bodis-Wollner, M. Yahr, L. Mylin, and J. Thornton, 1982, Dopaminergic deficiency and delayed visual evoked potentials in humans, <u>Ann. Neurol.</u>, 11:478-483.

9. I. Bodis-Wollner, M.S. Marx, S. Mitra, P. Bobak, L. Mylin, and M. Yahr, 1987, Visual dysfunction in Parkinson's disease, <u>Brain</u>, 110:1675-1698.

10. A.P. Mariani and J.N. Hokoc, 1988, Two types of tyrosine hydroxylase-immunoreactive amacrine cell in the rhesus monkey retina, <u>J. Comp. Neurol.</u>, 276:81-91.

11. D. Regan and C. Maxner, 1987, Orientation-selection visual loss in patients with Parkinson's disease, 1987, <u>Brain</u>, 110:415-432.

12. C. Bulens, J.D. Meerwaldt, G.J. van der Wildt, and C.K. Keemink, 1986, Contrast sensitivity in Parkinson's disease, <u>Neurology</u>, 36:1121-1125.

13. A. Benton, K. Hamsher, N. Varney, and O. Spreen, 1983, "Contributions to Neuropsychological Assessment: A Clinical Manual," Oxford University Press, New York.

14. F. Bowen, M. Burus, E. Brady, and M.D. Yahr, 1972, A note on alterations of personal orientation in parkinsonism, <u>Neuropsychologia</u>, 14:425-429.

15. S. Sutton, M. Braren, and J. Zubin, 1965, Evoked potential correlates of stimulus uncertainty, <u>Science</u>, 150:1187-1188.

16. M.M. Hoehn and M.D. Yahr, 1967, Parkinsonism: onset, progression and mortality, <u>Neurology</u>, 17:427-442.

17. I. Bodis-Wollner, L. Mylin, and S. Frkovic, 1989, The topography of the N70 component of the visual evoked potential in humans, <u>in</u>: "Topographic Brain Mapping of EEG and Evoked Potentials," K. Maurer, ed., Springer-Verlag, Berlin.

18. T. Malmfors, 1963, Evidence of adrenergic neurons with synaptic terminals in the retina of rats demonstrated with fluorescence and electron microscopy, <u>Acta Physiol. Scand.</u>, 64:58-66.

19. R.S. Dyer, W.E. Howell, and R.C. McPhail, 1981, Dopamine depletion slows retinal transmission, <u>Exp. Neurol.</u>, 71:326-340.

20. M. Onofrj and I. Bodis-Wollner, 1982, Dopaminergic deficiency causes delayed visual evoked potentials in rats, <u>Ann. Neurol.</u>, 11:484-490.

21. M.F. Ghilardi, M.S. Marx, I. Bodis-Wollner, C.B. Camras, and A. Glover, 1989, The effect of intraocular 6-hydroxydopamine on retinal processing of primates, <u>Ann. Neurol.</u>, 25:357-364.

22. M.F. Ghilardi, E. Chung, I. Bodis-Wollner, M. Dvorzniak, A. Glover, and M. Onofrj, 1988, Systemic 1-methyl, 4-phenyl, 1-2-3-6-tetrahydropyridine (MPTP) administration decreases retinal dopamine content in primates, <u>Life Sci.</u>, 43:255-262.

23. M.S. Marx, I. Bodis-Wollner, P. Bobak, C. Harnois, L. Mylin, and M. Yahr, 1986, Temporal frequency-dependent VEP changes in Parkinson's Disease, <u>Vision Res.</u>, 26:185-193.

24. S.A. Raskin, J.C. Borod, J. Wasserstein, I. Bodis-Wollner, L. Coscia, and M.D. Yahr, 1990, Visuospatial orientation in Parkinson's Disease, <u>Int. J. Neurosci.</u>, 51:9-18.

25. A.M. Gotham, R.G. Brown, and C.D. Marsden, 1988, 'Frontal' cognitive function in patients with Parkinson's Disease 'On' and 'Off' levodopa, <u>Brain</u>, 111:299-321.

26. S. Funahashi, C.J. Bruce, and P.S. Goldman-Rakic, 1989, Mnemonic coding of visual space in the monkey's dorsolateral prefrontal cortex, J. Neurophysiology, 61,331-349.

27. L.D. Selemon and P.S. Goldman-Rakic, 1988, Common cortical and subcortical targets of the dorsolateral prefrontal and posterior parietal cortices in the rhesus monkey: evidence for a distributed neural network subserving spatially guided behavior, J. Neurosci., 8(11):4049-4068.

28. E. Halgren, N.K. Squires, C.L. Wilson, J.R. Rohrbaugh, T.L. Babb, and P.H. Crandall, 1980, Endogenous potentials generated in the human hippocampal formation and amygdala by infrequent events, Science, 210:803-805.

29. E.C. Hansch, K. Syndulko, S.S.N. Cohen, Z.I. Goldberg, A.R. Potvin, and W.W. Tourellotte, 1982, Cognition in Parkinson's disease: An event-related potential perspective, Ann. Neurol., 11:599-607.

30. D.S. Goodin and M.J. Aminoff, 1987, Electrophysiological differences between demented and nondemented patients with Parkinson's disease, Ann. Neurol., 21(1):90-94.

31. I. Bodis-Wollner, M.D. Yahr, and L.H. Mylin, 1984, Nonmotor functions of the basal ganglia, in:"Advances in Neurology", R.G. Hassler and J.R. Christ, eds., Raven Press, New York.

32. S. Pang, J.C. Borod, I. Bodis-Wollner, S. Raskin, L. Mylin, L. Coscia, and M.D. Yahr, 1990, The auditory P300 correlates with specific cognitive deficits in Parkinson's disease, J. Neural Transm., in press.

33. G. Kass, 1988, Electrophysiological and psychophysical investigation of spatial contrast discrimination in human vision, doctoral dissertation, I. Bodis-Wollner, supervisor, University of Michigan, Ann Arbor, Microfilm file.

Acknowledgement: This work was partially supported by EY01708, NS-11631, and NIH-ST35DK07420.

Abercrombie E.D.
Departments of
Behavioural Neuro-
Science and Psychiatry
University of
Pittsburg, U.S.A.

Albanese A.
Institute of Neurology
Università Cattolica
del Sacro Cuore
Rome, Italy

Aldrigde J.W.
Department of Neurology
University of Michigan
U.S.A.

Altavista M.C.
Institute of Neurology
Università Cattolica
del Sacro Cuore
Rome, Italy

Alvarez F.J.
Preclinical Research
Sandoz Ltd.
Basel, Switzerland

Amato G.
Institute of Human
Physiology
University of Palermo
Palermo, Italy

Apicella P.
LNF 3 - CNRS
31 Chemin Joseph Aiguier
Marseille, France

Apicella P.
Institut de Physiologie
Université de Fribourg
Fribourg, Switzerland

Arbuthnott G.W.
Department of
Preclinical
Veterinary Sciences
University of
Edinburgh, U.K.

Armstrong D.M.
FIDIA
Georgetown Institute
of Neurosciences
Washington D.C.

Barbeito L.
Collège de France
INSERM U114
Paris, France

Barone P.
Institute of Neurology
II Faculty of Medicine
University of Naples
Naples, Italy

Bentivolgio A.R.
Institute of Neurology
Università Cattolica
del Sacro Cuore
Rome, Italy

Berendse H.W.
Department of Anatomy
and Embryology
Vrije Universiteit
Amsterdam

Berger T.W.
Departments of
Behavioural
Neuroscience and
Psychiatry
University of Pittsburg
U.S.A.

Bernardi G.
Neurological Clinic
II University of Rome
Rome, Italy

Berrard S.
Laboratoire di
Neurobiologie
Cellulaire e Moléculaire
C.N.R.S.
Gif-sur-Yvette, France

Berretta S.
Institute of Human
Physiology
University of Catania
Catania, Italy

Besson M.J.
Laboratoire de
Neurochimie-Anatomie
IDN-CNRS, Paris

Block F.
Max Plank Institute
for Experimental Medicine
Gottingen, FRG

Bodis-Wollner I.
Department of Neurology
The Mount Sinai School
of Medicine
New York
U.S.A.

Boeijinga P.H.
Department of Exp.
Zoology
University of
Amsterdam
Amsterdam, Holland

Bolam P.
MRC Anatomical
Neuropharmacology
Unit
University Department
of Pharmacology
Oxford, England

Bonuccelli U.
Institute of Clinical
Neurology
University of Pisa
Pisa, Italy

Boucher B.
Oregon Regional
Primate
Research Center

Beaverton, Oregon
U.S.A.

Boyce S.
Merck Sharp and Dohme
Research Laboratories
Neuroscience Research
Center
Harlow, Essex, England

Brault E.
Laboratoire di
Neurobiologie
Cellulaire e Moléculaire
C.N.R.S., Gif-sur-Yvette
France

Brotchie J.M.
Department of Cell
and Structural Biology
University of Manchester
Manchester, England

Bruce K.
Department of Anatomy
Michigan State University
E. Lansing, MI, USA

Buchwald N.A.
Mental Retardation
Research Center
University of California
Los Angeles, U.S.A.

Caboche J.
Laboratoire de
Neurochimie-Anatomie
IDN-CNRS
Paris

Cagnacci A.
Institute of Obstetrics and
Gynecology
University of Pisa
Pisa, Italy

Calabresi P.
Neurological Clinic
II University of Rome
Rome, Italy

Caltagirone C.
Neurological Clinic
II University of Rome
Rome, Italy

Calzà L.
Institue of Human
Physiology
University of Cagliari
Cagliari, Italy

Campanella G.
Institute of Neurology
II Faculty of Medicine
University of Naples
Naples, Italy

Campbell K.J.
Department of Anatomy
University of Toronto
Ontario, Canada

Caraceni T.
Center for the Study
of Parkinson's Disease
and Extrapyramidal
Diseases
Istituto Neurologico
Nazionale "C. Besta",
Milan, Italy

Carboni S.
"B.B.Brodie" Department
of Neuroscience
Chair of Toxicology
Medical School
Cagliari, Italy

Carlesimo A.G.
Neurological Clinic
II University of Rome
Rome, Italy

Carpenter M.B.
Uniformed Services
University
Bethesda, MD
U.S.A.

Cepeda C.
Mental Retardation
Research Center
University of California
Los Angeles
U.S.A.

Chase T.N.
Experimental
Therapeutics Branch
NINCDS, Bethesda, MD
U.S.A.

Cheramy A.
Collège de France
INSERM U114
Paris, France

Cherubini E.
INSERM U-029
Paris, France

Chinaglia G.

Department of
Pathology
University of Basel
Basel, Switzerland

Chockkan V.
Centre de Recherche
en Neurobiologie
Université Laval et
Hopital de
l'Enfant-Jésus
Quebec, Canada

Colosimo C.
Institute of
Neurology
Università Cattolica
del Sacro Cuore
Rome, Italy

Colzi A.
Pharmaceutical Research
Department
F.Hoffmann-La Roche Ltd.
Switzerland

Cools A.R.
Psychoneuropharmacol.
Research Unit
Cath. University
of Nijmegen
The Netherlands

Corsini G.U.
Institute of Obstetrics and
Gynecology
University of Pisa
Pisa, Italy

Cozzolino A.
Institute of
Experimental
Pharmacology and
Toxicology
University of
Cagliari
Cagliari, Italy

Crossman A.R.
Department of Cell
and Structural
Biology
University of
Manchester
Manchester, England

Dall'Olio R.
Institute of
Pharmacology
University of Bologna
Bologna, Italy

Da Prada M.
Pharmaceutical Research
Department
F.Hoffmann-La Roche Ltd.
Basel, Switzerland

De Montis G.
Institute of
Pharmacology
and Biochemical
Pathology
University of
Cagliari
Cagliari, Italy

De Murtas M.
Neurological Clinic
II University of Rome
Rome, Italy

Desban M.
INSERM U114
Chaire de
Neuropharmacologie
Collège de France
Paris, France

Diana M.
Department of
Neuroscience
University of Cagliari
Cagliari, Italy

Di Chiara G.
Institute of
Experimental
Pharmacology and
Toxicology
University of Cagliari
Cagliari, Italy

Divac I.
Institute of
Neurophysiology
University of
Copenhagen
Copenhagen,
Denmark

Ellenbroek B.A.
Psychoneuropharmacol
Research Unit
Cath. University
of Nijmegen
The Netherlands

Falk A.
Department of Neurology
The Mount Sinai School
of Medicine

New York
U.S.A.

Feger J.
Laboratoire de
Pharmacologie
Université
R. Descartes
Paris

Fenelon G.
Laboratoire de
Neuromorphologie
et de Neurologie
Expérimentale du
mouvement
INSERM, Hopital de la
Salpetrière
Paris, France

Fenu S.
Institute of
Experimental
Pharmacology
and Toxicology
University of Cagliari
Cagliari, Italy

Fetoni V.
Istituto Neurologico
C. Besta
Milan, Italy

Filion M.
Centre de Recerche en
Neurobiologie
Université
Laval et Hopital
de l'Enfant-Jésus
Quebec, Canada

Fishell G.
Neurobiology
Research Group
Department of Anatomy
University of Toronto
Canada

Flaherty A.W.
Laboratory of Brain &
Cognitive Sciences
M.I.T.
Cambridge, U.S.A.

Francois C.
Laboratoire de
Neuromorphologie
et de Neurologie
Expérimentale du
mouvement

INSERM, Hopital de la
Salpetrière
Paris, France

Gandolfi O.
Institute of
Pharmacology
University of
Bologna
Bologna, Italy

Gauchy C.
INSERM U114
Chaire de
Neuropharmacologie
Collège de France
Paris, France

Geminiani G.
Center for
the Study of
Parkinson's Disease
and Extrapyramidal
Disease
Istituto Neurologico
Nazionale "C. Besta",
Milan, Italy

Genitrini S.
Center for the Study of
Parkinson's Disease
and Extrapyramidal
Disease
Istituto Neurologico
Nazionale "C. Besta",
Milan, Italy

Gessa G.L.
"B.B.Brodie" Department
of Neuroscience
Chair of Toxicology,
Medical School
Cagliari, Italy

Ghilardi M.F.
Department of Neurology
The Mount Sinai School
of Medicine
New York
U.S.A.

Giardino L.
Institute of Human
Physiology
University of Cagliari
Cagliari, Italy

Gilman S.
Department of Neurology
University of Michigan
U.S.A.

Gimenez-Amaya J.M.
Department of Morphology
University of Madrid
Madrid, Spain

Giovanni A.
Department of Neurology
University of Medicine
and Dentistry of
New Jersey
Piscataway
New Jersey, U.S.A.

Giovannini P.
Center for the Study of
Parkinson's Disease
and Extrapyramidal
Disease
Istituto Neurologico
Nazionale "C. Besta"
Milan, Italy

Girotti F.
Center for the Study
of Parkinson's Disease
and Extrapyramidal
Disease
Istituto Neurologico

Nazionale "C. Besta",
Milan, Italy

Gliessman P.
Oregon Regional
Primate
Research Center
Beaverton, Oregon
U.S.A.

Glowinski J.
INSERM U114
Chaire de
Neuropharmacologie
Collège de France
Paris, France

Gobert A.
Laboratoire de
Physiologie Nerveuse
CNRS, Gif-sur-Yvette
France

Grace A.A.
Departments of
Behavioural
Neuroscience
and Psychiatry
Center for Neuroscience
University of Pittsburg
U.S.A.

Graham W.C.
Department of Cell and
Structural Biology
University of Manchester
Manchester, England

Graybiel A.M.
Laboratory of Neuroanatomy
Department of Brain &
Cognitive Sciences
M.I.T. Cambridge, Mass.
U.S.A.

Groenewegen H.J.
Department of Anatomy
and Embryology
Vrije Universiteit
Amsterdam, Holland

Grofova I.
Department of
Anatomy
Michigan State
University
MI, USA

Groves P.M.
Departments of
Psychiatry
and Neuroscience
University of
California
San Diego, U.S.A.

Guibert B.
Laboratoire de
Physiologie Nerveuse
CNRS, Gif-sur-Yvette
France

Hammerstad J.
Oregon Health
Sciences University
Portland, Oregon
U.S.A.

Hantraye P.
CNRS
Service Hospitalier
Frédéric Joliot
Orsay, France

Haracz J.L.
Program in Neural
Science
Department of
Psychology
Indiana University
Indiana, U.S.A.

Hasegawa Y.
Department of Anatomy
Kagoshima University
Kagoshima, Japan

Hattori T.
Department of Anatomy
University of Toronto
Ontario, Canada

Hazrati L.N.
Centre de Recerche en
Neurobiologie
Hopital de
l'Enfant-Jèsus
Quebec, Canada

Heikkila R.E.
Department of
Neurology
University of Medicine
and Dentistry
of New Jersey
Piscataway, New Jersey
U.S.A.

Heim C.
Max Plank Institute for
Experimental Medicine
Gottingen, FRG

Herrera-Marschitz M.
Department of
Pharmacology
Karolinska Institutet
Stockholm, Sweden

Hirato M.
Department of
Neurosurgery
Gunma University School
of Medicine
Maebashi, Gunma
Japan

Holstein G.R.
Departments of Neurology and
Cell Biology
Mount Sinai School of
Medicine
New York

Hood S.H.
Department of
Preclinical
Veterinary Sciences
University of
Edinburgh
Edinburgh, U.K.

Hossmann K.A.
Max Planck Institut
fur Neurologische
Forschung
Dept. Experimental
Neurology
Koln, FRG

Hull C.D.
Mental Retardation
Research Center
University of California
Los Angeles, California
U.S.A.

Hunt M.A.
Neuropsychiatric
Research Institute
Fargo, North Dakota
U.S.A.

Imai H.
Department of Neurology
Juntendo University
School of Medicine
Tokyo, Japan

Ingham C.A.
Department of Preclinical
Veterinary Sciences
University of Edinburgh
Edinburgh, U.K.

Isacson O.
The Regeneration
Laboratory
Harvard Medical School
Belmont, MA
U.S.A.

Iversen S.D.
Merck Sharp and Dohme
Research Laboratories
Neuroscience Research
Center
Harlow, Essex, England

Jaeger D.
Department of Neurology
University of Michigan
Michigan, U.S.A.

Jaspers R.
Max Plank Institute for
Experimental Medicine
Gottingen, FRG

Jayaraman A.
Louisiana
State University

New Orleans
Louisiana
U.S.A.

Johnston J.G.
Neurobiology Research
Group
Department of Anatomy
University of Toronto
Canada

Julien J.F.
Departement de
Génétique
Moléculaire
Laboratoire de
Neurobiologie
C.N.R.S.,
Gif-sur-Yvette
France

Kawashima Y.
Department of
Neurosurgery
Gunma University School
of Medicine
Maebashi, Gunma, Japan

Kayahara T.
Department of Anatomy
Mie University
Tsu, Japan

Kemel M.L.
INSERM U114
Chaire de
Neuropharmacologie
Collège de France
Paris, France

Kerkerian Le Goff L.
Unité de Neurochemie
CNRS, Marseille
France

Kettler R.
Pharmaceutical Research
Department
F.Hoffmann-La Roche Ltd.
Basel, Switzerland

Khalil R.
Service de Neurologie
CHU La Timone
Marseille, France

Kito S.
Third Department of
Internal Medicine
Hiroshima University

Hiroshima, Japan

Koller W.C.
University of Kansas
Medical School
Kansas City, Kansas
U.S.A.

Krushel L.A.
Neurobiology Research
Group
Department of Anatomy
University of Toronto
Canada

Kuga Y.
Department of Anatomy
Mie University
Tsu, Japan

Lacey M.G.
Department of Pharmacology
SKF Research Ltd.
Herts
U.K.

Lavoie B.
Centre de Recerche
en Neurobiologie
Hopital de
l'Enfant-Jèsus
Quebec, Canada

Legallet E.
LNF 3 - CNRS
Marseille
France

Leviel V.
Laboratoire de
Physiologie Nerveuse
CNRS, Gif-sur-Yvette
France

Levine M.S.
Mental Retardation
Research Center
University of California
Los Angeles, California
U.S.A.

Ljungberg T.
Institut de Physiologie
Université de Fribourg
Fribourg, Switzerland

Lopes da Silva F.H.
Department of Exp.
Zoology
University of Amsterdam

Amsterdam, Holland

Loschmann P.A.
Research Laboratories
of Schering AG
Berlin and Bergkamen
F.R.G.

Macchi G.
Institute of Neurology
Università Cattolica
del Sacro Cuore
Rome, Italy

Mallet J.
Departement de Génétique
Moléculaire
Laboratoire de
Neurobiologie
C.N.R.S., Gif-sur-Yvette
France

Martone M.E.
Departments of Psychiatry
and Neuroscience
University of California
San Diego, U.S.A.

Matsumura M.
Department of
Neurosurgery

Gunma University
School
of Medicine
Maebashi, Gunma
Japan

Maziere M.
CNRS
Service Hospitalier
Frédéric Joliot
Orsay, France

Melis G.B.
Institute of Obstetrics and
Gynecology
University of Pisa
Pisa, Italy

Mengod G.
Preclinical Research
Sandoz Pharma AG
Basel

Mercuri N.B.
Neurological Clinic
II University of Rome
Rome, Italy

Meredith G.E.
Department of Anatomy
and Embryology
Vrije Universiteit
Amsterdam

Mitchell L.J.
Department of Cell
and Structural Biology
University of Manchester
Manchester, England

Miyashita N.
Department of Neurology
Juntendo University
School of Medicine
Tokyo, Japan

Miyoshi R.
Third Department of
Internal Medicine
Hiroshima University
Hiroshima, Japan

Morelli M.
Institute of Experimental
Pharmacology and
Toxicology
University of Cagliari
Cagliari, Italy

Muramatsu S.
Department of Neurology
Jichi Medical School
Tochigi, Japan

Muratorio A.
Institute of Clinical
Neurology
University of Pisa
Pisa, Italy

Mylin L.
Department of Neurology
The Mount Sinai School
of Medicine
New York
U.S.A.

Nakamura S.
Kanazawa University
Faculty of Medicine
Kanazawa, Japan

Nakamura S.
Department of Neurology
Jichi Medical School
Tochigi, Japan

Nakamura T.

Department of Neurology
Juntendo University
School
of Medicine
Tokyo, Japan

Nakano K.
Department of Anatomy
Mie University
Tsu, Japan

Napolitano A.
Institute of Clinical
Neurology
University of Pisa
Pisa, Italy

Narabayashi H.
Department of Neurology
Juntendo University
School of Medicine
Tokyo, Japan

Nieoullon A.
Unité de Neurochemie
CNRS, Marseille, France

Nisenbaum E.S.
Departments of
Behavioural Neuroscience
and Psychiatry
Center for Neuroscience
University of Pittsburg
U.S.A.

Nishi K.
Department of Neurology
Juntendo University
School of Medicine
Tokyo, Japan

Nocentini U.
Institute of Neurology
II University of Rome
Rome, Italy

Nomura Y.
Segawa Neurological
Clinic for Children
Tokyo, Japan

North R.A.
Vollum Institute
Oregon Health Sciences
University
Portland, Oregon, U.S.A.

Nutt J.G.
Oregon Regional Primate
Research Center

Beaverton, Oregon, U.S.A.

Ohye C.
Department of
Neurosurgery
Gunma University School
of Medicine
Maebashi, Gunma, Japan

Oliva D.
Center for the Study of
Parkinson's Disease
and Extrapyramidal
Disease
Istituto Neurologico
Nazionale "C. Besta"
Milan, Italy

Onn S.P.
Departments of
Behavioural
Neuroscience and
Psychiatry
Center for Neuroscience
University of Pittsburg
U.S.A.

Page R.D.
Department of Cell and
Structural Biology
University of Manchester
Manchester, England

Palacios J.M.
Preclinical Research
Sandoz Pharma AG
Basel, Switzerland

Palma V.
Institute of Neurology
II Faculty of Medicine
University of Naples
Naples, Italy

Pani L.
"B.B.Brodie" Department
of Neuroscience
Chair of Toxicology
Medical School
Cagliari, Italy

Pang S.
The VEP Laboratory
Department of Neurology
The Mount Sinai School of
Medicine
New York, U.S.A.

Parashos S.A.
Experimental Therapeutics
Branch

NINCDS, Bethesda, Md
U.S.A.

Parent A.
Centre de Recherches en
Neurobiologie
Hopital de l'Enfant-Jèsus
Quebec, Canada

Paschen W.
Max Planck Institut fur
Neurologische Forschung
Dept. Experimental
Neurology
Koln, FRG

Pasik P.
Departments of Neurology
and Cell Biology
Mount Sinai School of
Medicine
New York, U.S.A.

Pasik T.
Department of Neurology
Mount Sinai School of
Medicine
New York , U.S.A.

Paul M.L.
Department of
Pharmacology
Dalhousie University
Halifax, Canada

Pennartz C.M.A.
Department of Exp.
Zoology
University of Amsterdam
Amsterdam

Peppe A.
Institute of
Nervous and Mental
Diseases
University of Rome
"La Sapienza"
Rome, Italy

Percheron G.
Laboratoire de
Neuromorphologie et
de Neurologie
Expérimentale du mouvement
INSERM, Hopital de la
Salpetrière, Paris
France

Perciavalle V.
Institute of Human
Physiology

University of Catania
Catania, Italy

Piazza P.V.
Institute of Human
Physiology
University of Palermo
Palermo, Italy

Piccini P.
Institute of Clinical
Neurology
University of Pisa
Pisa, Italy

Pierelli F.
Institute of Nervous
and Mental Diseases
University of Rome
"La Sapienza"
Rome, Italy

Portas C.
"B.B.Brodie" Department
of Neuroscience
Chair of Toxicology
Medical School
Cagliari, Italy

Probst A.
Department of Pathology
University of Basel
Basel, Switzerland

Quinn B.
Laboratory of Neuroanatomy
Department of Brain &
Cognitive Sciences
M.I.T., Cambridge, Mass.
U.S.A.

Rataboul P.
Departement de Génétique
Moléculaire
Laboratoire de
Neurobiologie
C.N.R.S., Gif-sur-Yvette
France

Rebec G.V.
Program in Neural Science
Department of Psychology
Indiana University
Indiana, U.S.A.

Renwart N.
Laboratoire de
Pharmacologie
Université
R. Descartes

Paris, France

Rettig K.J.
Research Laboratories
of Schering AG
Berlin and Bergkamen
F.R.G.

Riche D.
Laboratoire de
Physiologie Nerveuse
CNRS, Gif-sur-Yvette
France

Robertson G.S.
Department of
Pharmacology
Dalhousie University
Halifax, Canada

Robertson H.A.
Department of
Pharmacology
Dalhousie University
Halifax, Canada

Robertson R.G.
Department of Cell and
Structural Biology
University of Manchester
Manchester, England

Robledo P.
Laboratoire de
Pharmacologie
Université
R. Descartes
Paris, France

Rogard M.
Laboratoire de
Neurochimie-Anatomie
IDN-CNRS
Paris, France

Rossetti Z.L.
"B.B.Brodie" Department
of Neuroscience
Chair of Toxicology
Medical School
Cagliari, Italy

Rossi P.
Institute of Neurology
Università Cattolica
del Sacro Cuore
Rome, Italy

Ruberte E.
Preclinical Research

Sandoz Pharma AG
Basel, Switzerland

Rupniak N.M.
Merck Sharp and Dohme
Research Laboratories
Neuroscience Research
Center
Harlow, Essex, England

Ryan L.J.
Department of Psychology
Oregon State University
Oregon, U.S.A.

Sadikot A.F.
Centre de Recherches en
Neurobiologie
Hopital de l'Enfant-Jésus
Quebec, Canada

Salin P.
Unité de Neurochemie
CNRS, Marseille, France

Sambrook M.A.
Department of Cell and
Structural Biology
University of Manchester
Manchester, England

Sancesario G.
Neurological Clinic
II University of Rome
Rome, Italy

Sauer G.
Research Laboratories
of Schering AG
Berlin and Bergkamen
F.R.G.

Schmidt-Kastner R.
Max Planck Institut fur
Neurologische Forschung
Dept. Experimental
Neurology
Koln, FRG

Schultz W.
Institut de Physiologie
Université de Fribourg
Fribourg, Switzerland

Segawa M.
Segawa Neurological
Clinic for Children
Tokyo, Japan

Shibasaki T.
Department of

Neurosurgery
Gunma University
School of Medicine
Maebashi, Gunma, Japan

Shibazaki T.
Department of
Neurosurgery
Gunma University
School of Medicine
Maebashi, Gunma, Japan

Smith Y.
MRC Anatomical
Neuropharmacology Unit
University Department
of Pharmacology
Oxford, England

Soliveri P.
Istituto Neurologico
C. Besta
Milan, Italy

Sonsalla P.K.
Department of Neurology
University of Medicine
and Dentistry of
New Jersey, Piscataway
New Jersey, U.S.A.

Sontag K.H.
Max Plank Institute
for Experimental Medicine
Gottingen, FRG

Spann B.M.
Department of Anatomy
Michigan State University
E. Lansing, MI, USA

Stanzione P.
Neurological Clinic
II University of Rome
Rome, Italy

Stefani A.
Neurological Clinic
II University of Rome
Rome, Italy

Steventon M.J.
Merck Sharp and Dohme
Research Laboratories
Neuroscience Research
Center
Harlow, Essex, England

Tagliati M.
Neurological Clinic
II University of Rome

Rome, Italy

Takada M.
Department of Anatomy
University of Tenessee
Tenessee, Memphis
USA

Tamma F.
Center for the Study of
Parkinson's Disease
and Extrapyramidal
Disease
Istituto Neurologico
Nazionale "C. Besta"
Milan, Italy

Tanaka A.
Third Department of
Internal Medicine
Hiroshima University
Hiroshima, Japan

Tepper J.M.
Center for Molecular and
Behavioural Neuroscience
State University of
New Jersey
U.S.A.

Testa D.
Istituto Neurologico
C. Besta
Milan, Italy

Tremblay L.
Centre de Recherche
en Neurobiologie
Université Laval et
Hopital de l'Enfant-Jésus
Quebec, Canada

Trent F.
Department of Biological
Sciences
State University of
New Jersey
U.S.A.

Trouche E.
LNF 3 - CNRS
Marseille, France

Trugman J.M.
Department of Neurology
University of Virginia
Health Sciences Center
Charlottesville, Virginia
U.S.A.

Tschanz J.T.

Program in Neural Science
Department of Psychology
Indiana University
Indiana, U.S.A.

Uchimura N.
Departments of Physiology
and Neuropsychiatry
Kurume University
Kurume, Japan

Vaccari A.
"B.B.Brodie" Department
of Neuroscience
Chair of Toxicology
Medical School
Cagliari, Italy

Van der Kooy D.
Neurobiology Research
Group
Department of Anatomy
University of Toronto
Toronto, Canada

Vernier P.
Departement de Génétique
Moléculaire
Laboratoire de
Neurobiologie
C.N.R.S., Gif-sur-Yvette
France

Viallet F.
Service de Neurologie
CHU La Timone
Marseille, France

Vuillet J.
Université Aix-Marseille II
Faculté de Médicine
Marseilles, France

Wachtel H.
Research Laboratories
of Schering AG
Berlin and Bergkamen
F.R.G.

Walsh J.P.
Mental Retardation
Research Center
University of California
Los Angeles, California
U.S.A.

Wamsley J.K.
Neuropsychiatric
Research Institute
Fargo, North Dakota
U.S.A.

Wang Z.
Program in Neural Science
Department of Psychology
Indiana University
Indiana, U.S.A.

White I.
Program in Neural Science
Department of Psychology
Indiana University
Indiana, U.S.A.

Woodward W.R.
Oregon Health
Sciences University
Portland, Oregon, U.S.A.

Wooten F.G.
Department of Neurology
University of Virginia
Health Sciences Center
Charlottesville, Virginia
U.S.A.

Wouterlood F.G.
Department of Anatomy
and Embryology
Vrije Universiteit
Amsterdam, Holland

Yelnik J.
Laboratoire de
Neuromorphologie

et de Neurologie
Expérimentale du
mouvement
INSERM, Hopital de la

Salpetrière, Paris
France

Yoshida M.
Department of Neurology
Jichi Medical School
Tochigi, Japan

Young S.J.
Departments of Psychiatry
and Neuroscience
University of California
San Diego, U.S.A.

Zigmond M.J.
Departments of
Behavioural
Neuroscience and
Psychiatry
University of Pittsburg
U.S.A.

Zurcher G.
Pharmaceutical Research
Department
F. Hoffmann-La Roche Ltd.
Basel, Switzerland